The Combined Works of Charles Fort

The Combined Works of Charles Fort

Charles Fort

ADULTBRAIN publishing

CONTENTS

Adultbrain Publishing is dedicated to breathing new life into timeless literary works by resurrecting old classics for the modern age. We meticulously curate and convert these masterpieces into high-quality digital and audio formats, making them accessible to a new generation of readers and listeners. Our commitment to preserving the essence of these works, while enhancing them with today's technology, allows us to offer immersive experiences that retain the authenticity of the original texts. Whether rediscovering a beloved classic or experiencing it for the first time, our editions invite readers to start using their Adultbrain today.
Published by Adultbrain Publishing.
ISBN: 978-1-0690495-7-5
eISBN: 978-1-0690495-6-8
Title: The Combined Works of Charles Fort
Start using your adultbrain today.
For more information, visit: www.adultbrain.ca

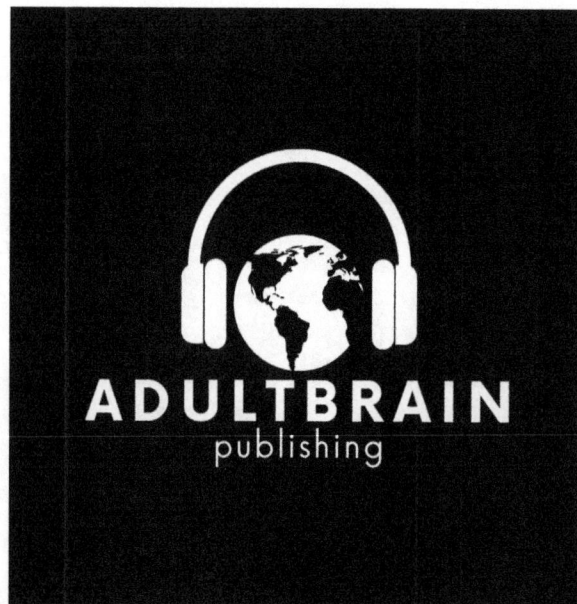

Book of the Damned (1919)

The Book of The Damned

Chapter I

A procession of the damned.

By the damned, I mean the excluded.

We shall have a procession of data that Science has excluded.

Battalions of the accursed, captained by pallid data that I have exhumed, will march. You'll read them – or they'll march. Some of them livid and some of them fiery and some of them rotten.

Some of them are corpses, skeletons, mummies, twitching, tottering, animated by companions that have been damned alive. There are giants that will walk by, though sound asleep. There are things that are theorems and things that are rags: they'll go by like Euclid arm in arm with the spirit of anarchy. Here and there will flit little harlots. Many are clowns. But many are of the highest respectability. Some are assassins. There are pale stenches and gaunt superstitions and mere shadows and lively malices: whims and amiabilities. The naive and the pedantic and the bizarre and the grotesque and the sincere and the insincere, the profound and the puerile.

A stab and a laugh and the patiently folded hands of hopeless propriety.

The ultra-respectable, but the condemned, anyway.

The aggregate appearance is of dignity and dissoluteness: the aggregate voice is a defiant prayer: but the spirit of the whole is processional.

The power that has said to all these things that they are damned, is Dogmatic Science.

But they'll march.

The little harlots will caper, and freaks will distract attention, and the clowns will break the rhythm of the whole with their buffooneries – but the solidity of the procession as a whole: the impressiveness of things that pass and pass and pass, and keep on and keep on and keep on coming.

The irresistibleness of things that neither threaten nor jeer nor defy, but arrange themselves in mass-formations that pass and pass keep on passing.

So, by the damned, I mean the excluded.

But by the excluded I mean that which will some day be the excluding.

Or everything that is, won't be.

And everything that isn't, will be –

But, of course, will be that which won't be –

It is our expression that the flux between that which isn't and that which won't be, or the state that is commonly and absurdly called "existence," is a rhythm of heavens and hells: that the damned won't stay damned; that salvation only precedes perdition.

The inference is that some day our accursed tatterdemalions will be sleek angels. Then the sub-inference is that some later day, back they'll go whence they came.

It is our expression that nothing can attempt to be, except by attempting to exclude something else: that that which is commonly called "being" is a state that is wrought more or less definitely proportionately to the appearance of positive difference between that which is included and that which is excluded.

But it is our expression that there are no positive differences:

that all things are like a mouse and a bug in the heart of a cheese.

Mouse and a bug: no two things could seem more unlike. They're there a week, or they stay there a month: both are then only transmutations of cheese. I think we're all bugs and mice, and are only different expressions of an all-inclusive cheese.

Or that red is not positively different from yellow: is only another degree of whatever vibrancy yellow is a degree of: that red and yellow are continuous, or that they merge in orange.

So then that, if, upon the basis of yellowness and redness, Science should attempt to classify all phenomena, including all red things as veritable, and excluding all yellow things as false or illusory, the demarcation would have to be false and arbitrary, because things colored orange, constituting continuity, would belong on both sides of the attempted border-line.

As we go along, we shall be impressed with this:

That no basis for classification, or inclusion and exclusion, more reasonable than that of redness and yellowness has ever been conceived of.

Science has, by appeal to various bases, included a multitude of data. Had it not done so, there would be nothing with which to seem to be. Science has, by appeal to various bases, excluded a multitude of data. Then, if redness is continuous with yellowness:

if every basis of admission is continuous with every basis of exclusion, Science must have excluded some things that are continuous with the accepted. In redness and yellowness, which merge in orangeness, we typify all tests, all standards, all means of forming an opinion –

Or that any positive opinion upon any subject is illusion built – upon the fallacy that there are positive differences to judge by –

That the quest of all intellection has been for something – a fact, a basis, a generalization, law, formula, a major premise that is positive:

that the best that has ever been done has been to say that some things are self-evident – whereas, by evidence we mean the support of something else –

That this is the quest; but that it has never been attained; but that Science has acted, ruled, pronounced, and condemned as if it had been attained.

What is a house?

It is not possible to say what anything is, as positively distinguished from anything else, if there are no positive differences.

A barn is a house, if one lives in it. If residence constitutes houseness, because style of architecture does not, then a bird's nest is a house: and human occupancy is not the standard to judge by, because we speak of dogs' houses; nor material, because we speak of snow houses of Eskimos – or a shell is a house to a hermit crab – or was to the mollusk that made it – or things seemingly so positively different as the White House at Washington and a shell on the seashore are seen to be continuous.

So no one has ever been able to say what electricity is, for instance.

It isn't anything, as positively distinguished from heat or magnetism or life. Metaphysicians and theologians and biologists have tried to define life. They have failed, because, in a positive sense, there is nothing to define: there is no phenomenon of life that is not, to some degree, manifest in chemism, magnetism, astronomic motions.

White coral islands in a dark blue sea.

Their seeming of distinctness: the seeming of individuality, or of positive difference one from another but all are only projections from the same sea bottom. The difference between sea and land is not positive. In all water there is some earth: in all earth there is some water.

So then that all seeming things are not things at all, if all are inter-continuous, any more than is the leg of a table a thing in itself, if it is only a projection from something else: that not one of us is a real person, if, physically, we're continuous with environment; if, psychically, there is nothing to us but expression of relation to environment.

Our general expression has two aspects:

Conventional monism, or that all "things" that seem to have identity of their own are only islands that are projections from something underlying, and have no real outlines of their own. But that all "things," though only projections, are projections that are striving to break away from the underlying that denies them identity of their own.

I conceive of one inter-continuous nexus, in which and of which, all seeming things are only different expressions, but in which all things are localizations of one attempt to break away and become real things, or to establish entity or positive difference or final demarcation or unmodified independence or personality or soul, as it is called in human phenomena –

That anything that tries to establish itself as a real, or positive, or absolute system, government, organization, self, soul, entity, individuality, can so attempt only by drawing a line about itself, or about the inclusions that constitute itself, and damning or excluding, or breaking away from, all other "things":

That, if it does not so act, it can not seem to be;

That, if it does so act, it falsely and arbitrarily and futilely and disastrously acts, just as would one who draws a circle in the sea, including a few waves, saying that the other waves, with which the included are continuous, are positively different, and stakes his life upon maintaining that the admitted and the damned are positively different.

Our expression is that our whole existence is animation of the local by an ideal that is realizable only in the universal:

That, if all exclusions are false, because always are included and excluded continuous: that if all seeming of existence perceptible to us is the product of exclusion, there is nothing that is perceptible to us that really is: that only the universal can really be. Our especial interest is in modem science as a manifestation of this one ideal or purpose or process:

That it has falsely excluded, because there are no positive standards to judge by : that it has excluded things that, by its own pseudostandards, have as much right to come in as have the chosen.

Our general expression:

That the state that is commonly and absurdly called "existence," is a flow, or a current, or an attempt, from negativeness to positiveness, and is intermediate to both.

By positiveness we mean:

Harmony, equilibrium, order, regularity, stability, consistency, unity, realness, system, government, organization, liberty, independence, soul, self, personality, entity, individuality, truth, beauty, justice, perfection, definiteness –

That all that is called development, progress, or evolution is movement toward, or attempt toward, this state for which, or for aspects of which, there are so many names, all of which are summed up in the one word "positiveness."

At first this summing up may not be very readily acceptable.

At first it may seem that all these words are not synonyms: that "harmony" may mean "order," but that by "independence," for instance, we do not mean "truth," or that by "stability" we do not mean "beauty," or "system," or "justice."

I conceive of one inter-continuous nexus, which expresses itself in astronomic phenomena, and chemic, biologic, psychic, sociologic:

that it is everywhere striving to localize positiveness: that to this attempt in various fields of phenomena – which are only quasi-different – we give different names. We speak of the "system" of the planets, and not of their "government": but in considering a store, for instance, and its management, we see that the words are interchangeable.

It used to be customary to speak of chemic equilibrium, but not of social equilibrium: that false demarcation has been broken down. We shall see that by all these words we mean the same state. As every-day conveniences, or in terms of common illusions, of course, they are not synonyms. To a child an earth worm is not an animal. It is to the biologist.

By "beauty," I mean that which seems complete.

Obversely, that the incomplete, or the mutilated, is the ugly.

Venus de Milo.

To a child she is ugly.

When a mind adjusts to thinking of her as a completeness, even though, by physiologic standards, incomplete, she is beautiful.

A hand thought of only as a hand, may seem beautiful.

Found on a battlefield – obviously a part – not beautiful.

But everything in our experience is only a part of something else that in turn is only a part of still something else – or that there is nothing beautiful in our experience: only appearances that are intermediate to beauty and ugliness – that only universality is complete:

that only the complete is the beautiful: that every attempt to achieve beauty is an attempt to give to the local the attribute of the universal.

By stability, we mean the immovable and the unaffected. But all seeming things are only reactions to something else. Stability, too, then, can be only the universal, or that besides which there is nothing else. Though some things seem to have – or have – higher approximations to stability than have others, there are, in our experience, only various degrees of intermediateness to stability and instability. Every man, then, who works for stability under its various names of "permanency," "survival," "duration," is striving to localize in something the state that is realizable only in the universal.

By independence, entity, and individuality, I can mean only that besides which there is nothing else, if given only two things, they must be continuous and mutually affective, if everything is only a reaction to something else, and any two things would be destructive of each other's independence, entity, or individuality.

All attempted organizations and systems and consistencies, some approximating far higher than others, but all only intermediate to Order and Disorder, fail eventually because of their relations with outside forces. All are attempted completenesses. If to all local phenomena there are always outside forces, these attempts, too, are realizable only in the state of completeness, or that to which there are no outside forces.

Or that all these words are synonyms, all meaning the state that we call the positive state –

That our whole "existence" is a striving for the positive state.

The amazing paradox of it all:

That all things are trying to become the universal by excluding other things.

That there is only this one process, and that it does animate all expressions, in all fields of phenomena, of that which we think of as one inter-continuous nexus:

The religious and their idea or ideal of the soul. They mean distinct, stable entity, or a state that is independent, and not a mere flux of vibrations or complex of reactions to environment, continuous with environment, merging away with an infinitude of other interdependent complexes.

But the only thing that would not merge away into something else would be that besides which there is nothing else.

That Truth is only another name for the positive state, or that the quest for Truth, is the attempt to achieve positiveness:

Scientists who have thought that they were seeking Truth, but who were trying to find out astronomic, or chemic, or biologic truths.

But Truth is that besides which there is nothing: nothing to modify it, nothing to question it, nothing to form an exception: the all-inclusive, the complete –

By Truth I mean the Universal.

So chemists have sought the true, or the real, and have always failed in their endeavors, because of the outside relations of chemical phenomena: have failed in the sense that never has a chemical law, without exceptions, been discovered: because chemistry is continuous with astronomy, physics, biology – For instance, if the sun should greatly change its distance from this earth, and if human life could survive, the familiar chemic formulas would no longer work out: a new science of chemistry would have to be learned –

Or that all attempts to find Truth in the special are attempts to find the universal in the local.

And artists and their striving for positiveness, under the name of "harmony" – but their pigments that are oxidizing, or are responding to a deranging environment – or the strings of musical instruments that are differently and disturbingly adjusting to outside chemic and thermal and gravitational forces – again and again this oneness of all ideals, and that it is the attempt to be, or to achieve, locally, that which is realizable only universally. In our experience there is only intermediateness to harmony and discord. Harmony is that besides which there are no outside forces.

And nations that have fought with only one motive: for individuality, or entity, or to be real, final nations, not subordinate to, or parts of, other nations. And that nothing but intermediateness has ever been attained, and that history is record of failures of this one attempt, because there always have been outside forces, or other nations contending for the same goal.

As to physical things, chemic, mineralogic, astronomic, it is not customary to say that they act to achieve Truth or Entity, but it is understood that all motions are toward Equilibrium: that there is no motion except toward Equilibrium, of course always away from some other approximation to Equilibrium.

All biologic phenomena act to adjust: there are no biologic actions other than adjustments.

Adjustment is another name for Equilibrium. Equilibrium is the Universal, or that which has nothing external to derange it.

But that all that we call "being" is motion: and that all motion is the expression, not of equilibrium, but of equilibrating, or of equilibrium unattained: that life-motions are expressions of equilibrium unattained: that all thought relates to the unattained: that to have what is called being in our quasi-state, is not to be in the positive sense, or is to be intermediate to Equilibrium and Inequilibrium.

So then:

That all phenomena in our intermediate state, or quasi-state, represent this one attempt to organize, stabilize, harmonize, individualize – or to positivize, or to become real:

That only to have seeming is to express failure or intermediateness
to final failure and final success;

That every attempt – that is observable – is defeated by Continuity, or by outside forces – or by the excluded that are continuous with the included:

That our whole "existence" is an attempt by the relative to be the absolute, or by the local to be the universal.

In this book, my interest is in this attempt as manifested in modern science:

That it has attempted to be real, true, final, complete, absolute:

That, if the seeming of being, here, in our quasi-state, is the product of exclusion that is always false and arbitrary, if always are included and excluded continuous, the whole seeming system, or entity, of modern science is only quasi-system, or quasi-entity, wrought by the same false and arbitrary process as that by which the still less positive system that preceded it, or the theological system, wrought the illusion of its being.

In this book, I assemble some of the data that I think are of the falsely and arbitrarily excluded.

The data of the damned.

I have gone into the outer darkness of scientific and philosophical transactions and proceedings, ultra-respectable, but covered with the dust of disregard. I have descended into journalism. I have come back with the quasi-souls of lost data.

They will march.

As to the logic of our expressions to come –

That there is only quasi-logic in our mode of seeming:

That nothing ever has been proved –

Because there is nothing to prove.

When I say that there is nothing to prove, I mean that to those who accept Continuity, or the merging away of all phenomena into other phenomena, without positive demarcations one from another, there is, in a positive sense, no one thing. There is nothing to prove.

For instance nothing can be proved to be an animal – because animalness and vegetableness are not positively different. There are some expressions of life that are as much vegetable as animal, or that represent the merging of animalness and vegetableness.

There is then no positive test, standard, criterion, means of forming an opinion. As distinct from vegetables, animals do not exist.

There is nothing to prove. Nothing could be proved to be good, for instance. There is nothing in our "existence" that is good, in a positive sense, or as really outlined from evil. If to forgive be good in times of peace, it is evil in wartime. There is nothing to prove:

good in our experience is continuous with, or is only another aspect of evil.

As to what I'm trying to do now – I accept only. If I can't see universally, I only localize.

So, of course then, that nothing ever has been proved:

That theological pronouncements are as much open to doubt as ever they were, but that, by a hypnotizing process, they became dominant over the majority of minds in their era;

That, in a succeeding era, the laws, dogmas, formulas, principles, of materialistic science never were proved, because they are only localizations simulating the universal; but that the leading minds of their era of dominance were hypnotized into more or less firmly believing them.

Newton's three laws, and that they are attempts to achieve positiveness, or to defy and break Continuity, and are as unreal as are all other attempts to localize the universal:

That, if every observable body is continuous, mediately or immediately, with all other bodies, it can not be influenced only by its own inertia, so that there is no way of knowing what the phenomena of inertia may be; that, if all things are reacting to an infinitude of forces, there is no way of knowing what the effects of only one impressed force would be; that if every reaction is continuous with its action, it can not be conceived of as a whole, and that there is no way of conceiving what it might be equal and opposite to –

Or that Newton's three laws are three articles of faith;

Or that demons and angels and inertias and reactions are all mythological characters;

But that, in their eras of dominance, they were almost as firmly believed in as if they had been proved.

Enormities and preposterousnesses will march.

They will be "proved" as well as Moses or Darwin or Lyell ever "proved" anything.

We substitute acceptance for belief.

Cells of an embryo take on different appearances in different eras.

The more firmly established, the more difficult to change.

That social organism is embryonic.

That firmly to believe is to impede development.

That only temporarily to accept is to facilitate.

But:

Except that we substitute acceptance for belief, our methods will be the conventional methods; the means by which every belief has been formulated and supported: or our methods will be the methods of theologians and savages and scientists and children. Because, if all phenomena are continuous, there can be no positively different methods. By the inconclusive means and methods of cardinals and fortune tellers and evolutionists and peasants, methods which must be inconclusive, if they relate always to the local, and if there is nothing local to conclude, we shall write this book.

If it function as an expression of its era, it will prevail.

All sciences begin with attempts to define.

Nothing ever has been defined.

Because there is nothing to define.

Darwin wrote "The Origin of Species."

He was never able to tell what he meant by a "species."

It is not possible to define.

Nothing has ever been finally found out.

Because there is nothing final to find out.

It's like looking for a needle that no one ever lost in a haystack

that never was –

But that all scientific attempts really to find out something, whereas really there is nothing to find out, are attempts, themselves, really to be something.

A seeker of Truth. He will never find it. But the dimmest of Possibilities – he may himself become Truth.

Or that science is more than an inquiry:

That is it a pseudo-construction, or a quasi-organization: that it is an attempt to break away and locally establish harmony, stability, equilibrium, consistency, entity –

Dimmest of possibilities – that it may succeed.

That ours is a pseudo-existence, and that all appearances in it partake of its essential fictitiousness –

But that some appearances approximate far more highly to the positive state than do others.

We conceive of all "things" as occupying gradations, or steps in series between positiveness and negativeness, or realness and unrealness: that some seeming things are more nearly consistent, just, beautiful, unified, individual, harmonious, stable – than others.

We are not realists. We are not idealists. We are intermediatists that nothing is real, but that nothing is unreal: that all phenomena are approximations one way or the other between realness and unrealness.

So then:

That our whole quasi-existence is an intermediate stage between positiveness and negativeness or realness and unrealness.

Like purgatory, I think.

But in our summing up, which was very sketchily done, we omitted to make clear that Realness – is an aspect of the positive state.

By Realness, I mean that which does not merge away into something else, and that which is not partly something else: that which is not a reaction, to, or an imitation of, something else. By a real hero, we mean one who is not partly a coward, or whose actions and motives do not merge away into cowardice. But, if in Continuity, all things do merge, by Realness, I mean the Universal, besides which there is nothing with which to merge.

That, though the local might be universalized, it is not conceivable that the universal can be localized: but that high approximations there may be, and that these approximate successes may be translated out of Intermediateness into Realness – quite as, in a relative sense, the industrial world recruits itself by translating out of unrealness, or out of the seemingly less real imaginings of inventors, machines which seem, when set up in factories, to have more of Realness than they had when only imagined.

That all progress, if all progress is toward stability, organization, harmony, consistency, or positiveness, is the attempt to become real.

So, then, in general metaphysical terms, our expression is that, like a purgatory, all that is commonly called "existence," which we call Intermediateness, is quasi-existence, neither real nor unreal, but expression of attempt to become real, or to generate for or recruit a real existence.

Our acceptance is that Science, though usually thought of so specifically, or in its own local terms, usually supposed to be a prying into old bones, bugs, unsavory messes, is an expression of this one spirit animating all Intermediateness: that, if Science could absolutely exclude all data but its own present data, or that which is assimilable with the present quasi-organization, it would be a real system, with positively definite outlines – it would be real.

Its seeming approximation to consistency, stability, system – positiveness or realness – is sustained by damning the irreconcilable or the unassimilable –

All would be well.

All would be heavenly –

If the damned would only stay damned.

Chapter II

   In the autumn of 1883, and for years afterward, occurred brilliant- colored sunsets, such as had never been seen before, within the memory of all observers. Also there were blue moons.

I think that one is likely to smile incredulously at the notion of blue moons. Nevertheless they were as common as were green suns in 1883.

Science had to account for these unconventionalities. Such publications as Nature and Knowledge were besieged with inquiries.

I suppose, in Alaska and in the South Sea Islands, all the medicine men were similarly upon trial. Something had to be thought of.

Upon the 28th of August, 1883, the volcano of Krakatoa, of the Straits of Sunda, had blown up. Terrific.

We're told that the sound was heard 2,000 miles, and that 36,380 persons were killed. Seems just a little unscientific, or impositive, to me: marvel to me we're not told 2,163 miles and 36,387 persons. The volume of smoke that went up must have been visible to other Planets – or, tormented with our crawlings and scurryings, the earth complained to Mars; swore a vast black oath at us.

In all text-books that mention this occurrence – no exception so far as I have read – it is said that the extraordinary atmospheric effects of 1883 were first noticed in the last of August or the first of September.

That makes a difficulty for us.

It is said that these phenomena were caused by particles of volcanic dust that were cast high in the air by Krakatoa.

This is the explanation that was agreed upon in 1883 –

But for seven years the atmospheric phenomena continued –

Except that, in the seven, there was a lapse of several years – and where was the volcanic dust all that time?

You'd think that such a question as that would make trouble?

Then you haven't studied hypnosis. You have never tried to demonstrate to a hypnotic that a table is not a hippopotamus.

According to our general acceptance, it would be impossible to demonstrate such a thing. Point out a hundred reasons for saying that a hippopotamus is not a table: you'll have to end up agreeing that neither is a table a table – it only seems to be a table. Well, that's what the hippopotamus seems to be. So how can you prove that something is not something else, when neither is something else some other thing? There's nothing to prove.

This is one of the profundities that we advertised in advance.

You can oppose an absurdity only with some other absurdity.

But Science is established preposterousness. We divide all intellection:

the obviously preposterous and the established.

But Krakatoa: that's the explanation that the scientists gave. I don't know what whopper the medicine men told.

We see, from the start, the very strong inclination of science to deny, as much as it can, external relations of this earth.

This book is an assemblage of data of external relations of this earth. We take the position that our data have been damned, upon no consideration for individual merits or demerits, but in conformity with a general attempt to hold out for isolation of this earth. This is attempted positiveness. We take the position that science can no more succeed than, in a similar endeavor, could the Chinese, or than could the United States. So then, with only pseudo-consideration of the phenomena of 1883, or as an expression of positivism in its aspect of isolation, or unrelatedness, scientists have perpetrated such an enormity as suspension of volcanic dust seven years in the air – disregarding the lapse of several years – rather than to admit the arrival of dust from somewhere beyond this earth. Not that scientists themselves have ever achieved positiveness, in its aspect of unitedness, among themselves – because Nordenskiold, before 1883, wrote a great deal upon his theory of cosmic dust, and Prof. Cleveland Abbe contended against the Krakatoan explanation – but that this is the orthodoxy of the main body of scientists. My own chief reason for indignation here:

That this preposterous explanation interferes with some of my own enormities.

It would cost me too much explaining, if I should have to admit that this earth's atmosphere has such sustaining power.

Later, we shall have data of things that have gone up in the air and that have stayed up – somewhere – weeks – months – but not by the sustaining power of this earth's atmosphere. For instance, the turtle of Vicksburg. It seems to me that it would be ridiculous to think of a good-sized turtle hanging, for three or four months, upheld only by the air, over the town of Vicksburg. When it comes to the horse and the barn – I think that they'll be classics some day, but I can never accept that a horse and a barn could float several months in this earth's atmosphere.

The orthodox explanation:

See the Report of the Krakatoa Committee of the Royal Society.

It comes out absolutely for the orthodox explanation – absolutely and beautifully, also expensively. There are 492 pages in the "Report," and 40 plates, some of them marvellously colored.

It was issued after an investigation that took five years. You couldn't think of anything done more efficiently, artistically, authoritatively.

The mathematical parts are especially impressive: distribution of the dust of Krakatoa; velocity of translation and rates of subsidence; altitudes and persistences –

Annual Register, 1883-105:

That the atmospheric effects that have been attributed to Krakatoa were seen in Trinidad before the eruption occurred;

Knowledge, 5-418:

That they were seen in Natal, South Africa, six months before the eruption.

Inertia and its inhospitality.

Or raw meat should not be fed to babies.

We shall have a few data initiatorily.

I fear me that the horse and the barn were a little extreme for our budding liberalities.

The outrageous is the reasonable, if introduced politely.

Hailstones, for instance. One reads in the newspapers of hailstones the size of hens' eggs. One smiles. Nevertheless I will engage to list one hundred instances, from the Monthly Weather Review, of hailstones the size of hens' eggs. There is an account in Nature, Nov. I, 1894, of hailstones that weighed almost two pounds each. See Chambers' Encyclopedia for three-pounders. Report of the Smithsonian Institute, 1870-479 – two-pounders authenticated, and six-pounders reported. At Seringapatam, India, about the year 1800, fell a hailstone –

I fear me, I fear me: this is one of the profoundly damned. I blurt out something that should, perhaps, be withheld for several hundred pages – but that damned thing was the size of an elephant. We laugh.

Or snowflakes. Size of saucers. Said to have fallen at Nashville, Tenn., Jan. 24, 1891. One smiles.

"In Montana, in the winter of 1887, fell snowflakes 15 inches across, and 8 inches thick." (Monthly Weather Review, 1915-73.)

In the topography of intellection, I should say that what we call knowledge is ignorance surrounded by laughter.

Black rains – red rains – the fall of a thousand tons of butter.

Jet-black snow – pink snow – blue hailstones – hailstones flavored like oranges.

Punk and silk and charcoal.

About one hundred years ago, if anyone was so credulous as to think that stones had ever fallen from the sky, he was reasoned with:

In the first place there are no stones in the sky:

Therefore no stones can fall from the sky.

Or nothing more reasonable or scientific or logical than that could be said upon any subject. The only trouble is the universal trouble:

that the major premise is not real, or is intermediate somewhere between realness and unrealness.

In 1772, a committee, of whom Lavoisier was a member, was appointed by the French Academy, to investigate a report that a stone had fallen from the sky at Luce, France. Of all attempts at positiveness, in its aspect of isolation, I don't know of anything that has been fought harder for than the notion of this earth's unrelatedness. Lavoisier analyzed the stone of Luce. The exclusionists' explanation at that time was that stones do not fall from the sky: that luminous objects may seem to fall, and that hot stones may be picked up where a luminous object seemingly had landed – only lightning striking a stone, heating, even melting it.

The stone of Luce showed signs of fusion.

Lavoisier's analysis "absolutely proved" that this stone had not fallen: that it had been struck by lightning.

So, authoritatively, falling stones were damned. The stock means of exclusion remained the explanation of lightning that was seen to strike something – that had been upon the ground in the first place.

But positiveness and the fate of every positive statement. It is not customary to think of damned stones raising an outcry against a sentence of exclusion, but, subjectively, aerolites did or data of them bombarded the walls raised against them –

Monthly Review, 1796-426:

"The phenomenon which is the subject of the remarks before us will seem to most persons as little worthy of credit as any that could be offered. The falling of large stones from the sky, without any assignable cause of their previous ascent, seems to partake so much of the marvellous as almost entirely to exclude the operation of known and natural agents. Yet a body of evidence is here brought to prove that such events have actually taken place, and we ought not to withhold from it a proper degree of attention."

The writer abandons the first, or absolute, exclusion, and modifies it with the explanation that the day before a reported fall of stones in Tuscany, June 16, 1794, there had been an eruption of Vesuvius –

Or that stones do fall from the sky, but that they are stones that have been raised to the sky from some other part of the earth's surface by whirlwinds or by volcanic action.

It's more than one hundred and twenty years later. I know of no aerolite that has ever been acceptably traced to terrestrial origin.

Falling stones had to be undamned though still with a reservation that held out for exclusion of outside forces.

One may have the knowledge of a Lavoisier, and still not be able to analyze, not be able even to see, except conformably with the hypnoses, or the conventional reactions against hypnoses, of one's era. We believe no more.

We accept.

Little by little the whirlwind and volcano explanations had to be abandoned, but so powerful was this exclusion-hypnosis, sentence of damnation, or this attempt at positiveness, that far into our own times some scientists, notably Prof. Lawrence Smith and Sir Robert Ball, continued to hold out against all external origins, asserting that nothing could fall to this earth, unless it had been cast up or whirled up from some other part of this earth's surface.

It's as commendable as anything ever has been – by which I mean it's intermediate to the commendable and the censurable.

It's virginal.

Meteorites, data of which were once of the damned, have been admitted, but the common impression of them is only a retreat of attempted exclusion: that only two kinds of substance fall from the sky: metallic and stony: that the metallic objects are of iron and nickel –

Butter and paper and wool and silk and resin.

We see, to start with, that the virgins of science have fought and wept and screamed against external relations upon two grounds: There in the first place;

Or up from one part of this earth's surface and down to another.

At late as November, 1902, in Nature Notes, 13-231, a member of the Selborne Society still argued that meteorites do not fall from the sky; that they are masses of iron upon the ground "in the first place," that attract lightning; that the lightning is seen, and is mistaken for a falling, luminous object –

By progress we mean rape.

Butter and beef and blood and a stone with strange inscriptions upon it.

Chapter III

So then it is our expression that Science relates to real knowledge no more than does the growth of a plant, or the organization of a department store, or the development of a nation: that all are assimilative, or organizing, or systematizing processes that represent different attempts to attain the positive state – the state commonly called heaven, I suppose I mean.

There can be no real science where there are indeterminate variables, but every variable is, in finer terms, indeterminate, or irregular, if only to have the appearance of being in Intermediateness – is to express regularity unattained. The invariable, or the real and stable, would be nothing at all in Intermediateness – rather as, but in relative terms, an undistorted interpretation of external sounds in the mind of a dreamer could not continue to exist in a dreaming mind, because that touch of relative realness would be of awakening and not of dreaming. Science is the attempt to awaken to realness, wherein it is attempt to find regularity and uniformity. Or the regular and uniform would be that which has nothing external to disturb it. By the universal we mean the real. Or the notion is that the underlying super-attempt, as expressed in Science, is indifferent to the subject-matter of Science: that the attempt to regularize is the vital spirit. Bugs and stars and chemical messes: that they are only quasi-real, and that of them there is nothing real to know; but that systematization of pseudo-data is approximation to realness or final awakening –

Or a dreaming mind and its centaurs and canary birds that turn into giraffes - there could be no real biology upon such subjects, but attempt, in a dreaming mind, to systematize such appearances would be movement toward awakening if better mental co-ordination is all that we mean by the state of being awake – relatively awake.

So it is, that having attempted to systematize, by ignoring externality to the greatest possible degree, the notion of things dropping in upon this earth, from externality, is as unsettling and as unwelcome to Science as tin horns blowing in upon a musician's relatively symmetric composition - flies alighting upon a painter's attempted harmony, and tracking colors one into another – suffragist getting up and making a political speech at a prayer meeting.

If all things are of a oneness, which is a state intermediate to unrealness and realness, and if nothing has succeeded in breaking away and establishing entity for itself, and could not continue to "exist" in intermediateness, if it should succeed, any more than could the bora still at the same time be the uterine, I of course know of no positive difference between Science and Christian Science – and the attitude of both toward the unwelcome is the same – "it does not exist."

A Lord Kelvin and a Mrs. Eddy, and something not to their liking - it does not exist.

Of course not, we Intermediates say: but, also, that, in Intermediateness, neither is there absolute non-existence.

Or a Christian Scientist and a toothache – neither exists in the final sense: also neither is absolutely non-existent, and, according to our therapeutics, the one that more highly approximates to realness will win.

A secret of power –

I think it's another profundity.

Do you want power over something?

Be more nearly real than it.

We'll begin with yellow substances that have fallen upon this earth: we'll see whether our data of them have a higher approximation to realness than have the dogmas of those who deny their existence – that is, as products from somewhere external to this earth.

In mere impressionism we take our stand. We have no positive tests nor standards. Realism in art: realism in science they pass away. In 1859, the thing to do was to accept Darwinism; now many biologists are revolting and trying to conceive of something else. The thing to do was to accept it in its day, but Darwinism of course was never proved:

The fittest survive.

What is meant by the fittest?

Not the strongest; not the cleverest –

Weakness and stupidity everywhere survive.

There is no way of determining fitness except in that a thing does survive.

"Fitness," then, is only another name for "survival."

Darwinism:

That survivors survive.

Although Darwinism, then, seems positively baseless, or absolutely irrational, its massing of supposed data, and its attempted coherence approximate more highly to Organization and Consistency than did the inchoate speculations that preceded it.

Or that Columbus never proved that the earth is round. Shadow of the earth on the moon?

No one has ever seen it in its entirety. The earth's shadow is much larger than the moon. If the periphery of the shadow is curved – but the convex moon – a straight-edged object will cast a curved shadow upon a surface that is convex. All the other so-called proofs may be taken up in the same way.

It was impossible for Columbus to prove that the earth is round. It was not required: only that with a higher seeming of positiveness than that of his opponents, he should attempt. The thing to do, in 1492, was nevertheless to accept that beyond Europe, to the west, were other lands. I offer for acceptance, as something concordant with the spirit of this first quarter of the 20th century, the expression that beyond this earth are – other lands – from which come things as, from America, float things to Europe.

As to yellow substances that have fallen upon this earth, the endeavor to exclude extra-mundane origins is the dogma that all yellow rains and yellow snows are colored with pollen from this earth's pine trees. Symons' Meteorological Magazine is especially prudish in this respect and regards as highly improper all advances made by other explainers.

Nevertheless, the Monthly Weather Review, May, 1877, reports a golden-yellow fall, of Feb. 27, 1877, at Peckloh, Germany, in which four kinds of organisms, not pollen, were the coloring matter. There were minute things shaped like arrows, coffee beans, horns, and disks.

They may have been symbols. They may have been objective Hieroglyphics –

Mere passing fancy – let it go –

In the Annales de Chimie, 85-288, there is a list of rains said to have contained sulphur. I have thirty or forty other notes. I'll not use one of them. I'll admit that every one of them is upon a fall of pollen. I said, to begin with, that our methods would be the methods of theologians and scientists, and they always begin with an appearance of liberality. I grant thirty or forty points to start with. I'm as liberal as any of them – or that my liberality won't cost me anything – the enormousness of the data that we shall have.

Or just to look over a typical instance of this dogma, and the way it works out:

In the American Journal of Science, 1-42-196, we are told of a yellow substance that fell by the bucketful upon a vessel, one "windless" night in June, in Pictou Harbor, Nova Scotia. The writer analyzed the substance, and it was found to "give off nitrogen and ammonia and an animal odor."

Now, one of our Intermediatist principles, to start with, is that so far from positive, in the aspect of Homogeneousness, are all substances, that, at least in what is called an elementary sense, anything can be found anywhere. Mahogany logs on the coast of Greenland; bugs of a valley on the top of Mt. Blanc; atheists at a prayer meeting; ice in India. For instance, chemical analysis can reveal that almost any dead man was poisoned with arsenic, we'll say, because there is no stomach without some iron, lead, tin, gold, arsenic in it and of it – which, of course, in a broader sense, doesn't matter much, because a certain number of persons must, as a restraining influence, be executed 'for murder every year; and, if detectives aren't able really to detect anything, illusion of their success is all that is necessary, and it is very honorable to give up one's life for society as a whole.

The chemist who analyzed the substance of Pictou sent a sample to the Editor of the Journal. The Editor of course found pollen in it.

My own acceptance is that there'd have to be some pollen in it:

that nothing could very well fall through the air, in June, near the pine forests of Nova Scotia, and escape all floating spores of pollen.

But the Editor does not say that this substance "contained" pollen.

He disregards "nitrogen, ammonia, and an animal odor," and says that the substance was pollen. For the sake of our thirty or forty tokens of liberality, or pseudo-liberality, if we can't be really liberal, we grant that the chemist of the first examination probably wouldn't know an animal odor if he were janitor of a menagerie.

As we go along, however, there can be no such sweeping ignoring of this phenomenon:

The fall of animal-matter from the sky.

I'd suggest, to start with, that we'd put ourselves in the place of deep-sea fishes:

How would they account for the fall of animal-matter from above?

They wouldn't try –

Or it's easy enough to think of most of us as deep-sea fishes of a kind.

Jour. Franklin Inst., 90-11:

That, upon the 4th of February, 1870, there fell, at Genoa, Italy, according to Director Boccardo, of the Technical Institute of Genoa, and Prof. Castellani, a yellow substance. But the microscope revealed numerous globules of cobalt blue, also corpuscles of a pearly color that resembled starch. See Nature, 2-166.

Comptes Rendus, 56-972:

M. Bouis says of a substance, reddish varying to yellowish, that fell enormously and successively, or upon April 30, May 1 and May 2, in France and Spain, that it carbonized and spread the odor of charred animal matter – that it was not pollen – that in alcohol it left a residue of resinous matter. Hundreds of thousands of tons of this matter must have fallen.

"Odor of charred animal matter."

Or an aerial battle that occurred in inter-planetary space several hundred years ago – effect of time in making diverse remains uniform in appearance –

It's all very absurd because, even though we are told of a prodigious quantity of animal matter that fell from the sky – three days – France and Spain – we're not ready yet: that's all. M. Bouis says that this substance was not pollen; the vastness of the fall makes acceptable that it was not pollen; still, the resinous residue does suggest pollen of pine trees. We shall hear a great deal of a substance with a resinous residue that has fallen from the sky:

finally we shall divorce it from all suggestion of pollen.

Blackwood's Magazine, 3-338:

A yellow powder that fell at Gerace, Calabria, March 14, 1813.

Some of this substance was collected by Sig. Simenini, Professor of Chemistry, at Naples. It had an earthy, insipid taste, and is described as "unctuous." When heated, this matter turned brown, then black, then red. According to the Annals of Philosophy, 11-466, one of the components was a greenish-yellow substance, which, when dried, was found to be resinous.

But concomitants of this fall:

Loud noises were heard in the sky.

Stones fell from the sky.

According to Chladni, these concomitants occurred, and to me they seem – rather brutal? – or not associable with something so soft and gentle as a fall of pollen?

Black rains and black snows rains as black as a deluge of ink – jet-black snowflakes.

Such a rain as that which fell in Ireland, May 14, 1849, described in the Annals of Scientific Discovery, 1850, and the Annual Register, 1849. It fell upon a district of 400 square miles, and was the color of ink, and of a fetid odor and very disagreeable taste. The rain at Castlecommon, Ireland, April 30, 1887 "thick, black ram." (Amer. Met. Jour., 4-193.)

A black rain fell in Ireland, Oct. 8 and 9, 1907. (Symons' Met. Mag. 43-2.) "It left a most peculiar and disagreeable smell in the air."

The orthodox explanation of this rain occurs in Nature, March 2, 1908 – cloud of soot that had come from South Wales, crossing the Irish Channel and all of Ireland.

So the black rain of Ireland, of March, 1898: ascribed in Symons' Met. Mag. 33-40, to clouds of soot from the manufacturing towns of North England and South Scotland.

Our Intermediatist principle of pseudo-logic, or our principle of Continuity is, of course, that nothing is unique, or individual: that all phenomena merge away into all other phenomena: that, for instance - suppose there should be vast celestial super-oceanic, or inter-planetary vessels that come near this earth and discharge volumes of smoke at times. We're only supposing such a thing as that now, because, conventionally, we are beginning modestly and tentatively. But if it were so, there would necessarily be some phenomenon upon this earth, with which that phenomenon would merge.

Extra-mundane smoke and smoke from cities merge, or both would manifest in black precipitations in rain.

In Continuity, it is impossible to distinguish phenomena at their merging-points, so we look for them at their extremes. Impossible to distinguish between animal and vegetable in some infusoria – but hippopotamus and violet. For all practical purposes they're distinguishable enough. No one but a Barnum or a Bailey would send one a bunch of hippopotami as a token of regard.

So away from the great manufacturing centers:

Black rain in Switzerland, Jan. 20, 1911. Switzerland is so remote, and so ill at ease is the conventional explanation here, that Nature, 85-451, says of this rain that in certain conditions of weather, snow may take on an appearance of blackness that is quite deceptive.

May be so. Or at night, if dark enough, snow may look black.

This is simply denying that a black rain fell in Switzerland, Jan.

20, 1911.

Extreme remoteness from great manufacturing centers:

La Nature, 1888, 2-406:

That Aug. 14, 1888, there fell at the Cape of Good Hope, a rain so black as to be described as a "shower of ink."

Continuity dogs us. Continuity rules us and pulls us back. We seemed to have a little hope that by the method of extremes we could get away from things that merge indistinguishably into other things. We find that every departure from one merger is entrance upon another. At the Cape of Good Hope, vast volumes of smoke from great manufacturing centers, as an explanation, can not very acceptably merge with the explanation of extra-mundane origin – but smoke from a terrestrial volcano can, and that is the suggestion that is made in La Nature.

There is, in human intellection, no real standard to judge by, but our acceptance, for the present, is that the more nearly positive will prevail. By the more nearly positive we mean the more nearly Organized. Everything merges away into everything else, but proportionately to its complexity, if unified, a thing seems strong, real, and distinct: so, in aesthetics, it is recognized that diversity in unity is higher beauty, or approximation to Beauty, than is simpler unity; so the logicians feel that agreement of diverse data constitute greater convincingness, or strength, than that of mere parallel instances: so to Herbert Spencer the more highly differentiated and integrated is the more fully evolved. Our opponents hold out for mundane origin of all black rains. Our method will be the presenting of diverse phenomena in agreement with the notion of some other origin. We take up not only black rains but black rains and their accompanying phenomena.

A correspondent to Knowledge, 5-190, writes of a black rain that fell in the Clyde Valley, March 1, 1884: of another black rain that fell two days later. 'According to the correspondent, a black rain had fallen in the Clyde Valley, March 20, 1828: then again March 22, 1828. According to Nature, 9-43, a. black rain fell at Marlsford, England, Sept. 4, 1873; more than twenty-four hours later another black rain fell in the same small town.

The black rains of Slains:

According to Rev. James Rust (Scottish Showers):

A black rain at Slains, Jan. 14, 1862 – another at Carluke, 140 miles from Slains, May 1, 1862 – at Slains, May 20, 1862 – Slains, Oct. 28, 1863.

But after two of these showers, vast quantities of a substance described sometimes as "pumice stone," but sometimes as "slag," were washed upon the sea coast near Slains. A chemist's opinion is given that this substance was slag: that it was not a volcanic product: slag from smelting works. We now have, for black rains, a concomitant that is irreconcilable with origin from factory chimneys. Whatever it may have been the quantity of this substance was so enormous that, in Mr. Rust's opinion, to have produced so much of it would have required the united output of all the smelting works in the world. If slag it were, we accept that an artificial product has, in enormous quantities, fallen from the sky. If you don't think that such occurrences are damned by Science, read Scottish Showers and see how impossible it was for the author to have this matter taken up by the scientific world.

The first and second rains corresponded, in time, with ordinary ebullitions of Vesuvius.

The third and fourth, according to Mr. Rust, corresponded with no known volcanic activities upon this earth.

La Science Pour Tons, 11-26:

That, between October, 1863, and January, 1866, four more black rains fell at Slains, Scotland. The writer of this supplementary account tells us, with a better, or more unscrupulous, orthodoxy than Mr. Rust's, that of the eight black rains, five coincided with eruptions of Vesuvius and three with eruptions of Etna.

The fate of all explanation is to close one door only to have another fly wide open. I should say that my own notions upon this subject will be considered irrational, but at least my gregariousness is satisfied in associating here with the preposterous or this writer, and those who think in his rut, have to say that they can think of four discharges from one far-distant volcano, passing over a great part of Europe, precipitating nowhere else, discharging precisely over one small northern parish –

But also of three other discharges, from another far-distant volcano, showing the same precise preference, if not marksmanship, for one small parish in Scotland.

Nor would orthodoxy be any better off in thinking of exploding meteorites and their debris: preciseness and recurrence would be just as difficult to explain. My own notion is of an island near an oceanic trade-route: it might receive debris from passing vessels seven times in four years.

Other concomitants of black rains:

In Timb's Year Book, 1851-270, there is an account of "a sort of rumbling, as of wagons, heard for upward of an hour without ceasing," July 16, 1850, Bulwick Rectory, Northampton, England.

On the 1 9th, a black rain fell.

In Nature, 30-6, a correspondent writes of an intense darkness at Preston, England, April 26, 1884: page 32, another correspondent writes of black rain at Crowle, near Worcester, April 26: that a week later, or May 3, it had fallen again: another account of black rain, upon the 28th of April, near Church Shetton, so intense that the following day brooks were still dyed with it. According to four accounts by correspondents to Nature there were earthquakes in England at this time.

Or the black rain of Canada, Nov. 9, 1819. This time it is orthodoxy to attribute the black precipitate to smoke of forest fires south of the Ohio River –

Zurcher, Meteors, p. 238:

That this black rain was accompanied by "shocks like those of an earthquake."

Edinburgh Philosophical Journal, 2-381:

That the earthquake had occurred at the climax of intense darkness and the fall of black rain.

Red rains.

Orthodoxy:

Sand blown by the sirocco, from the Sahara to Europe.

Especially in the earthquake regions of Europe, there have been many falls of red substance, usually, but not always, precipitated in rain. Upon many occasions, these substances have been "absolutely identified" as sand from the Sahara. When I first took this matter up, I came across assurance after assurance, so positive to this effect, that, had I not been an Intermediatist, I'd have looked n further. Samples collected from a rain at Genoa - samples of sand forwarded from the Sahara "absolute agreement" some writers said: same color, same particles of quartz, even the same shells of diatoms mixed in. Then the chemical analyses: not a disagreement worth mentioning.

Our intermediatist means of expression will be that, with proper exclusions, after the scientific or theological method, anything can be identified with anything else, if all things are only different expressions of an underlying oneness. To many minds there's rest and there's satisfaction in that expression, "absolutely identified." Absoluteness, or the illusion of it - the universal quest. If chemists have identified substances that have fallen in Europe as sand from African deserts, swept up in African whirlwinds, that's assuasive to all the irritations that occur to those cloistered minds that must repose in the concept of a snug, isolated, little world, free from contact with cosmic wicked-nesses, safe from stellar guile, undisturbed by inter-planetary prowlings and invasions. The only trouble is that a chemist's analysis, which seems so final and authoritative to some minds, is no more nearly absolute than is identification by a child or description by an imbecile – I take some of that back: I accept that the approximation is higher –

But that it's based upon delusion, because there is no definiteness, no homogeneity, no stability, only different stages somewhere between them and indefiniteness, heterogeneity, and instability. There are no chemical elements. It seems acceptable that Ramsay and others have settled that. The chem-ical elements are only another disappointment in the quest for the positive, as the definite, the ho-mogeneous, and the stable. If there were real elements, there could be a real science of chemistry.

Upon Nov. 12 and 13, 1902, occurred the greatest fall of matter in the history of Australia. Upon the 14th of November, it "rained mud," in Tasmania. It was of course attributed to Australian whirl-winds, but, according to the Monthly Weather Review, 32- 365, there was a haze all the way to the Philippines, also as far as Hong Kong. It may be that this phenomenon had no especial relation with the even more tremendous fall of matter that occurred in Europe, February, 1903.

For several days, the south of England was a dumping ground - from somewhere.

If you'd like to have a chemist's opinion, even though it's only a chemist's opinion, see the report of the meeting of the Royal Chemical Society, April 2, 1903. Mr. E. G. Clayton read a paper upon some of the substance that had fallen from the sky, collected by him. The Sahara explanation applies mostly to falls that occur in southern Europe. Farther away, the conventionalists are a little uneasy: for instance, the editor of the Monthly Weather Review, 29-121, says of a red rain that fell near the coast of Newfoundland, early in 1890: "It would be very remarkable if this was Sahara dust." Mr. Clayton said that the matter examined by him was "merely wind-borne dust from the roads and lanes of Wessex." This opinion is typical of all scientific opinion - or theological opinion - or femi-nine opinion - all very well except for what it disregards.

The most charitable thing I can think of - because I think it gives us a broader tone to relieve our malices with occasional charities - is that Mr. Clayton had not heard of the astonishing extent of this fall - had covered the Canary Islands, on the 19th, for instance.

I think, myself, that in 1903, we passed through the remains of a powdered world - left over from an ancient inter-planetary dispute, brooding in space like a red resentment ever since. Or, like every other opinion, the notion of dust from Wessex turns into a provincial thing when we look it over.

To think is to conceive incompletely, because all thought relates only to the local. We metaphysi-cians, of course, like to have the notion that we think of the unthinkable. As to opinions, or pro-nouncements, I should say, because they always have such an authoritative air, of other chemists, there is an analysis in Nature, 68-54, giving water and organic matter at 9.08 per cent. It's that car-rying out of fractions that's so convincing.

The substance is identified as sand from the Sahara. The vastness of this fall. In Nature, 68-65, we are told that it had occurred in Ireland, too. The Sahara, of course - because, prior to Feb. 19, there had been dust storms in the Sahara disregarding that in that great region there's always, in some part of it, a dust storm. However, just at present, it does look reasonable that dust had come from Africa, via the Canaries.

The great difficulty that authoritativeness has to contend with is some other authoritativeness. When an infallibility clashes with a pontification –

They explain. Nature, March 5, 1903:

Another analysis - 36 per cent organic matter.

Such disagreements don't look very well, so, in Nature, 68-109, one of the differing chemists explains. He says that his analysis was of muddy rain, and the other was of sediment of rain –

We're quite ready to accept excuses from the most high, though I do wonder whether we're quite so damned as we were, if we find ourselves in a gracious and tolerant mood toward the powers that condemn but the tax that now comes upon our good manners and unwillingness to be too severe –

Nature, 68-223:

Another chemist. He says it was 23.49 per cent water and organic matter.

He "identifies" this matter as sand from an African desert - but after deducting organic matter –

But you and I could be "identified" as sand from an African desert, after deducting all there is to us except sand –

Why we can not accept that this fall was of sand from the Sahara, omitting the obvious objection that in most parts the Sahara is not red at all, but is usually described as "dazzling white" –

The enormousness of it: that a whirlwind might have carried it, but that, in that case it would be no supposititious, or doubtfully identified whirlwind, but the greatest atmospheric cataclysm in the history of this earth:

Jour. Roy. Met. Soc., 30-56:

That, up to the 27th of February, this fall had continued in Belgium, Holland, Germany and Austria; that in some instances it was not sand, or that almost all the matter was organic: that a vessel had reported the fall as occurring in the Atlantic Ocean, midway between Southampton and the Barbados. The calculation is given that, in England alone, 10,000,000 tons of matter had fallen. It had fallen in Switzerland (Symons' Met. Mag., March, 1903). It had fallen in Russia (Bull. Com. Geolog., 22-48). Not only had a vast quantity of matter fallen several months before, in Australia, but it was at this time falling in Australia (Victorian Naturalist, June, 1903) - enormously - red mud - fifty tons per square mile.

The Wessex explanation -

Or that every explanation is a Wessex explanation: by that I mean an attempt to interpret the enormous in terms of the minute but that nothing can be finally explained, because by Truth we mean the Universal; and that even if we could think as wide as Universality, that would not be requital to the cosmic quest which is not for Truth, but for the local that is true - not to universalize the local but to localize the universal - or to give to a cosmic cloud absolute interpretation in terms of the little dusty roads and lanes of Wessex. I can not conceive that this can be done: I think of high approximation.

Our Intermediatist concept is that, because of the continuity of all "things," which are not separate, positive, or real things, all pseudo-things partake of the underlying, or are only different expressions, degrees, or aspects of the underlying: so then that a sample from somewhere in anything must correspond with a sample from somewhere in anything else.

That, by due care in selection, and disregard for everything else, or the scientific and theological method, the substance that fell, February, 1903, could be identified with anything, or with some part or aspect of anything that could be conceived of –

With sand from the Sahara, sand from a barrel of sugar, or dust of your great, great grandfather.

Different samples are described and listed hi the Journal of the Royal Meteorological Society, 30-57 - or we'll see whether my notion that a chemist could have identified some one of these samples as from anywhere conceivable, is extreme or not:

"Similar to brick dust," in one place; "buff or light brown, in another place; "chocolate-colored and silky to the touch and slightly iridescent"; "gray"; "red-rust color"; "reddish raindrops and gray sand"; "dirty gray"; "quite red"; "yellow-brown, with a tinge of pink"; "deep yellow-clay color."

In Nature, it is described as of a peculiar yellowish cast in one place, reddish somewhere else, and salmon-colored in another place.

Or there could be real science if there were really anything to be scientific about.

Or the science of chemistry is like a science of sociology, prejudiced in advance, because only to see is to see with a prejudice, setting out to "prove" that all inhabitants of New York came from Africa. Very easy matter. Samples from one part of town. Disregard for all the rest.

There is no science but Wessex-science.

According to our acceptance, there should be no other, but that approximation should be higher: that metaphysics is super-evil: that the scientific spirit is of the cosmic quest.

Our notion is that, in a real existence, such a quasi-system of fables as the science of chemistry could not deceive for a moment: but that in an "existence" endeavoring to become real, it represents that endeavor, and will continue to impose its pseudo-positiveness until it be driven out by a higher approximation to realness;

That the science of chemistry is as impositive as fortune-telling -

Or no -

That, though it represents a higher approximation to realness than does alchemy, for instance, and so drove out alchemy, it is still only somewhere between myth and positiveness.

The attempt at realness, or to state a real and unmodified fact here, is the statement:

All red rains are colored by sands from the Sahara desert.

My own impositivist acceptances are:

That some red rains are colored by sands from the Sahara desert;

Some by sands from other terrestrial sources;

Some by sands from other worlds, or from their deserts also from aerial regions too indefinite or amorphous to be thought of as "worlds" or planets –

That no supposititious whirlwind can account for the hundreds of millions of tons of matter that fell upon Australia, Pacific Ocean and Atlantic Ocean and Europe in 1902 and 1903 - that a whirlwind that could do that would not be supposititious.

But now we shall cast off some of our own wessicality by accepting that there have been falls of red substance other than sand.

We regard every science as an expression of the attempt to be real. But to be real is to localize the universal - or to make some one thing as wide as all things - successful accomplishment of which I cannot conceive of. The prime resistance to this endeavor is the refusal of the rest of the universe to be damned, excluded, disregarded, to receive Christian Science treatment, by something else so attempting.

Although all phenomena are striving for the Absolute - or have surrendered to and have incorporated themselves in higher attempts, simply to be phenomenal, or to have seeming in Intermediateness is to express relations.

A river.

It is water expressing the gravitational relation of different levels.

The water of the river.

Expression of chemic relations of hydrogen and oxygen which are not final.

A city.

Manifestation of commercial and social relations.

How could a mountain be without base in a greater body?

Storekeeper live without customers?

The prime resistance to the positivist attempt by Science is its relations with other phenomena, or that it only expresses those relations in the first place. Or that a Science can have seeming, or survive in Intermediateness, as something pure, isolated, positively different, no more than could a river or a city or a mountain or a store.

This Intermediateness-wide attempt by parts to be wholes which can not be realized in our quasi-state, if we accept that in it the co-existence of two or more wholes or universals is impossible high approximation to which, however, may be thinkable -

Scientists and their dream of "pure science."

Artists and their dream of "art for art's sake."

It is our notion that if they could almost realize, that would be almost realness: that they would instantly be translated into real existence. Such thinkers are good positivists, but they are evil in an economic and sociologic sense, if, in that sense, nothing has justification for being, unless it serve, or function for, or express the relations, of some higher aggregate. So Science functions for and serves society at large, and would, from society at large, receive no support, unless it did so divert itself or dissipate and prostitute itself. It seems that by prostitution I mean usefulness.

There have been red rains that, in the middle ages, were called "rains of blood." Such rains terrified many persons, and were so unsettling to large populations, that Science, in its sociologic relations, has sought, by Mrs. Eddy's method, to remove an evil -

That "rains of blood" do not exist;

That rains so called are only of water colored by sand from the

Sahara desert.

My own acceptance is that such assurances, whether fictitious or not, whether the Sahara is a "dazzling white" desert or not, have wrought such good effects, in a sociologic sense, even though prostitutional in the positivist sense, that, in the sociologic sense, they were well justified;

But that we've gone on: that this is the twentieth century; that most of us have grown up so that such soporifics of the past are no longer necessary:

That if gushes of blood should fall from the sky upon New York City, business would go on as usual.

We began with rains that we accepted ourselves were, most likely, only of sand. In my own still immature hereticalness - and by heresy, or progress, I mean, very largely, a return, though with many modifications, to the superstitions of the past, I think I feel considerable aloofness to the idea of rains of blood. Just at present, it is my conservative, or timid purpose, to express only that there have been red rains that very strongly suggest blood or finely divided animal matter -

Debris from inter-planetary disasters.

Aerial battles.

Food-supplies from cargoes of super-vessels, wrecked in interplanetary traffic.

There was a red rain in the Mediterranean region, March 6, 1888.

Twelve days later, it fell again. Whatever this substance may have been, when burned, the odor of animal matter from it was strong and persistent. (L' Astronomie, 1888-205.)

But - infinite heterogeneity - or debris from many different kinds of aerial cargoes - there have been red rains that have been colored by neither sand nor animal matter.

Annals of Philosophy, 16-226:

That, Nov. 2, 1819 - week before the black rain and earthquake of Canada - there fell, at Blanken-berge, Holland, a red rain. As to sand, two chemists of Bruges concentrated 144 ounces of the rain to 4 ounces - "no precipitate fell." But the color was so marked that had there been sand, it would have been deposited, if the substance had been diluted instead of concentrated. Experiments were made, and various reagents did cast precipitates, but other than sand. The chemists concluded that the rain-water contained muriate of cobalt - which is not very enlightening: that could be said of many substances carried in vessels upon the Atlantic Ocean. Whatever it may have been, in the An-nales de Chimie, 2-12-432, its color is said to have been red-violet. For various chemic reactions, see Quar. Jour. Roy. Inst., 9-202, and Edin. Phil. Jour., 2-381.

Something that fell with dust said to have been meteoric, March 9, 10, n, 1872: described in the Chemical News, 25-300, as a "peculiar substance," consisted of red iron ochre, carbonate of lime, and organic matter.

Orange-red hail, March 14, 1873, in Tuscany. (Notes and Queries, 9-5-16.)

Rain of lavender-colored substance, at Oudon, France, Dec. 19, 1903. (Bull. Soc. Met. de France, 1904-124.) La Nature, 1885-2-351:

That, according to Prof. Schwedoff, there fell, in Russia, June 14, 1880, red hailstones, also blue hailstones, also gray hailstones.

Nature, 34-123:

A correspondent writes that he had been told by a resident of a small town in Venezuela, that there, April 17, 1886, had fallen hailstones, some red, some blue, some whitish: informant said to have been one unlikely ever to have heard of the Russian phenomenon; described as an "honest, plain countryman."

Nature, July 5, 1877, quotes a Roman correspondent to the London Times who sent a translation from an Italian newspaper: that a red rain had fallen in Italy, June 23, 1877, containing "microscop-ically small particles of sand."

Or, according to our acceptance, any other story would have been an evil thing, in the sociologic sense, in Italy, in 1877. But the English correspondent, from a land where terrifying red rains are uncommon, does not feel this necessity. He writes: "I am by no means satisfied that the rain was of sand and water." His observations are that drops of this rain left stains "such as sandy water could not leave." He notes that when the water evaporated, no sand was left behind.

L'Annee Scientifique, 1888-75: That, Dec. 13, 1887, there fell, in Cochin China, a substance like blood, somewhat coagulated.

Annales de Chimie, 85-266:

That a thick, viscous, red matter fell at Ulm, in 1812.

We now have a datum with a factor that has been foreshadowed;

which will recur and recur and recur throughout this book. It is a factor that makes for speculation so revolutionary that it will have to be re-enforced many times before we can take it into full acceptance.

Year Book of Facts, 1861-273:

Quotation from a letter from Prof. Campini to Prof. Matteucci:

That, upon Dec. 28, 1860, at about 7 a. m., in the northwestern part of Siena, a reddish rain fell copiously for two hours.

A second red shower fell at n o'clock.

Three days later, the red rain fell again.

The next day another red rain fell.

Still more extraordinarily:

Each fall occurred in "exactly the same quarter of town."

Chapter IV

It is in the records of the French Academy that, upon March 17, 1669, in the town of Chatillon-sur-Seine, fell a reddish substance that was "thick, viscous, and putrid."

American Journal of Science, 1-41-404:

Story of a highly unpleasant substance that had fallen from the sky, in Wilson County, Tennessee. We read that Dr. Troost visited the place and investigated. Later we're going to investigate some investigations - but never mind that now. Dr. Troost reported that the substance was clear blood and portions of flesh scattered upon tobacco fields. He argued that a whirlwind might have taken an animal up from one place, mauled it around, and have precipitated its remains somewhere else.

But, in volume 44, page 216, of the Journal, there is an apology.

The whole matter is, upon newspaper authority, said to have been a hoax by negroes, who had pretended to have seen the shower, for the sake of practicing upon the credulity of their masters : that they had scattered the decaying flesh of a dead hog over the tobacco fields.

If we don't accept this datum, at least we see the sociologically necessary determination to have all falls accredited to earthly origins - even when they're falls that don't fall.

Annual Register, 1821-687:

That, upon the 13th of August, 1819, something had fallen from the sky at Amherst, Mass. It had been examined and described by Prof. Graves, formerly lecturer at Dartmouth College. It was an object that had upon it a nap, similar to that of milled cloth. Upon removing this nap, a buff-colored, pulpy substance was found. It had an offensive odor, and, upon exposure to the air, turned to a vivid red. This thing was said to have fallen with a brilliant light.

Also see the Edinburgh Philosophical Journal, 5-295. In the Annales de Chimie, 1821-67, M. Arago accepts the datum, and gives four instances of similar objects or substances said to have fallen from the sky, two of which we shall have with our data of gelatinous, or viscous matter, and two of which I omit, because it.

seems to me that the dates given are too far back.

In the American Journal of Science, 1-2-335, is Professor Graves' account, communicated by Professor Dewey:

That, upon the evening of August 13, 1819, a light was seen in Amherst - a falling object - sound as if of an explosion.

In the home of Prof. Dewey, this light was reflected upon a wall of a room in which were several members of Prof. Dewey's family.

The next morning, in Prof. Dewey's front yard, in what is said to have been the only position from which the light that had been seen in the room, the night before, could have been reflected, was found a substance "unlike anything before observed by anyone who saw it." It was a bowl-shaped object, about 8 inches in diameter, and one inch thick. Bright buff-colored, and having upon it a "fine nap." Upon removing this covering, a buff-colored, pulpy substance of the consistency of soft-soap, was found "of an offensive, suffocating smell."

A few minutes of exposure to the air changed the buff color to "a livid color resembling venous blood." It absorbed moisture quickly from the air and liquefied. For some of the chemic reactions, see the Journal.

There's another lost quasi-soul of a datum that seems to me to belong here:

London Times, April 19, 1836:

Fall of fish that had occurred in the neighborhood of Allahabad, India, It is said that the fish were of the chalwa species, about a span in length and a seer in weight - you know.

They were dead and dry.

Or they had been such a long time out of water that we can't accept that they had been scooped out of a pond, by a whirlwind - even though they were so definitely identified as of a known local species -

Or they were not fish at all.

I incline, myself, to the acceptance that they were not fish, but slender, fish-shaped objects of the same substance as that which fell at Amherst - it is said that, whatever they were, they could not be eaten: that "in the pan, they turned to blood."

For details of this story see the Journal of the Asiatic Society of Bengal, 1834-307. May 16 or 17, 1834, is the date given in the Journal.

In the American Journal of Science, 1-25-362, occurs the inevitable damnation of the Amherst object:

Prof. Edward Hitchcock went to live in Amherst. He says that years later, another object, like the one said to have fallen in 1819, had been found at "nearly the same place." Prof. Hitchcock was invited by Prof. Graves to examine it. Exactly like the first one.

Corresponded in size and color and consistency. The chemic reactions were the same.

Prof. Hitchcock recognized it in a moment.

It was a gelatinous fungus.

He did not satisfy himself as to just the exact species it belonged to, but he predicted that similar fungi might spring up within twenty four hours -

But, before evening, two others sprang up.

Or we've arrived at one of the oldest of the exclusionists' conventions - or nostoc. We shall have many data of gelatinous substance said to have fallen from the sky: almost always the exclusionists argue that it was only nostoc, an Alga, or, in some respects, a fungous growth. The rival convention is "spawn of frogs or of fishes."

These two conventions have made a strong combination. In instances where testimony was not convincing that gelatinous matter had been seen to fall, it was said that the gelatinous substance was nostoc, and had been upon the ground in the first place: when the testimony was too good that it had fallen, it was said to be spawn that had been carried from one place to another in a whirlwind.

Now, I can't say that nostock is always greenish, any more than I can say that blackbirds are always black, having seen a white one: we shall quote a scientist who knew of flesh-colored nostoc, when so to know was convenient. When we come to reported falls of gelatinous substances, I'd like it to be noticed how often they are described as whitish or grayish. In looking up the subject, myself, I have read only of greenish nostoc. Said to be greenish, in Webster's Dictionary - said to be "blue-green" in the New International Encyclopedia "from bright green to olive-green" (Science Gossip, 10-114); "green" (Science Gossip, 7-260); "greenish" (Notes and Queries, 1-11-219). It would seem acceptable that, if many reports of white birds should occur, the birds are not blackbirds, even though there have been white blackbirds. Or that, if often reported, grayish or whitish gelatinous substance is not nostoc, and is not spawn if occurring in times unseasonable for spawn.

"The Kentucky Phenomenon."

So it was called, in its day, and now we have an occurrence that attracted a great deal of attention in its own time. Usually these things of the accursed have been hushed up or disregarded suppressed like the seven black rains of Slains - but, upon March 3, 1876, something occurred, in Bath County, Kentucky, that brought many newspaper correspondents to the scene.

The substance that looked like beef that fell from the sky.

Upon March 3, 1876, at Olympian Springs, Bath County, Kentucky, flakes of a substance that looked like beef fell from the sky - "from a clear sky." We'd like to emphasize that it was said that nothing but this falling substance was visible in the sky. It fell in flakes of various sizes; some two inches square, one, three or four inches square. The flake-formation is interesting: later we shall think of it as signifying pressure - somewhere. It was a thick shower, on the ground, on trees, on fences, but it was narrowly localized: or upon a strip of land about 100 yards long and about 50 yards wide. For the first account, see the Scientific American, 34-197, and the New York Times, March 10, 1876.

Then the exclusionists.

Something that looked like beef: one flake of it the size of a square envelope.

If we think of how hard the exclusionists have fought to reject the coming of ordinary-looking dust from this earth's externality, we can sympathize with them in this sensational instance, perhaps.

^Newspaper correspondents Wrote broadcast and witnesses were quoted, and this time there is no mention of a hoax, and, except by one scientist, there is no denial that the fall did take place.

It seems to me that the exclusionists are still more emphatically conservators. It is not so much that they are inimical to all data of externally derived substances that fall upon this earth, as that they are inimical to all data discordant with a system that does not include such phenomena -

Or the spirit or hope or ambition of the cosmos, which we call attempted positivism: not to find out the new; not to add to what is called knowledge, but to systematize.

Scientific American Supplement, 2-426:

That the substance reported from Kentucky had been examined by Leopold Brandeis.

"At last we have a proper explanation of this much talked of phenomenon."

"It has been comparatively easy to identify the substance and to fix its status. The Kentucky 'wonder' is no more or less than nostoc."

Or that it had not fallen; that it had been upon the ground in the fast place, and had swollen in rain, and, attracting attention by greatly increased volume, had been supposed by unscientific observers to have fallen in rain -

What rain, I don't know.

Also it is spoken of as "dried" several times. That's one of the most important of the details.

But die relief of outraged propriety, expressed in the Supplement, is amusing to some of us, who, I fear, may be a little improper at times. Very spirit of the Salvation Army, when some third-rate scientist comes out with an explanation of the vermiform appendix or the os cocyx that would have been acceptable to Moses.

To give completeness to "the proper explanation," it is said that Mr. Brandeis had identified the substance as "flesh-colored" nostoc.

Prof. Lawrence Smith, of Kentucky, one of the most resolute of the exclusionists:

New York Times, March 12, 1876:

That the substance had been examined and analyzed by Prof. Smith, according to whom, it gave every indication of being the "dried" spawn of some reptile, "doubtless of the frog" or up from one place and down in another. As to "dried," that may refer to condition when Prof. Smith received it.

In the Scientific American Supplement, 2-473, Dr. A. Mead Edwards, President of the Newark Scientific Association, writes that, when he saw Mr. Brandeis' communication, his feeling was of conviction that propriety had been re-established, or that the problem had been solved, as he expresses it: knowing Mr. Brandeis well, he had called upon that upholder of respectability, to see the substance that had been identified as nostoc. But he had also called upon Dr. Hamilton, who had a specimen, and Dr. Hamilton had declared it to be lung-tissue. Dr. Edwards writes of the substance that had so completely, or beautifully - if beauty is completeness - been identified as nostoc "It turned out to be lung tissue also." He wrote to other persons who had specimens, and identified other specimens as masses of cartilage or muscular fibres. "As to whence it came, I have no theory." Nevertheless he endorses the local explanation - and a bizarre thing it is:

A flock of gorged, heavy-weighted buzzards, but far up and invisible in the clear sky -

They had disgorged.

Prof. Fassig lists the substance, in his "Bibliography," as fish spawn. McAtee (Monthly Weather Review, May, 1918), lists it as a jelly-like material, supposed to have been the "dried" spawn either of fishes or of some batrachian.

Or this is why, against the seemingly insuperable odds against all things new, there can be what is called progress -

That nothing is positive, in the aspects of homogeneity and unity:

If the whole world should seem to combine against you, it is only unreal combination, or interme-diateness to unity and disunity. Every resistance is itself divided into parts resisting one another. The simplest strategy seems to be - never bother to fight a thing: set its own parts fighting one another.

We are merging away from carnal to gelatinous substance, and here there is an abundance of in-stances or reports of instances. These data are so improper they're obscene to the science of to-day, but we shall see that science, before it became so rigorous, was not so prudish. Chladni was not, and Greg was not.

I shall have to accept, myself, that gelatinous substance has often fallen from the sky -

Or that, far up, or far away, the whole sky is gelatinous?

That meteors tear through and detach fragments?

That fragments are brought down by storms?

That the twinkling of stars is penetration of light through something that quivers?

I think, myself, that it would be absurd to say that the whole sky is gelatinous: it seems more accept-able that only certain areas are.

Humboldt (Cosmos, 1-119) says that all our data in this respect must be "classed amongst the myth-ical fables of mythology." He is very sure, but just a little redundant.

We shall be opposed by the standard resistances:

There in the first place;

Up from one place, in a whirlwind, and down in another.

We shall not bother to be very convincing one way or another, because of the over-shadowing of the datum with which we shall end up. It will mean that something had been in a stationary position for several days over a small part of a small town in England: this is the revolutionary thing that we have alluded to before; whether the substance were nostoc, or spawn, or some kind of a larval nexus, doesn't matter so much. If it stood in the sky for several days, we rank with Moses as a chronicler of improprieties or was that story, or datum, we mean, told by Moses? Then we shall have so many records of gelatinous substance said to have fallen with meteorites, that, between the two phenom-ena, some of us will have to accept connection or that there are at least vast gelatinous areas aloft, and that meteorites tear through, carrying down some of the substance.

Comptes Rendus, 3-554:

That, in 1836, M. Vallot, member of the French Academy, placed before the Academy some frag-ments of a gelatinous substance, said to have fallen from the sky, and asked that they be analyzed. There is no further allusion to this subject.

Comptes Rendus, 23-542:

That, in Wilna, Lithuania, April 4, 1846, in a rainstorm, fell nut-sized masses of a substance that is described as both resinous and gelatinous. It was odorless until burned: then it spread a very pro-nounced sweetish odor. It is described as like gelatine, but much firmer: but, having been in water 24 hours, it swelled out, and looked altogether gelatinous -

It was grayish.

We are told that, in 1841 and 1846, a similar substance had fallen in Asia Minor.

In Notes and Queries, 8-6-190, it is said that, early in August, 1894, thousands of jelly fish, about the size of a shilling, had fallen at Bath, England. I think it is not acceptable that they were jelly fish: but it does look as if this time frog spawn did fall from the sky, and may have been translated by a whirlwind because, at the same time, small frogs fell at Wigan, England.

Nature, 87-10:

That, June 24, 1911, at Eton, Bucks, England, the ground was found covered with masses of jelly, the size of peas, after a heavy rainfall. We are not told of nostoc, this time: it is said that the object contained numerous eggs of "some species of Chironomus, from which larvae soon emerged."

I incline, then, to think that the objects that fell at Bath were neither jelly fish nor masses of frog spawn, but something of a larval kind -

This is what had occurred at Bath, England, 23 years before.

London Times, April 24, 1871:

That, upon the 22nd of April, 1871, a storm of glutinous drops neither jelly fish nor masses of frog spawn, but something of a railroad station, at Bath. "Many soon developed into a wormlike chrysalis, about an inch in length." The account of this occurrence in the Zoologist, 2-6-2686, is more like the Eton-datum: of minute forms, said to have been infusoria; not forms about an inch in length.

Trans. Ent. Soc. of London, 1871-proc. xxii:

That the phenomenon had been investigated by the Rev. L. Jenyns, of Bath. His description is of minute worms in filmy envelopes. He tries to account for their segregation. The mystery of it is: what could have brought so many of them together? Many other falls we shall have record of, and in most of them segregation is the great mystery. A whirlwind seems anything but a segregative force. Segregation of things that have fallen from the sky has been avoided as most deep-dyed of the damned. Mr. Jenyns conceives of a large pool, in which were many of these spherical masses: of the pool drying up and concentrating all in a small area; of a whirlwind then scooping all up together - But several days later, more of these objects fell in the same place.

That such marksmanship is not attributable to whirlwinds seems to me to be what we think we mean by common sense: It may not look like common sense to say that these things had been stationary over the town of Bath, several days -

The seven black rains of Slains;

The four red rains of Siena.

An interesting sidelight on the mechanics of orthodoxy is that Mr. Jenyns dutifully records the second fall, but ignores it in his explanation.

R. P. Greg, one of the most notable of cataloguers of meteoritic phenomena, records (Phil. Mag.: 4-8-463) falls of viscid substance in the years 1652, 1686, 1718, 1796, 1811, 1819, 1844. He gives earlier dates, but I practice exclusions, myself. In the Report of the British Association, 1860-63, Greg records a meteor that seemed to pass near the ground, between Barsdorf and Freiburg, Germany: the next day a jelly-like mass was found in the snow -

Unseasonableness for either spawn or nostoc.

Greg's comment in this instance is: "curious if true." But he records without modification the fall of a meteorite at Gotha, Germany, Sept. 6, 1835, "leaving a jelly-like mass on the ground."

We are told that this substance fell only three feet away from an observer. In the Report of the British Association, 1855-94, according to a letter from Greg to Prof. Baden-Powell, at night, Oct. 8, 1844, near Coblentz, a German, who was known to Greg, and another person, saw a luminous body fall close to them. They returned next morning and found a gelatinous mass of grayish color.

According to Chladni's account (Annals of Philosophy, n.s., 12-94) a. viscous mass fell with a luminous meteorite between Siena and Rome, May, 1652; viscous matter found after the fall of a fire ball, in Lusatia, March, 1796; fall of a gelatinous substance, after the explosion of a meteorite, near Heidelburg, July, 1811. In the Edinburgh Philosophical Journal, 1-234, the substance that fell at Lusatia is said to have been of the "color and odor of dried, brown varnish." In the Amer. Jour. Sci., 1-26-133, it is said that gelatinous matter fell with a globe of fire, upon the island of Lethy, India, 1718.

In the Amer. Jour. Sci., 1-26-396, in many observations upon the meteors of November, 1833, are reports of falls of gelatinous substance:

That, according to newspaper reports, "lumps of jelly" were found on the ground at Rahway, N. J. The substance was whitish, or resembled the coagulated white of an egg;

That Mr. H. H. Garland, of Nelson County, Virginia, had found a jelly-like substance of about the circumference of a twenty-five cent piece;

That, according to a communication from A. C. Twining to Prof. Olmstead, a woman at West Point, N. Y., had seen a mass the size of a tea cup. It looked like boiled starch;

That, according to a newspaper, of Newark, N. J., a mass of gelatinous substance, like soft soap, had been found. "It possessed little elasticity, and, on the application of heat, it evaporated as readily as water."

It seems incredible that a scientist would have such hardihood, or infidelity, as to accept that these things had fallen from the sky: nevertheless, Prof. Olmstead, who collected these lost souls says:

"The fact that the supposed deposits were so uniformly described as gelatinous substance forms a presumption in favor of the supposition that they had the origin ascribed to them."

In contemporaneous scientific publications considerable attention was given to Prof. Olmstead's series of papers upon this subject of the November meteors. You will not find one mention of the part that treats of gelatinous matter.

Chapter V

I shall attempt not much of correlation of dates. A mathematic–minded positivist, with his delusion that in an intermediate state twice two are four, whereas, if we accept Continuity, we can not accept that there are anywhere two things to start with, would search our data for periodicities. It is so obvious to me that the mathematic, or the regular, is the attribute of the Universal, that I have not much inclination to look for it in the local. Still, in this solar system, "as a whole," there is considerable approximation to regularity; or the mathematic is so nearly localized that eclipses, for instance, can, with rather high approximation, be foretold, though I have notes that would deflate a little the astronomers' vainglory in this respect - or would if that were possible. An astronomer is poorly paid, uncheered by crowds, considerably isolated: he lives upon his own inflations: deflate a bear and it couldn't hibernate. This solar system is like every other phenomenon that can be regarded "as a whole" - or the affairs of a ward are interfered with by the affairs of the city of which it is a part; city by county; county by state; state by nation; nation by other nations; all nations by climatic conditions; climatic conditions by solar circumstances; sun by general planetary circumstances; solar system "as a whole" by other solar systems - so the hopelessness of finding the phenomena of entirety in the ward of a city. But positivists are those who try to find the unrelated in the ward of a city. In our acceptance this is the spirit of cosmic religion. Objectively the state is not realizable in the ward of a city. But, if a positivist could bring himself to absolute belief that he had found it, that would be a subjective realization of that which is unrealizable objectively.

Of course we do not draw a positive line between the objective and the subjective - or that all phenomena called things or persons are subjective within one all-inclusive nexus, and that thoughts within those that are commonly called "persons" are sub-subjective.

It is rather as if Intermediateness strove for Regularity in this solar system and failed: then generated the mentality of astronomers, and, in that secondary expression, strove for conviction that failure had been success.

I have tabulated all the data of this book, and a great deal besides card system and several proximities, thus emphasized, have been revelations to me: nevertheless, it is only the method of theologians and scientists - worst of all, of statisticians.

For instance, by the statistic method, I could "prove" that a black rain has fallen "regularly" every seven months, somewhere upon this earth. To do this, I'd have to include red rains and yellow rains, but, conventionally, I'd pick out the black particles in red substances and in yellow substances, and disregard the rest.

Then, too, if here and there, a black rain should be a week early or a month late - that would be "acceleration" or "retardation."

This is supposed to be legitimate in working out the periodicities of comets. If black rains, or red or yellow rains with black particles in them, should not appear at all near some dates we have not read Darwin in vain - "the records are not complete." As to other, interfering black rains, they'd be either gray or brown, or for them we'd find other periodicities.

Still, I have had to notice the year 1819, for instance. I shall not note them all in this book, but I have records of 31 extraordinary events in 1883. Some one should write a book upon the phenomena of this one year - that is if books should be written.

1849 is notable for extraordinary falls, so far apart that a local explanation seems inadequate not only the black rain of Ireland, May, 1849, but a red rain in Sicily and a red rain in Wales. Also, it is said (Timb's Year Book, 1850-241) that, upon April 18 or 20, 1849, shepherds near Mt. Araat, found a substance that was not indigenous, upon areas measuring 8 to 10 miles in circumference. Presumably it had fallen there.

We have already gone into the subject of Science and its attempted positiveness, and its resistances in that it must have relations of service. It is very easy to see that most of the theoretic science of the 1 9th century was only a relation of reaction against theologic dogma, and has no more to do with Truth than has a wave that bounds back from a shore. Or, if a shop girl, or you or I, should pull out a piece of chewing gum about a yard long, that would be quite as scientific a performance as was the stretching of this earth's age several hundred million of years.

All "things" are not things, but only relations, or expressions of relations: but all relations are striving to be the unrelated, or have surrendered to, and subordinated to, higher attempts. So there is a positivist aspect to this reaction that is itself only a relation, and that is the attempt to assimilate all phenomena under the materialist explanation, or to formulate a final, all-inclusive system, upon the materialist basis. If this attempt could be realized, that would be the attaining of realness; but this attempt can be made only by disregarding psychic phenomena, for instance - or, if science shall eventually give in to the psychic, it would be no more legitimate to explain the immaterial in terms of the material, than to explain the material in terms of the immaterial. Our own acceptance is that material and immaterial are of a oneness, merging, for instance, in a thought that is continuous with a physical action: that oneness can not be explained, because the process of explaining is the interpreting of something in terms of something else. All explanation is assimilation of something in terms of something else that has been taken as a basis: but, in Continuity, there is nothing that is any more basic than anything else - unless we think that delusion built upon delusion is less real than its pseudo-foundation.

In 1829 (Timb's Year Book, 1848-235) in Persia, fell a substance that the people said they had never seen before. As to what it was, they had not a notion, but they saw that the sheep ate it.

They ground it into flour and made bread, said to have been passable enough, though insipid.

That was a chance that science did not neglect. Manna was placed upon a reasonable basis, or was assimilated and reconciled with the system that had ousted the older - and less nearly real - system. It was said that, likely enough, manna had fallen in ancient times because it was still falling but that there was no tutelary influence behind it that it was a lichen from the steppes of Asia Minor " - up from one place in a whirlwind and down in another place." In the American Almanac, 1833-71, it is said that this substance " - unknown to the inhabitants of the region" - was "immediately recognized" by scientists who examined it: and that "the chemical analysis also identified it as a lichen."

This was back in the days when Chemical Analysis was a god.

Since then his devotees have been shocked and disillusioned. Just how a chemical analysis could so botanize, I don't know - but it was Chemical Analysis who spoke, and spoke dogmatically. It seems to me that the ignorance of inhabitants, contrasting with the local knowledge of foreign scientists, is overdone: if there's anything good to eat, within any distance conveniently covered by a whirlwind - inhabitants know it. I have data of other falls, in Persia and Asiatic Turkey, of edible substances. They are all dogmatically said to be "manna"; and "manna" is dogmatically said to be a species of lichens from the steppes of Asia Minor. The position that I take is that this explanation was evolved in ignorance of the fall of vegetable substances, or edible substances, in other parts of the world: that it is the familiar attempt to explain the general in terms of the local ; that, if we shall have data of falls of vegetable substance, in, say, Canada, or India, they were not of lichens from the steppes of Asia Minor; that, though all falls in Asiatic Turkey and Persia are sweepingly and conveniently called showers of "manna," they have not been even all of the same substance. In one instance the particles are said to have been "seeds." Though, in Comptes Rendus, the substance that fell in 1841 and 1846, is said to have been gelatinous, in the Bull. Sci. Nat. de Neuchatel, it is said to have been of something, in lumps the size of a filbert, that had been ground into flour; that of this flour had been made bread, very attractive-looking, but flavorless.

The great difficulty is to explain segregation in these showers -

But deep sea fishes and occasional falls down to them, of edible substances; bags of grain, barrels of sugar; things that had not been whirled up from one part of the ocean-bottom, in storms or submarine disturbances, and dropped somewhere else -

I suppose one thinks but grain in bags never has fallen -

Object of Amherst its covering like "milled cloth" -

Or barrels of corn lost from a vessel would not sink but a host of them clashing together, after a wreck they burst open ; the corn sinks, or does when saturated; the barrel staves float longer -

If there be not an overhead traffic in commodities similar to our own commodities carried over this earth's oceans - I'm not the deep-sea fish I think I am.

I have no data other than the mere suggestion of the Amherst object of bags or barrels, but my notion is that bags and barrels from a wreck on one of this earth's oceans, would, by the time they reached the bottom, no longer be recognizable as bags or barrels; that, if we can have data of the fall of fibrous material that may have been cloth or paper or wood, we shall be satisfactory and grotesque enough.

Proc. Roy. Irish Acad., 1-379:

"In the year 1686, some workmen, who had been fetching water from a pond, seven German miles from Memel, on returning to their work, after dinner (during which there had been a snow storm) found the flat ground around the pond covered with a coal-black, leafy mass; and a person who lived near said he had seen it fall like flakes with the snow."

Some of these flake-like formations were as large as a table-top. "The mass was damp and smelt disagreeably, like rotten seaweed, but, when dried, the smell went off."

"It tore fibrously, like paper."

Classic explanation:

"Up from one place, and down in another."

But what went up, from one place, in a whirlwind? Of course, our Intermediatist acceptance is that had this been the strangest substance conceivable, from the strangest other world that could be thought of; somewhere upon this earth there must be a substance similar to it, or from which it would, at least subjectively, or according to description, not be easily distinguishable. Or that everything in New York City is only another degree or aspect of something, or combination of things, in a village of Central Africa.

The novel is a challenge to vulgarization: write something that looks new to you: some one will point out that the thrice-accursed Greeks said it long ago. Existence is Appetite: the gnaw of being; the one attempt of all things to assimilate all other things, if they have not surrendered and submitted to some higher attempt. It was cosmic that these scientists, who had surrendered to and submitted to the Scientific System, should, consistently with the principles of that system, attempt to assimilate the substance that fell at Memel with some known terrestrial product. At the meeting of the Royal Irish Academy it was brought out that there is a substance, of rather rare occurrence, that has been known to form in thin sheets upon marsh land.

It looks like greenish felt.

The substance of Memel:

Damp, coal-black, leafy mass.

But, if broken up, the marsh-substance is flake-like, and it tears fibrously.

An elephant can be identified as a sunflower - both have long stems. A camel is indistinguishable from a peanut - if only their humps be considered.

Trouble with this book is that we'll end up a lot of intellectual roues: we'll be incapable of being astonished with anything. We knew, to start with, that science and imbecility are continuous; nevertheless so many expressions of the merging-point are at first startling. We did think that Prof. Hitchcock's performance in identifying the Amherst phenomenon as a fungus was rather notable as scientific vaudeville, if we acquit him of the charge of seriousness - or that, in a place where fungi were so common that, before a given evening two of them sprang up, only he, a stranger in this very fungiferous place, knew a fungus when he saw something like a fungus - if we disregard its quick liquefaction, for instance.

It was only a monologue, however: now we have an all-star cast: and they're not only Irish; they're royal Irish.

The royal Irishmen excluded "coal-blackness" and included fibrousness: so then that this substance was "marsh-paper," which "had been raised into the air by storms of wind, and had again fallen."

Second act:

It was said that, according to M. Ehrenberg, "the meteor-paper was found to consist partly of vegetable matter, chiefly of conifervae."

Third act:

Meeting of the royal Irishmen: chairs, tables, Irishmen:

Some flakes of marsh-paper were exhibited.

Their composition was chiefly of conifervse.

This was a double inclusion: or it's the method of agreement that logicians make so much of. So no logician would be satisfied with identifying a peanut as a camel, because both have humps: he demands accessory agreement - that both can live a long time without water, for instance.

Now, it's not so very unreasonable, at least to the free and easy vaudeville standards that, throughout this book, we are considering, to think that a green substance could be snatched up from one place in a whirlwind, and fall as a black substance somewhere else: but the royal Irishmen excluded something else, and it is a datum that was as accessible to them as it is to me:

That, according to Chladni, this was no little, local deposition that was seen to occur by some indefinite person living near a pond somewhere.

It was a tremendous fall from a vast sky-area. Likely enough all the marsh paper in the world could not have supplied it.

At the same time, this substance was falling "in great quantities," in Norway and Pomerania. Or see Kirkwood, Meteoric Astronomy, p. 66: "Substance like charred paper fell in Norway and other parts of northern Europe, Jan. 31, 1686."

Or a whirlwind, with a distribution as wide as that, would not acceptably, I should say, have so specialized in the rare substance called "marsh paper." There'd have been falls of fence rails, roofs of houses, parts of trees. Nothing is said of the occurrence of a tornado in northern Europe, in January, 1686. There is record only of this one substance having fallen in various places.

Time went on, but the conventional determination to exclude data of all falls to this earth, except of substances of this earth, and of ordinary meteoric matter, strengthened.

Annals of Philosophy, 16-68:

The substance that fell in January, 1686, is described as "a mass of black leaves, having the appearance of burnt paper, but harder, and cohering, and brittle."

"Marsh paper" is not mentioned, and there is nothing said of the "conifervae," which seemed so convincing to the royal Irishmen.

Vegetable composition is disregarded, quite as it might be by some one who might find it convenient to identify a crook-necked squash as a big fish hook.

Meteorites are usually covered with a black crust, more or less scale-like. The substance of 1686 is black and scale-like. If so be convenience, "leaf-likeness" is "scale-likeness." In this attempt to assimilate with the conventional, we are told that the substance is a mineral mass: that it is like the black scales that cover meteorites.

The scientist who made this "identification" was Von Grotthus. He had appealed to the god Chemical Analysis. Or the power and glory of mankind with which we're not always so impressed but the gods must tell us what we want them to tell us. We see again that, though nothing has identity of its own, anything can be "identified" as anything. Or there's nothing that's not reasonable, if one snoopeth not into its exclusions. But here the conflict did not end. Berzelius examined the substance. He could not find nickel in it. At that time, the presence of nickel was the "positive" test of meteoritic matter. Whereupon, with a supposititious "positive" standard of judgment against him, Von Grotthus revoked his "identification."(Annals and Mag. of Nat. Hist., 1-3-185.)

This equalization of eminences, permits us to project with our own expression, which, otherwise, would be subdued into invisibility:

That it's too bad that no one ever looked to see - hieroglyphics? - something written upon these sheets of paper?

If we have no very great variety of substances that have fallen to this earth; if, upon this earth's surface there is infinite variety of substances detachable by whirlwinds, two falls of such a rare substance as marsh paper would be remarkable.

A writer in the Edinburgh Review, 87-194, says that, at the time of writing, he had before him a portion of a sheet of 200 square feet, of a substance that had fallen at Carolath, Silesia, in 1839 - exactly similar to cotton-felt, of which clothing might have been made. The god Microscopic Examination had spoken. The substance consisted chiefly of conifervae.

Jour. Asiatic Soc. of Bengal, i84y-pt. 1-193:

That March 16, 1846 - about the time of a fall of edible substance in Asia Minor an olive-gray powder fell at Shanghai. Under the microscope, it was seen to be an aggregation of hairs of two kinds, black ones and rather thick white ones. They were supposed to be mineral fibres, but, when burned, they gave out "the common ammonical smell and smoke of burnt hair or feathers." The writer described the phenomenon as "a cloud of 3800 square miles of fibers, alkali, and sand." In a postscript, he says that other investigators, with more powerful microscopes, gave opinion that the fibres were not hairs; that the substance consisted chiefly of conifervae.

Or the pathos of it, perhaps; or the dull and uninspired, but courageous persistence of the scientific: everything seemingly found out is doomed to be subverted by more powerful microscopes and telescopes; by more refined, precise, searching means and methods the new pronouncements irrepressibly bobbing up; their reception always as Truth at last; always the illusion of the final; very little of the Intermediatist spirit -

That the new that has displaced the old will itself some day be displaced; that it, too, will be recognized as myth-stuff -

But that if phantoms climb, spooks of ladders are good enough for them.

Annual Register, 1821-681:

That, according to a report by M. Laine, French Consul at Pernambuco, early in October, 1821, there was a shower of a substance resembling silk. The quantity was as tremendous as might be a whole cargo, lost somewhere between Jupiter and Mars, having drifted around perhaps for centuries, the original fabrics slowly disintegrating.

In Annales de Chimie, 2-15-427, it is said that samples of this substance were sent to France by M. Laine, and that they proved to have some resemblances to silky filaments which, at certain times of the year, are carried by the wind near Paris.

In the Annals of Philosophy, n.s., 12-93, there is mention of a fibrous substance like blue silk that fell near Naumberg, March 23, 1665. According to Chladni (Annales de Chimie, 2-31-264), the quantity was great. He places a question mark before the date.

One of the advantages of Intermediatism is that, in the oneness of quasiness, there can be no mixed metaphors. Whatever is acceptable of anything, is, in some degree or aspect, acceptable of everything. So it is quite proper to speak, for instance, of something that is as firm as a rock and that sails in a majestic march.

The Irish are good monists: they have of course been laughed at for their keener perceptions. So it's a book we're writing, or it's a procession, or it's a museum, with the Chamber of Horrors rather over-emphasized. A rather horrible correlation occurs in the Scientific American, 1859-178. What interests us is that a correspondent saw a silky substance fall from the sky - there was an aurora borealis at the time he attributes the substance to the aurora.

Since the time of Darwin, the classic explanation has been that all silky substances that fall from the sky are spider webs. In 1832, aboard the Beagle, at the mouth of La Plata River, 60 miles from land, Darwin saw an enormous number of spiders, of the kind usually known as "gossamer" spiders, little aeronauts that cast out filaments by which the wind carries them.

It's difficult to express that silky substances that have fallen to this earth were not spider webs. My own acceptance is that spider webs are the merger; that there have been falls of an externally derived silky substance, and also of the webs, or strands, rather, of aeronautic spiders indigenous to this earth ; that in some instances it is impossible to distinguish one from the other. Of course, our expression upon silky substances will merge away into expressions upon other seeming textile substances, and I don't know how much better off we'll be -

Except that, if fabricable materials have fallen from the sky -

Simply to establish acceptance of that may be doing well enough in this book of first and tentative explorations.

In All the Year Round, 8-254, is described a fall that took place in England, Sept. 21, 1741, in the towns of Bradly, Selbourae, and Alresford, and in a triangular space included by these three towns. The substance is described as "cobwebs" - but it fell in flake-formation, or in "flakes or rags about one inch broad and five or six inches long." Also these flakes were of a relatively heavy substance - "they fell with some velocity." The quantity was great - the shortest side of the triangular space is eight miles long. In the Wernerian Nat. Hist. Soc. Trans., 5-386, it is said that there were two falls that they were some hours apart a datum that is becoming familiar to us - a datum that can not be taken into the fold, unless we find it repeated over and over and over again. It is said that the second fall lasted from nine o'clock in the morning until night.

Now the hypnosis of the classic - that what we call intelligence is only an expression of inequilibrium; that when mental adjustments are made, intelligence ceases - or, of course, that intelligence is the confession of ignorance. If you have intelligence upon any subject, that is something you're still learning if we agree that that which is learned is always mechanically done in quasi-terms, of course, because nothing is ever finally learned.

It was decided that this substance was spiders' web. That was adjustment. But it's not adjustment to me; so I'm afraid I shall have some intelligence in this matter. If I ever arrive at adjustment upon this subject, then, upon this subject, I shall be able to have no thoughts, except routine-thoughts. I haven't yet quite decided absolutely everything, so I am able to point out:

That this substance was of quantity so enormous that it attracted wide attention when it came down -

That it would have been equally noteworthy when it went up -

That there is no record of any one, in England or elsewhere, having seen tons of "spider webs" going up, September, 1741.

Further confession of intelligence upon my part:

That, if it be contested, then, that the place of origin may have been far away, but still terrestrial -

Then it's that other familiar matter of incredible "marksmanship" - again hitting a small, triangular space for hours interval of hours - then from nine in the morning until night: same small triangular space.

These are the disregards of the classic explanation. There is no mention of spiders having been seen to fall, but a good inclusion is that, though this substance fell in good-sized flakes of considerable weight, it was viscous. In this respect it was like cobwebs: dogs nosing it on grass, were blindfolded with it. This circumstance does strongly suggest cobwebs -

Unless we can accept that, in regions aloft, there are vast viscous or gelatinous areas, and that things passing through become daubed.

Or perhaps we dear up the confusion in the descriptions of the substance that fell in 1841 and 1846, in Asia Minor, described in one publication as gelatinous, and in another as a cereal that it was a cereal that had passed through a gelatinous region. That the paper-like substance of Memel may have had such an experience may be indicated in that Ehrenberg found in it gelatinous matter, which he called "nostoc." (Annals and Mag. of Nat. Hist., 1-3-185.)

Scientific American, 45-337:

Fall of a substance described as "cobwebs," latter part of October, 1881, in Milwaukee, Wis., and other towns: other towns mentioned are Green Bay, Vesburge, Fort Howard, Sheboygan, and Ozaukee.

The aeronautic spiders are known as "gossamer" spiders, because of the extreme lightness of the filaments that they cast out to the wind. Of the substance that fell in Wisconsin, it is said:

"In all instances the webs were strong in texture and very white."

The Editor says:

"Curiously enough, there is no mention in any of the reports that we have seen, of the presence of spiders."

So our attempt to divorce a possible external product from its terrestrial merger: then our joy of the prospector who thinks he's found something:

The Monthly Weather Review, 26-566, quotes the Montgomery (Ala.) Advertiser:

That, upon Nov. 21, 1898, numerous batches of spider-web-like substance fell in Montgomery, in strands and in occasional masses several inches long and several inches broad. According to the writer, it was not spiders' web, but something like asbestos; also that it was phosphorescent.

The Editor of the Review says that he sees no reason for doubting that these masses were cobwebs.

La Nature, 1883-342:

A correspondent writes that he sends a sample of a substance said to have fallen at Montussan (Gironde), Oct. 16, 1883. According to a witness, quoted by the correspondent, a thick cloud, accompanied by rain and a violent wind, had appeared. This cloud was composed of a woolly substance in lumps the size of a fist, which fell to the ground. The Editor (Tissandier) says of this substance that it was white, but was something that had been burned. It was fibrous. M. Tissandier astonishes us by saying that he can not identify this substance. We thought that anything could be 'identified" as anything. He can say only that the cloud in question must have been an extraordinary conglomeration.

Annual Register, 1832-447:

That, March, 1832, there fell, in the fields of Kourianof, Russia, a combustible yellowish substance, covering, at least two inches thick, an area of 600 or 700 square feet. It was resinous and yellowish: so one inclines to the conventional explanation that it was pollen from pine trees - but, when torn, it had the tenacity of cotton. When placed in water, it had the consistency of resin.

"This resin had the color of amber, was elastic, like India rubber, and smelled like prepared oil mixed with wax."

So in general our notion of cargoes - and our notion of cargoes of food supplies:

In Philosophical Transactions, 19-224, is an extract from a letter by Mr. Robert Vans, of Kilkenny, Ireland, dated Nov. 15, 1695: that there had been "of late," in the counties of Limerick and Tipperary, showers of a sort of matter like butter or grease . . . having "a very stinking smell."

There follows an extract from a letter by the Bishop of Cloyne, upon "a very odd phenomenon," which was observed in Munster and Leinster: that for a good part of the spring of 1695 there fell a substance which the country people called "butter" - "soft, clammy, and of a dark yellow" - that cattle fed "indifferently" in fields where this substance lay.

"It fell in lumps as big as the end of one's finger." It had a "strong ill scent." His Grace calls it a "stinking dew."

In Mr. Van's letter, it is said that the "butter" was supposed to have medicinal properties, and "was gathered in pots and other vessels by some of the inhabitants of this place."

And:

In all the following volumes of Philosophical Transactions there is no speculation upon this extraordinary subject. Ostracism.

The fate of this datum is a good instance of damnation, not by denial, and not by explaining away, but by simple disregard. The fall is listed by Chladni, and is mentioned in other catalogs, but, from the absence of all inquiry, and of all but formal mention, we see that it has been under excommunication as much as was ever anything by the preceding system. The datum has been buried alive. It is as irreconcilable with the modern system of dogmas as ever were geologic strata and vermiform appendix with the preceding system -

If, intermittently, or "for a good part of the spring," this substance fell in two Irish provinces, and nowhere else, we have, stronger than before, a sense of a stationary region overhead, or a region that receives products like this earth's products, but from external sources, a region in which this earth's gravitational and meteorological forces are relatively inert if for many weeks a good part of this substance did hover before finally falling. We suppose that, in 1685, Mr. Vans and the Bishop of Cloyne could describe what they saw as well as could witnesses in 1885: nevertheless, it is going far back; we shall have to have many modern instances before we can accept.

As to other falls, or another fall, it is said in the Amer. Jour. Set., 1-28-361, that, April n, 1832 about a month after the fall of the substance of Kourianof fell a substance that was wineyellow, transparent, soft, and smelling like rancid oil. M. Herman, a chemist who examined it, named it "sky oil." For analysis and chemic reactions, see the Journal. The Edinburgh New Philosophical Journal, 13-368, mentions an "unctuous" substance that fell near Rotterdam, in 1832. In Comptes Rendus, 13-215, there is an account of an oily, reddish matter that fell at Genoa, February, 1841.

Whatever it may have been -

Altogether, most of our difficulties are problems that we should leave to later developers of supergeography, I think. A discoverer of America should leave Long Island to some one else. If there be, plying back and forth from Jupiter and Mars and Venus, superconstructions that are sometimes wrecked, we think of fuel as well as cargoes. Of course the most convincing data would be of coal falling from the sky: nevertheless, one does suspect that oil-burning engines were discovered ages ago in more advanced worlds but, as I say, we should leave something to our disciples so we'll not especially wonder whether these butter-like, or oily substances were food or fuel. So we merely note that in the Scientific American, 24-323, is an account of hail that fell, in the middle of April, 1871, in Mississippi, in which was a substance described as turpentine.

Something that tasted like orange water, in hailstones, about the first of June, 1842, near Nimes, France; identified as nitric acid (Jour. de Pharmacie, 1845-273).

Hail and ashes, in Ireland, 1755 (Sci. Amer., 5-168).

That, at Elizabeth, N. J., June 9, 1874, fell hail in which was a substance, said, by Prof. Leeds, of Stevens Institute, to be carbonate of soda (Sci. Amer., 30-262).

We are getting a little away from the lines of our composition, but it will be an important point later that so many extraordinary falls have occurred with hail. Or - if they were of substances that had had origin upon some other part of this earth's surface - had the hail, too, that origin? Our acceptance here will depend upon the number of instances. Reasonably enough, some of the things' that fall to this earth should coincide with falls of hail.

As to vegetable substances in quantities so great as to suggest lost cargoes, we have a note in the Intellectual Observer, 3-468: that, upon the first of May, 1863, a rain fell at Perpignan, "bringing down with it a red substance, which proved on examination to be a red meal mixed with fine sand." At various points along the Mediterranean, this substance fell.

There is, in Philosophical Transactions, 16-281, an account of a seeming cereal, said to have fallen in Wiltshire, in 1686 said that some of the "wheat" fell "enclosed in hailstones" but the writer in Transactions, says that he had examined the grains, and that they were nothing but seeds of ivy berries dislodged from holes and chinks where birds had hidden them. If birds still hide ivy seeds, and if winds still blow, I don't see why the phenomenon has not repeated in more than two hundred years since.

Or the red matter in rain, at Siena, Italy, May, 1830; said, by Arago, to have been vegetable matter (Arago, (Euvres, 12-468).

Somebody should collect data of falls at Siena alone.

In the Monthly Weather Review, 29-465, a correspondent writes that, upon Feb. 16, 1901, at Pawpaw, Michigan, upon a day that was so calm that his windmill did not run, fell a brown dust that looked like vegetable matter. The Editor of the Review concludes that this was no widespread fall from a tornado, because it had been reported from nowhere else.

Rancidness – putridity – decomposition - a note that has been struck many times. In a positive sense, of course, nothing means anything, or every meaning is continuous with all other meanings: or that all evidences of guilt, for instance, are just as good evidences of innocence - but this condition seems to mean things lying around among the stars a long time. Horrible disaster in the time of Julius Caesar; remains from it not reaching this earth till the time of the Bishop of Cloyne: we leave to later research the discussion of bacterial action and decomposition, and whether bacteria could survive in what we call space, of which we know nothing -

Chemical News, 35-183:

Dr. A. T. Machattie, F.C.S., writes that, at London, Ontario, Feb. 24, 1868, in a violent storm, fell, with snow, a dark-colored substance, estimated at 500 tons, over a belt 50 miles by 10 miles. It was examined under a microscope, by Dr. Machattie, who found it to consist mainly of vegetable matter "far advanced in decomposition."

The substance was examined by Dr. James Adams, of Glasgow, who gave his opinion that it was the remains of cereals.

Dr. Machattie points out that for months before this fall the ground of Canada had been frozen, so that in this case a more than ordinarily remote origin has to be thought of. Dr. Machattie thinks of origin to the south. "However," he says, "this is mere conjecture."

Amer. Jour. Sci., 1841-40:

That, March 24, 1840 - during a thunderstorm - at Rajkit, India, occurred a fall of grain. It was reported by Col. Sykes, of the British Association.

The natives were greatly excited - because it was grain of a kind unknown to them.

Usually comes forward a scientist who knows more of the things that natives know best than the natives know - but it so happens that the usual thing was not done definitely in this instance:

"The grain was shown to some botanists, who did not immediately recognize it, but thought it to be either a spartium or a vicia."

Chapter VI

    Lead, silver, diamonds, glass.

They sound like the accursed, but they're not: they're now of the chosen - that is, when they occur in metallic or stony masses that Science has recognized as meteorites. We find that resistance is to substances not so mixed in or incorporated.

Of accursed data, it seems to me that punk is pretty damnable.

In the Report of the British Association, 1878-376, there is mention of a light chocolate-brown substance that has fallen with meteorites. No particulars given; not another mention anywhere else that I can find. In this English publication, the word "punk" is not used; the substance is called "amadou." I suppose, if the datum has anywhere been admitted to French publications, the word "amadou" has been avoided, and "punk" used.

Or oneness of allness: scientific works and social registers: a Goldstein who can't get in as Goldstein, gets in as Jackson.

The fall of sulphur from the sky has been especially repulsive to the modern orthodoxy - largely because of its associations with the superstitions or principles of the preceding orthodoxy - stories of devils: sulphurous exhalations. Several writers have said that they have had this feeling. So the scientific reactionists, who have rabidly fought the preceding, because it was the preceding: and the scientific prudes, who, in sheer exclusionism, have held lean hands over pale eyes, denying falls of sulphur. I have many notes upon the sulphurous odor of meteorites, and many notes upon phosphorescence of things that come from externality. Some day I shall look over old stories of demons that have appeared sulphurously upon this earth, with the idea of expressing that we have often had undesirable visitors from other worlds; or that an indication of external derivation is sulphurousness. I expect some day to rationalize demonology, but just at present we are scarcely far enough advanced to go so far back.

For a circumstantial account of a mass of burning sulphur, about the size of a man's fist, that fell at Pultusk, Poland, Jan. 30, 1868, upon a road, where it was stamped out by a crowd of villagers, see Rept. Brit. Assoc., 1874-272.

The power of the exclusionists lies in that in their stand are combined both modern and archaic systematists. Falls of sandstone and limestone are repulsive to both theologians and scientists.

Sandstone and limestone suggest other worlds upon which occur processes like geological processes; but limestone, as a fossiliferous substance, is of course especially of the unchosen.

In Science, March 9, 1888, we read of a block of limestone, said to have fallen near Middleburgh, Florida. It was exhibited at the Sub-tropical Exposition, at Jacksonville. The writer, in Science, denies that it fell from the sky. His reasoning is: There is no limestone in the sky;

Therefore this limestone did not fall from the sky.

Better reasoning I can not conceive of because we see that a final major premise - universal true - would include all things: that, then, would leave nothing to reason about so then that all reasoning must be based upon "something" not universal, or only a phantom intermediate to the two finalities of nothingness and allness, or negativeness and positiveness.

La Nature, 1890-2-127:

Fall, at Pel-et-Der (L' Aube) France, June 6, 1890, of limestone pebbles. Identified with limestone at Chateau Landon - or up and down in a whirlwind. But they fell with hail - which, in June, could not very well be identified with ice from Chateau-Landon.

Coincidence, perhaps.

Upon page 70, Science Gossip, 1887, the Editor says, of a stone that was reported to have fallen at Little Lever, England, that a sample had been sent to him. It was sandstone. Therefore it had not fallen, but had been on the ground in the first place. But, upon page 140, Science Gossip, 1887, is an account of "a large, smooth, waterworn, gritty sandstone pebble" that had been found in the wood of a full-grown beech tree. Looks to me as if it had fallen red-hot, and had penetrated the tree with high velocity.

But I have never heard of anything falling red-hot from a whirlwind -

The wood around this sandstone pebble was black, as if charred.

Dr. Farrington, for instance, in his books, does not even mention sandstone. However, the British Association, though reluctant, is less exclusive: Report of 1860, p. 197: substance about the size of a duck's egg, that fell at Raphoe, Ireland, June 9, 1860 date questioned. It is not definitely said that this substance was sandstone, but that it "resembled" friable sandstone.

Falls of salt have occurred often. They have been avoided by scientific writers, because of the dictum that only water and not substances held in solution, can be raised by evaporation. However, falls of salty water have received attention from Dalton and others, and have been attributed to whirlwinds from the sea. This is so reasonably contested - quasi-reasonably - as to places not far from the sea -

But the fall of salt that occurred high in the mountains of Switzerland -

We could have predicted that that datum could be found somewhere.

Let anything be explained in local terms of the coast of England - but also has it occurred high in the mountains of Switzerland.

Large crystals of salt fell in a hail storm Aug. 20, 1870, in Switzerland. The orthodox explanation is a crime: whoever made it, should have had his finger-prints taken. We are told (An. Rec. Sci., 1872) that these objects of salt "came over the Mediterranean from some part of Africa."

Or the hypnosis of the conventional - provided it be glib. One reads such an assertion, and provided it be suave and brief and conventional, one seldom questions - or thinks "very strange" and then forgets. One has an impression from geography lessons : Mediterranean not more than three inches wide, on the map ; Switzerland only a few more inches away. These sizable masses of salt are described in the Amer. Jour. Sci., 3-3-239, as "essentially imperfect cubic crystals of common salt." As to occurrence with hail - that can in one, or ten, or twenty, instances be called a coincidence.

Another datum: extraordinary year 1883:

London Times, Dec. 25, 1883:

Translation from a Turkish newspaper; a substance that fell at Scutari, Dec. 2, 1883; described as an unknown substance, in particles or flakes? like snow. "It was found to be saltish to the taste, and to dissolve readily in water."

Miscellaneous:

"Black, capillary matter" that fell, Nov. 16, 1857, at Charleston, S. C. (Amer. Jour. Sci., 2-31-459).

Fall of small, friable, vesicular masses, from size of a pea to size of a walnut, at Lobau, Jan. 18, 1835 (Rept. Brit. Assoc., 1860-85).

Objects that fell at Peshawur, India, June, 1893, during a storm: substance that looked like crystallized nitre, and that tasted like sugar (Nature, July 13, 1893). I suppose sometimes deep-sea fishes have their noses bumped by cinders. If their regions be subjacent to Cunard or White Star routes, they're especially likely to be bumped. I conceive of no inquiry: they're deep-sea fishes. Or the slag of Slains. That it was a furnace-product. The Rev. James Rust seemed to feel bumped. He tried in vain to arouse inquiry.

As to a report, from Chicago, April 9, 1879, that slag had fallen from the sky, Prof. E. S. Bastian (Amer. Jour. Sci., 3-18-78) says that the slag "had been on the ground in the first place." It was furnace-slag. "A chemical examination of the specimens has shown that they possess none of the characteristics of true meteorites.'

Over and over and over again, the universal delusion; hope and despair of attempted positivism; that there can be real criteria, or distinct characteristics of anything. If anybody can define - not merely suppose, like Prof. Bastian, that he can define - the true characteristics of anything, or so localize trueness anywhere, he makes the discovery for which the cosmos is laboring. He will .be instantly translated, like Elijah, into the Positive Absolute. My own notion is that, in a moment of super-concentration, Elijah became so nearly a real prophet that he was translated to heaven or to the Positive Absolute, with such velocity that he left an incandescent train behind him. As we go along, we shall find the "true test of meteoritic material," which in the past has been taken as an absolute, dissolving into almost utmost nebulosity. Prof. Bastian explains mechanically, or in terms of the usual reflexes to all reports of unwelcome substances: that near where the slag had been found, telegraph wires had been struck by lightning; that particles of melted wire had been seen to fall near the slag which had been on the ground in the first place. But, according to the N. Y. Times, April 14, 1879, about two bushels of this substance had fallen.

Something that was said to have fallen at Darmstadt, June 7, 1846; listed by Greg (Rept. Brit. Assoc., 1867-416) as "only slag."

Philosophical Magazine, 4-10-381:

That, in 1855, a large stone was found far in the interior of a tree, in Battersea Fields.

Sometimes cannon balls are found embedded in trees. Doesn't seem to be anything to discuss; doesn't seem discussable that any one would cut a hole in a tree and hide a cannon ball, which one could take to bed, and hide under one's pillow, just as easily. So with the stone of Battersea Fields. What is there to say, except that it fell with high velocity and embedded in the tree? Nevertheless, there was a great deal of discussion -

Because, at the foot of the tree, as if broken off the stone, fragments of slag were found.

I have nine other instances.

Slag and cinders and ashes, and you won't believe, and neither will I, that they came from the furnaces of vast aerial superconstructions.

We'll see what looks acceptable.

As to ashes, the difficulties are great, because we'd expect many falls of terrestrially derived - ashes volcanoes and forest fires. In some of our acceptances, I have felt a little radical -

I suppose that one of our main motives is to show that there is, in quasi-existence, nothing but the preposterous or something intermediate to absolute preposterousness and final reasonableness that the new is the obviously preposterous; that it becomes the established and disguisedly preposterous; that it is displaced, after a while, and is again seen to be the preposterous. Or that all progress is from the outrageous to the academic or sanctified, and back to the outrageous modified, however, by a trend of higher and higher approximation to the impreposterous. Sometimes I feel a little more uninspired than at other times, but I think we're pretty well accustomed now to the oneness of all-ness; or that the methods of science in maintaining its system are as outrageous as the attempts of the damned to break in. In the Annual Record of Science, 1875-241, Prof. Daubree is quoted: that ashes that had fallen in the Azores had come from the Chicago fire -

Or the damned and the saved, and there's little to choose between them; and angels are beings that have not obviously barbed tails to them - or never have such bad manners as to stroke an angel below the waist-line.

However this especial outrage was challenged: the Editor of the Record returns to it, in the issue of 1876: considers it "in the highest degree improper to say that the ashes of Chicago were landed in the Azores."

Bull. Soc. Astro, de France, 22-245:

Account of a white substance, like ashes, that fell at Annoy, France, March 27, 1908: simply called a curious phenomenon; no attempt to trace to a terrestrial source.

Flake formations, which may signify passage through a region of pressure, are common; but spherical formations as if of things that have rolled and rolled along planar regions somewhere are commoner:

Nature, Jan. 10, 1884, quotes a Kimberly newspaper:

That, toward the close of November, 1883, a thick shower of ashy matter fell at Queenstown, South Africa. The matter was in marble-sized balls, which were soft and pulpy, but which, upon drying, crumbled at touch. The shower was confined to one narrow streak of land. It would be only ordinarily preposterous to attribute this substance to Krakatoa -

But, with the fall, loud noises were heard -

But I'll omit many notes upon ashes: if ashes should sift down upon deep-sea fishes, that is not to say that they came from steamships.

Data of falls of cinders have been especially damned by Mr. Symons, the meteorologist, some of whose investigations we'll investigate Later – nevertheless -

Notice of a fall, in Victoria, Australia, April 14, 1875 (Rept. Brit. Assoc., 1875-242) at least we are told, in the reluctant way, that some one "thought" he saw matter fall near him at night, and the next day found something that looked like cinders. In the Proc. of the London Roy. Soc., 19-122, there is an account of cinders that fell on the deck of a lightship, Jan. 9, 1873. In the Amer. Jour. Sci.,

2-24-449, there is a notice that the Editor had received a specimen of cinders said to have fallen in showery weather - upon a farm, near Ottowa, Ill, Jan. 17, 1857.

But after all, ambiguous things they are, cinders or ashes or slag or clinkers, the high priest of the accursed that must speak aloud for us is - coal that has fallen from the sky.

Or coke:

The person who thought he saw something like cinders, also thought he saw something like coke, we are told.

Nature, 36-119:

Something that "looked exactly like coke" that fell - during a thunder storm - in the Orne, France, April 24, 1887.

Or charcoal:

Dr. Angus Smith, in the Lit. and Phil. Soc. of Manchester Memoirs, 2-9-146, says that, about 1827 - like a great deal in Lyell's Principles and Darwin's Origin, this account is from Hearsay - something fell from the sky, near Allport, England. It fell luminously, with a loud report, and scattered in a field. A fragment that was seen by Dr. Smith, is described by him as having "the appearance of a piece of common wood charcoal." Nevertheless, the reassured feeling of the faithful, upon reading this, is burdened with data of differences: the substance was so uncommonly heavy that it seemed as if it had iron in it; also there was "a sprinkling of sulphur." This material is said, by Prof. Baden-Powell, to be "totally unlike that of any other meteorite." Greg, in his catalogue (Rept. Brit. Assoc., 1860-73) calls it "a more than doubtful substance" - but again, against reassurance, that is not doubt of authenticity. Greg says that it is like compact charcoal, with particles of sulphur and iron pyrites embedded.

Reassurance rises again:

Prof. Baden-Powell says: "It contains also charcoal, which might perhaps be acquired from matter among which it fell." This is a common reflex with the exclusionists: that substances not "truly meteoritic" did not fall from the sky, but were picked up by "truly meteoritic" things, of course only on their surfaces, by impact with this earth.

Rhythm of reassurances and their declines:

According to Dr. Smith, this substance was not merely coated with charcoal; his analysis gives 43.59 per cent carbon.

Our acceptance that coal has fallen from the sky will be via data of resinous substances and bituminous substances, which merge so that they can not be told apart.

Resinous substance said to have fallen at Kaba, Hungary, April 15, 1887 (Rept. Brit. Assoc., 1860-94).

A resinous substance that fell after a fireball? at Neuhaus, Bohemia, Dec. 17, 1824 (Rept. Brit. Assoc., 1860-70).

Fall, July 28, 1885, at Luchon, during a storm, of a brownish substance; very friable, carbonaceous matter; when burned it gave out a resinous odor (Comptes Rendus, 103-837).

Substance that fell, Feb. 17, 18, 19, 1841, at Genoa, Italy, said to have been resinous; said by Arago ((Euvres, 12-469) to have been bituminous matter and sand.

Fall - during a thunderstorm - July, 1681, near Cape Cod, upon the deck of an English vessel, the

Albemarle, of "burning, bituminous matter" (Edin. New Phil. Jour., 26-86); a fall, at Christiania, Norway, June 13, 1822, of bituminous matter, listed by Greg as doubtful; fall of bituminous matter, in Germany, March 8, 1798, listed by Greg. Lockyer (The Meteoric Hypothesis, p. 24) says that the substance that fell at the Cape of Good Hope, Oct. 13, 1838 - about five cubic feet of it: substance so soft that it was cuttable with a knife " - after being experimented upon, it left a residue, which gave out a very bituminous smell."

And this inclusion of Lockyer's - so far as findable in all books that I have read - is, in books, about as close as we can get to our desideratum - that coal has fallen from the sky. Dr. Farrington, except with a brief mention, ignores the whole subject of the fall of carbonaceous matter from the sky. Proctor, in all of his books that I have read - is, in books, about as close as we can get to duction to the Study of Meteorites," p. 53) excommunicates with the admission that carbonaceous matter has been found in meteorites "in very minute quantities" - or my own suspicion is that it is possible to damn something else only by losing one's own soul - quasi-soul, of course.

Sci. Amer., 35-120:

That the substance that fell at the Cape of Good Hope "resembled a piece of anthracite coal more than anything else."

It's a mistake, I think: the resemblance is to bituminous coal - but it is from the periodicals that we must get our data. To the writers of books upon meteorites, it would be as wicked - by which we mean departure from the characters of an established species - quasi-established, of course - to say that coal has fallen from the sky, as would be, to something in a barnyard, a temptation that it climb a tree and catch a bird. Domestic things in a barnyard: and how wild things from forests outside seem to them. Or the homeopathist but we shall shovel data of coal.

And, if over and over, we shall learn of masses of soft coal that have fallen upon this earth, if in no instance has it been asserted that the masses did not fall, but were upon the ground in the first place; if we have many instances, this time we turn down good and hard the mechanical reflex that these masses were carried from one place to another in whirlwinds, because we find it too difficult to accept that whirlwinds could so select, or so specialize in a peculiar substance. Among writers of books, the only one I know of who makes more than brief mention is Sir Robert Ball. He represents a still more antique orthodoxy, or is an exclusionist of the old type, still holding out against even meteorites. He cites several falls of carbonaceous matter, but with disregards that make for reasonableness that earthy matter may have been caught up by whirlwinds and flung down somewhere else. If he had given a full list, he would be called upon to explain the special affinity of whirlwinds for a special kind of coal. He does not give a full list. We shall have all that's findable, and we shall see that against this disease we're writing, the homeopathist's prescription availeth not. Another exclusionist was Prof. Lawrence Smith. His psycho-tropism was to respond to all reports of carbonaceous matter falling from the sky, by saying that this damned matter had been deposited upon things of the chosen by impact with this earth. Most of our data antedate him, or were contemporaneous with him, or were as accessible to him as to us. In his attempted positivism it is simply - and beautifully - disregarded that, according to Berthelot, Berzelius, Cloez, Wohler and others these masses are not merely coated with carbonaceous matter, but are carbonaceous throughout, or are permeated throughout. How any one could so resolutely and dogmatically and beautifully and blindly hold out, would puzzle us were it not for our acceptance that only to think is to exclude and include; and to exclude some things that have as much right to come in as have the included that to have an opinion upon any subject is to be a Lawrence Smith because there is no definite subject.

Dr. Walter Flight (Eclectic Magazine, 89-71) says, of the substance that fell near Alais, France, March 15, 1806, that it "emits a faint bituminous substance" when heated, according to the observations of Berzelius and a commission appointed by the French Academy. This time we have not the reluctances expressed in such words as "like" and "resembling." We are told that this substance is "an earthy kind of coal."

As to "minute quantities" we are told that the substance that fell at the Cape of Good Hope has in it a little more than a quarter of organic matter, which, in alcohol, gives the familiar reaction of yellow, resinous matter. Other instances given by Dr. Flight are:

Carbonaceous matter that fell in 1840, in Tennessee; Cranbourne, Australia, 1861; Montauban, France, May 14, 1864 (twenty masses, some of them as large as a human head; of a substance that "resembled a dull-colored earthy lignite") ; Goalpara, India, about 1867 (about 8 per cent of a hydrocarbon); at Ornans, France, July n, 1868; substance with "an organic, combustible ingredient," at Hessle, Sweden, Jan. 1, 1860.

Knowledge, 4-134:

That, according to M. Daubree, the substance that had fallen in the Argentine Republic, "resembled certain kinds of lignite and boghead coal." In Comptes Rendus, 96-1764, it is said that this mass fell, June 30, 1880, in the province Entre Rios, Argentina: that it is "like" brown coal; that it resembles all the other carbonaceous masses that have fallen from the sky.

Something that fell at Grazac, France, Aug. 10, 1885: when burned, it gave out a bituminous odor (Comptes Rendus, 104-1771).

Carbonaceous substance that fell at Rajpunta, India, Jan. 22, 1911: very friable: 50 per cent of it soluble in water (Records Geol.

Survey of India, 44-pt. 1-41).

A combustible carbonaceous substance that fell with sand at Naples, March 14, 1818 (Amer. Jour. Set., 1-1-309).

Sci. Amer. Sup., 29-11798:

That, June 9, 1889, a very friable substance, of a deep, greenish black, fell at Mighei, Russia. It contained 5 per cent organic matter, which, when powdered and digested in alcohol, yielded, after evaporation, a bright yellow resin. In this mass was 2 per cent of an unknown mineral.

Cinders and ashes and slag and coke and charcoal and coal.

And the things that sometimes deep-sea fishes are bumped by.

Reluctances and the disguises or covered retreats of such words as "like" and "resemble" - or that conditions of Intermediateness forbid abrupt transitions but that the spirit animating all Intermediateness is to achieve abrupt transitions because, if anything could finally break away from its origin and environment, that would be a real thing - something not merging away indistinguishably with the surrounding. So all attempt to be original; all attempt to invent something that is more than mere extension or modification of the preceding, is positivism or that if one could conceive of a device to catch flies, positively different from, or unrelated to, all other devices up - he'd shoot to heaven, or the Positive Absolute - leaving behind such an incandescent train that in one age it would be said that he had gone aloft in a fiery chariot, and in another age that he had been struck by lightning - I'm collecting notes upon persons supposed to have been struck by lightning. I think that high approximation to positivism has often been achieved - instantaneous translation - residue of negativeness left behind, looking much like effects of a stroke of lightning.

Some day I shall tell the story of the Marie Celeste "properly," as the Scientific American Supplement would say mysterious disappearance of a sea captain, his family, and the crew -

Of positivists, by the route of Abrupt Transition, I think that Manet was notable but that his approximation was held down by his intense relativity to the public - or that it is quite as impositive to flout and insult and defy as it is to crawl and placate. Of course, Manet began with continuity with Courbet and others, and then, between him and Monet there were mutual influences but the spirit of abrupt difference is the spirit of positivism, and Manet's stand was against the dictum that all lights and shades must merge away suavely into one another and prepare for one another. So a biologist like De Vries represents positivism, or the breaking of Continuity, by trying to conceive of evolution by mutation against the dogma of indistinguishable gradations by "minute variations." A Copernicus conceives of helio-centricity. Continuity is against him. He is not permitted to break abruptly with the past. He is permitted to publish his work, but only as "an interesting hypothesis."

Continuity - and that all that we call evolution or progress is attempt to break away from it -

That our whole solar system was at one time attempt by planets to break away from a parental nexus and set up as individualities, and, failing, move in quasi-regular orbits that are expressions of relations with the sun and with one another, all having surrendered, being now quasi-incorporated in a higher approximation to system;

Intermediateness in its mineralogic aspect of positivism - or Iron that strove to break away from Sulphur and Oxygen, and be real, homogeneous Iron failing, inasmuch as elemental iron exists only in text-book chemistry;

Intermediateness in its biologic aspect of positivism or the wild, fantastic, grotesque, monstrous things it conceived of, sometimes in a frenzy of effort to break away abruptly from all preceding types - but failing, in the giraffe-effort, for instance, or only caricaturing an antelope -

All things break one relation only by the establishing of some other relation -

All things cut an umbilical cord only to clutch a breast.

So the fight of the exclusionists to maintain the traditional - or to prevent abrupt transition from the quasi-established - fighting so" that here, more than a century after meteorites were included, no other notable inclusion has been made, except that of cosmic dust, data of which Nordenskiold made more nearly real than data in opposition.

So Proctor, for instance, fought and expressed his feeling of the preposterous, against Sir W. H. Thomson's notions of arrival upon this earth of organisms on meteorites -

"I can only regard it as a jest" (Knowledge, 1-302).

Or that there is nothing but jest or something intermediate to jest and tragedy; That ours is not an existence but an utterance;

That Momus is imagining us for the amusement of the gods, often with such success that some of us seem almost alive like characters in something a novelist is writing; which often to considerable degree take their affairs away from the novelist -

That Momus is imagining us and our arts and sciences and religions, and is narrating or picturing us as a satire upon the gods' real existence.

Because - with many of our data of coal that has fallen from the sky as accessible then as they are now, and with the scientific pronouncement that coal is fossil, how, in a real existence, by which we mean a consistent existence, or a state in which there is real intelligence, or a form of thinking that does not indistinguishably merge away with imbecility, could there have been such a row as that which was raised about forty years ago over Dr. Hahn's announcement that he had found fossils in meteorites?

Accessible to anybody at that time:

Philosophical Magazine, 4-17-425:

That the substance that fell at Kaba, Hungary, April 15, 1857, contained organic matter "analogous to fossil waxes."

Or limestone:

Of the block of limestone which was reported to have fallen at Midleburgh, Florida, it is said (Science, 11-118) that, though something had been seen to fall in "an old cultivated field," the witnesses who ran to it picked up something that "had been upon the ground in the first place." The writer who tells us this, with the usual exclusion-imagination, known as stupidity, but unjustly, because there is no real stupidity, thinks he can think of a good-sized stone that had for many years been in a cultivated field, but that had never been seen before - had never interfered with plowing, for instance. He is earnest and unjarred when he writes that this stone weighs 200 pounds. My own notion, founded upon my own experience in seeing, is that a block of stone weighing 500 pounds might be in one's parlor twenty years, virtually unseen - but not in an old cultivated field, where it interfered with plowing - not anywhere - if it interfered.

Dr. Hahn said that he had found fossils in meteorites. There is a description of the corals, sponges, shells, and crinoids, all of them microscopic, which he photographed, in Popular Science, 20-83.

Dr. Hahn was a well-known scientist. He was better known after that.

Anybody may theorize upon other worlds and conditions upon them that are similar to our own conditions: if his notions be presented undisguisedly as fiction, or only as an "interesting hypothesis," he'll stir up no prude rages.

But Dr. Hahn said definitely that he had found fossils in specified meteorites: also he published photographs of them. His book is in the New York Public Library. In the reproductions every feature of some of the little shells is plainly marked. If they're not shells, neither are things under an oyster-counter. The striations are very plain: one sees even the hinges where bivalves are joined.

Prof. Lawrence Smith (Knowledge, 1-258):

"Dr. Hahn is a kind of half-insane man, whose imagination has run away with him."

Conservation of Continuity.

Then Dr. Weinland examined Dr. Hahn's specimens. He gave his opinion that they are fossils and that they are not crystals of enstatite, as asserted by Prof. Smith, who had never seen them.

The damnation of denial and the damnation of disregard: After the publication of Dr. Weinland's findings - silence.

Chapter VII

    The living things that have come down to this earth:

Attempts to preserve the system:

That small frogs and toads, for instance, never have fallen from the sky, but were - "on the ground, in the first place"; or that there have been such falls - "up from one place in a whirlwind, and down in another."

Were there some especially froggy place near Europe, as there is an especially sandy place, the scientific explanation would of course be that all small frogs falling from the sky in Europe, come from that center of frogeity.

To start with, I'd like to emphasize something that I am permitted to see because I am still primitive or intelligent or in a state of maladjustment:

That there is not one report findable of a fall of tadpoles from the sky.

As to "there in the first place":

See Leisure Hours, 3-779, for accounts of small frogs, or toads, said to have been seen to fall from the sky. The writer says that all observers were mistaken: that the frogs or toads must have fallen from trees or other places overhead.

Tremendous number of little toads, one or two months old, that were seen to fall from a great thick cloud that appeared suddenly in a sky that had been cloudless, August, 1804, near Toulouse, France, according to a letter from Prof. Pontus to M. Arrago.

(Comptes Rendus, 3-54.)

Many instances of frogs that were seen to fall from the sky.

("Notes and Queries," 8-6-104) ; accounts of such falls, signed by witnesses. ("Notes and Queries," 8-6-190.)

Scientific American, July 12, 1873:

"A shower of frogs which darkened the air and covered the ground for a long distance is the reported result of a recent rainstorm at Kansas City, Mo."

As to having been there "in the first place":

Little frogs found in London, after a heavy storm, July 30, 1838.

(Notes and Queries, 8-7-437) ;

Little toads found in a desert, after a rainfall (Notes and Queries, 8-8-493).

To start with I do not deny – positively - the conventional explanation of "up and down." I think that there may have been such occurrences. I omit many notes that I have upon indistinguishables.

In the London Times, July 4, 1883, there is an account of a shower of twigs and leaves and tiny toads in a storm upon the slopes of the Apennines. These may have been the ejectamenta of a whirlwind. I add, however, that I have notes upon two other falls of tiny toads, in 1883, one in France and one in Tahiti; also of fish in Scotland. But in the phenomenon of the Apennines, the mixture seems to me to be typical of the products of a whirlwind. The other instances seem to me to be typical of something like migration? Their great numbers and their homogeneity.

Over and over in these annals of the damned occurs the datum of segregation. But a whirlwind is thought of as a condition of chaos - quasi-chaos: not final negativeness, of course -

Monthly Weather Review, July, 1881:

"A small pond in the track of the cloud was sucked dry, the water being carried over the adjoining fields together with a large quantity of soft mud, which was scattered over the ground for half a mile around."

It is so easy to say that small frogs that have fallen from the sky had been scooped up by a whirlwind; but here are the circumstances of a scoop; in the exclusionist-imagination there is no regard for mud, debris from the bottom of a pond, floating vegetation, loose things from the shores - but a precise picking out of frogs only. Of all instances I have that attribute the fall of small frogs or toads to whirlwinds, only one definitely identifies or places the whirlwind. Also, as has been said before, a pond going up would be quite as interesting as frogs coming down. Whirlwinds we read of over and over but where and what whirlwind? It seems to me that anybody who had lost a pond would be heard from. In Symons' Meteorological Magazine, 32-106, a fall of small frogs, near Birmingham, June 30, 1892, is attributed to a specific whirlwind - but not a word as to any special pond that had contributed. And something that strikes my attention here is that these frogs are described as almost white.

I'm afraid there is no escape for us: we shall have to give to civilization upon this earth - some new worlds.

Places with white frogs in them.

Upon several occasions we have had data of unknown things that have fallen from - somewhere. But something not to be overlooked is that if living things have landed alive upon this earth - in spite of all we think we know of the accelerative velocity of falling bodies - and have propagated - why the exotic becomes the indigenous, or from the strangest of places we'd expect the familiar. Or if hosts of living frogs have come here - from somewhere else - every living thing upon this earth may, ancestrally, have come from - somewhere else.

I find that I have another note upon a specific hurricane:

Annals and Mag. of Nat. Hist., 1-3-185:

After one of the greatest hurricanes in the history of Ireland, some fish were found "as far as 15 yards from the edge of a lake."

Have another: this is a good one for the exclusionists:

Fall of fish in Paris: said that a neighboring pond had been blown dry. (Living Age, 52-186.) Date not given, but I have seen it recorded somewhere else.

The best known fall of fishes from the sky is that which occurred at Mountain Ash, in the Valley of Abedare, Glamorganshire, Feb. n, 1859.

The Editor of the Zoologist, 2-677, having published a report of a fall of fishes, writes: "I am continually receiving similar accounts of frogs and fishes." But, in all the volumes of the Zoologist, I can find only two reports of such falls. There is nothing to conclude other than that hosts of data have been lost because orthodoxy does not look favorably upon such reports. The Monthly Weather Review records several falls of fishes in the United States; but accounts of these reported occurrences are not findable in other American publications. Nevertheless, the treatment by the Zoologist, of the fall reported from Mountain Ash is fair. First appears, in the issue of 1859-6493, a letter from the Rev. John Griffith, Vicar of Abedare, asserting that the fall had occurred, chiefly upon the property of Mr. Nixon, of Mountain Ash. Upon page 6540, Dr. Gray, of the British Museum, bridling with exclusionism, writes that some of these fishes, which ad been sent to him alive, were "very young minnows." He said: "On reading the evidence, it seems to me most probably only a practical joke: that one of Mr. Nixon's employees had thrown a pailful of water upon another, who had thought fish in it had fallen from the sky" - had dipped up a pailful from a brook.

Those fishes - still alive - were exhibited at the Zoological Gardens, Regent's Park. The Editor says that one was a minnow and that the rest were sticklebacks.

He says that Dr. Gray's explanation is no doubt right.

But, upon page 6564, he publishes a letter from another correspondent, who apologizes for opposing "so high an authority as Dr. Gray/' but says that he had obtained some of these fishes from persons who lived at a considerable distance apart, or considerably out of range of the playful pail of water.

According to the Annual Register, 1859-14, the fishes themselves had fallen by pailfuls.

If these fishes were not upon the ground in the first place, we base our objections to the whirlwind explanation, upon two data:

That they fell in no such distribution as one could attribute to the discharge of a whirlwind, but upon a narrow strip of land: about 80 yards long and 12 yards wide -

The other datum is again the suggestion that at first seemed so incredible, but for which support is piling up, a suggestion of a stationary source overhead -

That ten minutes later another fall of fishes occurred upon this same narrow strip of land.

Even arguing that a whirlwind may stand still axially, it discharges tangentially. Wherever the fishes came from it does not seem thinkable that some could have fallen and that others could have whirled even a tenth of a minute, then falling directly after the first to fall. Because of these evil circumstances the best adaptation was to laugh the whole thing off and say that some one had soused some one else with a pailful of water, in which a few "very young" minnows had been caught up.

In the London Times, March 2, 1859, is a letter from Mr. Aaron Roberts, curate of St. Peter's, Carmathon. In this letter the fishes are said to have been about four inches long, but there is some question of species. I think, myself, that they were minnows and sticklebacks. Some persons, thinking them to be sea fishes, placed them in salt water, according to Mr. Roberts. "The effect is stated to have been almost instantaneous death." "Some were placed in fresh water. These seemed to thrive well." As to narrow distribution, we are told that the fishes fell "in and about the premises of Mr. Nixon." "It was not observed at the time that any fish fell in any other part of the neighborhood, save in the particular spot mentioned."

In the London Times, March 10, 1859, Vicar Griffith writes an account:

"The roofs of some houses were covered with them."

In this letter it is said that the largest fishes were five inches long, and that these did not survive the fall.

Report of the British Association, 1859-158:

"The evidence of the fall of fish on this occasion was very conclusive."

A specimen of the fish was exhibited and was found to be the Gasterosteus leirus.

Gasterosteus is the stickleback.

Altogether I think we have not a sense of total perdition, when we're damned with the explanation that some one soused some one else with a pailful of water, in which were thousands of fishes four or five inches long, some of which covered roofs of houses, and some of which remained ten minutes in the air. By way of contrast we offer our own acceptance:

That the bottom of a super-geographical pond had dropped out.

I have a great many notes upon the fall of fishes, despite the difficulty these records have in getting themselves published, but I pick out the instances that especially relate to our super-geographical acceptances, or to the Principles of Super-Geography: or data of things that have been in the air longer than acceptably could a whirlwind carry them ; that have fallen with a distribution narrower than is attributable to a whirlwind; that have fallen for a considerable length of time upon the same narrow area of land.

These three factors indicate, somewhere not far aloft, a region of inertness to this earth's gravitation, of course, however, a region that, by the flux and variation of all things, must at times be susceptible - but, afterward, our heresy will bifurcate -

In amiable accommodation to the crucifixion it'll get, I think -

But so impressed are we with the datum that, though there have been many reports of small frogs that have fallen from the sky, not one report upon a fall of tadpoles is findable, that to these circumstances another adjustment must be made.

Apart from our three factors of indication, an extraordinary observation is the fall of living things without injury to them. The devotees of St. Isaac explain that they fall upon thick grass and so survive: but Sir James Emerson Tennant, in his "History of Ceylon," tells of a fall of fishes upon gravel, by which they were seemingly uninjured. Something else apart from our three main interests is a phenomenon that looks like what one might call an alternating series of falls of fishes, whatever the significance may be:

Meerut, India, July, 1824 (Living Age, 52-186); Fifeshire, Scotland, summer of 1824 (Wernerian Nat. Hist. Soc. Trans., 5-575); Moradabad, India, July, 1826 (Living Age, 52-186); Ross-shire, Scotland, 1828 (Living Age, 52-186); Moradabad, India, July 20, 1829 (Lin. Soc. Trans., 16-764); Perthshire, Scotland, (Living Age, 52-186); Argyleshire, Scotland, 1830, March 9, 1830 (Recreative Science, 3-339); Feridpoor, India, Feb. 19, 1830 (Jour. Asiatic Soc. of Bengal, 2-650).

A psycho-tropism that arises here disregarding serial significance - or mechanical, unintelligent, repulsive reflex - is that the fishes of India did not fall from the sky; that they were found upon the ground after torrential rains, because streams had overflowed and had then receded.

In the region of Inertness that we think we can conceive of, or a zone that is to this earth's gravitation very much like the neutral zone of a magnet's attraction, we accept that there are bodies of water and also clear spaces bottoms of ponds dropping out - very interesting ponds, having no earth at bottom vast drops of water afloat in what is called space fishes and deluges of water falling -

But also other areas, in which fishes however they got there: a matter that we'll consider remain and dry, or even putrefy, then sometimes falling by atmospheric dislodgment.

After a "tremendous deluge of rain, one of the heaviest falls on record" (All the Year Round, 8-255) at Rajkote, India, July 25, 1850, "the ground was found literally covered with fishes."

The word 'found" is agreeable to the repulsion of the conventionalists and their concept of an overflowing stream - but, according to Dr. Buist, some of these fishes were "found" on the tops of haystacks.

Ferrel (A Popular Treatise, p. 414) tells of a fall of living fishes - some of them having been placed in a tank, where they survived - that occurred in India, about 20 miles south of Calcutta, Sept. 20, 1839. A witness of this fall says:

"The most strange thing which ever struck me was that the fish did not fall helter-skelter, or here and there, but they fell in a straight line, not more than a cubit in breadth." See Living Age, 52-186.

Amer. Jour. Sci., 1-32-199:

That, according to testimony taken before a magistrate, a fall occurred, Feb. 19, 1830, near Ferid-poor, India, of many fishes, of various sizes - some whole and fresh and others "mutilated and pu-trefying." Our reflex to those who would say that, in the climate of India, it would not take long for fishes to putrefy, is - that high in the air, the climate of India is not torrid. Another peculiarity of this fall is that some of the fishes were much larger than others.

Or to those who hold out for segregation in a whirlwind, or that objects, say, twice as heavy as others would be separated from the lighter, we point out that some of these fishes were twice as heavy as others.

In the Journal of the Asiatic Society of Bengal, 2-650, depositions of witnesses are given:

"Some of the fish were fresh, but others were rotten and without heads."

'Among the number which I got, five were fresh and the rest stinking and headless."

They remind us of His Grace's observation of some pages back.

According to Dr. Buist, some of these fishes weighed one and a half pounds each and others three pounds.

A fall of fishes, at Futtepoor, India, May 16, 1833:

"They were all dead and dry." (Dr. Buist, Living Age, 52-186.)

India is far away: about 1830 was long ago:

Nature, Sept. 19, 1918-46:

A correspondent writes, from the Dove Marine Laboratory, Cuttercoats, England, that, at Hindon, a suburb of Sunderland, Aug. 24, 1918, hundreds of small fishes, identified as sand eels, had fallen - Again the small area: about 60 by 30 yards.

The fall occurred during a heavy rain that was accompanied by Thunder - or indications of distur-bance aloft but by no visible lightning. The sea is close to Hindon, but if you try to think of these fishes having described a trajectory in a whirlwind from the ocean, consider this remarkable datum:

That, according to witnesses, the fall upon this small area, occupied ten minutes.

I can not think of a clearer indication of a direct fall from a stationary source.

And:

"The fish were all dead, and indeed stiff and hard, when picked up, immediately after the occur-rence."

By all of which I mean that we have only begun to pile up our data of things that fall from a station-ary source overhead: we'll have to take up the subject from many approaches before our acceptance, which seems quite as rigorously arrived at as ever has been a belief, can emerge from the accursed.

I don't know how much the horse and the barn will help us to emerge: but, if ever anything did go up from this earth's surface and stay up - those damned things - may have:

Monthly Weather Review, May, 1878:

In a tornado, in Wisconsin, May 23, 1878, "a barn and a horse were carried completely away, and neither horse nor barn, nor any , portion of either have since been found." After that, which would be a little strong were it not for a steady improvement in our digestions that I note as we go along, there is little of the bizarre or the unassimilable, in the turtle that hovered six months or so over a small town in Mississippi:

Monthly Weather Review, May, 1894:

That, May n, 1894, at Vicksburg, Miss, fell a small piece of alabaster; that, at Bovina, eight miles from Vicksburg, fell a gopher turtle.

They fell in a hailstorm.

This item was widely copied at the time: for instance, Nature, one of the volumes of 1894, page 430, and Jour. Roy. Met. Soc., 20-273. As to discussion - not a word. Or Science and its continuity with Presbyterianism - data like this are damned at birth. The Weather Review does sprinkle, or baptize, or attempt to save, this infant but in all the meteorological literature that I have gone through, after that date not a word, except mention once or twice.

The Editor of the Review says:

"An examination of the weather map shows that these hailstorms occur on the south side of a region of cold northerly winds, and were but a small part of a series of similar storms ; apparently some special local whirls or gusts carried heavy objects from this earth's surface up to the cloud regions."

Of all incredibilities that we have to choose from, I give first place to a notion of a whirlwind pouncing upon a region and scrupulously selecting a turtle and a piece of alabaster. This time, the other mechanical thing "there in the first place" can not rise in response to its stimulus: it is resisted in that these objects were coated with ice - month of May in a southern state. If a whirlwind at all, there must have been very limited selection: there is no record of the fall of other objects. But there is no attempt in the Review to specify a whirlwind.

These strangely associated things were remarkably separated.

They fell eight miles apart.

Then - as if there were real reasoning they must have been high to fall with such divergence, or one of them must have been carried partly horizontally eight miles farther than the other. But either supposition argues for power more than that of a local whirl or gust, or argues for a great, specific disturbance, of which there is no record for the month of May, 1894.

Nevertheless - as if I really were reasonable - I do feel that I have to accept that this turtle had been raised from this earth's surface, somewhere near Vicksburg - because the gopher turtle is common in the southern states.

Then I think of a hurricane that occurred in the state of Mississippi weeks or months before May n, 1894.

No - I don't look for it - and inevitably find it.

Or that things can go up so high in hurricanes that they stay up indefinitely - but may, after a while, be shaken down by storms. Over and over have we noted the occurrence of strange falls in storms. So then that the turtle and the piece of alabaster may have had far different origins - from different worlds, perhaps - have entered a region of suspension over this earth wafting near each other long duration - final precipitation by atmospheric disturbance - with hail - or that hailstones, too, when large, are phenomena of suspension of long duration : that it is highly unacceptable that the very large ones could become so great only in falling from the clouds.

Over and over has the note of disagreeableness, or of putrefaction, been struck - long duration. Other indications of long duration. I think of a region somewhere above this earth's surface, in which gravitation is inoperative, and is not governed by the square of the distance quite as magnet-ism is negligible at a very short distance from a magnet. Theoretically the attraction of a magnet should decrease with the square of the distance, but the falling-off is found to be almost abrupt at a short distance.

I think that things raised from this earth's surface to that region have been held there until shaken down by storms - The Super-Sargasso Sea.

Derelicts, rubbish, old cargoes from inter-planetary wrecks; things cast out into what is called space by convulsions of other planets, things from the times of the Alexanders, Caesars and Napoleons of Mars and Jupiter and Neptune; things raised by this earth's cyclones: horses and barns and ele-phants and flies and dodoes, moas, and pterodactyls; leaves from modern trees, and leaves of the Car-boniferous era all, however, tending to disintegrate into homogeneous-looking muds or dusts, red or black or yellow - treasure-troves for the palaeontologists and for the archaeologists - accumula-tions of centuries - cyclones of Egypt, Greece, and Assyria - fishes dried and hard, there a short time: others there long enough to putrefy - But the omnipresence of Heterogeneity - or living fishes, also ponds of fresh water: oceans of salt water.

As to the Law of Gravitation, I prefer to take one simple stand:

Orthodoxy accepts the correlation and equivalence of forces:

Gravitation is one of these forces.

All other forces have phenomena of repulsion and of inertness irrespective of distance, as well as of attraction.

But Newtonian Gravitation admits attraction only:

Then Newtonian Gravitation can be only one-third acceptable even to the orthodox, or there is de-nial of the correlation and equivalence of forces.

Or still simpler:

Here are the data.

Make what you will, yourself, of them.

In our Intermediatist revolt against homogeneous, or positive, explanations, or our acceptance that the all-sufficing cannot be less than universality, besides which, however, there would be nothing to suffice, our expression upon the Super-Sargasso Sea, though it harmonizes with data of fishes that fall as if from a stationary source and, of course, with other data, too is inadequate to account for two peculiarities of the falls of frogs:

That never has a fall of tadpoles been reported;

That never has a fall of full-grown frogs been reported -

Always frogs a few months old.

It sounds positive, bat, if there be such reports they are somewhere out of my range of reading.

But tadpoles would be more likely to fall from the sky, than would frogs, little or big, if such falls be attributed to whirlwinds; and more likely to fall from the Super-Sargasso Sea, if, though very tentatively and provisionally, we accept the Super-Sargasso Sea. Before we take up an especial expression upon the fall of immature and larval forms of life to this earth, and the necessity then of conceiving of some factor besides mere stationariness or suspension or stagnation, there are other data that are similar to data of falls of fishes.

Science Gossip, 1886-238:

That small snails, of a land species, had fallen near Redruth, Cornwall, July 8, 1886, "during a heavy thunderstorm:" roads and fields strewn with them, so that they were gathered up by the hatful: none seen to fall by the writer of this account: snails said to be "quite different to any previously known in this district."

But, upon page 282, we have better orthodoxy. Another correspondent writes that he had heard of the supposed fall of snails: that he had supposed that all such stories had gone the way of witch stories; that, to his astonishment, he had read an account of this absurd story in a local newspaper of "great and deserved repute."

"I thought I should for once like to trace the origin of one of these fabulous tales."

Our own acceptance is that justice can not be in an intermediate existence, in which there can be approximation only to justice or to injustice; that to be fair is to have no opinion at all; that to be honest is to be uninterested; that to investigate is to admit prejudice; that nobody has ever really investigated anything, but has always sought positively to prove or to disprove something that was conceived of, or suspected, in advance.

"As I suspected," says this correspondent, "I found that the snails were of a familiar land-species" - that they had been upon the ground "in the first place."

He found that the snails had appeared after the rain: that "astonished rustics had jumped to the conclusion that they had fallen." He met one person who said that he had seen the snails fall.

"This was his error," says the investigator.

In the Philosophical Magazine, 58-310, there is an account of snails said to have fallen at Bristol, in a field of three acres, in such quantities that they were shovelled up. It is said that the snails "may be considered as a local species." Upon page 457, another correspondent says that the numbers had been exaggerated, and that in his opinion they had been upon the ground in the first place.

But that there had been some unusual condition aloft comes out in his observation upon "the curious azure-blue appearance of the sun, at the time."

Nature, 47-278:

That, according to Das Wetter, Dec., 1892, upon August 9, 1892, a yellow cloud appeared over Paderborn, Germany. From this cloud, fell a torrential rain, in which were hundreds of mussels. There is no mention of whatever may have been upon the ground in the first place, nor of a whirlwind.

Lizards - said to have fallen on the sidewalks of Montreal, Canada, Dec. 28, 1857. (Notes and Queries, 8-6-104.) In the Scientific American, 3-112, a correspondent writes, from South Granville, N. Y., that, during a heavy shower, July 3, 1860, he heard a peculiar sound at his feet, and looking down, saw a snake lying as if stunned by a fall. It then came to life. Gray snake, about a foot long.

These data have any meaning or lack of meaning or degree of damnation you please: but, in the matter of the fall that occurred at Memphis, Tennessee, occur some strong significances. Our quasireasoning upon this subject applies to all segregations so far considered.

Monthly Weather Review, Jan. 15, 1877:

That, in Memphis, Tenn., Jan. 15, 1877, rather strictly localized, or "in a space of two blocks," and after a violent storm in which the rain "fell in torrents," snakes were found. They were crawling on sidewalks, in yards, and in streets, and in masses - but "none were found on roofs or any other elevation above ground" and "none were seen to fall."

* If you prefer to believe that the snakes had always been there, or had been upon the ground in the first place, and that it was only that something occurred to call special attention to them, in the streets of Memphis, Jan. 15, 1877 - why that's sensible: that's the common sense that has been against us from the first.

It is not said whether the snakes were of a known species or not, but that "when first seen, they were of a dark brown, almost black." Blacksnakes, I suppose.

If we accept that these snakes did fall, even though not seen to fall by all the persons who were out sight-seeing in a violent storm, and had not been in the streets crawling loose or in thick tangled masses, in the first place;

If we try to accept that these snakes had been raised from some other part of this earth's surface in a whirlwind;

If we try to accept that a whirlwind could segregate them -

We accept the segregation of other objects raised in that whirlwind.

Then, near the place of origin, there would have been a fall of heavier objects that had been snatched up with the snakes - stones, fence rails, limbs of trees. Say that the snakes occupied the next gradation, and would be the next to fall. Still farther would there have been separate falls of lightest objects: leaves, twigs, tufts of grass.

In the Monthly Weather Review there is no mention of other falls said to have occurred anywhere in January, 1877.

Again ours is the objection against such selectiveness by a whirlwind.

Conceivably a whirlwind could scoop out a den of hibernating snakes, with stones and earth and an infinitude of other debris, snatching up dozens of snakes, - I don't know how many to a den - hundreds may be - but, according to the account of this occurrence in the New York Times, there were thousands of them; alive; from one foot to eighteen inches in length. The Scientific American, 36-86, records the fall, and says that there were thousands of them.

The usual whirlwind-explanation is given - "but in what locality snakes exist in such abundance is yet a mystery."

This matter of enormousness of numbers suggests to me something of a migratory nature - but that snakes in the United States do not migrate in the month of January, if ever.

As to falls or flutterings of winged insects from the sky, prevailing notions of swarming would seem explanatory enough: nevertheless, in instances of ants, there are some peculiar circumstances.

L'Astronomie, 1 889-3 53 :

Fall of fishes, June 13, 1889, in Holland; ants, Aug. 1, 1889, Strasbourg; little toads, Aug. 2, 1889,

Savoy. Fall of ants, Cambridge, England, summer of 1874 - "some were wingless." (Scientific American, 30-193.) Enormous fall of ants, Nancy, France, July 21, 1887 - "most of them were wingless." (Nature, 36-349.) Fall of enormous, unknown ants - size of wasps - Manitoba, June, 1895. (Sci. Amer., 72-385.)

However, our expression will be:

That wingless, larval forms of life, in numbers so enormous that migration from some place external to this earth is suggested, have fallen from the sky.

That these "migrations" - if such can be our acceptance - have occurred at a time of hibernation and burial far in the ground of larvae in the northern latitudes of this earth; that there is significance in recurrence of these falls in the last of January - or that we have the square of an incredibility in such a notion as that of selection of larvae by whirlwinds, compounded with selection of the last of January.

I accept that there are "snow worms" upon this earth - whatever their origin may have been. In the Proc. Acad. Nat. Sci. of Philadelphia, 1899-125, there is a description of yellow worms and black worms that have been found together on glaciers in Alaska. Almost positively were there no other forms of insect-life upon these glaciers, and there was no vegetation to support insect-life, except microscopic organisms. Nevertheless the description of this probably polymorphic species fits a description of larvae said to have fallen in Switzerland, and less definitely fits another description. There is no opposition here, if our data of falls are clear. Frogs of every-day ponds look like frogs said to have fallen from the sky - except the whitish frogs of Birmingham. However, all falls of larvae have not positively occurred in the last of January:

London Times, April 14, 1837:

That, in the parish of Bramford Speke, Devonshire, a large number of black worms, about three-quarters of an inch in length, had fallen in a snow storm.

In Timb's Year Book, 1877-26, it is said that, in the winter of 1876, at Christiania, Norway, worms were found crawling upon the ground. The occurrence is considered a great mystery, because the worms could not have come up from the ground, inasmuch as the ground was frozen at the time, and because they were reported from other places, also, in Norway.

Immense number of black insects in a snowstorm, in 1827, at Pakroff, Russia. (Scientific American, 30-193.)

Fall, with snow, at Orenburg, Russia, Dec. 14, 1830, of a multitude of small, black insects, said to have been gnats, but also said to have had flea-like motions. (Amer. Jour. Sci., 1-22-375.)

Large number of worms found in a snowstorm, upon the surface of snow about four inches thick, near Sangerfield, N. Y., Nov. 18, 1850 (Scientific American, 6-96). The writer thinks that the worms had been brought to the surface of the ground by rain, which had fallen previously.

Scientific American, Feb. 21, 1891:

"A puzzling phenomenon has been noted frequently in some parts of the Valley Bend District, Randolph County, Va., this winter. The crust of the snow has been covered two or three times with worms resembling the ordinary cut worms. Where they come from, unless they fall with the snow is inexplicable." In the Scientific American, Mar. 7, 1891, the Editor says that similar worms had been seen upon the snow near Utica, N. Y., and in Oheida and Herkimer Counties; that some of the worms had been sent to the Department of Agriculture at Washington. Again two species, or polymorphism. According to Prof. Riley, it was not polymorphism, "but two distinct species" - which, because of our data, we doubt.

One kind was larger than the other: color-differences not distinctly stated. One is called the larvae of the common soldier beetle and the other "seems to be a variety of the bronze cut worm." No attempt to explain the occurrence in snow.

Fall of great numbers of larvae of beetles, near Mortagne, France, May, 1858. The larvae were inanimate as if with cold. (Annales Society Entomologique de France, 1858.)

Trans. Ent. Soc. of London, 1871-183, records "snowing of larvae," in Silesia, 1806; "appearance of many larvae on the snow," in Saxony, 1811; "larvae found alive on the snow," 1828; larvae and snow which "fell together," in the Eifel, Jan. 30, 1847; "fall of insects," Jan. 24, 1849, m Lithuania; occurrence of larvae estimated at 300,000 on the snow in Switzerland, in 1856. The compiler says that most of these larvae live underground, or at the roots of trees j- that whirlwinds uproot trees, and carry away the larvae - conceiving of them as not held in masses of frozen earth - all as neatly detachable as currants in something. In the Revue et Magasin de Zoologie, 1849-72, there is an account of the fall in Lithuania, Jan. 24, 1849 - that black larvae had fallen in enormous numbers.

Larvae thought to have been of beetles, but described as "caterpillars," not seen to fall, but found crawling on the snow, after a snowstorm, at Warsaw, Jan. 20, 1850. (All the Year Round, 8-253.)

Flammarion (The Atmosphere, p. 414) tells of a fall of larvae that occurred Jan. 30, 1869, in a snowstorm, in Upper Savoy:

"They could not have been hatched in the neighborhood, for, during the days preceding, the temperature had been very low"; said to have been of a species common in the south of France. In La Science Pour Tons, 14-183, it is said that with these larvae there were developed insects.

L'Astronomie, 1890-313:

That, upon the last of January, 1890, there fell, in a great tempest, in Switzerland, incalculable numbers of larvae: some black and some yellow; numbers so great that hosts of birds were attracted.

Altogether we regard this as one of our neatest expressions for external origins and against the whirlwind-explanation. If an exclusionist says that, in January, larvae were precisely and painstakingly picked out of frozen ground, in incalculable numbers, he thinks of a tremendous force - disregarding its refinements: then if origin and precipitation be not far apart, what becomes of an infinitude of other debris, conceiving of no time for segregation?

If he thinks of a long translation - all the way from the south of France to Upper Savoy, he may think then of a very fine sorting over by differences of specific gravity but in such a fine selection, larvae would be separated from developed insects.

As to differences in specific gravity - the yellow larvae that fell in Switzerland Jan., 1890, were three times the size of the black larvae that fell with them. In accounts of this occurrence, there is no denial of the fall.

Or that a whirlwind never brought them together and held them together and precipitated them and only them together -

That they came from Genesistrine.

There's no escape from it. We'll be persecuted for it. Take it or leave it -

Genesistrine.

The notion is that there is somewhere aloft a place of origin of life relatively to this earth. Whether it's the planet Genesistrine, or the moon, or a vast amorphous region super-jacent to this earth, or an island in the Super-Sargasso Sea, should perhaps be left to the researches of other super - or extra - geographers. That the first unicellular organisms may have come here from Genesistrine or that men or anthropomorphic beings may have come here before amoebae: that, upon Genesistrine, there may have been an evolution expressible in conventional biologic terms, but that evolution upon this earth has been - like evolution in modern Japan - induced by external influences; that evolution, as a whole, upon this earth, has been a process of population by immigration or by bombardment.

Some notes I have upon remains of men and animals encysted, or covered with clay or stone, as if fired here as projectiles, I omit now, because it seems best to regard the whole phenomenon as a tropism - as a geotropism - probably atavistic, or vestigial, as it were, or something still continuing long after expiration of necessity; that, once upon a time, all kinds of things came here from Genesistrine, but that now only a few kinds of bugs and things, at long intervals, feel the inspiration.

Not one instance have we of tadpoles that have fallen to this earth. It seems reasonable that a whirlwind could scoop up a pond, frogs and all and cast down the frogs somewhere else: but, then, more reasonable that a whirlwind could scoop up a pond, tadpoles and all because tadpoles are more numerous in their season than are the frogs in theirs: but the tadpole-season is earlier in the spring, or in a time that is more tempestuous. Thinking in terms of causation - as if there were real causes - our notion is that, if X is likely to cause Y, but is more likely to cause Z, but does not cause Z, X is not the cause of Y. Upon this quasi-sorites, we base our acceptance that the little frogs that have fallen to this earth are not products of whirlwinds: that they came from externality, or from Genesistrine.

I think of Genesistrine in terms of biologic mechanics: not that somewhere there are persons who collect bugs in or about the last of January and frogs in July and August, and bombard this earth, any more than do persons go through northern regions, catching and collecting birds, every autumn, then casting them southward.

But atavistic, or vestigial, geotropism in Genesistrine - or a million larvae start crawling, and a million little frogs start hopping - knowing no more what it's all about than we do when we crawl to work in the morning and hop away at night.

I should say, myself, that Genesistrine is a region in the Super – Sargasso Sea, and that parts of the Super-Sargasso Sea have rhythms of susceptibility to this earth's attraction.

Chapter VIII

I accept that, when there are storms, the dam-dest of excluded, excommunicated things - things that are leprous to the faithful f - are brought down from the Super-Sargasso Sea - or from what for convenience we call the Super-Sargasso Sea - which by no means has been taken into full acceptance yet.

That things are brought down by storms, just as, from the depths of the sea things are brought up by storms. To be sure it is orthodoxy that storms have little, if any, effect, below the waves of the ocean – but - of course - only to have an opinion is to be ignorant of, or to disregard a contradiction, or something else that modifies an opinion out of distinguishability.

Symon's Meteorological Magazine, 47-180:

That, along the coast of New Zealand, in regions not subject to submarine volcanic action, deep-sea fishes are often brought up by storms.

Iron and stones that fall from the sky; and atmospheric disturbances:

"There is absolutely no connection between the two phenomena." (Symons.)

The orthodox belief is that objects moving at planetary velocity would, upon entering this earth's atmosphere, be virtually unaffected by hurricanes; might as well think of a bullet swerved by some one fanning himself. The only trouble with the orthodox reasoning is the usual trouble - its phantom-dominant - its basing upon a myth - data we've had, and more we'll have, of things in the sky having no independent velocity. There are so many storms and so many meteors and meteorites that it would be extraordinary if there were no concurrences. Nevertheless so many of these concurrences are listed by Prof. Baden-Powell (Rept. Brit. Assoc., 1850-54) that one notices.

See Rept. Brit. Assoc., 1860 - other instances.

The famous fall of stones at Siena, Italy, 1794 - "in a violent thunderstorm."

See Greg's Catalogues many instances. One that stands out is - "bright ball of fire and light in a hurricane in England, Sept. 2, 1786." The remarkable datum here is that this phenomenon was invisible forty minutes. That's about 800 times the duration that the orthodox give to meteors and meteorites.

See the Annual Register - many instances.

In Nature, Oct. 25, 1877, and the London Times, Oct. 15, 1877, something that fell in a gale of Oct. 14, 1877, is described as a "huge ball of green fire." This phenomenon is described by another correspondent, in Nature, 17-10, and an account of it by another correspondent was forwarded to Nature by W. F. Denning.

There are so many instances that some of us will revolt against the insistence of the faithful that it is only coincidence, and accept that there is connection of the kind called causal. If it is too difficult to think of stones and metallic masses swerved from their courses by storms, if they move at high velocity, we think of low velocity, or of things having no velocity at all, hovering a few miles above this earth, dislodged by storms, and falling luminously.

But the resistance is so great here, and "coincidence" so insisted upon that we'd better have some more instances:

Aerolite in a storm at St. Leonards-on-sea, England, Sept. 17, 1885 - no trace of it found (Annual Register, 1885) ; meteorite in a gale, March 1, 1886, described in the Monthly Weather Review, March, 1886; meteorite in a thunderstorm, off coast of Greece, Nov. 19, 1899 (Nature, 61-111); fall of a meteorite in a storm, July 7, 1883, near Lachine, Quebec (Monthly Weather Review, July, 1883); same phenomenon noted in Nature, 28-319; meteorite in a whirlwind, Sweden, Sept. 24, 1883 (Nature, 29-15).

London Roy. Soc. Proc., 6-276:

A triangular cloud that appeared in a storm, Dec. 17, 1852; a red nucleus, about half the apparent diameter of the moon, and a long tail; visible 13 minutes; explosion of the nucleus.

Nevertheless, in Science Gossip, n.s., 6-65, it is said that, though meteorites have fallen in storms, no connection is supposed to exist between the two phenomena, except by the ignorant peasantry.

But some of us peasants have gone through the Report of the British Association, 1852. Upon page 239, Dr. Buist, who had never heard of the Super-Sargasso Sea, says that, though it is difficult to trace connection between the phenomena, three aerolites had fallen in five months, in India, during thunderstorms, in 1851 (may have been 1852). For accounts by witnesses, see page 229, of the Report.

Or - we are on our way to account for "thunderstones." It seems to me that, very strikingly here, is borne out the general acceptance that ours is only an intermediate existence, in which there is nothing fundamental, or nothing final to take as a positive standard to judge by.

Peasants believed in meteorites.

Scientists excluded meteorites.

Peasants believe in "thunderstones."

Scientists exclude "thunderstones."

It is useless to argue that peasants are out in the fields, and that scientists are shut up in laboratories and lecture rooms. We can not take for a real base that, as to phenomena with which they are more familiar, peasants are more likely to be right than are scientists:

a host of biologic and meteorologic fallacies of peasants rises against us.

I should say that our "existence" is like a bridge - except that that comparison is in static terms - but like the Brooklyn Bridge, upon which multitudes of bugs are seeking a fundamental coming to a girder that seems firm and final - but the girder is built upon supports. A support then seems final. But it is built upon underlying structures. Nothing final can be found in all the bridge, because the bridge itself is not a final thing in itself, but is a relationship between Manhattan and Brooklyn. If our "existence" is a relationship between the Positive Absolute and the Negative Absolute, the quest for finality in it is hopeless: everything in it must be relative, if the "whole" is not a whole, but is, itself, a relation.

In the attitude of Acceptance, our pseudo-base is:

Cells of an embryo are in the reptilian era of the embryo;

Some cells feel stimuli to take on new appearances.

If it be of the design of the whole that the next era be mammalian, those cells that turn mammalian will be sustained against resistance, by inertia, of all the rest, and will be relatively right, though not finally right, because they, too, in time will have to give away to characters of other eras of higher development.

If we are upon the verge of a new era, in which Exclusionism must be overthrown, it will avail thee not to call us base-born and frowsy peasants.

In our crude, bucolic way, we now offer an outrage upon common sense that we think will some day be an unquestioned commonplace:

That manufactured objects of stone and iron have fallen from the sky:

That they have been brought down from a state of suspension, in a region of inertness to this earth's attraction, by atmospheric disturbances.

The "thunderstone" is usually "a beautifully polished, wedge-shaped piece of greenstone," says a writer in the Cornhill Magazine, 50-517. It isn't: it's likely to be of almost any kind of stone, but we call attention to the skill with which some of them have been made. Of course this writer says it's all superstition. Otherwise he'd be one of us crude and simple sons of the soil.

Conventional damnation is that stone implements, already on the ground - "on the ground in the first place" - are found near where lightning was seen to strike: that are supposed by astonished rustics, or by intelligence of a low order, to have fallen in or with lightning.

Throughout this book, we class a great deal of science with bad fiction. When is fiction bad, cheap, low? If coincidence is overworked.

That's one way of deciding. But with single writers coincidence seldom is overworked: we find the excess in the subject at large. Such a writer as the one of the Cornhill Magazine tells us vaguely of beliefs of peasants: there is no massing of instance after instance after instance. Here ours will be the method of mass-formation.

Conceivably lightning may strike the ground near where there was a wedge-shaped object in the first place: again and again and again: lightning striking ground near wedge-shaped object in China; lightning striking ground near wedge-shaped object in Scotland; lightning striking ground near wedge-shaped object in Central Africa: coincidence in France; coincidence in Java; coincidence in South America -

We grant a great deal but note a tendency to restlessness. Nevertheless this is the psycho-tropism of science to all "thunderstones" said to have fallen luminously.

As to greenstone, it is in the island of Jamaica, where the notion is general that axes of a hard greenstone fall from the sky "during the rains." (Jour. Inst. Jamaica, 2-4.) Some other time we shall inquire into this localization of objects of a specific material. "'They are of a stone nowhere else to be found in Jamaica." (Notes and Queries, 2-8-24.)

In my own tendency to exclude, or in the attitude of one peasant or savage who thinks he is not to be classed with other peasants or savages, I am not very much impressed with what natives think. It would be hard to tell why. If the word of a Lord Kelvin carries no more weight, upon scientific subjects, than the word of a Sitting Bull, unless it be in agreement with conventional opinion - I think it must be because savages have bad table manners. However, my snobbishness, in this respect, loosens up somewhat before very widespread belief by savages and peasants. And the notion of "thunderstones" is as wide as geography itself.

The natives of Burmah, China, Japan, according to Blinkenberg (Thunder Weapons, p. 100) - not, of course, that Blinkenberg accepts one word of it - think that carved stone objects have fallen from the sky, because they think they have seen such objects fall from the sky. Such objects are called "thunderbolts" in these countries.

They are called "thunderstones" in Moravia, Holland, Belgium, France, Cambodia, Sumatra, and Siberia. They're called "storm stones" in Lausitz; "sky arrows" in Slavonia; "thunder axes" in England and Scotland; "lightning stones" in Spain and Portugal; "sky axes" in Greece; "lightning flashes" in Brazil; "thunder teeth" in Amboina.

The belief is as widespread as is belief in ghosts and witches, which only the superstitious deny today.

As to beliefs by North American Indians, Tyler gives a list of references (Primitive Culture, 2-237).

As to South American Indians -

"Certain stone hatchets are said to have fallen from the heavens. (Jour. Amer. Folk Lore, 17-203.)

If you, too, revolt against coincidence after coincidence after coincidence, but find our interpretation of "thunderstones" just a little too strong or rich for digestion, we recommend the explanation of one, Tallius, written in 1649:

"The naturalists say they are generated in the sky by fulgurous exhalation conglobed in a cloud by the circumfused humor."

Of course the paper in the Cornhill Magazine was written with no intention of trying really to investigate this subject, but to deride the notion that worked-stone objects have ever fallen from the sky.

A writer in the Amer. Jour. Sci., 1-21-325, read this paper and thinks it remarkable "that any man of ordinary reasoning powers should write a paper to prove that thunderbolts do not exist."

I confess that we're a little flattered by that.

Over and over:

"It is scarcely necessary to suggest to the intelligent reader that thunderstones are a myth."

We contend that there is a misuse of a word here: we admit that only we are intelligent upon this subject, if by intelligence is meant the inquiry of inequilibrium, and that all other intellection is only mechanical reflex - of course that intelligence, too, is mechanical, but less orderly and confined: less obviously mechanical - that as an acceptance of ours becomes firmer and firmer-established, we pass from the state of intelligence to reflexes in ruts. An odd thing is that intelligence is usually supposed to be creditable. It may be in the sense that it is mental activity trying to find out, but it is confession of ignorance. The bees, the theologians, the dogmatic scientists are the intellectual aristocrats. The rest f us are plebeians, not yet graduated to Nirvana, or to the instinctive and suave as differentiated from the intelligent and crude.

Blinkenberg gives many instances of the superstition of "thunderstones" which flourishes only where mentality is in a lamentable state - or universally. In Malacca, Sumatra, and Java, natives say that stone axes have often been found under trees that have been struck by lightning. Blinkenberg does not dispute this, but says it is coincidence: that the axes were of course upon the ground in the first place: that the natives jumped to the conclusion that these carved stones had fallen in or with lightning. In Central Africa, it is said that often have wedge-shaped, highly polished objects of stone, described as "axes," been found sticking in trees that have been struck by lightning - or by what seemed to be lightning.

The natives, rather like the unscientific persons of Memphis, Tenn., when they saw snakes after a storm, jumped to the conclusion that the "axes" had not always been sticking in the trees. Livingstone (Last Journal, pages 83, 89, 442, 448) says that he had never heard of stone implements used by natives of Africa, A writer in the Report of the Smithsonian Institute, 1877-308, says that there are a few.

That they are said, by the natives, to have fallen in thunderstorms.

As to luminosity, it is my lamentable acceptance that bodies falling through this earth's atmosphere, if not warmed even, often fall with a brilliant light, looking like flashes of lightning. This matter seems important: we'll take it up later, with data.

In Prussia, two stone axes were found in the trunks of trees, one under the bark. (Blinkenberg, Thunder Weapons, p. 100.)

The finders jumped to the conclusion that the axes had fallen there.

Another stone ax - or wedge-shaped object of worked stone - said to have been found in a tree that had been struck by something that looked like lightning. (Thunder Weapons, p. 71.)

The finder jumped to the conclusion.

Story told by Blinkenberg, of a woman, who lived near Kulsbjaergene, Sweden, who found a flint near an old willow - "near her house." I emphasize "near her house" because that means familiar ground. The willow had been split by something. –

She jumped.

Cow killed by lightning, or by what looked like lightning (Isle of Sark, near Guernsey). The peasant who owned the cow dug up the ground at the spot and found a small greenstone "ax." Blinkenberg says that he jumped to the conclusion that it was this object that had fallen luminously, killing the cow.

Reliquary, 1867-208:

A flint ax found by a farmer, after a severe storm - described as a "fearful storm" - by a signal staff, which had been split by something. I should say that nearness to a signal staff may be considered familiar ground.

Whether he jumped, or arrived at the conclusion by a more leisurely process, the farmer thought that the flint object had fallen in the storm.

In this instance we have a lamentable scientist with us. It's impossible to have positive difference between orthodoxy and heresy : somewhere there must be a merging into each other, or an overlapping.

Nevertheless, upon such a subject as this, it does seem a little shocking. In most works upon meteorites, the peculiar, sulphurous odor of things that fall from the sky is mentioned. Sir John Evans ("Stone Implements," p. 57) says - with extraordinary reasoning powers, if he could never have thought such a thing with ordinary reasoning powers - that this flint object "proved to have been the bolt, by its peculiar smell when broken."

If it did so prove to be, that settles the whole subject. If we prove that only one object of worked stone has fallen from the sky, all piling up of further reports is unnecessary. However, we have already taken the stand that nothing settles anything; that the disputes of ancient Greece are no nearer solution now than they were several thousand years ago - all because, in a positive sense, there is nothing to prove or solve or settle. Our object is to be more nearly real than our opponents. Wideness is an aspect of the Universal. We go on widely. According to us the fat man is nearer godliness than is the thin man. Eat, drink, and approximate to the Positive Absolute. Beware of negativeness, by which we mean indigestion.

The vast majority of "thunderstones" are described as "axes," but Meunier (La Nature, 1892-2-381) tells of one that was in his possession; said to have fallen at Ghardia, Algeria, contrasting "profoundment" (pear-shaped) with the angular outlines of ordinary meteorites. The conventional explanation that it had been formed as a drop of molten matter from a larger body seems reasonable to me; but with less agreeableness I note its fall in a thunderstorm, the datum that turns the orthodox meteorologist pale with rage, or induces a slight elevation of his eyebrows, if you mention it to him.

Meunier tells of another "thunderstone" said to have fallen in North Africa. Meunier, too, is a little lamentable here: he quotes a soldier of experience that such objects fall most frequently in the deserts of Africa.

Rather miscellaneous now:

"Thunderstone" said to have fallen in London, April, 1876:

weight about 8 pounds: no particulars as to shape (Timb's Year Book, 1877-246).

"Thunderstone" said to have fallen at Cardiff, Sept. 26, 1916 (London Times, Sept. 28, 1916). According to Nature, 98-95, it was coincidence; only a lightning flash had been seen. Stone that fell in a storm, near St. Albans, England: accepted by the Museum of St. Albans; said, at the British Museum, not to be of "true meteoritic material." (Nature, 80-34.)

London Times, April 26, 1876:

That, April 20, 1876, near Wolverhampton, fell a mass of meteoritic iron during heavy fall of rain. An account of this phenomenon in Nature, 14 – 272, by H. S. Maskelyne, who accepts it as authentic.

For three . other instances see the Scientific American, 47-194; 52-83; 68-325.

As to wedge-shaped larger than could very well be called an "ax":

Nature, 30-300

That, May 27, 1884, at Tysnas, Norway, a meteorite had fallen: that the turf was torn up at the spot where the object had been [103/104] supposed to have fallen; that two days later "a very peculiar stone" was found near by. The description is -- "in shape and size very like the fourth part of a large Stilton cheese." It is our acceptance that many objects and different substances have been brought down, by atmospheric disturbance from what - only as a matter of convenience now, and until we have more data - we call the Super-Sargasso Sea; however, our chief interest is in objects that have been shaped by means similar to human handicraft.

Description of the "thunderstones" of Burmah (Proc. Asiatic Soc. of Bengal, 1869-183): said to be of a kind of stone unlike any other found in Burmah; called "thunderbolts" by the natives.

I think there's a good deal of meaning in such expressions as "unlike any other found in Burmah" - but that if they had said anything more definite, there would have been unpleasant consequences to writers in the 19th century.

More about the "thunderstones" of Burmah, in the Proc. Soc. Antiq. of London, 2-3-97. One f them, described as an "adze," was exhibited by Captain Duff, who wrote that there was no stone like it in its neighborhood.

Of course it may not be very convincing to say that because a stone is unlike neighboring stones it had foreign origin - also we fear it is a kind of plagiarism: we got it from the geologists, who demonstrate by this reasoning the foreign origin of erratics. We fear we're a little gross and scientific at times.

But it's my acceptance that a great deal of scientific literature must be read between the lines. It's not every one who has the lamentableness of a Sir John Evans. Just as a great deal of Voltaire's meaning was inter-linear, we suspect that a Captain Duff merely hints rather than to risk having a Prof. Lawrence Smith fly at him and call him "a half-insane man." Whatever Captain Duff's meaning may have been, and whether he smiled like a Voltaire when he wrote it, Captain Duff writes of "the extremely soft nature of the stone, rendering it equally useless as an offensive or defensive weapon."

Story, by a correspondent, in Nature, 34-53, of a Malay, of "considerable social standing" and one thing about our data is that, damned though they be, they do so often bring us into awful good company who knew of a tree that had been struck, about a month before, by something in a thunderstorm. He searched among the roots of this tree and found a "thunderstone." Not said whether he jumped or leaped to the conclusion that it had fallen: process likely to be more leisurely in tropical countries. Also I'm afraid his way of reasoning was not very original: just so were fragments of the Bath-furnace meteorite, accepted by orthodoxy, discovered. We shall now have an unusual experience. We shall read of some reports of extraordinary circumstances that were investigated by a man of science - not, of course that they were really investigated by him, but that his phenomena occupied a position approximating higher to real investigation than to utter neglect. Over and over we read of extraordinary occurrences - no discussion; not even a comment afterward findable; mere mention occasionally - burial and damnation.

The extraordinary and how quickly it is hidden away.

Burial and damnation, or the obscurity of the conspicuous.

We did read of a man who, in the matter of snails, did travel some distance to assure himself of something that he had suspected in advance; and we remember Prof. Hitchcock, who had only to smite Amherst with the wand of his botanical knowledge, and lo! two fungi sprang up before night; and we did read of Dr. Gray and his thousands of fishes from one pailful of water - but these instances stand out; more frequently there was no "investigation." We now have a good many reported occurrences that were "investigated."

Of things said to have fallen from the sky, we make, in the usual scientific way, two divisions: miscellaneous objects and substances, and symmetric objects attributable to beings like human beings, sub-dividing into wedges, spheres, and disks.

Jour. Roy. Met. Soc., 14-207:

That, July 2, 1866, a correspondent to a London newspaper wrote that something had fallen from the sky, during a thunderstorm of June 30, 1866, at Netting Hill. Mr. G. T. Symons, of Symons' Meteorological Magazine, investigated, about as fairly, and with about as unprejudiced a mind, as anything ever has been investigated.

He says that the object was nothing but a lump of coal: that, next door to the home of the correspondent coal had been unloaded the day before. With the uncanny wisdom of the stranger upon unfamiliar ground that we have noted before, Mr. Symons saw that the coal reported to have fallen from the sky, and the coal unloaded more prosaically the day before, were identical. Persons in the neighborhood, unable to make this simple identification, had bought from the correspondent pieces of the object reported to have fallen from the sky. As to credulity, I know of no limits for it - but when it comes to paying out money for credulity - oh, no standards to judge by, of course - just the same -

The trouble with efficiency is that it will merge away into excess.

With what seems to me to be super-abundance of convincingness, Mr. Symons then lugs another character into his little comedy:

That it was all a hoax by a chemist's pupil, who had filled a capsule with an explosive, and "during the storm had thrown the burning mass into the gutter, so making an artificial thunderbolt."

Or even Shakespeare, with all his inartistry, did not lug in King Lear to make Hamlet complete. Whether I'm lugging in something that has no special meaning, myself, or not, I find that this storm of June 30, 1866, was peculiar.

It is described in the London Times, July 2, 1866: that "during the storm, the sky, in many places remained partially clear while hail and rain were falling." That may have more meaning when we take up the possible extra-mundane origin of some hailstones, especially if they fall from a cloudless sky. Mere suggestion, not worth much, that there may have been falls of extra-mundane substances, in London, June 30, 1866.

Clinkers, said to have fallen, during a storm, at Kilburn, July 5, 1877:

According to the Kilburn Times, July 7, 1877, quoted by Mr. Symons, a street had been "literally strewn," during the storm, with a mass of clinkers, estimated at about two bushels: sizes from that of a walnut to that of a man's hand "pieces of the clinkers can be seen at the Kilburn Times office."

If these clinkers, or cinders, were refuse from one of the supermercantile constructions from which coke and coal and ashes occasionally fall to this earth, or, rather, to the Super-Sargasso Sea, from which dislodgment by tempests occurs, it is intermediatistic to accept that they must merge away somewhere with local phenomena of the scene of precipitation. If a red-hot stove should drop from a cloud into Broadway, some one would find that at about the time of the occurrence, a moving van had passed, and that the moving men had tired of the stove, or something - that it had not been really red-hot, but had been rouged instead of blacked, by some absent-minded housekeeper. Compared with some of the scientific explanations that we have encountered, there's considerable restraint, I think, in that one.

Mr. Symons learned that in the same street - he emphasizes that it was a short street - there was a fire-engine station. I had such an impression of him hustling and bustling around at Notting Hill, searching cellars until he found one with newly arrived coal in it; ringing door bells, exciting a whole neighborhood, calling up to second-story windows, stopping people in the streets, hotter and hotter on the trail of a wretched impostor of a chemist's pupil. After his efficiency at Notting Hill, we'd expect to hear that he went to the station, and - something like this:

"It is said that clinkers fell, in your street, at about ten minutes past four o'clock, afternoon of July fifth. Will you look over your records and tell me where your engine was at about ten minutes past four, July fifth?"

Mr. Symons says:

"I think that most probably they had been raked out of the steam fire-engine."

June 20, 1880, it was reported that a "thunderstone" had struck the house at 180 Oakley Street, Chelsea, falling down the chimney, into the kitchen grate.

Mr. Symons investigated.

He describes the "thunderstone" as an "agglomeration of brick, soot, unburned coal, and cinder."

He says that, in his opinion, lightning had flashed down the chimney, and had fused some of the brick of it.

He does think it remarkable that the lightning did not then scatter the contents of the grate, which were disturbed only as if a heavy body had fallen. If we admit that climbing up the chimney to find out, is too rigorous a requirement for a man who may have been large, dignified and subject to expansions, the only unreasonableness we find in what he says - as judged by our more modern outlook, is:

"I suppose that no one would suggest that bricks are manufactured in the atmosphere."

Sounds a little unreasonable to us, because it is so of the positivistic spirit of former times, when it was not so obvious that the highest incredibility and laughability must merge away with the "proper" - as the Sci. Am. Sup. would say. The preposterous is always interpretable in terms of the "proper," with which it must be continuous - or - clay-like masses such as have fallen from the sky - tremendous heat generated by their velocity - they bake - bricks.

We begin to suspect that Mr. Symons exhausted himself at Netting Hill. It's a warning to efficiency-fanatics.

Then the instance of three lumps of earthy matter, found upon a well-frequented path, after a thunderstorm, at Reading, July 3, 1883.

There are so many records of the fall of earthy matter from the sky that it would seem almost uncanny to find resistance here, were we not so accustomed to the uncompromising stands of orthodoxy - which, in our metaphysics, represent good, as attempts, but evil in their insufficiency. If I thought it necessary, I'd list one hundred and fifty instances of earthy matter said to have fallen from the sky. It is his antagonism to atmospheric disturbance associated with the fall of things from the sky that blinds and hypnotizes a Mr. Symons here. This especial Mr. Symons rejects the Reading substance because it was not "of true meteoritic material." It's uncanny or it's not uncanny at all, but universal - if you don't take something for a standard of opinion, you can't have any opinion at all: but, if you do take a standard, in some of its applications it must be preposterous. The carbonaceous meteorites, which are unquestioned though avoided, as we have seen - by orthodoxy, are more glaringly of untrue meteoritic material than was this substance of Reading. Mr. Symons says that these three lumps were upon the ground "in the first place."

Whether these data are worth preserving or not, I think that the appeal that this especial Mr. Symons makes is worthy of a place in the museum we're writing. He argues against belief in all external origins "for our credit as Englishmen." He is a patriot, but I think that these foreigners had a small chance "in the first place" for hospitality from him.

Then comes a "small lump of iron (two inches in diameter)" said to have fallen, during a thunderstorm, at Brixton, Aug. 17, 1887. Mr. Symons says: "At present I can not trace it."

He was at his best at Netting Hill: there's been a marked falling off in his later manner:

In the London Times, Feb. 1, 1888, it is said that a roundish object of iron had been found, "after a violent thunderstorm," in a garden at Brixton, Aug. 17, 1887. It was analyzed by a chemist, who could not identify it as true meteoritic material. Whether a product of workmanship like human workmanship or not, this object is described as an oblate spheroid, about two inches across its major diameter. The chemist's name and address are given: Mr. J. James Morgan: Ebbw Vale.

Garden - familiar ground - I suppose that, in Mr. Symons' opinion this symmetric object had been upon the ground "in the first place," though he neglects to say this. But we do note that he described this object as a "lump," which does not suggest the spheroidal or symmetric. It is our notion that the word "lump" was, because of its meaning of amorphousness, used purposely to have the next datum stand alone, remote, without similars. If Mr. Symons had said that there had been a report of another round object that had fallen from the sky, his readers would be attracted by an agreement. He distracts his readers by describing in terms of the unprecedented -

"Iron cannon ball."

It was found in a manure heap, in Sussex, after a thunderstorm.

However, Mr. Symons argues pretty reasonably, it seems to me, that, given a cannon ball in a manure heap, in the first place, lightning might be attracted by it, and, if seen to strike there, the untutored mind, or mentality below the average, would leap or jump, or proceed with less celerity, to the conclusion that the iron object had fallen.

Except that - if every farmer isn't upon very familiar ground or if every farmer doesn't know his own manure heap as well as Mr. Symons knew his writing desk -

Then comes the instance of a man, his wife, and his three daughters, at Casterton, Westmoreland, who were looking out at their lawn, during a thunderstorm, when they "considered," as Mr. Symons expresses it, that they saw a stone fall from the sky, kill a sheep, and bury itself in the ground.

They dug.

They found a stone ball.

Symons:

Coincidence. It had been there in the first place.

This object was exhibited at a meeting of the Royal Meteorological Society by Mr. C. Carus-Wilson. It is described in the Journal's list of exhibits as a "sandstone" ball. It is described as "sandstone" by Mr. Symons.

Now a round piece of sandstone may be almost anywhere in the ground - in the first place - but, by our more or less discreditable habit of prying and snooping, we find that this object was rather more complex and of material less commonplace. In snooping through Knowledge, Oct. 9, 1885, we read that this "thunderstone" was in the possession of Mr. C. Carus-Wilson, who tells the story of the witness and his family - the sheep killed, the burial of something in the earth, the digging, and the finding. Mr. C. Carus-Wilson describes the object as a ball of hard, ferruginous quartzite, about the size of a cocoanut, weight about twelve pounds, Whether we're feeling around for significance or not, there is a suggestion not only of symmetry but of structure in this object: it had an external shell, separated from a loose nucleus. Mr. Carus – Wilson attributes this cleavage to unequal cooling of the mass.

My own notion is that there is very little deliberate misrepresentation in the writings of scientific men: that they are quite as guiltiness in intent as are other hypnotic subjects. Such a victim of induced belief reads of a stone ball said to have fallen from the sky. Mechanically in his mind arise impressions of globular lumps, or nodules, of sandstone, which are common almost everywhere.

He assimilates the reported fall with his impressions of objects in the ground, in the first place. To an intermediatist, the phenomena of intellection are only phenomena of universal process localized in human minds. The process called "explanation" is only a local aspect of universal assimilation. It looks like materialism: but the intermediatist holds that interpretation of the immaterial, as it is called, in terms of the material, as it is called, is no more rational than interpretation of the "material" in terms of the "immaterial": that there is in quasi-existence neither the material nor the immaterial, but approximations one way or the other. But so hypnotic quasi-reasons : that globular lumps of sandstone are common. Whether he jumps or leaps, or whether only the frowsy and baseborn are so athletic, his is the impression, by assimilation, that this especial object is a ball of sandstone. Or human mentality: its inhabitants are conveniences. It may be that Mr. Symons' paper was written before this object was exhibited to the members of the Society, and with the charity with which, for the sake of diversity, we intersperse our malices, we are willing to accept that he "investigated" something that he had never seen. But whoever listed this object was uncareful: it is listed as "sandstone." We're making excuses for them.

Really - as it were - you know we're not quite so damned as we were.

One does not apologize for the gods and at the same time feel quite utterly prostrate before them.

If this were a real existence, and all of us real persons, with real standards to judge by, I'm afraid we'd have to be a little severe with some of these Mr. Symonses. As it is, of course, seriousness seems out of place.

We note an amusing little touch in the indefinite allusion to "a, man," who with his un-named family, had "considered" that he had seen a stone fall. The "man" was the Rev. W. Carus-Wilson, who was well-known in his day.

The next instance was reported by W. B. Tripp, F. R. M. S. - that, during a thunderstorm, a farmer had seen the ground in front of him plowed up by something that was luminous.

Dug.

Bronze ax.

My own notion is that an expedition to the north pole could not be so urgent as that representative scientists should have gone to that farmer and there spend a summer studying this one reported occurrence. As it is - un-named farmer somewhere - no date. The thing must stay damned.

Another specimen for our museum is a comment in Nature, upon these objects: that they are "of an amusing character, thus clearly showing that they were of terrestrial, and not a celestial, character." Just why celestiality, or that of it which, too, is only of Intermediateness should not be quite as amusing as terrestriality is beyond our reasoning powers, which we have agreed are not ordinary.

Of course there is nothing amusing about wedges and spheres at all - or Archimedes and Euclid are humorists. It is that they were described derisively. If you'd like a little specimen of the standardization of orthodox opinion -

Amer. Met. Jour., 4-589:

"They are of an amusing character, thus clearly showing that they were of a terrestrial and not a celestial character." I'm sure - not positively, of course - that we've tried to be as easy-going and lenient with Mr. Symons, as his obviously scientific performance would permit. Of course it may be that sub-consciously we were prejudiced against him, instinctively classing him with St. Augustine, Darwin, St. Jerome, and Lyell. As to the "thunderstones," I think that he investigated them mostly "for the credit of Englishmen," or in the spirit of the Royal Krakatoa Committee, or about as the commission from the French Academy investigated meteorites. According to a writer in Knowledge, 5-418, the Krakatoa Committee attempted not in the least to prove what had caused the atmospheric effects of 1883, but to prove - that Krakatoa did it.

Altogether I should think that the following quotation should be enlightening to any one who still thinks that these occurrences were investigated not to support an opinion formed in advance:

In opening his paper, Mr. Symons says that he undertook his investigation as to the existence of "thunderstones," or "thunderbolts" as he calls them - "feeling certain that there was a weak point somewhere, inasmuch as 'thunderbolts' have no existence."

We have another instance of the reported fall of a "cannon ball."

It occurred prior to Mr. Symons' investigations, but is not mentioned by him. It was investigated, however. In the Proc. Roy.

Soc. Edin.f 3-147, is the report of a "thunderstone," "supposed to have fallen in Hampshire, Sept., 1852." It was an iron cannon ball, or it was a "large nodule of iron pyrites or bisulphuret of iron." No one had seen it fall. It had been noticed, upon a garden path, for the first time, after a thunderstorm. It was only a "supposed" thing, because - "It had not the character of any known meteorite."

In the London Times, Sept. 16, 1852, appears a letter from Mr. George E. Bailey, a chemist of Andover, Hants. He says that, in a very heavy thunderstorm, of the first week of September, 1852, this iron object had fallen in the garden of Mr. Robert Bowling, of Andover; that it had fallen upon a path "within six yards of the house." It had been picked up "immediately" after the storm by Mrs. Dowling. It was about the size of a cricket ball: weight four pounds. No one had seen it fall. In the Times, Sept. 15, 1852, there is an account of this thunderstorm, which was of unusual violence.

There are some other data relative to the ball of quartz of Westmoreland.

They're poor things. There's so little to them that they look like ghosts of the damned. However, ghosts, when multiplied, take on what is called substantiality - if the solidest thing conceivable, in quasi-existence, is only concentrated phantomosity. It is not only that there have been other reports of quartz that has fallen from the sky; there is another agreement. The round quartz object of Westmoreland, if broken open and separated from its loose nucleus would be a round, hollow, quartz object. My pseudo-position is that two reports of similar extraordinary occurrences, one from England and one from Canada are interesting.

Proc. Canadian Institute, 3-7-8:

That, at the meeting of the Institute, of Dec. 1, 1888, one of the members, Mr. J. A. Livingstone, exhibited a globular quartz body which he asserted had fallen from the sky. It had been split open. It was hollow.

But the other members of the Institute decided that the object
was spurious, because it was not of "true meteoritic material."

No date; no place mentioned; we note the suggestion that it was only a geode, which had been upon the ground in the first place. It's crystalline lining was geode-like.

Quartz is upon the "index prohibitory" of Science. A monk who would read Darwin would sin no more than would a scientist who would admit that, except by the "up and down" process, quartz has ever fallen from the sky but Continuity: it is not excommunicated if part of or incorporated in a baptized meteorite St. Catherine's of Mexico, I think. It's as epicurean a distinction as any ever made by theologians. Fassig lists a quartz pebble, found in a hailstone (Bibliography, part 2-355). "Up and down," of course. Another object of quartzite was reported to have fallen, in the autumn of 1880, at Schroon Lake, N. Y. said in the Scientific American, 43-272 to be a fraud it was not the usual. About the first of May, 1899, the newspapers published a story of a "snow-white" meteorite that had fallen, at Vincennes, Indiana. The Editor of the Monthly Weather Review ("M. W. R." April, 1899) requested the local observer, at Vincennes, to investigate. The Editor says that the thing was only a fragment of a quartz bowlder. He says that any one with at least a public school education should know better than to write that quartz has ever fallen from the sky.

Notes and Queries, 2-8-92:

That, in the Leyden Museum of Antiquities, there is a disk of quartz: 6 centimeters by 5 millimeters by about 5 centimeters; said to have fallen upon a plantation in the Dutch West Indies, after a meteoric explosion.

Bricks.

I think this is a vice we're writing. I recommend it to those who have hankered for a new sin. At first some of our data were of so frightful or ridiculous mien, as to be hated, or eyebrowed, was only to be seen. Then some pity crept in? I think that we can now embrace bricks.

The baked-clay-idea was all right in its place, but it rather lacks distinction, I think. With our minds upon the concrete boats that have been building terrestrially lately, and thinking of wrecks that may occur to some of them, and of a new material for the deep-sea fishes to disregard -

Object that fell at Richland, South Carolina - yellow to gray - said to look like a piece of brick. (Amer. Jour. Sci., 2-34-298.)

Pieces of "furnace-made brick" said to have fallen - in a hailstorm - at Padua, Aug., 1834. (Edin. New Phil. Jour., 19-87.)

The writer offered an explanation that started another convention:

that the fragments of brick had been knocked from buildings by the hailstones. But there is here a concomitant that will be disagreeable to any one who may have been inclined to smile at the now digestible-enough notion that furnace-made bricks have fallen from the sky. It is that in some of the hailstones - two per cent of them - that were found with the pieces of brick, was a light grayish powder.

Monthly Notices of the Royal Astronomical Society, 337-365:

Padre Sechi explains that a stone said to have fallen, in a thunderstorm, at Supino, Italy, Sept., 1875, had been knocked from a roof.

Nature, 33-153 :

That it had been reported that a good-sized stone, of form clearly artificial, had fallen at Naples, Nov., 1885. The stone was described by two professors of Naples, who had accepted it as inexplicable but veritable. They were visited by Dr. H. Johnstone-Lavis, the correspondent to Nature, whose investigations had convinced him that the object was a "shoemaker's lapstone."

Now to us of the initiated, or to us of the wider outlook, there is nothing incredible in the thought of shoemakers in other worlds - but I suspect that this characterization is tactical. This object of worked stone, or this shoemaker's lapstone, was made of Vesuvian lava, Dr. Johnstone-Lavis thinks: most probably of lava of the flow of 1631, from the La Scala quarries. We condemn "most probably" as bad positivism. As to the "men of position,"

who had accepted that this thing had fallen from the sky - "I have now obliged them to admit their mistake," says Dr. Johnstone-Lavis - or it's always the stranger in Naples who knows La Scala lava better than the natives know it.

Explanation:

That the thing had been knocked from, or thrown from, a roof. As to attempt to trace the oc-currence to any special roof - nothing said upon that subject. Or that Dr. Johnstone-Lavis called a carved stone a "lapstone," quite as Mr. Symons called a spherical object a "cannon ball": bent upon a discrediting incongruity:

Shoemaking and celestiality.

It is so easy to say that axes, or wedge-shaped stones found on the ground, were there in the first place, and that it is only coincidence that lightning should strike near one - but the credibility of coincidences decreases as the square root of their volume, I think. Our massed instances speak too much of coincidences of coincidences.

But the axes, or wedge-shaped objects that have been found in trees are more difficult for orthodoxy. For instance, Arago accepts that such finds have occurred, but he argues that, if wedge-shaped stones have been found in tree trunks, so have toads been found in tree trunks - did the toads fall there? Not at all bad for a hypnotic.

Of course, in our acceptance, the Irish are the Chosen People. It's because they are characteristically best in accord with the underlying essence of quasi-existence. M. Arago answers a question by asking another question. That's the only way a question can be answered in our Hibernian kind of an exis-tence.

Dr. Bodding argued with the natives of the Santal Parganas, India, who said that cut and shaped stones had fallen from the sky, some of them lodging in tree trunks. Dr. Bodding, with orthodox no-tions of velocity of falling bodies, having missed, I suppose, some of the notes I have upon large hail-stones, which, for size, have fallen with astonishingly low velocity, argued that anything falling from the sky would be "smashed to atoms." He accepts that objects of worked stone have been found in tree trunks, but he explains :

That the Santals often steal trees, but do not chop them down in the usual way, because that would be to make too much noise: they insert stone wedges, and hammer them instead: then, if they should be caught, wedges would not be the evidence against them that axes would be.

Or that a scientific man can't be desperate and reasonable too.

Or that a pickpocket, for instance, is safe, though caught with his hand in one's pocket, if he's gloved, say: because no court in the land would regard a gloved hand in the same way in which a bare hand would be regarded.

That there's nothing but intermediateness to the rational and the preposterous: that this status of our own ratiocinations is perceptible wherein they are upon the unfamiliar.

Dr. Bodding collected 50 of these shaped stones, said to have fallen from the sky, in the course of many years. He says that the Santals are a highly developed race, and for ages have not used stone implements - except in this one nefarious convenience to him.

All explanations are localizations. They fade away before the universal. It is difficult to express that black rains in England do not originate in the smoke of factories - less difficult to express that black rains of South Africa do not. We utter little stress upon the absurdity of Dr. Bodding's explanation, because, if anything's absurd everything's absurd, or, rather, has in it some degree or aspect of absur-dity, and we've never had experience with any state except something somewhere between ultimate absurdity and final reasonableness.

Our acceptance is that Dr. Bodding's elaborate explanation does not apply to cut-stone objects found in tree trunks in other lands: we accept that for the general, a local explanation is inadequate. As to "thunderstones" not said to have fallen luminously, and not said to have been found sticking in trees, we are told by faithful hypnotics that astonished rustics come upon prehistoric axes that have been washed into sight by rains, and jump to the conclusion that the things had fallen from the sky. But simple rustics come upon many prehistoric things: scrapers, pottery, knives, hammers.

We have no record of rusticity coming upon old pottery after a rain, reporting the fall of a bowl from the sky.

Just now, my own acceptance is that wedge-shaped stone objects, formed by means similar to human workmanship, have often fallen from the sky. Maybe there are messages upon them. My acceptance is that they have been called "axes" to discredit them: or the more familiar a term, the higher the incongruity with vague concepts of the vast, remote, tremendous, unknown. In Notes and Queries, 2-8-92, a writer says that he had a "thunderstone," which he had brought from Jamaica. The description is of a wedge-shaped object; not of an ax:

"It shows no mark of having been attached to a handle."

Of ten "thunderstones," figured upon different pages in Blinkenberg's book, nine show no sign of ever having been attached to a handle: one is perforated.

But in a report by Dr. C. Leemans, Director of the Leyden Museum of Antiquities, objects, said by the Japanese to have fallen from the sky, are alluded to throughout as "wedges." In the Archaeologic Journal, 11-118, in a paper upon the "thunderstones" of Java, the objects are called "wedges" and not "axes."

Our notion is that rustics and savages call wedge-shaped objects that fall from the sky, "axes": that scientific men, when it suits their purposes, can resist temptations to prolixity and pedantry, and adopt the simple: that they can be intelligible when derisive. All of which lands us in a confusion, worse, I think than we were in before we so satisfactorily emerged from the distresses of - butter and blood and ink and paper and punk and silk. Now it's cannon balls and axes and disks - if a "lapstone" be a disk - it's a flat stone, at any rate.

A great many scientists are good impressionists: they snub the impertinences of details. Had he been of a coarse, grubbing nature, I think Dr. Bodding could never have so simply and beautifully explained the occurrence of stone wedges in tree trunks. But to a realist, the story would be something like this:

A man who needed a tree, in a land of jungles, where, for some unknown reason, every one's very selfish with his trees, conceives that hammering stone wedges makes less noise than does the chopping of wood: he and his descendants, in a course of many years, cut down trees with wedges, and escape penalty, because it never occurs to a prosecutor that the head of an ax is a wedge.

The story is like every other attempted positivism - beautiful and complete, until we see what it excludes or disregards; whereupon it becomes the ugly and incomplete - but not absolutely, because there is probably something of what is called foundation for it. Perhaps a mentally incomplete Santal did once do something of the kind. Story told to Dr. Bodding: in the usual scientific way, he makes a dogma of an aberration.

Or we did have to utter a little stress upon this matter, after all. They're so hairy and attractive, these scientists of the 19th century. We feel the zeal of a Sitting Bull when we think of their scalps. We shall have to have an expression of our own upon this confusing subject. We have expressions: we don't call them explanations: we've discarded explanations with beliefs. Though every one who scalps is, in the oneness of allness, himself likely to be scalped, there is such a discourtesy to an enemy as the wearing of wigs.

Cannon balls and wedges, and what may they mean?

Bombardments of this earth -

Attempts to communicate -

Or visitors to this earth, long ago explorers from the moon -

taking back with them, as curiosities, perhaps, implements of this earth's prehistoric inhabitants - a wreck - a cargo of such things held for ages in suspension in the Super-Sargasso Sea - falling, or shaken, down occasionally by storms -

But, by preponderance of description, we can not accept that "thunderstones" ever were attached to handles, or are prehistoric axes -

As to attempts to communicate with this earth, by means of wedge-shaped objects especially adapted to the penetration of vast, gelatinous areas spread around this earth -

In the Proc. Roy. Irish Acad., 9-337, there is an account of a stone wedge that fell from the sky, near Cashel, Tipperary, Aug. 2, 1865. The phenomenon is not questioned, but the orthodox preference is to call it, not ax-like, nor wedge-shaped, but "pyramidal."

For data of other pyramidal stones said to have fallen from the sky, see Rept. Brit. Assoc., 1861-34. One fell at Segowolee, India, March 6, 1853. Of the object that fell at Cashel, Dr. Haughton says in the Proceedings: "A singular feature is observable in this stone, that I have never seen in any other: the rounded edges of the pyramid are sharply marked by lines on the black crust, as perfect as if made by a ruler." Dr. Haughton's idea is that the marks may have been made by "some peculiar tension in the cooling." It must have been very peculiar, if in all aerolites not wedge-shaped, no such phenomenon had ever been observed. It merges away with one or two instances, known, after Dr. Haughton's time, of seeming stratification in meteorites. Stratification in meteorites, however, is denied by the faithful.

I begin to suspect something else.

A whopper is coming.

Later it will be as reasonable, by familiarity, as anything else

ever said.

If some one should study the stone of Cashel, as Champollion studied the Rosetta stone, he might - or, rather, would inevitably find meaning in those lines, and translate them into English -

Nevertheless I begin to suspect something else: something more subtle and esoteric than graven characters upon stones that have fallen from the sky, in attempts to communicate. The notion that other worlds are attempting to communicate with this world is widespread: my own notion is that it is not attempt at all that it was achievement centuries ago.

I should like to send out a report that a "thunderstone" had fallen, say, somewhere in New Hampshire - And keep track of every person who came to examine that stone trace down his affiliations - keep track of him -

Then send out a report that a "thunderstone" had fallen at Stockholm, say -

Would one of the persons who had gone to New Hampshire, be met again in Stockholm? But what if he had no anthropological, lapidarian, or meteorological affiliations but did belong to a secret society -

It is only a dawning credulity.

Of the three forms of symmetric objects that have, or haven't, fallen from the sky, it seems to me that the disk is the most striking.

So far, in this respect, we have been at our worst - possibly that's pretty bad - but "lapstones" are likely to be of considerable variety of form, and something that is said to have fallen at sometime somewhere in the Dutch West Indies is profoundly of the unchosen.

Now we shall have something that is high up in the castes of the accursed:

Comptes Rendus, 1887-182:

That, upon June 20, 1887, in a "violent storm" - two months before the reported fall of the symmetric iron object of Brixton - a small stone had fallen from the sky at Tarbes, France: 13 millimeters in diameter; 5 millimeters thick; weight 2 grammes.

Reported to the French Academy by M. Sudre, professor of the Normal School, Tarbes.

This time the old convenience "there in the first place" is too greatly resisted the stone was covered with ice.

This object had been cut and shaped by means similar to human hands and human mentality. It was a disk of worked stone - "tres regulier." "II a ete assurement travaille."

There's not a word as to any known whirlwind anywhere: nothing of other objects or debris that fell at or near this date, in France. The thing had fallen alone. But as mechanically as any part of a machine responds to its stimulus, the explanation appears in Comptes Rendus, that this stone had been raised by a whirlwind and then flung down.

It may be that in the whole nineteenth century no event more important than this occurred. In La Nature, 1887, and in L'Annee Scientifique, 1887, this occurrence is noted. It is mentioned in one of the summer numbers of Nature, 1887. Fassig lists a paper upon it in the Annuaire de Soc. Met., 1887.

Not a word of discussion.

Not a subsequent mention can I find.

Our own expression:

What matters it how we, the French Academy, or the Salvation

Army may explain?

A disk of worked stone fell from the sky, at Tarbes, France, June 20, 1887.

Chapter IX

My own pseudo-conclusion:

That we've been damned by giants sound asleep, or by great scientific principles and abstractions that cannot realize themselves: that little harlots have visited their caprices upon us; that clowns, with buckets of water from which they pretend to cast thousands of good-sized fishes have anathematized us for laughing disrespectfully, because, as with all clowns, underlying buffoonery is the desire to be taken seriously; that pale ignorances, presiding over microscopes by which they cannot distinguish flesh from nostoc or fishes' spawn or frogs' spawn, have visited upon us their wan solemnities. We've been damned by corpses and skeletons and mummies, which twitch and totter with pseudo-life derived from conveniences.

Or there is only hypnosis. The accursed are those who admit they're the accursed.

If we be more nearly real we are reasons arraigned before a jury of dream-phantasms.

Of all meteorites in museums, very few were seen to fall. It is considered sufficient grounds for admission if specimens can't be accounted for in any way other than that they fell from the sky - as if in the haze of uncertainty that surrounds all things, or that is the essence of everything, or in the merging away of everything into something else, there could be anything that could be accounted for in only one way. The scientist and the theologian reason that if something can be accounted for in only one way, it is accounted for in that way or logic would be logical, if the conditions that it imposes, but, of course, does not insist upon, could anywhere be found in quasi-existence. In our acceptance, logic, science, art, religion are, in our "existence," premonitions of a coming awakening, like dawning awarenesses of surroundings in the mind of a dreamer.

Any old chunk of metal that measures up to the standard of "true meteoritic material" is admitted by the museums. It may seem incredible that modern curators still have this delusion, but we suspect that the date on one's morning newspaper hasn't much to do with one's modernity all day long. In reading Fletcher's catalogue, for instance, we learn that some of the best-known meteorites were "found in draining a field" - "found in making a road" - "turned up by the plow" occurs a dozen times. Some one fishing in Lake Okeechobee, brought up an object in his fishing net. No meteorite had ever been seen to fall near it. The U. S. National Museum accepts it. *

If we have accepted only one of the data of "untrue meteoritic material" - one instance of "carbonaceous" matter if it be too difficult to utter the word "coal" - we see that in this inclusion-exclusion, as in every other means of forming an opinion, false inclusion and false exclusion have been practiced by curators of museums. There is something of ultra-pathos - of cosmic sadness in this universal search for a standard, and in belief that one has been revealed by either inspiration or analysis, then the dogged clinging to a poor sham of a thing long after its insufficiency has been shown - or renewed hope and search for the special that can be true, or for something local that could also be universal. It's as if "true meteoritic material" were a "rock of ages" to some scientific men.

They cling. But clingers cannot hold out welcoming arms.

The only seemingly conclusive utterance, or seemingly substantial thing to cling to, is a product of dishonesty, ignorance, or fatigue.

All sciences go back and back, until they're worn out with the process, or until mechanical reaction occurs: then they move forward as it were. Then they become dogmatic, and take for bases, positions that were only points of exhaustion. So chemistry divided and sub-divided down to atoms; then, in the essential insecurity of all quasi-constructions, it built up a system, which, to any one so obsessed by his own hypnoses that he is exempt to the chemist's hypnoses, is perceptibly enough an intellectual anaemia built upon infinitesimal debilities.

In Science, n.s., 31-298, E. D. Hovey, of the American Museum of Natural History, asserts or confesses, that often have objects of material such as fossiliferous limestone and slag been sent to him.

He says that these things have been accompanied by assurances that they have been seen to fall on lawns, on roads, in front of houses.

They are all excluded. They are not of true meteoritic material.

They were on the ground in the first place. It is only by coincidence that lightning has struck, or that a real meteorite, which was unfindable, has struck near objects of slag and limestone.

Mr. Hovey says that the list might be extended indefinitely.

That's a tantalizing suggestion of some very interesting stuff -

He says:

"But it is not worth while."

I'd like to know what strange, damned, excommunicated things have been sent to museums by persons who have felt convinced that they had seen what they may have seen, strongly enough to risk ridicule, to make up bundles, go to express offices, and write letters.

I accept that over the door of every museum, into which such things enter, is written:

"Abandon Hope."

If a Mr. Symons mentions one instance of coal, or of slag or cinders, said to have fallen from the sky, we are not - except by association with the "carbonaceous" meteorites strong in our impression that coal sometimes falls to this earth from coal-burning super-constructions, up somewhere -

In Comptes Rendus, 91-197, M. Daubree tells the same story.

Our acceptance, then, is that other curators could tell this same story. Then the phantomosity of our impression substantiates proportionately to its multiplicity. ,M. Daubree says that often have strange damned things been sent to the French museums, accompanied by assurances that they had been seen to fall from the sky.

Especially to our interest, he mentions coal and slag.

Excluded.

Buried un-named and undated in Science's potter's field.

I do not say that the data of the damned should have the same rights as the data of the saved. That would be justice. That would be of the Positive Absolute, and, though the ideal of, a violation of, the very essence of quasi-existence, wherein only to have the appearance of being is to express a preponderance of force one way or another or inequilibrium, or inconsistency, or injustice.

Our acceptance is that the passing away of exclusionism is a phenomenon of the twentieth century: that gods of the twentieth century will sustain our notions be they ever so unwashed and frowsy.

But, in our own expressions, we are limited, by the oneness of quasiness, to the very same methods by which orthodoxy established and maintains its now sleek, suave preposterousnesses.

At any rate, though we are inspired by an especial subtle essence or imponderable, I think that pervades the twentieth century, we have not the superstition that we are offering anything as a positive fact. Rather often we have not the delusion that we're any less superstitious and credulous than any logician, savage, curator, or rustic.

An orthodox demonstration, in terms of which we shall have some heresies is that if things found in coal could have got there only by falling there - they fell there.

So, in the Manchester Lit. and Phil. Soc. Mems., 2-9-306, it is argued that certain roundish stones that have been found in coal are "fossil aerolites": that they had fallen from the sky, ages ago, when the coal was soft, because the coal had closed around them, showing no sign of entrance.

Proc. Soc. of Antiq. of Scotland, 1-1-121:

That, in a lump of coal, from a mine in Scotland, an iron instrument had been found -

"The interest attaching to this singular relic arises from the fact of its having been found in the heart of a piece of coal, seven feet under the surface."

If we accept that this object of iron was of workmanship beyond the means and skill of the primitive men who may have lived in Scotland when coal was forming there -

"The instrument was considered to be modern."

That our expression has more of realness, or higher approximation to realness, than has the attempt to explain that is made in the Proceedings:

That in modern times some one may have bored for coal, and that his drill may have broken off in the coal it had penetrated.

Why he should have abandoned such easily accessible coal, I don't know. The important point is that there was no sign of boring: that this instrument was in a lump of coal that had closed around it so that its presence was not suspected, until the lump of coal was broken.

No mention can I find of this damned thing in any other publication.

Of course there is an alternative here: the thing may not have fallen from the sky: if in coal-forming times, in Scotland, there were, indigenous to this earth, no men capable of making such an iron instrument, it may have been left behind by visitors from other worlds.

In an extraordinary approximation to fairness and justice, which is permitted to us, because we are quite as desirous to make acceptable that nothing can be proved as we are to sustain our own expressions, we note:

That in Notes and Queries, 11-1-408, there is an account of an ancient copper seal, about the size of a penny, found in chalk, at a depth of from five to six feet, near Bredenstone, England. The design upon it is said to be of a monk kneeling before a virgin and child: a legend upon the margin is said to be: "St. Jordanis Monachi Spaldingie."

I don't know about that. It looks very desirable - undesirable to us.

There's a wretch of an ultra-frowsy thing in the Scientific American, 7-298, which we condemn ourselves, if somewhere, because of the oneness of allness, the damned must also be the damning. It's a newspaper story: that about the first of June, 1851, a powerful blast, near Dorchester, Mass., cast out from a bed of solid rock a bell-shaped vessel of an unknown metal: floral designs inlaid with silver; "art of some cunning workman." The opinion of the Editor of the Scientific American is that the thing had been made by Tubal Cain, who was the first inhabitant of Dorchester. Though I fear that this is a little arbitrary, I am not disposed to fly rabidly at every scientific opinion.

Nature, 35-36:

A block of metal found in coal, in Austria, 1885. It is now in the Salsburg museum.

This time we have another expression. Usually our intermediatist attack upon provincial positivism is: Science, in its attempted positivism takes something such as "true meteoritic material" as a standard of judgment; but carbonaceous matter, except for its relative infrequency, is just as veritable a standard of judgment; carbonaceous matter merges away into such a variety of organic substances, that all standards are reduced to indistinguishability: if, then, there is no real standard against us, there is no real resistance to our own acceptances. Now our intermediatism is: Science takes "true meteoritic material" as a standard of admission; but now we have an instance that quite as truly makes "true meteoritic material" a standard of exclusion; or, then, a thing that denies itself is no real resistance to our own acceptances this depending upon whether we have a datum of something of "true meteoritic material" that orthodoxy can never accept fell from the sky.

We're a little involved here. Our own acceptance is upon a carved, geometric thing that, if found in a very old deposit, antedates human life, except, perhaps, very primitive human life, as an indigenous product of this earth: but we're quite as much interested in the dilemma it made for the faithful.

It is of "true meteoritic material." In L'Astronomie, 1887-114, it is said that, though so geometric, its phenomena so characteristic of meteorites exclude the idea that it was the work of man.

As to the deposit - Tertiary coal.

Composition - iron, carbon, and a small quantity of nickel.

It has the pitted surface that is supposed by the faithful to be characteristic of meteorites.

For a full account of this object, see Comptes Rendus, 103-702.

The scientists who examined it could reach no agreement. They bifurcated: then a compromise was suggested; but the compromise is a product of disregard:

That it was of true meteoritic material, and had not been shaped by man;

That it was not of true meteoritic material, but telluric iron that had been shaped by man;

That it was true meteoritic material that had fallen from the sky, but had been shaped by man, after its fall.

The data, one or more of which must be disregarded by each of these three explanations are: "true meteoritic material" and surface markings of meteorites; geometric form; presence in an ancient deposit; material as hard as steel; absence upon this earth, in Tertiary times, of men who could work in material as hard as steel.

It is said that, though of "true meteoritic material," this object is virtually a steel object.

St. Augustine, with his orthodoxy, was never in - well, very much worse - difficulties than are the faithful here. By due disregard of a datum or so, our own acceptance that it was a steel object that had fallen from the sky to this earth, in Tertiary times, is not forced upon one. We offer ours as the only synthetic expression. For instance, in Science Gossip, 1887-58, it is described as a meteorite: in this account there is nothing alarming to the pious, because, though everything else is told, its geometric form is not mentioned.

It's a cube. There is a deep incision all around it. Of its faces, two that are opposite are rounded.

Though I accept that our own expression can only rather approximate to Truth, by the wideness of its inclusions, and because it seems, of four attempts, to represent the only complete synthesis, and can be nullified or greatly modified, by data that we, too, have somewhere disregarded, the only means of nullification that I can think of would be demonstration that this object is a mass of iron pyrites, which sometimes forms geometrically. But the analysis mentions not a trace of sulphur. Of course our weakness, or impositiveness, lies in that, by any one to whom it would be agreeable to find sulphur in this thing, sulphur would be found in it by our own intermediatism there is some sulphur in everything, or sulphur is only a localization or emphasis of something that, unemphasized, is in all things.

So there have, or haven't, been found upon this earth, things that fell from the sky, or that were left behind by extra-mundane visitors to this earth -

A yarn in the London Times, June 22, 1844: that some workmen, quarrying rock, close to the Tweed, about a quarter of a mile below Rutherford Mills, discovered a gold thread embedded in the stone, at a depth of 8 feet: that a piece of the gold thread had been sent to the office of the Kelso Chronicle.

Pretty little thing; not at all frowsy; rather damnable.

London Times, Dec. 24, 1851:

That Hiram De Witt, of Springfield, Mass., returning from California, had brought with him a piece of auriferous quartz about the size of a man's fist. It was accidentally dropped - split open - nail in it. There was a cut-iron nail, size of a six-penny nail, slightly corroded. "It was entirely straight and had a perfect head."

Or - California ages ago, when auriferous quartz was forming super-carpenter, million of miles or so up in the air - drops a nail.

To one not an intermediatist, it would seem incredible that this datum, not only of the damned, but of the lowest of the damned, or of the journalistic caste of the accursed, could merge away with something else damned only by disregard, and backed by what is called "highest scientific authority" -

Communication by Sir David Brewster (Rept. Brit. Assoc., 1845-51):

That a nail had been found in a block of stone, from Kingoodie Quarry, North Britain. The block in which the nail was found was nine inches thick, but as to what part of the quarry it had come from, there is no evidence except that it could not have been from the surface. The quarry had been worked about twenty years. It consisted of alternate layers of hard stone and a substance called "till." The point of the nail, quite eaten with rust, projected into some "till," upon the surface of the block of stone. The rest of the nail lay upon the surface of the stone, to within an inch of the head - that inch of it was embedded in the stone.

Although its caste is high, this is a thing profoundly of the damned - sort of a Brahmin as regarded by a Baptist. Its case was stated fairly; Brewster related all circumstances available to him but there was no discussion at the meeting of the British Association: no explanation was offered Nevertheless the thing can be nullified -

But the nullification that we find is as much against orthodoxy, in one respect as it is against our own expression that inclusion in quartz or sandstone indicates antiquity or there would have to be a re-vision of prevailing dogmas upon quartz and sandstone and age indicated by them, if the opposing data should be accepted. Of course it may be contended by both the orthodox and us heretics that the opposition is only a yarn from a newspaper. By an odd combination, we find our two lost souls that have tried to emerge, chucked back to perdition by one blow:

Pop. Sci. News, 1884-41:

That, according to the Carson Appeal, there had been found in a mine, quartz crystals that could have had only 15 years in which to form: that, where a mill had been built, sandstone had been found, when the mill was torn down, that had hardened in 12 years: that in this sandstone was a piece of wood, "with a nail in it."

Annals of Scientific Discovery, 1853-71:

That, at the meeting of the British Association, 1853, Sir David Brewster had announced that he had to bring before the meeting, an object "of so incredible a nature that nothing short of the strongest evidence was necessary to render the statement at all probable."

A crystal lens had been found in the treasure-house at Ninevah.

In many of the temples and treasure houses of old civilizations upon this earth have been preserved things that have fallen from the sky or meteorites.

Again we have a Brahmin. This thing is buried alive in the heart of propriety: it is in the British Museum.

Carpenter, in The Microscope and Its Revelations, gives two drawings of it. Carpenter argues that it is impossible to accept that optical lenses had ever been made by the ancients. Never occurred to him - some one a million miles or so up in the air - looking through his telescope - lens drops out.

This does not appeal to Carpenter: he says that this object must have been an ornament.

According to Brewster, it was not an ornament, but "a true optical lens," In that case, in ruins of an old civilization upon this earth, has been found an accursed thing that was, acceptably, not a product of any old civilization indigenous to this earth.

Chapter X

Early explorers have Florida mixed up with Newfoundland.

But the confusion is worse than that still earlier. It arises from simplicity. Very early explorers think that all land westward is one land, India: awareness of other lands as well as India comes as a slow process. I do not now think of things arriving upon this earth from some especial other world. That was my notion when I started to collect our data. Or, as is a commonplace of observation, all intellection begins with the illusion of homogeneity. It's one of Spencer's data: we see homogeneousness in all things distant, or with which we have small acquaintance. Advance from the relatively homogeneous to the relatively heterogeneous is Spencerian Philosophy like everything else, so-called: not that it was really Spencer's discovery, but was taken from von Baer, who, in turn, was continuous with preceding evolutionary speculation. Our own expression is that all things are acting to advance to the homogeneous, or are trying to localize Homogeneousness. Homogeneousness is an aspect of the Universal, wherein it is a state that does not merge away into something else. We regard homogeneousness as an aspect of positiveness, but it is our acceptance that infinite frustrations of attempts to positivize manifest themselves in infinite heterogeneity: so that though things try to localize homogeneousness they end up in heterogeneity so great that it amounts to infinite dispersion or indistinguishability.

So all concepts are little attempted positivenesses, but soon have to give in to compromise, modification, nullification, merging away into indistinguishability - unless, here and there, in the world's history, there may have been a super-dogmatist, who, for only an infinitesimal of time, has been able to hold out against heterogeneity or modification or doubt or "listening to reason," or loss of identity - in which case - instant translation to heaven or the Positive Absolute.

Odd thing about Spencer is that he never recognized that "homogeneity," "integration," and "definiteness" are all words for the same state, or the state that we call "positiveness." What we call his mistake is in that he regarded "homogeneousness" as negative.

I began with a notion of some one other world, from which objects and substances have fallen to this earth; which had, or which, to less degree, has a tutelary interest in this earth; which is now attempting to communicate with this earth - modifying, because of data which will pile up later, into acceptance that some other world is not attempting but has been, for centuries, in communication with a sect, perhaps, or & secret society, or certain esoteric ones of this earth's inhabitants.

I lose a great deal of hypnotic power in not being able to concentrate attention upon some one other world.

As I have admitted before I'm intelligent, as contrasted with the orthodox. I haven't the aristocratic disregard of a New York curator or an Eskimo medicine-man.

I have to dissipate myself in acceptance of a host of other worlds: size of the moon, some of them: one of them, at least, - tremendous thing: we'll take that up later. Vast, amorphous aerial regions, to which such definite words as "worlds" and "planets" seem inapplicable.

And artificial constructions that I have called "super-constructions": one of them about the size of Brooklyn, I should say, off hand. And one or more of them wheel-shaped things, a goodly number of square miles in area.

I think that earlier in this book, before we liberalized into embracing everything that comes along, your indignation, or indigestion would have expressed in the notion that, if this were so, astronomers would have seen these other worlds and regions and vast geometric constructions. You'd have had that notion: you'd have stopped there.

But the attempt to stop is saying "enough" to the insatiable. In cosmic punctuation there are no periods: illusion of periods is incomplete view of colons and semi-colons.

We can't stop with the notion that if there were such phenomena, astronomers would have seen them. Because of our experience with suppression and disregard, we suspect, before we go into the subject at all, that astronomers have seen them; that navigators and meteorologists have seen them; that individual scientists and other trained observers have seen them many times -

That it is the System that has excluded data of them.

As to the Law of Gravitation, and astronomers' formulas, remember that these formulas worked out in the time of La Place as well as they do now. But there are hundreds of planetary bodies now known that were then not known. So a few hundred worlds more of ours won't make any difference. La Place knew of about only thirty bodies in this solar system: about six hundred are recognized now -

What are the discoveries of geology and biology to a theologian?

His formulas still work out as well as they ever did.

If the Law of Gravitation could be stated as a real utterance, it might be a real resistance to us. But we are told only that gravitation is gravitation. Of course to an intermediatist, nothing can be defined except in terms of itself - but even the orthodox, in what seems to me to be the innate premonitions of realness, not founded upon experience, agree that to define a thing in terms of itself is not real definition. It is said that by gravitation is meant the attraction of all things proportionately to mass and inversely as the square of the distance. Mass would mean inter-attraction holding together final particles, if there were final particles. Then, until final particles be discovered, only one term of this expression survives, or mass is attraction. But distance is only extent of mass, unless one holds out for absolute vacuum among planets, a position against which we could bring a host of data. But there is no possible means of expressing that gravitation is anything other than attraction.

So there is nothing to resist us but such a phantom as - that gravitation is the gravitation of all gravitations proportionately to gravitation and inversely as the square of gravitation. In a quasi-existence, nothing more sensible than this can be said upon any so-called subject - perhaps there are higher approximations to ultimate sensibleness.

Nevertheless we seem to have a feeling that with the System against us we have a kind of resistance here. We'd have felt so formerly, at any rate: I think the Dr. Grays and Prof. Hitchcocks have modified our trustfulness toward indistinguishability. As to the perfection of this System that quasi-opposes us and the infallibility of its mathematics as if there could be real mathematics in a mode of seeming where twice two are not four - we've been told over and over of their vindication in the discovery of Neptune.

I'm afraid that the course we're taking will turn out like every other development. We began humbly, admitting that we're of the damned -

But our eyebrows -

Just a faint flicker in them, or in one of them, every time we hear of the "triumphal discovery of Neptune" - this "monumental achievement of theoretical astronomy," as the text books call it.

The whole trouble is that we've looked it up.

The text-books omit this:

That, instead of the orbit of Neptune agreeing with the calculations of Adams and Leverrier, it was so different - that Leverrier said that it was not the planet of his calculations.

Later it was thought best to say no more upon that subject.

The text-books omit this:

That, in 1846, every one who knew a sine from a cosine was out sining and cosining for a planet beyond Uranus.

Two of them guessed right.

To some minds, even after Leverrier's own rejection of Neptune, the word "guessed" may be objectionable - but, according to Prof.

Peirce, of Harvard, the calculations of Adams and Leverrier would have applied quite as well to positions many degrees from the position of Neptune.

Or for Prof. Peirce's demonstration that the discovery of Neptune was only a "happy accident," see Proc. Amer. Acad. Sciences, 1-65.

For references, see Lowell's Evolution of Worlds.

Or comets: another nebulous resistance to our own notions. As to eclipses, I have notes upon several of them that did not occur upon scheduled time, though with differences only of seconds - and one delightful lost soul, deep-buried, but buried in the ultra-respectable records of the Royal Astronomical Society, upon an eclipse that did not occur at all. That delightful, ultra-sponsored thing of perdition is too good and malicious to be dismissed with passing notice: we'll have him later. Throughout the history of astronomy, every comet that has come back upon predicted time - not that, essentially, there was anything" more abstruse about it than is a prediction that you can make of a postman's periodicities to-morrow - was advertised for all it was worth. It's the way reputations are worked up for fortune-tellers by the faithful. The comets that didn't come back - omitted or explained. Or Encke's comet. It came back slower and slower.

But the astronomers explained. Be almost absolutely sure of that: they explained. They had it all worked out and formulated and "proved" why that comet was coming back slower and slower - and there the dam thing began coming faster and faster.

Halley's comet.

Astronomy - "the perfect science, as we astronomers like to call it." (Jacoby.)

It's my own notion that if, in a real existence, an astronomer could not tell one longitude from another, he'd be sent back to this purgatory of ours until he could meet that simple requirement.

Halley was sent to the Cape of Good Hope to determine its longitude. He got it degrees wrong. He gave to Africa's noble Roman promontory a retrousse twist that would take the pride out of any Kaffir.

We hear everlastingly of Halley's comet. It came back - maybe.

But, unless we look the matter up in contemporaneous records, we hear nothing of - the Leonids, for instance. By the same methods as those by which Halley's comet was predicted, the Leonids were predicted. Nov., 1898 - no Leonids. It was explained. They had been perturbed. They would appear in November, 1899. Nov., 1899 - Nov., 1900 - no Leonids.

My notion of astronomic accuracy:

Who could not be a prize marksman, if only his hits be recorded?

As to Halley's comet, of 1910 - everybody now swears he saw it. He has to perjure himself: otherwise he'd be accused of having no interest in great, inspiring things that he's never given any attention to. Regard this:

That there never is a moment when there is not some comet in the sky. Virtually there is no year in which several new comets are not discovered, so plentiful are they. Luminous fleas on a vast black dog in popular impressions, there is no realization of the extent to which this solar system is flea-bitten.

If a comet have not the orbit that astronomers have predicted - perturbed. If - like Halley's comet - it be late - even a year late - perturbed. When a train is an hour late, we have small opinion of the predictions of time tables. When a comet's a year late, all we ask is that it be explained. We hear of the inflation and arrogance of astronomers. My own acceptance is not that they are imposing upon us: that they are requiting us. For many of us priests no longer function to give us seeming rapport with Perfection, Infallibility - the Positive Absolute. Astronomers have stepped forward to fill a vacancy - with quasi-phantomosity - but, in our acceptance, with a higher approximation to substantiality than had the attenuations that preceded them. I should say, myself, that all that we call progress is not so much response to "urge" as it is response to a hiatus - or if you want something to grow somewhere, dig out everything else in its area. So I have to accept that the positive assurances of astronomers are necessary to us, or the blunderings, evasions and disguises of astronomers would never be tolerated: that, given such latitude as they are permitted to take, they could not be very disastrously mistaken. Suppose the comet called Halley's had not appeared -

Early in 1910, a far more important comet than the anaemic luminosity said to be Halley's, appeared. It was so brilliant that it was visible in daylight. The astronomers would have been saved anyway. If this other comet did not have the predicted orbit - perturbation. If you're going to Coney Island, and predict there'll be a special kind of a pebble on the beach, I don't see how you can disgrace yourself, if some other pebble will do just as well - because the feeble thing said to have been seen in 1910 was no more in accord with the sensational descriptions given out by astronomers in advance than is a pale pebble with a brick-red bowlder.

I predict that next Wednesday, a large Chinaman, in evening clothes, will cross Broadway, at 42nd Street, at 9 p. m. He doesn't, but a tubercular Jap, in a sailor's uniform does cross Broadway, at 35th Street, Friday, at noon. Well, a Jap is a perturbed Chinaman, and clothes are clothes.

I remember the terrifying predictions made by the honest and credulous astronomers, who must have been themselves hypnotized, or they could not have hypnotized the rest of us, in 1909. Wills were made. Human life might be swept from this planet. In quasiexistence, which is essentially Hibernian, that would be no reason why wills should not be made. The less excitable of us did expect at least some pretty good fireworks.

I have to admit that it is said that, in New York, a light was seen in the sky.

It was about as terrifying as the scratch of a match on the seat of some breeches half a mile away.

It was not on time.

Though I have heard that a faint nebulosity, which I did not see, myself, though I looked when I was told to look, was seen in the sky, it appeared several days after the time predicted.

A hypnotized host of imbeciles of us: told to look up at the sky:

we did - like a lot of pointers hypnotized by a partridge.

The effect:

Almost everybody now swears that he saw Halley's comet, and that it was a glorious spectacle.

An interesting circumstance here is that seemingly we are trying to discredit astronomers because astronomers oppose us that's not my impression. We shall be in the Brahmin caste of the hell of the Baptists. Almost all our data, in some regiments of this procession, are observations by astronomers, few of them mere amateur astronomers. It is the System that opposes us. - It is the System that is suppressing astronomers. I think we pity them in their captivity. Ours is not malice in a positive sense. It's chivalry - somewhat. Unhappy astronomers looking out from high towers in which they are imprisoned we appear upon the horizon. But, as I have said, our data do not relate to some especial other world. I mean very much what a savage upon an ocean island might vaguely think of in his speculations not upon some other land, but complexes of continents and their phenomena: cities, factories in cities, means of communication -

Now all the other savages would know of a few vessels sailing in their regular routes, passing this island in regularized periodicities.

The tendency in these minds would be expression of the universal tendency toward positivism - or Completeness - or conviction that these few regularized vessels constituted all. Now I think of some especial savage who suspects otherwise - because he's very backward and unimaginative and insensible to the beautiful ideals of the others: not piously occupied, like the others, in bowing before impressive – looking sticks of wood; dishonestly taking time for his speculations, while the others are patriotically witch-finding. So the other higher and nobler savages know about the few regularized vessels: know when to expect them; have their periodicities all worked out ; just about when vessels will pass, or eclipse each other - explaining that all vagaries were due to atmospheric conditions. They'd come out strong in explaining.

You can't read a book upon savages without noting what resolute explainers they are.

They'd say that all this mechanism was founded upon the mutual attraction of the vessels deduced from the fall of a monkey from a palm tree - or, if not that, that devils were pushing the vessels - something of the kind.

Storms.

Debris, not from these vessels, cast up by the waves.

Disregarded.

How can one think of something and something else, too?

I'm in the state of mind of a savage who might find upon a shore, washed up by the same storm, buoyant parts of a piano and a paddle that was carved by cruder hands than his own: something light and summery from India, and a fur overcoat from Russia - or all science, though approximating wider and wider, is attempt to conceive of India in terms of an ocean island, and of Russia in terms of India so interpreted. Though I am trying to think of Russia and India in world-wide terms, I cannot think that that, or the universalizing of the local, is cosmic purpose. The higher idealist is the positivist who tries to localize the universal, and is in accord with cosmic purpose: the super-dogmatist of a local savage who can hold out, without a flurry of doubt, that a piano washed up on a beach is the trunk of a palm tree that a shark has bitten, leaving his teeth in it. So we fear for the soul of Dr. Gray, because, he did not devote his whole life to that one stand that, whether possible or inconceivable, thousands of fishes had been cast from one bucket.

So, unfortunately for myself, if salvation be desirable, I look out widely but amorphously, indefinitely and heterogeneously. If I say I conceive of another world that is now in secret communication with certain esoteric inhabitants of this earth, I say I conceive of still other worlds that are trying to establish communication with all the inhabitants of this earth. I fit my notions to the data I find.

That is supposed to be the right and logical and scientific thing to do; but it is no way to approximate to form, system, organization.

Then I think I conceive of other worlds and vast structures that pass us by, within a few miles, without the slightest desire to communicate, quite as tramp vessels pass many islands without particularizing one from another. Then I think I have data of a vast construction that has often come to this earth, dipped into an ocean, submerged there a while, then going away - Why? I'm not absolutely sure. How would an Eskimo explain a vessel, sending ashore for coal, which is plentiful upon some Arctic beaches, though of unknown use to the natives, then sailing away, with no interest in the natives?

A great difficulty in trying to understand vast constructions that show no interest in us:

The notion that we must be interesting.

I accept that, though we're usually avoided, probably for moral reasons, sometimes this earth has been visited by explorers. I think that the notion that there have been extra-mundane visitors to China, within what we call the historic period, will be only ordinarily absurd, when we come to that datum.

I accept that some of the other worlds are of conditions very similar to our own. I think of others that are very different so that visitors from them could not live here without artificial adaptations. How some of them could breathe our attenuated air, if they came from a gelatinous atmosphere - Masks.

The masks that have been found in ancient deposits.

Most of them are of stone, and are said to have been ceremonial regalia of savages -

But the mask that was found in Sullivan County, Missouri, in 1879 (American Antiquarian, 3-336). It is made of iron and silver.

Chapter XI

One of the dam-dest in our whole saturnalia of the accursed -

Because it is hopeless to try to shake off an excommunication only by saying that we're damned by blacker things than ourselves; and that the damned are those who admit they're of the damned. Inertia and hypnosis are too strong for us. We say that: then we go right on admitting we're of the damned. It is only by being more nearly real that we can sweep away the quasi-things that oppose us. Of course, as a whole, we have considerable amorphousness, but we are thinking now of "individual" acceptances. Wideness is an aspect of Universalness or Realness. If our syntheses disregard fewer data than do opposing syntheses which are often not syntheses at all, but mere consideration of some one circumstance less widely synthetic things fade away before us. Harmony is an aspect of the Universal, by which we mean Realness. If we approximate more highly to harmony among the parts of an expression and to all available circumstances of an occurrence, the self-contradictors turn hazy. Solidity is an aspect of realness. We pile them up, and we pile them up, or they pass and pass and pass: things that bulk large as they march by, supporting and solidifying one another -

And still, and for regiments to come, hypnosis and inertia rule us -

One of the dam-dest of our data:

In the Scientific American, Sept. 10, 1910, Charles F. Holder writes:

"Many years ago, a strange stone resembling a meteorite, fell into the Valley of the Yaqui, Mexico, and the sensational story went from one end to the other of the country that a stone bearing human inscriptions had descended to the earth."

The bewildering observation here is Mr. Holder's assertion that this stone did fall. It seems to me that he must mean that it fell by dislodgment from a mountain side into a valley - but we shall see that it was such a marked stone that very unlikely would it have been unknown to dwellers in a valley, if it had been reposing upon a mountain side above them. It may have been carelessness: intent may have been to say that a sensational story of a strange stone said to have fallen, etc.

This stone was reported, by Major Frederick Burnham, of the British Army. Later Major Burnham re-visited it, and Mr. Holder accompanied him, their purpose to decipher the inscriptions upon it, if possible.

"This stone was a brown, igneous rock, its longest axis about eight feet, and on the eastern face, which had an angle of about forty-five degrees, was the deep-cut inscription."

Mr. Holder says that he recognized familiar Mayan symbols in the inscription. His method was the usual method by which anything can be "identified" as anything else: that is to pick out whatever is agreeable and disregard the rest. He says that he has demonstrated that most of the symbols are Mayan. One of our intermediatist pseudo-principles is that any way of demonstrating anything is just as good a way of demonstrating anything else. By Mr. Holder's method we could demonstrate that we're Mayan if that should be a source of pride to us. One of the characters upon this stone is a circle within a circle - similar character found by Mr. Holder in a Mayan manuscript. There are two 6's. 6's can be found in Mayan manuscripts. A double scroll. There are dots and there are dashes. Well, then, we, in turn, disregard the circle within a circle and the double scroll and emphasize that 6's occur in this book, and that dots are plentiful, and would be more plentiful if it were customary to use the small "I" for the first personal pronoun - that when it comes to dashes - that's demonstrated: we're Mayan.

I suppose the tendency is to feel that we're sneering at some valuable archaeologic work, and that Mr. Holder did make a veritable identification.

He writes:

"I submitted the photographs to the Field Museum and the Smithsonian and one or two others, and, to my surprise, the reply was that they could make nothing out of it."

Our indefinite acceptance, by preponderance of three or four groups of museum-experts against one person, is that a stone bearing inscriptions unassimilable with any known language upon this earth, is said to have fallen from the sky. Another poor wretch of an outcast belonging here is noted in the Scientific American, 48-261: that, of an object, or a meteorite, that fell Feb. 16, 1883, near Brescia, Italy, a false report was circulated that one of the fragments bore the impress of a hand. That's all that is findable by me upon this mere gasp of a thing. Intermediatistically, my acceptance is that, though in the course of human history, there have been some notable approximations, there never has been a real liar: that he could not survive in intermediateness, where everything merges away or has its pseudo-base in something else - would be instantly translated to the Negative Absolute. So my acceptance is that, though curtly dismissed, there was something to base upon in this report; that there were unusual markings upon this object. Of course that is not to jump to the conclusion that they were cuneiform characters that looked like finger prints.

Altogether, I think that in some of our past expressions, we must have been very efficient, if the experience of Mr. Symons be typical, so indefinite are we becoming here. Just here we are interested in many things that have been found, especially in the United States, which speak of a civilization, or of many civilizations not indigenous to this earth. One trouble is in trying to decide whether they fell here from the sky, or were left behind by visitors from other worlds.

We have a notion that there have been disasters aloft, and that coins have dropped here: that inhabitants of this earth found them or saw them fall, and then made coins imitatively: it may be that coins were showered here by something of a tutelary nature that undertook to advance us from the stage of barter to the use of a medium.

If coins should be identified as Roman coins, we've had so much experience with "identifications" that we know a phantom when we see one - but, even so, how could Roman coins have got to North America - far in the interior of North America - or buried under the accumulation, of centuries, of soil - unless they did drop from wherever the first Romans came from? Ignatius Donnelly, in "Atlantis," gives a list of objects that have been found in mounds that are supposed to antedate all European influence in America: lathe-made articles, such as traders - from somewhere would supply to savages marks of the lathe said to be unmistakable. Said to be: of course we can't accept that anything is unmistakable. In the Rept. Smithson. Inst., 1881-619, there is an account, by Charles C. Jones, of two silver crosses that were found in Georgia.

They are skillfully made, highly ornamented crosses, but are not conventional crucifixes: all arms of equal length. Mr. Jones is a good positivist - that De Sota had halted at the "precise" spot where these crosses were found. But the spirit of negativeness that lurks in all things said to be "precise" shows itself in that upon one of these crosses is an inscription that has no meaning in Spanish or any other known, terrestrial language:

"IYNKICIDU," according to Mr. Jones. He thinks that this is a name, and that there is an aboriginal ring to it, though I should say, myself, that he was thinking of the far-distant Incas: that the Spanish donor cut on the cross the name of an Indian to whom it was presented. But we look at the inscription ourselves and see that the letters said to be "C" and "D" are turned the wrong way, and that the letter said to be "K" is not only turned the wrong way, but is upside down.

It is difficult to accept that the remarkable, the very extensive, copper mines in the region of Lake Superior, were ever the works of American aborigines. Despite the astonishing extent of these mines, nothing has ever been found to indicate that the region was ever inhabited by permanent dwellers - "... not a vestige of a dwelling, a skeleton, or a bone has been found." The Indians have no traditions relating to the mines. (Amer. Antiquarian, 25-258.)

I think that we've had visitors: that they have come here for copper, for instance. As to other relics of them but we now come upon frequency of a merger that has not so often appeared before: Fraudulency.

Hair called real hair - then there are wigs. Teeth called real teeth - then there are false teeth. Official money - counterfeit money.

It's the bane of psychic research. If there be psychic phenomena, there must be fraudulent psychic phenomena. So desperate is the situation here that Carrington argues that, even if Palladino be caught cheating, that is not to say that all her phenomena are fraudulent. My own version is: that nothing indicates anything, in a positive sense, because, in a positive sense, there is nothing to be indicated. Everything that is called true must merge away indistinguishably into something called false. Both are expressions of the same underlying quasiness, and are continuous. Fraudulent antiquarian relics are very common, but they are not more common than are fraudulent paintings.

W. S. Forest, "Historical Sketches of Norfolk, Virginia":

That, in Sept., 1833, when some workmen, near Norfolk, were boring for water, a coin was drawn up from a depth of about 30 feet.

It was about the size of an English shilling, but oval an oval disk, if not a coin. The figures upon it were distinct, and represented "a warrior or hunter and other characters, apparently of Roman origin."

The means of exclusion would probably be - men digging a hole no one else looking: one of them drops a coin into the hole - as to where he got a strange coin, remarkable in shape even that's disregarded.

Up comes the coin - expressions of astonishment from the evil one who had dropped it.

However the antiquarians have missed this coin. I can find no other mention of it.

Another coin. Also a little study in the genesis of a prophet.

In the American Antiquarian, 16-313, is copied a story by a correspondent to the Detroit News, of a copper coin about the size of a two-cent piece, said to have been found in a Michigan mound. The Editor says merely that he does not endorse the find.

Upon this slender basis, he buds out, in the next number of the Antiquarian:

"The coin turns out, as we predicted, to be a fraud."

You can imagine the scorn of Elijah, or any of the old more nearly real prophets.

Or all things are tried by the only kind of jurisprudence we have in quasi-existence:

Presumed to be innocent until convicted - but they're guilty.

The Editor's reasoning is as phantom-like as my own, or St. Paul's, or Darwin's. The coin is condemned because it came from the same region from which, a few years before, had come pottery that had been called fraudulent. The pottery had been condemned because it was condemnable.

Scientific American, June 17, 1882:

That a farmer, in Cass Co., Ill., had picked up, on his farm, a bronze coin, which was sent to Prof. F. F. Hilder, of St. Louis, who identified it as a coin of Antiochus IV. Inscription said to be in ancient Greek characters: translated as "King Antiochus Epiphanes (Illustrious) the Victorious." Sounds quite definite and convincing - but we have some more translations coming. In the American Pioneer, 2-169, are shown two faces of a copper coin, with characters very much like those upon the Grave Creek stone - which, with translations, we'll take up soon. This coin is said to have been found in Connecticut, in 1843.

"Records of the Past," 12-182:

That, early in 1913, a coin, said to be a Roman coin, was reported as discovered in an Illinois mound. It was sent to Dr. Emerson, of the Art Institute, of Chicago. His opinion was that the coin is "of the rare mintage of Domitius Domitianus, Emperor in Egypt." As to its discovery in an Illinois mound, Dr. Emerson disclaims responsibility. But what strikes me here is that a joker should not have been satisfied with an ordinary Roman coin. Where did he get a rare coin, and why was it not missed from some collection?

I have looked over numismatic journals enough to accept that the whereabouts of every rare coin in any one's possession is known to coin-collectors. Seems to me nothing left but to call this another "identification."

Proc. Amer. Phil. Soc., 12-224:

That, in July, 1871, a letter was received from Mr. Jacob W. Moffit, of Chillicothe, Ill., enclosing a photograph of a coin, which he said had been brought up, by him, while boring, from a depth of 120 feet.

Of course, by conventional scientific standards, such depth has some extraordinary meaning. Paleontologists, geologists, and archaeologists consider themselves reasonable in arguing ancient origin of the far-buried. We only accept: depth is a pseudo-standard with us; one earthquake could bury a coin of recent mintage 120 feet below the surface.

According to a writer in the Proceedings, the coin is uniform in thickness, and had never been hammered out by savages - "there are other tokens of the machine shop."

But according to Prof. Leslie, it is an astrologic amulet. "There are upon it the signs of Pisces and Leo."

Or, with due disregard, you can find signs of your great grandmother, or of the Crusades, or of the Mayans, upon anything that ever came from Chillicothe or from a five and ten cent store.

Anything that looks like a cat and a gold fish looks like Leo and Pisces: but, by due suppressions and distortions there's nothing that can't be made to look like a cat and a gold fish. I fear me we're turning a little irritable here. To be damned by slumbering giants and interesting little harlots and clowns who rank high in their profession is at least supportable to our vanity; but, we find that the anthropologists are of the slums of the divine, or of an archaic kindergarten of intellectuality, and it is very unflattering to find a mess of moldy infants sitting in judgment upon us.

Prof. Leslie then finds, as arbitrarily as one might find that some joker put the Brooklyn Bridge where it is, that "the piece was placed there as a practical joke, though not by its present owner; and is a modern fabrication, perhaps of the sixteenth century, possibly Hispano – American or French-American origin."

It's sheer, brutal attempt to assimilate a thing that may or may not have fallen from the sky, with phenomena admitted by the anthropologic system: or with the early French or Spanish explorers of Illinois. Though it is ridiculous in a positive sense, to give reasons, it is more acceptable to attempt reasons more nearly real than opposing reasons. Of course, in his favor, we note that Prof. Leslie qualifies his notions. But his disregards are that there is nothing either French or Spanish about this coin. A legend upon it is said to be "somewhere between Arabic and Phoenician, without being either."

Prof. Winchell (Sparks from a Geologist's Hammer, p. 170) says of the crude designs upon this coin, which was in his possession scrawls of an animal and of a warrior, or of a cat and a gold fish, whichever be convenient - that they had been neither stamped nor engraved, but "looked as if etched with an acid." That is a method unknown in numismatics of this earth. As to the crudity of design upon this coin, and something else - that, though the "warrior" may be, by due disregards, either a cat or a gold fish, we have to note that his head dress is typical of the American Indian - could be explained, of course, but for fear that we might be instantly translated to the Positive Absolute, which may not be absolutely desirable, we prefer to have some flaws or negativeness in our own expressions.

Data of more than the thrice-accursed:

Tablets of stone, with the ten commandments engraved upon them, in Hebrew, said to have been found in mounds in the United States;

Masonic emblems said to have been found in mounds in the United States.

We're upon the borderline of our acceptances, and we're amorphous in the uncertainties and mergings of our outline. Conventionally, or, with no real reason for so doing, we exclude these things, and then, as grossly and arbitrarily and irrationally - though our attempt is always to approximate away from these negative states - as ever a Kepler, Newton, or Darwin, made his selections, without which he could not have seemed to be, at all, because every one of them is now seen to be an illusion, we accept that other lettered things have been found in mounds in the United States. Of course we do what we can to make the selection seem not gross and arbitrary and irrational.

Then, if we accept that inscribed things of ancient origin have been found in the United States; that can not be attributed to any race indigenous to the western hemisphere; that are not in any language ever heard of in the eastern hemisphere - there's nothing to it but to turn non-Euclidian and try to conceive of a third "hemisphere," or to accept that there has been intercourse between the western hemisphere and some other world.

But there is a peculiarity to these inscribed objects. They remind me of the records left, by Sir John Franklin, in the Arctic; but, also, of attempts made by relief expeditions to communicate with the Franklin expedition. The lost explorers cached their records - or concealed them conspicuously in mounds. The relief expeditions sent up balloons, from which messages were dropped broadcast.

Our data are of things that have been cached, and of things that seem to have been dropped - Or a Lost Expedition from - Somewhere.

Explorers from somewhere, and their inability to return then, a long, sentimental, persistent attempt, in the spirit of our own Arctic relief-expeditions at least to establish communication -
What if it may have succeeded?

We think of India - the millions of natives who are ruled by a small band of esoterics - only because they receive support and direction from somewhere else - or from England.

In 1838, Mr. A. B. Tomlinson, owner of the great mound at Grave Creek, West Virginia, excavated the mound. He said that, in the presence of witnesses, he had found a small, flat, oval stone - or disk - upon which were engraved alphabetic characters.

Col. Whittelsey, an expert in these matters, says that the stone is now "universally regarded by archaeologists as a fraud": that, in his opinion, Mr. Tomlinson had been imposed upon.

Avebury, Prehistoric Times, p. 271:

"I mention it because it has been the subject of much discussion, but it is now generally admitted to be a fraud. It is inscribed with Hebrew characters, but the forger has copied the modern instead of the ancient form of the letters."

As I have said, we're as irritable here, under the oppressions of the anthropologists as ever were slaves in the south toward superiorities from "poor white trash." When we finally reverse our relative positions we shall give lowest place to the anthropologists.

A Dr. Gray does at least look at a fish before he conceives of a miraculous origin for it. We shall have to submerge Lord Avebury far below him - if we accept that the stone from Grave Creek is generally regarded as a fraud by eminent authorities who did not know it from some other object - or, in general, that so decided an opinion must be the product of either deliberate disregard or ignorance or fatigue. The stone belongs to a class of phenomena that is repulsive to the System. It will not assimilate with the System. Let such an object be heard of by such a systematist as Avebury, and the mere mention of it is as nearly certainly the stimulus to a conventional reaction as is a charged body to an electroscope or a glass of beer to a prohibitionist. It is of the ideals of Science to know one object from another before expressing an opinion upon a thing, but that is not the spirit of universal mechanics:

A thing. It is attractive or repulsive. Its conventional reaction follows.

Because it is not the stone from Grave Creek that is in Hebrew characters, either ancient or modern: it is a stone from Newark, Ohio, of which the story is told that a forger made this mistake of using modern instead of ancient Hebrew characters. We shall see that the inscription upon the Grave Creek stone is not in Hebrew. Or all things are presumed to be innocent, but are supposed to be guilty - unless they assimilate.

Col. Whittelsey (Western Reserve Historical Tracts, No. 33) says that the Grave Creek stone was considered a fraud by Wilson, Squires, and Davis. Then he comes to the Congress of Archaeologists at Nancy, France, 1875. It is hard for Col. Whittelsey to admit that, at this meeting, which sounds important, the stone was endorsed. He reminds us of Mr. Symons, and "the man" who "considered" that he saw something. Co). Whittelsey's somewhat tortuous expression is that the finder of the stone "so imposed his views" upon the congress that it pronounced the stone genuine.

Also the stone was examined by Schoolcraft. He gave his opinion for genuineness.

Or there's only one process, and "see-saw" is one of its aspects.

Three or four fat experts on the side against us. We find four or five plump ones on our side. Or all that we call logic and reasoning ends up as sheer preponderance of avoirdupois.

Then several philologists came out in favor of genuineness. Some of them translated the inscription. Of course, as we have said, it is our method - or the method of orthodoxy - way in which all conclusions are reached to have some awfully eminent, or preponderantly plump, authorities with us whenever we can - in this case, however, we feel just a little apprehensive in being caught in such excellently obese, but somewhat negativized company:

Translation by M. Jombard:

"Thy orders are laws: thou shinest in impetuous elan and rapid chamois."

M. Maurice Schwab:

"The chief of Emigration who reached these places (or this island) has fixed these characters forever."

M. Oppert:

"The grave of one who was assassinated here. May God, to revenge him, strike his murderer, cutting off the hand of his existence."

I like the first one best. I have such a vivid impression from it of some one polishing up brass or something, and in an awful hurry. Of course the third is more dramatic - still they're all very good. They are perturbations of one another, I suppose.

In Tract 44, Col. Whittelsey returns to the subject. He gives the conclusion of Major De Helward, at the Congress of Luxembourg, 1877:

"If Prof. Read and myself are right in the conclusion that the figures are neither of the Runic, Phoenician, Canaanite, Hebrew, Lybian, Celtic, or any other alphabet-language, its importance has been greatly over-rated."

Obvious to a child; obvious to any mentality not helplessly subjected to a system:

That just therein lies the importance of this object.

It is said that an ideal of science is to find out the new - but, unless a thing be of the old, it is "unimportant."

"It is not worth while." (Hovey.)

Then the inscribed ax, or wedge, which, according to Dr. John C. Evans, in a communication to the American Ethnological Society, was plowed up, near Pemberton, N. J., 1859. The characters upon this ax, or wedge, are strikingly similar to the characters on the Grave Creek stone. Also, with a little disregard here and a little more there, they look like tracks in the snow by some one's who's been out celebrating, or like your handwriting, or mine, when we think there's a certain distinction in illegibility. Method of disregard: anything's anything.

Dr. Abbott describes this object in the Report of the Smithsonian Institution, 1875-260.

He says he has no faith in it.

All progress is from the outrageous to the commonplace. Or quasi-existence proceeds from rape to the crooning of lullabies.

It's been interesting to me to go over various long-established periodicals, and note controversies between attempting positivists, and then intermediatistic issues. Bold, bad intruders of theories; ruffians with dishonorable intentions - the alarms of Science; her attempts to preserve that which is dearer than life itself – submission - then a fidelity like Mrs. Micawber's. So many of these ruffians, or wandering comedians that were hated, or scorned, pitied, embraced, conventionalized. There's not a notion in this book that has a more frightful, or ridiculous, mien than had the notion of human footprints in rocks, when that now respectabilized ruffian, or clown, was first heard from. It seems bewildering to one whose interests are not scientific that such rows should be raised over such trifles: but the feeling of a systematist toward such an intruder is just about what any one's would be if a tramp from the street should come in, sit at one's dinner table, and say he belonged there. We know what hypnosis can do: let him insist with all his might that he does belong there, and one begins to suspect that he may be right; that he may have higher perceptions of what's right. The prohibitionists had this worked out very skillfully.

So the row that was raised over the stone from Grave Creek - but time and cumulativeness, and the very factor we make so much of - or the power of massed data. There were other reports of inscribed stones, and then, half a century later, some mounds - or caches, as we call them were opened by the Rev. Mr. Gass, near the city of Davenport. (American Antiquarian, 15-73.) Several stone tablets were found. Upon one of them, the letters "TFTOWNS" may easily be made out. In this instance we hear nothing of fraudulency - time, cumulativeness, the power of massed data. The attempt to assimilate this datum is:

That the tablet was probably of Mormon origin.

Why?

Because, at Mendon, Ill., was found a brass plate, upon which were similar characters.

Why that?

Because that was found "near a house once occupied by a Mormon."

In a real existence, a real meteorologist, suspecting that cinders had come from a fire engine - would have asked a fireman.

Tablets of Davenport - there's not a record findable that it ever occurred to any antiquarian - to ask a Mormon.

Other tablets were found. Upon one of them are two "F's" and two "8's." Also a large tablet, twelve inches by eight to ten inches "with Roman numerals and Arabic." It is said that the figure "8" occurs three times, and the figure, or letter "O" seven times. "With these familiar characters are others that resemble ancient alphabets, either Phoenecian or Hebrew."

It may be that the discovery of Australia, for instance, will turn out to be less important than the discovery and the meaning of these tablets -

But where will you read of them in anything subsequently published; what antiquarian has ever since tried to understand them, and their presence, and indications of antiquity, in a land that we're told was inhabited only by unlettered savages?

These things that are exhumed only to be buried in some other way.

Another tablet was found, at Davenport, by Mr. Charles Harrison, president of the American Antiquarian Society. "... 8 and other hieroglyphics are upon this tablet." This time, also, fraud is not mentioned. My own notion is that it is very unsportsmanlike ever to mention fraud. Accept anything. Then explain it your way. Anything that assimilates with one explanation, must have assimilable relations, to some degree, with all other explanations, if all explanations are somewhere continuous. Mormons are lugged in again, but the attempt is faint and helpless - "because general circumstances make it difficult to explain the presence of these tablets."

Altogether our phantom resistance is mere attribution to the Mormons, without the slightest attempt to find base for the attribution.

We think of messages that were showered upon this earth, and of messages that were cached in mounds upon this earth. The similarity to the Franklin situation is striking. Conceivably centuries from now, objects dropped from relief-expedition-balloons may be found in the Arctic, and conceivably there are still undiscovered caches left by Franklin, in the hope that relief expeditions would find them. It would be as incongruous to attribute these things to the Eskimos as to attribute tablets and lettered stones to the aborigines of America. Some time I shall take up an expression that the queer-shaped mounds upon this earth were built by explorers from Somewhere, unable to get back, designed to attract attention from some other world, and that a vast sword-shaped mound has been discovered upon the moon - Just now we think of lettered things and their two possible significances.

A bizarre little lost soul, rescued from one of the morgues of the American Journal of Science:

An account, sent by a correspondent, to Prof. Silliman, of something that was found in a block of marble, taken Nov., 1829, from a quarry, near Philadelphia (Am. J. Sci., 1-19-361). The block was cut into slabs. By this process, it is said, was exposed an indentation in the stone, about one and a half inches by five-eighths of an inch. A geometric indentation: in it were two definite-looking raised letters, like "I U": only difference is that the corners of the "U" are not rounded, but are right angles. We are told that this block of stone came from a depth of seventy or eighty feet - or that, if acceptable, this lettering was done long, long ago. To some persons, not sated with the commonness of the incredible that has to be accepted, it may seem grotesque to think that an indentation in sand could have tons of other sand piled upon it and hardening into stone, without being pressed out - but the famous Nicaraguan footprints were found in a quarry under eleven strata of solid rock.

There was no discussion of this datum. We only take it out for an airing, As to lettered stones that may once upon a time have been showered upon Europe, if we cannot accept that the stones were inscribed by indigenous inhabitants of Europe, many have been found in caves - whence they were carried as curiosities by prehistoric men, or as ornaments, I suppose. About the size and shape of the Grave Creek stone, or disk: "flat and oval and about two inches wide." (Sollas.) Characters painted upon them: found first by M. Piette, in the cave of Mas d' Azil, Ariege. According to Sollas, they are marked in various directions with red and black lines. "But on not a few of them, more complex characters occur, which in a few instances simulate some of the capital letters of the Roman alphabet."

In one instance the letters "F El" accompanied by no other markings to modify them, are as plain as they could be. According to Sollas ("Ancient Hunters," p. 95) M. Cartailhac has confirmed the observations of Piette, and M. Boule has found additional examples.

"They offer one of the darkest problems of prehistoric times." (Sollas.)

As to caches in general, I should say that they are made with two purposes: to proclaim and to conceal; or that caches documents are hidden, or covered over, in conspicuous structures; at least, so are designed the cairns in the Arctic.

Trans N. Y. Acad. of Sciences, 11-27:

That Mr. J. H. Hooper, Bradley Co., Term., having come upon a curious stone, in some woods upon his farm, investigated. He dug.

He unearthed a long wall. Upon this wall were inscribed many alphabetic characters. "872 characters have been examined, many of them duplicates, and a few imitations of animal forms, the moon, and other objects. Accidental imitations of oriental alphabets are numerous.

The part that seems significant:

That these letters had been hidden under a layer of cement.

And still, in our own heterogeneity, or unwillingness, or inability, to concentrate upon single concepts, we shall - or we shan't - accept that, though there may have been a Lost Colony or Lost Expedition from Somewhere, upon this earth, and extra-mundane visitors Who could never get back, there have been other extra-mundane visitors, who have gone away again altogether quite in analogy with the Franklin Expedition and Peary's Sittings in the Arctic -

And a wreck that occurred to one group of them -

And the loot that was lost overboard -

The Chinese seals of Ireland.

Not the things with the big, wistful eyes; that lie on ice, and that are taught to balance objects on their noses but inscribed stamps, with which to make impressions.

Proc. Roy. Irish Acad., 1-381:

A paper was read by Mr. J. Huband Smith, descriptive of about a dozen Chinese seals that had been found in Ireland. They are all alike: each a cube with an animal seated upon it. "It is said that the inscriptions upon them are of a very ancient class of Chinese characters."

The three points that have made a leper and an outcast of this datum - but only in the sense of disregard, because nowhere that I know of is it questioned - :

Agreement among archaeologists that there were no relations, in the remote past, between China and Ireland;

That no other objects, from ancient China - virtually, I suppose - have ever been found in Ireland;

The great distances at which these seals have been found apart. After Mr. Smith's investigations - if he did investigate, or do more than record - many more Chinese seals were found in Ireland, and, with one exception, only in Ireland. In 1852, about 60 had been found. Of all archaeologic finds in Ireland, "none is enveloped in greater mystery." (Chamber's Journal, 16-364.) According to the writer in Chamber's Journal, one of these seals was found in a curiosity shop in London. When questioned, the shopkeeper said that it had come from Ireland.

In this instance, if you don't take instinctively to our expression, there is no orthodox explanation for your preference. It is the astonishing scattering of them, over field and forest, that has hushed the explainers. In the Proceedings of the Royal Irish Academy, 10-171, Dr. Frazer says that they "appear to have been sown broadcast over the country in some strange way that I cannot offer solution of."

The struggle for expression of a notion that did not belong to Dr. Frazer's era: "The invariable story of their find is what we might expect if they had been accidentally dropped. . . ."

Three were found in Tipperary; six in Cork; three in Down; four in Waterford; all the rest one or two to a county.

But one of these Chinese seals was found in the bed of the River Boyne, near Clonard, Meath, when workmen were raising gravel.

That one, at least, had been dropped there.

Chapter XII

Astronomy.

And a watchman looking at half a dozen lanterns, where a street's been torn up.

There are gas lights and kerosene lamps and electric lights in the neighborhood: matches flaring, fires in stoves, bonfires, house afire somewhere; lights of automobiles, illuminated signs -

The watchman and his one little system.

Ethics.

And some young ladies and the dear old professor of a very "select" seminary.

Drugs and divorce and rape: venereal diseases, drunkenness, murder -

Excluded.

The prim and the precise, or the exact, the homogeneous, the single, the puritanic, the mathematic, the pure, the perfect. We can have illusion of this state but only by disregarding its infinite denials. It's a drop of milk afloat in acid that's eating it. The positive swamped by the negative. So it is in intermediateness, where only to "be" positive is to generate corresponding and, perhaps, equal negativeness. In our acceptance, it is, in quasi-existence, premonitory, or pre-natal, or pre-awakening consciousness of a real existence.

But this consciousness of realness is the greatest resistance to efforts to realize or to become real - because it is feeling that realness has been attained. Our antagonism is not to Science, but to the attitude of the sciences that they have finally realized; or to belief, instead of acceptance; to the insufficiency, which, as we have seen over and over, amounts to paltriness and puerility, of scientific dogmas and standards. Or, if several persons start out to Chicago, and get to Buffalo, and one be under the delusion that Buffalo is Chicago, that one will be a resistance to the progress of the others. So astronomy and its seemingly exact, little system -

But data we shall have of round worlds and spindle-shaped worlds, and worlds shaped like a wheel; worlds like titanic pruning hooks; world linked together by streaming filaments; solitary worlds, and worlds in hordes: tremendous worlds and tiny worlds: some of them made of material like the material of this earth; and worlds that are geometric super-constructions made of iron and steel -

Or not only fall from the sky of ashes and cinders and coke and charcoal and oily substances that suggest fuel - but the masses of iron that have fallen upon this earth.

Wrecks and flotsam and fragments of vast iron constructions - Or steel. Sooner or later we shall have to take up an expression that fragments of steel have fallen from the sky. If fragments, not of iron, but of steel, have fallen upon this earth -

But what would a deep-sea fish learn even if a steel plate of a wrecked vessel above him should drop and bump him on the nose?

Our submergence in a sea of conventionality of almost impenetrable density.

Sometimes I'm a savage who has found something on the beach of his island. Sometimes I'm a deep-sea fish with a sore nose.

The greatest of mysteries:

Why don't they ever come here, or send here, openly?

Of course there's nothing to that mystery if we don't take so seriously the notion - that we must be interesting. It's probably for moral reasons that they stay away but even so, there must be some degraded ones among them.

Or physical reasons:

When we can specially take up that subject, one of our leading ideas, or credulities, will be that near approach by another world to this world would be catastrophic: that navigable worlds would avoid proximity; that others that have survived have organized into protective remotenesses, or orbits which approximate to regularity, though by no means to the degree of popular supposition.

But the persistence of the notion that we must be interesting.

Bugs and germs and things like that: they're interesting to us:

some of them are too interesting.

Dangers of near approach - nevertheless our own ships that dare not venture close to a rocky shore can send rowboats ashore -

Why not diplomatic relations established between the United States and Cyclorea - which, in our advanced astronomy, is the name of a remarkable wheel-shaped world or super-construction?

Why not missionaries sent here openly to convert us from our barbarous prohibitions and other taboos, and to prepare the way for a good trade in ultra-bibles and super-whiskeys; fortunes made in selling us cast-off super-fineries, which we'd take to like an African chief to some one's old silk hat from New York or London? The answer that occurs to me is so simple that it seems immediately acceptable, if we accept that the obvious is the solution of all problems, or if most of our perplexities consist in laboriously and painfully conceiving of the unanswerable, and then looking for answers using such words as "obvious" and "solution" conventionally -

Or:

Would we, if we could, educate and sophisticate pigs, geese, cattle?

Would it be wise to establish diplomatic relation with the hen that now functions, satisfied with mere sense of achievement by way of compensation?

I think we're property.

I should say we belong to something:

That once upon a time, this earth was No-man's Land, that other worlds explored and colonized here, and fought among themselves for possession, but that now it's owned by something:

That something owns this earth - all others warned off.

Nothing in our own times - perhaps - because I am thinking of certain notes I have has ever appeared upon this earth, from somewhere else, so openly as Columbus landed upon San Salvador, or as Hudson sailed up his river. But as to surreptitious visits to this earth, in recent times, or as to emissaries, perhaps, from other worlds, or voyagers who have shown every indication of intent to evade and avoid, we shall have data as convincing as our data of oil or coal-burning aerial super-constructions.

But, in this vast subject, I shall have to do considerable neglecting or disregarding, myself. I don't see how I can, in this book, take up at all the subject of possible use of humanity to some other mode of existence, or the flattering notion that we can possibly be worth something.

Pigs, geese, and cattle.

First find out that they are owned.

Then find out the whyness of it.

I suspect that, after all, we're useful - that among contesting claimants, adjustment has occurred, or that something now has a legal right to us, by force, or by having paid out analogues of beads for us to former, more primitive, owners of us - all others warned off - that all this has been known, perhaps for ages, to certain ones upon this earth, a cult or order, members of which function like bellwethers to the rest of us, or as superior slaves or overseers, directing us in accordance with instructions received - from Somewhere else - in our mysterious usefulness.

But I accept that, in the past, before proprietorship was established, inhabitants of a host of other worlds have dropped here, hopped here, wafted, sailed, flown, motored - walked here, for all I know - been pulled here, been pushed; have come singly, have come in enormous numbers; have visited occasionally, have visited periodically for hunting, trading, replenishing harems, mining: have been unable to stay here, have established colonies here, have been lost here; far-advanced peoples, or things, and primitive peoples or whatever they were: white ones, black ones, yellow ones -

I have a very convincing datum that the ancient Britons were blue ones.

Of course we are told by conventional anthropologists that they only painted themselves blue, but in our own advanced anthropology, they were veritable blue ones -

Annals of Philosophy, 14-51:

Note of a blue child born in England.

That's atavism.

Giants and fairies. We accept them, of course. Or, if we pride ourselves upon being awfully far-advanced, I don't know how to sustain our conceit except by very largely going far back. Science of to-day the superstition of to-morrow. Science of to-morrow the superstition of to-day.

Notice of a stone ax, 17 inches long: 9 inches across broad end.

(Proc. Soc. of Ants, of Scotland, 1-9-184.)

Amer. Antiquarian, 18-60:

Copper ax from an Ohio mound: 22 inches long; weight 38 pounds.

Amer. Anthropologist, n. s., 8-229:

Stone ax found at Birchwood, Wisconsin exhibited in the collection of the Missouri Historical Society - found with "the pointed end embedded in the soil" - for all I know, may have dropped there - 28 inches long, 14 wide, 11 thick - weight 300 pounds.

Or the footprints, in sandstone, near Carson, Nevada each print 1 8 to 20 inches long. (Amer. Jour. Sci., 3-26-139.)

These footprints are very clear and well-defined: reproduction of them in the Journal - but they assimilate with the System, like sour apples to other systems: so Prof. Marsh, a loyal and unscrupulous systematist, argues:

"The size of these footprints and specially the width between the right and left series, are strong evidence that they were not made by men, as has been so generally supposed."

So these excluders. Stranglers of Minerva. Desperadoes of disregard.

Above all, or below all, the anthropologists. I'm inspired with a new insult - some one offends me: I wish to express almost absolute contempt for him - he's a systematistic anthropologist.

Simply to read something of this kind is not so impressive as to see for one's self: if any one will take the trouble to look up these footprints, as pictured in the Journal, he will either agree with Prof.

Marsh or feel that to deny them is to indicate a mind as profoundly enslaved by a system as was ever the humble intellect of a medieval monk. The reasoning of this representative phantom of the chosen, or of the spectral appearances who sit in judgment, or condemnation, upon us of the more nearly real:

That there never were giants upon this earth, because gigantic footprints are more gigantic than prints made by men who are not giants.

We think of giants as occasional visitors to this earth. Of course -

Stonehenge, for instance. It may be that, as time goes on, we shall have to admit that there are remains of many tremendous habitation of giants upon this earth, and that their appearances here were more than casual - but their bones - or the absence of their bones -

Except that, no matter how cheerful and unsuspicious my disposition may be, when I go to the American Museum of Natural History, dark cynicisms arise the moment I come to the fossils or old bones that have been found upon this earth gigantic things that have been reconstructed into terrifying but "proper" dinosaurs - but my uncheerfulness -

The dodo did it.

On one of the floors below the fossils, they have a reconstructed dodo. It's frankly a fiction: it's labeled as such but it's been reconstructed so cleverly and so convincingly -

Fairies.

"Fairy crosses."

Harper's Weekly, 50-715:

That, near the point where the Blue Ridge and the Allegheny Mountains unite, north of Patrick County, Virginia, many little stone crosses have been found. A race of tiny beings.

They crucified cockroaches.

Exquisite beings - but the cruelty of the exquisite. In their diminutive way they were human beings. They crucified. The "fairy crosses," we are told in Harper's Weekly, range in weight from one-quarter of an ounce to an ounce: but it is said, in the Scientific American, 79-395, that some of them are no larger than the head of a pin.

They have been found in two other states, but all in Virginia are strictly localized on and along Bull Mountain.

We are reminded of the Chinese seals in Ireland.

I suppose they fell there.

Some are Roman crosses, some St. Andrew's, some Maltese. This time we are spared contact with the anthropologists and have geologists instead, but I am afraid that the relief to our finer, or more nearly real, sensibilities will not be very great. The geologists were called upon to explain the "fairy crosses." Their response was the usual scientific tropism - "Geologists say that they are crystals."

The writer in Harper's Weekly points out that this "hold up," or this anaesthetic, if theoretic science be little but attempt to assuage pangs of the unexplained, fails to account for the localized distributions of these objects - which make me think of both aggregation and separation at the bottom of the sea, if from a wrecked ship, similar objects should fall in large numbers but at different times.

But some are Roman crosses, some St. Andrew's, some Maltese.

Conceivably there might be a mineral that would have a diversity of geometric forms, at the same time restricted to some expression of the cross, because snowflakes, for instance, have diversity but restriction to the hexagon, but the guilty geologists, cold-blooded as astronomers and chemists and all the other deep-sea fishes - though less profoundly of the pseudo-saved than the wretched anthropologists - disregarded the very datum that it was wise to disregard :

That the "fairy crosses" are not all made of the same material.

It's the same old disregard, or it's the same old psycho-tropism, or process of assimilation. Crystals are geometric forms. Crystals are included in the System. So then "fairy crosses" are crystals.

But that different minerals should, in a few different regions, be inspired to turn into different forms of the cross - is the kind of resistance that we call less nearly real than our own acceptances.

We now come to some "cursed" little things that are of the "lost," but for the "salvation" of which scientific missionaries have done their dam-dest.

"Pigmy flints."

They can't very well be denied.

They're lost and well known.

"Pigmy flints" are tiny, prehistoric implements. Some of them are a quarter of an inch in size. England, India, France, South Africa - they've been found in many parts of the world - whether showered there or not. They belong high up in the froth of the accursed: they are not denied, and they have not been disregarded ; there is an abundant literature upon this subject. One attempt to rationalize them, or assimilate them, or take them into the scientific fold, has been the notion that they were toys of prehistoric children.

It sounds reasonable. But, of course, by the reasonable we mean that for which the equally reasonable, but opposing, has not been found out - except that we modify that by saying that, though nothing's finally reasonable, some phenomena have higher approximations to Reasonableness than have others. Against the notion of toys, the higher approximation is that where "pigmy flints" are found, all flints are pigmies - at least so in India, where, when larger implements have been found in the same place, there are separations by strata. (Wilson.)

The datum that, just at present, leads me to accept that these flints were made by beings about the size of pickles, is a point brought out by Prof. Wilson (Rept. National Museum, 1892-455):

Not only that the flints are tiny but that the chipping upon them is "minute."

Struggle for expression, in the mind of a 19th-century-ite, of an idea that did not belong to his era:

In Science Gossip, 1896-36, R. A. Galty says:

"So fine is the chipping that to see the workmanship a magnifying glass is necessary.

I think that would be absolutely convincing, if there were anything - absolutely anything - either that tiny beings, from pickle to cucumber-stature made these things, or that ordinary savages made them under magnifying glasses.

The idea that we are now going to develop, or perpetrate, is rather intensely of the accursed, or the advanced. It's a lost soul, I admit - or boast - but it fits in. Or, as conventional as ever, our own method is the scientific method of assimilating. It assimilates, if we think of the inhabitants of Elvera -

By the way, I forgot to tell the name of the giant's world:

Monstrator.

Spindle-shaped world - about 100,000 miles along its major axis more details to be published later.

But our coming inspiration fits in, if we think of the inhabitants of Elvera as having only visited here: having, in hordes as dense as clouds of bats, come here, upon hunting excursions - for mice, I should say: for bees, very likely or most likely of all, or inevitably, to convert the heathen here - horrified with any one who would gorge himself with more than a bean at a time; fearful for the souls of beings who would guzzle more than a dew drop at a time - hordes of tiny missionaries, determined that right should prevail, determining right by their own minutenesses. They must have been missionaries.

Only to be is motion to convert or assimilate something else.

The idea now is that tiny creatures coming here from their own little world, which may be Eros, though I call it Elvera, would flit from the exquisite to the enormous - gulp of a fair-sized terrestrial animal - half a dozen of them gone and soon digested. One falls into a brook - torn away in a mighty torrent -

Or never anything but conventional, we adopt from Darwin:

"The geological records are incomplete."

Their flints would survive, but, as to their fragile bodies one - might as well search for prehistoric frost-traceries. A little whirlwind - Elverean carried away a hundred yards - body never found by his companions. They'd mourn for the departed. Conventional emotion to have: they'd mourn. There'd have to be a funeral: there's no getting away from funerals. So I adopt an explanation that I take from the anthropologists: burial in effigy. Perhaps the Elvereans would not come to this earth again until many years later - another distressing occurrence - one little mausoleum for all burials in effigy.

.London Times, July 20, 1836:

That, early in July, 1836, some boys were searching for rabbits' burrows in the rocky formation, near Edinburgh, known as Arthur's Seat. In the side of a cliff, they came upon some thin sheets of slate, which they pulled out.

Little cave.

Seventeen tiny coffins.

Three or four inches long.

In the coffins were miniature wooden figures. They were dressed differently both in style and material. There were two tiers of eight coffins each, and a third tier begun, with one coffin.

The extraordinary datum, which has especially made mystery here:

That the coffins had been deposited singly, in the little cave, and at intervals of many years. In the first tier, the coffins were quite decayed, and the wrappings had moldered away. In the second tier, the effects of age had not advanced so far. And the top coffin was quite recent-looking.

In the Proceedings of the Society of Antiquarians of Scotland, 3-12-460, there is a full account of this find. Three of the coffins and three of the figures are pictured.

So Elvera with its downy forests and its microscopic oyster shells - and if the Elvereans be not very far-advanced, they take baths - with sponges the size of pin heads -

Or that catastrophes have occurred: that fragments of Elvera have fallen to this earth:

In Popular Science, 20-83, Francis Bingham, writing of the corals and sponges and shells and crinoids that Dr. Hahn had asserted that he had found in meteorites, says, judging by the photographs of them, that their "notable peculiarity" is their "extreme smallness." The corals, for instance, are about one-twentieth the size of terrestrial corals. "They represent a veritable pigmy animal world," says Bingham.

The inhabitants of Monstrator and Elvera were primitives, I think, at the time of their occasional visits to this earth - though, of course, in a quasi-existence, anything that we semi-phantoms call evidence of anything may be just as good evidence of anything else. Logicians and detectives and jurymen and suspicious wives and members of the Royal Astronomic Society recognize this indeterminateness, but have the delusion that in the method of agreement there is final, or real evidence. The method is good enough for an "existence" that is only semi-real, but also it is the method of reasoning by which witches were burned, and by which ghosts have been feared. I'd not like to be so unadvanced as to deny witches and ghosts, but I do think that there never have been witches and ghosts like those of popular supposition. But stories of them have been supported by astonishing fabrications of details and of different accounts in agreement.

So, if a giant left impressions of his bare feet in the ground, that is not to say that he was a primitive - bulk of culture out taking the Kneipp cure. So, if Stonehenge is a large, but only roughly geometric construction, the inattention to details by its builders - signifies anything you please - ambitious dwarfs or giants - if giants, that they were little more than cave men, or that they were postimpressionist architects from a very far-advanced civilization.

If there are other worlds, there are tutelary worlds - or that Kepler, for instance, could not have been absolutely wrong: that his notion of an angel assigned to push along and guide each planet may not be very acceptable, but that, abstractedly, or in the notion of a tutelary relation, we may find acceptance.

Only to be is to be tutelary.

Our general expression:

That "everything*' in Intermediateness is not a thing, but is an endeavor to become something - by breaking away from its continuity, or merging away, with all other phenomena is an attempt to break away from the very essence of a relative existence and become absolute - if it have not surrendered to, or become part of, some higher attempt:

That to this process there are two aspects:

Attraction, or the spirit of everything to assimilate all other things if it have not given in and subordinated to - or have not been assimilated by - some higher attempted system, unity, organization, entity, harmony, equilibrium -

And repulsion, or the attempt of everything to exclude or disregard the unassimilable.

Universality of the process:

Anything conceivable:

A tree. It is doing all it can to assimilate substances of the soil and substances of the air, and sunshine, too, into tree-substance: obversely it is rejecting or excluding or disregarding that which it can not assimilate.

Cow grazing, pig rooting, tiger stalking: planets trying, or acting, to capture comets; rag pickers and the Christian religion, and a cat down headfirst in a garbage can; nations fighting for more territory, sciences correlating the data they can, trust magnates organizing, chorus girl out for a little late supper - all of them stopped somewhere by the unassimilable. Chorus girl and the broiled lobster.

If she eats not shell and all she represents universal failure to positivize. Also, if she does she represents universal failure to positivize: her ensuing disorders will translate her to the Negative Absolute. Or Science and some of our cursed hard-shelled data.

One speaks of the tutelarian as if it were something distinct in itself. So one speaks of a tree, a saint, a barrel of pork, the Rocky Mountains. One speaks of missionaries, as if they were positively different, or had identity of their own, or were a species by themselves.

To the Intermediatist, everything that seems to have identity is only attempted identity, and every species is continuous with all other species, or that which is called the specific is only emphasis upon some aspect of the general. If there are cats, they're only emphasis upon universal felinity. There is nothing that does not partake of that of which the missionary, or the tutelary, is the special. Every conversation is a conflict of missionaries, each trying to convert the other, to assimilate, or to make the other similar to himself. If no progress be made, mutual repulsion will follow.

If other worlds have ever in the past had relations with this earth, they were attempted positivizations: to extend themselves, by colonies, upon this earth; to convert, or assimilate, indigenous inhabitants of this earth.

Or parent-worlds and their colonies here -

Super-Romanimus -

Or where the first Romans came from.

It's as good as the Romulus and Remus story.

Super-Israelimus -

Or that, despite modern reasoning upon this subject, there was once something that was super-parental or tutelary to early orientals.

Azuria, which was tutelary to the early Britons:

Azuria, whence came the blue Britons, whose descendants gradually diluting, like blueing in a wash-tub, where a faucet's turned on, have been most emphasized of sub-tutelarians, or assimilators ever since.

Worlds that were once tutelarian worlds - before this earth became sole property of one of them their attempts to convert or assimilate - but then the state that comes to all things in their missionary-frustrations unacceptance by all stomachs of some things; rejection by all societies of some units; glaciers that sort over and cast out stones -

Repulsion. Wrath of the baffled missionary. There is no other wrath. All repulsion is reaction to the unassimilable.

So then the wrath of Azuria -

Because surrounding peoples of this earth would not assimilate with her own colonists in the part of the earth that we now call England.

I don't know that there has ever been more nearly just, reasonable, or logical wrath, in this earth's history - if there is no other wrath.

The wrath of Azuria, because the other peoples of this earth would not turn blue to suit her.

History is a department of human delusion that interests us. We are able to give a little advancement to history. In the vitrified forts of a few parts of Europe, we find data that the Humes and Gibbons have disregarded. The vitrified forts surrounding England, but not in England.

The vitrified forts of Scotland, Ireland, Brittany, and Bohemia. Or that, once upon a time, with electric blasts, Azuria tried to swipe this earth clear of the peoples who resisted her.

The vast blue bulk of Azuria appeared in the sky. Clouds turned green. The sun was formless and purple in the vibrations of wrath that were emanating from Azuria. The whitish, or yellowish, or brownish peoples of Scotland, Ireland, Brittany, and Bohemia fled to hill tops and built forts. In a real existence, hill tops, or easiest accessibility to an aerial enemy, would be the last choice in refuges. But here, in quasi-existence, if we're accustomed to run to hill tops, in times of danger, we run to them just the same, even with danger closest to hill tops. Very common in .quasi-existence: attempt to escape by running closer to the pursuing.

They built forts, or already had forts, on hill tops.

Something poured electricity upon them.

The stones of these forts exist to this day, vitrified, or melted and turned to glass.

The archaeologists have jumped from one conclusion to another, like the "rapid chamois" we read of a while ago, to account for vitrified forts, always restricted by the commandment that unless their conclusions conformed to such tenets as Exclusionism, of the System, they would be excommunicated. So archaeologists, in their medieval dread of excommunication, have tried to explain vitrified forts in terms of terrestrial experience. We find in their insufficiencies the same old assimilating of all that could be assimilated, and disregard for the unassimilable, conventionalizing into the explanation that vitrified forts were made by prehistoric peoples who built vast fires - often remote from wood-supply - to melt externally, and to cement together, the stones of their constructions. But negativeness always: so within itself a science can never be homogeneous or unified or harmonious. So Miss Russel, in the Journal of the B. A. A., has pointed out that it is seldom that single stones, to say nothing of long walls, of large houses that are burned to the ground, are vitrified.

If we pay a little attention to this subject, ourselves, before starting to write upon it, which is one of the ways of being more nearly real than oppositions so far encountered by us, we find:

That the stones of these forts are vitrified in no reference to cementing them: that they are cemented here and there, in streaks, as if special blasts had struck, or played, upon them.

Then one thinks of lightning?

Once upon a time something melted, in streaks, the stones of forts on the tops of hills in Scotland, Ireland, Brittany, and Bohemia.

Lightning selects the isolated and conspicuous.

But some of the vitrified forts are not upon tops of hills: some are very inconspicuous: their walls too are vitrified in streaks.

Something once had effect, similar to lightning, upon forts, mostly on hills, in Scotland, Ireland, Brittany, and Bohemia.

But upon hills, all over the rest of the world, are remains of forts that are not vitrified.

There is only one crime, in the local sense, and that is not to turn blue, if the gods are blue: but, in the universal sense, the one crime is not to turn the gods themselves green, if you're green.

Chapter XIII

One of the most extraordinary of phenomena, or alleged phenomena, of psychic research, or alleged research - if in quasiexistence there never has been real research, but only approximations to research that merge away, or that are continuous with, prejudice and convenience -

"Stone-throwing."

It's attributed to poltergeists. They're mischievous spirits.

Poltergeists do not assimilate with our own present quasi-system, which is an attempt to correlate denied or disregarded data as phenomena of extra-telluric forces, expressed in physical terms. Therefore I regard poltergeists as evil or false or discordant or absurd - names that we give to various degrees or aspects of the unassimilable, or that which resists attempts to organize, harmonize, systematize, or, in short, to positivize - names that we give to our recognitions of the negative state. I don't care to deny poltergeists, because I suspect that later, when we're more enlightened, or when we widen the range of our credulities, or take on more of that increase of ignorance that is called knowledge, poltergeists may become assimilable. Then they'll be as reasonable as trees. By reasonableness I mean that which assimilates with a dominant force, or system, or a major body of thought which is, itself, of course, hypnosis and delusion - developing, however, in our acceptance, to higher and higher approximations to realness. The poltergeists are now evil or absurd to me, proportionately to their present unassimilableness, compounded, however, with the factor of their possible future assimilableness.

We lug in the poltergeists, because some of our own data, or alleged data, merge away indistinguishably with data, or alleged data, of them: Instances of stones that have been thrown, or that have fallen, upon a small area, from an unseen and undetectable source.

London Times, April 27, 1872:

"From 4 o'clock, Thursday afternoon, until half past eleven, Thursday night, the houses, 56 and 58 Reverdy Road, Bermondsey, were assailed with stones and other missiles coming from an unseen quarter. Two children were injured, every window broken, and several articles of furniture were destroyed. Although there was a strong body of policemen scattered in the neighborhood, they could not trace the direction whence the stones were thrown."

"Other missiles" make a complication here. But if the expression means tin cans and old shoes, and if we accept that the direction could not be traced because it never occurred to any one to look upward - why we've lost a good deal of our provincialism by this time.

London Times, Sept. 16, 1841:

That, in the home of Mrs. Charton, at Sutton Courthouse, Sutton Lane, Chiswick, windows had been broken "by some unseen agent."

Every attempt to detect the perpetrator failed. The mansion was detached and surrounded by high walls. No other building was near it.

The police were called. Two constables, assisted by members of the household, guarded the house, but the windows continued to be broken "both in front and behind the house."

Or the floating islands that are often stationary in the Super –

Sargasso Sea; and atmospheric disturbances that sometimes affect them, and bring things down within small areas, upon this earth, from temporarily stationary sources.

Super-Sargasso Sea and the beaches of its floating islands from which I think, or at least accept, pebbles have fallen:

Wolverhampton, England, June, 1860 - violent storm - fall of so many little black pebbles that they were cleared away by shoveling (La Sci. Pour Tous, 5-264); great number of small black stones that fell at Birmingham, England, Aug., 1858 - violent storm -said to be similar to some basalt a few leagues from Birmingham (Rept. Brit. Assoc., 1864-37); pebbles described as "common water-worn pebbles" that fell at Palestine, Texas, July 6, 1888 - "of a formation not found near Palestine" (W. H. Perry, Sergeant, Signal Corps), Monthly Weather Review, July, 1888); round, smooth pebbles at Kandahor, 1834 (Am. J. Sci., 1-26-161); "a number of stones of peculiar formation and shapes, unknown in this neighborhood, fell in a tornado at Hillsboro, 111., May 18, 1883." (Monthly Weather Review, May, 1883.)

Pebbles from aerial beaches and terrestrial pebbles as products of whirlwinds, so merge in these instances that, though it's interesting to hear of things of peculiar shape that have fallen from the sky, it seems best to pay little attention here, and to find phenomena of the Super-Sargasso Sea remote from the merger:

To this requirement we have three adaptations:

Pebbles that fell where no whirlwind to which to attribute them could be learned of;

Pebbles which fell in hail so large that incredibly could that hail have been formed in this earth's atmosphere;

Pebbles which fell and were, long afterward, followed by more pebbles, as if from some aerial, stationary source, in the same place.

In September, 1898, there was a story in a New York newspaper, of lightning - or an appearance of luminosity? - in Jamaica - something had struck a tree: near the tree were found some small pebbles. It was said that the pebbles had fallen from the sky, with the lightning. But the insult to orthodoxy was that they were not angular fragments such as might have been broken from a stony meteorite: that they were "water-worn pebbles."

In the geographical vagueness of a mainland, the explanation "up from one place and down in another" is always good, and is never overworked, until the instances are massed as they are in this book: but, upon this occasion, in the relatively small area of Jamaica, there was no whirlwind findable - however "there in the first place' bobs up.

Monthly Weather Review, Aug., 1898-363:

That the government meteorologist had investigated: had reported that a tree had been struck by lightning, and that small water-worn pebbles had been found near the tree: but that similar pebbles could be found all over Jamaica.

Monthly Weather Review, Sept., 1915-446:

Prof. Fassig gives an account of a fall of hail that occurred in Maryland, June 22, 1915: hailstones the size of baseballs "not at all uncommon."

"An interesting, but unconfirmed, account stated that small pebbles were found at the center of some of the larger hail gathered at Annapolis. The young man who related the story offered to produce the pebbles, but has not done so."

A footnote:

"Since writing this, the author states that he has received some of the pebbles."

When a young man "produces" pebbles, that's as convincing as anything else I've ever heard of, though no more convincing than, if having told of ham sandwiches falling from the sky, he should "produce" ham sandwiches. If this "reluctance" be admitted by us, we correlate it with a datum reported by a Weather Bureau observer, signifying that, whether the pebbles had been somewhere aloft a long time or not, some of the hailstones that fell with them, had been. The datum is that some of these hailstones were composed of from twenty to twenty-five layers alternately of clear ice and snow-ice. In orthodox terms I argue that a fair-sized hailstone falls from the clouds with velocity sufficient to warm it so that it would not take on even one layer of ice. To put on twenty layers of ice, I conceive of something that had not fallen at all, but had rolled somewhere, at a leisurely rate, for a long time.

We now have a commonplace datum that is familiar in two respects:

Little, symmetric objects of metal that fell at Orenburg, Russia, Sept., 1824 (Phil. Mag., 4-8-463).

A second fall of these objects, at Orenburg, Russia, Jan. 25, 1825 (Quar. Jour. Roy. Inst., 1828-1-447).

I now think of the disk of Tarbes, but when first I came upon these data I was impressed only with recurrence, because the objects of Orenburg were described as crystals of pyrites, or sulphate of iron. I had no notion of metallic objects that might have been shaped or molded by means other than crystallization, until I came to Arago's account of these occurrences (Euvres, 11-644). Here the analysis gives 70 per cent, red oxide of iron, and sulphur and loss by ignition 5 per cent. It seems to me acceptable that iron with considerably less than 5 per cent, sulphur in it is not iron pyrites - then little, rusty iron objects, shaped by some other means, have fallen, four months apart, at the same place. M. Arago expresses astonishment at this phenomenon of recurrence so familiar to us.

Altogether, I find opening before us, vistas of heresies to which I, for one, must shut my eyes. I have always been in sympathy with the dogmatists and exclusionists: that is plain in our opening lines: that to seem to be is falsely and arbitrarily and dogmatically to exclude. It is only that exclusionists who are good in the nineteenth century are evil in the twentieth century. Constantly we feel a merging away into infinitude; but that this book shall approximate to form, or that our data shall approximate to organization, or that we shall approximate to intelligibility, we have to call ourselves back constantly from wandering off into infinitude. The thing that we do, however, is to make our own outline, or the difference between what we include and what we exclude, vague.

The crux here, and the limit beyond which we may not go - very much - is:

Acceptance that there is a region that we call the Super-Sargasso Sea not yet fully accepted, but a provisional position that has received a great deal of support -

But is it a part of this earth, and does it revolve with and over this earth -

Or does it flatly overlie this earth, not revolving with and over this earth -

That this earth does not revolve, and is not round, or roundish, at all, but is continuous with the rest of its system, so that, if one could break away from the traditions of the geographers, one might walk and walk, and come to Mars, and then find Mars continuous with Jupiter?

I suppose some day such queries will sound absurd - the thing will be so obvious -

Because it is very difficult to me to conceive of little metallic objects hanging precisely over a small town in Russia, for four months, if revolving, unattached, with a revolving earth -

It may be that something aimed at that town, and then later took another shot.

These are speculations that seem to me to be evil relatively to these early years in the twentieth century -

Just now, I accept that this earth is not round, of course: that is very old-fashioned - but roundish, or, at least, that it has what is called form of its own, and does revolve upon its axis, and in an orbit around the sun. I only accept these old traditional notions -

And that above it are regions of suspension that revolve with it: from which objects fall, by disturbances of various kinds, and then, later, fall again, in the same place:

Monthly Weather Review, May, 1884-134:

Report from the Signal Service observer, at Bismarck, Dakota:

That, at 9 o'clock, in the evening of May 22, 1884, sharp sounds were heard throughout the city, caused by a fall of flinty stones striking against windows.

Fifteen hours later another fall of flinty stones occurred at Bismarck. There is no report of stones having fallen anywhere else.

This is a thing of the ultra-damned. All Editors of scientific publications read the Monthly Weather Review and frequently copy from it. The noise made by the stones of Bismarck, rattling against those windows, may be in a language that aviators will some day interpret: but it was a noise entirely surrounded by silences. Of this ultra-damned thing, there is no mention, findable by me, in any other publication.

The size of some hailstones has worried many meteorologists - but not text-book meteorologists. I know of no more serene occupation than that of writing text-books - though writing for the War Cry, of the Salvation Army, may be equally unadventurous. In the drowsy tranquillity of a text-book, we easily and unintelligently read of dust particles around which icy rain forms, hailstones, in their fall, then increasing by accretion - but in the meteorological journals, we read often of air-spaces nucleating hailstones - But it's the size of the things. Dip a marble in icy water. Dip and dip and dip it. If you're a resolute dipper, you will, after a while, have an object the size of a baseball - but

I think a thing could fall from the moon in that length of time. Also the strata of them. The Maryland hailstones are unusual, but a dozen strata have often been counted. Ferrel gives an instance of thirteen strata.

Such considerations led Prof. Schwedoff to argue that some hailstones are not, and can not, be generated in this earth's atmosphere that they come from somewhere else. Now, in a relative existence, nothing can of itself be either attractive or repulsive: its effects are functions of its associations or implications. Many of our data have been taken from very conservative scientific sources: it was not until their discordant implications, or irreconcilabilities with the System, were perceived, that ex-communication was pronounced against them.

Prof. Schwedoff's paper was read before the British Association (Rept, of 1882, p. 453).

The implication, and the repulsiveness of the implication to the snug and tight little exclusionists of 1882 - though we hold out that they were functioning well and ably relatively to 1882 -

That there is* water oceans or lakes and ponds, or rivers of it - that there is water away from, and yet not far-remote from, this earth's atmosphere and gravitation -

The pain of it:

That the snug little system of 1882 would be ousted from its Reposefulness -

A whole new science to learn:

The Science of Super-Geography -

And Science is a turtle that says that its own shell encloses all things.

So the members of the British Association. To some of them Prof. Schwedoff's ideas were like slaps on the back of an environment – denying turtle: to some of them his heresy was like an offering of meat, raw and dripping, to milk-fed lambs. Some of them bleated like lambs, and some of them turled like turtles. We used to crucify, but now we ridicule: or, in the loss of vigor of all progress, the spike has etherealized into the laugh.

Sir William Thomson ridiculed the heresy, with the phantomosities of his era:

That all bodies, such as hailstones, if away from this earth's atmosphere, would have to move at planetary velocity - which would be positively reasonable if the pronouncements of St. Isaac were anything but articles of faith - that a hailstone falling through this earth's atmosphere, with planetary velocity, would perform 13,000 times as much work as would raise an equal weight of water one degree centigrade, and therefore never fall as a hailstone at all; be more than melted - super-volatalized -

These turls and these bleats of pedantry - though we insist that, relatively to 1882, these turls and bleats should be regarded as respectfully as we regard rag dolls that keep infants occupied and noise-less - it is the survival of rag dolls into maturity that we object to - so these pious and naive ones who believed that 13,000 times something could have - that is, in quasi-existence an exact and calcu-lable resultant, whereas there is - in quasi-existence - nothing that can, except by delusion and con-venience, be called a unit, in the first place - whose devotions to St. Isaac required blind belief in formulas of falling bodies -

Against data that were piling up, in their own time, of slow-falling meteorites; "milk warm" ones admitted even by Farrington and Merrill; at least one icy meteorite nowhere denied by the present or-thodoxy, a datum as accessible to Thomson, in 1882, as it is now to us, because it was an occurrence of 1860. Beans and needles and tacks and a magnet. Needles and tacks adhere to and systematize relatively to a magnet, but, if some beans, too, be caught up, they are irreconcilables to this system and drop right out of it. A member of the Salvation Army may hear over and over data that seem so memorable to an evolutionist. It seems remarkable that they do not influence him - one finds that he can not remember them. It is incredible that Sir William Thomson had never heard of slow-falling, cold meteorites. It is simply that he had no power to remember such irreconcilabilities.

And then Mr. Symons again. Mr. Symons was a man who probably did more for the science of meteorology than did any other man of his time: therefore he probably did more to hold back the science of meteorology than did any other man of his time. In Nature, 41-135, Mr. Symons says that Prof. Schwedoff's ideas are "very droll."

I think that even more amusing is our own acceptance that, not very far above this earth's surface, is a region that will be the subject of a whole new science - super-geography - with which we shall immortalize ourselves in the resentments of the schoolboys of the future -

Pebbles and fragments of meteors and things from Mars and Jupiter and Azuria: wedges, delayed messages, cannon balls, bricks, nails, coal and coke and charcoal and offensive old cargoes - things that coat in ice in some regions and things that get into areas so warm that they putrefy - or that there are all the climates of geography in super-geography. I shall have to accept that, floating in the sky of this earth, there often are fields of ice as extensive as those on the Arctic Ocean - volumes of water in which are many fishes and frogs tracts of land covered with caterpillars -

Aviators of the future. They fly up and up. Then they get out and walk. The fishing's good: the bait's right there. They find messages from other worlds - and within three weeks there's a big trade worked up in forged messages. Sometime I shall write a guide book to the Super-Sagasso Sea, for aviators, but just at present there wouldn't be much call for it.

We now have more of our expression upon hail as a concomitant, or more data of things that have fallen from the sky, with hail.

In general, the expression is:

These things may have been raised from some other part of the earth's surface, in whirlwinds, or may not have fallen, and may have been upon the ground, in the first place - but were the hailstones found with them, raised from some other part of the earth's surface, or were the hailstones upon the ground, in the first place?

As I said before, this expression is meaningless as to a few instances; it is reasonable to think of some coincidence between the fall of hail and the fall of other things: but, inasmuch as there have been a good many instances, - we begin to suspect that this is not so much a book we're writing as a sanitarium for overworked coincidences. If not conceivably could very large hailstones and lumps of ice form in this earth's atmosphere, and so then had to come from external regions, then other things in or accompanying very large hailstones and lumps of ice came from external regions - which worries us a little: we may be instantly translated to the Positive Absolute.

Cosmos, 13-120, quotes a Virginia newspaper, that fishes said to have been catfishes, a foot long, some of them, had fallen, in 1853, at Norfolk, Virginia, with hail.

Vegetable debris, not only nuclear, but frozen upon the surfaces of large hailstones, at Toulouse, France, July 28, 1874. (La Science Pour Tous, 1874-270.)

Description of a storm, at Pontiac, Canada, July n, 1864, in which it is said that it was not hailstones that fell, but "pieces of ice, from half an inch to over two inches in diameter." (Canadian Naturalist, 2-1-308):

"But the most extraordinary thing is that a respectable farmer, of undoubted veracity, says he picked up a piece of hail, or ice, in the center of which was a small green frog."

Storm at Dubuque, Iowa, June 16, 1882, in which fell hailstones and pieces of ice (Monthly Weather Review, June, 1882):

"The foreman of the Novelty Iron Works, of this city, states that in two large hailstones melted by him were found small living frogs."

But the pieces of ice that fell upon this occasion had a peculiarity that indicates though by as bizarre an indication as any we've had yet - that they had been for a long time motionless or floating somewhere.

We'll take that up soon.

Living Age, 52-186:

That, June 30, 1841, fishes, one of which was ten inches long, fell at Boston; that, eight days later, fishes and ice fell at Derby.

In Timb's Year Book, 1842-275, it is said that, at Derby, the fishes had fallen in enormous numbers; from half an inch to two inches long, and some considerably larger. In the Athenaeum, 1841 – 542, copied from the Sheffield Patriot, it is said that one of the fishes weighed three ounces. In several accounts, it is said that, with the fishes, fell many small frogs and "pieces of half-melted ice."

We are told that the frogs and the fishes had been raised from some other part of the earth's surface, in a whirlwind; no whirlwind specified; nothing said as to what part of the earth's surface comes ice, in the month of July - interests us that the ice is described as "half melted."

In the London Times, July 15, 1841, it is said that the fishes were sticklebacks; that they had fallen with ice and small frogs, many of which had survived the fall. We note that, at Dumferline, three months later (Oct. 7, 1841) fell many fishes, several inches in length, in a thunderstorm. (London Times, Oct. 12, 1841.)

Hailstones, we don't care so much about. The matter of stratification seems significant, but we think more of the fall of lumps of ice from the sky, as possible data of the Super-Sargasso Sea:

Lumps of ice, a foot in circumference, Derbyshire, England, May 12, 1811 (Annual Register, 1811-54); cuboidal mass, six inches in diameter, that fell at Birmingham, 26 days later (Thomson "Intro, to Meteorology," p. 179); size of pumpkins, Bungalore, India, May 22, 1851 (Rept. Brit. Assoc., 1855-35); masses of ice of a pound and a half each, New Hampshire, Aug. 13, 1851 (Lummis, "Meteorology," p. 129) ; masses of ice, size of a man's head, in the Delphos tornado (Ferrel, "Popular Treatise," p. 428) ; large as a man's hand, killing thousands of sheep, Texas, May 3, 1877 (Monthly Weather Review, May, 1877); "pieces of ice so large that they could not be grasped in one hand," in a tornado, in Colorado, June 24, 1877 (Monthly Weather Review, June, 1877); lumps of ice four and a half inches long, Richmond, England, Aug. 2, 1879 (Symons' Met. Mag., 14-100); mass of ice, 21 inches in circumference that fell with hail, Iowa, June, 1881 (Monthly Weather Review, June, 1881) ; "pieces of ice," eight inches long, and an inch and a half thick, Davenport, Iowa, Aug. 30, 1882 (Monthly Weather Review, Aug., 1882); lump of ice size of a brick; weight two pounds, Chicago,

July 12, 1883 (Monthly Weather Review, July, 1883); lumps of ice that weighed one pound and a half each, India, May (?), 1888 (Nature, 37-42); lump of ice weighing four pounds, Texas, Dec. 6, 1893 (Sc. Am., 68-58); lumps of ice one pound in weight, Nov. 14, 1901, in a tornado, Victoria (Meteorology of Australia, p. 34).

Of course it is our acceptance that these masses not only accompanied tornadoes, but were brought down to this earth by tornadoes.

Flammarion, "The Atmosphere," p. 34: Block of ice, weighing four and a half pounds that fell at Cazorta, Spain, June 15, 1829; block of ice, weighing eleven pounds, at Cette, France, Oct., 1844; mass of ice three feet long, three feet wide, and more than two feet thick, that fell, in a storm, in Hungary, May 8, 1802.

Scientific American, 47-119:

That, according to the Salina Journal, a mass of ice, weighing about 80 pounds had fallen from the sky, near Salina, Kansas, Aug., 1882. We are told that Mr. W. J. Hagler, the North Santa Fe" merchant became possessor of it, and packed it in sawdust in his store.

London Times, April 7, 1860:

That, upon the 16th of March, 1860, in a snowstorm, in Upper Wasdale, blocks of ice, so large that at a distance they looked like a flock of sheep, had fallen.

Rept. Brit. Assoc., 1851-32:

That a mass of ice about a cubic yard in size had fallen at Candeish, India, 1828.

Against these data, though, so far as I know, so many of them have never been assembled together before, there is a silence upon the part of scientific men that is unusual. Our Super-Sargasso Sea may not be an unavoidable conclusion, but arrival upon this earth of ice from external regions does seem to be -except that there must be, be it ever so faint, a merger. It is in the notion that these masses of ice are only congealed hailstones. We have data against this notion, as applied to all our instances, but the explanation has been offered, and, it seems to me, may apply in some instances. In the Bull. Soc. Astro, de France, 20-245, it is said of blocks of ice the size of decanters that had fallen at Tunis that they were only masses of congealed hailstones.

London Times, Aug. 4, 1857:

That a block of ice, described as "pure" ice, weighing 25 pounds, had been found in the meadow of Mr. Warner, of Cricklewood. There had been a storm the day before. As in some of our other instances, no one had seen this object fall from the sky. It was found after the storm: that's all that can be said about it. Letter from Capt. Blakiston, communicated by Gen. Sabine, to the Royal Society (London Roy. Soc. Proc., 10-468) :

That, Jan. 14, 1860, in a thunderstorm, pieces of ice had fallen upon Capt. Blakiston's vessel - that it was not hail. "It was not hail, but irregular shaped pieces of solid ice of different dimensions, up to the size of half a brick."

According to the Advertiser-Scotsman, quoted by the Edinburgh New Philosophical Magazine, 47-371, an irregular-shaped mass of ice fell at Ord, Scotland, Aug., 1849, after "an extraordinary peal of thunder."

It is said that this was homogeneous ice, except in a small part, which looked like congealed hailstones.

The mass was about 20 feet in circumference.

The story, as told in the London Times, Aug. 14, 1849, is that, upon the evening of the 13th of August, 1849, after a loud peal of thunder, a mass of ice said to have been 20 feet in circumference, had fallen upon the estate of Mr. Moffat, of Balvullich, Rossshire.

It is said that this object fell alone, or without hailstones.

Altogether, though it is not so strong for the Super-Sargasso Sea, I think this is one of our best expressions upon external origins. That large blocks of ice could form in the moisture of this earth's atmosphere is about as likely as that blocks of stone could form in a dust whirl. Of course, if ice or water comes to this earth from external sources, we think of at least minute organisms in it, and on, with our data, to frogs, fishes; on to anything that's thinkable, coming from external sources. It's of great importance to us to accept that large lumps of ice have fallen from the sky, but what we desire most - perhaps because of our interest in its archaeologic and palaeontologic treasures - is now to be through with tentativeness and probation, and to take the Super-Sargasso Sea into full acceptance in our more advanced fold of the chosen of this twentieth century.

In the Report of the British Association, 1855-37, it is said that, at Poorhundur, India, Dec. n, 1854, flat pieces of ice, many of them weighing several pounds - each, I suppose - had fallen from the sky. They are described as "large ice-flakes."

Vast fields of ice in the Super-Arctic regions, or strata, of the Super-Sargasso Sea. When they break up, their fragments are flake-like. In our acceptance, there are aerial ice-fields that are remote from this earth; that break up, fragments grinding against one another, rolling in vapor and water, of different constituency in different regions, forming slowly as stratified hailstones - but that there are ice-fields near this earth, that break up into just such flat pieces of ice as cover any pond or river when ice of a pond or river is broken, and are sometimes soon precipitated to the earth, in this familiar flat formation.

Symons' Met. Mag., 43-154:

A correspondent writes that, at Braemar, July 2, 1908, when the sky was clear overhead, and the sun shining, flat pieces of ice fell - from somewhere. The sun was shining, but something was going on somewhere: thunder was heard.

Until I saw the reproduction of a photograph in the Scientific American, Feb. 21, 1914, I had supposed that these ice-fields must be, say, at least ten or twenty miles away from this earth, and invisible, to terrestrial observers, except as the blurs that have so often been reported by astronomers and meteorologists. The photograph published by the Scientific American is of an aggregation supposed to be clouds, presumably not very high, so clearly detailed are they. The writer says that they looked to him like "a field of broken ice." Beneath is a picture of a conventional field of ice, floating ordinarily in water. The resemblance between the two pictures is striking - nevertheless, it seems to me incredible that the first of the photographs could be of an aerial ice-field, or that gravitation could cease to act at only a mile or so from this earth's surface –

Unless:

The exceptional: the flux and vagary of all things.

Or that normally this earth's gravitation extends, say, ten or fifteen miles outward - but that gravitation must be rhythmic. Of course, in the pseudo-formulas of astronomers, gravitation as a fixed quantity is essential. Accept that gravitation is a variable force, and astronomers deflate, with a perceptible hissing sound, into the punctured condition of economists, biologists, meteorologists, and all the others of the humbler divinities, who can admittedly offer only insecure approximations.

We refer all who would not like to hear the hiss of escaping arrogance, to Herbert Spencer's chapters upon the rhythm of all phenomena.

If everything else - light from the stars, heat from the sun, the winds and the tides; forms and colors and sizes of animals; demands and supplies and prices; political opinions and chemic reactions and religious doctrines and magnetic intensities and the ticking of clocks ; and arrival and departure of the seasons - if everything else is variable, we accept that the notion of gravitation as fixed and formulable is only another attempted positivism, doomed, like all other illusions of realness in quasi-existence. So it is intermediatism to accept that, though gravitation may approximate higher to invariability than do the winds, for instance, it must be somewhere between the Absolutes of Stability and Instability. Here then we are not much impressed with the opposition of physicists and astronomers, fearing, a little mournfully, that their language is of expiring sibilations.

So then the fields of ice in the sky, and that, though usually so far away as to be mere blurs, at times they come close enough to be seen in detail. For description of what I call a "blur," see Pop. Sci. News, Feb., 1884 - sky, in general, unusually clear, but, near the sun, "a white, slightly curdled haze, which was dazzlingly bright."

We accept that sometimes fields of ice pass between the sun and the earth: that many strata of ice, or very thick fields of ice, or superimposed fields would obscure the sun - that there have been occasions when the sun was eclipsed by fields of ice:

Flammarion, "The Atmosphere," p. 394:

That a profound darkness came upon the city of Brussels, June 18, 1839:

There fell flat pieces of ice, an inch long.

Intense darkness at Aitkin, Minn., April 2, 1889: sand and "solid chunks of ice" reported to have fallen (Science, April 19, 1889).

In Symons' Meteorological Magazine, 32-172, are outlined roughedged but smooth-surfaced pieces of ice that fell at Manassas, Virginia, Aug. 10, 1897. They look as much like the roughly broken fragments of a smooth sheet of ice - as ever have roughly broken fragments of a smooth sheet of ice looked. About two inches across, and one inch thick. In Cosmos, 3-116, it is said that, at Rouen, July 5, 1853, fell irregular-shaped pieces of ice, about the size of a hand, described as looking as if all had been broken from one enormous block of ice. That I think was an aerial iceberg. In the awful density, or almost absolute stupidity of the 19th century, it never occurred to anybody to look for traces of polar bears or of seals upon these fragments.

Of course, seeing what we want to see, having been able to gather these data only because they are in agreement with notions formed in advance, we are not so respectful to our own notions as to a similar impression forced upon an observer who had no theory or acceptance to support. In general, our prejudices see and our prejudices investigate, but this should not be taken as an absolute.

Monthly Weather Review, July, 1894:

That, from the Weather Bureau, of Portland, Oregon, a tornado, of June 3, 1894, was reported. Fragments of ice fell from the sky.

They averaged three to four inches square, and about an inch thick. In length and breadth they had the smooth surfaces required by our acceptance: and, according to the writer in the Review, "gave the impression of a vast field of ice suspended in the atmosphere, and suddenly broken into fragments about the size of the palm of the hand."

This datum, profoundly of what we used to call the "damned," or before we could no longer accept judgment, or cut and dried condemnation by infants, turtles, and lambs, was copied - but without comment - in the Scientific American, 71-371.

Our theology is something like this:

Of course we ought to be damned - but we revolt against adjudication by infants, turtles, and lambs. We now come to some remarkable data in a rather difficult department of super-geography. Vast fields of aerial ice. There's a lesson to me in the treachery of the imaginable. Most of our opposition is in the clearness with which the conventional, but impossible, becomes the imaginable, and then the resistant to modifications.

After it had become the conventional with me, I conceived clearly of vast sheets of ice, a few miles above this earth - then the shining of the sun, and the ice partly melting that note upon the ice that fell at Derby - water trickling and forming icicles upon the lower surface of the ice sheet. I seemed to look up and so clearly visualized those icicles hanging like stalactites from a flat-roofed cave, in white calcite. Or I looked up at the under side of an aerial ice-lump, and seemed to see a papillation similar to that observed by a calf at times. But then - but then - if icicles should form upon the under side of a sheet of aerial ice, that would be by the falling of water toward this earth; an icicle is of course an expression of gravitation - and, if water melting from ice should fall toward this earth, why not the ice itself fall before an icicle could have time to form? Of course, in quasi-existence, where everything is a paradox, one might argue that the water falls, but the ice does not, because the ice is heavier that - is, in masses. That notion, I think, belongs in a more advanced course than we are taking at present.

Our expression upon icicles:

A vast field of aerial ice - it is inert to this earth's gravitation but by universal flux and variation, part of it sags closer to this earth, and is susceptible to gravitation - by cohesion with the main mass, this part does not fall, but water melting from it does fall, and forms icicles - then, by various disturbances, this part sometimes falls in fragments that are protrusive with icicles.

Of the ice that fell, some of it enclosing living frogs, at Dubuque, Iowa, June 16, 1882, it is said (Monthly Weather Review, June, 1882) that there were pieces from one to seventeen inches in circumference, the largest weighing one pound and three-quarters - that upon some of them were icicles half an inch in length. We emphasize that these objects were not hailstones.

The only merger is that of knobby hailstones, or of large hailstones with protuberances wrought by crystallization: but that is no merger with terrestrial phenomena, and such formations are unaccountable to orthodoxy; or it is incredible that hail could so crystallize - not forming by accretion - in the fall of a few seconds.

For an account of such hailstones, see Nature, 61-594. Note the size "some of them the size of turkeys' eggs."

It is our expression that sometimes the icicles themselves have fallen, as if by concussion, or as if something had swept against the under side of an aerial ice floe, detaching its papillations.

Monthly Weather Review, June, 1889:

That, at Oswego, N. Y., June n, 1889, according to the Turin (N. Y.) Leader, there fell, in a thunder-storm, pieces of ice that "resembled the fragments of icicles."

Monthly Weather Review, 29-506:

That on Florence Island, St. Lawrence River, Aug. 8, 1901, with ordinary hail, fell pieces of ice "formed like icicles, the size and shape of lead pencils that had been cut into sections about three-eighths of an inch in length."

So our data of the Super-Sargasso Sea, and its Arctic region: and, for weeks at a time, an ice field may hang motionless over a part of this earth's surface - the sun has some effect upon it, but not much until late in the afternoon, I should say - part of it has sagged, but is held up by cohesion with the main mass - whereupon we have such an occurrence as would have been a little uncanny to us once upon a time - or fall of water from a cloudless sky, day after day, in one small part of this earth's surface, late in the afternoon, when the sun's rays had had time for their effects:

Monthly Weather Review, Oct., 1886:

That, according to the Charlotte Chronicle, Oct. 21, 1886, for three weeks there had been a fall of water from the sky, in Charlotte, N. C., localized in one particular spot, every afternoon, about three o'clock; that, whether the sky was cloudy or cloudless, the water or rain fell upon a small patch of land between two trees and nowhere else.

This is the newspaper account, and, as such, it seems in the depths of the unchosen, either by me or any other expression of the Salvation Army. The account by the Signal Service observer, at Charlotte, published in the Review, follows:

"An unusual phenomenon was witnessed on the 21st: having been informed that, for some weeks prior to date, rain had been falling daily, after 3 p. m., on a particular spot, near two trees, corner of 9th and D streets, I visited the place, and saw precipitation in the form of rain drops at 4.47 and 4.55 p. m., while the sun was shining brightly. On the 22nd, I again visited the place, and from 4.05 to 4.25 p. m., a light shower of rain fell from a cloudless sky. . . . Sometimes the precipitation falls over an area of half an acre, but always appears to center at these two trees, and when lightest occurs there only."

Chapter XIV

We see conventionally. It is not only that we think and act and speak and dress alike, because of our surrender to social attempt at Entity, in which we are only super-cellular. We see what it is "proper" that we should see. It is orthodox enough to say that a horse is not a horse, to an infant - any more than is an orange an orange to the unsophisticated. It's interesting to walk along a street sometimes and look at things and wonder what they'd look like, if we hadn't been taught to see horses and trees and houses as horses and trees and houses. I think that to super-sight they are local stresses merging indistinguishably into one another, in an all-inclusive nexus.

I think that it would be credible enough to say that many times have Monstrator and Elvera and Azuria crossed telescopic fields of vision, and were not even seen - because it wouldn't be proper to see them; it wouldn't be respectable, and it wouldn't be respectful: it would be insulting to old bones to see them: it would bring on evil influences from the relics of St. Isaac to see them.

But our data:

Of vast worlds that are orbitless, or that are navigable, or that are adrift in inter-planetary tides and currents: the data that we shall have of their approach, in modern times, within five or six miles of this earth -

But then their visits, or approaches, to other planets, or to other of the few regularized bodies that have surrendered to the attempted Entity of this solar system as a whole -

The question that we can't very well evade:

Have these other worlds, or super-constructions, ever been seen by astronomers?

I think there would not be much approximation to realness in taking refuge in the notion of astronomers who stare and squint and see only that which it is respectable and respectful to see. It is all very well to say that astronomers are hypnotics, and that an astronomer looking at the moon is hypnotized by the moon, but our acceptance is that the bodies of this present expression often visit the moon, or cross it, or are held in temporary suspension near - it then some of them must often have been within the diameter of an astronomer's hypnosis.

Our general expression:

That, upon the oceans of this earth, there are regularized vessels, but also that there are tramp vessels:

That, upon the super-ocean, there are regularized planets, but also that there are tramp worlds:

That astronomers are like mercantile purists who would deny commercial vagabondage.

Our acceptance is that vast celestial vagabonds have been excluded by astronomers, primarily because their irresponsibilities are an affront to the pure and the precise, or to attempted positivism; and secondarily because they have not been seen so very often.

The planets steadily reflect the light of the sun: upon this uniformity a system that we call Primary Astronomy has been built up; but now the subject-matter of Advanced Astronomy is data of celestial phenomena that are sometimes light and sometimes dark, varying like some of the satellites of Jupiter, but with a wider range. However, light or dark, they have been seen and reported so often that the only important reason for their exclusion is - that they don't fit in.

With dark bodies that are probably external to our own solar system, I have, in the provincialism that no one can escape, not much concern. Dark bodies afloat in outer space would have been damned a few years ago, but now they're sanctioned by Prof. Barnard - and, if he says they're all right, you may think of them without the fear of doing something wrong or ridiculous - the close kinship we note so often between the evil and the absurd - I suppose by the ridiculous I mean the froth of evil. The dark companion of Algol, for instance. Though that's a clear case of celestial miscegenation, the purists, or positivists, admit that's so. In the Proceedings of the National Academy of Science, 1915-394, Prof.

Barnard writes of an object - he calls it an "object" in Cephus.

His idea is that there are dark, opaque bodies outside this solar system. But in the Astrophysical Journal, 1916-1, he modifies into regarding them as "dark nebulae." That's not so interesting.

We accept that Venus, for instance, has often been visited by other worlds, or by super-constructions, from which come cinders and coke and coal; that sometimes these things have reflected light and have been seen from this earth - by professional astronomers.

It will be noted that throughout this chapter our data are accursed Brahmins - as, by hypnosis and inertia, we keep on and keep on saying, just as a good many of the scientists of the 19th century kept on and kept on admitting the power of the system that preceded them - or Continuity would be smashed. There's a big chance here for us to be instantaneously translated to the Positive Absolute - oh, well -

What I emphasize here is that our damned data are observations by astronomers of the highest standing, excommunicated by astronomers of similar standing but backed up by the dominant spirit of their era - to which all minds had to equilibrate or be negligible, unheard, submerged. It would seem sometimes, in this book, as if our revolts were against the dogmatisms and pontifications of single scientists of eminence. This is only a convenience," because it seems necessary to personify. If we look over Philosophical Transactions, or the publications of the Royal Astronomical Society, for instance, we see that Herschel, for instance, was as powerless as any boy star-gazer, to enforce acceptance of any observation of his that did not harmonize with the system that was growing up as independently of him and all other astronomers, as a phase in the development of an embryo compels all cells to take on appearances concordantly with the design and the predetermined progress and schedule of the whole.

Visitors to Venus:

Evans, "Ways of the Planets," p. 140:

That, in 1645, a body large enough to look like a satellite was seen near Venus. Four times in the first half of the 18th century, a similar observation was reported. The last report occurred in 1767.

A large body has been seen seven times, according to Science Gossip, 1886-178 - near Venus. At least one astronomer, Houzeau, accepted these observations and named the - world, planet, superconstruction - "Neith." His views are mentioned "in passing, but without endorsement," in the Trans. N. Y. Acad., 5-249.

Houzeau or some one writing for the magazine-section of a Sunday newspaper - outer darkness for both alike. A new satellite in this solar system might be a little disturbing - though the formulas of La Place, which were considered final in his day, have survived the admittance of five or six hundred bodies not included in those formulas - a satellite to Venus might be a little disturbing, but would be explained - but a large body approaching a planet - staying a while - going away - coming back some other time - anchoring, as it were -

Azuria is pretty bad, but Azuria is no worse than Neith.

Astrophysical Journal, 1-127:

A light-reflecting body, or a bright spot near Mars: seen Nov. 25, 1894, by Prof. Pickering and others, at the Lowell Observatory, above an unilluminated part of Mars - self-luminous, it would seem thought - to have been a cloud - but estimated to have been about twenty miles away from the planet.

Luminous spot seen moving across the disk of Mercury, in 1799, by Harding and Schroeter. (Monthly Notices of the R. A. S., 38-338.) In the first Bulletin issued by the Lowell Observatory, in 1903, Prof. Lowell describes a body that was seen on the terminator of Mars, May 20, 1903. On May 27, it was "suspected." If still there, it had moved, we are told, about 300 miles - "probably a dust cloud."

Very conspicuous and brilliant spots seen on the disk of Mars, Oct. and Nov., 1911. (Popular Astronomy, Vol. 19, No. 10.)

So one of them accepted six or seven observations that were in agreement, except that they could not be regularized, upon a world - planet – satellite - and he gave it a name. He named it "Neith."

Monstrator and Elvera and Azuria and Super-Romanimus -

Or heresy and orthodoxy and the oneness of all quasiness, and our ways and means and methods are the very same. Or, if we name things that may not be, we are not of lonely guilt in the nomenclature of absences -

But now Leverrier and "Vulcan."

Leverrier again.

Or to demonstrate the collapsibility of a froth, stick a pin in the largest bubble of it. Astronomy and inflation: and by inflation we mean expansion of the attenuated. Or that the science of Astronomy is a phantom-film distended with myth-stuff - but always our acceptance that it approximates higher to substantiality than did the system that preceded it.

So Leverrier and the "planet Vulcan."

And we repeat, and it will do us small good to repeat. If you be of the masses that the astronomers have hypnotized - being themselves hypnotized, or they could not hypnotize others - or that the hypnotist's control is not the masterful power that it is popularly supposed to be, but only transference of state from one hypnotic to another -

If you be of the masses that the astronomers have hypnotized, you will not be able even to remember. Ten pages from here, and Leverrier and the "planet Vulcan" will have fallen from your mind, like beans from a magnet, or like data of cold meteorites from the mind of a Thomson.

Leverrier and the "planet Vulcan."

And much the good it will do us to repeat.

But at least temporarily we shall have an impression of a historic fiasco, such as, in our acceptance, could occur only in a quasiexistence.

In 1859, Dr. Lescarbault, an amateur astronomer, of Orgères, France, announced that, upon March 26, of that year, he had seen a body of planetary size cross the sun. We are in a subject that is now as unholy to the present system as ever were its own subjects to the system that preceded it, or as ever were slanders against miracles to the preceding system. Nevertheless few text-books go so far as quite to disregard this tragedy. The method of the systematists is slightingly to give a few instances of the unholy, and dispose of the few. If it were desirable to them to deny that there are mountains upon this earth, they would record a few observations upon some slight eminences near Orange, N. J., but say that commuters, though estimable persons in several ways, are likely to have their observations mixed. The text-books casually mention a few of the "supposed" observations upon "Vulcan," and then pass on.

Dr. Lescarbault wrote to Leverrier, who hastened to Orgères -

Because this announcement assimilated with his own calculations upon a planet between Mercury and the sun -

Because this solar system itself has never attained positiveness in the aspect of Regularity: there are to Mercury, as there are to Neptune, phenomena irreconcilible with the formulas, or motions that betray influence by something else.

We are told that Leverrier "satisfied himself as to the substantial accuracy of the reported observation." The story of this investigation is told in Monthly Notices, 20-98. It seems too bad to threaten the naive little thing with our rude sophistications, but it is amusingly of the ingenuousness of the age from which present dogmas have survived. Lescarbault wrote to Leverrier. Leverrier hastened to Orgères. But he was careful not to tell Lescarbault who he was. Went right in and "subjected Dr. Lescarbault to a very severe cross-examination" - just the way you or I may feel at liberty to go into anybody's home and be severe with people "pressing him hard step by step" - just as any one might go into some one else's house and press him hard, though unknown to the hard-pressed one. Not until he was satisfied, did Leverrier reveal his identity. I suppose Dr. Lescarbault expressed astonishment.

I think there's something Utopian about this: it's so unlike the stand-offishness of New York life.

Leverrier gave the name "Vulcan" to the object that Dr. Lescarbault had reported.

By the same means by which he is, even to this day, supposed by the faithful - to have discovered Neptune, he had already announced the probable existence of an Intra-Mercurial body, or group of bodies. He had five observations besides Lescarbault's upon something that had been seen to cross the sun. In accordance with the mathematical hypnoses of his era, he studied these six transits.

Out of them he computed elements giving "Vulcan" a period of about 20 days, or a formula for heliocentric longitude at any time.

But he placed the time of best observation away up in 1877.

But even so, or considering that he still had probably a good many years to live, it may strike one that he was a little rash - that is if one have not gone very deep into the study of hypnoses - that, having "discovered" Neptune by a method which, in our acceptance, had no more to recommend it than had once equally well-thought-of methods of witch-finding, he should not have taken such chances: that if he was right as to Neptune, but should be wrong as to' "Vulcan," his average would be away below that of most fortunetellers, who could scarcely hope to do business upon a fifty per cent basis - all that the reasoning of a tyro in hypnoses.

The date:

March 22, 1877.

The scientific world was up on its hind legs nosing the sky. The thing had been done so authoritatively. Never a pope had said a thing with more of the seeming of finality. If six observations correlated, what more could be asked? The Editor of Nature, a week before the predicted event, though cautious, said that it is difficult to explain how six observers, unknown to one another, could have data that could be formulated, if they were not related phenomena.

In a way, at this point occurs the crisis of our whole book.

Formulas are against us.

But can astronomic formulas, backed up by observations in agreement, taken many years apart, calculated by a Leverrier, be as meaningless, in a positive sense, as all other quasi-things that we have encountered so far?

The preparations they made, before March 22, 1877. In England, the Astronomer Royal made it the expectation of his life: notified observers at Madras, Melbourne, Sydney, and New Zealand, and arranged with observers in Chili and the United States. M. Struve had prepared for observations in

Siberia and Japan -

March 22, 1877 -

Not absolutely hypocritically, I think it's pathetic, myself. If any one should doubt the sincerity of Leverrier, in this matter, we note, whether it has meaning or not, that a few months later he died.

I think we'll take up Monstrator, though there's so much to this subject that we'll have to come back. According to the Annual Register, 9-120, upon the 9th of August, 1762, M. de Rostan, of Basle, France, was taking altitudes of the sun, at Lausanne. He saw a vast, spindle-shaped body, about three of the sun's digits in breadth and nine in length, advancing slowly across the disk of the sun, or "at no more than half the velocity with which the ordinary solar spots move." It did not disappear until the 7th of September, when it reached the sun's limb. Because of the spindle-like form, I incline to think of a super-Zeppelin, but another observation, which seems to indicate that it was a world, is that, though it was opaque, and "eclipsed the sun," it had around it a kind of nebulosity or atmosphere? A penumbra would ordinarily be a datum of a sun spot, but there are observations that indicate that this object was at a considerable distance from the sun:

It is recorded that another observer, at Paris, watching the sun, at this time, had not seen this object; But that M. Croste, at Sole, about forty-five German leagues northward from Lausanne, had seen it, describing the same spindleform, but disagreeing a little as to breadth. Then comes the important point: that he and M. de Rostan did not see it upon the same part of the sun. This, then, is parallax, and, compounded with invisibility at Paris, is great parallax - or that, in the course of a month, in the summer of 1762, a large, opaque, spindle-shaped body traversed the disk of the sun, but at a great distance from the sun. The writer in the Register says: "In a word, we know of nothing to have recourse to, in the heavens, by which to explain this phenomenon." I suppose he was not a hopeless addict to explaining. Extraordinary - we fear he must have been a man of loose habits in some other respects.

As to us -

Monstrator.

In the Monthly Notices of the R. A. S., Feb., 1877, Zeverrier, who never lost faith, up to the last day, gives the six observations upon an unknown body of planetary size, that he had formulated: Fritsche, Oct. 10, 1802; Stark, Oct. 9, 1819; De Cuppis, Oct. 30, 1839; Sidebotham, Nov. 12, 1849; Lescarbault, March 26, 1859; Lummis, March 20, 1862.

If we weren't so accustomed to Science in its essential aspect of Disregard, we'd be mystified and impressed, like the Editor of Nature, with the formulation of these data: agreement of so many instances would seem incredible as a coincidence: but our acceptance is that, with just enough disregard, astronomers and fortune-tellers can formulate anything - or we'd engage, ourselves, to formulate periodicities in the crowds in Broadway - say that every Wednesday morning, a tall man, with one leg and a black eye, carrying a rubber plant, passes the Singer Building, at quarter past ten o'clock. Of course it couldn't really be done, unless such a man did have such periodicity, but if some Wednesday mornings it should be a small child lugging a barrel, or a fat negress with a week's wash, by ordinary disregard that would be prediction good enough for the kind of quasi-existence we're in.

So whether we accuse, or whether we think that the word "accuse" over-dignifies an attitude toward a quasi-astronomer, or mere figment in a super-dream, our acceptance is that Leverrier never did formulate observations -

That he picked out observations that could be formulated -

That of this type are all formulas -

That, if Leverrier had not been himself helplessly hypnotized, or if he had had in him more than a tincture of realness, never could he have been beguiled by such a quasi-process: but that he was hypnotized, and so extended, or transferred, his condition to others, that upon March 22, 1877, he had this earth bristling with telescopes, with the rigid and almost inanimate forms of astronomers behind them -

And not a blessed thing of any unusuality was seen upon that day or succeeding days.

But that the science of Astronomy suffered the slightest in prestige?

It couldn't. The spirit of 1877 was behind it. If, in an embryo, some cells should not live up to the phenomena of their era, the others will sustain the scheduled appearances. Not until an embryo enters the mammalian stage are cells of the reptilian stage false cells.

It is our acceptance that there were many equally authentic reports upon large planetary bodies that had been seen near the sun; that, of many, Leverrier picked out six; not then deciding that all the other observations related to still other large, planetary bodies, but arbitrarily, or hypnotically, disregarding - or heroically disregarding - every one of them - that to formulate at all he had to exclude falsely. The denouement killed him, I think. I'm not at all inclined to place him with the Grays and Hitchcocks and Symonses. I'm not, because, though it was rather unsportsmanlike to put the date so far ahead, he did give a date, and he did stick to it with such a high approximation -

I think Leverrier was translated to the Positive Absolute.

The disregarded:

Observation, of July 26, 1819, by Gruthinson but that was of two bodies that crossed the sun together -

Nature, 14-469:

That, according to the astronomer, J. R. Hind, Benjamin Scott, City Chamberlain of London, and Mr. Wray, had, in 1847, seen a body similar to "Vulcan" cross the sun.

Similar observation by Hind and Lowe, March 12, 1849 (L'Année Scientifique, 1876-9). Nature, 14-505:

Body of apparent size of Mercury, seen, Jan. 29, 1860, by F. A. R. Russell and four other observers, crossing the sun.

De Vico's observation of July 12, 1837 ("Observatory," 2-424).

L'Année Scientifique, 1865-16:

That another amateur astronomer, M. Coumbray, of Constantinople, had written to Leverrier, that, upon the 8th of March, 1865, he had seen a black point, sharply outlined, traverse the disk of the sun. It detached itself from a group of sun spots near the limb of the sun, and took 48 minutes to reach the other limb. Figuring upon the diagram sent by M. Coumbray, a central passage would have taken a little more than an hour. This observation was disregarded by Leverrier, because his formula required about four times that velocity. The point here is that these other observations are as authentic as those that Leverrier included; that, then, upon data as good as the data of "Vulcan," there must be other "Vulcans" - the heroic and defiant disregard, then, of trying to formulate one, omitting the others, which, by orthodox doctrine, must have influenced it greatly, if all were in the relatively narrow space between Mercury and the sun.

Observation upon another such body, of April 4, 1876, by M. Weber, of Berlin. As to this observation, Leverrier was informed by Wolf, in Aug., 1876 (L'Année Scientifique, 1876-7). It made no difference, so far as can be known, to this notable positivist. Two other observations noted by Hind and Denning - London Times, Nov. 3, 1871, and March 26, 1873.

Monthly Notices of the R. A. S., 20-100:

Standacher, Feb., 1762; Lichtenberg, Nov. 19, 1762; Hoffman, May. 1764; Dangos, Jan. 18, 1798; Stark, Feb. 12, 1820. An observation by Schmidt, Oct. 11, 1847, is said to be doubtful: but, upon page 192, it is said that this doubt had arisen because of a mistaken translation, and two other observations by Schmidt are given: Oct. 14, 1849, and Feb. 18, 1850 - also an observation by Lofft, Jan. 6, 1818. Observation by Steinheibel, at Vienna, April 27, 1820 (Monthly Notices, 1862).

Haase had collected reports of twenty observations like Lescarbault's.

The list was published in 1872, by Wolf. Also there are other instances like Gruthinsen's:

Amer. Jour. Sci., 2-28-446:

Report by Pastorff that he had seen twice in 1836, and once in 1837, two round spots of unequal size, moving across the sun, changing position relatively to each other, and taking a different course, if not orbit, each time: that, in 1834, he had seen similar bodies pass six times across the disk of the sun, looking very much like Mercury in his transits.

March 22, 1876 -

But to point out Leverrier's poverty-stricken average - or discovering planets upon a fifty per cent, basis - would be to point out the low percentage of realness in the quasi-myth-stuff of which the whole system is composed. We do not accuse the text-books of omitting this fiasco, but we do note that theirs is the conventional adaptation here of all beguilers who are in difficulties -

The diverting of attention.

It wouldn't be possible in a real existence, with real mentality, to deal with, but I suppose it's good enough for the quasi-intellects that stupefy themselves with text-books. The trick here is to gloss over Leverrier's mistake, and blame Lescarbault - he was only an amateur - had delusions. The reader's attention is led against Lescarbault by a report from M. Lias, director of the Brazilian Coast Survey, who, at the time of Lescarbault's "supposed" observation had been watching the sun in Brazil, and, instead of seeing even ordinary sun spots, had noted that the region of the "supposed transit" was of "uniform intensity."

But the meaninglessness of all utterances in quasi-existence -

"Uniform intensity" turns our way as much as against us - or some day some brain will conceive a way of beating Newton's third law - if every reaction, or resistance, is, or can be, interpretable as stimulus instead of resistance - if this could be done in mechanics, there's a way open here for some one to own the world - specifically in this matter, "uniform intensity" means that Lescarbault saw no ordinary sun spot, just as much as it means that no spot at all was seen upon the sun. Continuing the interpretation of a resistance as an assistance, which can always be done with mental forces - making us wonder what applications could be made with steam and electric forces - we point out that invisibility in Brazil means parallax quite as truly as it means absence, and, inasmuch as "Vulcan" was supposed to be distant from the sun, we interpret denial as corroboration - method of course of every scientist, politician, theologian, high-school debater.

So the text-books, with no especial cleverness, because no especial cleverness is needed, lead the reader into contempt for the amateur of Orgères, and forgetfulness of Leverrier - and some other subject is taken up.

But our own acceptance:

That these data are as good as ever they were;

That, if some one of eminence, should predict an earthquake, and if there should be no earthquake at the predicted time, that would discredit the prophet, but data of past earthquakes would remain as good as ever they had been. It is easy enough to smile at the illusion of a single amateur -

The mass-formation:

Fritsche, Stark, De Cuppis, Sidebotham, Lescarbault, Lummis, Gruthinson, De Vico, Scott, Wray, Russell, Hind, Lowe, Coumbray, Weber, Standacher, Lichtenberg, Dangos, Hoffman, Schmidt, Lofft, Steinheibel, Pastorff -

These are only the observations conventionally listed relatively to an Intra-Mercurial planet. They are formidable enough to prevent our being diverted, as if it were all the dream of a lonely amateur - but they're a mere advance-guard. From now on other data of large celestial bodies, some dark and some reflecting light, will pass and pass and keep on passing -

So that some of us will remember a thing or two, after the procession's over - possibly.

Taking up only one of the listed observations -

Or our impression that the discrediting of Leverrier has nothing to do with the acceptability of these data:

In the London Times, Jan. 10, 1860, is Benjamin Scott's account of his observation:

That, in the summer of 1847, he had seen a body that had seemed to be the size of Venus, crossing the sun. He says that, hardly believing the evidence of his sense of sight, he had looked for some one, whose hopes or ambitions would not make him so subject to illusion. He had told his little son, aged five years, to look through the telescope. The child had exclaimed that he had seen "a little balloon" crossing the sun. Scott says that he had not had sufficient self-reliance to make public announcement of his remarkable observation at the time, but that, in the evening of the same day, he had told Dr. Dick, F.R.A.S., who had cited other instances. In the Times, Jan. 12, 1860, is published a letter from Richard Abbott, F.R.A.S.: that he remembered Mr. Scott's letter to him upon this observation, at the time of the occurrence.

I suppose that, at the beginning of this chapter, one had the notion that, by hard scratching through musty old records we might rake up vague, more than doubtful data, distortable into what's called evidence of unrecognized worlds or constructions of planetary size -

But the high authenticity and the support and the modernity of these of the accursed that we are now considering -

And our acceptance that ours is a quasi-existence, in which above all other things, hopes, ambitions, emotions, motivations, stands Attempt to Positivize: that we are here considering an attempt to systematize that is sheer fanaticism in its disregard of the unsystematizable - that it represented the highest good in the 19th century that it is mono-mania, but heroic mono-mania that was quasidivine in the 19th century -

But that this isn't the 19th century.

As a doubly sponsored Brahmin in the regard of Baptists - the objects of July 29, 1878, stand out and proclaim themselves so that nothing but disregard of the intensity of mono-mania can account for their reception by the system:

Or the total eclipse of July 29, 1878, and the reports by Prof. Watson, from Rawlins, Wyoming, and by Prof. Swift, from Denver, Colorado: that they had seen two shining objects at a considerable distance from the sun.

It's quite in accord with our general expression: not that there is an Intra-Mercurial planet, but that there are different bodies, many vast things; near this earth sometimes, near the sun sometimes; orbitless worlds, which, because of scarcely any data of collisions, we think of as under navigable control - or dirigible super-constructions.

Prof. Watson and Prof. Swift published their observations. Then the disregard that we can not think of in terms of ordinary, sane exclusions.

The text-book systematists begin by telling us that the trouble with these observations is that they disagree widely: there is considerable respectfulness, especially for Prof. Swift, but we are told that by coincidence these two astronomers, hundreds of miles apart, were illuded: their observations were so different -

Prof. Swift (Nature, Sept. 19, 1878):

That his own observation was "in close approximation to that given by Prof. Watson."

In the Observatory, 2-161, Swift says that his observations and Watson's were "confirmatory of each other."

The faithful try again:

That Watson and Swift mistook stars for other bodies.

In the Observatory, 2-193, Prof. Watson says that he had previously committed to memory all stars near the sun, down to the seventh magnitude - "

And he's damned anyway.

How such exclusions work out is shown by Lockyer (Nature, Aug. 20, 1878). He says: "There is little doubt that an Intra – Mercurial planet has been discovered by Prof. Watson." That was before excommunication was pronounced.

He says:

"If it will fit one of Leverrier's orbits" -

It didn't fit.

In Nature, 21-301, Prof. Swift says:

"I have never made a more valid observation, nor one more free from doubt."

He's damned anyway.

We shall have some data that will not live up to most rigorous requirements, but, if any one would like to read how carefully and minutely these two sets of observations were made, see Prof. Swift's detailed description in the Am. Jour. Sci., 116-313; and the technicalities of Prof. Watson's observations in Monthly Notices, 38-525.

Our own acceptance upon dirigible worlds, which is assuredly enough, more nearly real than attempted concepts of large planets relatively near this earth, moving in orbits, but visible only occasionally; which more nearly approximates to reasonableness than does wholesale slaughter of Swift and Watson and Fritsche and Stark and De Cuppis - but our own acceptance is so painful to so many minds that, in another of the charitable moments that we have now and then for the sake of contrast, we offer relief:

The things seen high in the sky by Swift and Watson -

Well, only two months before the horse and the barn -

We go on with more observations by astronomers, recognizing that it is the very thing that has given them life, sustained them, held them together, that has crushed all but the quasi-gleam of independent life out of them. Were they not systematized, they could not be at all, except sporadically and without sustenance. They are systematized: they must not vary from the conditions of the system: they must not break away for themselves.

The two great commandments:

Thou shalt not break Continuity;

Thou shalt try.

We go on with these disregarded data, some of which, many of which, are of the highest degree of acceptability. It is the System that pulls back its variations, as this earth is pulling back the Matterhorn. It is the System that nourishes and rewards, and also freezes out life with the chill of disregard. We do note that, before excommunication is pronounced, orthodox journals do liberally enough record unassimilable observations.

All things merge away into everything else.

That is Continuity.

So the System merges away and evades us when we try to focus against it.

We have complained a great deal. At least we are not so dull as to have the delusion that we know just exactly what it is that we are complaining about. We speak seemingly definitely enough of "the System," but we're building upon observations by members of that very system. Or what we are doing - gathering up the loose heresies of the orthodox. Of course "the System" fringes and ravels away, having no real outline. A Swift will antagonize "the System," and a Lockyer will call him back; but, then, a Lockyer will vary with a "meteoric hypothesis," and a Swift will, in turn, represent "the System." This state is to us typical of all intermediatist phenomena; or that not conceivably is anything really anything, if its parts are likely to be their own opposites at any time. We speak of astronomers - as if there were real astronomers - but who have lost their identity in a System - as if it were a real System - but behind that System is plainly a rapport, or loss of identity in the Spirit of an Era.

Bodies that have looked like dark bodies, and lights that may have been sunlight reflected from interplanetary - objects, masses, constructions -

Lights that have been seen upon - or near? - the moon:

In Philosophical Transactions, 82-27, is Herschel's report upon many luminous points, which he saw upon - or near? - the moon, during an eclipse. Why they should be luminous, whereas the moon itself was dark, would get us into a lot of trouble - except that later we shall, or we sha'n't, accept that many times have luminous objects been seen close to this earth - at night But numerousness is a new factor, or new disturbance, to our explorations -

A new aspect of inter-planetary inhabitancy or occupancy -

Worlds in hordes or beings winged beings perhaps - wouldn't astonish me if we should end up by discovering angels or beings in machines - argosies of celestial voyagers -

In 1783 and 1787, Herschel reported more lights on or near the moon, which he supposed were volcanic.

The word of a Herschel has had no more weight, in divergences from the orthodox, than has had the word of a Lescarbault. These observations are of the disregarded.

Bright spots seen on the moon, Nov., 1821 (Proc. London Roy. Soc., 2-167).

For four other instances, see Loomis ("Treatise on Astronomy," p. 174).

A moving light is reported in Phil. Trans., 84-429. To the writer, it looked like a star passing over the moon - "which, on the next moment's consideration I knew to be impossible." "It was a fixed, steady light upon the dark part of the moon." I suppose "fixed" applies to luster.

In the Report of the Brit. Assoc., 1847-18, there is an observation by Rankin, upon luminous points seen on the shaded part of the moon, during an eclipse. They seemed to this observer like reflections of stars. That's not very reasonable: however, we have, in the Annual Register, 1821-687, a light not referable to a star because it moved with the moon: was seen three nights in succession; reported by Capt. Kater. See Quart. Jour. Roy. Inst., 12-133.

Phil. Trans., 112-237:

Report from the Cape Town Observatory: a whitish spot on the dark part of the moon's limb. Three smaller lights were seen.

The call of positiveness, in its aspects of singleness, or homogeneity, or oneness, or completeness. In data now coming, I feel it myself. A Leverrier studies more than twenty observations. The inclination is irresistible to think that they all relate to one phenomenon.

It is an expression of cosmic inclination. Most of the observations are so irreconcilable with any acceptance other than of orbitless, dirigible worlds that he shuts his eyes to more than two-thirds of them; he picks out six that can give him the illusion of completeness, or of all relating to one planet. Or let it be that we have data of many dark bodies - still do we incline almost irresistibly to think of one of them as the dark-bodyin – chief. Dark bodies, floating, or navigating, in inter-planetary space - and I conceive of one that's the Prince of Dark Bodies:

Melanicus.

Vast dark thing with the wings of a super-bat, or jet-black superconstruction; most likely one of the spores of the Evil One.

The extraordinary year, 1883:

London Times, Dec. 17, 1883:

Extract from a letter by Hicks Pashaw: that, in Egypt, Sept. 24, 1883, he had seen, through glasses, "an immense black spot upon the lower part of the sun."

Sun spot, may be.

One night an astronomer was looking up at the sky, when something obscured a star, for three and a half seconds. A meteor had been seen nearby, but its train had been only momentarily visible.

Dr. Wolf was the astronomer (Nature, 86-528).

The next datum is one of the most sensational we have, except that there is very little to it. A dark object that was seen by Prof. Heis, for eleven degrees of arc, moving slowly across the Milky Way. (Greg's Catalogue, Rept. Brit. Assoc., 1867-426.)

One of our quasi-reasons for accepting that orbitless worlds are dirigible is the almost complete absence of data of collisions: of course, though in defiance of gravitation, they may, without direction like human direction, adjust to one another in the way of vortex rings of smoke - a very human-like way that is. But in Knowledge, Feb., 1894, are two photographs of Brooks' comet that are shown as evidence of its seeming collision with a dark object, Oct., 1893. Our own wording is that it "struck against something": Prof. Barnard's is that it had "entered some dense medium, which shattered it." For all I know it had knocked against merely a field of ice.

Melanicus.

That upon the wings of a super-bat, he broods over this earth and over other worlds, perhaps deriving something from them:

hovers on wings, or wing-like appendages, or planes that are hundreds of miles from tip to tip - a super-evil thing that is exploiting us. By Evil I mean that which makes us useful.

He obscures a star. He shoves a comet. I think he's a vast, black, brooding vampire.

Science, July 31, 1896:

That, according to a newspaper account, Mr. W. R. Brooks, director of the Smith Observatory, had seen a dark round object pass rather slowly across the moon, in a horizontal direction. In Mr. Brooks' opinion it was a dark meteor. In Science, Sept. 14, 1896, a correspondent writes that, in his opinion, it may have been a bird. We shall have no trouble with the meteor and bird mergers, if we have observations of long duration and estimates of size up to hundreds of miles. As to the body that was seen by Brooks, there is a note from the Dutch astronomer, Muller, in the Scientific American, 75-251, that, upon April 4, 1892, he had seen a similar phenomenon. In Science Gossip, n. s., 3-135, are more details of the Brooks object - apparent diameter about one-thirtieth of the moon's - moon's disk crossed in three or four seconds. The writer, in Science Gossip, says that, on June 27, 1896, at one o'clock in the morning, he was looking at the moon with a 2 -inch achromatic, power 44, when a long black object sailed past, from west to east, the transit occupying 3 or 4 seconds. He believed this object to be a bird - there was, however, no fluttering motion observable in it.

In the Astronomische Nachrichten, No. 3477, Dr. Brendel, of Griefswald, Pomerania, writes that Postmaster Ziegler and other observers had seen a body about 6 feet in diameter crossing the sun's disk. The duration here indicates something far from the earth, and also far from the sun. This thing was seen a quarter of an hour before it reached the sun. Time in crossing the sun was about an hour. After leaving the sun it was visible an hour.

I think he's a vast, black vampire that sometimes broods over this earth and other bodies.

Communication from Dr. F. B. Harris (Popular Astronomy, 20 – 398):

That, upon the evening of January 27, 1912, Dr. Harris saw, upon the moon, "an intensely black object." He estimated it to be 2 50 miles long and 50 miles wide. "The object resembled a crow poised, as near as anything." Clouds then cut off observation.

Dr. Harris writes:

"I cannot but think that a very interesting and curious phenomenon happened."

Chapter XV

Short chapter coming now, and it's the worst of them all. I think it's speculative. It's a lapse from our usual pseudostandards.

I think it must mean that the preceding chapter was very efficiently done, and that now by the rhythm of all quasi-things - which can't be real things, if they're rhythms, because a rhythm is an appearance that turns into its own opposite and then back again - but now, to pay up, we're what we weren't. Short chapter, and I think we'll fill in with several points in Intermediatism.

A puzzle:

If it is our acceptance that, out of the Negative Absolute, the Positive Absolute is generating itself, recruiting, or maintaining, itself, via a third state, or our own quasi-state, it would seem that we're trying to conceive of Universalness manufacturing more Universalness from Nothingness. Take that up yourself, if you're willing to run the risk of disappearing with such velocity that you'll leave an incandescent train behind, and risk being infinitely happy forever, whereas you probably don't want to be happy -

I'll sidestep that myself, and try to be intelligible by regarding the Positive Absolute from the aspect of Realness instead of Universalness, recalling that by both Realness and Universalness we mean the same state, or that which does not merge away into something else, because there is nothing else. So the idea is that out of Unrealness, instead of Nothingness, Realness, instead of Universalness, is, via our own quasi-state, manufacturing more Realness.

Just so, but in relative terms, of course, all imaginings that materialize into machines or statues, buildings, dollars, paintings or books in paper and ink are graduations from unrealness to realness - in relative terms. It would seem then that Intermediateness is a relation between the Positive Absolute and the Negative Absolute.

But the absolute cannot be the related - of course a confession that we can't really think of it at all, if here we think of a limit to the unlimited. Doing the best we can, and encouraged by the reflection that we can't do worse than has been done by metaphysicians in the past, we accept that the absolute can't be the related. So then that our quasi-state is not a real relation, if nothing in it is real. On the other hand, it is not an unreal relation, if nothing in it is unreal. It seems thinkable that the Positive Absolute can, by means of Intermediateness, have a quasi-relation, or be only quasi-related, or be the unrelated, in final terms, or, at least, not be the related, in final terms.

As to free will and Intermediatism - same answer as to everything else. By free will we mean Independence - or that which does not merge away into something else so, in Intermediateness, neither free-will nor slave-will - but a different approximation for every so-called person toward one or the other of the extremes. The hackneyed way of expressing this seems to me to be the acceptable way, if in Intermediateness, there is only the paradoxical : that we're free to do what we have to do.

I am not convinced that we make a fetich of the preposterous.

I think our feeling is that in first gropings there's no knowing what will afterward be the acceptable. I think that if an early biologist heard of birds that grow on trees, he should record that he had heard of birds that grow on trees: then let sorting over of data occur afterward. The one thing that we try to tone down, but that is to a great degree unavoidable is having our data all mixed up like Long Island and Florida in the minds of early American explorers. My own notion is that this whole book is very much like a map of North America in which the Hudson River is set down as a passage leading to Siberia. We think of Monstrator and Melanicus and of a world that is now in communication with this earth: if so, secretly, with certain esoteric ones upon this earth. Whether that world's Monstrator and Monstrator's Melanicus - must be the subject of later inquiry. It would be a gross thing to do: solve up everything now and leave nothing to our disciples. I have been very much struck with phenomena of "cup marks." They look to me like symbols of communication.

But they do not look to me like means of communication between some of the inhabitants of this earth and other inhabitants of this earth.

My own impression is that some external force has marked, with symbols, rocks of this earth, from far away.

I do not think that cup marks are inscribed communications among different inhabitants of this earth, because it seems too unacceptable that inhabitants of China, Scotland, and America should all have conceived of the same system.

Cup marks are strings of cup-like impressions in rocks. Sometimes there are rings around them, and sometimes they have only semi-circles. Great Britain, America, France, Algeria, Circassia, Palestine: they're virtually everywhere - except in the far north, I think. In China, cliffs are dotted with them. Upon a cliff, near Lake Como, there is a maze of these markings. In Italy and Spain and India they occur in enormous numbers.

Given that a force, say like electric force, could, from a distance, mark such a substance as rocks, as, from a distance of hundreds of miles, selenium can be marked by telephotographers - but I am of two minds -

The Lost Explorers from Somewhere, and an attempt, from Somewhere, to communicate with them: so a frenzy of showering of messages toward this earth, in the hope that some of them would mark rocks near the lost explorers -

Or that somewhere upon this earth, there is an especial rocky surface, or receptor, or polar construction, or a steep, conical hill, upon which for ages have been received messages from some other world; but that at times messages go astray and mark substances perhaps thousands of miles from the receptor;

That perhaps forces behind the history of this earth have left upon the rocks of Palestine and England and India and China records that may some day be deciphered, of their misdirected instructions to certain esoteric ones - Order of the Freemasons - the Jesuits - I emphasize the row-formation of cup marks:

Prof. Douglas (Saturday Review, Nov. 24, 1883):

"Whatever may have been their motive, the cup-markers showed a decided liking for arranging their sculpturings in regularly spaced rows."

That cup marks are an archaic form of inscription was first suggested by Canon Greenwell many years ago. But more specifically adumbratory to our own expression are the observations of Rivett – Carnac (Jour. Roy. Asiatic Soc., 1903-515):

That the Braille system of raised dots is an inverted arrangement of cup marks: also that there are strong resemblances to the Morse code. But no tame and systematized archaeologist can do more than casually point out resemblances, and merely suggest that strings of cup marks look like messages, because - China, Switzerland, Algeria, America - if messages they be, there seems to be no escape from attributing one origin to them - then, if messages they be, I accept one external origin, to which the whole surface of this earth was accessible, for them.

Something else that we emphasize:

That rows of cup marks have often been likened to foot prints.

But, in this similitude, their uni-linear arrangement must be disregarded - of course often they're mixed up in every way, but arrangement in single lines is very common. It is odd that they should so often be likened to footprints: I suppose there are exceptional cases, but unless it's something that hops on one foot, or a cat going along a narrow fence-top, I don't think of anything that makes footprints one directly ahead of another - Cop, in a station house, walking a chalk line, perhaps.

Upon the Witch's Stone, near Ratho, Scotland, there are twenty four cups, varying in size from one and a half to three inches in diameter, arranged in approximately straight lines. Locally it is explained that these are tracks of dogs' feet (Proc. Soc. Antiq. Scotland, 2-4-79). Similar marks are scattered bewilderingly all around the Witch's Stone like a frenzy of telegraphing, or like messages repeating and repeating, trying to localize differently.

In Inverness-shire, cup marks are called "fairies' footmarks." At Valna's church, Norway, and St. Peter's, Ambleteuse, there are such marks, said to be horses' hoofprints. The rocks of Clare, Ireland, are marked with prints supposed to have been made by a mythical cow ("Folklore," 21-184).

We now have such a ghost of a thing that I'd not like to be interpreted as offering it as a datum: it simply illustrates what I mean by the notion of symbols, like cups, or like footprints, which, if like those of horses or cows, are the reverse of, or the negatives of, cups - of symbols that are regularly received somewhere upon this earth - steep, conical hill, somewhere, I think - but that have often alighted in wrong places considerably to the mystification of persons waking up some morning to find them upon formerly blank spaces.

An ancient record still worse, an ancient Chinese record - of a courtyard of a palace dwellers of the palace waking up one morning, finding the courtyard marked with tracks like the footprints of an ox supposed that the devil did it. (Notes and Queries, 9-6-225.)

Chapter XVI

Angels.

Hordes upon hordes of them.

Beings massed like the clouds of souls, or the commingling whiffs of spirituality, or the exhalations of souls that Dore pictured so often.

It may be that the Milky Way is a composition of stiff, frozen, finally - static, absolute angels. We shall have data of little Milky Ways, moving swiftly; or data of hosts of angels, not absolute, or still dynamic. I suspect, myself, that the fixed stars are really fixed, and that the minute motions said to have been detected in them are illusions. I think that the fixed stars are absolutes. Their twinkling is only the interpretation by an intermediatist state of them. I think that soon after Leverrier died, a new fixed star was discovered - that, if Dr. Gray had stuck to his story of the thousands of fishes from one pail of water, had written upon it, lectured upon it, taken to street corners, to convince the world that, whether conceivable or not, his explanation was the only true explanation: had thought of nothing but this last thing at night and first thing in the morning - his obituary - another "nova" reported in Monthly Notices.

I think that Milky Ways, of an inferior, or dynamic, order, have often been seen by astronomers. Of course it may be that the phenomena that we shall now consider are not angels at all. We are simply feeling around, trying to find out what we can accept. Some of our data indicate hosts of rotund and complacent tourists in inter-planetary space but then data of long, lean, hungry ones. I think that there are, out in inter-planetary space Super Tamerlanes, at the head of hosts of celestial ravagers which have come here and pounced upon civilizations of the past, cleaning them up all but their bones, or temples and monuments for which later historians have invented exclusionist histories. But if something now has a legal right to us, and can enforce its proprietorship, they've been warned off. It's the way of all exploitation. I should say that we're now under cultivation: that we're conscious of it, but have the impertinence to attribute it all to our own nobler and higher instincts. Against these notions is the same sense of finality that opposes all advance. It's why we rate acceptance as a better adaptation than belief. Opposing us is the strong belief that, as to interplanetary phenomena, virtually everything has been found out. Sense of finality and illusion of homogeneity. But that what is called advancing knowledge is violation of the sense of blankness.

A drop of water. Once upon a time water was considered so homogeneous that it was thought of as an element. The microscope - and not only that the supposititiously elementary was seen to be of infinite diversity, but that in its protoplasmic life there were new orders of beings.

Or the year 1491 - and a European looking westward over the ocean - his feeling that that suave western droop was unbreakable; that gods of regularity would not permit that smooth horizon to be disturbed by coasts or spotted with islands. The unpleasantness of even contemplating such a state wide, smooth west, so clean against the sky - spotted with islands geographic leprosy.

But coasts and islands and Indians and bison, in the seemingly vacant west: lakes, mountains, rivers –

One looks up at the sky: the relative homogeneity of the relatively unexplored: one thinks of only a few kinds of phenomena. But the acceptance is forced upon me that there are modes and modes and modes of inter-planetary existence: things as different from planets and comets and meteors as Indians are from bison and prairie dogs: a super-geography- or celestiography - of vast stagnant regions, but also of Super-Niagaras and Ultra-Mississippis: and a super-sociology - voyagers and tourists and ravagers: the hunted and the hunting: the super-mercantile, the super-piratic, the super-evangelical. Sense of homogeneity, or our positivist illusion of the unknown - and the fate of all positivism.

Astronomy and the academic.

Ethics and the abstract.

The universal attempt to formulate or to regularize - an attempt that can be made only by disregarding or denying.

Or all things disregard or deny that which will eventually invade and destroy them –

Until comes the day when some one thing shall say, and enforce upon Infinitude:

"Thus far shalt thou go: here is absolute demarcation."

The final utterance:

"There is only I."

In the Monthly Notices of the R. A. S., 11-48, there is a letter from the Rev. W. Read:

That, upon the 4th of September, 1851, at 9.30 a. m., he had seen a host of self-luminous bodies, passing the field of his telescope, some slowly and some rapidly. They appeared to occupy a zone several degrees in breadth. The direction of most of them was due east to west, but some moved from north to south. The numbers were tremendous. They were observed for six hours.

Editor's note:

"May not these appearances be attributed to an abnormal state of the optic nerves of the observer?"

In Monthly Notices, 12-38, Mr. Read answers that he had been a diligent observer, with instruments of a superior order, for about 28 years - "but I have never witnessed such an appearance before."

As to illusion he says that two other members of his family had seen the objects.

The Editor withdraws his suggestion.

We know what to expect. Almost absolutely - in an existence that is essentially Hibernian - we can predict the past - that is, look over something of this kind, written in 1851, and know what to expect from the Exclusionists later. If Mr. Read saw a migration of dissatisfied angels, numbering millions, they must merge away, at least subjectively, with commonplace terrestrial phenomena - of course disregarding Mr. Read's probable familiarity, of 28 years' duration, with the commonplaces of terrestrial phenomena.

Monthly Notices, 12-183:

Letter from Rev. W. R. Dawes: That he had seen similar objects - and in the month of September - that they were nothing but seeds floating in the air.

In the Report of the British Association, 1852-235, there is a communication from Mr. Read to Prof. Baden-Powell:

That the objects that had been seen by him and by Mr. Dawes were not similar. He denies that he had seen seeds floating in the air. There had been little wind, and that had come from the sea, where seeds would not be likely to have origin. The objects that he had seen were round and sharply defined, and with none of the feathery appearance of thistle down. He then quotes from a letter from C. B. Chalmers, F. R. A. S., who had seen a similar stream, a procession, or migration, except that some of the bodies were more elongated - or lean and hungry - than globular.

He might have argued for sixty-five years. He'd have impressed nobody - of importance. The supermotif, or dominant, of his era, was Exclusionism, and the notion of seeds in the air assimilates - with due disregards - with that dominant.

Or pageantries here upon our earth, and things looking down upon us and the Crusades were only dust clouds, and glints of the sun on shining armor were only particles of mica in dust clouds. I think it was a Crusade that Read saw - but that it was right, relatively to the year 1851, to say that it was only seeds in the wind, whether the wind blew from the sea or not. I think of things that were luminous with religious zeal, mixed up, like everything else in Intermediateness, with black marauders and from gray to brown beings of little personal ambitions. There may have been a Richard Coeur de Lion, on his way to right wrongs in Jupiter. It was right, relatively to 1851, to say that he was a seed of a cabbage.

Prof. Coffin, U. S. N. (Jour. Frank. Inst., 88-151):

That, during the eclipse of Aug., 1869, he had noted the passage, across his telescope, of several bright flakes resembling thistleblows, floating in the sunlight. But the telescope was so focussed that, if these things were distinct, they must have been so far away from this earth that the difficulties of orthodoxy remain as great, one way or another, no matter what we think they were –

They were "well-defined," says Prof. Coffin.

Henry Waldner (Nature, 5-304) :

That, April 27, 1863, he had seen great numbers of small, shining bodies passing from west to east. He had notified Dr. Wolf, of the Observatory of Zurich, who "had convinced himself of this strange phenomenon." Dr. Wolf had told him that similar bodies had been seen by Sig. Capocci, of the Capodimonte Observatory, at Naples, May n, 1845.

The shapes were of great diversity - or different aspects of similar shapes?

Appendages were seen upon some of them.

We are told that some were star-shaped, with transparent appendages.

I think, myself, it was a Mohammed and his Hegira. May have been only his harem. Astonishing sensation: afloat in space with ten million wives around one. Anyway, it would seem that we have considerable advantage here, inasmuch as seeds are not in season in April - but the pulling back to earth, the bedraggling by those sincere but dull ones of some time ago. ,We have the same stupidity necessary, functioning stupidity - of attribution of something that was so rare that an astronomer notes only one instance between 1845 and 1863, to an everyday occurrence—

 Or Mr. Waldner's assimilative opinion that he had seen only ice crystals.

Whether they were not very exclusive veils of a super-harem, or planes of a very light material, we have an impression of star-shaped things with transparent appendages that have been seen in the sky. Hosts of small bodies - black, this time - that were seen by the astronomers Herrick, Buys-Ballot, and De Cuppis (L'Année Scientifique, 1860-25); vast number of bodies that were seen by M. Lamey, to cross the moon ((L'Année Scientifique, 1874-62); another instance of dark ones; prodigious number of dark, spherical bodies reported by Messier, June 17, 1777 (Arago, (Euvres, 9-38); considerable number of luminous bodies which appeared to move out from the sun, in diverse directions; seen at Havana, during eclipse of the sun, May 15, 1836, by Prof. Auber (Poey); M. Poey cites a similar instance, of Aug. 3, 1886; M. Lotard's opinion that they were birds (L'Astronomie, 1886-391); large number of small bodies crossing disk of the sun, some swiftly, some slowly; most of them globular, but some seemingly triangular, and some of more complicated structure; seen by M. Trouvelet, who, whether seeds, insects, birds, or other commonplace things, had never seen anything resembling these forms (L'Année Scientifique, 1885-8); report from the Rio de Janeiro Observatory, of vast numbers of bodies crossing the sun, some of them luminous and some of them dark, from some time in December, 1875, until Jan- 22, 1876 (La Nature, 1876-384).

Of course, at a distance, any form is likely to look round or roundish: but we point out that we have notes upon the seeming of more complex forms. In L'Astronomie, 1886-70, is recorded M. Briguiere's observation, at Marseilles, April 15 and April 25, 1883, upon the crossing of the sun by bodies that were irregular in form.

Some of them moved as if in alignment.

Letter from Sir Robert Inglis to Col. Sabine (Rept. Brit. Assoc., 1849-17):

That, at 3 p. m., Aug. 8, 1849, at Gais, Switzerland, Inglis had seen thousands and thousands of brilliant white objects, like snowflakes in a cloudless sky. Though this display lasted about twentyfive minutes, not one of these seeming snowflakes was seen to fall.

Inglis says that his servant "fancied" that he had seen something like wings on these - whatever they were. Upon page 18, of the Report, Sir John Herschel says that, in 1845 or 1846, his attention had been attracted by objects of considerable size, in the air, seemingly not far away. He had looked at them through a telescope. He says that they were masses of hay, not less than a yard or two in diameter. Still there are some circumstances that interest me. He says that, though no less than a whirlwind could have sustained these masses, the air about him was calm. "No doubt wind prevailed at the spot, but there was no roaring noise." None of these masses fell within his observation or knowledge. To walk a few fields away and find out more would seem not much to expect from a man of science, but it is one of our superstitions, that such a seeming trifle is just what - by the Spirit of an Era, we'll call it - one is not permitted to do. If those things were not masses of hay, and if Herschel had walked a little and found out, and had reported that he had seen strange objects in the air - that report, in 1846, would have been as misplaced as the appearance of a tail upon an embryo still in its gastrula era. I have noticed this inhibition in my own case many times. Looking back - why didn't I do this or that little thing that would have cost so little and have meant so much? Didn't belong to that era of my own development.

Nature, 22-64:

That, at Kattenau, Germany, about half an hour before sunrise, March 22, 1880, "an enormous number of luminous bodies rose from the horizon, and passed in a horizontal direction from east to west." They are described as having appeared in a zone or belt. "They shone with a remarkably brilliant light."

So they've thrown lassos over our data to bring them back to earth. But they're lassos that cannot tighten. We can't pull out of them: we may step out of them, or lift them off. Some of us used to have an impression of Science sitting in calm, just judgment: some of us now feel that a good many of our data have been lynched. If a Crusade, perhaps from Mars to Jupiter, occur in the autumn - "seeds." If a Crusade or outpouring of celestial vandals is seen from this earth in the spring - "ice crystals." If we have record of a race of aerial beings, perhaps with no substantial habitat, seen by some one in India - "locusts."

This will be disregarded:

If locusts fly high, they freeze and fall in thousands.

Nature, 47-581:

Locusts that were seen in the mountains of India, at a height of 12,750 feet - "in swarms and dying by thousands."

But no matter whether they fly high or fly low, no one ever wonders what's in the air when locusts are passing overhead, because of the failing of stragglers. I have especially looked this matter up - no mystery when locusts are flying overhead - constant falling of stragglers.

Monthly Notices, 30-135:

"An unusual phenomenon noticed by Lieut. Herschel, Oct. 17 and 1 8, 1870, while observing the sun, at Bangalore, India."

Lieut. Herschel had noticed dark shadows crossing the sun - but away from the sun there were luminous, moving images. For two days bodies passed in a continuous stream, varying in size and velocity.

The Lieutenant tries to explain, as we shall see, but he says: "As it was, the continuous flight, for two whole days, in such numbers, in the upper regions of the air, of beasts that left no stragglers, is a wonder of natural history, if not of astronomy."

He tried different focussing - he saw wings - perhaps he saw planes. He says that he saw upon the objects either wings or phantom-like appendages.

Then he saw something that was so bizarre that, in the fullness of his nineteenth-centuriness, he writes:

"There was no longer doubt: they were locusts or flies of some sort."

One of them had paused.

It had hovered.

Then it had whisked off.

The Editor says that at that time "countless locusts had descended upon certain parts of India."

We now have an instance that is extraordinary in several respects - super-voyagers or super-ravagers; angels, ragamuffins, crusaders, emigrants, aeronauts, or aerial elephants, or bison or dinosaurs - except that I think the thing had planes or wings - one of them has been photographed. It may be that in the history of photography no more extraordinary picture than this has ever been taken.

L'Astronomie, 1885-347:

That, at the Observatory of Zacatecas, Mexico, Aug. 12, 1883, about 2,500 meters above sea level, were seen a large number of small luminous bodies, entering upon the disk of the sun. M. Bonilla telegraphed to the Observatories of the City of Mexico and of Puebla. Word came back that the bodies were not visible there. Because of this parallax, M. Bonilla placed the bodies "relatively near the earth." But when we find out what he called "relatively near the earth" - birds or bugs or hosts of a Super-Tamerlane or army of a celestial Richard Coeur de Lion - our heresies rejoice anyway. His estimate is "less distance than the moon."

One of them was photographed. See L'Astronomie, 1885-349.

The photograph shows a long body surrounded by indefinite structures, or by the haze of wings or planes in motion.

L'Astronomie, 1887-66:

Signor Ricco, of the Observatory of Palermo, writes that, Nov. 30, 1880, at 8.30 o'clock in the morning, he was watching the sun, when he saw, slowly traversing its disk, bodies in two long, parallel lines, and a shorter, parallel line. The bodies looked winged to him. But so large were they that he had to think of large birds. He thought of cranes.

He consulted ornithologists, and learned that the configuration of parallel lines agrees with the flight-formation of cranes. This was in 1880: anybody now living in New York City, for instance, would tell him that also it is a familiar formation of aeroplanes.

But, because of data of focus and subtended angles, these beings or objects must have been high.

Sig. Ricco argues that condors have been known to fly 3 or 4 miles high, and that heights reached by other birds have been estimated at 2 or 3 miles. He says that cranes have been known to fly so high that they have been lost to view.

Our own acceptance, in conventional terms, is that there is not a bird of this earth that would not freeze to death at a height of more than four miles: that if condors fly three or four miles high, they are birds that are especially adapted to such altitudes.

Sig. Ricco's estimate is that these objects or beings or cranes must have been at least five and a half miles high.

Chapter XVII

The vast dark thing that looked like a poised crow of unholy dimensions. Assuming that I shall ever have any readers, let him, or both of them, if I shall ever have such popularity as that, note how dim that bold black datum is at the distance of only two chapters.

The question:

Was it a thing or the shadow of a thing? Acceptance either way calls not for mere revision but revolution in the science of astronomy. But the dimness of the datum of only two chapters ago. The carved stone disk of Tarbes, and the rain that fell every afternoon for twenty - if I haven't forgotten, myself, whether it was twenty-three or twenty-five days! - upon one small area. We are all Thomsons, with brains that have smooth and slippery, though corrugated, surfaces - or that all intellection is associative - or that we remember that which correlates with a dominant and a few chapters go by, and there's scarcely an impression that hasn't slid off our smooth and slippery brains, of Leverrier and the "planet Vulcan." There are two ways by which irreconcilables can be remembered - if they can be correlated in a system more nearly real than the system that rejects them - and by repetition and repetition and repetition.

Vast black thing like a crow poised over the moon.

The datum is so important to us, because it enforces, in another field, our acceptance that dark bodies of planetary size traverse this solar system.

Our position:

That the things have been seen:

Also that their shadows have been seen.

Vast black thing poised like a crow over the moon. So far it is a single instance. By a single instance, we mean the negligible.

In Popular Science, 34-158, Serviss tells of a shadow that Schroeter saw, in 1788, in the lunar Alps. First he saw a light. But then, when this region was illuminated, he saw a round shadow where the light had been.

Our own expression:

That he saw a luminous object near the moon: that that part of the moon became illuminated, and the object was lost to view; but that then its shadow underneath was seen.

Serviss explains, of course. Otherwise he'd not be Prof. Serviss. It's a little contest in relative approximations to realness. Prof. Serviss thinks that what Schroeter saw was the "round" shadow of a mountain - in the region that had become lighted. He assumes that Schroeter never looked again to see whether the shadow could be attributed to a mountain. That's the crux: conceivably a mountain could cast a round - and that means detached - shadow, in the lighted part of the moon. Prof. Serviss could, of course, explain why he disregards the light in the first place - maybe it had always been there "in the first place." If he couldn't explain, he'd still be an amateur.

We have another datum. I think it is more extraordinary than –

Vast thing, black and poised, like a crow, over the moon.

But only because it's more circumstantial, and because it has corroboration, do I think it more extraordinary than –

Vast poised thing, black as a crow, over the moon.

Mr. H. C. Russell, who was usually as orthodox as anybody, I suppose - at least, he wrote "F. R. A. S." after his name - tells in the Observatory, 2-374, one of the wickedest, or most preposterous, stories that we have so far exhumed:

That he and another astronomer, G. D. Hirst, were in the Blue Mountains, near Sydney, N. S. W., and Mr. Hirst was looking at the moon—

He saw on the moon what Russell calls "one of those remarkable facts, which being seen should be recorded, although no explanation can at present be offered."

That may be so. It is very rarely done. Our own expression upon evolution by successive dominants and their correlates is against it. On the other hand, we express that every era records a few observations out of harmony with it, but adumbratory or preparatory to the spirit of eras still to come. It's very rarely done. Lashed by the phantom-scourge of a now passing era, the world of astronomers is in a state of terrorism, though of a highly attenuated, modernized, devitalized kind. Let an astronomer see something that is not of the conventional, celestial sights, or something that it is "improper" to see - his very dignity is in danger. Some one of the corralled and scourged may stick a smile into his back. He'll be thought of unkindly.

With a hardihood that is unusual in his world of ethereal sensitivenesses, Russell says, of Hirst's observation:

"He found a large part of it covered with a dark shade, quite as dark as the shadow of the earth during an eclipse of the moon."

But the climax of hardihood or impropriety or wickedness, preposterousness or enlightenment:

"One could hardly resist the conviction that it was a shadow, yet it would not be the shadow of any known body."

Richard Proctor was a man of some liberality. After a while we shall have a letter, which once upon a time we'd have called delirious - don't know that we could read such a thing now, for the first time, without incredulous laughter - which Mr. Proctor permitted to be published in Knowledge. But a dark, unknown world that could cast a shadow upon a large part of the moon, perhaps extending far beyond the limb of the moon; a shadow as deep as the shadow of this earth—

Too much for Mr. Proctor's politeness.

I haven't read what he said, but it seems to have been a little coarse. Russell says that Proctor "freely used" his name in the Echo, of March 14, 1879, ridiculing this observation which had been made by Russell as well as Hirst. If it hadn't been Proctor, it would have been some one else - but one notes that the attack came out in a newspaper. There is no discussion of this remarkable subject, no mention in any other astronomic journal. The disregard was almost complete - but we do note that the columns of the Observatory were open to Russell to answer Proctor.

In the answer, I note considerable intermediateness. Far back in 1879, it would have been a beautiful positivism, if Russell had said—

"There was a shadow on the moon. Absolutely it was cast by an unknown body."

According to our religion, if he had then given all his time to the maintaining of this one stand, of course breaking all friendships, all ties with his fellow astronomers, his apotheosis would have occurred, greatly assisted by means well known to quasi-existence when its compromises and evasions, and phenomena that are partly this and partly that, are flouted by the definite and uncompromising. It would be impossible in a real existence, but Mr. Russell, of quasi-existence, says that he did resist the conviction; that he had that one could "hardly resist"; and most of his resentment is against Mr. Proctor's thinking that he had not resisted. It seems too bad - if apotheosis be desirable.

The point in Intermediatism here is:

Not that to adapt to the conditions of quasi-existence is to have what is called success in quasi-existence, but is to lose one's soul—

But is to lose "one's" chance of attaining soul, self, or entity.

One indignation quoted from Proctor interests us:

"What happens on the moon may at any time happen to this earth."

Or:

That is just the teaching of this department of Advanced Astronomy:

That Russell and Hirst saw the sun eclipsed relatively to the moon by a vast dark body;

That many times have eclipses occurred relatively to this earth, by vast, dark bodies;

That there have been many eclipses that have not been recognized as eclipses by scientific kindergartens.

There is a merger, of course. We'll take a look at it first - that, after all, it may have been a shadow that Hirst and Russell saw, but the only significance is that the sun was eclipsed relatively to the moon by a cosmic haze of some kind, or a swarm of meteors close together, or a gaseous discharge left behind by a comet. My own acceptance is that vagueness of shadow is a function of vagueness of intervention; that a shadow as dense as the shadow of this earth is cast by a body denser than hazes and swarms. The information seems definite enough in this respect - "quite as dark as the shadow of this earth during the eclipse of the moon."

Though we may not always be as patient toward them as we should be, it is our acceptance that the astronomic primitives have done a great deal of good work: for instance, in the allaying of fears upon this earth. Sometimes it may seem as if all science were to us very much like what a red flag is to bulls and anti-socialists. It's not that: it's more like what unsquare meals are to bulls and anti-socialists - not the scientific, but the insufficient. Our acceptance is that Evil is the negative state, by which we mean the state of maladjustment, discord, ugliness, disorganization, inconsistency, injustice, and so on - as determined in Intermediateness, not by real standards, but only by higher approximations to adjustment, harmony, beauty, organization, consistency, justice, and so on. Evil is outlived virtue, or incipient virtue that has not yet established itself, or any other phenomenon that is not in seeming adjustment, harmony, consistency with a dominant. The astronomers have functioned bravely in the past. They've been good for business: the big interests think kindly, if at all, of them. It's bad for trade to have an intense darkness come upon an unaware community and frighten people out of their purchasing values. But if an obscuration be foretold, and if it then occur - may seem a little uncanny only a shadow - and no one who was about to buy a pair of shoes runs home panic-stricken and saves the money.

Upon general principles we accept that astronomers have quasi-systematized data of eclipses - or have included some and disregarded others.

They have done well.

They have functioned.

But now they're negatives, or they're out of harmony—

If we are in harmony with a new dominant, or the spirit of a new era, in which Exclusionism must be overthrown; if we have data of many obscurations that have occurred, not only upon the moon, but upon our own earth, as convincing of vast intervening bodies, usually invisible, as is any regularized, predicted eclipse.

One looks up at the sky.

It seems incredible that, say, at the distance of the moon, there could be, but be invisible, a solid body, say the size of the moon.

One looks up at the moon, at a time when only a crescent of it is visible. The tendency is to build up the rest of it in one's mind; but the unillumined part looks as vacant as the rest of the sky, and it's of the same blueness as the rest of the sky. There's a vast area of solid substance before one's eyes. It's indistinguishable from the sky.

In some of our little lessons upon the beauties of modesty and humility, we have picked out basic arrogances - tail of a peacock, horns of a stag, dollars of a capitalist - eclipses of astronomers. Though I have no desire for the job, I'd engage to list hundreds of instances in which the report upon an expected eclipse has been "sky overcast" or "weather unfavorable." In our Super-Hibernia, the unfavorable has been construed as the favorable. Some time ago, when we were lost, because we had not recognized our own dominant, when we were still of the unchosen and likely to be more malicious than we now are - because we have noted a steady tolerance creeping into our attitude - if astronomers are not to blame, but are only correlates to a dominant - we advertised^ predicted eclipse that did not occur at all. Now, without any especial feeling, except that of recognition of the fate of all attempted absolutism, we give the instance, noting that, though such an evil thing to orthodoxy, it was orthodoxy that recorded the non-event.

Monthly Notices of the R. A. S., 8-132:

"Remarkable appearances during the total eclipse of the moon on March 19, 1848":

In an extract from a letter from Mr. Forster, of Bruges, it is said that, according to the writer's observations at the time of the predicted total eclipse, the moon shone with about three times the intensity of the mean illumination of an eclipsed lunar disk: that the British Consul, at Ghent, who did not know of the predicted eclipse, had written enquiring as to the "blood-red" color of the moon.

This is not very satisfactory to what used to be our malices. But there follows another letter, from another astronomer, Walkey, who had made observations at Clyst St. Lawrence: that, instead of an eclipse, the moon became - as is printed in italics "most beautifully illuminated" . . . "rather tinged with a deep red" . . . "the moon being as perfect with light as if there had been no eclipse whatever."

I note that Chambers, in his work upon eclipses, gives Forster's letter in full - and not a mention of Walkey's letter.

There is no attempt in Monthly Notices to explain upon the notion of greater distance of the moon, and the earth's shadow falling short, which would make as much trouble for astronomers, if that were not foreseen, as no eclipse at all. Also there is no refuge in saying that virtually never, even in total eclipses, is the moon totally dark - "as perfect with light as if there had been no eclipse whatever." It is said that at the time there had been an aurora borealis, which might have caused the luminosity, without a datum that such an effect, by an aurora, had ever been observed upon the moon.

But single instances - so an observation by Scott, in the Antarctic.

The force of this datum lies in my own acceptance, based upon especially looking up this point, that an eclipse nine-tenths of totality has great effect, even though the sky be clouded.

Scott (Voyage of the Discovery, vol.II, p. 215):

"There may have been an eclipse of the sun, Sept. 21, 1903, as the almanac said, but we should, none of us, have liked to swear to the fact."

This eclipse had been set down at nine-tenths of totality. The sky was overcast at the time.

So it is not only that many eclipses unrecognized by astronomers as eclipses have occurred, but that intermediatism, or impositivism, breaks into their own seemingly regularized eclipses.

Our data of unregularized eclipses, as profound as those that are conventionally - or officially? - recognized, that have occurred relatively to this earth:

In Notes and Queries there are several allusions to intense darknesses that have occurred upon this earth, quite as eclipses occur, but that are not referable to any known eclipsing body. Of course there is no suggestion here that these darknesses may have been eclipses. My own acceptance is that, if in the nineteenth century any one had uttered such a thought as that, he'd have felt the blight of a Dominant; that Materialistic Science was a jealous god, excluding, as works of the devil, all utterances against the seemingly uniform, regular, periodic; that to defy him would have brought on - withering by ridicule - shrinking away by publishers - contempt of friends and family - justifiable grounds for divorce - that one who would so defy would feel what unbelievers in relics of saints felt in an earlier age; what befell virgins who forgot to keep fires burning, in a still earlier age - but that, if he'd almost absolutely hold out, just the same - new fixed star reported in Monthly Notices. Altogether, the point in Positivism here is that by Dominants and their correlates, quasi-existence strives for the positive state, aggregating, around a nucleus, or dominant, systematized members of a religion, a science, a society - but that "individuals" who do not surrender and submerge may of themselves highly approximate to positiveness - the fixed, the real, the absolute.

In Notes and Queries, 2-4-139, there is an account of a darkness in Holland, in the midst of a bright day, so intense and terrifying that many panic-stricken persons lost their lives stumbling into the canals.

Gentleman's Magazine, 33-414:

A darkness that came upon London, August 19, 1763, "greater than at the great eclipse of 1748."

However, our preference is not to go so far back for data. For a list of historic "dark days," see Humboldt, Cosmos, 1-120.

Monthly Weather Review, March, 1886-79:

That, according to the La Crosse Daily Republican, of March 20.

1886, darkness suddenly settled upon the city of Oshkosh, Wis., at 3 p. m., March 19. In five minutes the darkness equaled that of midnight.

Consternation.

I think that some of us are likely to overdo our own superiority and the absurd fears of the Middle Ages—

Oshkosh.

People in the streets rushing in all directions - horses running away women and children running into cellars - little modern touch after all: gas meters instead of images and relics of saints.

This darkness, which lasted from eight to ten minutes, occurred in a day that had been "light but cloudy." It passed from west to east, and brightness followed: then came reports from towns to the west of Oshkosh: that the same phenomenon had already occurred there. A "wave of total darkness" had passed from west to east.

Other instances are recorded in the Monthly Weather Review, but, as to all of them, we have a sense of being pretty well-eclipsed, ourselves, by the conventional explanation that the obscuring body was only a very dense mass of clouds. But some of the instances are interesting - intense darkness at Memphis, Tenn., for about fifteen minutes, at 10 a. m., Dec. 2, 1904 - "We are told that in some quarters a panic prevailed, and that some were shouting and praying and imagining that the end of the world had come." (M. W. R., 32-522.) At Louisville, Ky., March 7, 1911, at about 8 a. m.: duration about half an hour; had been raining moderately, and then hail had fallen. "The intense blackness and general ominous appearance of the storm spread terror throughout the city." (M. W. R., 39-345-)

However, this merger between possible eclipses by unknown dark bodies and commonplace terrestrial phenomena is formidable.

As to darknesses that have fallen upon vast areas, conventionality is - smoke from forest fires. In the U. S. Forest Service Bulletin, No. 117, F. G. Plummer gives a list of eighteen darknesses that have occurred in the United States and Canada. He is one of the primitives, but I should say that his dogmatism is shaken by vibrations from the new Dominant. His difficulty, which he acknowledges, but which he would have disregarded had he written a decade or so earlier, is the profundity of some of these obscurations. He says that mere smokiness can not account for such "awe-inspiring dark days." So he conceives of eddies in the air, concentrating the smoke from forest fires. Then, in the inconsistency or discord of all quasi-intellection that is striving for consistency or harmony, he tells of the vastness of some of these darknesses. Of course Mr. Plummer did not really think upon this subject, but one does feel that he might have approximated higher to real thinking than by speaking of concentration and then listing data of enormous area, or the opposite of circumstances of concentration - because, of his nineteen instances, nine are set down as covering all New England. In quasi-existence, everything generates or is part of its own opposite. Every attempt at peace prepares the way for war; all attempts at justice result in injustice in some other respect: so Mr. Plummer's attempt to bring order into his data, with the explanation of darkness caused by smoke from forest fires, results in such confusion that he ends up by saying that these daytime darknesses have occurred "often with little or no turbidity of the air near the earth's surface" - or with no evidence at all of smoke - except that there is almost always a forest fire somewhere.

However, of the eighteen instances, the only one that I'd bother to contest is the profound darkness in Canada and northern parts of the United States, Nov. 19, 1819 - which we have already considered.

Its concomitants:

Lights in the sky;

Fall of a black substance;

Shocks like those of an earthquake.

In this instance, the only available forest fire was one to the south of the Ohio River. For all I know, soot from a very great fire south of the Ohio might fall in Montreal, Canada, and conceivably, by some freak of reflection, light from it might be seen in Montreal, but the earthquake is not assimilable with a forest fire. On the other hand, it will soon be our expression that profound darkness, fall of matter from the sky, lights in the sky, and earthquakes are phenomena of the near approach of other worlds to this world. It is such comprehensiveness, as contrasted with inclusion of a few factors and disregard for the rest, that we call higher approximation to realness - or universalness.

A darkness, of April 17, 1904, at Wimbledon, England (Symons' Met. Mag., 39-69). It came from a smokeless region: no rain, no thunder; lasted 10 minutes; too dark to go "even out in the open."

As to darknesses in Great Britain, one thinks of fogs - but in Nature, 25-289, there are some observations by Major J. Herschel, upon an obscuration in London, Jan. 22, 1882, at 10.30 a. m., so great that he could hear persons upon the opposite side of the street, but could not see them - "It was obvious that there was no fog to speak of."

Annual Register, 1857-132:

An account by Charles A. Murray, British Envoy to Persia, of a darkness of May 20, 1857, that came upon Bagdad - "a darkness more intense than ordinary midnight, when neither stars nor moon are visible. . . ." "After a short time the black darkness was succeeded by a red, lurid gloom, such as I never saw in any part of the world."

"Panic seized the whole city."

"A dense volume of red sand fell."

This matter of sand falling seems to suggest conventional explanation enough, or that a simoon, heavily charged with terrestrial sand, had obscured the sun, but Mr. Murray, who says that he had had experience with simoons, gives his opinion that "it can not have been a simoon."

It is our comprehensiveness now, or this matter of concomitants of darknesses that we are going to capitalize. It is all very complicated and tremendous, and our own treatment can be but impressionistic, but a few of the rudiments of Advanced Seismology we shall now take up - or the four principle phenomena of another world's close approach to this world.

If a large substantial mass, or super-construction, should enter this earth's atmosphere, it is our acceptance that it would sometimes - depending upon velocity - appear luminous or look like a cloud, or like a cloud with a luminous nucleus. Later we shall have an expression upon luminosity different from the luminosity of incandescence - that comes upon objects falling from the sky, or entering this earth's atmosphere. Now our expression is that worlds have often come close to this earth, and that smaller objects size of a haystack or size of several dozen skyscrapers lumped, have often hurtled through this earth's atmosphere, and have been mistaken for clouds, because they were enveloped in clouds—

Or that around something coming from the intense cold of interplanetary space that is of some regions: our own suspicion is that other regions are tropical - the moisture of this earth's atmosphere would condense into a cloud-like appearance around it. In Nature, 20-121, there is an account by Mr. S. W. Clifton, Collector of Customs, at Freemantle, Western Australia, sent to the Melbourne Observatory - a clear day appearance of a small black cloud, moving not very swiftly - bursting into a ball of fire, of the apparent size of the moon—

Or that something with the velocity of an ordinary meteorite could not collect vapor around it, but that slower-moving objects - speed of a railway train, say - may.

The clouds of tornadoes have so often been described as if they were solid objects that I now accept that sometimes they are: that some so-called tornadoes are objects hurtling through this earth's atmosphere, not only generating disturbances by their suctions, but crushing, with their bulk, all things in their way, rising and falling and finally disappearing, demonstrating that gravitation is not the power that the primitives think it is, if an object moving at relatively low velocity be not pulled to this earth, or being so momentarily affected, bounds away.

In Finley's Reports on the Character of 600 Tornadoes very suggestive bits of description occur:

"Cloud bounded along the earth like a ball"—

Or that it was no meteorological phenomenon, but something very much like a huge solid ball that was bounding along, crushing and carrying with it everything within its field—

"Cloud bounded along, coming to the earth every eight hundred or one thousand yards."

Here's an interesting bit that I got somewhere else. I offer it as a datum in super-biology, which, however, is a branch of advanced science that I'll not take up, restricting to things indefinitely called "objects"—

"The tornado came wriggling, jumping, whirling like a great green snake, darting out a score of glistening fangs."

Though it's interesting, I think that's sensational, myself. It may be that vast green snakes sometimes rush past this earth, taking a swift bite wherever they can, but, as I say, that's a super-biologic phenomenon. Finley gives dozens of instances of tornado clouds that seem to me more like solid things swathed in clouds, than clouds. He notes that, in the tornado at Americus, Georgia, July 18, 1881, "a strange sulphurous vapor was emitted from the cloud."

In many instances, objects, or meteoritic stones, that have come from this earth's externality, have had a sulphurous odor. Why a wind effect should be sulphurous is not clear. That a vast object from external regions should be sulphurous is in line with many data. This phenomenon is described in the Monthly Weather view, July, 1881, as "a strange sulphurous vapor . . . burning and sickening all who approached close enough to breathe it."

The conventional explanation of tornadoes as wind-effects - which we do not deny in some instances - is so strong in the United States that it is better to look elsewhere for an account of an object that has hurtled through this earth's atmosphere, rising and falling and defying this earth's gravitation.

Nature, 7-112:

That, according to a correspondent to the Birmingham Morning News, the people living near

King's Sutton, Banbury, saw, about one o'clock, Dec. 7, 1872, something like a haycock hurtling through the air. Like a meteor it was accompanied by fire and a dense smoke and made a noise like that of a railway train. "It was sometimes high in the air and sometimes near the ground." The effect #as tornado-like: trees and walls were knocked down. It's a late day now to try to verify this story, but a list is given of persons whose property was injured. We are told that this thing then disappeared "all at once."

These are the smaller objects, which may be derailed railway trains or big green snakes, for all I know - but our expression upon approach to this earth by vast dark bodies—

That likely they'd be made luminous: would envelope in clouds, perhaps, or would have their own clouds—

But that they'd quake, and that they'd affect this earth with quakes—

And that then would occur a fall of matter from such a world, or rise of matter from this earth to a nearby world, or both fall and rise, or exchange of matter-process known to Advanced Seismology as celestio-metathesis—

Except that - if matter from some other world - and it would be like some one to get it into his head that we absolutely deny gravitation, just because we can not accept orthodox dogmas - except that, if matter from another world, filling the sky of this earth, generally, as to a hemisphere, or locally, should be attracted to this earth, it would seem thinkable that the whole thing should drop here, and not merely its surface-materials.

Objects upon a ship's bottom. From time to time they drop to the bottom of the ocean. The ship does not.

Or, like our acceptance upon dripping from aerial ice-fields, we think of only a part of a nearby world succumbing, except in being caught in suspension, to this earth's gravitation, and surface materials falling from that part—

Explain or express or accept, and what does it matter? Our attitude is:

Here are the data.

See for yourself.

What does it matter what my notions may be?

Here are the data.

But think for yourself, or think for myself, all mixed up we must be. A long time must go by before we can know Florida from Long Island. So we've had data of fishes that have fallen from our now established and respectabilized Super-Sargasso Sea - which we've almost forgotten, it's now so respectable - but we shall have data of fishes that have fallen during earthquakes. These we accept were dragged down from ponds or other worlds that have been quaked, when only a few miles away, by this earth, some other world also quaking this earth.

In a way, or in its principle, our subject is orthodox enough. Only grant proximity of other worlds - which, however, will not be a matter of granting, but will be a matter of data - and one conventionally conceives of their surfaces quaked - even of a whole lake full of fishes being quaked and dragged down from one of them. The lake full of fishes may cause a little pain to some minds, but the fall of sand and stones is pleasantly enough thought of. More scientific persons, or more faithful hypnotics than we, have taken up this subject, unpainfully, relatively to the moon. For instance, Perrey has gone over 15,000 records of earthquakes, and he has correlated many with proximities of the moon, or has attributed many to the pull of the moon when nearest this earth. Also there is a paper upon this subject in the Proc. Roy. Soc. of Cornwall, 1845. Or, theoretically, when at its closest to this earth, the moon quakes the face of this earth, and is itself quaked but does not itself fall to this earth. As to showers of matter that may have come from the moon at such times - one can go over old records and find what one pleases.

That is what we now shall do.

Our expressions are for acceptance only.

Our data:

We take them from four classes of phenomena that have preceded or accompanied earthquakes: Unusual clouds, darkness profound, luminous appearances in the sky, and falls of substances and objects whether commonly called meteoritic or not.

Not one of these occurrences fits in with principles of primitive, or primary, seismology, and every one of them is a datum of a quaked body passing close to this earth or suspended over it. To the primitives there is not a reason in the world why a convulsion of this earth's surface should be accompanied by unusual sights in the sky, by darkness, or by the fall of substances or objects from the sky. As to phenomena like these, or storms, preceding earthquakes, the irreconcilability is still greater.

It was before 1860 that Perrey made his great compilation. We take most of our data from lists compiled long ago. Only the safe and unpainful have been published in recent years - at least in ambitious, voluminous form. The restraining hand of the "System" - as we call it, whether it has any real existence or not - is tight upon the sciences of to-day. The uncanniest aspect of our quasi-existence that I know of is that everything that seems to have one identity has also as high a seeming of everything else. In this oneness of allness, or continuity, the protecting hand strangles; the parental stifles; love is inseparable from phenomena of hate. There is only Continuity - that is in quasi-existence. Nature, at least in its correspondents' columns, still evades this protective strangulation, and the Monthly Weather Review is still a rich field of unfaithful observation: but, in looking over other long-established periodicals, I have noted their glimmers of quasi-individuality fade gradually, after about 1860, and the surrender of their attempted identities to a higher attempted organization. Some of them, expressing Intermediateness-wide endeavor to localize the universal, or to localize self, soul, identity, entity - or positiveness or realness - held out until as far as 1880; traces findable up to 1890 and then, expressing the universal process - except that here and there in the world's history there may have been successful approximations to positiveness by "individuals" - who only then became individuals and attained to selves or souls of their own - surrendered, submitted, became parts of a higher organization's attempt to individualize or systematize into a complete thing, or to localize the universal or the attributes of the universal. After the death of Richard Proctor, whose occasional illiberalities I'd not like to emphasize too much, all succeeding volumes of Knowledge have yielded scarcely an unconventionality. Note the great number of times that the American Journal of Science and the Report of the British Association are quoted: note that, after, say, 1885, they're scarcely mentioned in these inspired but illicit pages - as by hypnosis and inertia, we keep on saying. About 1880.

Throttle and disregard.

But the coercion could not be positive, and many of the excommunicated continued to creep in; or, even to this day, some of the strangled are faintly breathing.

Some of our data have been hard to find. We could tell stories of great labor and fruitless quests that would, though perhaps imperceptibly, stir the sympathy of a Mr. Symons. But, in this matter of concurrence of earthquakes with aerial phenomena, which are as unassociable with earthquakes, if internally caused, as falls of sand on convulsed small boys full of sour apples, the abundance of so-called evidence is so great that we can only sketchily go over the data, beginning with Robert Mallet's Catalogue (Rept. Brit. Assoc., 1852), omitting some extraordinary instances, because they occurred before the eighteenth century:

Earthquake "preceded" by a violent tempest, England, Jan. 8, 1704 - "preceded" by a brilliant meteor, Switzerland, Nov. 4, 1704 - "luminous cloud, moving at high velocity, disappearing behind the horizon," Florence, Dec. 9, 1731 - "thick mists in the air, through which a dim light was seen: several weeks before the shock, globes of light had been seen in the air," Swabia, May 22, 1732 - rain of earth, Carpentras, France, Oct. 18, 1737 - a black cloud, London, March 19, 1750 - violent storm and a strange star of octagonal shape, Slavange, Norway, April 15, 1752 - balls of fire from a streak in the sky, Augermannland, 1752 - numerous meteorites, Lisbon, Oct. 15, 1755 "terrible tempests" over and over "falls of hail" and "brilliant meteors," instance after instance "an immense globe," Switzerland, Nov. 2, 1761 - oblong, sulphurous cloud, Germany, April, 1767 - extraordinary mass of vapor, Bologne, April, 1780 - heavens obscured by a dark mist, Grenada, Aug. 7, 1804 - "strange, howling noises in the air, and large spots obscuring the sun," Palermo, Italy, April 16, 1817 "luminous meteor moving in the same direction as the shock," Naples, Nov. 22, 1821 fire ball appearing in the sky: apparent size of the moon, Thuringerwald, Nov. 29, 1831.

And, unless you be polarized by the New Dominant, which is calling for recognition of multiplicities of external things, as a Dominant, dawning new over Europe in 1492, called for recognition of terrestrial externality to Europe - unless you have this contact with the new, you have no affinity for these data - beans that drop from a magnet - irreconcilables that glide from the mind of a Thomson—

Or my own acceptance that we do not really think at all; that we correlate around super-magnets that I call Dominants - a Spiritual Dominant in one age, and responsively to it up spring monasteries, and the stake and the cross are its symbols: a Materialist Dominant, and up spring laboratories, and microscopes and telescopes and crucibles are its ikons - that we're nothing but iron filings relatively to a succession of magnets that displace preceding magnets.

With no soul of your own, and with no soul of my own - except that some day some of us may no longer be Intermediatisms, but may hold out against the cosmos that once upon a time thousands of fishes were cast from one pail of water - we have psycho-valency for these data, if we're obedient slaves to the New Dominant, and repulsion to them, if we're mere correlates to the Old Dominant. I'm a soulless and selfless correlate to the New Dominant, myself: I see what I have to see. The only inducement I can hold out, in my attempt to rake up disciples, is that some day the New will be fashionable: the new correlates will sneer at the old correlates. After all, there is some inducement to that - and I'm not altogether sure it's desirable to end up as a fixed star.

As a correlate to the New Dominant, I am very much impressed with some of these data - the luminous object that moved in the same direction as an earthquake - it seems very acceptable that a quake followed this thing as it passed near this earth's surface. The streak that was seen in the sky - or only a streak that was visible of another world - and objects, or meteorites, that were shaken down from it. The quake at Carpentras, France: and that, above Carpentras, was a smaller world, more violently quaked, so that earth was shaken down from it.

But I like best the super-wolves that were seen to cross the sun, during the earthquake at Palermo. They howled.

Or the loves of the worlds. The call they feel for one another.

They try to move closer and howl when they get there.

The howls of the planets.

I have discovered a new unintelligibility.

In the Edinburgh New Philosophical Journal - have to go away back to 1841 - days of less efficient strangulation - Sir David Milne lists phenomena of quakes in Great Britain. I pick out a few that indicate to me that other worlds were near this earth's surface:

Violent storm before a shock of 1703 - ball of fire "preceding," 1750 - a large ball of fire seen upon day following a quake, 1755 - "uncommon phenomenon in the air: a large luminous body, bent like a crescent, which stretched itself over the heavens, 1816 - vast ball of fire, 1750 - black rains and black snows, 1755 - numerous instances of upward projection or upward attraction? during quakes "preceded by a cloud, very black and lowering," 1795 - fall of black powder, preceding a quake, by six hours, 1837.

Some of these instances seem to me to be very striking - a smaller world: it is greatly racked by the attraction of this earth - black substance is torn down from it not until six hours later, after an approach still closer, does this earth suffer perturbation. As to the extraordinary spectacle of a thing, world, super-construction, that was seen in the sky, in 1816, I have not yet been able to find out more. I think that here our acceptance is relatively sound: that this occurrence was tremendously of more importance than such occurrence as, say, transits of Venus, upon which hundreds of papers have been written - that not another mention have I found, though I have not looked so especially as I shall look for more data - that all but undetailed record of this occurrence was suppressed.

Altogether we have considerable agreement here between data of vast masses that do not fall to this earth, but from which substances fall, and data of fields of ice from which ice may not fall, but from which water may drip. I'm beginning to modify: that, at a distance from this earth, gravitation has more effect than we have supposed, though less effect than the dogmatists suppose and "prove." I'm coming out stronger for the acceptance of a Neutral Zone - that this earth, like other magnets, has a neutral zone, in which is the Super – Sargasso Sea, and in which other worlds may be buoyed up, though projecting parts may be subject to this earth's attraction-

But my preference:

Here are the data.

I now have one of the most interesting of the new correlates. I think I should have brought it in before, but, whether out of place here, because not accompanied by earthquake, or not, we'll have it. I offer it as an instance of an eclipse, by a vast, dark body, that has been seen and reported by an astronomer. The astronomer is M. Lias: the phenomenon was seen by him, at Pernambuco, April 11, 1860.

Comptes Rendus, 50-1197:

It was about noon - sky cloudless - suddenly the light of the sun was diminished. The darkness increased, and, to illustrate its intensity, we are told that the planet Venus shone brilliant. But Venus was of low visibility at this time. The observation that burns incense to the New Dominant is:

That around the sun appeared a corona.

There are many other instances that indicate proximity of other worlds during earthquakes. I note a few - quake and an object in the sky, called "a large, luminous meteor" (Quar. Jour. Roy. Inst., 5-132); luminous body in the sky, earthquake, and fall of sand, Italy, Feb. 12 and 13, 1870 (La Science Pour Tous, 15-159); many reports upon luminous object in the sky and earthquake, Connecticut, Feb. 27, 1883 (Monthly Weather Review, Feb., 1883); luminous object, or meteor, in the sky, fall of stones from the sky, and earthquake, Italy, Jan. 20, 1891 (L'Astronomie, 1891-154); earthquake and prodigious number of luminous bodies, or globes, in the air, Boulogne, France, June 7, 1779 (Sestier, "La Foudre" 1-169) ; earthquake at Manila, 1863, and "curious luminous appearance in the sky," (Ponton, "Earthquakes," p. 124).

The most notable appearance of fishes during an earthquake is that of Riobamba. Humboldt sketched one of them, and it's an uncanny-looking thing. Thousands of them appeared upon the ground during this tremendous earthquake. Humboldt says that they were cast up from subterranean sources. I think not myself, and have data for thinking not, but there'd be such a row arguing back and forth that it's simpler to consider a clearer instance of the fall of living fishes from the sky, during an earthquake. I can't quite accept, myself, whether a large lake, and all the fishes in it, was torn down from some other world, or a lake in the Super-Sargasso Sea, distracted between two pulling worlds, was dragged down to this earth –

Here are the data:

La Science Pour Tous, 6-191:

Feb. 16, 1861. An earthquake at Singapore. Then came an extraordinary downpour of rain - or as much water as any good-sized lake would consist of. For three days this rain or this fall of water came down in torrents. In pools on the ground, formed by this deluge, great numbers of fishes were found. The writer says that he had, himself, seen nothing but water fall from the sky. Whether I'm emphasizing what a deluge it was or not, he says that so terrific had been the downpour that he had not been able to see three steps away from him. The natives said that the fishes had fallen from the sky. Three days later the pools dried up and many dead fishes were found, but, in the first place - though that's an expression for which we have an instinctive dislike - the fishes had been active and uninjured. Then follows material for another of our little studies in the phenomena of disregard. A psycho-tropism here is mechanically to take pen in hand and mechanically write that fishes found on the ground after a heavy rainfall came from overflowing streams. The writer of the account says that some of the fishes had been found in his courtyard, which was surrounded by high walls - paying no attention to this, a correspondent (La Science Pour Tous, 6-317) explains that in the heavy rain a body of water had probably overflowed, carrying fishes with it. We are told by the first writer that these fishes of Singapore were of a species that was very abundant near Singapore. So I think, myself, that a whole lakeful of them had been shaken down from the Super – Sargasso Sea, under the circumstances we have thought of. However, if appearance of strange fishes after an earthquake be more pleasing in the sight, or to the nostrils, of the New Dominant, we faithfully and piously supply that incense - An account of the occurrence at Singapore was read by M. de Castelnau, before the French Academy. M. de Castelnau recalled that, upon a former occasion, he had submitted to the Academy the circumstance that fishes of a new species had appeared at the Cape of Good Hope, after an earthquake.

It seems proper, and it will give luster to the new orthodoxy, now to have an instance in which, not merely quake and fall of rocks, or meteorites, or quake and either eclipse or luminous appearances in the sky have occurred, but in which are combined all the phenomena, one or more of which, when accompanying earthquake, indicate, in our acceptance, the proximity of another world. This time a longer duration is indicated than in other instances.

In the Canadian Institute Proceedings, 2-7-198, there is an account, by the Deputy Commissioner at Dhurmsalla, of the extraordinary Dhurmsalla meteorite - coated with ice. But the combination of events related by him is still more extraordinary:

That within a few months of the fall of this meteorite there had been a fall of live fishes at Benares, a shower of red substance at Furruckabad, a dark spot observed on the disk of the sun, an earthquake, "an unnatural darkness of some duration," and a luminous appearance in the sky that looked like an aurora borealis –

But there's more to this climax:

We are introduced to a new order of phenomena:

Visitors.

The Deputy Commissioner writes that, in the evening, after the fall of the Dhurmsalla meteorite, or mass of stone covered with ice, he saw lights. Some of them were not very high. They appeared and went out and reappeared. I have read many accounts of the Dhurmsalla meteorite - July 28, 1860 - but never in any other of them a mention of this new correlate - something as out of place in the nineteenth century as would have been an aeroplane - the invention of which would not, in our acceptance, have been permitted, in the nineteenth century, though adumbrations to it were permitted. This writer says that the lights moved like fire balloons, but:

"I am sure that they were neither fire balloons, lanterns, nor bonfires, or any other thing of that sort, but bona fide lights in the heavens."

It's a subject for which we shall have to have a separate expression- trespassers upon territory to which something else has a legal right perhaps someone lost a rock, and he and his friends came down looking for it, in the evening - or secret agents, or emissaries, who had an appointment with certain esoteric ones near Dhurmsalla - things or beings coming down to explore, and unable to stay down long –

In a way, another strange occurrence during an earthquake is suggested. The ancient Chinese tradition the marks like hoof marks in the ground. We have thought - with a low degree of acceptance - of another world that may be in secret communication with certain esoteric ones of this earth's inhabitants - and of messages in symbols like hoof marks that are sent to some receptor, or special hill, upon this earth - and of messages that at times miscarry.

This other world comes close to this world - there are quakes - but advantage of proximity is taken, to send a message the message, designed for a receptor, in India, perhaps, or in Central Europe, miscarries all the way to England - marks like the marks of the Chinese tradition are found upon a beach, in Cornwall, after an earthquake –

Phil. Trans., 50-500:

After the quake of July 15, 1757, upon the sands of Penzance, Cornwall, in an area of more than 100 square yards, were found marks like hoof prints, except that they were not crescentic. We feel a similarity, but note an arbitrary disregard of our own, this time. It seems to us that marks described as "little cones surrounded by basins of equal diameter" would be like hoof prints, if hoofs printed complete circles. Other disregards are that there were black specks on the tops of cones, as if something, perhaps gaseous, had issued from them; that from one of these formations came a gush of water as thick as a man's wrist. Of course the opening of springs is common in earthquakes - but we suspect, myself, that the Negative Absolute is compelling us to put in this datum and its disorders.

There's another matter in which the Negative Absolute seems to work against us. Though to super-chemistry, we have introduced the principle of celestio-metathesis, we have no good data of exchange of substances during proximities. The data are all of falls and not of upward translations. Of course upward impulses are common during earthquakes, but I haven't a datum upon a tree or a fish or a brick or a man that ever did go up and stay up and that never did come down again. Our classic of the horse and barn occurred in what was called a whirlwind.

It is said that, in an earthquake in Calabria, paving stones shot up far in the air.

The writer doesn't specifically say that they came down again, but something seems to tell me they did.

The corpses of Riobamba.

Humboldt reported that, in the quake of Riobamba, "bodies were torn upward from graves"; that "the vertical motion was so strong that bodies were tossed several hundred feet in the air."

I explain.

I explain that, if in the center of greatest violence of an earthquake, anything ever has gone up, and has kept on going up, the thoughts of the nearest observers were very likely upon other subjects.

The quay of Lisbon.

We are told that it went down.

A vast throng of persons ran to the quay for refuge. The city of Lisbon was in profound darkness. The quay and all the people on it disappeared. If it and they went down not a single corpse, not a shred of clothing, not a plank of the quay, nor so much as a splinter of it ever floated to the surface.

Chapter XVIII

    The New Dominant.

I mean "primarily" all that opposes Exclusionism –

That Development or Progress or Evolution is Attempt to Positivize, and is a mechanism by which a positive existence is recruited - that what we call existence is the womb of infinitude, and is itself only incubatory - that eventually all attempts are broken down by the falsely excluded. Subjectively, the breaking down is aided by our own sense of false and narrow limitations. So the classic and academic artists wrought positivist paintings, and expressed the only ideal that I am conscious of, though we so often hear of "ideals" instead of different manifestations, artistically, scientifically, theologically, politically, of the One Ideal. They sought to satisfy, in its artistic aspect, cosmic craving for unity or completeness, sometimes called harmony, called beauty in some aspects. By disregard they sought completeness. But the light-effects that they disregarded, and their narrow confinement to standardized subjects brought on the revolt of the Impressionists. So the Puritans tried to systematize, and they disregarded physical needs, or vices, or relaxations: they were invaded and overthrown when their narrowness became obvious and intolerable. All things strive for positiveness, for themselves, or for quasi-systems of which they are parts. Formality and the mathematic, the regular and the uniform are aspects of the positive state - but the Positive is the Universal - so all attempted positiveness that seems to satisfy in the aspects of formality and regularity, sooner or later disqualifies in the aspect of wideness or universalness. So there is revolt against the science of to-day, because the formulated utterances that were regarded as final truths in a past generation, are now seen to be insufficiencies. Every pronouncement that has opposed our own acceptances has been found to be a composition like any academic painting: something that is arbitrarily cut off from relations with environment, or framed off from interfering and disturbing data, or outlined with disregards. Our own attempt has been to take in the included, but also to take in the excluded into wider expressions. We accept, however, that for every one of our expressions there are irreconcilables somewhere - that final utterance would include all things. However, of such is the gossip of angels. The final is unutterable in quasi-existence, where to think is to include but also to exclude, or be not final. If we admit that for every opinion we have expressed, there must somewhere be an irreconcilable, we are Intermediatists and not positivists; not even higher positivists. Of course it may be that some day we shall systematize and dogmatize and refuse to think of anything that we may be accused of disregarding, and believe instead of merely accepting: then, if we could have a wider system, which would acknowledge no irreconcilables we'd be higher positivists. So long as we only accept, we are not higher positivists, but our feeling is that the New Dominant, even though we have thought of it only as another enslavement, will be the nucleus for higher positivism - and that it will be the means of elevating into infinitude a new batch of fixed stars - until, as a recruiting instrument, it, too, will play out, and will give way to some new medium for generating absoluteness. It is our acceptance that all astronomers of to-day have lost their souls, or, rather, all chance of attaining Entity, but that Copernicus and Kepler and Galileo and Newton, and, conceivably, Leverrier are now fixed stars. Some day I shall attempt to identify them. In all this, I think we're quite a Moses. We point out the Promised Land, but, unless we be cured of our Intermediatism, will never be reported in Monthly Notices, ourself.

In our acceptance, Dominants, in their succession, displace preceding Dominants not only because they are more nearly positive, but because the old Dominants, as recruiting mediums, play out. Our expression is that the New Dominant, of Wider Inclusions, is now manifesting throughout the world, and that the old Exclusionism is everywhere breaking down. In physics Exclusionism is breaking down by its own researches in radium, for instance, and in its speculations upon electrons, or its merging away into metaphysics, and by the desertion that has been going on for many years, by such men as Gurney, Crookes, Wallace, Flammarion, Lodge, to formerly disregarded phenomena

- no longer called "spiritualism" but now "psychic research." Biology is in chaos: conventional Dar-winites mixed up with mutationists and orthogenesists and followers of Wisemann, who take from Darwinism one of its pseudo-bases, and nevertheless try to reconcile their heresies with orthodoxy. The painters are metaphysicians and psychologists. The breaking down Exclusionism in China and Japan and in the United States has astonished History. The science of astronomy is going downward so that, though Pickering, for instance, did speculate upon a Trans – Neptunian planet, and Low-ell did try to have accepted heretical ideas as to marks on Mars, attention is now minutely focussed upon such technicalities as variations in shades of Jupiter's fourth satellite. I think that, in general acceptance, over-refinement indicates decadence.

I think that the stronghold of Inclusionism is in aeronautics. I think that the stronghold of the Old Dominant, when it was new, was in the invention of the telescope. Or that, coincidentally with the breakdown of Exclusionism appears the means of finding out - whether there are vast aerial fields of ice and floating lakes full of frogs and fishes or not - where carved stones and black substances and great quantities of vegetable matter and flesh, which may be dragons' flesh, come from whether there are inter-planetary trade routes and vast areas devastated by Super-Tamerlanes - whether some-times there are visitors to this earth who - might be pursued and captured and questioned.

Chapter XIX

Have industriously sought data for an expression upon birds, but the prospecting has not been very quasi-satisfactory. I think I rather emphasize our industriousness, because a charge likely to be brought against the attitude of Acceptance is that one who only accepts must be one of languid in-terest and little application of energy. It doesn't seem to work out: we are very industrious, suggest to some of our disciples that they look into the matter of , messages upon pigeons, of course attrib-uted to earthly owners, but I said to be undecipherable. I'd do it, ourselves, only that would be self-ish. That's more of the Intermediatism that will keep us out of the firmament: Positivism is absolute egoism. But look back in the time of Andree's Polar Expedition. Pigeons that would have no public-ity ordinarily, were often reported at that time.

In the Zoologist, 3-18-21, is recorded an instance of a bird (puffin) that had fallen to the ground with a fractured head. Interesting, but mere speculation - but what solid object, high in the air, had that bird struck against?

Tremendous red rain in France, Oct. 16 and 17, 1846; great storm at the time, and red rain supposed to have been colored by matter swept up from this earth's surface, and then precipitated (Comptes Rendus, 23-832). But in Comptes Rendus, 24-625, the description of this red rain differs from one's impression of red, sandy or muddy water. It is said that this rain was so vividly red and so blood-like that many persons in France were terrified. Two analyses are given (Comptes Rendus, 24-812). One chemist notes a great quantity of corpuscles whether blood-like corpuscles - or not - in the matter. The other chemist sets down organic matter at 35 per cent. It may be that an inter-planetary dragon had been slain somewhere, or that this red fluid, in which were many corpuscles, came from some-thing not altogether pleasant to contemplate, about the size of the Catskill Mountains, perhaps - but the present datum is that with this substance, larks, quail, ducks, and water hens, some of them alive, fell at Lyons and Grenoble and other places.

I have notes upon other birds that have fallen from the sky, unaccompanied by the red rain that makes the fall of birds in France peculiar, and very peculiar, if it be accepted that the red substance was extra-mundane. The other notes are upon birds that have fallen from the sky, in the midst of storms, or of exhausted, but living, birds, falling not far from a storm-area. But now we shall have an instance for which I can find no parallel: fall of dead birds, from a clear sky, far-distant from any storm to which they could be attributed - so remote from any discoverable storm that –

My own notion is that, in the summer of 1896, something, or some beings, came as near to this earth as they could, upon a hunting expedition; that, in the summer of 1896, an expedition of super-scientists passed over this earth, and let down a dragnet - and what would it catch, sweeping through the air, supposing it to have reached not quite to this earth?

In the Monthly Weather Review, May, 1917, W. L. McAtee quotes from the Baton Rouge correspondence to the Philadelphia Times:

That, in the summer of 1896, into the streets of Baton Rouge, La., and from a "clear sky," fell hundreds of dead birds. There were wild ducks and cat birds, woodpeckers, and "many birds of strange plumage," some of them resembling canaries.

Usually one does not have to look very far from any place to learn of a storm. But the best that could be done in this instance was to say:

"There had been a storm on the coast of Florida."

And, unless he have psycho-chemic repulsion for the explanation, the reader feels only momentary astonishment that dead birds from a storm in Florida should fall from an unstormy sky in Louisiana, and with his intellect greased like the plumage of a wild duck, the datum then drops oft.

Our greasy, shiny brains. That they may be of some use after all: that other modes of existence place a high value upon them as lubricants; that we're hunted for them; a hunting expedition to this earth - the newspapers report a tornado.

If from a clear sky, or a sky in which there were no driven clouds, or other evidences of still-continuing wind-power - or, if from a storm in Florida, it could be accepted that hundreds of birds had fallen far away, in Louisiana, I conceive, conventionally, of heavier objects having fallen in Alabama, say, and of the fall of still heavier objects still nearer the origin in Florida.

The sources of information of the Weather Bureau are widespread.

It has no records of such falls.

So a drag net that was let down from above somewhere –

Or something that I learned from the more scientific of the investigators
of psychic phenomena:

The reader begins their works with prejudice against telepathy and everything else of psychic phenomena. The writers deny spirit-communication, and say that the seeming data are data of "only telepathy." Astonishing instances of seeming clairvoyance - "only telepathy." After a while the reader finds himself agreeing that it's only telepathy - which, at first, had been intolerable to him.

So maybe, in 1896, a super-dragnet did not sweep through this earth's atmosphere, gathering up all the birds within its field, the meshes then suddenly breaking –

Or that the birds of Baton Rouge were only from the Super-Sargasso Sea –

Upon which we shall have another expression. We thought we'd settled that, and we thought we'd established that, but nothing's ever settled, and nothing's ever established, in a real sense, if, in a real sense, there is nothing in quasiness.

I suppose there had been a storm somewhere, the storm in Florida, perhaps, and many birds had been swept upward into the Super-Sargasso Sea. It has frigid regions and it has tropical regions - that birds of diverse species had been swept upward, into an icy region, where, huddling together for warmth, they had died. Then, later, they had been dislodged - meteor coming along – boat – bicycle – dragon - don't know what did come along something dislodged them.

So leaves of trees, carried up there in whirlwinds, staying there years, ages, perhaps only a few months, but then falling to this earth at an unseasonable time for dead leaves - fishes carried up there, some of them dying and drying, some of them living in volumes of water that are in abundance up there, or that fall sometimes in the deluges that we call "cloudbursts."

The astronomers won't think kindly of us, and we haven't done anything to endear ourselves to the meteorologists but we're weak and mawkish Intermediatists - several times we've tried to get the aeronauts with us - extraordinary things up there: things that curators of museums would give up all hope of ever being fixed stars, to obtain: things left over from whirlwinds of the time of the Pharaohs, perhaps: or that Elijah did go up in the sky in something like a chariot, and may not be Vega, after all, and that there may be a wheel or so left of whatever he went up in. We basely suggest that it would bring a high price - but sell soon, because after a while there'd be thousands of them hawked around –

We weakly drop a hint to the aeronauts.

In the Scientific American, 33-197, there is an account of some hay that fell from the sky. From the circumstances we incline to accept that this hay went up, in a whirlwind, from this earth, in the first place, reached the Super-Sargasso Sea, and remained there a long time before falling. An interesting point in this expression is the usual attribution to a local and coinciding whirlwind, and identifica-tion of it - and then data that make that local whirlwind unacceptable –

That, upon July 27, 1875, small masses of damp hay had fallen at Monkstown, Ireland. In the Dublin Daily Express, Dr. J. W. Moore had explained: he had found a nearby whirlwind, to the south of Monkstown, that coincided. But, according to the Scientific American, a similar fall had occurred near Wrexham, England, two days before.

In November, 1918, I made some studies upon light objects thrown into the air. Armistice-day. I suppose I should have been more emotionally occupied, but I made notes upon torn-up papers thrown high in the air from windows of office buildings. Scraps of paper did stay together for a while. Several minutes, sometimes.

Cosmos, 3-4-574:

That, upon the loth of April, 1869, at Autriche (Indre-et-Loire) a great number of oak leaves enor-mous - segregation of them - fell from the sky. Very calm day. So little wind that the leaves fell almost vertically. Fall lasted about ten minutes.

Flammarion, in "The Atmosphere" p. 412, tells this story.

He has to find a storm.

He does find a squall - but it had occurred upon April 3rd.

Flammarion's two incredibilities are that leaves could remain a week in the air: that they could stay together a week in the air.

Think of some of your own observations upon papers .thrown from an aeroplane.

Our one incredibility:

That these leaves had been whirled up six months before, when they were common on the ground, and had been sustained, of course not in the air, but in a region gravitationally inert; and had been precipitated by the disturbances of April rains.

I have no records of leaves that have so fallen from the sky, in October or November, the season when one might expect dead leaves to be raised from one place and precipitated somewhere else. I emphasize that this occurred in April.

La Nature, 1889-2-94:

That, upon April 19, 1889, dried leaves, of different species, oak, elm, etc., fell from the sky. This day, too, was a calm day. The fall was tremendous. The leaves were seen to fall fifteen minutes, but, judging from the quantity on the ground, it is the writer's opinion that they had already been falling half an hour. I think that the geyser of corpses that sprang from Riobamba toward the sky must have been an interesting sight. If I were a painter, I'd like that subject. But this cataract of dried leaves, too, is a study in the rhythms of the dead. In this datum, the point most agreeable to us is the very point that the writer in La Nature emphasizes. Windlessness. He says that the surface of the Loire was "absolutely smooth." The river was strewn with leaves as far as he could see.

L'Astronomie, 1894-194:

That, upon the 7th of April, 1894, dried leaves fell at Clairvaux and Outre-Aube, France. The fall is described as prodigious. Half an hour. Then, upon the 11th, a fall of dried leaves occurred at Pont-carré.

It is in this recurrence that we found some of our opposition to the conventional explanation. The Editor (Flammarion) explains.

He says that the leaves had been caught up in a cyclone which had expended its force; that the heavier leaves had fallen first. We think that that was all right for 1894, and that it was quite good enough for 1894. But, in these more exacting days, we want to know how wind-power insufficient to hold some leaves in the air could sustain others four days.

The factors in this expression are unseasonableness, not for dried leaves, but for prodigious numbers of dried leaves; direct fall, windlessness, month of April, and localization in France. The factor of localization is interesting. Not a note have I upon fall of leaves from the sky, except these notes. Were the conventional explanation, or "old correlate" acceptable, it would seem that similar occurrences in other regions should be as frequent as in France. The indication is that there may be quasi-permanent undulations in the Super-Sargasso Sea, or a pronounced inclination toward France –

Inspiration:

That there may be a nearby world complementary to this world, where autumn occurs at the time that is springtime here.

Let some disciple have that.

But there may be a dip toward France, so that leaves that are borne high there, are more likely to be held in suspension than highflying leaves elsewhere. Some other time I shall take up Super-geography, and be guilty of charts. I think, now, that the Super-Sargasso Sea is an oblique belt, with changing ramifications, over Great Britain, France, Italy, and on to India. Relatively to the United States I am not very clear, but think especially of the Southern States.

The preponderance of our data indicates frigid regions aloft.

Nevertheless such phenomena as putrefaction have occurred often enough to make super-tropical regions, also, acceptable. We shall have one more datum upon the Super-Sargasso Sea. It seems to me that, by this time, our requirements of support and reinforcement and agreement have been quite as rigorous for acceptance as ever for belief: at least for full acceptance. By virtue of mere acceptance, we may, in some later book, deny the Super-Sargasso Sea, and find that our data relate to some other complementary world instead - or the moon - and have abundant data for accepting that the moon is not more than twenty or thirty miles away. However, the Super-Sargasso Sea functions very well as a nucleus around which to gather data that oppose Exclusionism. That is our main motive: to oppose Exclusionism.

Or our agreement with cosmic processes. The climax of our general expression upon the Super-Sargasso Sea. Coincidentally appears something else that may overthrow it later.

Notes and Queries, 8-12-228:

That in the province of Macerata, Italy (summer of 1897?) an immense number of small, blood-colored clouds covered the sky. About an hour later a storm broke, and myriad seeds fell to the ground. It is said that they were identified as products of a tree found only in Central Africa and the Antilles. If - in terms of conventional reasoning - these seeds had been high in the air, they had been in a cold region. But it is our acceptance that these seeds had, for a considerable time, been in a warm region, and for a time longer than is attributable to suspension by wind-power:

"It is said that a great number of the seeds were in the first stage of germination."

Chapter XX

 The New Dominant.

Inclusionism.

In it we have a pseudo-standard.

We have a datum, and we give it an interpretation, in accordance with our pseudo-standard. At present we have not the delusions of Absolutism that may have translated some of the positivists of the nineteenth century to heaven. We are Intermediatists - but feel a lurking suspicion that we may some day solidify and dogmatize and illiberalize into higher positivists. At present we do not ask whether something be reasonable or preposterous, because we recognize that by reasonableness and preposterousness are meant agreement and disagreement with a standard - which must be a delusion -though not absolutely, of course - and must some day be displaced by a more advanced quasi-delusion. Scientists in the past have taken the positivist attitude - is this or that reasonable or unreasonable? Analyze them and we find that they meant relatively to a standard, such as Newtonism, Daltonism, Darwinism, or Lyellism. But they have written and spoken and thought as if they could mean real reasonableness and real unreasonableness.

So our pseudo-standard is Inclusionism, and, if a datum be a correlate to a more widely inclusive outlook as to this earth and its externality and relations with externality, its harmony with Inclusionism admits it. Such was the process, and such was the requirement for admission in the days of the

Old Dominant: our difference is in underlying Intermediatism, or consciousness that though we're more nearly real, we and our standards are only quasi –

Or that all things - in our intermediate state - are phantoms in a super-mind in a dreaming state - but striving to awaken to realness.

Though in some respects our own Intermediatism is unsatisfactory, our underlying feeling is –

That in a dreaming mind awakening is accelerated - if phantoms in that mind know that they're only phantoms in a dream. Of course, they too are quasi, or - but in a relative sense - they have an essence of what is called realness. They are derived from experience or from sense-relations, even though grotesque distortions. It seems acceptable that a table that is seen when one is awake is more nearly real than a dreamed table, which, with fifteen or twenty legs, chases one.

So now, in the twentieth century, with a change of terms, and a change in underlying consciousness, our attitude toward the New Dominant is the attitude of the scientists of the nineteenth century to the Old Dominant. We do not insist that our data and interpretations shall be as shocking, grotesque, evil, ridiculous, childish, insincere, laughable, ignorant to nineteenth-centuryites as were their data and interpretations to the medieval-minded. We ask only whether data and interpretations correlate. If they do, they are acceptable, perhaps only for a short time, or as nuclei, or scaffolding, or preliminary sketches, or as gropings and tentativenesses.

Later, of course, when we cool off and harden and radiate into space most of our present mobility, which expresses in modesty and plasticity, we shall acknowledge no scaffoldings, gropings or tentativenesses, but think we utter absolute facts. A point in Intermediatism here is opposed to most current speculations upon Development. Usually one thinks of the spiritual as higher than the material, but, in our acceptance, quasi-existence is a means by which the absolutely immaterial materializes absolutely, and, being intermediate, is a state in which nothing is finally either immaterial or material, all objects, substances, thoughts, occupying some grade of approximation one way or the other. Final solidification of the ethereal is, to us, the goal of cosmic ambition. Positivism is Puritanism. Heat is Evil. Final Good is Absolute Frigidity. An Arctic winter is very beautiful, but I think that an interest in monkeys chattering in palm trees accounts for our own Intermediatism.

Visitors.

Our confusion here, out of which we are attempting to make quasi-order is as great as it has been throughout this book, because we have not the positivist's delusion of homogeneity. A positivist would gather all data that seem to relate to one kind of visitors and coldly disregard all other data. I think of as many different kinds of visitors to this earth as there are visitors to New York, to a jail, to a church - some persons go to church to pick pockets, for instance.

My own acceptance is that either a world or a vast super-construction - or a world, if red substances and fishes fell from it - hovered over India in the summer of 1860. Something then fell from somewhere, July 17, 1860, at Dhurmsalla. Whatever "it" was, "it" is so persistently alluded to as "a meteorite" that I look back and see that I adopted this convention myself. But in the London Times, Dec. 26, 1860, Syed Abdoolah, Professor of Hindustani, University College, London, writes that he had sent to a friend in Dhurmsalla, for an account of the stones that had fallen at that place. The answer: "... divers forms and sizes, many of which bore great resemblance to ordinary cannon balls just discharged from engines of war."

It's an addition to our data of spherical objects that have arrived upon this earth. Note that they are spherical stone objects. And, in the evening of this same day that something - took a shot at Dhurmsalla - or sent objects upon which there may be decipherable markings - lights were seen in the air – I think, myself, of a number of things, beings, whatever they were, trying to get down, but resisted, like balloonists, at a certain altitude, trying to get farther up, but resisted.

Not in the least except to good positivists, or the homogeneousminded, does this- speculation interfere with the concept of some other world that is in successful communication with certain esoteric ones upon this earth, by a code of symbols that print in rock, like symbols of telephotographers in selenium.

I think that sometimes, in favorable circumstances, emissaries have come to this earth - secret meetings –

Of course it sounds –

But:

Secret meetings – emissaries - esoteric ones in Europe, before the war broke out –

And those who suggested that such phenomena could be.

However, as to most of our data, I think of super-things that have passed close to this earth with no more interest in this earth than have passengers upon a steamship in the bottom of the sea - or passengers may have a keen interest, but circumstances of schedules and commercial requirements forbid investigation of the bottom of the sea.

Then, on the other hand, we may have data of super-scientific attempts to investigate phenomena of this earth from - perhaps by beings from so far away that they had never even heard that something, somewhere, asserts a legal right to this earth.

Altogether, we're good intermediatists, but we can't be very good hypnotists.

Still another source of the merging away of our data:

That, upon general principles of Continuity, if super-vessels, or super-vehicles, have traversed this earth's atmosphere, there must be mergers between them and terrestrial phenomena: observations upon them must merge away into observations upon clouds and balloons and meteors. We shall begin with data that we can not distinguish ourselves and work our way out of mergers into extremes.

In the Observatory, 35-168, it is said that, according to a newspaper, March 6, 1912, residents of Warmley, England, were greatly excited by something that was supposed to be "a splendidly illuminated aeroplane, passing over the village." "The machine was apparently traveling at a tremendous rate, and came from the direction of Bath, and went on toward Gloucester." The Editor says that it was a large, triple-headed fireball. "Tremendous indeed!" he says. "But we are prepared for anything nowadays."

That is satisfactory. We'd not like to creep up stealthily and then jump out of a corner with our data.

This Editor, at least, is prepared to read –

Nature, Oct. 27, 1898:

A correspondent writes that, in the County Wicklow, Ireland, at about 6 o'clock in the evening, he had seen, in the sky, an object that looked like the moon in its three-quarter aspect. We note the shape which approximates to triangularity, and we note that in color it is said to have been golden yellow. It moved slowly, and in about five minutes disappeared behind a mountain.

The Editor gives his opinion that the object may have been an escaped balloon.

In Nature, Aug. u, 1898, there is a story, taken from the July number of the Canadian Weather Review, by the meteorologist, F. F. Payne: that he had seen, in the Canadian sky, a large, pear-shaped object, sailing rapidly. At first he supposed that the object was a balloon, "its outline being sharply defined." "But, as no cage was seen, it was concluded that it must be a mass of cloud." In about six minutes this object became less definite - whether because of increasing distance or not - "the mass became less dense, and finally it disappeared." As to cyclonic formation - "no whirling motion could be seen."

Nature, 58-294:

That, upon July 8, 1898, a correspondent had seen, at Kiel, an object in the sky, colored red by the sun, which had set. It was about as broad as a rainbow, and about twelve degrees high. "It remained in its original brightness about five minutes, and then faded rapidly, and then remained almost stationary again, finally disappearing about eight minutes after I first saw it."

In an intermediate existence, we quasi-persons have nothing to judge by because everything is its own opposite. If a hundred dollars a week be a standard of luxurious living to some persons, it is poverty to others. We have instances of three objects that were seen in the sky in a space of three months, and this concurrence seems to me to be something to judge by. Science has been built upon concurrence: so have been most of the fallacies and fanaticisms.

I feel the positivism of a Leverrier, or instinctively take to the notion that all three of these observations relate to the same object. However, I don't formulate them and predict the next transit. Here's another chance for me to become a fixed star - but as usual - oh, well –

A point in Intermediatism:

That the Intermediatist is likely to be a flaccid compromiser.

Our own attitude:

Ours is a partly positive and partly negative state, or a state in which nothing is finally positive or finally negative –

But, if positivism attract you, go ahead and try: you will be in harmony with cosmic endeavor - but Continuity will resist you.

Only to have appearance in quasiness is to be proportionately positive, but beyond a degree of attempted positivism, Continuity will rise to pull you back. Success, as it is called though there is only success-failure in Intermediateness - will, in Intermediateness, be yours proportionately as you are in adjustment with its own state, or some positivism mixed with compromise and retreat. To be very positive is to be a Napoleon Bonaparte, against whom the rest of civilization will sooner or later combine. For interesting data, see newspaper accounts of fate of one Dowie, of Chicago.

Intermediatism, then, is recognition that our state is only a quasi-state: it is no bar to one who desires to be positive: it is recognition that he can not be positive and remain in a state that is positive-negative. Or that a great positivist – isolated - with no system to support him - will be crucified, or will starve to death, or will be put in jail and beaten to death - that these are the birth-pangs of translation to the Positive Absolute.

So, though positive-negative, myself, I feel the attraction of the positive pole of our intermediate state, and attempt to correlate these three data: to see them homogeneously; to think that they relate to one object.

In the aeronautic journals, and in the London Times there is no mention of escaped balloons, in the summer or fall of 1898. In the New York Times there is no mention of ballooning in Canada or the United States, in the summer of 1898.

London Times, Sept. 29, 1885:

A clipping from the Royal Gazette, of Bermuda, of Sept. 8, 1885, sent to the Times by General Lefroy:

That, upon Aug. 27, 1885, at about 8:30 a. m., there was observed by Mrs. Adelina D. Bassett, "a strange object in the clouds, coming from the north." She called the attention of Mrs. L. Lowell to it, and they were both somewhat alarmed. However, they continued to watch the object steadily for some time. It drew nearer. It was of triangular shape, and seemed to be about the size of a pilot-boat mainsail, with chains attached to the bottom of it. While crossing the land it had appeared to descend, but, as it went out to sea, it ascended, and continued to ascend, until it was lost to sight high in the clouds.

Or with such power to ascend, I don't think much myself of the notion that it was an escaped balloon, partly deflated. Nevertheless, General Lefroy, correlating with Exclusionism, attempts to give a terrestrial interpretation to this occurrence. He argues that the thing may have been a balloon that had escaped from France or England - or the only aerial thing of terrestrial origin that, even to this date of about thirty-five years later, has been thought to have crossed the Atlantic Ocean. He accounts for the triangular form by deflation - "a shapeless bag, barely able to float." My own acceptance is that great deflation does not accord with observations upon its power to ascend.

In the Times, Oct. 1, 1885, Charles Harding, of the R. M. S., argues that if it had been a balloon from Europe, surely it would have been seen and reported by many vessels. Whether he was as good a Briton as the General or not, he shows awareness of the United States - or that the thing may have been a partly collapsed balloon that had escaped from the United States.

General Lefroy wrote to Nature about it (Nature, 33-99), saying - whatever his sensitivenesses may have been - that the columns of the Times were "hardly suitable" for such a discussion. If, in the past, there had been more persons like General Lefroy, we'd have better than the mere fragments of data that in most cases are too broken up very well to piece together. He took the trouble to write to a friend of his, W. H. Gosling, of Bermuda - who also was an extraordinary person. He went to the trouble of interviewing Mrs. Bassett and Mrs. Lowell. Their description to him was somewhat different:

An object from which nets were suspended –

Deflated balloon, with its network hanging from it –

A super-dragnet?

That something was trawling overhead?

The birds of Baton Rouge.

Mr. Gosling wrote that the item of chains, or suggestion of a basket that had been attached, had originated with Mr. Bassett, who had not seen the object. Mr. Gosling mentioned a balloon that had escaped from Paris in July. He tells of a balloon that fell in Chicago, Sept. 17, or three weeks later than the Bermuda object.

It's one incredibility against another, with disregards and convictions governed by whichever of the two Dominants looms stronger in each reader's mind. That he can't think for himself any more than I can is understood.

My own correlates:

I think that we're fished for. It may be that we're highly esteemed by super-epicures somewhere. It makes me more cheerful when I think that we may be of some use after all. I think that dragnets have often come down and have been mistaken for whirlwinds and waterspouts. Some accounts of seeming structure in whirlwinds and waterspouts are astonishing. And I have data that, in this book, I can't take up at all - mysterious disappearances. I think we're fished for. But this is a little expression on the side: relates to trespassers; has nothing to do with the subject that I shall take up at some other time - or our use to some other mode of seeming that has a legal right to us.

Nature, 33-137:

"Our Paris correspondent writes that in relation to the balloon which is said to have been seen over Bermuda, in September, no ascent took place in France which can account for it."

Last of August: not September. In the London Times there is no mention of balloon ascents in Great Britain, in the summer of 1885, but mention of two ascents in France. Both balloons had escaped. In L'Aeronauts, Aug., 1885, it is said that these balloons had been sent up from fetes of the fourteenth of July - 44 days before the observation at Bermuda. The aeronauts were Gower and Eloy. Gower's balloon was found floating on the ocean, but Eloy's balloon was not found. Upon the 17th of July it was reported by a sea captain: still in the air; still inflated.

But this balloon of Eloy's was a small exhibition balloon, made for short ascents from fetes and fair grounds. In La Nature, 1885 – 2-131, it is said that it was a very small balloon, incapable of remaining long in the air.

As to contemporaneous ballooning in the United States, I find only one account: an ascent in Connecticut, July 29, 1885. Upon leaving this balloon, the aeronauts had pulled the "rip cord," "turning it inside out." (N. Y. Times, Aug. 10, 1885.)

To the Intermediatist, the accusation of "anthropomorphism" is meaningless. There is nothing in anything that is unique or positively different. We'd be materialists were it not quite as rational to express the material in terms of the immaterial as to express the immaterial in terms of the material. Oneness of allness in quasiness. I will engage to write the formula of any novel in psychochemic terms, or draw its graph in psycho-mechanic terms: or write, in romantic terms, the circumstances and sequences of any chemic or electric or magnetic reaction: or express any historic event in algebraic terms - or see Boole and Jevons for economic situations expressed algebraically.

I think of the Dominants as I think of persons - not meaning that they are real persons - not meaning that we are real persons –

Or the Old Dominant and its jealousy, and its suppression of all things and thoughts that endangered its supremacy. In reading discussions of papers, by scientific societies, I have often noted how, when they approached forbidden - or irreconcilable - subjects, the discussions were thrown into confusion and ramification. It's as if scientific discussions have often been led astray - as if purposefully - as if by something directive, hovering over them. Of course I mean only the Spirit of all Development. Just so, in any embryo, cells that would tend to vary from the appearances of their era are compelled to correlate.

In Nature, 90-169, Charles Tilden Smith writes that, at Chisbury, Wiltshire, England, April 8, 1912, he saw something in the sky –

"-unlike anything that I had ever seen before."

"Although I have studied the skies for many years, I have never seen anything like it."

He saw two stationary dark patches upon clouds.

The extraordinary part:

They were stationary upon clouds that were rapidly moving.

They were fan-shaped - or triangular - and varied in size, but kept the same position upon different clouds as cloud after cloud came along. For more than half an hour Mr. Smith watched these dark patches –

His impression as to the one that appeared first:

That it was "really a heavy shadow cast upon a thin veil of clouds by some unseen object away in the west, which was intercepting the sun's rays."

Upon page 244, of this volume of Nature, is a letter from another correspondent, to the effect that similar shadows are cast by mountains upon clouds, and that no doubt Mr. Smith was right in attributing the appearance to "some unseen object, which was intercepting the sun's rays." But the Old Dominant that was a jealous Dominant, and the wrath of the Old Dominant against such an irreconcilability as large, opaque objects in the sky, casting down shadows upon clouds. Still the Dominants are suave very often, or are not absolute gods, and the way attention was led away from this subject is an interesting study in quasi-divine bamboozlement.

Upon page 268, Charles J. P. Cave, the meteorologist, writes that, upon April 5 and 8, at Ditcham Park, Petersfield, he had observed a similar appearance, while watching some pilot balloons but he describes something not in the least like a shadow on clouds, but a stationary cloud the inference seems to be that the shadows at Chisbury may have been shadows of pilot balloons. Upon page 322, another correspondent writes upon shadows cast by mountains; upon page 348 some one else carries on the divergence by discussing this third letter: then some one takes up the third letter mathematically; and then there is a correction of error in this mathematic demonstration - I think it looks very much like what I think it looks like.

But the mystery here:

That the dark patches at Chisbury could not have been cast by stationary pilot balloons that were to the west, or that were between clouds and the setting sun. If, to the west of Chisbury, a stationary object were high in the air, intercepting the sun's rays, the shadow of the stationary object would not have been stationary, but would have moved higher and higher with the setting of the sun.

I have to think of something that is in accord with no other data whatsoever:

A luminous body - not the sun - in the sky but, because of some unknown principle or atmospheric condition, its light extended down only about to the clouds; that from it were suspended two triangular objects, like the object that was seen in Bermuda; that it was this light that fell short of the earth that these objects intercepted; that the objects were drawn up and lowered from something overhead, so that, in its light, their shadows changed size.

If my grope seem to have no grasp in it, and, if a stationary balloon will, in half an hour, not cast a stationary shadow from the setting sun, we have to think of two triangular objects that accurately maintained positions in a line between sun and clouds, and at the same time approached and receded from clouds. Whatever it may have been, it's enough to make the devout make the sign of the crucible, or whatever the devotees of the Old Dominant do in the presence of a new correlate.

Vast, black thing poised like a crow over the moon.

It is our acceptance that these two shadows of Chisbury looked, from the moon, like vast things, black as crows, poised over the earth. It is our acceptance that two triangular luminosities and then two triangular patches, like vast black things, poised like crows over the moon, and, like the triangularites at Chisbury, have been seen upon, or over, the moon:

Scientific American, 46-49:

Two triangular, luminous appearances reported by several observers in Lebanon, Conn., evening of July 3, 1882, on the moon's upper limb. They disappeared, and two dark triangular appearances that looked like notches were seen three minutes later upon the lower limb. They approached each other, met and instantly disappeared.

The merger here is notches that have at times been seen upon the moon's limb: thought to be cross sections of craters (Monthly Notices, R. A. S., 37-432). But these appearances of July 3, 1882, were vast upon the moon - "seemed to be cutting off or obliterating nearly a quarter of its surface."

Something else that may have looked like a vast black crow poised over this earth from the moon:

Monthly Weather Review, 41-599:

Description of a shadow in the sky, of some unseen body, April 8, 1913, Fort Worth, Texas - supposed to have been cast by an unseen cloud - this patch or shade moved with the declining sun.

Rept. Brit. Assoc., 1854-410:

Account by two observers of a faint but distinctly triangular object, visible for six nights in the sky. It was observed from two stations that were not far apart. But the parallax was considerable. Whatever it was, it was, acceptably, relatively close to this earth.

I should say that relatively to phenomena of light we are in confusion as great as some of the discords that orthodoxy is in relatively to light. Broadly and intermediatistically, our position is:

That light is not really and necessarily light - any more than is anything else really and necessarily anything - but an interpretation of a mode of force, as I suppose we have to call it, as light. At sea level, the earth's atmosphere interprets sunlight as red or orange or yellow. High up on mountains the sun is blue. Very high up on mountains the zenith is black. Or it is orthodoxy to say that in inter-planetary space, where there is no air, there is no light. So then the sun and comets are black, but this earth's atmosphere, or, rather, dust particles in it, interpret radiations from these black objects as light.

We look up at the moon.

The jet-black moon is so silvery white.

I have about fifty notes indicating that the moon has atmosphere: nevertheless most astronomers hold out that the moon has no atmosphere. They have to: the theory of eclipses would not work out otherwise. So, arguing in conventional terms, the moon is black. Rather astonishing - explorers upon the moon - stumbling and groping in intense darkness - with telescopes powerful enough, we could see them stumbling and groping in brilliant light.

Or, just because of familiarity, it is not now obvious to us how the preposterousnesses of the old system must have seemed to the correlates of the system preceding it.

Ye jet-black silvery moon.

Altogether, then, it may be conceivable that there are phenomena of force that are interpretable as light as far down as the clouds, but not in denser strata of air, or just the opposite of familiar interpretations.

I now have some notes upon an occurrence that suggests a force not interpreted by air as light, but interpreted, or reflected by the ground as light. I think of something that, for a week, was suspended over London: of an emanation that was not interpreted as light until it reached the ground.

Lancet, June 1, 1867:

That every night for a week, a light had appeared in Woburn Square, London, upon the grass of a small park, enclosed by railings. Crowds gathering - police called out "for the special service of maintaining order and making the populace move on." The Editor of the Lancet went to the Square. He says that he saw nothing but a patch of light falling upon an arbor at the northeast corner of the enclosure. Seems to me that that was interesting enough.

In this Editor we have a companion for Mr. Symons and Dr. Gray. He suggests that the light came from a street lamp - does not say that he could trace it to any such origin himself - but recommends that the police investigate neighboring street lamps.

I'd not say that such a commonplace as light from a street lamp would not attract and excite and deceive great crowds for a week - but I do accept that any cop who was called upon for extra work would have needed nobody's suggestion to settle that point the very first thing.

Or that something in the sky hung suspended over a London Square for a week.

Chapter XXI

Knowledge, Dec. 28, 1883:

Seeing so many meteorological phenomena in your excellent paper, Knowledge, I am tempted to ask for an explanation of the following, which I saw when on board the British India Company's steamer Patna, while on a voyage up the Persian Gulf. In May, 1880, on a dark night, about 11:30 p. m., there suddenly appeared on each side of the ship an enormous luminous wheel, whirling around, the spokes of which seemed to brush the ship along. The spokes would be 200 or 300 yards long, and resembled the birch rods of the dames' schools. Each wheel contained about sixteen spokes, and, although the wheels must have been some 500 or 600 yards in diameter, the spokes could be distinctly seen all the way round. The phosphorescent gleam seemed to glide along flat on the surface of the sea, no light being visible in the air above the water. The appearance of the spokes could be almost exactly represented by standing in a boat and flashing a bull's eye lantern horizontally along the surface of the water, round and round. I may mention that the phenomenon was also seen by Captain Avern, of the Patna, and Mr. Manning, third officer.

"Lee Fore Brace.

"P. S. The wheels advanced along with the ship for about twenty minutes. L. F. B."

Knowledge, Jan. u, 1884:

Letter from "A. Mc. D.":

That "Lee Fore Brace," "who sees 'so many meteorological phenomena in your excellent paper/ should have signed himself 'The Modern Ezekiel,' for his vision of wheels is quite as wonderful as the prophet's." The writer then takes up the measurements that were given, and calculates a velocity at the circumference of a wheel, of about 166 yards per second, apparently considering that especially incredible. He then says: "From the nom de plume he assumes, it might be inferred that your correspondent is in the habit of 'sailing close to the wind/ " He asks permission to suggest an explanation of his own. It is that before 11:30 p. m. there had been numerous accidents to the "main brace," and that it had required splicing so often that almost any ray of light would have taken on a rotary motion.

In Knowledge, Jan. 25, 1884, Mr. "Brace" answers and signs himself "J.W. Robertson":

"I don't suppose A. Mc. D. means any harm, but I do think it's rather unjust to say a man is drunk because he sees something out of the common. If there's one thing I pride myself upon, it's being able to say that never in my life have I indulged in anything stronger than water." From this curiosity of pride, he goes on to say that he had not intended to be exact, but to give his impressions of dimensions and velocity. He ends amiably: "However, 'no offense taken, where I suppose none is meant.' "

To this letter Mr. Proctor adds a note, apologizing for the publication of "A. Me. D's." letter, which had come about by a misunderstood instruction. Then Mr. Proctor wrote disagreeable letters, himself, about other persons - what else would you expect in a quasi-existence?

The obvious explanation of this phenomenon is that, under the surface of the sea, in the Persian Gulf, was a vast luminous wheel: that it was the light from its submerged spokes that Mr. Robertson saw, shining upward. It seems clear that this light did shine upward from origin below the surface of the sea. But at first it is not so clear how vast luminous wheels, each the size of a village, ever got under the surface of the Persian Gulf: also there may be some misunderstanding as to what they were doing there.

A deep-sea fish, and its adaptation to a dense medium –

That, at least in some regions aloft, there is a medium dense even to gelatinousness –

A deep-sea fish, brought to the surface of the ocean: in a relatively attenuated medium, it disintegrates –

Super-constructions adapted to a dense medium in inter-planetary space - sometimes, by stresses of various kinds, they are driven into this earth's thin atmosphere –

Later we shall have data to support just this: that things entering this earth's atmosphere disintegrate and shine with a light that is not the light of incandescence: shine brilliantly, even if cold –

Vast wheel-like super-constructions - they enter this earth's atmosphere, and, threatened with disintegration, plunge for relief into an ocean, or into a denser medium.

Of course the requirements now facing us are:

Not only data of vast wheel-like super-constructions that have relieved their distresses in the ocean, but data of enormous wheels that have been seen in the air, or entering the ocean, or rising from the ocean and continuing their voyages.

Very largely we shall concern ourselves with enormous fiery obi jects that have either plunged into the ocean or risen from the ocean. Our acceptance is that, though disruption may intensify into incandescence, apart from disruption and its probable fieriness, things that enter this earth's atmosphere have a cold light which would not, like light from molten matter, be instantly quenched by water. Also it seems acceptable that a revolving wheel would, from a distance, look like a globe; that a revolving wheel, seen relatively close by, looks like a wheel in few aspects. The mergers of ball-lightning and meteorites are not resistances to us: our data are of enormous bodies.

So we shall interpret - and what does it matter?

Our attitude throughout this book:

That here are extraordinary data that they never would be exhumed, and never would be massed together, unless –

Here are the data:

Our first datum is of something that was once seen to enter an ocean. It's from the puritanic publication, Science, which has yielded us little material, or which, like most puritans, does not go upon a spree very often. Whatever the thing could have been, my impression is of tremendousness, or of bulk many times that of all meteorites in all museums combined: also of relative slowness, or of long warning of approach. The story, in Science, 5-242, is from an account sent to the Hydrographic Office, at Washington, from the branch office, at San Francisco:

That, at midnight, Feb. 24, 1885, Lat. 37° N., and Long. 170° E., or somewhere between Yokohama and Victoria, the captain of the bark Innerwich was aroused by his mate, who had seen something unusual in the sky. This must have taken appreciable time. The captain went on deck and saw the sky turning fiery red. "All at once, a large mass of fire appeared over the vessel, completely blinding the spectators." The fiery mass fell into the sea. Its size may be judged by the volume of water cast up by it, said to have rushed toward the vessel with a noise that was "deafening." The bark was struck flat aback, and "a roaring, white sea passed ahead." "The master, an old, experienced mariner, declared that the awfulness of the sight was beyond description."

In Nature, 37-187, and L'Astronomie, 1887-76, we are told that an object, described as "a large ball of fire," was seen to rise from the sea, near Cape Race. We are told that it rose to a height of fifty feet, and then advanced close to the ship, then moving away, remaining visible about five minutes. The supposition in Nature is that it was "ball lightning," but Flammarion, "Thunder and Lightning," p. 68, says that it was enormous. Details in the American Meteorological Journal, 6-443 - Nov. 12, 1887 - British steamer Siberian - that the object had moved "against the wind" before retreating - that Captain Moore said that at about the same place he had seen such appearances before.

Report of the British Association, 1861-30:

That, upon June 18, 1845, according to the Malta Times, from the brig Victoria, about 900 miles east of Adalia, Asia Minor (36° 40' 56", N. Lat: 13° 44' 36" E. Long.) three luminous bodies were seen to issue from the sea, at about half a mile from the vessel. They were visible about ten minutes. The story was never investigated, but other accounts that seem acceptably to be other observations upon this same sensational spectacle came in, as if of their own accord, and were published by Prof. Baden-Powell. One is a letter from a correspondent at Mt. Lebanon. He describes only two luminous bodies. Apparently they were five times the size of the moon: each had appendages, or they were connected by parts that are described as "sail-like or streamer-like," looking like "large flags blown out by a gentle breeze."

The important point here is not only suggestion of structure, but duration. The duration of meteors is a few seconds: duration of fifteen seconds is remarkable, but I think there are records up to half a minute. This object, if it were all one object, was visible at Mt. Lebanon about one hour. An interesting circumstance is that the appendages did not look like trains of meteors, which shine by their own light, but "seemed to shine by light from the main bodies."

About 900 miles west of the position of the Victoria is the town of Adalia, Asia Minor. At about the time of the observation reported by the captain of the Victoria, the Rev. F. Hawlett, F. R. A. S., was in Adalia. He, too, saw this spectacle, and sent an account to Prof. Baden-Powell. In his view it was a body that appeared and then broke up. He places duration at twenty minutes to half an hour. In the Report of the British Association, 1860-82, the phenomenon was reported from Syria and Malta, as two very large bodies "nearly joined."

Rept. Brit. Assoc., 1860-77:

That, at Cherbourg, France, Jan. 12, 1836, was seen a luminous body, seemingly two-thirds the size of the moon. It seemed to rotate on an axis. Central to it there seemed to be a dark cavity. For other accounts, all indefinite, but distortable into data of wheel-like objects in the sky, see Nature, 22-617; London Times, Oct. 15, 1859; Nature, 21-225; Monthly Weather Review, 1883-264.

L'Astronomie, 1894-157:

That, upon the morning of Dec. 20, 1893, an appearance in the sky was seen by many persons in Virginia, North Carolina, and South Carolina. A luminous body passed overhead, from west to east, until at about 15 degrees in the eastern horizon, it appeared to stand still for fifteen or twenty minutes. According to some descriptions it was the size of a table. To some observers it looked like an enormous wheel. The light was a brilliant white. Acceptably it was not an optical illusion - the noise of its passage through the air was heard. Having been stationary, or having seemed to stand still fifteen or twenty minutes, it disappeared, or exploded. No sound of explosion was heard.

Vast wheel-like constructions. They're especially adapted to roll through a gelatinous medium from planet to planet. Sometimes, because of miscalculations, or because of stresses of various kinds, they enter this earth's atmosphere. They're likely to explode. They have to submerge in the sea. They stay in the sea a while, revolving with relative leisureliness, until relieved, and then emerge, sometimes close to vessels. Seamen tell of what they see: their reports are interred in scientific morgues. I should say that the general route of these constructions is along latitudes not far from the latitudes of the Persian Gulf.

Journal of the Royal Meteorological Society, 28-29:

That, upon April 4, 1901, about 8:30, in the Persian Gulf, Captain Hoseason, of the steamship Kilwa, according to a paper read before the Society by Captain Hoseason, was sailing in a sea in which there was no phosphorescence - "there being no phosphorescence in the water."

I suppose I'll have to repeat that:

"... there being no phosphorescence in the water."

Vast shafts of light - though the Captain uses the word "ripples" - suddenly appeared. Shaft followed shaft, upon the surface of the sea. But it was only a faint light, and, in about fifteen minutes, died out: having appeared suddenly; having died out gradually. The shafts revolved at a velocity of about 60 miles an hour.

Phosphorescent jelly fish correlate with the Old Dominant: in one of the most heroic compositions of disregards in our experience, it was agreed, in the discussion of Capt. Hoseason's paper, that the phenomenon was probably pulsations of long strings of jelly fish.

Nature, 21-410:

Reprint of a letter from R. E. Harris, Commander of the A. H. N. Co.'s steamship Shahjehan, to the Calcutta Englishman, Jan. 21, 1880:

That upon the 5th of June, 1880, off the coast of Malabar, at 10 p.m., water calm, sky cloudless, he had seen something that was so foreign to anything that he had ever seen before, that he had stopped his ship. He saw what he describes as waves of brilliant light, with spaces between. Upon the water were floating patches of a substance that was not identified. Thinking in terms of the conventional explanation of all phosphorescence at sea, the Captain at first suspected this substance. However, he gives his opinion that it did no illuminating but was, with the rest of the sea, illuminated by tremendous shafts of light. Whether it was a thick and oily discharge from the engine of a submerged construction or not, I think that I shall have to accept this substance as a concomitant, because of another note. "As wave succeeded wave, one of the most grand and brilliant, yet solemn, spectacles that one could think of, was here witnessed."

Jour. Roy. Met. Soc., 32-280:

Extract from a letter from Mr. Douglas Carnegie, Blackheath, England. Date some time in 1906 –

"This last voyage we witnessed a weird and most extraordinary electric display." In the Gulf of Oman, he saw a bank of apparently quiescent phosphorescence: but, when within twenty yards of it, "shafts of brilliant light came sweeping across the ship's bows at a prodigious speed, which might be put down as anything between 60 and 200 miles an hour." "These light bars were about 20 feet apart and most regular." As to phosphorescence - "I collected a bucketful of water, and examined it under the microscope, but could not detect anything abnormal." That the shafts of light came up from something beneath the surface - "They first struck us on our broadside, and I noticed that an intervening ship had no effect on the light beams: they started away from the lee side of the ship, just as if they had traveled right through it."

The Gulf of Oman is at the entrance to the Persian Gulf.

Jour. Roy. Met. Soc., 33-294:

Extract from a letter by Mr. S. C. Patterson, second officer of the P. and O. steamship Delta: a spectacle which the Journal continues to call phosphorescent:

Malacca Strait, 2 a.m., March 14, 1907:

"... shafts which seemed to move round a center - like the spokes of a wheel - and appeared to be about 300 yards long." The phenomenon lasted about half an hour, during which time the ship had traveled six or seven miles. It stopped suddenly,"

L'Astronomie, 1891-312:

A correspondent writes that, in October, 1891, in the China Sea, he had seen shafts or lances of light that had had the appearance of rays of a searchlight, and that had moved like such rays.

Nature, 20-291:

Report to the Admiralty by Capt. Evans, the Hydrographer of the British Navy:

That Commander J.E. Pringle, of H.M.S. Vulture, had reported that, at Lat. 26° 26' N., and Long. 53° 11' E.- in the Persian Gulf May 15, 1879, he had noticed luminous waves or pulsations in the water, moving at great speed. This time we have a definite datum upon origin somewhere below the surface. It is said that these waves of light passed under the Vulture. "On looking toward the east, the appearance was that of a revolving wheel with a center on that bearing, and whose spokes were illuminated, and, looking toward the west, a similar wheel appeared to be revolving, but in the opposite direction. Or finally as to submergence - "These waves of light extended from the surface well under the water." It is Commander Pringle's opinion that the shafts constituted one wheel, and that doubling was an illusion. He judges the shafts to have been about 25 feet broad, and the spaces about 100. Velocity about 84 miles an hour. Duration about 35 minutes. Time 9:40 p.m. Before and after this display the ship had passed through patches of floating substance described as "oily-looking fish spawn."

Upon page 428 of this number of Nature, E.L. Moss says that, in April, 1875, when upon H.M.S. Bulldog, a few miles north of Vera Cruz, he had seen a series of swift lines of light. He had dipped up some of the water, finding in it animalcule, which would, however, not account for phenomena of geometric formation and high velocity. If he means Vera Cruz, Mexico, this is the only instance we have out of oriental waters.

Scientific American, 106-51:

That, in the Nautical Meteorological Annual, published by the Danish Meteorological Institute, appears a report upon a "singular phenomenon" that was seen by Capt. Gabe, of the Danish East Asiatic Co.'s steamship Bintang. At 3 a.m., June 10, 1909, while sailing through the Straits of Malacca, Captain Gabe saw a vast revolving wheel of light, flat upon the water "long arms issuing from a center around which the whole system appeared to rotate." So vast was the appearance that only half of it could be seen at a time, the center lying near the horizon. This display lasted about fifteen minutes. Heretofore we have not been clear upon the important point that forward motions of these wheels do not synchronize with a vessel's motions, and freaks of disregard, or, rather, commonplaces of disregard, might attempt to assimilate with lights of a vessel. This time we are told that the vast wheel moved forward, decreasing in brilliancy, and also in speed of rotation, disappearing when the center was right ahead of the vessel - or my own interpretation would be that the source of light was submerging deeper and deeper and slowing down because meeting more and more resistance.

The Danish Meteorological Institute reports another instance:

That, when Capt. Breyer, of the Dutch steamer Valentijn, was in the South China Sea, midnight, Aug. 12, 1910, he saw a rotation of flashes. "It looked like a horizontal wheel, turning rapidly." This time it is said that the appearance was above water. "The phenomenon was observed by the captain, the first and second mates, and the first engineer, and upon all of them it made a somewhat uncomfortable impression."

In general, if our expression be not immediately acceptable, we recommend to rival interpreters that they consider the localization - with one exception - of this phenomenon, to the Indian Ocean and adjacent waters, or Persian Gulf on one side and China Sea on the other side. Though we're Intermediatists, the call of attempted Positivism, in the aspect of Completeness, is irresistible. We have expressed that from few aspects would wheels of fire in the air look like wheels of fire, but, if we can get it, we must have observation upon vast luminous wheels, not interpretable as optical illusions, 266 BOOK OF THE DAMNED but enormous, substantial things that have smashed down material resistances, and have been seen to plunge into the ocean:

Athenceum, 1848-833:

That at the meeting of the British Association, 1848, Sir W. S. Harris said that he had recorded an account sent to him of a vessel toward which had whirled "two wheels of fire, which the men described as rolling millstones of fire." "When they came near, an awful crash took place: the topmasts were shivered to pieces." It is said that there was a strong sulphurous odor.

Chapter XXII

　　Journal of the Royal Meteorological Society, 1-157:

Extract from the log of the barque Lady of the Lake, by Capt. F.W. Banner:

Communicated by R.H. Scott, F.R.S.:

That, upon the 22nd of March, 1870, at Lat. 5° 47' N., Long. 27 52' W., the sailors of the Lady of the Lake saw a remarkable object, or "cloud," in the sky. They reported to the Captain.

According to Capt. Banner, it was a cloud of circular form, with an included semicircle divided into four parts, the central dividing shaft beginning at the center of the circle and extending far outward, and then curving backward.

Geometricity and complexity and stability of form: and the small likelihood of a cloud maintaining such diversity of features, to say nothing of appearance of organic form.

The thing traveled from a point at about 20 degrees above the horizon to a point about 80 degrees above. Then it settled down to the northeast, having appeared from the south, southeast.

Light gray in color, or it was cloud-color.

"It was much lower than the other clouds."

And this datum stands out:

That, whatever it may have been, it traveled against the wind.

"It came up obliquely against the wind, and finally settled down right in the wind's eye."

For half an hour this form was visible. When it did finally disappear that was not because it disintegrated like a cloud, but because it was lost to sight in the evening darkness.

Capt. Banner draws the following diagram:

Chapter XXIII

　　Text-books tell us that the Dhurmsalla meteorites were picked up "soon," or "within half an hour." Given a little time the conventionalists may argue that these stones were hot when they fell, but that their great interior coldness had overcome the molten state of their surfaces.

According to the Deputy Commissioner of Dhurmsalla, these stones had been picked up "immediately" by passing coolies. These stones were so cold that they benumbed the fingers. But they had fallen with a great light. It is described as "a flame of fire about two feet in depth and nine feet in length." Acceptably this light was not the light of molten matter.

In this chapter we are very intermediatistic - and unsatisfactory.

To the intermediatist there is but one answer to all questions:

Sometimes and sometimes not.

Another form of this intermediatist "solution" of all problems is:

Yes and no.

Everything that is, also isn't.

A positivist attempts to formulate: so does the intermediatist, but with less rigorousness: he accepts but also denies: he may seem to accept in one respect and deny in some other respect, but no real line can be drawn between any two aspects of anything. The intermediatist accepts that which seems to correlate with something that he has accepted as a dominant. The positivist correlates with a belief. In the Dhurmsalla meteorites we have support for our expression that things entering this earth's atmosphere sometimes shine with a light that is not the light of incandescence - or so we account, or offer an expression upon, "thunderstones," or carved stones that have fallen luminously to this earth, in streaks that have looked like strokes of lightning - but we accept, also, that some things that have entered this earth's atmosphere, disintegrate with the intensity of flame and molten matter - but some things, we accept, enter this earth's atmosphere and collapse non-luminously, quite like deep-sea fishes brought to the surface of the ocean. Whatever 268 BOOK OF THE DAMNED 269 agreement we have is an indication that somewhere aloft there is a medium denser than this earth's atmosphere. I suppose our stronghold is in that such is not popular belief –

Or the rhythm of all phenomena:

Air dense at sea level upon this earth - less and less dense as one ascends - then denser and denser. A good many bothersome questions arise –

Our attitude:

Here are the data:

Luminous rains sometimes fall (Nature, March 9, 1882; Nature, 25-437). This is light that is not the light of incandescence, but no one can say that these occasional, or rare, rains come from this earth's externality. We simply note cold light of falling bodies.

For luminous rain, snow, and dust, see Hartwig, "Aerial World," p. 319. As to luminous clouds, we have more nearly definite observations and opinions: they mark transition between the Old Dominant and the New Dominant. We have already noted the transition in Prof. Schwedoff's theory of external origin of some hailstones - and the implications that, to a former generation, seemed so preposterous - "droll" was the word - that there are in interplanetary regions volumes of water whether they have fishes and frogs in them or not. Now our acceptance is that clouds sometimes come from external regions, having had origin from super-geographical lakes and oceans that we shall not attempt to chart, just at present - only suggesting to enterprising aviators - and we note that we put it all up to them, and show no inclination to go Columbussing on our own account - that they take bathing suits, or, rather, deep-sea-diving-suits along. So then that some clouds come from inter-planetary oceans - of the Super-Sargasso Sea if we still accept the Super-Sargasso Sea and shine, upon entering this earth's atmosphere. In Himmel und Erde, Feb., 1889 a phenomenon of transition of thirty years ago - Herr O. Jesse, in his observations upon luminous night-clouds, notes the great height of them, and drolly or sensibly suggests that some of them may have come from regions external to this earth. I suppose he means only from other planets. But it's a very droll and sensible idea either way.

In general I am accounting for a great deal of this earth's isolation: that it is relatively isolated by circumstances that are similar to the circumstances that make for relative isolation of the bottom of the ocean - except that there is a clumsiness of analogy now. To call ourselves deep-sea fishes has been convenient, but, in a quasi-existence, there is no convenience that will not sooner or later turn awkward - so, if there be denser regions aloft, these regions should now be regarded as analogues of far-submerged oceanic regions, and things coming to this earth would be like things rising to an attenuated medium - and exploding - sometimes incandescently, sometimes with cold light - sometimes non-luminously, like deep-sea fishes brought to the surface - altogether conditions of inhospitality. I have a suspicion that, in their own depths, deep-sea fishes are not luminous. If they are, Darwinism is mere Jesuitism, in attempting to correlate them. Such advertising would so attract attention that all advantages would be more than offset. Darwinism is largely a doctrine of concealment: here we have brazen proclamation - if accepted. Fishes in the Mammoth Cave need no light to see by. We might have an expression that deep-sea fishes turn luminous upon entering a less dense medium - but models in the American Museum of Natural History: specialized organs of luminosity upon these models. Of course we do remember that awfully convincing "dodo," and some of our sophistications we trace to him - at any rate disruption is regarded as a phenomenon of coming from a dense to a less dense medium.

An account by M. Acharius, in the Transactions of the Swedish Academy of Sciences, 1808-215, translated for the North American Review, 3-319:

That M. Acharius, having heard of "an extraordinary and probably hitherto unseen phenomenon," reported from near the town of Skeninge, Sweden, investigated:

That, upon the 16th of May, 1808, at about 4 p.m., the sun suddenly turned dull brick-red. At the same time there appeared, upon the western horizon, a great number of round bodies, dark brown, and seemingly the size of a hat crown. They passed overhead and disappeared in the eastern horizon. Tremendous procession. It lasted two hours. Occasionally one fell to the ground. When the place of a fall was examined, there was found a film, which soon dried and vanished. Often, when approaching the sun, these bodies seemed to link together, or were then seen to be linked together, in groups not exceeding eight, and, under the sun, they were seen to have tails three or four fathoms long. Away from the sun the tails were invisible. Whatever their substance may have been, it is described as gelatinous - "soapy and jellied."

I place this datum here for several reasons. It would have been a good climax to our expression upon hordes of small bodies that, in our acceptance, were not seeds, nor birds, nor ice-crystals: but the tendency would have been to jump to the homogeneous conclusion that all our data in that expression related to this one kind of phenomena, whereas we conceive of infinite heterogeneity of the external: of crusaders and rabbles and emigrants and tourists and dragons and things like gelatinous hat crowns. Or that all things, here, upon this earth, that flock together, are not necessarily sheep, Presbyterians, gangsters, or porpoises. The datum is important to us, here, as indication of disruption in this earth's atmosphere - dangers in entering this earth's atmosphere.

I think, myself, that thousands of objects have been seen to fall from aloft, and have exploded luminously, and have been called "ball lightning."

"As to what ball lightning is, we have not yet begun to make intelligent guesses." (Monthly Weather Review, 34-17.)

In general, it seems to me that when we encounter the opposition "ball lightning" we should pay little attention, but confine ourselves to guesses that are at least intelligent, that stand phantomlike in our way. We note here that in some of our acceptances upon intelligence we should more clearly have pointed out that they were upon the intelligent as opposed to the instinctive. In the Monthly Weather Review, 33-409, there is an account of "ball lightning" that struck a tree. It made a dent such as a falling object would make. Some other time I shall collect instances of "ball lightning," to express that they are instances of objects that have fallen from the sky, luminously, exploding terrifically. So bewildered is the old orthodoxy by these phenomena that many scientists have either denied "ball lightning" or have considered it very doubtful. I refer to Dr. Sestier's list of one hundred and fifty instances, which he considered authentic.

In accord with our disaccord is an instance related in the Monthly Weather Review, March, 1887 something that fell luminously from the sky, accompanied by something that was not so affected, or that was dark:

That, according to Capt. C.D. Sweet, of the Dutch bark, J.P.A., upon March 19, 1887, N. 37° 39', W. 57° 00', he encountered a severe storm. He saw two objects in the air above the ship. One was luminous, and might be explained in several ways, but the other was dark. One or both fell into the sea, with a roar and the casting up of billows. It is our acceptance that these things had entered this earth's atmosphere, having first crashed through a field of ice - "immediately afterward lumps of ice fell."

One of the most astonishing of the phenomena of "ball lightning" is a phenomenon of many meteorites: violence of explosion out of all proportion to size and velocity. We accept that the icy meteorites of Dhurmsalla could have fallen with no great velocity, but the sound from them was tremendous. The soft substance that fell at the Cape of Good Hope was carbonaceous, but was unburned, or had fallen with velocity insufficient to ignite it. The tremendous report that it made was heard over an area more than seventy miles in diameter.

That some hailstones have been formed in a dense medium, and violently disintegrate in this earth's relatively thin atmosphere:

Nature, 88-350:

Large hailstones noted at the University of Missouri, Nov. n, 1911: they exploded with sounds like pistol shots. The writer says that he had noticed a similar phenomenon, eighteen years before, at Lexington, Kentucky. Hailstones that seemed to have been formed in a denser medium: when melted under water they gave out bubbles larger than their central air spaces. (Monthly Weather Review, 33-445.)

Our acceptance is that many objects have fallen from the sky, but that many of them have disintegrated violently. This acceptance will coordinate with data still to come, but, also, we make it easy for ourselves in our expressions upon super-constructions, if we're asked why, from thinkable wrecks of them, girders, plates, or parts recognizably of manufactured metal have not fallen from the sky. However, as to composition, we have not this refuge, so it is our expression that there have been reported instances of the fall of manufactured metal from the sky.

The meteorite of Rutherford, North Carolina, is of artificial material: mass of pig iron. It is said to be fraudulent. (Amer. Jour. Sci., 2-34-298.)

The object that was said to have fallen at Marblehead, Mass., in 1858, is described in the Amer. Jour.

Sci., 2-34-135, as "a furnace product, formed in smelting copper ores, or iron ores containing copper." It is said to be fraudulent.

According to Ehrenberg, the substance reported by Capt. Callam to have fallen upon his vessel, near Java, "offered complete resemblance to the residue resulting from combustion of a steel wire in a flask of oxygen." (Zurcher, "Meteors" p. 239.) Nature, Nov. 21, 1878, publishes a notice that, according to the Yuma Sentinel, a meteorite that "resembles steel" had been found in the Mohave Desert. In Nature, Feb. 15, 1894, we read that one of the meteorites brought to the United States by Peary, from Greenland, is of tempered steel. The opinion is that meteoric iron had fallen in water or snow, quickly cooling and hardening. This does not apply to composition. Nov. 5, 1898, Nature publishes a notice of a paper by Prof. Berwerth, of Vienna, upon "the close connection between meteoric iron and steel-works' steel.

At the meeting of Nov. 24, 1906, of the Essex Field Club, was exhibited a piece of metal said to have fallen from the sky, Oct. 9, 1906, at Braintree. According to the Essex Naturalist, Dr. Fletcher, of the British Museum, had declared this metal to be smelted iron - so that the mystery of its reported 'fall' remained unexplained."

Chapter XXIV

We shall have an outcry of silences. If a single instance of anything be disregarded by a System - our own attitude is that a single instance is a powerless thing. Of course our own method of agreement of many instances is not a real method. In Continuity, all things must have resemblances with all other things. Anything has any quasi-identity you please. Some time ago conscription was assimilated with either autocracy or democracy with equal facility. Note the need for a dominant to correlate to. Scarcely anybody said simply that we must have conscription: but that we must have conscription, which correlates with democracy/, which was taken as a base, or something basically desirable. Of course between autocracy and democracy nothing but false demarcation can be drawn. So I can conceive of no subject upon which there should be such poverty as a single instance, if anything one pleases can be whipped into line. However, we shall try to be more nearly real than the Darwinites who advance concealing coloration as Darwinism, and then drag in proclaiming luminosity, too, as Darwinism. I think the Darwinites had better come in with us as to the deep-sea fishes - and be sorry later, I suppose. It will be amazing or negligible to read all the instances now to come of things that have been seen in the sky, and to think that all have been disregarded. My own opinion is that it is not possible, or very easy, to disregard them, now that they have been brought together - but that, if prior to about this time we had attempted such an assemblage, the Old Dominant would have withered our typewriter - as it is the letter "e" has gone back on us, and the "s" is temperamental. "Most extraordinary and singular phenomenon," North Wales, Aug. 26, 1894; a disk from which projected an orange-colored body that looked like "an elongated flatfish," reported by Admiral Ommanney (Nature, 50-524) ; disk from which projected a hook-like form, India, about 1838; diagram of it given; disk about size of the moon, but brighter than the moon; visible about twenty minutes; by G. Pettit, in Prof. Baden-Powell's Catalogue (Rept. Brit. Assoc., 1849); very brilliant hook-like form, seen in the sky at Poland, Trumbull Co., Ohio, during the stream of meteors, of 1833; visible more than an hour: large luminous body, almost stationary "for a time"; shaped like a square table; Niagara Falls, Nov. 13, 1833 (Amer. Jour. Sci., 1-25-391); something described as a bright white cloud, at night, Nov. 3, 1886, at Hamar, Norway; from it were emitted brilliant rays of light; drifted across the sky; "retained throughout its original form" (Nature, Dec. 16, 1886-158); thing with an oval nucleus, and streamers with dark bands and lines very suggestive of structure; New Zealand, May 4, 1888 (Nature, 42-402); luminous object, size of full moon, visible an hour and a half, Chili, Nov. 5, 1883 (Comptes Rendus, 103-682); bright object near sun, Dec. 21, 1882 (Knowledge, 3-13); light that looked like a great flame, far out at sea, off Ryook Phyoo, Dec. 2, 1845 (London Roy. Soc. Proc., 5-627); something like a gigantic trumpet, suspended, vertical, oscillating gently, visible five or six minutes, length estimated at 425 feet, at Oaxaca, Mexico, July 6, 1874 (Sci. Am. Sup., 6-2365); two luminous bodies, seemingly united, visible five or six minutes, June 3, 1898 (La Nature, 1898-1-127); thing with a tail, crossing moon, transit half a minute, Sept. 26, 1870 (London Times, Sept. 30, 1870); object four or five times size of moon, moving slowly across sky, Nov. 1, 1885, near Adrianople (L'Astronomie, 1886-309) ; large body, colored red, moving slowly, visible 15 minutes, reported by Coggia, Marseilles, Aug. 1, 1871 (Chem. News, 24-193); details of this observation, and similar observation by Guillemin, and other instances by de Fonville (Comptes Rendus, 73-297, 755); thing that was large and that was stationary twice in seven minutes, Oxford, Nov. 19, 1847; listed by Lowe (Rec. Sci., 1-136) ; grayish object that looked to be about three and a half feet long, rapidly approaching the earth at Saarbruck, April 1, 1826; sound like thunder; object expanding like a sheet (Am. Jour. Sci., 1-26-133; Quar. Jour. Roy. Inst., 24-488); report by an astronomer, N. S. Drayton upon an object duration of which seemed to him extraordinary; duration three-quarters of a minute, Jersey City, July 6, 1882 (Sci. Amer., 47-53); object like a comet, but with proper motion of 10 degrees an hour; visible one hour; reported by Purine and Clancy from the Cordoba Observatory, Argentine, March 14, 1916 (Sci. Amer., 115-493); something like a signal light, reported by Glaisher, Oct. 4, 1844; bright as Jupiter, "sending out quick flickering waves of light" (Year Book of Facts, 1845-278).

I think that with the object known as Eddie's "comet," passes away the last of our susceptibility to the common fallacy of personifying. It is one of the most deep-rooted of positivist illusions - that people are persons. We have been guilty too often of spleens and spites and ridicules against astronomers, as if they were persons, or final unities, individuals, completenesses, or selves - instead of indeterminate parts. But, so long as we remain in quasi-existence, we can cast out illusion only with some other illusion, though the other illusion may approximate higher to reality. So we personify no more - but we super-personify. We now take into full acceptance our expression that Development is an Autocracy of Successive Dominants - which are not final - but which approximate higher to individuality or self-ness, than do the human tropisms that irresponsibly correlate to them.

Eddie reported a celestial object, from the Observatory at Grahamstown, South Africa. It was in 1890. The New Dominant was only heir presumptive then, or heir apparent but not obvious. The thing that Eddie reported might as well have been reported by a night watchman, who had looked up through an unplaced sewer pipe.

It did not correlate.

The thing was not admitted to Monthly Notices. I think myself that if the Editor had attempted to let it in – earthquake - or a mysterious fire in his publishing house.

The Dominants are jealous gods.

In Nature, presumably a vassal of the new god, though of course also plausibly rendering homage to the old, is reported a cometlike body, of Oct. 27, 1890, observed at Grahamstown, by Eddie. It may have looked comet-like, but it moved 100 degrees while visible, or one hundred degrees in three-quarters of an hour. See Nature, 43-89, 90.

In Nature, 44-519, Prof. Copeland describes a similar appearance that he had seen, Sept. 10, 1891. Dreyer says (Nature, 44 – 541) that he had seen this object at the Armagh Observatory. He likens it to the object that was reported by Eddie. It was seen by Dr. Alexander Graham Bell, Sept. 11, 1891, in Nova Scotia.

But the Old Dominant was a jealous god.

So there were different observations upon something that was seen in November, 1883. These observations were Philistines in 1883. In the Amer. Met. Jour., 1 - 110, a correspondent reports having seen an object like a comet, with two tails, one up and one down, Nov. 10 or 12, 1883. Very likely this phenomenon should be placed in our expression upon torpedo-shaped bodies that have been seen in the sky our data upon dirigibles, or super-Zeppelins - but our attempted classifications are far from rigorous - or are mere gropes. In the Scientific American, 50-40, a correspondent writes from Humacao, Porto Rico, that, Nov. 21, 1883, he and several other – persons - or persons, as it were - had seen a majestic appearance, like a comet. Visible three successive nights: disappeared then. The Editor says that he can offer no explanation. If accepted, this thing must have been close to the earth. If it had been a comet, it would have been seen widely, and the news would have been telegraphed over the world, says the Editor. Upon page 07 of this volume of the Scientific American, a correspondent writes that, at Sulphur Springs, Ohio, he had seen "a wonder in the sky," at about the same date. It was torpedo-shaped, or something with a nucleus, at each end of which was a tail. Again the Editor says that he can offer no explanation: that the object was not a comet. He associates it with the atmospheric effects general in 1883. But it will be our expression that, in England and Holland, a similar object was seen in November, 1882.

In the Scientific American, 40-294, is published a letter from Henry Harrison, of Jersey City, copied from the N.Y. Tribune: that upon the evening of April 13, 1879, Mr. Harrison was searching for Brorsen's comet, when he saw an object that was moving so rapidly that it could not have been a comet. He called a friend to look, and his observation was confirmed. At two o'clock in the morning this object was still visible. In the Scientific American Supplement, 7-2885, Mr. Harrison disclaims sensationalism, which he seems to think unworthy, and gives technical details: he says that the object was seen by Mr. J. Spencer Devoe, of Manhattanville.

Chapter XXV

A formation having the shape of a dirigible." It was reported from Huntington, West Virginia (Sci. Amer., 115-241). Luminous object that was seen July 19, 1916, at about eleven p. m. Observed through "rather powerful field glasses," it looked to be about two degrees long and half a degree wide. It gradually dimmed, disappeared, reappeared, and then faded out of sight. Another person - as we say: it would be too inconvenient to hold to our intermediatist recognitions - another person who observed this phenomenon suggested to the writer of the account that the object was a dirigible, but the writer says that faint stars could be seen behind it. This would seem really to oppose our notion of a dirigible visitor to this earth - except for the inconclusiveness of all things in a mode of seeming that is not final - or we suggest that behind some parts of the object, thing, construction, faint stars were seen. We find a slight discussion here. Prof. H. M. Russell thinks that the phenomenon was a detached cloud of aurora borealis. Upon page 369 of this volume of the Scientific American, another correlator suggests that it was a light from a blast furnace - disregarding that, if there be blast furnaces in or near Huntington, their reflections would be commonplaces there.

We now have several observations upon cylindrical-shaped bodies that have appeared in this earth's atmosphere: cylindrical, but pointed at both ends, or torpedo-shaped. Some of the accounts are not very detailed, but out of the bits of description my own acceptance is that super-geographical routes are traversed by torpedo-shaped super-constructions that have occasionally visited, or that have occasionally been driven into this earth's atmosphere. From data, the acceptance is that upon entering this earth's atmosphere, these vessels have been so racked that had they not sailed away, disintegration would have occurred: that, before leaving this earth, they have, whether in attempted communication or not, or in mere wantonness or not, dropped objects, which did almost immediately violently disintegrate or explode. Upon general principles we think that explosives have not been purposely dropped, but that parts have been racked off, and have fallen, exploding like the things called "ball lightning." May have been objects of stone or metal with inscriptions upon them, for all we know, at present. In all instances, estimates of dimensions are valueless, but ratios of dimensions are more acceptable. A thing said to have been six feet long may have been six hundred feet long; but shape is not so subject to the illusions of distance.

Nature, 40-415:

That, Aug. 5, 1889, during a violent storm, an object that looked to be about 15 inches long and 5 inches wide, fell, rather slowly, at East Twickenham, England. It exploded. No substance from it was found.

L'Année Scientifique, 1864-54:

That, Oct. 10, 1864, M. Leverrier had sent to the Academy three letters from witnesses of a long luminous body, tapering at both ends, that had been seen in the sky.

In Thunder and Lightning, p. 87, Flammarion says that on Aug. 20, 1880, during a rather violent storm, M. A. Trécul, of the French Academy, saw a very brilliant yellowish-white body, apparently 35 to 40 centimeters long, and about 25 centimeters wide. Torpedo-shaped., Or a cylindrical body, "with slightly conical ends." It dropped something, and disappeared in the clouds. Whatever it may have been that was dropped, it fell vertically, like a heavy object, and left a luminous train. The scene of this occurrence may have been far from the observer. No sound was heard. For M. Trécul's account, see Comptes Rendus, 103-849.

Monthly Weather Review, 1907-310:

That, July 2, 1907, in the town of Burlington, Vermont, a terrific explosion had been heard throughout the city. A ball of light, or a luminous object, had been seen to fall from the sky or from a torpedo-shaped thing, or construction, in the sky. No one had seen this thing that had exploded fall from a larger body that was in the sky but if we accept that at the same time there was a larger body in the sky –

My own acceptance is that a dirigible in the sky, or a construction that showed every sign of disrupting, bad barely time to drop - whatever it did drop - and to speed away to safety above.

The following story is told, in the Review, by Bishop John S. Michaud:

"I was standing on the corner of Church and College Streets, just in front of the Howard Bank, and facing east, engaged in conversation with Ex-Governor Woodbury and Mr. A. A. Buell, when, without the slightest indication, or warning, we were startled by what sounded like a most unusual and terrific explosion, evidently very nearby. Raising my eyes, and looking eastward along College street, I observed a torpedo-shaped body, some 300 feet away, stationary in appearance, and suspended in the air, about 50 feet above the tops of the buildings. In size it was about 6 feet long by 8 inches in diameter, the shell, or covering, having a dark appearance, with here and there tongues of fire issuing from spots on the surface, resembling red-hot, unburnished copper. Although stationary when first noticed, this object soon began to move, rather slowly, and disappeared over Dolan Brothers' store, southward. As it moved, the covering seemed rupturing in places, and through these the intensely red flames issued."

Bishop Michaud attempts to correlate it with meteorological observations.

Because of the nearby view this is perhaps the most remarkable of the new correlates, but the correlate now coming is extraordinary because of the great number of recorded observations upon it. My own acceptance is that, upon Nov. 17, 1882, a vast dirigible crossed England, but by the definiteness-indefiniteness of all things quasi-real, some observations upon it can be correlated with anything one pleases.

E. W. Maunder, invited by the Editors of the Observatory to write some reminiscences for the sooth number of their magazine, gives one that he says stands out (Observatory, 39-214). It is upon something that he terms "a strange celestial visitor." Maunder was at the Royal Observatory, Greenwich, Nov. 17, 1882, at night. There was an aurora, without features of special interest. In the midst of the aurora, a great circular disk of greenish light appeared and moved smoothly across the sky. But the circularity was evidently the effect of foreshortening. The thing passed above the moon, and was, by other observers, described as "cigar-shaped," "like a torpedo," "a spindle," "a shuttle." The idea of foreshortening is not mine: Maunder says this. He says: "Had the incident occurred a third of a century later, beyond doubt every one would have selected the same simile - it would have been 'just like a Zeppelin.' " The duration was about two minutes. Color said to have been the same as that of the auroral glow in the north. Nevertheless, Maunder says that this thing had no relation to auroral phenomena. "It appeared to be a definite body." Motion too fast for a cloud, but "nothing could be more unlike the rush of a meteor." In the Philosophical Magazine, 5-15-318, J. Rand Capron, in a lengthy paper, alludes throughout to this phenomenon as an "auroral beam," but he lists many observations upon its "torpedo-shape," and one observation upon a "dark nucleus" in it host of most confusing observations - estimates of height between 40 and 200 miles - observations in Holland and Belgium. We are told that according to Capron's spectroscopic observations the phenomenon was nothing but a beam of auroral light. In the Observatory, 6-192, is Maunder's contemporaneous account. He gives apparent approximate length and breadth at twenty-seven degrees and three degrees and a half. He gives other observations seeming to indicate structure - "remarkable dark marking down the center."

In Nature, 27-84, Capron says that because of the moonlight he had been able to do little with the spectroscope.

Color white, but aurora rosy (Nature, 27-87).

Bright stars seen through it, but not at the zenith, where it looked opaque. This is the only assertion of transparency (Nature, 27-87). Too slow for a meteor, but too fast for a cloud (Nature, 27-86). "Surface had a mottled appearance" (Nature, 27-87).

"Very definite in form, like a torpedo" (Nature, 27-100). "Probably a meteoric object" (Dr. Groneman, Nature, 27-296). Technical demonstration by Dr. Groneman, that it was a cloud of meteoric matter (Nature, 28-105). See Nature, 27-315, 338, 365, 388, 412, 434 –

"Very little doubt it was an electric phenomenon" (Proctor, Knowledge, 2-419).

In the London Times, Nov. 20, 1882, the Editor says that he had received a great number of letters upon this phenomenon. He publishes two. One correspondent describes it as "well-defined and shaped like a fish . . . extraordinary and alarming." The other correspondent writes of it as "a most magnificent luminous mass, shaped somewhat like a torpedo."

Chapter XXVI

Notes and Queries, 5-3-306:

About 8 lights that were seen in Wales, over an area of about 8 miles, all keeping their own ground, whether moving together perpendicularly, horizontally, or over a zigzag course. They looked like electric lights - disappearing, reappearing dimly, then shining as bright as ever. "We have seen them three or four at a time afterward, on four or five occasions."

London Times, Oct. 5, 1877:

"From time to time the west coast of Wales seems to have been the scene of mysterious lights. . . . And now we have a statement from Towyn that within the last few weeks lights of various colors have been seen moving over the estuary of the Dysynni River, and out to sea. They are generally in a northerly direction, but sometimes they hug the shore, and move at high velocity for miles toward Aberdovey, and suddenly disappear.

L'Annee Scientifique, 1877-45:

Lights that appeared in the sky, above Vence, France, March 23, 1877; described as balls of fire of dazzling brightness; appeared from a cloud about a degree in diameter; moved relatively slowly.

They were visible more than an hour, moving northward. It is said that eight or ten years before similar lights or objects had been seen in the sky, at Vence.

London Times, Sept. 19, 1848:

That, at Inverness, Scotland, two large, bright lights that looked like stars had been seen in the sky: sometimes stationary, but occasionally moving at high velocity.

L'Année Scientifique, 1888-66:

Observed near St. Petersburg, July 30, 1880, in the evening: a large spherical light and two smaller ones, moving along a ravine: visible three minutes; disappearing without noise.

Nature, 35-173 –

That, at Yloilo, Sept. 30, 1886, was seen a luminous object the size of the full moon. It "floated" slowly "northward," followed by smaller ones close to it. 282

"The False Lights of Durham."

Every now and then in the English newspapers, in the middle of the nineteenth century, there is something about lights that were seen against the sky, but as if not far above land, oftenest upon the coast of Durham. They were mistaken for beacons by sailors. Wreck after wreck occurred. The fishermen were accused of displaying false lights and profiting by wreckage. The fishermen answered that mostly only old vessels, worthless except for insurance, were so wrecked.

In 1866 (London Times, Jan. 9, 1866) popular excitement became intense. There was an investigation. Before a commission, headed by Admiral Collinson, testimony was taken. One witness described the light that had deceived him as "considerably elevated above ground." No conclusion was reached: the lights were called "the mysterious lights." But whatever the "false lights of Durham" may have been, they were unaffected by the investigation. In 1867, the Tyne Pilotage Board took the matter up. Opinion of the Mayor of Tyne - "a mysterious affair."

In the Report of the British Association, 1877-152, there is a description of a group of "meteors" that traveled with "remarkable slowness." They were in sight about three minutes. "Remarkable," it seems, is scarcely strong enough: one reads of "remarkable" as applied to a duration of three seconds. These "meteors" had another peculiarity; they left no train. They are described as "seemingly huddled together like a flock of wild geese, and moving with the same velocity and grace of regularity."

Jour. Roy. Astro. Soc. of Canada, Nov. and Dec., 1913:

That, according to many observations collected by Prof. Chant, of Toronto, there appeared, upon the night of Feb. 9, 1913, a spectacle that was seen in Canada, the United States, and at sea, and in Bermuda. A luminous body was seen. To it there was a long tail. The body grew rapidly larger. "Observers differ as to whether the body was single, or was composed of three or four parts, with a tail to each part." The group, or complex structure, moved with "a peculiar, majestic deliberation." "It disappeared in the distance, and another group emerged from its place of origin. Onward they moved, at the same deliberate pace, in twos or threes or fours." They disappeared. A third group, or a third structure, followed. Some observers compared the spectacle to a fleet of airships: others to battleships attended by cruisers and destroyers.

According to one writer:

"There were probably 30 or 32 bodies, and the peculiar thing about them was their moving in fours and threes and twos, abreast of one another; and so perfect was the lining up that you would have thought it was an aerial fleet maneuvering after rigid drilling."

Nature, May 25, 1893:

A letter from Capt. Charles J. Norcock, of H.M.S. Caroline:

That, upon the 24th of February, 1893, at 10 p.m., between Shanghai and Japan, the officer of the watch had reported "some unusual lights."

They were between the ship and a mountain. The mountain was about 6,000 feet high. The lights seemed to be globular. They moved sometimes massed, but sometimes strung out in an irregular line. They bore "northward," until lost to sight. Duration two hours.

The next night the lights were seen again.

They were, for a time, eclipsed by a small island. They bore north at about the same speed and in about the same direction as speed and direction of the Caroline. But they were lights that cast a reflection: there was a glare upon the horizon under them. A telescope brought out but few details: that they were reddish, and seemed to emit a faint smoke. This time the duration was seven and a half hours.

Then Capt. Norcock says that, in the same general locality, and at about the same time, Capt. Castle, of H.M.S. Leander, had seen lights. He had altered his course and had made toward them. The lights had fled from him. At least, they had moved higher in the sky.

Monthly Weather Review, March, 1904-115:

Report from the observations of three members of his crew by Lieut. Frank H. Schofield, U.S.N., of the U.S.S. Supply:

Feb. 24, 1904. Three luminous objects, of different sizes, the largest having an apparent area of about six suns. When first sighted, they were not very high. They were below clouds of an estimated height of about one mile.

They fled, or they evaded, or they turned.

They went up into the clouds below which they had, at first, been sighted. Their unison of movement.

But they were of different sizes, and of different susceptibilities to all forces of this earth and of the air.

Monthly Weather Review, Aug., 1898-358:

Two letters from C.N. Crotsenburg, Crow Agency, Montana:

That, in the summer of 1896, when this writer was a railroad postal clerk - or one who was experienced in train-phenomena - while his train was going "northward," from Trenton, Mo., he and another clerk saw, in the darkness of a heavy rain, a light that appeared to be round, and of a dull-rose color, and seemed to be about a foot in diameter. It seemed to float within a hundred feet of the earth, but soon rose high, or "midway between horizon and zenith." The wind was quite strong from the east, but the light held a course almost due north.

Its speed varied. Sometimes it seemed to outrun the train "considerably." At other times it seemed to fall behind. The mail clerks watched until the town of Linville, Iowa, was reached. Behind the depot of this town, the light disappeared, and was not seen again. All this time there had been rain, but very little lightning, but Mr. Crotsenburg offers the explanation that it was "ball lightning."

The Editor of the Review disagrees. He thinks that the light may have been a reflection from the rain, or fog, or from leaves of trees, glistening with rain, or the train's light - not lights.

In the December number of the Review is a letter from Edward M. Boggs - that the light was a reflection, perhaps, from the glare - one light, this time - from the locomotive's fire-box, upon wet telegraph wires - an appearance that might not be striated by the wires, but consolidated into one rotundity - that it had seemed to oscillate with the undulations of the wires, and had seemed to change horizontal distance with the varying angles of reflection, and had seemed to advance or fall behind, when the train had rounded curves.

All of which is typical of the best of quasi-reasoning. It includes and assimilates diverse data: but it excludes that which will destroy it:

That, acceptably, the telegraph wires were alongside the track beyond, as well as leading to Linville.

Mr. Crotsenburg thinks of "ball lightning," which, though a sore bewilderment to most speculation, is usually supposed to be a correlate with the old system of thought: but his awareness of "something else" is expressed in other parts of his letters, when he says that he has something to tell that is "so strange that I should never have mentioned it, even to my friends, had it not been corroborated ... so unreal that I hesitated to speak of it, fearing that it was some freak of the imagination."

Chapter XXVII

Vast and black. The thing that was poised, like a crow over the moon.

Round and smooth. Cannon balls. Things that have fallen from the sky to this earth.

Our slippery brains.

Things like cannon balls have fallen, in storms, upon this earth.

Like cannon balls are things that, in storms, have fallen to this earth.

Showers of blood.

Showers of blood.

Showers of blood.

Whatever it may have been, something like red-brick dust, or a red substance in a dried state, fell at Piedmont, Italy, Oct. 27, 1814 (Electric Magazine, 68-437). A red powder fell, in Switzerland, winter of 1867 (Pop. Sci. Rev., 10-112) –

That something, far from this earth had bled super-dragon that had rammed a comet –

Or that there are oceans of blood somewhere in the sky - substance that dries, and falls in a powder - wafts for ages in powdered form - that there is a vast area that will some day be known to aviators as the Desert of Blood. We attempt little of super-topography, at present, but Ocean of Blood, or Desert of Blood - or both - Italy is nearest to it - or to them.

I suspect that there were corpuscles in the substance that fell in Switzerland, but all that could be published in 1867 was that in this substance there was a high proportion of "variously shaped organic matter."

At Giessen, Germany, in 1821, according to the Report of the British Association, 5-2, fell a rain of a peach-red color. In this rain were flakes of a hyacinthine tint. It is said that this substance was organic: we are told that it was pyrrhine.

But distinctly enough, we are told of one red rain that it was of corpuscular composition - red snow, rather. It fell, March 12, 1876, near the Crystal Palace, London (Year Book of Facts 1876 - 89; Nature, 13-414). As to the "red snow" of polar and mountainous regions, we have no opposition, because that "snow" has never been seen to fall from the sky: it is a growth of microorganisms, or of a "protococcus," that spreads over snow that is on the ground. This time nothing is said of "sand from the Sahara."

It is said of the red matter that fell in London, March 12, 1876, that it was composed of corpuscles

–

Of course:

That they looked like "vegetable cells."

A note:

That nine days before had fallen the red substance - flesh - whatever it may have been - of Bath County, Kentucky. I think that a super-egotist, vast, but not so vast as it had supposed, had refused to move to one side for a comet.

We summarize our general super-geographical expressions:

Gelatinous regions, sulphurous regions, frigid and tropical regions: a region that has been Source of Life relatively to this earth: regions wherein there is density so great that things from them, entering this earth's thin atmosphere, explode. We have had a datum of explosive hailstones. We now have support to the acceptance that they had been formed in a medium far denser than air of this earth at sea-level. In the Popular Science News, 22-38, is an account of ice that had been formed, under great pressure, in the laboratory of the University of Virginia. When released, and brought into contact with ordinary air, this ice exploded.

And again the flesh-like substance that fell in Kentucky: its flake-like formation. Here is a phenomenon that is familiar to us: it suggests flattening, under pressure. But the extraordinary inference is - pressure not equal on all sides. In the Annual Record of Science, 1873-350, it is said that, in 1873, after a heavy thunderstorm in Louisiana, a tremendous number of fish scales were found, for a distance of forty miles, along the banks of the Mississippi River: bushels of them picked up in single places: large scales that were said to be of the gar fish, a fish that weighs from five to fifty pounds. It seems impossible to accept this identification: one thinks of a substance that had been pressed into flakes or scales. And round hailstones with wide thin margins of ice irregularly around them - still, such hailstones seem to me more like things that had been stationary: had been held in a field of thin ice. In the Illustrated London News, 34-546, are drawings of hailstones so margined, as if they had been held in a sheet of ice.

Some day we shall have an expression which will be, to our advanced primitiveness, a great joy:

That devils have visited this earth: foreign devils: human-like beings, with pointed beards: good singers; one shoe ill-fitting - but with sulphurous exhalations, at any rate. I have been impressed with the frequent occurrence of sulphurousness with things that come from the sky. A fall of jagged pieces of ice, Orkney, July 24, 1818 (Trans. Roy. Soc. Edin., 9-187). They had a strong sulphurous odor. And the coke - or the substance that looked like coke - that fell at Mortrée, France, April 24, 1887: with it fell a sulphurous substance. The enormous round things that rose from the ocean, near the Victoria. Whether we still accept that they were super-constructions that had come from a denser atmosphere and, in danger of disruption, had plunged into the ocean for relief, then rising and continuing on their way to Jupiter or Uranus - it was reported that they spread a "stench of sulphur." At any rate, this datum of proximity is against the conventional explanation that these things did not rise from the ocean, but rose far away above the horizon, with illusion of nearness.

And the things that were seen in the sky July, 1898: I have another note. In Nature, 58-224, a correspondent writes that, upon July 1, 1898, at Sedberg, he had seen in the sky - a red object - or, in his own wording, something that looked like the red part of a rainbow, about 10 degrees long. But the sky was dark at the time. The sun had set. A heavy rain was falling.

Throughout this book, the datum that we are most impressed with:

Successive falls.

Or that, if upon one small area, things fall from the sky, and then, later, fall again upon the same small area, they are not products of a whirlwind, which though sometimes axially stationary, discharges tangentially –

So the frogs that fell at Wigan. I have looked that matter up again. Later more frogs fell.

As to our data of gelatinous substance said to have fallen to this earth with meteorites, it is our expression that meteorites, tearing through the shaky, protoplasmic seas of Genesistrine - against which we warn aviators, or they may find themselves suffocating in a reservoir of life, or stuck like currants in a blanc mange - that meteorites detach gelatinous, or protoplasmic, lumps that fall with them.

Now the element of positiveness in our composition yearns for the appearance of completeness. Super-geographical lakes with fishes in them. Meteorites that plunge through these lakes, on their way to this earth. The positiveness in our make-up must have egression in at least one record of a meteorite that has brought down a lot of fishes with it –

Nature, 3-512:

That, near the bank of a river, in Peru, Feb. 4, 1871, a meteorite fell. "On the spot, it is reported, several dead fishes were found, of different species." The attempt to correlate - is that the fishes "are supposed to have been lifted out of the river and dashed against the stones."

Whether this be imaginable or not depends upon each one's own hypnoses.

Nature, 4-169:

That the fishes had fallen among the fragments of the meteorite.

Popular Science Review, 4-126:

That one day, Mr. Le Gould, an Australian scientist, was traveling in Queensland. He saw a tree that had been broken off close to the ground. Where the tree had been broken was a great bruise. Near by was an object that "resembled a ten-inch shot."

A good many pages back there was an instance of overshadowing, I think. The little carved stone that fell at Tarbes is my own choice as the most impressive of our new correlates. It was coated with ice, remember. Suppose we should sift and sift and discard half the data in this book - suppose only that one datum should survive. To call attention to the stone of Tarbes would, in my opinion, be doing well enough, for whatever the spirit of this book is trying to do. Nevertheless, it seems to me that a datum that preceded it was slightingly treated.

The disk of quartz, said to have fallen from the sky, after a meteoric explosion:

Said to have fallen at the plantation Bleijendal, Dutch Guiana: sent to the Museum of Leyden by M. van Sypesteyn, adjutant to the Governor of Dutch Guiana (Notes and Queries, 2-8-92). And the fragments that fall from super-geographic ice fields: flat pieces of ice with icicles on them. I think that we did not emphasize enough that, if these structures were not icicles, but crystalline protuberances, such crystalline formations indicate long suspension quite as notably as would icicles. In the Popular Science News, 24-34, it is said that in 1869, near Tiflis, fell large hailstones with long protuberances. "The most remarkable point in connection with the hailstones is the fact that, judging from our present knowledge, a very long time must have been occupied in their formation." According to the Geological Magazine, 7-27, this fall occurred May 27, 1869. The writer in the Geological Magazine says that of all theories that he had ever heard of, not one could give him light as to this occurrence - "these growing crystalline forms must have been suspended a long time" –

Again and again this phenomenon:

Fourteen days later, at about the same place, more of these hailstones fell.

Rivers of blood that vein albuminous seas, or an egg-like composition, in the incubation of which this earth is a local center of development - that there are super-arteries of blood in Genesistrine: that sunsets are consciousness of them: that they flush the skies with northern lights sometimes: super-embryonic reservoirs from which life-forms emanate –

Or that our whole solar system is a living thing: that showers of blood upon this earth are its internal hemorrhages –

Or vast living things in the sky, as there are vast living things in the oceans –

Or some one especial thing: an especial time: an especial place.

A thing the size of the Brooklyn Bridge. It's alive in outer space - something the size of Central Park kills it –

It drips.

We think of the ice fields above this earth: which do not, themselves, fall to this earth, but from which water does fall –

Popular Science News, 35-104:

That, according to Prof. Luigi Palazzo, head of the Italian Meteorological Bureau, upon May 15, 1890, at Messignadi, Calabria, something the color of fresh blood fell from the sky.

This substance was examined in the public-health laboratories of Rome.

It was found to be blood.

"The most probable explanation of this terrifying phenomenon is that migratory birds (quails or swallows) were caught and torn in a violent wind."

So the substance was identified as birds' blood –

What matters it what the microscopists of Rome said - or had to say - and what matters it that we point out that there is no assertion that there was a violent wind at the time - and that such a substance would be almost infinitely dispersed in a violent wind - that no bird was said to have fallen from the sky or said to have been seen in, the sky that not a feather of a bird is said to have been seen-–

This one datum:

The fall of blood from the sky –

But later, in the same place, blood again fell from the sky.

Chapter XXVIII

Notes and Queries, 7-8-508:

A correspondent who had been to Devonshire writes for information as to a story that he had heard there: of an occurrence of about thirty-five years before the date of writing:

Of snow upon the ground - of all South Devonshire waking up one morning to find such tracks in the snow as had never before been heard of - "clawed footmarks" or "an unclassifiable form" - alternating at huge but regular intervals with what seemed to be the impression of the point of a stick but the scattering of the prints - amazing expanse of territory covered - obstacles, such as hedges, walls, houses, seemingly surmounted –

Intense excitement that the track had been followed by huntsmen and hounds, until they had come to a forest - from which the hounds had retreated, baying and terrified, so that no one had dared to enter the forest.

Notes and Queries, 7-9-18:

Whole occurrence well-remembered by a correspondent: a badger had left marks hi the snow: this was determined, and the excitement had "dropped to a dead calm in a single day."

Notes and Queries, 7-9-70:

That for years a correspondent had had a tracing of the prints, which his mother had taken from those in the snow in her garden, in Exmouth: that they were hoof-like marks but had been made by a biped.

Notes and Queries, 7-9-253:

Well remembered by another correspondent, who writes of the excitement and consternation of "some classes." He says that a kangaroo had escaped from a menagerie - "the footprints being so peculiar and far apart gave rise to a scare that the devil was loose."

We have had a story, and now we shall tell it over from contemporaneous sources. We have had the later accounts first very largely for an impression of the correlating effect that time brings about, by addition, disregard and distortion. For instance, the "dead calm in a single day." If I had found that the excitement did die out rather soon, I'd incline to accept that nothing extraordinary had occurred.

I found that the excitement had continued for weeks.

I recognize this as a well-adapted thing to say, to divert attention from a discorrelate.

All phenomena are "explained" in the terms of the Dominant of their era. This is why we give up trying really to explain, and content ourselves with expressing. Devils that might print marks in snow are correlates to the third Dominant back from this era. So it was an adjustment by nineteenth-century correlates, or human tropisms, to say that the marks in the snow were clawed. Hooflike marks are not only horsey but devilish. It had to be said in the nineteenth century that those prints showed claw-marks. We shall see that this was stated by Prof. Owen, one of the greatest biologists of his day - except that Darwin didn't think so. But I shall give reference to two representations of them that can be seen in the New York Public Library. In neither representation is there the faintest suggestion of a claw-mark. There never has been a Prof. Owen who has explained: he has correlated. Another adaptation, in the later accounts, is that of leading this discorrelate to the Old Dominant into the familiar scenery of a fairy story, and discredit it by assimilation to the conventionally fictitious - so the idea of the baying, terrified hounds, and forest like enchanted forests, which no one dared to enter. Hunting parties were organized, but the baying, terrified hounds do not appear in contemporaneous accounts.

The story of the kangaroo looks like adaptation to needs for an animal that could spring far, because marks were found in the snow on roofs of houses. But so astonishing is the extent of snow that was marked that after a while another kangaroo was added.

But the marks were in single lines.

My own acceptance is that not less than a thousand one-legged kangaroos, each shod with a very small horseshoe could have marked that snow of Devonshire.

London Times, Feb. 16, 1855:

"Considerable sensation has been caused in the towns of Topsham, Lymphstone, Exmouth, Teignmouth, and Dawlish, in Devonshire, in consequence of the discovery of a vast number of foot tracks of a most strange and mysterious description."

The story is of an incredible multiplicity of marks discovered in the morning of Feb. 8, 1855, in the snow, by the inhabitants of many towns and regions between towns. This great area must of course be disregarded by Prof. Owen and the other correlators. The tracks were in all kinds of unaccountable places: in gardens enclosed by high walls, and up on the tops of houses, as well as in the open fields. There was in Lymphstone scarcely one unmarked garden. We've had heroic disregards but I think that here disregard was titanic. And, because they occurred in single lines, the marks are said to have been "more like those of a biped than of a quadruped" - as if a biped would place one foot precisely ahead of another - unless it hopped but then we have to think of a thousand, or of thousands.

It is said that the marks were "generally 8 inches in advance of each other."

"The impression of the foot closely resembles that of a donkey's shoe, and measured from an inch and a half, in some instances, to two and a half inches across."

Or the impressions were cones in incomplete, or crescentic basins.

The diameters equaled diameters of very young colts' hoofs: too small to be compared with marks of donkey's hoofs.

"On Sunday last the Rev. Mr. Musgrave alluded to the subject in his sermon and suggested the possibility of the footprints being those of a kangaroo, but this could scarcely have been the case, as they were found on both sides of the Este. At present it remains a mystery, and many superstitious people in the above named towns are actually afraid to go outside their doors after night."

The Este is a body of water two miles wide.

London Times, March 6, 1855:

"The interest in this matter has scarcely yet subsided, many inquiries still being made into the origin of the footprints, which caused so much consternation upon the morning of the 8th ult. In addition to the circumstances mentioned in the Times a little while ago, it may be stated that at Dawlish a number of persons sallied out, armed with guns and other weapons, for the purpose, if possible, of discovering and destroying the animal which was supposed to have been so busy in multiplying its footprints. As might have been expected, the party returned as they went. Various speculations have been made as to the cause of the footprints. Some have asserted that they are those of a kangaroo, while others affirm that they are the impressions of claws of large birds driven ashore by stress of weather. On more than one occasion reports have been circulated that an animal from a menagerie had been caught, but the matter at present is as much involved in mystery as ever it was."

In the Illustrated London News, the occurrence is given a great deal of space. In the issue of Feb. 24, 1855, a sketch is given of the prints.

I call them cones in incomplete basins.

Except that they're a little longish, they look like prints of hoofs of horses - or, rather, of colts.

But they're in a single line.

It is said that the marks from which the sketch was made were 8 inches apart, and that this spacing was regular and invariable "in every parish." Also other towns besides those named in the Times are mentioned. The writer, who had spent a winter in Canada, and was familiar with tracks in snow, says that he had never seen "a more clearly defined track." Also he brings out the point that was so persistently disregarded by Prof. Owen and the other correlators - . that "no known animal walks in a line of single footsteps, not even man." With these wider inclusions, this writer concludes with us that the marks were not footprints. It may be that his following observation hits upon the crux of the whole occurrence: That whatever it may have been that had made the marks, it had removed, rather than pressed, the snow. According to his observations the snow looked "as if branded with a hot iron."

Illustrated London News, March 3, 1855-214:

Prof. Owen, to whom a friend had sent drawings of the prints, writes that there were claw-marks. He says that the "track" was made by "a" badger. Six other witnesses sent letters to this number of the News.

One mentioned, but not published, is a notion of a strayed swan.

Always this homogeneous-seeing "a" badger "a" swan "a" track. I should have listed the other towns as well as those mentioned in the Times.

A letter from Mr. Musgrave is published. He, too, sends a sketch of the prints. It, too, shows a single line. There are four prints, of which the third is a little out of line.

There is no sign of a claw-mark.

The prints look like prints of longish hoofs of a very young colt, but they are not so definitely outlined as in the sketch of Feb. 24th, as if drawn after disturbance by wind, or after thawing had set in. Measurements at places a mile and a half apart, gave the same Inter-spacing - "exactly eight inches and a half apart."

We now have a little study in the psychology and genesis of an attempted correlation. Mr. Musgrave says: "I found a very apt opportunity to mention the name 'kangaroo' in allusion to the report then current." He says that he had no faith in the kangaroo-story himself, but was glad "that a kangaroo was in the wind," because it opposed "a dangerous, degrading, and false impression that it was the devil."

"Mine was a word in season and did good."

Whether it's Jesuitical or not, and no matter what it is or isn't, that is our own acceptance: that, though we've often been carried away from this attitude controversially, that is our acceptance as to every correlate of the past that has been considered in this book - relatively to the Dominant of its era.

Another correspondent writes that, though the prints in all cases resembled hoof marks, there were indistinct traces of claws - that "an" otter had made the marks. After that many other witnesses wrote to the News. The correspondence was so great that, in the issue of March 10th, only a selection could be given. There's "a" jumping-rat solution and "a" hopping toad inspiration, and then some one came out strong with an idea of "a" hare that had galloped with pairs of feet held close together, so as to make impressions in a single line.

London Times, March 14, 1840:

"Among the high mountains of that elevated district where Glenorchy, Glenlyon and Glenochay are contiguous, there have been met with several times, during this and also the former winter upon the snow, the tracks of an animal seemingly unknown at present in Scotland. The print, in every respect, is an exact resemblance to that of a foal of considerable size, with this small difference, perhaps, that the sole seems a little longer, or not so round; but as no one has had the good fortune as yet to have obtained a glimpse of this creature, nothing more can be said of its shape or dimensions; only it has been remarked, from the depth to which the feet sank in the snow, that it must be a beast of considerable size. It has been observed also that its walk is not like that of the generality of quadrupeds, but that it is more like the bounding or leaping of a horse when scared or pursued. It is not in one locality that its tracks have been met with, but through a range of least twelve miles."

In the Illustrated London News, March 17, 1855, a correspondent from Heidelberg writes, "upon the authority of a Polish Doctor Medicine," that on the Piashowa-gora (Sand Hill) a small elevation on the border of Galicia, but in Russian Poland, such marks are to be seen in the snow every year, and sometimes in the sand of this hill, and "are attributed by the inhabitants to supernatural influences."

New Lands

1 LANDS in the sky-- That they are nearby-- That they do not move.
I take for a principle that all being is the infinitely serial, and that whatever has been will, with differences of particulars, be again--

The last quarter of the fifteenth century--land to the west!

This first quarter of the twentieth century--we shall have revelations.

There will be data. There will be many. Behind this book, unpublished collectively, or held as constituting its reserve forces, there are other hundreds of data, but independently I take for a principle that all existence is a flux and a re-flux, by which periods of expansion follow periods of contraction; that few men can even think widely when times are narrow times, but that human constrictions cannot repress extensions of thoughts and lives and enterprise and dominion when times are wider times--so then that the pageantry of foreign coasts that was revealed behind blank horizons after the year 1492, cannot be, in the course of development, the only astounding denial of seeming vacancy-- that the spirit, or the animation, and the stimulations and the needs of the fifteenth century are all appearing again, and that requital may appear again--

Aftermath of war, as in the year 1492: demands for readjustments; crowded and restless populations, revolts against limitations, intolerable restrictions against emigrations. The young man is no longer urged, or is no longer much inclined, to go westward. He will, or must, go somewhere. If directions alone no longer invite him, he may hear invitation in dimensions. There are many persons, who have not investigated for themselves, who think that both poles of this earth have been discovered. There are too many women traveling luxuriously in "Darkest Africa." Eskimos of Disco, Greenland, are publishing a newspaper. There must be outlet, or there will be explosion--

Outlet and invitation and opportunity--

San Salvadors of the Sky--a Plymouth Rock that hangs in the heavens of Servia--a foreign coast from which storms have brought materials to the city of Birmingham, England.

Or the mentally freezing, or dying, will tighten their prohibitions, and the chill of their censorships will contract, to extinction, our lives, which, without sin, represent matter deprived of motion. Their ideal is Death, or approximate death, warmed over occasionally only enough to fringe with uniform, decorous icicles-- from which there will be no escape, if, for the living and sinful and adventurous there be not San Salvadors somewhere else, a Plymouth Rock of reversed significance, coasts of sky-continents.

But every consciousness that we have of needs, and all hosts, departments, and sub-divisions of data that indicate the possible requital of needs are opposed-- not by the orthodoxy of the common Puritans, but by the Puritans of Science, and their austere, disheartening, dried or frozen orthodoxy. Islands of space--see Sci. Amer., vol. this and p. that--accounts from the Depts. of the Brit. Assoc. for the Ad. of Sci.--Nature, etc.--except for an occasional lapse, our sources of data will not be sneered at. As to our interpretations, I consider them, myself, more as suggestions and gropings and stimuli. Islands of space and the rivers and the oceans of an extra-geography--

Stay and let salvation damn you--or straddle an auroral beam and paddle from Rigel to Betelgeuse. If there be no accepting that there are such rivers and oceans beyond this earth, stay and travel upon steamships with schedules that can be depended upon, food so well cooked and well served, comfort looked after so carefully--or someday board the thing that was seen over the city of Marseilles, Aug. 19, 1887, and ride on that, bearing down upon the moon, giving up for lost, escaping collision by the swirl of a current that was never heard of before.

There are, or there are not, nearby cities of foreign existences. They have, or they have not, been seen, by reflection, in the skies, of Sweden and Alaska. As one will. Whether acceptable, or too preposterous to be thought of, our data are of rabbles of living things that have been seen in the sky; also of processions of military beings--monsters that live in the sky and die in the sky, and spatter this earth with their red life-fluids--ships from other worlds that have been seen by millions of the inhabitants of this earth, exploring, night after night, in the sky of France, England, New England, and Canada-- signals from the moon, which, according to notable indications, may not be so far from this earth as New York is from London--definitely reported and, in some instances, multitudinously witnessed, events that have been disregarded by our opposition--

A scientific priestcraft--

"Thou shalt not!" is crystallized in its frozen textbooks.

I have data upon data upon data of new lands that are not far away. I hold out expectations and the materials of new hopes and new despairs and new triumphs and new tragedies. I hold out my hands to point to the sky--there is a hierarchy that utters me manacles, I think--there is a dominant force that pronounces prisons that have dogmas for walls for such thoughts. It binds its formulas around all attempting extensions.

But sounds have been heard in the sky. They have been heard, and it is not possible to destroy the records of them. They have been heard. In their repetitions and regularities of series and intervals, we shall recognize perhaps interpretable language. Columns of clouds, different-colored by sunset, have vibrated to the artillery of other worlds like the strings of a cosmic harp, and I conceive of no buzzing of insects that can forever divert attention from such dramatic reverberations. Language has shone upon the dark parts of the moon: luminous exclamations that have fluttered in the lunar crater Copernicus; the eloquence of the starlike light in Aristarchus; hymns that have been chanted in lights and shades upon Linne; the wilder, luminous music in Plato--

But not a sound that has been heard in the sky, not a thing that has fallen from the sky, not a thing that "should not be," but that has nevertheless been seen in the sky can we, with any sense of freedom, investigate, until first we find out about the incubus that in the past has suffocated even speculation. I shall find out for myself: anybody who cares to may find out with me. A ship from a foreign world does, or does not, sail in the sky of this earth. It is in accordance with observations by hundreds of thousands of witnesses that this event has taken place, and, if the time be when aeronautics upon this earth is of small development, that is an important circumstance to consider--but there is suffocation upon the whole occurrence and every one of its circumstances. Nobody can give good attention to the data, if diverting his mind is consciousness, altogether respectful, of the scientists who say that there are no other physical worlds except planets, millions of miles away, distances that conceivable vessels could not traverse. I should like to let loose, in an opening bombardment, the data of the little black stones of Birmingham, which, time after time, in a period of eleven years, fell obviously from a fixed point in the sky, but such a release, now, would be wasted. It will have to be prepared for. Now each one would say to himself that there are no such fixed points in the sky. Why not? Because astronomers say that there are not.

But there is something else that is implied. Implied is the general supposition that the science of astronomy represents all that is most accurate, most exacting, painstaking, semi-religious in human thought, and is therefore authoritative.

Anybody who has not been through what I've been through, in investigating this subject, would ask what are the bases and what is the consistency of the science of astronomy. The miserable, though at times amusing, confusions of thought that I find in this field of supposed research word my inquiry differently--what of dignity, or even of decency, is in it?

Phantom dogmas, with their tails clutching at vacancies, are coiled around our data.

Serpents of pseudo-thought are stifling history.

They are squeezing "Thou shalt not!" upon Development.

New Lands--and the horrors and lights, explosions and music of them; rabbles of hellhounds and the march of military angels. But they are Promised Lands, and first must we traverse a desert. There is ahead of us a waste of parallaxes and spectrograms and triangulations. It may be weary going through a waste of astronomic determinations, but that depends--

If out of a dreary, academic zenith shower betrayals of frailty, folly, and falsification, they will be manna to our malices--

Or sterile demonstrations be warmed by our cheerful cynicisms into delicious little lies--blossoms and fruits of unexpected oases--

Rocks to strike with our suspicions--and the gush of exposures foaming with new implications.

Tyrants, dragons, giants--and, if all be dispatched with the skill and the might and the triumph over awful odds of the hero who himself tells his story--

I hear three yells from some hitherto undiscovered, grotesque critter at the very entrance of the desert.

2

"PREDICTION Confirmed!" "Another Verification!"

"A Third Verification of Prediction!"

Three times, in spite of its long-established sobriety, the Journal of the Franklin Institute, vols. 106 and 107, reels with an astronomer's exhilarations. He might exult and indulge himself, and that would be no affair of ours, and, in fact, we'd like to see everybody happy, perhaps, but it is out of these three chanticleerities by Prof. Pliny Chase that we materialize our opinion that, so far as methods and strategies are concerned, no particular differences can be noted between astrologers and astronomers, and that both represent engulfment in Dark Ages. Lord Bacon pointed out that the astrologers had squirmed into prestige and emolument by shooting at marks, disregarding their misses, and recording their hits with unseemly advertisement. When, in August, 1878, Prof. Swift and Prof. Watson said that, during an eclipse of the sun, they had seen two luminous bodies that might be planets between Mercury and the sun, Prof. Chase announced that, five years before, he had made a prediction, and that it had been confirmed by the positions of these bodies. Three times, in capital letters, he screamed, or announced, according to one's sensitiveness, or prejudices, that the "new planets" were in the exact positions of his calculations. Prof. Chase wrote that, before his time, there had been two great instances of astronomic calculation confirmed: the discovery of Neptune and the discovery of "the asteroidal belt," a claim that is disingenuously worded. If by mathematical principles, or by any other definite principles, there has ever been one great, or little, instance of astronomic discovery by means of calculations, confusion must destroy us, in the introductory position that we take, or expose our irresponsibility, and vitiate all that follows: that our data are oppressed by a tyranny of false announcements; that there never has been an astronomic discovery other than the observational or the accidental.

In The Story of the Heavens, Sir Robert Ball's opinion of the discovery of Neptune is that it is a triumph unparalleled in the annals of science. He lavishes--the great astronomer Leverrier, buried for months in profound meditations--the dramatic moment--Leverrier rises from his calculations and points to the sky--"Lo!" there a new planet is found.

My desire is not so much to agonize over the single fraudulencies or delusions, as to typify the means by which the science of Astronomy has established and maintained itself:

According to Leverrier, there was a planet external to Uranus; according to Hansen, there were two; according to Airy, "doubtful if there were one."

One planet was found--so calculated Leverrier, in his profound meditations. Suppose two had been found--confirmation of the brilliant computations by Hansen. None--the opinion of the great astronomer, Sir George Airy.

Leverrier calculated that the hypothetic planet was at a distance from the sun, within the limits of 35 and 37.9 times this earth's distance from the sun. The new planet was found in a position said to be 30 times this earth's distance from the sun. The discrepancy was so great that, in the United States, astronomers refused to accept that Neptune had been discovered by means of calculation: see such publications as the American Journal of Science, of the period.

Upon Aug. 29, 1849, Dr. Babinet read, to the French Academy, a paper in which he showed that, by the observations of three years, the revolution of Neptune would have to be placed at 165 years. Between the limits of 207 and 233 years was the period that Leverrier had calculated. Simultaneously, in England, Adams had calculated. Upon Sept. 2, 1846, after he had, for at least a month, been charting the stars in the region toward which Adams had pointed, Prof. Challis wrote to Sir George Airy that this work would occupy his time for three more months. This indicates the extent of the region toward which Adams had pointed. Discovery of the "asteroidal belt as deduced from Bode's Law":

We learn that Baron Von Zach had formed a society of twenty-four astronomers to search, in accordance with Bode's Law, for "a planet"--and not "a group," not "an asteroidal belt"--between Jupiter and Mars. The astronomers had organized, dividing the zodiac into twenty-four zones, assigning each zone to an astronomer. They searched. They found not one asteroid. Seven or eight hundred are now known.

Philosophical Magazine, 12-62:

That Piazzi, the discoverer of the first asteroid, had not been searching for a hypothetic body, as deduced from Bode's Law, but, upon an investigation of his own, had been charting stars in the constellation Taurus, night of Jan. 1, 1801. He noticed a light that he thought had moved, and, with his mind a blank, so far as asteroids and brilliant deductions were concerned, announced that he had discovered a comet.

As an instance of the crafty way in which some astronomers now tell the story, see Sir Robert Ball's Story of the Heavens, p. 230:

The organization of the astronomers of Lilienthal, but never a hint that Piazzi was not one of them-- "the search for a small planet was soon rewarded by a success that has rendered the evening of the first day of the nineteenth century memorable in astronomy." Ball tells of Piazzi's charting of the stars, and makes it appear that Piazzi had charted stars as a means of finding asteroids deductively, rewarded soon by success, whereas Piazzi had never heard of such a search, and did not know an asteroid when he saw one. "This laborious and accomplished astronomer had organized an ingenious system of exploring the heavens, which was eminently calculated to discriminate a planet among the starry host ... at length he was rewarded by a success which amply compensated him for all his toil."

Prof. Chase--these two great instances not of mere discovery, but of discovery by means of calculation, according to him--now the subject of his supposition that he, too, could calculate triumphantly--the verification depended upon the accuracy of Prof. Swift and Prof. Watson in recording the positions of the bodies that they had announced--

Sidereal Messenger, 6-84:

Prof. Colbert, Superintendent of Dearborn Observatory, leader of the party of which Prof. Swift was a member, says that the observations by Swift and Watson agreed, because Swift had made his observations agree with Watson's. The accusation is not that Swift had falsely announced a discovery of two unknown bodies, but that his precise determining of positions had occurred after Watson's determinations had been published.

Popular Astronomy, 7-13:

Prof. Asaph Hall writes that, several days after the eclipse, Prof. Watson told him that he had seen "a" luminous body near the sun, and that his declaration that he had seen two unknown bodies was not made until after Swift had been heard from.

Perched upon two delusions, Prof. Chase crowed his false raptures. The unknown never seen again. So it is our expression that hosts of astronomers calculate, and calculation-mad, calculate and calculate and calculate, and that, when one of them does point within 600,000,000 miles (by conventional measurements) of something that is found, he is the Leverrier of the text-books; that the others are the Prof. Chases not of the text-books.

As to most of us, the symbols of the infinitesimal calculus humble independent thinking into the conviction that used to be enforced by drops of blood from a statue. In the farrago and conflicts of daily lives, it is relief to feel such a rapport with finality, in a religious sense, or in a mathematical sense. So then, if the seeming of exactness in Astronomy be either infamously, or carelessly and laughingly, brought about by the connivances of which Swift and Watson were accused, and if the prestige of Astronomy be founded upon nothing but huge capital letters and exclamation points, or upon the disproportionality of balancing one Leverrier against hundreds of Chases, it may not be better that we should know this, if then to those of us who, in the religious sense, have nothing to depend upon, comes deprivation of even this last, lingering seeming of foundation, or seeming existence of exactness and realness, somewhere--

Except--that, if there be nearby lands in the sky and beings from foreign worlds that visit this earth, that is a great subject, and the trash that is clogging an epoch must be cleared away.

We have had a little sermon upon the insecurity of human triumphs, and, having brought it to a climax, now seems to be the time to stop; but there is still an involved "triumph" and I'd not like to have inefficiency, as well as probably everything else, charged against us--

The Discovery of Uranus.

We mention this stimulus to the text-book writers' ecstasies, because out of phenomena of the planet Uranus, the "Neptune-triumph" developed. For Richard Proctor's reasons for arguing that this discovery was not accidental, see Old and New Astronomy, p. 646. Philosophical Transactions, 71-492--a paper by Herschel--"An account of a comet discovered on March 13, 1781." A year went by, and not an astronomer in the world knew a new planet when he saw one: then Lexell did find out that the supposed comet was a planet.

Statues from which used to drip the life-blood of a parasitic cult-- Structures of parabolas from which bleed equations--

As we go along we shall develop the acceptance that astronomers might as well try to squeeze blood from images as to try to seduce symbols into conclusions, because applicable mathematics has no more to do with planetary inter-actions than have statues of saints. If this denial that the calculi have place in gravitational astronomy be accepted, the astronomers lose their supposed god; they become an unfocused simplest problem in celestial mechanics: that is, the formulation of the inter- actions of the sun and the moon and this earth. In the highest of mathematics, final, sacred mathematics, can this next to the simplest problem in so-called mathematical astronomy be solved?

It cannot be solved.

Every now and then, somebody announces that he has solved the Problem of the Three Bodies, but it is always an incomplete, or impressionistic, demonstration, compounded of abstractions, and ignoring the conditions of bodies in space. Over and over we shall find vacancy under supposed achievements; elaborate structures that are pretensions without foundation. Here we learn that astronomers cannot formulate the inter-actions of three bodies in space, but calculate anyway, and publish what they call the formula of a planet that is inter-acting with a thousand other bodies. They explain. It will be one of our most lasting impressions of astronomers: they explain and explain and explain. The astronomers explain that, though in finer terms, the mutual effects of three planets cannot be determined, so dominant is the power of the sun that all other effects are negligible.

Before the discovery of Uranus, there was no way by which the miracles of the astro-magicians could be tested. They said that their formulas worked out, and external inquiry was panic-stricken at the mention of a formula. But Uranus was discovered, and the magicians were called upon to calculate his path. They did calculate, and, if Uranus had moved in a regular path, I do not mean to say that astronomers or college boys have no mathematics by which to determine anything so simple.

They computed the orbit of Uranus. He went somewhere else.

They explained. They computed some more. They went on explaining and computing, year in and year out, and the planet Uranus kept on going somewhere else. Then they conceived of a powerful perturbing force beyond Uranus--so then that at the distance of Uranus the sun is not so dominant-- in which case effects of Saturn upon Uranus and Uranus upon Saturn are not so negligible--on through complexes of inter-actions that infinitely intensify by cumulativeness into a black outlook for the whole brilliant system. The palaeo-astronomers calculated, and for more than fifty years pointed variously at the sky. Finally two of them, of course agreeing upon the general background of Uranus, pointed within distances that are conventionally supposed to have been about six hundred millions of miles of Neptune, and now it is religiously, if not insolently, said that the discovery of Neptune was not accidental--

That the test of that which is not accidental is ability to do it again--

That it is within the power of anybody, who does not know a hyperbola from a cosine, to find out whether the astronomers are led by a cloud of rubbish by day and a pillar of bosh by night If, by the magic of his mathematics, any astronomer could have pointed to the position of Neptune, let him point to the planet past Neptune. According to the same reasoning by which a planet past Uranus was supposed to be, a Trans-Neptunian planet may be supposed to be. Neptune shows perturbations similar to those of Uranus.

According to Prof. Todd there is such a planet, and it revolves around the sun once in 375 years. There are two, according to Prof. Forbes, one revolving once in 1,000 years, and the other once in 5,000 years. See Macpherson's A Century's Progress in Astronomy. It exists, according to Dr. Eric Doolittle, and revolves once in 283 years (Sci. Amer., 122-641). According to Mr. Hind it revolves once in 1,600 years (Smithson. Miscell. Cols., 20-20).

So then we have found out some things, and, relatively to the oppressions that we felt from our opposition, they are reassuring. But also are they depressing.

Because, if, in this existence of ours, there is no prestige higher than that of astronomic science, and, if that seeming of substantial renown has been achieved by a composition of bubbles, what of anything like soundness must there be to all lesser reputes and achievements?

Let three bodies inter-act. There is no calculus by which their inter-actions can be formulated. But there are a thousand inter-acting bodies in this solar system-- or supposed solar system--and we find that the highest prestige in our existence is built upon the tangled assertions that there are magicians who can compute in a thousand quantities, though they cannot compute in three.

Then all other so-called human triumphs, or moderate successes, products of anybody's reasoning processes and labors--and what are they, if higher than them all, more academic, austere, rigorous, exact are the methods and the processes of the astronomers? What can be thought of our whole existence, its nature and its destiny?

That our existence, a thing within one solar system, or supposed solar system, is a stricken thing that is mewling through space, shocking able-minded, healthy systems with the sores on its sun, its ghastly moons, its civilizations that are all broken out with sciences; a celestial leper, holding out doddering expanses into which charitable systems drop golden comets? If it be the leprous thing that our findings seem to indicate, there is no encouragement for us to go on. We cannot discover: we can only betray new symptoms. If I be a part of such a stricken thing, I know of nothing but sickness and sores and rags to reason with: my data will be pustules; my interpretations will be inflammations--

3

SOUTHERN plantations and the woolly heads of Negroes pounding the ground--cries in northern regions and round white faces turned to the sky--fiery globes in the
sky--a study in black, white, and golden formations in one general glow. Upon the night of Nov. 13-14, 1833, occurred the most sensational celestial spectacle of the nineteenth century: for six hours fiery meteors gushed from the heavens, and were visible along the whole Atlantic coast of the United States.

One supposes that astronomers do not pound the ground with their heads, and presumably they do not screech, but they have feelings just the same. They itched. Here was something to formulate. When he hears of something new and unquestionable in the sky, an astronomer is diseased with ill-suppressed equations. Symbols persecute him for expression. His is the frenzy of someone who would stop automobiles, railroad trains, bicycles, all things, to measure them; run, with a yardstick, after sparrows, flies, all persons passing his door. This is supposed to be scientific, but it can be monomaniac. Very likely the distress and the necessity of Prof. Olmstead were keenest. He was the first to formulate. He "demonstrated" that these meteors, known as the Leonids, revolved around the sun once in six months.

They didn't.

Then Prof. Newton "demonstrated" that the "real" period was thirty-three and a quarter years. But this was done empirically, and that is not divine, nor even aristocratic, and the thing would have to be done rationally, or mathematically, by someone, because, if there be not mathematical treatment, in gravitational terms, of such phenomena, astronomers are in reduced circumstances. It was Dr. Adams, who, emboldened with his experience in not having to point anywhere near Neptune, but nevertheless being acclaimed by all patriotic Englishmen as the real discoverer of Neptune, mathematically "confirmed" Prof. Newton's "findings." Dr. Adams predicted that the Leonids would return in November, 1866, and in November, 1899, occupying several years, upon each occasion, in passing a point in this earth's orbit.

There were meteors upon the night of Nov. 13-14, 1866. They were plentiful. They often are in the middle of November. They no more resembled the spectacle of 1833 than an ordinary shower resembles a cloudburst. But the "demonstration" required that there should be an equal display, or, according to some aspects, a greater display, upon the corresponding night of the next year. There was a display, the next year; but it was in the sky of the United States, and was not seen in England. Another occurrence nothing like that of 1833 was reported from the United States.

By conventional theory, this earth was in a vast, wide stream of meteors, the earth revolving so as to expose successive parts to bombardment. So keenly did Richard Proctor visualize the earth so immersed and so bombarded, that, when nothing was seen in England, he explained. He spent most of his life explaining.

In the Student, 2-254, he wrote: "Had the morning of Nov. 14, 1867, been clear in England, we should have seen the commencement of the display, but not its more brilliant part."

We have had some experience with the "triumphs" of astronomers: we have some suspicions as to their greatly advertised accuracy. We shall find out for ourselves whether the morning of Nov. 14, 1867, was clear enough in England or not. We suspect that it was a charming morning, in England-- Monthly Notices, R. A. S. 28-32:

Report by E. J. Lowe, Highfield House, night of Nov. 13-14, 1867:

"Clear at 1.10 A.M.; high, thin cumuli, at 2 A.M., but sky not covered until 3.10 A.M., and the moon's place visible until 3.55 A.M.; sky not overcast until 5.50 A.M."

The determination of the orbital period of thirty-three years and a quarter, but with appearances of a period of thirty-three years, was arrived at by Prof. Newton by searching old records, finding that, in an intersection-period of thirty-three years, there had been extraordinary meteoric displays, from the year 902 A.D. to the year 1833 A.D. He reminds me of an investigator who searched old records for appearances of Halley's comet, and found something that he identified as Halley's comet, exactly on time, every seventy-five years, back to times of the Roman Empire. See the Edinburgh Review, vol. 66. It seems that he did not know that orthodoxy does not attribute exactly a seventy-five year period to Halley's comet. He got what he went looking for, anyway. I have no disposition for us to enjoy ourselves at Prof. Newton's expense, because, surely enough, his method, if regarded as only experimental, or tentative, is legitimate enough, though one does suspect him of very loose behavior in his picking and choosing. But Dr. Adams announced that, upon mathematical grounds, he had arrived at the same conclusion.

The test:

The next return of the Leonids was predicted for November, 1899. Memoirs of the British Astronomical Association, 9-6:

"No meteoric event ever before aroused such widespread interest, or so grievously disappointed anticipation."

There were no Leonids in November, 1899.

It was explained. They would be seen next year. There were no Leonids in November, 1900.

It was explained. They would be seen next year. No Leonids.

Vaunt and inflation and parade of the symbols of the infinitesimal calculus; the pomp of vectors, and the hush that surrounds quaternions: but when an axis of co- ordinates loses its rectitude, bin the service of a questionable selection, disciplined symbols become a rabble. The Most High of Mathematics--and one of his proposed prophets points to the sky. Nowhere near where he points, something is found. He points to a date--nothing happens.

Prof. Serviss, in Astronomy in a Nutshell, explains. He explains that the Leonids did not appear when they "should" have appeared, because Jupiter and Saturn had altered their orbits.

Back in the times of the Crusades, and nothing was disturbing the Leonids--and if you're stronger for dates than I am, think of some more dates, and nothing was altering the orbit of the Leonids-- discovery of America, and the Spanish Armada, in 1588, which, by some freak, I always remember, and no effects by Jupiter and Saturn--French Revolution and on to the year 1866, and still nothing the matter with the Leonids--but, once removed from "discovery" and "identification," and that's the end of their period, diverted by Jupiter and Saturn, old things that had been up in the sky at least as long as they had been. If we're going to accept the calculi at all, the calculus of probabilities must have a hearing. My own opinion, based upon reading many accounts of November meteors, is that decidedly the display of 1833 did not repeat in 1866: that a false priest sinned and that an equally false highpriest gave him sanction.

The tragedy goes comically on. I feel that, to all good Neo-astronomers, I can recommend the following serenity from an astronomer who was unperturbed by what happened to his science, in November, 1899, and some more Novembers

Bryant, A History of Astronomy, p. 252:

That the meteoric display of 1899 4 had failed to appear--"as had been predicted by Dr. Downing and Dr. Johnstone Stoney." One starts to enjoy this disguisement, thinking of virtually all the astronomers in the world who had predicted the return of the Leonids, and the finding, by Bryant, of two who had not, and his recording only the opinion of these two, coloring so as to look like another triumph--but we may thank our sorely stimulated suspiciousness for still richer enjoyment--

That even these two said no such saving thing-- Nature, Nov. 9, 1899:

Dr. Downing and Dr. Stoney, instead of predicting failure of the Leonids to appear, advise watch for them several hours later than had been calculated.

I conceive of the astronomers' fictitious paradise as malarchitectural with corrupted equations, and paved with rotten symbols. Seeming pure, white fountains of formal vanities--boasts that are gushing from decomposed triumphs. We shall find their furnishings shabby with tarnished comets. We turn expectantly to the subject of comets; or we turn cynically to the subject. We turn maliciously to the subject of comets. Nevertheless, threading the insecurities of our various feelings, is a motif that is the steady essence of Neo-astronomy:

That, in celestial phenomena, as well as in all other fields of research, the irregular, or the unformulable, or the uncapturable, is present in at least equal representation with the uniform: that, given any clear, definite, seemingly unvarying thing in the heavens, co-existently is something of wantonness or irresponsibility, bizarre and incredible, according to the standards of purists-- that the science of Astronomy concerns itself with only one aspect of existence, because of course there can be no science of the obverse phenomena--which is good excuse for so enormously disregarding, if we must have the idea that there are real sciences, but which shows the hopelessness of positively attempting. The story of the Comets, as not told in Mr. Chambers' book of that title, is almost unparalleled in the annals of humiliation. When a comet is predicted to return, that means faith in the Law of Gravitation. It is Newtonism that comets, as well as planets, obey the Law of Gravitation, and move in one of the conic sections. When a comet does not return when it "should," there is no refuge for an astronomer to say that planets perturbed it, because one will ask why he did not include such factors in his calculations, if these phenomena be subject to mathematical treatment. In his book, Mr. Chambers avoids, or indicates that he never heard of, a great deal that will receive cordiality from us, but he does publish a list of predicted comets that did not return. Writing, in 1909, he mentions others for which he had hopes:

Brooks' First Periodic Comet (1886, IV)--"We must see what 6 the years 1909 and 1910 bring forth."

This is pretty indefinite anticipation--however, nothing was brought forth, according to Monthly Notices, R. A. S., 1909 and 1910: the Brooks' comet that is recorded is Brooks', 1889. Giacobini's Second Periodic Comet (1900, III)--not seen in 1907--"so we shall not have a chance of knowing more about it until 1914." No more known about it in 1914. Borelly's Comet (1905, II)--"Its expected return, in 1911 or 1912, will be awaited with interest." This is pretty indefinite awaiting: it is now said that this comet did return upon Sept. 19, 1911. Denning's Second Periodic Comet (1894, I)--expected, in 1909, but not seen up to Mr. Chambers' time of writing--no mention in Monthly Notices. Swift's Comet, of Nov. 20, 1894--"must be regarded as lost, unless it should be found in December, 1912." No mention of it in Monthly Notices.

Three comets were predicted to return in 1913--not one of them returned (Monthly Notices, 74-326).

Once upon a time, armed with some of the best and latest cynicisms, I was hunting for prey in the Magazine of Science, and came upon an account of a comet that was expected in the year 1848. I supposed that the thing had been positively predicted, and very likely failed to appear, and, for such common game, had no interest. But I came upon the spoor of disgrace, in the word "triumph"--"If it does come, it will afford another astronomical triumph" (Mag. of Sci., 1848-107). The astronomers had predicted the return of a great comet in the year 1848. In Monthly Notices, April, 1847, Mr. Hind says that the result of his calculations had satisfied him that the identification had been complete, and that, in all probability, "the comet must be very near." Accepting Prof. Madler's determinations, he predicted that the comet would return to position nearest the sun, about the end of February, 1848.

No comet.

The astronomers explained. I don't know what the mind of an astronomer looks like, but I think of a fizzle with excuses revolving around it. A writer in the American Journal of Science, 2-9-442, explains excellently. It seems that, when the comet failed to return, Mr. Barber, of Etwell, again went over the calculations. He found that, between the years 1556 and 1592, the familiar attractions of Jupiter and Saturn had diminished the comet's period by 263 days, but that something else had wrought an effect that he set down positively at 751 days, with a resulting retardation of 488 days. This is magic that would petrify, with chagrin, the arteries of the hemorrhagicalest statue that ever convinced the faithful--reaching back through three centuries of inter-actions, which, without divine insight, are unimaginable when occurring in three seconds.

But there was no comet.

The astronomers explained. They went on calculating, and ten years later were still calculating. See Recreative Science, 1860-139. It would be heroic were it not mania. What was the matter with Mr. Barber, of Etwell, and the intellectual tentacles that he had thrust through centuries is not made clear in most of the contemporaneous accounts; but, in the year 1857, Mr. Hind published a pamphlet and explained. It seems that researches by Littrow had given new verification to a path that had been computed for the comet, and that nothing had been the matter with Mr. Barber, of Etwell, except his insufficiency of data, which had been corrected. Mr. Hind predicted. He pointed to the future, but he pointed like someone closing a thumb and spreading four fingers. Mr. Hind said that, according to Halley's calculations, the comet would arrive in the summer of 1865. However, an acceleration of five years had been discovered, so that the time should be set down for the middle of August, 1860. However, according to Mr. Hind's calculated orbit, the comet might return in the summer of 1864. However, allowing for acceleration, "the comet is found to be due early in August, 1858."

Then Bomme calculated. He predicted that the comet would return upon Aug. 2, 1858. There was no comet.

The astronomers went on calculating. They predicted that the comet would return upon Aug. 22, 1860.

No comet.

But I think that a touch of mercy is a luxury that we can afford; anyway, we'll have to be merciful or monotonous. For variety we shall switch from a comet that did not appear to one that did appear. Upon the night of June 30, 1861, a magnificent humiliator appeared in the heavens. One of the most brilliant luminosities of modern times appeared as suddenly as if it had dropped through the shell of our solar system--if it be a solar system. There were letters in the newspapers: correspondents wanted to know why this extraordinary object had not been seen coming, by astronomers. Mr. Hind explained. He wrote that the comet was a small object, and consequently had not been seen coming by astronomers. No one could deny the magnificence of the comet; nevertheless Mr. Hind declared that it was very small, looking so large because it was near this earth. This is not the later explanation: nowadays it is said that the comet had been in southern skies, where it had been observed. All contemporaneous astronomers agreed that the comet had come down from the north, and not one of them thought of explaining that it had been invisible because it had been in the south. A luminosity, with a mist around it, altogether the apparent size of the moon, had burst into view. In Recreative Science, 3-143, Webb says that nothing like it had been seen since the year 1680. Nevertheless the orthodox pronouncement was that the object was small and would fade away as quickly as it had appeared. See the Athenaeum, July 6, 1861--"So small an object will soon get beyond our view." (Hind)

Popular Science Review, 1-513:

That, in April, 1862, the thing was still visible.

Something else that was seen under circumstances that cannot be considered triumphant--upon Nov. 28, 1872, Prof. Klinkerfues, of Gottingen, looking for Biela's comet, saw meteors in the path of the expected comet. He telegraphed to Pogson, of Madras, to look near the star Theta Centauri, and he would see the comet. I'd not say that this was in the field of magic, but it does seem consummate. A dramatic telegram like this electrifies the faithful--an astronomer in the north telling an astronomer far in the south where to look, so definitely naming one special little star in skies invisible in the north. Pogson looked where he was told to look and announced that he saw what he was told to see. But at meetings of the R. A. S., Jan. to and March 14, 1873, Captain Tupman pointed out that, even if Biela's comet had appeared, it would have been nowhere near this star.

Among our later emotions will be indignation against all astronomers who say that they know whether stars are approaching or receding. When we arrive at that subject it will be the preciseness of the astronomers that will perhaps inflame us beyond endurance. We note here the far smaller difficulty of determining whether a relatively nearby comet is coming or going. Upon Nov. 6, 1892,

Edwin Holmes discovered a comet. In the Jour. B. A. A., 3-182, Holmes writes that different astronomers had calculated its distance from twenty million miles to two hundred million miles, and had determined its diameter to be all the way from twenty-seven thousand miles to three hundred thousand miles. Prof. Young said that the comet was approaching; Prof. Parkhurst wrote merely that the impression was that the comet was approaching the earth; but Prof. Berberich (Eng. Mec., 56-316) announced that, upon November 6, Holmes' comet had been 36,000,000 miles from this earth, and 6,000,000 miles away upon the 16th, and that the approach was so rapid that upon the 21st the comet would touch this earth.

The comet, which had been receding, kept on receding.

4

NEVERTHELESS I sometimes doubt that astronomers represent especial incompetence. They remind me too much of uplifters and grocers, philanthropists, expert accountants, makers of treaties, characters in international conferences, psychic researchers, biologists. The astronomers seem to me about as capitalists seem to socialists, and about as socialists seem to capitalists, or about as Presbyterians seem to Baptists; as Democrats seem to Republicans, or as artists of one school seem to artists of another school. If the basic fallacies, or the absence of base, in every specialization of thought can be seen by the units of its opposition, why then we see that all supposed foundations in our whole existence are myths, and that all discussion and supposed progress are the conflicts of phantoms and the overthrow of old delusions by new delusions. Nevertheless I am searching for some wider expression that will rationalize all of us--conceiving that what we call irrationality is our view of parts and functions out of relation to an underlying whole; an underlying something that is working out its development in terms of planets and acids and bugs, rivers and labor unions and cyclones, politicians and islands and astronomers. Perhaps we conceive of an underlying nexus in which all things, in our existence, are different manifestations--torn by its hurricanes and quaked by the struggles of Labor against Capital--and then, for the sake of balance, requiring relaxations. It has its rougher hoaxes, and some of the apes and some of the priests, and philosophers and wart hogs are nothing short of horse play; but the astronomers are the ironies of its less peasant-like moments--or the deliciousness of pretending to know whether a far-away star is approaching or receding, and at the same time exactly predicting when a nearby comet, which is receding, will complete its approach. This is cosmic playfulness; such pleasantries enable Existence to bear its catastrophes. Shattered comets and sickened nations and the hydrogenic anguishes of the sun--and there must be astronomers for the sake of relaxations.

It will be important to us that the astronomers shall not be less unfortunate in their pronouncements upon motions of the stars than they have turned out to be in other respects. Especially disagreeable to us is the doctrine that stars are variable because dark companies revolve around them; also we prefer to find that nothing fit for somewhat matured minds has been determined as to stars with light companions that encircle them, or revolve with them. If silence be the only true philosophy, and if every positive assertion be a myth, we should easily find requital for our negative preferences.

Prof. Otto Struve was one of the highest of astronomic authorities, and the faithful attribute triumphs to him. Upon March 19, 1873, Prof. Struve announced that he had discovered a companion to the star Procyon. That was an interesting observation, but the mere observation was not the triumph. Sometime before, Prof. Auwers, as credulous, if not jocular, as Newton and Leverrier and Adams, had computed the orbit of a hypothetic companion of Procyon's. Upon a chart of the stars, he had drawn a circle around Procyon. This orbit was calculated in gravitational terms, and a general theme of ours is that all such calculations are only ideal, and relate no more to stars and planets or anything else than do the spotless theories of uplifters to events that occur as spots in the one wide daub of existence. Specifically we wish to discredit this "triumph" of Struve's and Auwers', but in general we continue our expression that all uses of the calculus of celestial mechanics are false applications, and that this subject is for aesthetic enjoyment only, and has no place in the science of astronomy, if anybody can think that there is such a science. So, after great labor, or after considerable enjoyment, Auwers drew a circle around Procyon, and announced that that was the orbit of a companion-star. Exactly at the point in this circle where it "should" be, upon March 19, 1873, Struve saw the point of light which, it may be accepted, sooner or later someone would see. According to Agnes Clerke (System of the Stars, p. 173) over and over Struve watched the point of light, and convinced himself that it moved as it "should" move, exactly in the calculated orbit. In Reminiscences of an Astronomer, p, 138, Prof. Newcomb tells the story. According to him, an American astronomer then did more than confirm Struve's observations: he not only saw but exactly measured the supposed companion.

A defect was found between the lenses of Struve's telescope: it was found that this telescope showed a similar "companion," about 10" from every large star. It was found that the more than "confirmatory" determinations by the American astronomer had been upon "a long well-known star." (Newcomb)

Every astronomic triumph is a bright light accompanied by an imbecility, which may for a while make it variable with diminishments, and then be unnoticed.

Priestcrafts are not merely tyrannies: they're necessities. There must be more reassuring ways of telling this story. The good priest J. E. Gore (Studies in Astronomy, p. 104) tells it safely--not a thing except that, in the year 1873, a companion of Procyon's was, by Struve, "strongly suspected." Positive assurances of the sciences--they are islands of seeming stability in a cosmic jelly. We shall eclipse the story of Algol with some modern disclosures. In all minds not convinced that earnest and devoted falsifiers are holding back Development, the story, if remembered at all, will soon renew its fictitious luster. We are centers beaconed security.

Sir Robert Ball, in the Story of the Heavens, says that the period in which Algol blinks his magnitudes is 2 days, 20 hours, 48 minutes, and 55 seconds. He gives the details of Prof. Vogel's calculations upon a speck of light and an invisibility. It is a god-like command that out of the variations of light shall come the diameters of faint appearances and the distance and velocity of the unseeable-- that the diameter of the point of light is 1,054,000 miles, and that the diameter of the imperceptibility is 825,000 miles, and that their centers are 3,220,000 miles apart: orbital velocity of Algol, 26 miles a second, and the orbital velocity of the companion, 55 miles a second--should be stated 26.3 miles and 55.4 miles a second (Proctor, Old and New Astronomy, p. 773).

We come to a classic imposition like this, and at first we feel helpless. We are told that this thing is so. It is as if we were modes of motion and must go on, but are obstructed by an absolute bar of ultimate steel, shining, in our way, with an infinite polish. But all appearances are illusions.

No one with a microscope doubts this; no one who has gone specially from ordinary beliefs into minuter examination of any subject doubts this, as to his own specific experience--so then, broadly, that all appearances are illusions, and that, by this recognition, we shall dissipate resistances, monsters, dragons, oppressors that we shall meet in our pilgrimage. This bar-like calculation is itself a mode of motion. The static cannot absolutely resist the dynamic, because in the act of resisting it becomes itself proportionately the dynamic. We learn that modifications rusted into the steel of our opposition. The period of Algol, which Vogel carried out to a minute's 55th second, was, after all, so incompetently determined that the whole imposition was nullified--

Astronomical Journal, 11-553:

That, according to Chandler, Algol and his companion do not revolve around each other merely, but revolve together around some second imperceptibility--regularly.

Bull. Soc. Astro. de France, October, 5950:

That M. Mora has shown that in Algol's variations there were irregularities that neither Vogel nor Chandler had accounted for.

The Companion of Sirius looms up to our recognition that the story must be nonsense, or worse than nonsense--or that two light comedies will now disappear behind something darker. The story of the Companion of Sirius is that Prof.

Auwers, having observed, or in his mania for a pencil and something to scribble upon, having supposed he had observed, motions of the star Sirius, had deduced the existence of a companion, and had inevitably calculated its orbit. Early in the year 1862, Alvan Clark, Jr., turned his new telescope upon Sirius, and there, precisely where, according to Auwers' calculations, it should be, saw the companion. The story is told by Proctor, writing thirty years later: the finding of the companion, in the "precise position of the calculations"; Proctor's statement that, in the thirty years following, the companion had "conformed fairly well with the calculated orbit."

According to the Annual Record of Science and Industry, 1876-58, the companion, in half the time mentioned by Proctor, had not moved in the calculated orbit. In the Astronomical Register, 15-186, there are two diagrams by Flammarion: one is the orbit of the companion, as computed by Auwers; the other is the orbit, according to a mean of many observations. They do not conform fairly well. They do not conform at all.

I am now temporarily accepting that Flammarion and the other observing astronomers are right, and that the writers like Proctor, who do not say that they made observations of their own, are wrong, though I have data for thinking that there is no such companion-star. When Clark turned his telescope upon Sirius, the companion was found exactly where Auwers said it would be found. According to Flammarion and other astronomers, had he looked earlier or later it would not have been in this position. Then, in the name of the one calculus that astronomers seem never to have heard of, by what circumstances could that star have been precisely where it should be, when looked for, Jan. 31, 1862, if, upon all other occasions, it would not be where it should be?

Astronomical Register, 1-94:

A representation of Sirius--but with six small stars around him an account, by Dr. Dawes, of observations, by Goldschmidt, upon the "companion" and five other small stars near Sirius. Dr. Dawes' accusation, or opinion, is that it scarcely seems possible that some of these other stars were not seen by Clark. If Alvan Clark saw six stars, at various distances from Sirius, and picked out the one that was at the required distance, as if that were the only one, he dignifies our serials with a touch of something other than comedy. For Goldschmidt's own announcement, see Monthly Notices, R. A. S., 23-181, 243.

5

SMUGNESS and falseness and sequences of re-adjusting fatalities--and yet so great is the hypnotic power of astronomic science that it can outlive its "mortal" blows by the simple process of forgetting them, and, in general, simply by denying that it can make mistakes. Upon page 245, Old and New Astronomy, Richard Proctor says--"The ideas of astronomers in these questions of distance have not changed, and, in the present position of astronomy, based (in such respects) on absolute demonstration, they cannot change."

Sounds that have roared in the sky, and their vibrations have shaken down villages--if these be the voices of Development, commanding that opinions shall change, we shall learn what will become of the Proctors and their "absolute demonstrations." Lights that have appeared in the sky--that they are gleams upon the armament of Marching Organization. "There can be only one explanation of meteors"--I think it is that they are shining spear-points of slayers of dogmas. I point to the sky over a little town in Perthshire, Scotland--there may be a new San Salvador--it may be a new Plymouth Rock. I point to the crater Aristarchus, of the moon--there, for more than a century, a lighthouse may have been signaling.

Whether out of profound meditations, or farrago and bewilderment, I point, directly, or miscellaneously, and, if only a few of a multitude of data be our little horizons relax their constrictions.

I indicate that, in these pages, which are banners in a cosmic procession, I do feel a sense of responsibility, but how to maintain any great seriousness I do not know, because still is our subject astronomical "triumphs."

Once upon a time there was a young man, aged eighteen, whose name was Jeremiah Horrox. He was no astronomer. He was interested in astronomic subjects, but it may be that we shall agree that a young man of eighteen, who had not been heard of by one astronomer of his time, was an outsider. There was a transit of Venus in December, 1639, but not a grown-up astronomer in the world expected it, because the not always great and infallible Kepler had predicted the next transit of Venus for the year 1761. According to Kepler, Venus would pass below the sun in December, 1639. But there was another calculation: it was by the great, but sometimes not so great, Lansberg: that, in December, 1639, Venus would pass over the upper part of the sun. Jeremiah Horrox was an outsider. He was able to reason that, if Venus could not pass below the sun, and also over the upper part of the sun, she might take a middle course.

Venus did pass over the middle part of the sun's disc; and Horrox reported the occurrence, having watched it.

I suppose this was one of the most agreeable humiliations in the annals of busted inflations. One thinks sympathetically of the joy that went out from seventeenth- century Philistines. The story is told to this day by the Proctors and Balls and Newcombs: the way they tell this story of the boy who was able to conclude that something that could not occupy two extremes might be intermediate, and thereby see something that no professional observer of the time saw, is a triumph of absorption: That the transit of Venus, in December, 1639, was observed by Jeremiah Horrox, "the great astronomer."

We shall make some discoveries as we go along, and some of them will be worse thought of than others, but there is a discovery here that may be of interest: the secret of immortality--that there is a mortal resistance to everything; but that the thing that an keep on incorporating, or assimilating within itself, its own mortal resistances, will live forever. By its absorptions, the science of astronomy perpetuates its inflations, but there have been instances of indigestion. See the New York Herald, Sept. 16, 1909. Here Flammarion, who probably no longer asserts any such thing, claims Dr. Cook's "discovery of the north pole" as an "astronomical conquest." Also there are other ways. One suspects that the treatment that Dr. Lescarbault received from Flammarion illustrates other ways.

In the year 1859, it seems that Dr. Lescarbault was something of an astronomer. It seems that as far back as that he may have known a planet when he saw one, because, in an interview, he convinced Leverrier that he did know a planet when he saw one. He had at least heard of the planet Venus, because in the year 1882 he published a paper upon indications that Venus has an atmosphere. Largely because of an observation, or an announcement, of his, occurred the climax of Leverrier's fiascos: prediction of an intra-Mercurial planet that did not appear when it "should" appear. My suspicion is that astronomers pardonably, but frailly, had it in for Lescarbault, and that in the year 1891 came an occurrence that one of them made an opportunity. Early in the year 1891, Dr. Lescarbault announced that, upon the night of Jan. II, 1891, he had seen a new star. At the next meeting of the French Academy, Flammarion rose, spoke briefly, and sat down without over-doing. He said that Lescarbault had "discovered" Saturn.

If a navigator of at least thirty years' experience should announce that he had discovered an island, and if that island should turn out to be Bermuda, he would pair with Lescarbault--as Flammarion made Lescarbault appear. Even though I am a writer upon astronomical subjects, myself, I think that even I should know Saturn, if I should see him, at least in such a period as the year 1891, when the rings were visible. It is perhaps an incredible mistake. However. it will be agreeable to some of us to find that astronomers have committed just such almost incredible mistakes--

In Cosmos, n. s., 42-467, is a list of astronomers who reported "unknown" dark bodies that they had seen crossing the disc of the sun:

La Concha Montevideo Nov. 5, 1789;

Keyser Amsterdam Nov. 9, 1802;

Fisher Lisbon May 5, 1832;

Houzeau Brussels May 8, 1845.

According to the Nautical Almanac, the planet Mercury did cross the disc of the sun upon these dates.

It is either that the Flammarions do so punish those who see the new and the undesired, or that astronomers do "discover" Saturn, and do not know Mercury when they see him--and that Buckle overlooked something when he wrote that only the science of history attracts inferior minds often not fit even for clergymen.

Whatever we think of Flammarion, we admire his deftness. But we shall have an English instance of the ways in which Astronomy maintains itself and controls those who say that they see that which they "should" not see, which does seem beefy. One turns the not very attractive-looking pages of the English Mechanic, 1893, casually, perhaps, at any rate in no expectations of sensations--glaring at one, sketch of such a botanico-pathologic monstrosity as a muskmelon with rows of bunions on it (English Mechanic, Oct. 20, 1893). The reader is told, by Andrew Barclay, F.R.A.S., Kilmarnock, Scotland, that this enormity is the planet Jupiter, according to the speculum of his Gregorian telescope.

In the next issue of the English Mechanic, Capt. Noble, F.R.A.S., writes, gently enough, that, if he had such a telescope, he would dispose of the optical parts for whatever they would bring, and would make a chimney cowl of the tube.

English Mechanic, 1893-2-309--the planet Mars, by Andrew Barclay--a dark sphere, surrounded by a thick ring of lighter material; attached to it, another sphere, of half its diameter--a sketch as gross and repellent to a conventionalist as the museum-freak, in whose body the head of his dangling twin is embedded, its dwarfed body lopping out from his side. There is a description by Mr. Barclay, according to whom the main body is red, and the protuberance blue.

Capt. Noble--"Preposterous ... last straw that breaks the camel's back!"

Mr. Barclay comes back with some new observations upon Jupiter's lumps, and then in the rest of the volume is not heard from again. One reads on, interested in quieter matters, and gradually forgets the controversy.

English Mechanic, Aug. 23, 5897:

A gallery of monstrosities: Andrew Barclay, signing himself "F.R.A.S.," exhibiting:

The planet Jupiter, six times encircled with lumps; afflicted Mars, with his partly embedded twin reduced in size, but still a distress to all properly trained observers; the planet Saturn, shaped like a mushroom with a ring around it.

Capt. Noble--"Mr. Barclay is not a Fellow of the Royal Astronomical Society, and, were the game worth the candle, might be restrained by injunction from so describing himself!" And upon page 362, of this volume of the English Mechanic, Capt. Noble calls the whole matter "a pseudo F.R.A.S.'s crazy hallucinations."

Lists of the Fellows of the Royal Astronomical Society, from June, 1875, to June, 1896:

"Barclay, Andrew, Kilmarnock, Scotland; elected Feb. 8, 1856."

I cannot find the list for 1897 in the libraries. List for 1898--Andrew Barclay's name omitted. Thou shalt not see lumps on Jupiter.

Every one of Barclay's observations has something to support it. All conventional representations of Jupiter show encirclements by strings of rotundities that we are told are cloud-forms, but, in the Jour. B. A. A., December, 1910, is published a paper by Dr. Downing, entitled "Is Jupiter Humpy?" suggesting that various phenomena upon Jupiter agree with the idea that there are protuberances upon the planet. A common appearance, said to be an illusion, is Saturn as an oblong, if not mushroom-shaped: see any good index for observations upon the "square- shouldered aspect" of Saturn. In L'Astronomie, 1889-135, is a sketch of Mars, according to Fontana, in the year 1636--a sphere enclosed in a ring; in the center of the sphere a great protruding body, said, by Fontana, to have looked like a vast, black cone.

But, whether this or that should amuse or enrage us, should be accepted or rejected, is not to me the crux; but Andrew Bar-clay's own opening words are:

That, through a conventional telescope, conventional appearances are seen, and that a telescope is tested by the conventionality of its disclosures; but that there may be new optical principles, or applications, that may be, to the eye and the present telescope, what once the conventional telescope was to the eye--in times when scientists refused to look at the preposterous, enraging, impossible moons of Jupiter.

In the English Mechanic, 33-327, is a letter from the astronomer, A. Stanley Williams. He had written previously upon double stars, their colors and magnitudes. Another astronomer, Herbert Sadler, had pointed out some errors. Mr. Williams acknowledges the errors, saying that some were his own, and that some were from Smyth's Cycle of Celestial Objects. In the English Mechanic, 33-377, Sadler says that, earnestly, he would advise Williams not to use the new edition of Smyth's Cycle, because, with the exception of vol. 40, Memoirs of the Royal Astronomical Society, "a more disgracefully inaccurate" catalogue of double stars had never been published. "If," says one astronomer to the other astronomer, "you have a copy of this miserable production, sell it for waste paper. It is crammed with the most stupid errors."

A new character appears. He is George F. Chambers, F.R.A.S., author of a long list of astronomical works, and a tract, entitled, Where Are You Going, Sunday? He, too, is earnest. In this early correspondence, nothing ulterior is apparent, and we suppose that it is in the cause of Truth that he is so earnest. Says one astronomer that the other astronomer is "evidently one of those self-sufficient young men, who are nothing, if not abusive." But can Mr. Sadler have so soon forgotten what was done to him, on a former occasion, after he had slandered Admiral Smyth? Chambers challenges Sadler to publish a list of, say, fifty "stupid errors" in the book. He quotes the opinion of the Astronomer Royal: that the book was a work of "sterling merit." "Airy vs. Sadler," he says: "which is it to be?"

We began not very promisingly. Few excitements seemed to lurk in such a subject as double stars, their colors and magnitudes; but slander and abuse are livelier, and now enters curiosity: we'd like to know what was done to Herbert Sadler.

Late in the year 1876, Herbert Sadler was elected a Fellow of the Royal Astronomical Society. In Monthly Notices, R.A.S., January, 1879, appears his first paper that was read to the Society: Notes on the late Admiral Smyth's Cycle of Celestial Objects, volume second, known as the Bedford Catalogue. With no especial vehemence, at least according to our own standards of repression, Sadler expresses himself upon some "extraordinary mistakes" in this work.

At the meeting of the Society, May 9, 1879, there was an attack upon Sadler, and it was led by Chambers, or conducted by Chambers, who cried out that Sadler had slandered a great astronomer, and demanded that Sadler should resign. In the report of this meeting, published in the Observatory, there is not a trace of anybody's endeavors to find out whether there were errors in this book or not: Chambers ignored everything but his accusation of slander, and demanded again that Sadler should resign. In Monthly Notices, 39-389, the Council of the Society published regrets that it had permitted publication of Sadler's paper, "which was entirely unsupported by the citation of instances upon which his judgment was founded."

We find that it was Mr. Chambers who had revised and published the new edition of Smyth's Cycle. In the English Mechanic, Chambers challenged Sadler to publish, say, fifty "stupid errors." See page 451, vol. 33, English Mechanic--Sadler lists just fifty "stupid errors." He says that he could have listed, not 50, but 250, not trivial, but of the "grossest kind." He says that in one set of 167 observations, 117 were wrong.

The English Mechanic drops out of this comedy with the obvious title, but developments go on. Evidently withdrawing its "regrets," the Council permitted publication of a criticism of Chambers' edition of Smyth's Cycle, in Monthly Notices, 40-497, and the language in this criticism, by S. W. Burnham, was no less interpretable as slanderous than was Sadler's: that Smyth's data were "either roughly approximate or grossly incorrect, and so constantly recurring that it was impossible to explain that they were ordinary errors of observation." Burnham lists 30 pages of errors.

Following is a paper by E. B. Knobel, who published 17 pages of instances in which, in his opinion, Mr. Burnham had been too severe. Knowing of no objection by Burnham to this reduction, we have left 13 pages of errors in one standard astronomical work, which may fairly be considered as representative of astronomical work in general, inasmuch as it was, in the opinion of the Astronomer Royal, a book of "sterling merit."

I think that now we have accomplished something. After this we should all get along more familiarly and agreeably together. Thirteen pages of errors in one standard astronomical work are reassuring; there is a likeable fallibility here that should make for better relations. If the astronomers were what they think they are, we might as well make squeaks of disapproval against Alpine summits. As to astronomers who calculate positions of planets--of whom he was one--Newcomb, in Reminiscences of an Astronomer, says--"The men who have done it are therefore, in intellect, the select few of the human race--an aristocracy above all others in the scale of being." We could never get along comfortably with such awful selectness as that.

We are grateful to Mr. Sadler, in the cause of more comfortable relations.

6

English Mechanic, 56-184:

THAT, upon April 25, 1892, Archdeacon Nouri climbed Mt. Ararat. It was his hope that he should find something of archaeologic compensation for his clamberings. He found Noah's Ark.

About the same time, Dr. Holden, Director of the Lick Observatory, was watching one of the polished and mysterious-looking instruments that, in the new iconology, have replaced the images of saints. Dr. Holden was waiting for the appointed moment of the explosion of a large quantity of dynamite in San Francisco Bay. The moment came. The polished little "saint" revealed to the faithful scientist. He wrote an account of the record, and sent copies to the San Francisco newspapers.

Then he learned that the dynamite had not been fired off. He sent a second messenger after the first messenger, and, because messengers sometimes have velocities proportional to urgencies--"the Observatory escaped ridicule by a narrow margin." See the Observatory, 20-467. This revelation came from Prof. Colton, who, though probably faithful to all the "saints," did not like Dr. Holden.

The system that Archdeacon Nouri represented lost its power be. cause its claims exceeded all conceivableness, and because, in other respects, of its inertness to the obvious. The system that Dr. Holden represented is not different: there is the same seeing of whatever may be desirable, and the same profound meditations upon the remote, with the same inattention to fairly acceptable starting-points. The astronomers like to tell audiences of just what gases are burning in an unimaginably remote star, but have never reasonably made acceptable, for instance, that this earth is round, to start with. Of course I do not mean to say that this, or anything else, can be positively proved, but it is depressing to hear it said, so authoritatively, that the round shadow of this earth upon the moon proves that this earth is round, whereas records of angular shadows are common, and whereas, if this earth were a cube, its straight sides would cast a rounded shadow upon the convex moon. That the first part of a receding vessel to disappear should be the lower part may be only such an illusion of perspective as that by which railroad tracks seem to dip toward each other in the distance. Meteors sometimes appear over one part of the horizon and then seem to curve down behind the opposite part of the horizon, whereas they describe no such curve, because to a string of observers each observer is at the center of the seeming curve.

Once upon a time--about the year 1870--occurred an unusual sporting event. John Hampden, who was noted for his piety and his bad language, whose avowed purpose was to support the principles of this earth's earliest geodesist, offered to bet five hundred pounds that he could prove the flatness of this earth. Somewhere in England is the Bedford Canal, and along a part of it is a straight, unimpeded view, six miles in length. Orthodox doctrine--or the doctrine of the newer orthodoxy, because John Hampden considered that he was orthodox--is that the earth's curvature is expressible in the formula of 8 inches for the first mile, and then the square of the distance times 8 inches. For two miles, then, the square of 2, or 4, times 8. An object six miles away should be depressed 288 inches, or, allowing for refraction, according to Proctor (Old and New Astronomy) 216 inches. Hampden said that an object six miles away, upon this part of the Bedford Canal, was not depressed as it "should" be. Dr. Alfred Russell Wallace took up the bet. Mr. Walsh, Editor of the Field, was the stakeholder. A procession went to the Bedford Canal. Objects were looked at through telescopes, or looked for, and the decision was that Hampden had lost. There was rejoicing in the fold of the chosen, though Hampden, in one of his most furious bombardments of verses from the Bible, charged conspiracy and malfeasance and confiscation, and what else I don't know, piously and intemperately declaring that he had been defrauded.

In the English Mechanic, 80-40, someone writes to find out about the "Bedford Canal Experiment." We learn that the experiment had been made again. The correspondent writes that, if there were basis to the rumors that he had heard, there must be something wrong with established doctrine. Upon page 138, Lady Blount answers--that, upon May 11, 1904, she had gone to the Bedford Canal, accompanied by Mr. E. Clifton, a well-known photographer, who was himself uninfluenced by her motives, which were the familiar ones of attempting to restore the old gentleman who first took up the study of geodesy. However, she seethes with neither piety nor profanity. She says that, with his telescopic camera, Mr.

Clifton had photographed a sheet, six miles away, though by conventional theory the sheet should have been invisible. In a later number of the English Mechanic, a reproduction of this photograph is published. According to this evidence this earth is flat, or is a sphere enormously greater than is generally supposed. But at the 1901 meeting of the British Association for the Advancement of Science, Mr. H. Yule Oldham read a paper upon his investigations at the Bedford Canal. He, too, showed photographs. In his photographs, everything that should have been invisible was invisible.

I accept that anybody who is convinced that still are there relics upon Mt. Ararat, has only to climb Mt. Ararat, and he must find something that can be said to be part of Noah's Ark, petrified perhaps. If someone else should be convinced that a mistake has been made, and that the mountain is really Pike's Peak, he has only to climb Pike's Peak and prove that the most virtuous of all lands was once the Holy Land. The meaning that I read in the whole subject is that, in this Dark Age that we're living in, not even such rudimentary matters as the shape of this earth have ever been investigated except now and then to support somebody's theory, because astronomers have instinctively preferred the remote and the not so easily understandable and the safe from external inquiry. In Earth Features and Their Meaning, Prof. Hobbs says that this earth is top-shaped, quite as the sloping extremities of Africa and South America suggest. According to Prof. Hobbs, observations upon the pendulum suggest that this earth is shaped like a top. Some years ago, Dr. Gregory read a paper at a meeting of the Royal Geographical Society, giving data to support the theory of a top-shaped earth. In the records of the Society, one may read a report of the discussion that followed. There was no ridiculing. The President of the Society closed the discussion with virtual endorsement, recalling that it was Christopher Columbus who first said that this earth is top-shaped. For other expressions of this revolt against ancient dogmas, see Bull. Soc. Astro. de France, 17-315; 18-143; Pop. Sci. News, 31-234; Eng. Mec., 77-159; Sci. Amer.,100-441.

As to supposed motions of this earth, axial and orbital, circumstances are the same, despite the popular supposition that the existence of these motions has been established by syntheses of data and by unanswerable logic. All scientists, philosophers, religionists, are today looking back, wondering what could have been the matter with their predecessors to permit them to believe what they did believe. Granted that there will be posterity, we shall be predecessors. Then what is it that is conventionally taught today that will in the future seem as imbecilic as to all present orthodoxies seem the vaporings of preceding systems?

Well, for instance, that it is this earth that moves, though the sun seems to, by the same illusion by which to passengers on a boat, the shore seems to move, though it is the boat that is moving.

Apply this reasoning to the moon. The moon seems to move around the earth--but to passengers on a boat, the shore seems to move, whereas it is the boat that is moving--therefore the moon does not move.

As to the motions of the planets and stars that co-ordinate with the idea of a moving earth--they co-ordinate equally well with the idea of a stationary earth.

In the system that was conceived by Copernicus I find nothing that can be said to resemble foundation: nothing but the appeal of greater simplicity. An earth that rotates and revolves is simpler to conceive of than is a stationary earth with a rigid composition of stars, swinging around it, stars kept apart by some. unknown substance, or inter-repulsion. But all those who think that simplification is a standard to judge by are referred to Herbert Spencer's compilations of data indicating that advancing knowledge complicates, making, then, complexity, and not simplicity, the standard by which to judge the more advanced. My own acceptance is that there are fluxes one way and then the other way: that the Ptolemaic system was complex and was simplified; that, out of what was once a clarification, new complications have arisen, and that again will come flux toward simplification or clarification--that the simplification by Copernicus has now developed into an incubus of unintelligibilities revolving around a farrago of inconsistencies, to which the complexities of Ptolemy are clear geometry: miracles, incredibilities, puerilities; tottering deductions depending upon flimsy agreements; brutalized observations that are slaves to infatuated principles.

And one clear call that is heard above the rumble of readjusting collapses--the call for a Neo-astronomy--it may not be our Neo-astronomy.

Prof. Young, for instance, in his Manual of Astronomy, says that there are no common, obvious proofs that the earth moves around the sun, but that there are three abstrusities, all of modern determination. Then, if Copernicus founded the present system, he founded upon nothing. He had nothing to base upon. He either never heard of, or could not detect one of these abstrusities. All his logic is represented in his reasoning upon this earth's rotundity: that this earth is round, because of a general tendency to sphericity, manifesting, for instance, in fruits and in drops of water--showing that lie must have been unaware not only of abstrusities, but of icicles and bananas and oysters. It is not that I am snobbishly deriding the humble and more than questionable ancestry of modern astronomy. I am pointing out that a doctrine came into existence with nothing for a foundation: not a datum, not one observation to found upon; no astronomical principles, no mechanical principles to justify it. Our inquiry will be as to how, in the annals of false architecture, it could ever be said that--except miraculously, of course--a foundation was subsequently slipped under this baseless structure, dug under, rammed under, or God knows how devised and fashioned.

7

THE three abstrusities: The aberration of light, the annual parallax of the stars, the regular, annual shift of the lines of the stellar spectra. By the aberration of light is meant a displacement of all stars, during a year's observation, by which stars near the pole of the ecliptic describe circles, stars nearer the ecliptic describe ellipses, and the stars of the ecliptic, only little straight lines. It is supposed that light has velocity, and that these forms represent the ratio between the velocity of light and the supposed velocity of this earth in its orbit. In the year 1725, Bradley conceived of the present orthodox explanation of the aberration-forms of the stars: that they reflect or represent the path that this earth traverses around the sun, as it would look from the stars, appearing virtually circular from stars in the pole of the ecliptic, for instance. In Bradley's day there were no definite delusions as to the traversing by this earth of another path in space, as part of a whole moving system, so Bradley felt simple and satisfied. About a century later by some of the most amusing reasoning that one could be entertained with, astronomers decided that the whole supposed solar system is moving, at a rate of about 13 miles a second from the region of Sirius to a point near Vega, all this occurring in northern skies, because southern astronomers had not very much to say at that time. Now, then, if at one time in the year, and in one part of its orbit, this earth is moving in the direction in which the whole solar system is moving, there we have this earth traversing a distance that is the sum of its own motion and the general motion; then when the earth rounds about and retraces, there we have its own velocity minus the general velocity. The first abstrusity, then, is knocked flat on its technicalities, because the aberration-forms, then, do not reflect the annual motion of this earth: if, in conventional terms, though the path of this earth is circular or elliptic relatively to the sun, when compounding with solar motion it is not so formed relatively to stars; and there will have to be another explanation for the aberration-forms.

The second supposed proof that this earth moves around the sun is in the parallax of the stars. In conventional terms, it is said that opposite points in this earth's orbit are 185,000,000 miles apart. It is said that stars, so differently viewed, are minutely displaced against their backgrounds. Again solar-motion--if, in conventional terms, this earth has been traveling, as part of the solar system, from Sirius, toward Vega, in 2,000 years this earth has traveled 819,936,000,000 miles. This distance is 4,500 times the distance that is the base line for orbital parallax. Then displacement of the stars by solar-motion parallax in 2,000 years, should be 4,500 times the displacement by orbital parallax, in one year. Give to orbital parallax as minute a quantity as is consistent with the claims made for it, and 4,500 times that would dent the Great Dipper and nick the Sickle of Leo, and perhaps make the Dragon look like a dragon. But not a star in the heavens has changed more than doubtfully since the stars were catalogued by Hipparchus, 2,000 years ago. If, then, there be minute displacements of stars that are attributed to orbital parallax, they will have to be explained in some other way, if evidently the sun does not move from Sirius toward Vega, and if then, quite as reasonably, this earth may not move.

Prof. Young's third "proof" is spectroscopic.

To what degree can spectroscopy in astronomy be relied upon? Bryant, A History of Astronomy, p. 206:

That, according to Belopolsky, Venus rotates in about 24 hours, as determined by the spectroscope; that, according to Dr. Slipher, Venus rotates in about 224 days, as determined by the spectroscope.

According to observations too numerous to make it necessary to cite any, the seeming motions of stars, occulted by the moon, show that the moon has atmosphere. According to the spectroscope, there is no atmosphere upon the moon (Pubs. Astro. Soc. Pacific, vol. 6, no. 37).

The ring of light around Venus, during the transits of 1874 and 1882, indicated that Venus has atmosphere. Most astronomers say that Venus has an atmosphere of extreme density, obscuring the features of the planet. According to spectrum analysis, by Sir William Huggins, Venus has no atmosphere (Eng. Mec., 4-22).

In the English Mechanic, 89-439, are published results of spectroscopic examinations of Mars, by Director Campbell, of the Lick Observatory: that there is no oxygen, and that there is no water vapor on Mars. In Monthly Notices, R.A.S., 27-178, are published results of spectroscopic examinations of Mars by Huggins: abundance of oxygen; same vapors as the vapors of this earth.

These are the amusements of our Pilgrim's Progress, which has new San Salvadors for its goals, or new Plymouth Rocks for its expectations--but the experiences of pilgrims have variety--

In 1895, at the Allegheny Observatory, Prof. Keeler undertook to determine the rotation-period of Saturn's rings, by spectroscopy. It is gravitational gospel that particles upon the outside of the rings move at the rate of 10.69 miles a second; particles upon the inner edge, 13.01 miles a second. Prof. Keeler's determinations were what Sir Robert Ball calls "brilliant confirmation of the mathematical deduction." Prof. Keeler announced that according to the spectroscope, the outside particles of the rings of Saturn move at the rate of 10.1 miles a second, and that the inner particles move at the rate of 12.4 miles a second--"as they ought to," says Prof. Young, in his gospel, Elements of Astronomy. One reads of a miracle like this, the carrying out into decimals of different speeds of different particles in parts of a point of light, the parts of which cannot be seen at all without a telescope, whereby they seem to constitute a solid motionless structure, and one admires, or one worships, according to one's inexperience.

Or there comes upon one a sense of imposture and imposition that is not very bearable. Imposition or imposture or captivation--and it's as if we've been trapped and have been put into a revolving cage, some of the bars revolving at unthinkable speed, and other bars of it going around still faster, even though not conceivable. Disbelieve as we will, deride and accuse, and think of all the other false demonstrations that we have encountered, as we will--there's the buzz of the bars that encircle us. The concoction that has caged us is one the most brilliant harlots in modern prostitution: we're imprisoned at the pleasure of a favorite in the harem of the God of Gravitation. That's some relief: language always is--but how are we to determine" that the rings of Saturn do not move as they "ought" to, and thereby add more to the discrediting of spectroscopy in astronomy?

A gleam on a planet that's like shine on a sword to deliver us-- The White Spot of Saturn-- A bright and shining deliverer.

There's a gleam that will shatter concoctions and stop velocities. There's a shining thing on the planet Saturn, and the blow that it shines is lightning. Thus far has gone a revolution of 10.1 miles a second, but it stops by magic against magic; no farther buzzes a revolution of 12.4 miles a second-- that the rings of Saturn may not move as, to flatter one little god, they "ought" to, because, by the handiwork of Universality, they may be motionless.

Often has a white spot been seen upon the rings of Saturn: by Schmidt, Bond, Secchi, Schroeter, Harding, Schwabe, De Vico--a host of other astronomers.

It is stationary.

In the English Mechanic, 49-195, Thomas Gwyn Elger publishes a sketch of it as he saw it upon the nights of April 18 and 20, 1889. It occupied a position partly upon one ring and partly upon the other, showing no distortion. Let Prof. Keeler straddle two concentric merry-go-rounds, whirling at different velocities: there will be distortion. See vol. 49, English Mechanic, for observation after observation by astronomers upon this appearance, when seen for several months in the year 1889, the observers agreeing that, no matter what are the demands of theory, this fixed spot did indicate that the rings of Saturn do not move.

The White Spot on Saturn has blasted minor magic. He has little, black retainers who now function in the cause of completeness--the little, black spots of Saturn--

Nature, 53.109:

That, in July and August, 1895, Prof. Mascari, of the Catania Observatory, had seen dark spots upon the crepe ring of Saturn. The writer in Nature says that such duration is not easy to explain, if the rings of Saturn be formations of moving particles, because different parts of the discolored areas would have different velocities, so that soon would they distort and diffuse.

Certainly enough, relatively to my purpose, which is to find out for myself, and to find out with anybody else who may be equally impressed with a necessity, a brilliant, criminal thing has been slain by a gleam of higher intensity. Certainly enough, then, with the execution of one of its foremost exponents, the whole subject of spectroscopy in astronomy has been cast into rout and disgrace, of course only to ourselves, and not in the view of manufacturers of spectroscopes, for instance; but a phantom thing dies a phantom death, and must be slain over and over again.

I should say that just what is called the spectrum of a star is not commonly understood. It is one of the greatest uncertainties in science. The spectrum of a star is a ghost in the first place, but this ghost has to be further attenuated by a secondary process, and the whole appearance trembles so with the twinkling of a star that the stories told by spectra are gasps of palsied phantoms. So it is that, in one of the greatest indefinitenesses in science, an astronomer reads in a bewilderment that can be made to correspond with any desideratum. So it is our acceptance that when any faint, tremulous story told by a spectrum becomes standardized, the conventional astronomer is told, by the spectroscope, what he should be told, but that when anything new appears, for which there is no convention, the bewilderment of the astronomers is made apparent, and the worthlessness of spectroscopy in astronomy is shown to all except those who do not want to be shown. Upon the first of February, 1892, Dr. Thomas D. Anderson, of Edinburgh, discovered a new star that became known as Nova Aurigae. Here was something as to which there was no dogmatic "determination." Each astronomer had to see, not what he should, but what he could. We shall see that the astronomers might as well have gone, for information, to some of Mrs. Piper's "controls" as to think of depending upon their own ghosts.

In Monthly Notices, February, 1893, it is said that probably for seven weeks, up to the time of calculation, one part of this new star had been receding at a rate of 230 miles a second, and another part approaching at a rate of 320 miles a second, giving to these components a distance apart of 550 miles * 60 * 60 * 24 * 49, whatever that may be.

But there was another seance. This time Dr. Vogel was the medium. The ghosts told Dr. Vogel that the new star had three parts, one approaching this earth at the rate of about 420 miles a second, another approaching at a rate of 22 miles a second, a third part receding at a rate of 300 miles a second. See Jour. B. A. A., 2-258.

After that, the "controls" became hysterical. They flickered that there were six parts of this new star, according to Dr. Lowell's Evolution of Worlds, p. 9. The faithful will be sorry to read that Lowell revolted. He says: "There is not room for so many on the stage of the cosmic drama." For other reasons for repudiating spectroscopy, or spiritualism, in astronomy, read what else Lowell says upon this subject.

Nova Aurigae became fainter. Accordingly, Prof. Klinkerfues "found" that two bodies had passed, and had inflamed, each other, and that the light of their mutual disturbances would soon disappear (Jour. B. A. A., 2-365).

Nova Aurigae became brighter. Accordingly, Dr. Campbell "determined" that it was approaching this earth at a rate of 128 miles a second (Jour. B. A. A., 2-504).

Then Dr. Espin went into a trance. It was revealed to him that the object was a nebula (Eng. Mec., 56-61). Communication from Dr. and Mrs. Huggins, to the Royal Society--not a nebula, but a star (Eng. Mec., 57-397) . See Nature, 47-352, 425-- that, according to M. Eugen Gothard, the spectrum of N. A. agreed "perfectly" with the spectrum of a nebula: that, according to Dr. Huggins, no contrast could be more striking than the difference between the spectrum of N. A., and the spectrum of a nebula.

For an account of the revelations at Stonyhurst Observatory, see Mems. R. A. S., 51-129--that there never had been a composition of bodies moving at the rates that were so definitely announced, because N. A. was a single star.

Though I have read some of the communications from "Rector" and "Dr. Phinuit" to Mrs. Piper, I cannot think that they ever mouthed sillier babble than was flickered by the star-ghosts to the astronomers in the year 1892. We noted Prof. Klinkerfues' "finding" that two stars had passed each other, and that the illumination from their mutual perturbations would soon subside. There was no such disappearance. For observations upon N. A., ten years later, see Monthly Notices, 62-65. For Prof. Barnard's observations twenty years later, see Sci. Amer. Sup., 76-154.

The spectroscope is useful in a laboratory. Spoons are useful in a kitchen. If any other pilgrim should come upon a group of engineers trying to dig a canal with spoons, his experience and his temptation to linger would be like ours as to the astronomers and their attempted application of the spectroscope. I don't know what of remotest acceptability may survive in the third supposed proof that this earth moves around the sun, though we have not found it necessary to go into the technicalities of the supposed proof. I think we have killed the phantom thing, but I hope we have not quite succeeded, because we are moved more by the aesthetics of slaughter than by plain murderousness: we shall find unity in disposing of the third "proof" by the means by which the two others were disposed of--

Regular Annual Shift of Spectral Lines versus Solar Motion--

That, if this earth moves around the sun, the shift might be found by scientific Mrs. Pipers so to indicate--

But that if part of the time this earth, as a part of one traveling system, moves at a rate of 19 plus 13 miles a second and then part of the time at a rate of 19 minus 13 miles a second, compounding with great complexities at transverse times, that is the end of the regular annual shift that is supposed to apply to orbital motion.

We need not have admitted in the first place that the three abstrusities are resistances: however, we have a liking for revelations ourselves. Aberration and Parallax and Spectral Lines do not indicate only that this earth moves relatively to the stars: quite as convincingly they indicate that the stars in one composition gyrate relatively to a central and stationary earth, all of them in one concavity around this earth, some of them showing faintest of parallax, if this earth be not quite central to the revolving whole.

Something that I did not mention before, though I referred to Lowell's statements, is that astronomers now admit, or state, that the shift of spectral lines, which they say indicates that this earth moves around the sun, also indicates any one of three other circumstances, or sets of circumstances. Some persons will ask why I didn't say so at first, and quit the meaningless subject. Maybe it was a weakness of mine--something of a sporting instinct, I fear me, I have at times. I lingered, perhaps slightly intoxicated, with the deliciousness of Prof. Keeler and his decimals--like someone at a race track, determining that a horse is running at a rate of 2,653 feet and 4 inches a minute, by a method that means that no more than it means that the horse is brown, is making clattering sounds, or has a refreshing odor. For a study of a state of mind like that of many clergymen who try to believe in Moses, and in Darwin, too, see the works of Prof. Young, for instance. This astronomer teaches the conventional spectroscopic doctrine, and also mentions the other circumstances that make the doctrine meaningless. Such inconsistencies are phenomena of all transitions from the old to the new.

Three giants have appeared against us. Their hearts are bubbles. Their bones wilt. They are the limp caryatides that uphold the phantom structure of Palaeo- astronomy. By what miracle, we asked, could foundation be built subsequently under a baseless thing. But three ghosts can fit in anywhere. Sometimes astronomers cite the Foucault pendulum-experiment as "proof" of the motions of this earth. The circumstances of this demonstration are not easily mode clear: consequently one of normal suspiciousness is likely to let it impose upon him. But my practical and commonplace treatment is to disregard what the experiment and its complexities are, and to enquire whether it works out or not.

It does not. See Amer. Jour. Sci., 2-12-402; Eng. Mec., 93-293, 306; Astro. Reg., 2-265. Also we are told that experiments upon falling bodies have proved this earth's rotation. I get so tired of demonstrating that there never has been any Evolution mentally, except as to ourselves, that, if I could, I'd be glad to say that these experiments work out beautifully. Maybe they do. See Proctor's Old and New Astronomy, p. 229.

8

IT is supposed that astronomic subjects and principles and methods cannot be understood by the layman. I think this, myself. We shall take up some of the principles of astronomy, with the idea of expressing that of course they cannot be understood by the unhypnotized any more than can the stories of Noah's Ark and Jonah and the Whale be understood, but that our understanding, if we have any, will have some material for its exercises, just the same. The velocity of light is one of these principles. A great deal in the astronomic system depends upon this supposed velocity: determinations of distance, and amount of aberration depend. It will be our expression that these are ratios of impositions to mummeries, with such clownish products that formulas turn into antics, and we shall have scruples against taking up the subject at all, because we have much hard work to do, and we have qualms against stopping so often to amuse ourselves. But, then, sometimes in a more sentimental mood, I think that the pretty story of the velocity of light, and its "determination," will someday be of legitimate service; be rhymed someday, and told to children, in future kindergartens, replacing the story of Little Bopeep, with the tale of a planet that lost its satellites and sometimes didn't know where to find them, but that good magicians came along and formulated the indeterminable.

It was found by Roemer, a seventeenth-century astronomer, that, at times, the moons of Jupiter did not disappear behind him, and did not emerge from behind him, when they "should." He found that as distance between this earth and Jupiter increased, the delays increased. He concluded that these delays represented times consumed by the light of the moons in traveling greater distances. He found, or supposed he found, that when this earth is farthest from Jupiter, light from a satellite is seen 22 minutes later than when nearest Jupiter. Given measurement of the distance between opposite points in the earth's supposed orbit, and time consumed in traveling this distance--there you have the velocity of light.

I still say that it is a pretty story and should be rhymed; but we shall find that astronomers might as well try to formulate the gambols of the sheep of Little Bopeep, as to try to formulate anything depending upon the satellites of Jupiter.

In the Annals of Philosophy, 23-29, Col. Beaufoy writes that, upon Dec. 7, 1823, he looked for the emergence of Jupiter's third satellite, at the time set down in the National Almanac: for two hours he looked, and did not see the satellite emerge. In Monthly Notices, 44-8, an astronomer writes that, upon the night of Oct. 15, 1883, one of the satellites of Jupiter was forty-six minutes late. A paper was read at the meeting of the British Astronomical Association, Feb. 8, 1907, upon a satellite that was twenty minutes late. In Telescopic Work, p. 191,

W. F. Denning writes that, upon the night of Sept. 12, 1889, he and two other astronomers could not see satellite IV at all. See the Observatory, 9-237-- satellite IV disappeared 15 minutes before calculated time; about a minute later it re-appeared; disappeared again; re-appeared nine minutes later. For Todd's observations see the Observatory, 2-227--six times, between June 9 and July 2, 1878, a satellite was visible when, according to prediction, it should have been invisible. For some more instances of extreme vagaries of these satellites, see Monthly Notices, 43-427, and Jour. B. A. A., 14-27: observations by Noble, Turner, White, Holmes, Freeman, Goodacre, Ellis, and Molesworth. In periodical astronomical publications, there is no more easily findable material for heresy than such observations. We shall have other instances. They abound in the English Mechanic, for instance. But, in spite of a host of such observations, Prof. Young (The Sun, p. 35) says that the time occupied by light coming from these satellites is doubtful by "only a fraction of a second." It is of course another instance of the astronomers who know very little of astronomy.

It would have been undignified, if the astronomers had taken the sheep of Little Bopeep for their determinations. They took the satellites of Jupiter. They said that the velocity of light is about 190,000 miles a second.

So did the physicists. [p. 360]

Our own notion is that there is no velocity of light: that one sees a thing, or doesn't; that if the satellites of Jupiter behave differently according to proximity to this earth, that may be because this earth affects them, so affecting them, because the planets may not, as we may find, be at a thousandth part of the "demonstrated" distances. The notion of velocity of light finds support; we are told in the text-books, in the velocity of sound. If it does, it doesn't find support in gravitational effects, because, according to the same textbooks, gravitational effects have no velocity.

The physicists agreed with the astronomers. A beam of light is sent through, and is reflected back through, a revolving shutter--but it's complex, and we're simple: we shall find that there is no need to go into the details of this mechanism. It is not that a machine is supposed to register a velocity of 186,000 miles a second, or we'd have to be technical: it is that the eye is supposed to perceive--

And there is not a physicist in the world who can perceive when a parlor magician palms off playing-cards. Hearing, or feeling, or if one could smell light, some kind of a claim might be made--but the well-known limitations of seeing; common knowledge of little boys that a brand waved about in the dark cannot be followed by the eyes. The limit of the perceptible is said to be ten changes a second.

I think of the astronomers as occupying a little vortex of their own in the cosmic swoon in which wave all things, at least in this one supposed solar system. Call it swoon, or call it hypnosis--but that it is never absolute, and that all of us sometimes have awareness of our condition, and moments of wondering what it's all about and why we do and think the things that sometimes we wake up and find ourselves doing and thinking. Upon page 281, Old and New Astronomy, Richard Proctor awakens momentarily, and says: "The agreement between these results seems close enough, but those who know the actual difficulty of precise time- observations of the phenomena of Jupiter's satellites, to say nothing of the present condition of the theory of their motions, can place very little reliance on the velocity of light deduced from such observations." Upon pages 603-607, Proctor reviews some observations other than those that I have listed--satellites that have disappeared, come back, disappeared, returned again so bewilderingly that he wrote what we have quoted-- observations by Gorton, Wray, Gambart, Secchi, Main, Grover, Smyth-Maclear- Pearson, Hodgson, Carlisle, Siminton. And that is the last of his awareness: Proctor then swoons back into his hypnosis. He then takes up the determination of the velocity of light by the physicists, as if they could be relied upon, accepting every word, writing his gospel, glorying in this miracle of science. I call it a tainted agreement between the physicists and the astronomers. I prefer mild language. If by a method by which nothing could be found out, the astronomers determined that the velocity of light is about 190,000 miles a second, and if the physicists by another method found about the same result, what kind of harmony can that be other than the reekings of two consistent stenches? Proctor wrote that very little reliance could be placed upon anything depending upon Jupiter's satellites. It never occurred to him to wonder by what miracle the physicists agreed with these unreliable calculations. It is the situation that repeats in the annals of astronomy--a baseless thing that is supposed to have a foundation slipped under it, wedged in, or God knows how introduced or foisted. I prefer not to bother much with asking how the physicists could determine anything of a higher number of changes than ten per second. If it be accepted that the physicists are right, the question is--by what miracle were the astronomers right, if they had "very little" to rely upon?

Determinations of planetary distances and determinations of the velocity of light have squirmed together: they represent either an agreeable picture of co- operation, or a study in mutual support by writhing infamies. With most emphasis I have taken the position that the vagaries of the Jovian satellites are so great that extremely little reliance can be placed upon them, but now it seems to me that the emphasis should be upon the admission that, in addition to these factors of indeterminateness, it was, up to Proctor's day, not known with anything like accuracy when the satellites should appear and disappear. In that case one wonders as to the state of the theory in Roemer's day. It was in the mind of Roemer that the two "determinations" we are now considering first most notably satisfied affinity: mutual support by velocity of light and distances in this supposed solar system. Upon his Third Law, which, as we shall see later, he constructed upon at least three absences of anything to build upon, Kepler had, upon observations upon Mars, deduced 13,000,000 miles as this earth's distance from the sun. By the same method, which is the now discredited method of simultaneous observations, Roemer determined this distance to be 82,000,000 miles. I am not concerned with this great discrepancy so much as with the astronomers' reasons for starting off distances in millions instead of hundreds or thousands of miles.

In Kepler's day the strongest objection urged against the Copernican system was that, if this earth moves around the sun, the stars should show annual displacements--and it is only under modern "refinements" that the stars do so minutely vary, perhaps. The answer to this objection was that the stars are vastly farther away than was commonly supposed. Entailed by this answer was the necessity of enlarging upon common suppositions generally. Kepler determined or guessed, just as one pleases, and then Roemer outdid him. Roemer was followed by Huygens, with continued outdoing: 100,000,000 according to Huygens. Huygens took for his basis his belief that this earth is intermediate in size to Mars and Venus.

Astronomers, today, say that this earth is not so intermediate. We see that, in the secondary phase of development, the early astronomers, with no means of knowing whether the sun is a thousand or a million miles away, guessed or determined such distances as 82,000,000 miles and 100,000,000 miles, to account for the changelessness of the stars. If the mean of these extremes is about the distance of present dogmas, we'd like to know by what miracle a true distance so averages two products of wild methods. Our expression is that these developments had their origin in conspiracy and prostitution, if one has a fancy for such accusations; or, if everybody else has been so agreeable, we think so more amiably, ourselves, that it was all a matter of comfortably adjusting and being obliging all around. Our expression is that ever since the astronomers have seen and have calculated as they should see and should calculate. For instance, when this earth's distance from the sun was supposed to be 95,000,000 miles, all astronomers taking positions of Mars, calculated a distance of 95,000,000 miles; but then, when the distance was cut down to about 92,000,000 miles, all astronomers, taking positions of Mars, calculated about a distance of 92,000,000 miles. It may sound like a cynicism of mine, but in saying this I am quoting Richard Proctor, in one of his lucid suspicions (Old and New Astronomy, p. 280).

With nothing but monotony, and with nothing that looks like relief for us, the data of conspiracy, or of co-operation, continue. Upon worthless observations upon the transits of Venus, 1761 and 1769, this earth's orbit was found by Encke to be about 190,000,000 miles across (distance of the sun about 95,000,000 miles).

Altogether progress had been more toward the wild calculations of Huygens than toward the undomesticated calculations of Roemer. So, to agree with this change, if not progress, Delambre, taking worthless observations upon the satellites of Jupiter, cut down Roemer's worthless determinations, and announced that light crosses the plane of this earth's orbit in 16 minutes and 32 seconds--as it ought to, Prof. Young would say. It was then that the agreeably tainted physicists started spinning and squinting, calculating "independently," we are told, that Delambre was right. Everything settled--everybody comfortable--see Chambers' Handbook of Astronomy, published at this time--that the sun's distance had been ascertained, "with great accuracy," to be 95,298,260 miles.

But then occurred something that is badly, but protectively, explained, in most astronomical works. Foucault interfered with the deliciousness of those 95,298,260 miles. One may read many books that mention this subject, and one will always read that Foucault, the physicist, by an "independent" method, or by an "absolutely independent" method, disagreed somewhat. The "disagreement" is paraded so that one has an impression of painstaking, independent scientists not utterly slavishly supporting one another, but at the same time keeping well over the 90,000,000 mark, and so essentially agreeing, after all. But we find that there was no independence in Foucault's "experiments." We come across the same old disgusting connivance, or the same amiable complaisance, perhaps. See Clerke's History of Astronomy, p. 230. We learn that astronomers, to explain oscillations of the sun, had decided that the sun must be, not 95,298,260 miles away, but about 91,000,000. To oblige them, perhaps, or innocently, never having heard of them, perhaps, though for ten years they had been announcing that a new determination was needed, Foucault "found" that the velocity of light is less than had been necessary to suppose, when the sun was supposed to be about 95,000,000 miles away, and he "found" the velocity to be exactly what it should be, supposing the sun to be 91,000,000 miles away. Then it was that the astronomers announced, not that they had cut down the distance of the sun because of observations upon solar oscillations, but because they had been very much impressed by the "independent" observations upon the velocity of light, by Foucault, the physicist. This squirm occurred at the meeting of the Royal Astronomical Society, February, 1864. There would have to be more squirms. If, then, the distance across this earth's orbit was "found" to be less than Delambre had supposed, somebody would have to find that light comes from the satellites of Jupiter a little slower than Delambre had "proved." Whereupon, Glassenapp "found" that the time is 16 minutes and 40 seconds, which is what he should, or "ought to," find. Whereupon, there would have to be re-adjustment of Encke's calculations of distance of sun, upon worthless observations upon transits of Venus. And whereupon again, Newcomb went over the very same observations by which Encke had compelled agreement with the dogmas of his day, and Newcomb calculated, as was required, that the distance agreed with Foucault's reduction. Whether, in the first place, Encke ever did calculate, as he said he did, or not, his determination was mere agreement with Laplace's in the seventh book of the Mechanique Celeste. Of course he said that he had calculated independently, because his method was by triangulation, and Laplace's was the gravitational.

That the word "worthless" does apply to observations upon transits of Venus:

In Old and New Astronomy, Proctor says that the observations upon the transits of 1761 and 1769 were "altogether unsatisfactory." One supposes that anything that is altogether unsatisfactory can't be worth much. In the next transit, of 1874, various nations co-operated. The observations were so disappointing that the Russian, Italian, and Austrian Governments refused to participate in the expeditions of 1882. In Reminiscences of an Astronomer, p. 181, Newcomb says that the United States Commission, of which he was Secretary, had up to 1902 never published in full its observations, and probably never would, because by that time all other members were either dead or upon the retired list.

Method of Mars--more monotony--because of criticisms of the taking of parallax by simultaneous observations, Dr. David Gill went to the Island of Ascension, during the opposition of Mars of 1877, to determine alone, by the diurnal method, the distance of this earth from the sun, from positions of Mars. For particulars of Gill's method, see, for instance, Poor's Solar System, p. 86. Here Prof. Poor says that, of course, the orbital motion of Mars had to be allowed for, in Gill's calculations. If so, then of course this earth's orbital motion had to be allowed for. If Dr. Gill knew the space traversed by this earth in its orbit, and the curvature of its path, he knew the size and shape of the orbit, and consequently the distance from the sun. Then he took for the basis of his allowance that this earth is about 93,000,000 miles from the sun, and calculated that this earth is about 93,000,000 miles from the sun. For this classic deduction from the known to the same known, he received a gold medal.

In our earlier surveys, we were concerned with the false claim that there can be application of celestial mechanics to celestial phenomena; but, as to later subjects, the method is different. The method of all these calculations is triangulation.

One simple question:

To what degree can triangulation be relied upon?

To great degree in measuring the height of a building, or in the little distances of a surveyor's problems. It is clear enough that astronomers did not invent the telescope. They adopted the spectroscope from another science. Their primary mathematical principle of triangulation they have taken from the surveyors, to whom it is serviceable. The triangle is another emblem of the sterility of the science of astronomy. Upon the coat of arms of this great mule of the sciences, I would draw a prism within a triangle.

9

    ACCORDING to Prof. Newcomb, for instance, the distance of the sun is about 380 times the distance of the moon--as determined by triangulation. But, upon page 22, Popular Astronomy, Newcomb tells of another demonstration, with strikingly different results--as determined by triangulation.

A split god.

The god Triangulation is not one undivided deity.

The other method with strikingly different results is the method of Aristarchus. It cuts down the distance of the sun, from 380 to 20 times the distance of the moon. When an observer upon this earth sees the moon half-illumined, the angle at the moon, between observer and sun, is a right angle; a third line between observer and sun completes a triangle. According to Aristarchus, the tilt of the third line includes an angle of 86 degrees, making the sun-earth line 20 times longer than the moon-earth line.

"In principle," says Newcomb, "the method is quite correct and very ingenious, but it cannot be applied in practice." He says that Aristarchus measured wrong; that the angle between the moon-earth line and the earth-sun line is almost 90 degrees and not 86 degrees. Then he says that the method cannot be applied because no one can determine this angle that he had said is of almost 90 degrees. He says something that is so incongruous with the inflations of astronomers that they'd sizzle if their hypnotized readers could read and think at the same time. Newcomb says that the method of Aristarchus cannot be applied because no astronomer can determine when the moon is half-illumined.

We have had some experience.

Does anybody who has been through what we've been through suppose that there is a Prof. Keeler in the world who would not declare that trigonometrically and spectroscopically and micro-metrically he had determined the exact moment and exasperating, or delightful, decimal of a moment of semi-illumination of the moon, were it not that, according to at least as good a mathematician as he, determination based upon that demonstration does show that the sun is only 20 times as far away as the moon? But suppose we agree that this simple thing cannot be done.

Then instantly we think of some of the extravagant claims with which astronomers have stuffed supine credulities. Crawling in their unsightly confusion that sickens for simplification, is this offense to harmony:

That astronomers can tell under which Crusade, or its decimalated moment, a shine left a star, but cannot tell when a shine reaches a line on the moon--

Glory and triumph and selectness and inflation--or that we shall have renown as evangelists, spreading the homely and wholesome doctrine of humility. Hollis, in Chats on Astronomy, tells us that the diameter of this earth, at the equator, is 41,851,160 feet. But blessed be the meek, we tell him. In the Observatory, 19-118, is published the determination, by the astronomer Brenner, of the time of rotation of Venus, as to which other astronomers differ by hundreds of days. According to Brenner, the time is 23 hours, 57 minutes, and 7.5459 seconds. I do note that this especial refinement is a little too ethereal for the Editor of the Observatory: he hopes Brenner will pardon him, but is it necessary to carry out the finding to the fourth decimal of a second? However, I do not mean to say that all astronomers are as refined as Brenner, for instance. In the Jour. B. A. A., I-382, Edwin Holmes, perhaps coarsely, expresses some views. He says that such "exactness" as Capt. Noble's in writing that the diameter of Neptune is 38,133 miles and that of Uranus is 33,836 miles is bringing science into contempt, because very little is known of these planets; that, according to Neison, these diameters are 27,000 miles and 28,500 miles. Macpherson, in A Century's Progress in Science, quotes Prof. Serviss: that the average parallax of a star, which is an ordinary astronomic quantity, is "about equal to the apparent distance between two pins, placed one inch apart, and viewed from a distance of one hundred and eighty miles." Stick ins in a cushion, in New York--go to Saratoga and look at them --be overwhelmed with the more than human powers of the scientifically anointed--or ask them when shines half the moon.

The moon's surface is irregular. I do not say that anybody with brains enough to know when he has half a shoe polished should know when the sun has half the moon shined. I do say that if this simple thing cannot be known, the crowings of astronomers as to enormously more difficult determinations are mere barnyard disturbances.

Triangulation that, according to his little priests, straddles orbits and on his apex wears a star--that he's a false Colossus; shrinking, at the touch of data, back from the stars, deflating below the sun and moon; stubbing down below the clouds of this earth, so that the different stories that he told to Aristarchus and to Newcomb are the conflicting vainglories of an earth-tied squatter--

The blow that crumples a god:

That, by triangulation, there is not an astronomer in the world who can tell the distance of a thing only five miles away.

Humboldt, Cosmos, 5-138:

Height of Mauna Loa: 18,410 feet, according to Cook; 16,611, according to Marchand; 13,761, according to Wilkes--according to triangulation.

In the Scientific American, 119-31, a mountain climber calls the Editor to account for having written that Mt. Everest is 29,002 feet high. He says that, in his experience, there is always an error of at least ten per cent. in calculating the height of a mountain, so that all that can be said is that Mt.

Everest is between 26,100 and 31,900 feet high. In the Scientific American, 102-183, and 319, Miss Annie Peck cites two measurements of a mountain in India: they differ by 4,000 feet.

The most effective way of treating this subject is to find a list of measurements of a mountain's height before the mountain was climbed, and compare with the barometric determination, when the mountain was climbed. For a list of 8 measurements, by triangulation, of the height of Mt. St. Elias, see the Alpine Journal, 22-150: they vary from 12,672 to 19,500 feet. D'Abruzzi climbed Mt. St. Elias, Aug. 1, 1897. See a paper, in the Alpine Journal, 19-125 D'Abruzzi barometric determination-18,092 feet.

Suppose that, in measuring, by triangulation, the distance of anything five miles away, the error is, say, ten per cent. But, as to anything ten miles away, there is no knowing what the error would be. By triangulation, the moon has been "found" to be 240,000 miles away. It may be 240 or 240,000,000 miles away.

10

PSEUDO heart of a phantom thing--it is Keplerism, pulsating with Sir Isaac Newton's regularizations.

If triangulation cannot be depended upon accurately to measure distance greater than a mile or two between objects and observers, the aspects of Keplerism that depend upon triangulation should be of no more concern to us than two pins in a cushion 180 miles away: nevertheless so affected by something like seasickness are we by the wobbling deductions of the conventionalists that we shall have direct treatment, or independent expressions, whenever we can have, or seem to have, them. Kepler saw a planetary system, and he felt that, if that system could be formulated in terms of proportionality, by discovering one of the relations quantitatively, all its measurements could be deduced. I take from Newcomb, in Popular Astronomy, that, in Kepler's view, there was system in the arrangement and motions of the four little traitors that sneak around Jupiter; that Kepler, with no suspicions of these little betrayers, reasoned that this central body and its accompaniments were a representation, upon a small scale, of the solar system, as a whole. Kepler found that the cubes of mean distances of neighboring satellites of Jupiter, divided by the squares of their times, gave the same quotients. He reasoned that the same relations subsisted among planets, if the solar system be only an enlargement of the Jovian system.

Observatory, December, 1920: "The discordances between theory and observation (as to the motions of Jupiter's satellites) are of such magnitude that continued observations of their precise moments of

eclipses are very much to be desired." In the Report of the Jupiter Section of the British Astronomical Society (Mens. B. A. A., 8-83) is a comparison between observed times and calculated times of these satellites. 65 observations, in the year 1899, are listed. In one instance prediction and observation agree. Many differences of 3 or 4 minutes are noted, and there are differences of 5 or 6 minutes. Kepler formulated his law of proportionality between times and distances of Jupiter's satellites without knowing what the times are. It should be noted that the observations in the year 1899 took into consideration fluctuations that were discovered by Roemer, long after Kepler's time.

Just for the sake of having something that looks like opposition, let us try to think that Kepler was miraculously right anyway. Then, if something that may resemble Kepler's Third Law does subsist in the Jovian satellites that were known to Kepler, by what resemblance to logicality can that proportionality extend to the whole solar system, if a solar system can be supposed?

In the year 1892, a fifth satellite of Jupiter was discovered. Maybe it would conform to Kepler's law, if anybody could find out accurately in what time the faint speck does revolve. The sixth and the seventh satellites of Jupiter revolve so eccentrically that, in line of sight, their orbits intersect. Their distances are subject to very great variations; but, inasmuch as it might be said that their mean distances do conform to Kepler's Third Law, or would, if anybody could find out what their mean distances are, we go on to the others. The eighth and the ninth conform to nothing that can be asserted. If one of them goes around in one orbit at one time, the next time around it goes in some other orbit, and in some other plane. Inasmuch then as Kepler's Third Law, deduced from the system of Jupiter's satellites, cannot be thought to extend even within that minor system, one's thoughts stray into wondering what two pins in a cushion in Louisville, Ky., look like from somewhere up in the Bronx, rather than to dwell any more upon extension of any such pseudo-proportionality to the supposed solar system, as a whole.

It seems that in many of Kepler's demonstrations was this failure to have grounds for a starting-point, before extending his reasoning. He taught the doctrine of the music of the spheres, and assigned bass voices to Saturn and Jupiter, then tenor to Mars, contralto to the female planet, and soprano, or falsetto, rather, to little Mercury. And that is all very well and consistently worked out in detail, and it does seem reasonable that, if ponderous, if not lumpy, Jupiter does sing bass, the other planets join in, according to sex and huskiness--however, one does feel dissatisfied. We have dealt with Newcomb's account. But other conventionalists say that Kepler worked out his Third Law by triangulation upon Venus and Mercury when at greatest elongation, "finding" that the relation between Mercury and Venus is the same as the relation between Venus and this earth. If, according to conventionalists, there was no "proof" that this earth moves, in Kepler's time, Kepler started by assuming that this earth moves between "Venus and Mars; he assumed that the distance of Venus from the sun, at greatest elongation, represents mean distance; he assumed that observations upon Mercury indicated Mercury's orbit, an orbit that to this day defies analysis. However, for the sake of seeming to have opposition, we shall try to think that Kepler's data did give him material for the formulation of his law. His data were chiefly the observations of Tycho Brahe. But, by the very same data, Tycho had demonstrated that this earth does not move between Venus and Mars; that this earth is stationary. That stoutest of conventionalists, but at the same time seeming colleague of ours, Richard Proctor, says that Tycho Brahe's system was consistent with all data. I have never heard of an astronomer who denies this. Then the heart of modern astronomy is not Keplerism, but is one diversion of data that beat for such a monstrosity as something like Siamese Twins, serving both Keplerism and the Tychonic system. I fear that some of our attempts to find opposition are not very successful.

So far, this mediaeval doctrine, restricting to times and distances, though for all I know the planets sing proportionately as well as move proportionately, has data to interpret or to misinterpret. But, when it comes to extending Kepler's Third Law to the exterior planets, I have never read of any means that Kepler had of determining their proportional distances. He simply said that Mars and

Jupiter and Saturn were at distances that proportionalized with their times. He argued, reasonably enough, perhaps, that the slower-moving planets are the remoter, but that has nothing to do with proportional remoteness.

This is the pseudo heart of phantom astronomy.

To it Sir Isaac Newton gave a seeming of coherence.

I suspect that it was not by chance that the story of an apple should so importantly appear in two mythologies. The story of Newton and the apple was first told by Voltaire. One has suspicions of Voltaire's meanings. Suppose Newton did see an apple fall to the ground, and was so inspired, or victimized, into conceiving in terms of universal attraction. But had he tried to take a bone away from a dog, he would have had another impression, and would have been quite as well justified in explaining in terms of universal repulsion. If, as to all inter- acting things, electric, biologic, psychologic, economic, sociologic, magnetic, chemic, as well as canine, repulsion is as much of a determinant as is attraction, the Law of Gravitation, which is an attempt to explain in terms of attraction only, is as false as would be dogmas upon all other subjects if couched in terms of attraction only. So it is that the law of gravitation has been a rule of chagrin and fiasco. So, perhaps accepting, or passionately believing in every symbol of it, a Dr. Adams calculates that the Leonids will appear in November, 1899--but chagrin and fiasco--the Leonids do not appear. The planet Neptune was not discovered mathematically, because, though it was in the year 1846 somewhere near the position of the formula, in the year 1836 or 1856, it would have been nowhere near the orbit calculated by Leverrier and Adams. Some time ago, against the clamor that a Trans-Uranian planet had been discovered mathematically, it was our suggestion that, if this be not a myth, let the astronomer now discover the Trans-Neptunian planet mathematically. That there is no such mathematics, in the face of any number of learned treatises, is far more strikingly betrayed by those shining little misfortunes, the satellites of Jupiter. Satellite after satellite of Jupiter was discovered, but by accident or by observation, and not once by calculation: never were the perturbations of the earlier known satellites made the material for deducing the positions of other satellites. Astronomers have pointed to the sky, and there has been nothing; one of them pointed in four directions at once, and four times over, there was nothing; and many times when they have not pointed at all, there has been something.

Apples fall to the ground, and dogs growl, if their bones are taken away: also flowers bloom in the spring, and a trodden worm turns.

Nevertheless strong is the delusion that there is gravitational astronomy, and the great power of the Law of Gravitation, in popular respectfulness, is that it is mathematically expressed. According to my view, one might as well say that it is fetishly expressed. Descartes was as great a mathematician as Newton: veritably enough may it be said that he invented, or discovered, analytic geometry; only patriotically do Englishmen say that Newton invented, or discovered, the infinitesimal calculus. Descartes, too, formulated a law of the planets and not by a symbol was he less bewildering and convincing to the faithful, but his law was not in terms of gravitation, but in terms of vorticose motion. In the year 1732, the French Academy awarded a prize to John Bernouli, for his magnificent mathematical demonstration, which was as unintelligible as anybody's. Bernouli, too, formulated, or said he formulated, planetary inter-actions, as mathematically as any of his hypnotized admirers could have desired: it, too, was not gravitational.

The fault that I find with a great deal of mathematics in astronomy is the fault that I should find in architecture, if a temple, or a skyscraper, were supposed to prove something. Pure mathematics is architecture: it has no more place in astronomy than has the Parthenon. It is the arbitrary: it will not spoil a line nor dent a surface for a datum. There is a faint uniformity in every chaos: in discolorations on an old wall, anybody can see recognizable appearances; in such a mixture a mathematician will see squares and circles and triangles. If he would merely elaborate triangles and not apply his diagrams to theories upon the old wall itself, his constructions would be as harmless as poetry. In our metaphysics, unity cannot, of course, be the related. A mathematical expression of unity cannot, except approximately, apply to a planet, which is not final, but is part of something.

Sir Isaac Newton lived long ago. Every thought in his mind was a reflection of his era. To appraise his mind at all comprehensively, consider his works in general.

For some other instances of his love of numbers, see, in his book upon the Prophecies of Daniel, his determinations upon the eleventh horn of Daniel's fourth animal. If that demonstration be not very acceptable nowadays, some of his other works may now be archaic. For all I know Jupiter may sing bass, either smoothly or lumpily, and for all I know there may be some formulable ratio between an eleventh horn of a fourth animal and some other quantity: I complain against the dogmas that have solidified out of the vaporings of such minds, but I suppose I am not very substantial, myself. Upon general principles, I say that we take no ships of the time of Newton for models for the ships of to-day, and build and transport in ways that are magnificently, or perhaps disastrously, different, but that, at any rate, are not the same; and that the principles of biology and chemistry and all the other sciences, except astronomy, are not what they were in Newton's time, whether every one of them is a delusion or not. My complaint is that the still mediaeval science of astronomy holds back alone in a general appearance of advancement, even though there probably never has been real advancement.

There is something else to be said upon Keplerism and Newtonism. It is a squirm. I fear me that our experiences have sophisticated us. We have noted the division in Keplerism, by which, like everything else that we have examined, it is as truly interpretable one way as it is another way.

The squirm:

To lose all sense of decency and value of data, but to be agreeable; but to be like everybody else, and intend to turn our agreeableness to profit;

To agree with the astronomers that Kepler's three laws are not absolutely true, of course, but are approximations, and that the planets do move, as in Keplerian doctrine they are said to move but then to require only one demonstration that this earth is one of the planets;

To admire Newton's Principia from the beginning to the end of it, having, like almost all other admirers, never even seen a copy of it; to accept every theorem in it, without having the slightest notion what any one of them means; to accept that moving bodies do obey the laws of motion, and must move in one of the conic sections--but then to require only one demonstration that this earth is a moving body.

Kepler's three laws are popularly supposed to demonstrate that this earth moves around the sun. This is a mistake. There is something wrong with everything that is popular. As was said by us before, accept that this earth is stationary, and Kepler's doctrines apply equally well to a sun around which proportionately interspaced planets move in ellipses, the whole system moving around a central and stationary earth. All observations upon the motions of heavenly bodies are in accord with this interpretation of Kepler's laws. Then as to nothing but a quandary, which means that this earth is stationary, or which means that this earth is not stationary, just as one pleases, Sir Isaac Newton selected, or pleased himself and others. Without one datum, without one little indication more convincing one way than the other, he preferred to think that this earth is one of the moving planets. To this degree had he the "profundity" that we read about. He wrote no books upon the first and second horns of his dilemma: he simply disregarded the dilemma.

To anybody who may be controversially inclined, I offer simplification. He may feel at a disadvantage against batteries of integrals and bombardments of quaternions, transcendental functions, conics, and all the other stores of an astronomer's munitions--

Admire them. Accept that they do apply to the bodies that move around the sun. Require one demonstration that this earth is one of those bodies. For treatment of any such "demonstration," see our disquisition, or our ratiocinations upon the Three Abstrusities, or our intolerably painful attempts to write seriously upon the Three Abstrusities.

We began with three screams from an exhilarated mathematician. We have had some doubtful adventures, trying hard to pretend that monsters, or little difficulties, did really oppose us. We have reached, not the heart of a system, but the crotch of quandary.

11

WE have seen that some of the most brilliant inspirations of god-like intellects, or some of the most pestilential emanations from infected minds, have been attempts to account for the virtual changelessness of the stars. Above all other data of astronomy, that virtual changelessness of positions stands out as a crucial circumstance in my own mind. To account for constellations that have not changed in 2,000 years, astronomers say that they conceive of inconceivable distances. We shall have expressions of our own upon the virtually changeless positions of the stars; but there will be difficulties for us if the astronomers ever have found that some stars move around or with other stars. I shall take up the story of Prof. Struve and the "Companion of Procyon," with more detail, for the sake of some more light upon refinement, exactness, accuracy in astronomy, and for the sake of belittling, or for the sake of sneering, or anything else that anybody may choose to call it.

Prof. Struve's announcement of his discovery of the "Companion of Procyon" is published in Monthly Notices, 33-430--that, upon the 19th of March, 1873, Struve had discovered the companion of Procyon, having compared it micrometrically, having tested his observations with three determinations of position-angle, three measures of distance, and three additional determinations of position-angle, finding all in "excellent agreement." No optical illusion could be possible, it is said, because another astronomer, Lindemann, had seen the object. Technically, Struve publishes a table of his observations: sidereal time, distances, position- angles; from March 19 to April 2, 1873, after which his observations had to be discontinued until the following year. In Monthly Notices, 34-355, are published the resumed observations. Struve says that Auwers would not accept the discovery, unless, in the year that had elapsed, the "companion" had shown increase in position, consistent with theory. Struve writes--"This increase has really shown itself in the most remarkable manner." Therefore, he considers it "decisively established" that the object of his observations was the object of Auwers' calculations. He says that Ceraski, of Moscow, had seen the "companion," "without being warned of the place where it was to be looked for."

However--see back some chapters.

It may be said that, nevertheless, other stars have companions that do move as they should move. Later we shall consider this subject, thinking that it may be that lights have been seen to change position near some stars, but that never has a star revolved around another star, as to fit palaeo-astronomic theory it should. I take for a basis of analogy that never has one sat in a park and watched a tree revolve around one, but that given the affliction, or the endowment, of an astronomer, illusion of such a revolution one may have. We sit in a park. We notice a tree. Wherever we get the notion, we do have the notion that the tree has moved. Then, farther along, we notice another tree, and, as an indication of our vivid imagination or something else, we think it is the same tree, farther along. After that we pick out tree after tree, farther along, and, convinced that it is the same tree, of course conclude that the thing is revolving around us. Exactness and refinement develop: we compute the elements of its orbit. We close our eyes and predict where the tree will be when next we look; and there, by the same process of selection and identification, it is where it "should" be. And if we have something of almost everybody's mania for speed, we make that damn thing spin around with such velocity that we, too, reel in a chaos of very much unsettled botanic conventions. There is nothing far-fetched in this analogy, except the factor of velocity. Goldschmidt did announce that there were half a dozen faint points of light around Sirius, and it was Dawes' suspicion that Clark had arbitrarily picked out one of them. It is our expression that all around Sirius, at various distances

from Sirius, faint points of light were seen, and that at first, even for the first sixteen years, astronomers were not thoroughly hypnotized, and would not pick out the especial point of light that they should have picked out, so that there was nothing like agreement between the calculated and the observed orbit. Besides the irreconcilable observations noted by Flammarion, see the Intel. Obs., 1-482, for others. Then came standardized seeing. So, in the Observatory, 20-73, is published a set of observations, in the year 1896, upon the "Companion of Sirius," placing it exactly where it should be. Nevertheless, under this set of observations is published another set, so different that the Editor asks--"Does this mean that there are two companions?"

Dark Companions require a little more eliminative treatment. So the variable nebulae, then--and do dark nebulae revolve around light nebulae? For instances of variable nebulae, see Mems. R. A. S., 49-214; Comptes Rendus, 59-637; Monthly Notices, 38-104. It may be said that they are not of the Algol-type. Neither is Algol, we have shown.

According to the compulsions of data, our idea is that the stars that seem to be fixed in position are fixed in position, so now "proper motion" is as irreconcilable to us as relative motions.

As to "proper motion," the situation is this:

The stars that were catalogued 2,000 years ago have virtually not changed, or, if there be refinement in modern astronomy, have changed no more than a little more nearly exact charting would account for; but, in astronomic theory, the stars are said to be thought of as flying apart at unthinkable velocity; so then evidence of changed positions of stars is welcome to astronomers. As to well-known constellations, it cannot be said that there has been change; so, with several exceptions, "proper motion" is attributed to stars that are not well-known.

The result is an amusing trap. Great proper motion is said to indicate relative nearness to this earth. Of the twenty-five stars of supposed greatest proper motion, all but two are faintest of stars; so these twenty-three are said to be nearest this earth. But when astronomers take the relative parallax of a star, by reference to a fainter star, they agree that the fainter star, because fainter, is farther away. So one time faintness associates with nearness, and then conveniences change, and faintness associates with farness, and the whole subject so associates with humorousness, that if we're going to be serious at all in these expressions of ours we had better pass on.

Observatory, March, 1914:

A group of three stars that disappeared.

If three stars disappeared at once, they were acted upon by something that affected all in common. Try to think of someone force that would not tear the seeable into visible rags, that could blot out three stars, if they were trillions of miles apart. If they were close together that ends the explanation that only because stars are trillions of miles apart have they, for at least 2,000 years, seemed to hold the same relative positions.

In Agnes Clerke's System of the Stars, are cited many instances of stars that seem to be so closely related that it seems impossible to think that they are trillions, or billions, or millions of miles apart: such formations as "seven aligned stars appearing to be strung on a silvery filament." There are loops of stars in a cluster in Auriga; lines and arches in Opiuchus; zig-zag figures in Sagittarius. As to stars that not only seem close together but that are colored alike, Miss Clerke expresses her feeling that they are close together--"If these colors be inherent, it is difficult to believe that the stars distinguished by them are simply thrown together by perspective." As to figures in Sagittarius, Fison (Recent Advances in Astronomy) cites an instance of 30 small stars in the form of a forked twig, with dark rifts parallel. According to Fison, probability is overwhelmingly against the three uncommon stars in the belt of Orion falling into a straight line, by chance distribution, considering also that below this line is another of five faint stars parallel. There are dark lanes or rifts in the Milky Way that are like branches from main lanes or rifts, and the rifts sometimes have well-defined edges. In many regions where there are dark rifts there are lines of stars that are roughly parallel

That it is not distances apart that have held the stars from changing relatively to one another, because there are hosts of indications that some stars are close together, and are, or have been, affected, in common, by local formative forces.

For a detailed comparison, by J. E. Gore, of stars of today with stars catalogued by Al-Sufi about 1,000 years ago, see the Observatory, vol. 23. The stars have not changed in position, but it does seem that there have been many changes in magnitude.

Other changes--Pubs. Astro. Soc. Pacific, No. 185 (1920)--discovery of the seventeenth new star in one nebula (Andromeda). For lists of stars that have disappeared, see Monthly Notices, 8-16; 10-18; 11-47; Sidereal Messenger, 6-320; Jour. B. A. A., 14-255. Nebulae that have disappeared--see Amer. Jour. Sci., 2-33- 436; Clerke's System of the Stars, p. 293; Nature, 30-20.

In the Sidereal Messenger, 5-269, Prof. Colbert writes that, upon August 20, 1886, an astronomer, in Chicago, saw, for about half an hour, a small comet-like projection from the star Zeta, in Cassiopeia.

So, then, changes have been seen at the distance of the stars.

When the new star in Perseus appeared, in February, 1901, it was a point of light. Something went out from it, giving it in six months a diameter equal to half the apparent diameter of the moon. The appearances looked structural. To say loosely that they were light-effects, something like a halo, perhaps, is to ignore their complexity and duration and differences. According to Newcomb, who is occasionally quotable in our favor, these radiations were not mere light-rays, because they did not. go out uniformly from the star, but moved out variously and knotted and curved.

It was visible motion, at the distance of Nova Persei.

In Monthly Notices, 58-334, Dr. Espin writes that, upon the night of Jan. 16, 1898, he saw something that looked like a cloud in Perseus. It could have been nothing in the atmosphere of this earth, nor anything far from the constellation, because he saw it again in Perseus, upon January 24. He writes that, upon February 17, Mr. Heath and Dr. Halm saw it, like a cloud, dimming and discoloring stars shining through it. At the meeting of the British Astronomical Association, Feb. 23, 1898 (Jour. B. A. A., 8-216), Dr. Espin described this appearance and answered questions. "It was not a nebula, and was not like one." "Whatever it was it had the peculiar property of dimming and blotting out stars."

This thing moved into Perseus and then moved away.

Clerke, The System of the Stars, p. 295--a nebula that changed position abruptly, between the years 1833 and 1835, and then changed no more. According to Sir John Herschel, a star was central in this nebula, when observed in 1827, and in 1833, but, in August, 1835, the star was upon the eastern side of the nebula.

That it is not distance from this earth that has kept changes of position of the stars from being seen, for 2,000 years, because occasional, abrupt changes of position have been seen at the distance of the stars.

That, whether there be a shell-like, revolving composition, holding the stars in position, and in which the stars are openings, admitting light from an existence external to the shell, or not, all stars are at about the same distance from this earth as they would be if this earth were stationary and central to such a shell, revolving around it--

According to the aberration-forms of the stars.

All stars, at the pole of the ecliptic, describe circles annually; stars lower down describe ellipses that reduce more and more the farther down they are, until at the ecliptic they describe straight lines yearly.

Suppose all the stars to be openings, fixed in position relatively to one another, in some inter-spacing substance. Conceive of a gyration to the whole aggregation, and relatively to a central and stationary earth: then, as seen from this earth, all would describe circles, near the axis, ellipses lower down, and straight lines at the limit of transformation. If all were at the same distance from this earth, or if all were points in one gyrating concave formation, equidistant at all points from the central earth, all would have the same amplitude. All aberration- forms of the stars, whether of brilliant or faint stars, whether circles or ellipses or straight lines, have the same amplitude: about 41 seconds of arc.

If all stars are points of light admitted from externality, held fixed and apart in one shell-like composition that is opaque in some parts and translucent in some parts and perforated generally--

The Gegenschein--

That we have indication that there is such a shell around our existence.

The Gegenschein is a round patch of light in the sky. It seems to be reflected sunlight, at night, because it keeps position about opposite the sun's.

The crux:

Reflected sunlight--but reflecting from what?

That the sky is a matrix in which the stars are openings, and that, upon the inner, concave surface of this celestial shell, the sun casts its light, even if the earth is between, no more blotted out in the middle by the intervening earth than often to considerable degree is its light blotted out upon the moon during an eclipse of the moon, occupying no time in traveling the distance of the stars and back to this earth, because the stars are near, or because there is no velocity of light.

Suppose the Gegenschein could be a reflection of sunlight from anything at a distance less than the distance of the stars. It would have parallax against its background of stars.

Observatory, 17-47:

"The Gegenschein has no parallax."

At the meeting of the Royal Astronomical Society, Jan. 11, 1878, was read a paper by W. F. Denning. It was, by its implications, one of the most exciting documents in history. The subject was: "Suspected repetitions in meteor-showers." Mr. Denning listed twenty-two radiants that lasted from three to four months each.

In the year 1799, Humboldt noticed that the paths of meteors, when parts of one display, led back to one point of common origin, or one point from which all the meteors had radiated. This is the radiant-point, or the radiant. When a radiant occurs under a constellation, the meteors are named relatively. In the extraordinary meteoric display of Nov. 13-14, 1833, there was a circumstance that was as extraordinary as the display itself: that, though this earth is supposed to rotate upon its axis, giving to the stars the appearance of revolving nightly, and supposed to revolve around the sun, so affecting the seeming motions of the stars, these meteors of November, 1833, began under the constellation Leo, and six hours later, though Leo had changed position in the sky, had changed with, and seemed still coming from, Leo.

There was no parallax along the great base line from Canada to Florida.

Then these meteors did come from Leo, or parallax, or absence of parallax, is meaningless. The circumstance of precise position maintained under a moving constellation upon the night of Nov. 13-14, 1833, becomes insignificant relatively to Denning's data of such synchronization with a duration of months. When a radiant-point remains under Leo or Lyra, night after night, month after month, it is either that something is shifting it, without parallax, in exact coincidence with a doubly shifting constellation, which is so unthinkable that Denning says, "I cannot explain," or that the constellation is the radiant-point, in which case maintenance of precise position under it is unthinkable if it be far away--

That the stars are near.

Think of a ship, slowly sailing past a seacoast town, firing with smokeless powder, say. Shells from it burst before quite reaching the town, and all explosion-points are in line between the city and the ship, or are traceable to one such radiant. The bombardment continues. The ship moves slowly. Still all points of exploding shells are traceable to one point between the ship and the town. The bombardment goes on and goes on and goes on, and the ship is far from its first position. The point of exploding shells is still between the ship and the town. Wise men in the town say that the shells are not coming from the ship. They say this because formerly they had said that shells could not come from a ship. They reason: therefore shells are not coming from this ship. They are asked how, then, the point of explosion could so shift exactly in line with the moving ship. If there be a W. F. Denning among them, he will say, "I cannot explain." But the other wise men will be like Prof. Moulton, for instance. In his books, Prof. Moulton writes a great deal upon the subject of meteors, but he does not mention the meteors that, for months at a time, appear between observers and a shifting constellation.

There are other considerations. The shells are heard to explode. So then they explode near the town. But there is something the matter with that smokeless powder aboard ship: very feeble projectile-force, because also must the shells be exploding near the ship, or the radiant-point would not have the same background, as seen from different parts of the town. Then, in this town, inhabitants, provided they be not wise men, will conclude that, if the explosion-point is near the town, and is also near the ship, the ship is near the town--

Leo and Lyra and Andromeda--argosies that sail the sky and that bombard this earth--and that they are not far away.

And some of us there may be who, instead of trying to speculate upon an unthinkable remoteness, will suffer a sensitiveness to proximity instead; enter a new revolt against a black encompassment that glitters with a light beyond, and wonder what exists in a brilliant environment not far away-- and a new anguish for hyperaesthesia upon this earth: a suffocating consciousness of the pressure of the stars.

The Sickle of Leo, from which come the Leonids, gleams like a great question-mark in the sky.

The answer--

But God knows what the answer to anything is. Perhaps it is that the stars are very close indeed.

12

WE try to have independent expressions. Accept that it is not distance that has held the stars in unchanging position, if occasional, abrupt change of position has been seen at the distance of the stars, and it is implied that the not enormously distant stars are all about equally far away from this earth; or some would be greatly particularized, and that this earth does not move in an orbit, or stars would be seasonally particularized, but would not be, if the stars, in one composition revolve; also if this earth be relatively close to all stars, if many changes of magnitude and of appearance and disappearance have been seen at the distance of the stars, and, if, in the revolutions of the stars, they do not swirl in displacements as bewildering as a blizzard of luminous snowflakes, and if no state of inter-repulsion can be thought of, especially as many stars merge into others, this composition is a substantial, concave formation, or shell- like enclosure in which stars are points. So many of the expressions .in the preceding chapter imply others, or all others. However, we have tried to have independent expressions. Of course we realize that the supposed difference between inductive and deductive reasoning is a false demarcation; nevertheless we feel that deductions piled upon other de-ductions are only architecture, and a great deal in this book expresses the notion that architecture should be kept in its own place. Our general expression is not that there should be no architecture and no mathematics in astronomy, or neo-astronomy; not that there should be no poetry in biology; no chemistry in physiology--but that "pure" architecture or "pure" mathematics, biology, chemistry, has its own field, even though each is inextricably bound up with all the other aspects of being. So of course the very thing that we object to in its extreme manifestations is essential to us in some degree, and the deductive is findable somewhere in every one of our inductions, and we are not insensible to what we think is the gracefulness of some of the converging lines of our own constructions. We are not revolting against aspects, but against emphases and intrusions.

This first part of our work is what we consider neo-astronomic; and now to show that we have no rabidity against the mathematical except when over-emphasized, or misapplied, our language is that all expressions so far developed are to us of about 50% acceptability. A far greater attempted independence is coming, a second part of this work, considering phenomena so different that, if we term the first part of our explorations "neo-astronomic," even. some other term by which to designate the field of the second part will have to be thought of, and the word "extra-geographic" seems best for it. If in these two fields, our at least temporary conclusions be the same, we shall be impressed, in spite of all our cynicisms as to "agreements."

Neo-astronomy:

This supposed solar-system--an egg-like organism that is shelled away from external light and life--this central and stationary earth its nucleus--around it a revolving shell, in which the stars are pores, or functioning channels, through some of which spray irradiating fountains said to be "meteoric," but perhaps electric--in which the nebulae are translucent patches, and in which the many dark parts are areas of opaque, structural substance--and that the stars are not trillions nor even millions of miles away--with proportional reductions of all internal distances, so that the planets are not millions, nor even hundreds of thousands of miles away.

We conceive of the variability of the stars and the nebulae in terms of the incidence of external light upon a revolving shell and fluctuating passage through light-admitting points and parts. We conceive of all things being rhythmic, so, if stars be pores in a substance, that matrix must be subject to some changes, which may be of different periodicities in different regions. There may be local vortices in the most rigid substance, and so stars, or pores, might revolve around one another, but our tendency is to think that if light companions there be to some stars, they are reflections of light, passing through channels, upon surrounding substance, flickering from one position to another in the small undulations of this environment. So there may be other displacements, differences of magnitude, new openings and closings in a substance that is not absolutely rigid. So "proper motion" might be accounted for, but my own preference is to think, as to such stars as 1830 Groombridge and Barnard's "run-away star," that they are planets--also that some of the comets, especially the tailless comets, some of which have been seen to obscure stars, so that evidently they are not wisps of highly attenuated matter, are planets, all of them not conventionally recognized as planets, because of eccentricity and remoteness from the ecliptic, two departures, however, that many of the minor planets make to great degree. If some of these bodies be planets, the irregularities of some of them are consistent with the irregularities of Jupiter's satellites.

I suggest that a combination of the Ptolemaic and the Tychonic doctrines is in good accord with all the phenomena that we have considered, and with all planetary motions that we have had no occasion to pay much attention to--that the sun, carrying Mercury and Venus with him, revolves at a distance of a few thousand miles, or a few tens of thousands of miles, in a rising and falling spiral around this virtually, but not absolutely, stationary earth, which, according to modern investigations, is more top-shaped than spherical; moon, a few thousand miles away, revolving around this nucleus; and the exterior planets not only revolving around this whole central arrangement, but approaching and receding, in loops, also, quite as they seem, to the remotest of them preposterously near, according to conventional "determinations."

So all the phenomena of the skies may be explained. But all were explained in another way by Copernicus, in another way by Ptolemy, and in still another way by Tycho Brahe. One supposes that there are other ways. If there be a distant object, and, if one school of wise men can by their reasoning processes excellently demonstrate that it is a tree, another school positively determine that it is a house, and other investigators of the highest authoritativeness variously find and prove that it is a cloud or a buffalo or a geranium, why then, their reasoning processes may be admired but not trusted. Right at the heart of our opposition, and right at the heart of our own expressions, is the fatality that there is no reasoning, no logic, no explanation resembling the illusions in the vainglories of common suppositions. There is only the process of correlating to, or organizing or systematizing around, something that is arbitrarily taken for a base, or a dominant doctrine, or a major premise--the process of assimilating with something else, making agreement with something else, or interpreting in terms of something else, which supposed base is never itself final, but was originally an assimilation with still something else.

I typify the result of all examinations of all principles or laws or dominant thoughts, scientific, philosophic, or theologic, in what we find in examining the pronouncement that motion follows the least resistance: That motion follows least resistance.

How are we to identify least resistance? If motion follows it.

Then motion goes where motion goes.

If nothing can be positively distinguished from anything else there can be no positive logic, which is attempted positive distinguishment. Consider the popular "base" that Capital is tyranny, and almost utmost wickedness, and that Labor is pure and idealistic. But one's labor is one's capital, and capital that is not working is in no sense implicated in this conflict.

Nevertheless we now give up our early suspicion that our whole existence is a leper of the skies, quaking and cringing through space, having the isolation that astronomers suppose, because other celestial forms of being fly from infection--

That, if shelled away from external light and life, it is so surrounded and so protected in the same cause and functioning as that of similarly encompassed forms subsidiary to it--that our existence is super-embryonic.

Darkness of night and of lives and of thoughts--super-uterine entombment. Blackness of the unborn, quasi-illumined periodically by the little sun, which is not light, but less dark.

Then we think of an organism that needs no base, and needs nothing of finality, nor of special guidance to any part local to it, because all parts partake of the pre-determined development of the whole. Consequently our spleens subside, and our frequently unmannerly derisions are hushed by recognitions--that all organizations of thought must be baseless in themselves, and of course be not final, or they could not change, and must bear within themselves those elements that will, in time, destroy them--that seeming solidities that pass away, in phantom- successions, are functionaries relatively to their periods, and express the passage from phase to phase of all things embryonic.

So it is that one who searches for fundamentals comes to bifurcations; never to a base; only to a quandary. In our own field, let there be any acceptable finding.

It indicates that the earth moves around the sun. Just as truly it indicates that the sun moves around the earth. What is it that determines which will be accepted, hypnotically blinding the faithful to the other aspect? Our own expression is upon Development as serial reactions to successive Dominants. Let the dominant spirit of an era require that this earth be remote and isolated; Keplerism will support it: let the dominant change to a spirit of expansion, which would be impossible under such remoteness and isolation; Keplerism will support, or will not especially oppose, the new dominant. This is the essential process of embryonic growth, by which the same protoplasmic substance responds differently in different phases.

But I do not think that all data are so plastic. There are some that will not assimilate with a prevailing doctrine. They can have no effect upon an arbitrary system of thought, or a system subconsciously induced, in its time of dominance: they will simply be disregarded.

We have reached our catalogue of the sights and the sounds to which all that we have so far considered is merely introductory. For them there are either no conventional explanations or poor insufficiencies half-heartedly offered. Our data are glimpses of an epoch that is approaching with far-away explosions. It is vibrating on its edges with the tread of distant space-armies. Already it has pictured in the sky visions that signify new excitements, even now lapping over into the affairs of a self-disgusted, played-out hermitage.

We assemble the data. Unhappily, we shall be unable to resist the temptation to reason and theorize. May Super-embryology have mercy upon our own syllogisms. We consider that we are entitled to at least 13 pages of gross and stupid errors.

After that we shall have to explain.

Part II

13

JUNE, 1801--a mirage of an unknown city. It was seen, for more than an hour, at Youghal, Co. Cork, Ireland--a representation of mansions, surrounded by shrubbery and white palings--forests behind. In October, 1796, a mirage of a walled town had been seen distinctly for half an hour at Youghal. Upon March 9, 1797, had been seen a mirage of a walled town.

Feb. 7, 1802--an unknown body that was seen, by Fritsch, of Magdeburg, to cross the sun (Observatory, 3-136).

Oct. 10, 1802--an unknown dark body was seen, by Fritsch, rapidly crossing the sun (Comptes Rendus, 83-587). Between 10 and 11 o'clock, morning of Oct. 8, 1803, a stone fell from the sky, at the town of Apt, France. About eight hours later, "some persons believed that they felt an earthquake" (Rept. B. A., 1854-53).

Upon August 11, 1805, an explosive sound was heard at East Haddam, Connecticut. There are records of six prior sounds, as if of explosions, that were heard at East Haddam, beginning with the year 1791, but, unrecorded, the sounds had attracted attention for a century, and had been called the "Moodus" sounds, by the Indians. For the best account of the "Moodus" sounds, see the Amer. Jour. Sci., 39-339. Here a writer tries to show the phenomena were subterranean, but says that there was no satisfactory explanation.

Upon the 2nd of April, 1808, over the town of Pignerol, Piedmont, Italy, a loud sound was heard: in many places in Piedmont an earthquake was felt. In the Rept. B. A., 1854-68, it is said that aerial phenomena did occur; that, before the explosion, luminous objects had been seen in the sky over Pignerol, and that in several of the communes in the Alps aerial sounds, as if of innumerable stones colliding, had been heard, and that quakes had been felt. From April 2 to April 8, forty shocks were recorded at Pignerol; sounds like cannonading were heard at Barga. Upon the 18th of April, two detonations were heard at La Tour, and a luminous object was seen in the sky. The supposition, or almost absolute belief of most persons is that from the 2nd to the 18th of April this earth had moved far in its orbit and was rotating so that, if one should explain that probably meteors had exploded here, it could not very well be thought that more meteors were continuing to pick out this one point upon a doubly moving planet. But something was specially related to this one local sky. Upon the 19th of April, a stone fell from the sky at Borgo San Donnino, about 40 miles east of Piedmont (Rept. B. A., 1860). Sounds like cannonading were heard almost every day in this small region.

Upon the 13th of May, a red cloud such as marks the place of a meteoric explosion was seen in the sky. Throughout the rest of the year, phenomena that are now listed as "earthquakes" occurred in Piedmont. The last occurrence of which I have record was upon Jan. 22, 1810.

Feb. 9, 1812--two explosive sounds at East Haddam (Amer. Jour. Sci., 39-339). July 5, 1812--one explosive sound at East Haddam (Amer. Jour. Sci., 39-339).

Oct. 28, 1812--"phantom soldiers" at Havarah Park, near Ripley, England (Edinburgh Annual Register, 1812-II-124). When such appearances are explained by meteorologists, they are said to be displays of the aurora borealis. Psychic research explains variously. The physicists say that they are mirages of troops marching somewhere at a distance.

Night of July 31, 1813--flashes of light in the sky of Tottenham, near London (Year Book of Facts, 1853-272). The sky was clear. The flashes were attributed to a storm at Hastings, 65 miles away. We note not only that the planet Mars was in opposition at this time (July 30), but in one of the nearest of its oppositions in the 19th century.

Dec. 28, 1813--an explosive sound at East Haddam.

Feb. 2, 1816--a quake at Lisbon. There was something in the sky. Extraordinary sounds were heard, but were attributed to "flocks of birds." But six hours later something was seen in the sky: it is said to have been a meteor (Rept. B. A., 1854-106).

Since the year 1788, many earthquakes, or concussions that were listed as earthquakes, had occurred at the town of Comrie, Perthshire, Scotland. Seventeen instances were recorded in the year 1795. Almost all records of the phenomena of Comrie start with the year 1788, but, in Macara's Guide to Creifi, it is said that the disturbances were recorded as far back as the year 1597. They were slight shocks, and until the occurrence upon Aug. 13, 1816, conventional explanations, excluding all thought of relations with anything in the sky, seemed adequate enough. But, in an account in the London Times, Aug. 21, 1816, it is said that, at the time of the quake of August 13, a luminous object, or a "small meteor," had been seen at Dunkeld, near Comrie; and, according to David Milne (Edin. New Phil. Jour., 31-110), a resident of Comrie had reported "a large luminous body, bent like a crescent, which stretched itself over the heavens."

There was another quake in Scotland (Inverness) June 30, 1817. It is said that hot water fell from the sky (Rept. B. A., 1854-112).

Jan. 6, 1818--an unknown body that crossed the sun, according to Loft, of Ipswich; observed about three hours and a half (Quar. Jour. Roy. Inst., 5-117).

Five unknown bodies that were seen, upon June 26, 1819, crossing the sun, according to Gruithuisen (An. Sci. Disc., 1860-411). Also, upon this day, Pastorff saw something that he thought was a comet, which was then somewhere near the sun, but which, according to Olbers, could not have been the comet (Webb, Celestial Objects, p. 40).

Upon Aug. 28, 1819, there was a violent quake at Irkutsk, Siberia. There had been two shocks upon Aug. 22, 1813 (Rept. B. A., 1854-101). Upon April 6, 1805, or March 25, according to the Russian calendar, two stones had fallen from the sky at Irkutsk (Rept. B. A., 1860-12). One of these stones is now in the South Kensington Museum, London. Another violent shock at Irkutsk, April 7, 1820 (Rept. B. A., 1854-128).

Unknown bodies in the sky, in the year 1820, February 12 and April 27 (Comptes Rendus, 83-314). Things that marched in the sky--see Arago's Oeuvres, 11-576, or Annales de Chimie, 30-417--objects that were seen by many persons, in the streets of Embrun, during the eclipse of Sept. 7, 1820, moving in straight lines, turning and retracting in the same straight lines, all of them separated by uniform spaces.

Early in the year 1821--and a light shone out on the moon--a bright point of light in the lunar crater Aristarchus, which was in the dark at the time. It was seen, upon the 4th and the 7th of February, by Capt. Kater (An. Reg., 1821-689); and upon the 5th by Dr. Olbers (Mems. R. A. S., 1-159). It was a light like a star, and was seen again, May 4th and 6th, by the Rev. M. Ward and by Francis Bailey (Mems. R. A. S., 1-159). At Cape Town, nights of Nov. 28th and 29th, 1821, again a star-like light was seen upon the moon (Phil. Trans., 112-237).

Quar. Jour. Roy. Inst., 20-417:

That, early in the morning of March 20, 1822, detonations were heard at Melida, an island in the Adriatic. All day, at intervals, the sounds were heard. They were like cannonading, and it was supposed that they came from a vessel, or from Turkish artillery, practicing in some frontier village. For thirty days the detonations continued, sometimes thirty or forty, sometimes several hundred, a day. Upon April 13, 1822, it seems, according to description, that clearly enough was there an explosion in the sky of Comrie, and a concussion of the ground--"two loud reports, one apparently over our heads, and the other, which followed immediately, under our feet" (Edin. New Phil. Jour., 31-119).

July 15, 1822--the fall of perhaps unknown seeds from perhaps an unknown world--a great quantity of little round seeds that fell from the sky at Marienwerder, Germany. They were unknown to the inhabitants, who tried to cook them, but found that boiling seemed to have no effect upon them. Wherever they came from, they were brought down by a storm, and two days later, more of them fell, in a storm, in Silesia. It is said that these corpuscles were identified by some scientists as seeds of Galium spurium, but that other scientists disagreed. Later more of them fell at Posen, Mecklenburg. See Bull. des Sci. (math., astro., etc.) 1-1-298.

Aug. 19, 1822--a tremendous detonation at Melida--others continuing several days.

Oct. 23, 1822--two unknown dark bodies crossing the sun; observed by Pastorff (An. Sci. Disc., 1860-411).

An unknown, shining thing--it was seen, by Webb, May 22, 1823, near the planet Venus (Nature, 14-19).

More unknowns, in the year 1823--see Comptes Rendus, 49-811 and Webb's Celestial Objects, p. 43.

February, 1824--the sounds of Melida.

Upon Feb. II, 1824, a slight shock was felt at Irkutsk, Siberia (Rept. B. A., 1854-124). Upon February 18, or, according to other accounts, upon May 14, a stone that weighed five pounds, fell from the sky at Irkutsk (Rept. B. A., 1860-70).

Three severe shocks at Irkutsk, March 8, 1824 (Rept. B. A., 1854-124). September, 1824--the sounds of Melida.

At five o'clock, morning of Oct. 20, 1824, a light was seen upon the dark part of the moon, by Gruithuisen. It disappeared. Six minutes later it appeared again, disappeared again, and then flashed intermittently, until 5:30 A.M., when sunrise ended the observations (Sci. Amer. Sup., 7-2712). And, upon Jan. 22, 1825, again shone out the star-like light of Aristarchus, reported by the Rev. J. B. Emmett (Annals of Philosophy, 28-338).

The last sounds of Melida of which I have record, were heard in March, 1825. If these detonations did come from the sky, there was something that, for at least three years, was situated over, or was in some other way specially related to, this one small part of this earth's surface, subversively to all supposed principles of astronomy and geodesy. It is said that, to find out whether the sounds did come from the sky, or not, the Preteur of Melida went into underground caverns to listen. It is said that there the sounds could not be heard.

14

AND our own underground investigations--and whether there is something in the sky or not. We are in a hole in time. Cavern of Conventional Science--walls that are dogmas, from which drips ancient wisdom in a patter of slimy opinions--but we have heard a storm of data outside--

Of beings that march in the sky, and of a beacon on the moon--another dark body crosses the sun. Somewhere near Melida there is cannonading, and another stone falls from the sky, at Irkutsk, Siberia; and unknown grain falls from an unknown world, and there are flashes in the sky when the planet Mars is near.

In a farrago of lights and sounds and forms, I feel the presence of possible classifications that may thread a pattern of attempt to find out something. My attention is attracted by a streak of events that is beaded with little star-like points of light. First we shall find out what we can, as to the moon. In one of the numbers of the Observatory, an eminent authority, in some fields of research, is quoted as to the probable distance of the moon. According to his determinations, the moon is 37 miles away. He explains most reasonably: he is Mr.

G. B. Shaw. But by conventional doctrine, the moon is 240,000 miles away. My own idea is that somewhere between determinations by a Shaw and determinations by a Newcomb, we could find many acceptances.

I prefer questionable determinations, myself, or at any rate examinations that end up with questions or considerable latitude. It may be that as to the volcanoes of the moon we can find material for at least a seemingly intelligent question, if no statements are possible as to the size and the distance of the moon. The larger volcanoes of this earth are about three miles in diameter, though the craters of Haleakla, Hawaii, and Aso San, Japan, are seven miles across. But the larger volcanoes of the relatively little moon are said to be sixty miles across, though several are said to be twice that size. And I start off with just about the impression of disproportionality that I should have, if someone should tell me of a pygmy with ears five feet long.

Is there any somewhat good reason for thinking that the volcanic craters of the little moon are larger than, or particularly different in any other way from, the craters of this earth?

If not, we have a direct unit of measurement, according to which the moon is not 2,160, but about 100, miles in diameter.

How far away does one suppose to be an object with something like that diameter, and of the seeming size of the moon?

The astronomers explain. They argue that gravitation must be less powerful upon the moon than upon this earth, and that therefore larger volcanic formations could have been cast up on the moon. We explain. We argue that volcanic force must be less powerful upon the moon than upon this earth, and that therefore larger volcanic formations could not have been cast up on the moon.

The disproportionality that has impressed me has offended more conventional aesthetics than mine. Prof. See, for instance, has tried to explain that the lunar formations are not craters but are effects of bombardment by vast meteors, which spared this earth, for some reason not made clear. Viscid moon--meteor pops in--up splash walls and a central cone. If Prof. See will jump in swimming someday, and then go back some weeks later to see how big a splash he made, he will have other ideas upon such supposed persistences. The moon would have to have been virtually liquid to fit his theory, because there are no partly embedded, vast, round meteors protruding anywhere.

There have been lights like signals upon the moon. There are two conventional explanations: reflected sunlight and volcanic action. Of course, ultra-conventionalists do not admit that in our own times there has been even volcanic action upon the moon. Our instances will be of lights upon the dark part of the moon, and there are good reasons for thinking that our data do not relate to volcanic action. In volcanic eruptions upon this earth the glow is so accompanied by great volumes of smoke that a clear, definite point of light would seem not to be the appearance from a distance.

For Webb's account of a brilliant display of minute dots and streaks of light, in the Mare Crisium, July 4, 1832, see Astro. Reg.; 20-165. I have records of half a dozen similar illuminations here, in about 120 years, all of them when the Mare Crisium was in darkness. There can be no commonplace explanation for such spectacles, or they would have occurred oftener; nevertheless the Mare Crisium is a wide, open region, and at times there may have been uncommon percolations of sunlight, and I shall list no more of these interesting events that seem to me to have been like carnivals upon the moon.

Dec. 22, 1835--the star-like light in Aristarchus--reported by Francis Bailey--see Proctor's Myths and Marvels, p. 329.

Feb. 13, 1836--in the western crater of Messier--according to Gruithuisen (Sci. Amer. Sup., 7-2629)--two straight lines of light; between them a dark band that was covered with luminous points.

Upon the nights of March 18 and 19, 1847, large luminous spots were seen upon the dark part of the moon, and a general glow upon the upper limb, by the Rev. T. Rankin and Prof. Chevalier (Rept. B. A., 1847-18). The whole shaded part of the disc seemed to be a mixture of lights and shades. Upon the night of the 19th, there was a similar appearance upon this earth, an aurora, according to the London newspapers. It looks as if both the moon and this earth were affected by the same illumination, said to have been auroral. I offer this occurrence as indication that the moon is nearby, if moon and earth could be so affected in common.

But by signaling, I mean something like the appearance that was seen, by Hodgson, upon the dark part of the moon, night of Dec. 11, 1847--a bright light that flashed intermittently. Upon the next night it was seen again (Monthly Notices R. A. S., 8-55).

The oppositions of Mars occur once in about two years. and two months. In conventional terms, the eccentricity of the orbit of Mars is greater than the eccentricity of the orbit of this earth, and the part of its orbit that is traversed by this earth in August is nearest the orbit of Mars. When this earth is between Mars and the sun, Mars is said to be in opposition, and this is the position of nearest approach: when opposition occurs in August, that is the most favorable opposition. After that, every two years and about two months, the oppositions are less favorable, until the least favorable of all, in February, after which favorableness increases up to the climacteric opposition in August again. This is a cycle of changing proximities within a period of about fifteen years. In October, 1862, Lockyer saw a spot like a long train of clouds on Mars, and several days later Secchi saw a spot on Mars. And if that were signaling, it is very meager material upon which to suppose anything. And May 8-22, 1873--white spots on Mars. But, upon June 17, 1873, two months after nearest approach, but still in the period of opposition of Mars, there was either an extraordinary occurrence, or the extraordinariness is in our interpretation. See Rept. B. A., 1874-272. A luminous object came to this earth, and was seen and heard upon the night of June 17, 1873, to explode in the sky of Hungary, Austria, and Bohemia. In the words of various witnesses, termed according to their knowledge, the object was seen seemingly coming from Mars, or from "the red star in the south," where Mars was at the time. Our data were collected by Dr. Galle. The towns of Rybnik and Ratibor, Upper Silesia, are 15 miles apart. Without parallax, this luminous thing was seen from these points "to emerge and separate itself from the disc of the planet Mars." It so happens that we have a definite observation from one of these towns. At Rybnik, Dr. Sage was looking at Mars, at the time. He saw the luminous object "apparently issue from the planet." There is another circumstance, and for its reception our credulity, or our enlightenment, has been prepared. If this thing did come from Mars, it came from the planet to the point where it exploded in about 5 seconds: from the point of explosion, the sound traveled in several minutes. We have a description from Dr. Sage that indicates that a bolt of some kind, perhaps electric, did shoot from Mars, and that the planet quaked with the shock--"Dr.

Sage was looking attentively at the planet Mars, when he thus saw the meteor apparently issue from it, and the planet appear as if it was breaking up and dividing into two parts."

Some of the greatest surprises in commonplace experience are discoveries of the nearness of that which was supposed to be the inaccessibly remote.

It seems that the moon is close to this earth, because of the phenomenon of "earthshine." The same appearance has been seen upon the planet Venus. If upon the moon, it is light reflecting from this earth and back to this earth, what is it upon Venus? It is "some unexplained optical illusion" says Newcomb (Popular Astronomy, p. 296). For a list of more than twenty observations upon this illumination of Venus, see Rept. B. A., 1873-404. It is our expression that the phenomenon is "unexplained" because it does indicate that Venus is millions of miles closer to this earth than Venus "should" be.

Unknown objects have been seen near Venus. There were more than thirty such observations in the eighteenth century, not relating to so many different periods, however. Our own earliest datum is Webb's observation, of May 22, 1823. I know of only one astronomer who has supposed that these observations could relate to a Venusian satellite, pronouncedly visible sometimes, and then for many years being invisible: something else will have to be thought of. If these observations and others that we shall have, be accepted, they relate to unknown bulks that have, from outer space, gone to Venus, and have been in temporary suspension near the planet, even though the shade of Sir Isaac Newton would curdle at the suggestion. If, acceptably, from outer space, something could go to the planet Venus, one is not especially startled with the idea that something could sail out from the planet Venus--visit this earth, conceivably.

In the Rept. B. A., 1852-8, 35, it is said that, early in the morning of Sept. u, 1852, several persons at Fair Oaks, Staffordshire, had seen, in the eastern sky, a luminous object. It was first seen at 4:15 A.M. It appeared and disappeared several times, until 4:45 A.M., when it became finally invisible. Then, at almost the same place in the sky, Venus was seen, having risen above the eastern horizon.

These persons sent the records of their observations to Lord Wrottesley, an astronomer whose observatory was at Wolverhampton. There is published a letter from Lord Wrottesley, who says that at first he had thought that the supposititiously unknown object was Venus, with perhaps an extraordinary halo, but that he had received from one of the observers a diagram giving such a position relatively to the moon that he hesitated so to identify. It was in the period of nearest approach to this earth by Venus, and, since inferior conjunction (July 20, 1852) Venus had been a "morning star." If this thing in the sky were not Venus, the circumstances are that an object came close to this earth, perhaps, and for a while was stationary, as if waiting for the planet Venus to appear above the eastern horizon, then disappearing, whether to sail to Venus or not. We think that perhaps this thing did come close to this earth, because it was, it seems, seen only in the local sky of Fair Oaks. However, if, according to many of our data, professional astronomers have missed extraordinary appearances at reasonable hours, we can't conclude much from what was not reported by them, after 4 o'clock in the morning. I do not know whether this is the origin of the convention or not, but this is the first note I have upon the now standardized explanation that, when a luminous object is seen in the sky at the time of nearest approach by Venus, it is Venus, attracting attention by her great brilliance, exciting persons, unversed in astronomic matters, into thinking that a strange object had visited this earth. When reports are definite as to motions of a seemingly sailing or exploring, luminous thing, astronomers say that it was a fire-balloon.

In the Rept. B. A., 1856-54, it is said that, according to "Mrs. Ayling and friends," in a letter to Lord Wrottesley, a bright object had been seen in the sky of Petworth, Sussex, night of Aug. 11, 1855. According to description, it rose from behind hills, in the distance, at half past eleven o'clock. It was a red body, or it was a red-appearing construction, because from it were projections like spokes of a wheel; or they were "stationary rays," in the words of the description. "Like a red moon, it rose slowly, and diminished slowly, remaining visible one hour and a half." Upon Aug. 11, 1855, Venus was two weeks from primary greatest brilliance, inferior conjunction occurring upon September 30. The thing could not have been Venus, ascending in the sky, at this time of night. An astonishing thing, like a red moon, perhaps with spokes like a wheel's, might, if reported from nowhere else, be considered something that came from outer space so close to this earth that it was visible only in a local sky, except that it might have been visible in other places, and even half past eleven at night may be an unheard-of hour for astronomers, who specialize upon sunspots for a reason that is clearing up to us. Of course an ordinary fire-balloon could be extraordinarily described.

June 8, 1868--I have not the exact time, but one does suspect that it was early in the evening--an object that was reported from Radcliffe Observatory, Oxford. It looked like a comet, but inasmuch as it was reported only from Radcliffe, it may have been in the local sky of Oxford. It seemed to sail in the sky: it moved and changed its course. At first it was stationary; then it moved westward, then southward, then turning north, visible four minutes. See Eng. Mec., 7-351.

According to a correspondent to the Birmingham Gazette, May 28, 1868, there had been an extraordinary illumination upon Venus, some nights before: a red spot, visible for a few seconds, night of May 27. In the issue of the Gazette, of June 1st, someone else writes that he saw this light appearing and disappearing upon Venus. Upon March 15, Browning had seen something that looked like a little shaft of light from Venus (Eng. Mec., 40-130); and upon April 6, Webb had seen a similar appearance (Celestial Objects, p. 57). At the time of the appearance at Oxford, Venus was in the period of nearest approach (inferior conjunction July 16, 1868).

I think, myself, that there was one approximately great, wise astronomer. He was Tycho Brahe. For many years, he would not describe what he saw in the sky, because he considered it beneath his dignity to write a book. The undignified, or more or less literary, or sometimes altogether too literary, astronomers, who do write books, uncompromisingly say that when a luminous object is said to have moved to greater degree than could be considered illusory, in a local sky of this earth, it is a fire-balloon. It is not possible to find in the writings of astronomers who so explain, mention of the object that was seen by Coggia, night of Aug. 1, 1871. It seems that this thing was not far away, and did appear only in a local sky of this earth, and if it did come from outer space, how it could have "boarded" this earth, if this earth moves at a rate of 19 miles a second, or 1 mile a second, is so hard to explain that why Proctor and Hind, with their passionate itch for explaining, never took the matter up, I don't know. Upon Aug. 1, 1871, an unknown luminous object was seen in the sky of Marseilles, by Coggia (Comptes Rendus, 73- 398). According to description, it was a magnificent red object. It appeared at 10:43 P.M., and moved eastward, slowly, until 10:52:30. It stopped--moved northward, and again, at 10:59:30, was stationary. It turned eastward again, and, at 11:3:20, disappeared, or fell behind the horizon. Upon this night Venus was within three weeks of primary greatest brilliance, inferior conjunction occurring upon Sept. 25, 1871.

15

ONE repeating mystery--the mystery of the local sky.

How, if this earth be a moving earth, could anything sail to, fall to, or in any other way reach this earth, without being smashed into fine particles by the impact?

This earth is supposed to rip space at a rate of about 19 miles a second. Concepts smash when one tries to visualize such an accomplishment.

Now, three times over, we shall have other aspects of this one mystery of the local sky. First we shall take up data upon seeming relation between a region of this earth that is subject to earthquakes, or so-called earthquakes, and appearances in the sky of this especial region, and the repeating falls of objects and substances from this local sky and nowhere else at the times.

We have had records of quakes that occurred at Irkutsk, Siberia, and of stones that fell from the sky to Irkutsk. Upon March 8, 1829, a severe quake, preceded by clattering sounds, was felt at Irkutsk. There was something in the sky. Dr. Erman, the geologist, was in Irkutsk, at the time. In the Report of the British Association, 1854-20, it is said that, in Dr. Erman's opinion, the sounds that preceded the quake were in the sky.

The situation at Comrie, Perthshire, is similar. A stone fell, May 17, 1830, in the "earthquake region" around Comrie. It fell at Perth, 22 miles from Comrie. See Fletcher's List, p. 100. Upon Feb. 15, 1837, a black powder fell upon the Comrie region (Edin. New Phil. Jour., 31- 293). Oct. 12, 1839--a quake at Comrie. According to the Rev. M. Walker, of Comrie, the sky, at the time, was "peculiarly strange and alarming, and appeared as if hung with sackcloth." In Mallet's Catalogue (Rept. B. A., 1854-290) it is said that, throughout the month of October, shocks were felt at Comrie, sometimes slight and sometimes severe--"like distant thunder or reports of artillery"--"the noise sometimes seemed to be high in the air, and was often heard without any sensible shock." Upon the 23rd of October, occurred the most violent quake in the whole series of phenomena at Comrie. See the Edin. New Phil. Jour., vol. 32. All data in this publication were collected by David Milne. According to the Rev. M. Maxton, of Foulis Manse, ten miles from Comrie, rattling sounds were heard in the sky, preceding the shock that was felt. In vol. 33, p. 373, of the Journal, someone who lived seven miles from Comrie is quoted: "In every case, I am inclined to say that the sound proceeded not from . The sound seemed high in the air." Someone who lived at Gowrie, forty miles from Comrie, is quoted: "The most general opinion seems to be that the noise accompanying the concussion proceeded from above." See vol. 34, p. 87: another impression of explosion overhead and concussion underneath: "The noises heard first seemed to be in the air, and the rumbling sound in the earth." Milne's own conclusion--"It is plain that there are, connected with the earthquake shocks, sounds both in the earth and in the air, which are distinct and separate." If, upon the 23rd of October, 1839, there was a tremendous shock, not of subterranean origin, but from a great explosion in the sky of Comrie, and if this be accepted, there will be concussions somewhere else. The "faults" of dogmas will open; there will be seismic phenomena in science. I have a feeling of a conventional survey of this Scottish sky: vista of a fair, blue, vacant expanse--our suspicions daub the impression with black alarms--but also do we project detonating stimulations into the fair and blue, but unoccupied and meaningless. One cannot pass this single occurrence by, considering it only in itself: it is one of a long series of quakes of the earth at Comrie and phenomena in the sky at Comrie. We have stronger evidence than the mere supposition of many persons, in and near Comrie, that, upon Oct. 23, 1839, something had occurred in the sky, because sounds seemed to come from the sky. Milne says that clothes, bleaching on the grass, were entirely covered with black particles which presumably had fallen from the sky. The shocks were felt in November: in November, according to Milne, a powder like soot fell from the sky, upon Comrie and surrounding regions. In his report to the British Association, 1840, Milne, reviewing the phenomena from the year 1788, says: "Occasionally there was a fall of fine, black powder."

Jan. 8, 1840--sounds like cannonading, at Comrie, and a crackling sound in the air, according to some of the residents. Whether they were sounds of quakes or concussions that followed explosions, 247 occurrences, between Oct. 3, 1839, and Feb. 14, 1841, are listed in the Edin. New Phil. Jour., 32-107. It looks like bombardment, and like most persistent bombardment--from somewhere--and the frequent fall from the sky of the debris of explosions. Feb. 18, 1841 a shock and a fall of discolored rain at Comrie (Edin. New Phil. Jour., 35-148). See Roper's List of Earthquakes--year after year, and the continuance of this seeming bombardment in one small part of the sky of this earth, though I can find records only of dates and no details. However, I think I have found record of a fall from the sky of debris of an explosion, more substantial than finely powdered soot, at Crieff, which is several miles from Comrie. In the Amer. Jour. Sci., 2-28-275, Prof. Shepard tells a circumstantial story of an object that looked like a lump of slag, or cinders, reported to have fallen at Crieff. Scientists had refused to accept the story, upon the grounds that the substance was not of "true meteoric material." Prof. Shepard went to Crieff and investigated. He gives his opinion that possibly the object did fall from the sky. The story that he tells is that, upon the night of April 23, 1855, a young woman, in the home of Sir William Murray, Achterlyre House, Crieff, saw, or thought she saw, a luminous object falling, and picked it up, dropping it, because it was hot, or because she thought it was hot.

For a description, in a letter, presumably from Sir William Murray, or some member of his family, see Year Book of Facts, 1856-273. It is said that about 12 fragments of scorious matter, hot and emitting a sulphurous odor, had fallen.

In Ponton's Earthquakes, p. 118, it is said that, upon the 8th of October, 1857, there had been, in Illinois, an earthquake, preceded by "a luminous appearance, described by some as a meteor and by others as vivid flashes of lightning." Though felt in Illinois, the center of the disturbance was at St. Louis, Mo. One notes the misleading and the obscuring of such wording: in all contemporaneous accounts there is no such indefiniteness as one description by "some" and another notion by "others." Something exploded terrifically in the sky, at St. Louis, and shook the ground "severely" or "violently," at 4:20 A.M., Oct. 8, 1857. According to Timbs' Year Book of Facts, 1858-271, "a blinding meteoric ball from the heavens" was seen. "A large and brilliant meteor shot across the heavens" (St. Louis Intelligencer, October 8). Of course the supposed earthquake was concussion from an explosion in the sky, but our own interest is in a series that is similar to others that we have recorded. According to the New York Times, October 12, a slight shock was said to have been felt four hours before the great concussion, and another three days before. But see Milne's Catalog of Destructive Earthquakes--not a mention of anything that would lead one away from safe and standardized suppositions. See Bull. Seis. Soc. Amer., 3-68--here the "meteor" is mentioned, but there is no mention of the preceding concussions. Time after time, in a period of about three days, concussions were felt in and around St. Louis.

One of these concussions, with its "sound like thunder or the roar of artillery" (New York Times, October 8) was from an explosion in the sky. If the others were of the same origin--how could detonating meteors so repeat in one small local sky, and nowhere else, if this earth be a moving body? If it be said that only by coincidence did a meteor explode over a region where there had been other quakes, here is the question:

How many times can we accept that explanation as to similar series?

In the Proceedings of the Society for Psychical Research, 19-144, a correspondent writes that, in Herefordshire, Sept. 24, 1854, upon a day that was "perfectly still, sky cloudless," he had heard sounds like the discharges of heavy artillery, at intervals of about two minutes, continuing several hours. Again the "mystery of the local sky"--if these sounds did come from the sky. We have no data for thinking that they did.

In the London Times, Nov. 9, 1858, a correspondent writes that, in Cardiganshire, Wales, he had, in the autumn of 1855, often heard sounds like the discharges of heavy artillery, two or three reports rapidly, and then an interval of perhaps 20 minutes, also with long intervals, sometimes of days and sometimes of weeks, continuing throughout the winter of 1855-56. Upon the 3rd of November, 1858, he had heard the sounds again, repeatedly, and louder than they had been three years before. In the Times, November 12, someone else says that, at Dolgelly, he, too, had heard the "mysterious phenomenon," on the 3rd of November. Someone else--that, upon October 13, he had heard the sounds at Swansea. "The reports, as if of heavy artillery, came from the west, succeeding each other at apparently regular intervals, during the greater part of the afternoon of that day. My impression was that the sounds might have proceeded from practicing at Milford, but I ascertained, the following day, that there had been no firing of any kind there." Correspondent to the Times, November 20--that, with little doubt, the sounds were from artillery practice at Milford. He does not mention the investigation as to the sounds of October 13, but says that there had been cannon-firing, upon November 3rd, at Milford. Times, December 1--that most of the sounds could be accounted for as sounds of blasting in quarries. Daily News, November 16--that similar sounds had been heard, in 1848, in New Zealand, and were results of volcanic action. Standard, November 16--that the "mysterious noise" must have been from Devonport, where a sunken rock had been blown up. So, with at least variety these sounds were explained. But we learn that the series began before October 13. Upon the evening of September 28, in the Dartmoor District, at Crediton, a rumbling sound was heard. It was not supposed to be an earthquake, because no vibration of the ground was felt. It was thought that there had been an explosion of gunpowder. But there had been no such terrestrial explosion. About an hour later another explosive sound was heard. It was like all the other sounds, and in one place was thought to be distant cannonading--terrestrial cannonading. See Quar. Jour. Geolog. Soc. of London, vol. 15.

Somewhere near Barisal, Bengal, were occurring just such sounds as the sounds of Cardiganshire, which were like the sounds of Melida. In the Proc. Asiatic Soc. of Bengal, November, 1870, are published letters upon the Barisal Guns. One writer says that the sounds were probably booming of the surf. Someone else points out that the sounds, usually described as "explosive," were heard too far inland to be traced to such origin. A clear, calm day, in December, 1871--in Nature, 53-197, Mr. G. B. Scott writes that, in Bengal, he had heard "a dull, muffled boom, as if of distant cannon"--single detonations, and then two or three in quicker succession.

In the London Times, Jan. 20, 1860, several correspondents write as to a sound "resembling the discharge of a gun high in the air" that was heard near Reading, Berkshire, England, Jan. 17, 1860. See the Times, January 24th. To say that a meteor had exploded would, at present, well enough account for this phenomenon.

Sounds like those that were heard in Herefordshire, Sept. 24, 1854, were heard later. In the English Mechanic, 100-279, it is said that, upon Nov. 9, 1862, the Rev. T. Webb, the astronomer, of Hardwicke, fifteen miles west of Hereford, heard sounds that he attributed to gunfire at Milford Haven, about 85 miles from Hardwicke. Upon Aug. 1, 1865, Mr. Webb saw flashes upon the horizon, at Hardwicke, and attributed them to gunfire at Tenby, upon occasion of a visit by Prince Arthur. Tenby, too, is about 85 miles from Hardwicke. There were other phenomena in a region centering around Hereford and Worcester. Upon Oct. 6, 1863, there was a disturbance that is now listed as an earthquake; but in the London newspapers so many reports upon this occurrence state that a great explosion had been thought to occur, and that the quake was supposed to be an earthquake of subterranean origin only after no terrestrial explosion could be heard of, that the phenomenon is of questionable origin. There was a similar concussion in about the same region, Oct. 30, 1868. Again the shock was widely attributed to a great explosion, perhaps in London, and again was supposed to have been an earthquake when no terrestrial explosion could be heard of. Arcana of Science, 1829-196:

That, near Mhow, India, Feb. 27, 1828, fell a stone "perfectly similar" to the stone that fell near Allahabad, in 1802, and a stone that fell near Mooradabad, in 1808. These towns are in the Northwestern Provinces of India.

I have looked at specimens of these stones, and in my view they are similar. They are of brownish rock, streaked and spotted with a darker brown. A stone that fell at Chandakopur, in the same general region, June 6, 1838, is like them. All are as much alike as "erratics" that, because they are alike, geologists ascribe to the same derivation, stationary relatively to the places in which they are found.

It seems acceptable that, upon July 15 and 17, 1822, and then upon a later date, unknown seeds fell from the sky to this earth. If these seeds did come from some other world, there is another mystery as well as that of repetition in a local sky of this earth. How could a volume of seeds remain in one aggregation; how could the seeds be otherwise than scattered from Norway to Patagonia, if they met in space this earth, and if this earth be rushing through space at a rate of 19 miles a second? It may be that the seeds of 1822 fell again. According to Kaemtz (Meteorology, p. 465) yellowish brown corpuscles, some round, a few cylindrical, were found upon the ground, June, 1830, near Griesau, Silesia. Kaemtz says that they were tubercules from roots of a well-known Silesian plant--stalk of the plant dries up; heavy rain raises these tubercules to the ground--persons of a low order of mentality think that the things had fallen from the sky. Upon the night of March 24-25, 1852, a great quantity of seeds did fall from the sky, in Prussia, in Heinsberg, Erklenz, and Juliers, according to M. Schwann, of the University of Liege, in a communication to the Belgian Academy of Science (La Belgique Horticole, 2-319).

In Comptes Rendus, 5-549, is Dr. Wartmann's account of water that fell from the sky, at Geneva. At nine o'clock, morning of Aug. 9, 1837, there were clouds upon the horizon, but the zenith was clear. It is not remarkable that a little rain should fall now and then from a clear sky: we shall see wherein this account is remarkable. Large drops of warm water fell in such abundance that people were driven to shelter. The fall continued several minutes and then stopped. But then, several times during an hour, more of this warm water fell from the sky. Year Book of Facts, 1839-262-- that upon May 31, 1838, lukewarm water in large drops fell from the sky, at Geneva. Comptes Rendus, 15-290--no wind and not a cloud in the sky--at 10 o'clock, morning of May 11, 1842, warm water fell from the sky at Geneva, for about six minutes; five hours later, still no wind and no clouds, again fell warm water, in large drops; falling intermittently for several minutes.

In Comptes Rendus, 85-681, is noted a succession of falls of stones in Russia: June 12, 1863, at Buschof, Courland; Aug. 8, 1863, at Pillitsfer, Livonia; April 12, 1864, at Nerft, Courland. Also-- see Fletcher's List--a stone that fell at Dolgovdi, Volhynia, Russia, June 26, 1864. I have looked at specimens of all four of these stones, and have found them all very much alike, but not of uncommon meteoritic material: all gray stones, but Pillitsfer is darker than the others, and in a polished specimen of Nerft, brownish specks are visible.

In the Birmingham Daily Post, June 14, 1858, Dr. C. Mansfield Ingleby, a meteorologist, writes: "During the storm on Saturday (12th) morning, Birmingham was visited by a shower of aerolites. Many hundreds of thousands must have fallen, some of the streets being strewn with them." Someone else writes that many pounds of the stones had been gathered from awnings, and that they had damaged greenhouses, in the suburbs. In the Post, of the 15th, someone else writes that, according to his microscopic examinations, the supposed aerolites were only bits of the Rowley ragstone, with which Birmingham was paved, which had been washed loose by the rain. It is not often that sentiment is brought into meteorology, but in the Report of the British Association, 1864-37, Dr. Phipson explains the occurrence meteorologically, and with an unconscious tenderness. He says that the stones did fall from the sky, but that they had been carried in a whirlwind from Rowley, some miles from Birmingham. So we are to sentimentalize over the stones in Rowley that had been torn, by unfeeling paviers, from their companions of geologic ages, and exiled to the pavements of Birmingham, and then some of these little bereft companions, rising in a whirlwind and traveling, unerringly, if not miraculously, to rejoin the exiles. More dark companions. It is said that they were little black stones.

They fell again from the sky, two years later. In La Science Pour Tous, June 19, 1860, it is said that, according to the Wolverhampton Advertiser, a great number of little black stones had fallen, in a violent storm, at Wolverhampton. According to all records findable by me no such stones have ever fallen anywhere in Great Britain, except at Birmingham and Wolverhampton, which is 13 miles from Birmingham.

Eight years after the second occurrence, they fell again. English Mechanic, July 31, 1868--that stones "similar to, if not identical with the well-known Rowley ragstones" had fallen in Birmingham, having probably been carried from Rowley, in a whirlwind.

We were pleased with Dr. Phipson's story, but to tell of more of the little dark companions rising in a whirlwind and going unerringly from Rowley to rejoin the exiles in Birmingham is overdoing. That's not sentiment: that's mawkishness.

In the Birmingham Daily Post, May 30, 1868, is published a letter from Thomas Plant, a writer and lecturer upon meteorological subjects. Mr. Plant says, I think, that for one hour, morning of May 29, 1868, stones fell, in Birmingham, from the sky. His words may be interpretable in some other way, but it does not matter: the repeating falls are indication enough of what we're trying to find out--"From nine to ten, meteoric stones fell in immense quantities in various parts of town." "They resembled, in shape, broken pieces of Rowley ragstone ... in every respect they were like the stones that fell in 1858." In the Post, June 1, Mr. Plant says that the stones of 1858 did fall from the sky, and were not fragments washed out of the pavement by rain, because many pounds of them had been gathered from a platform that was 20 feet above the ground. It may be that for days before and after May 29, 1868, occasional stones fell from some unknown region stationary above Birmingham. In the Post, June 2, a correspondent writes that, upon the first of June, his niece, while walking in a field, was struck by a stone that injured her hand severely. He thinks that the stone had been thrown by some unknown person. In the Post, June 4, someone else writes that his wife, while walking down a lane, upon May 24th, had been cut on the head by a stone. He attributes this injury to stone-throwing by boys, but does not say that anyone had been seen to throw the stone.

Symons' Met. Mag., 4-137:

That, according to the Birmingham Gazette, a great number of small, black stones had been found in the streets of Wolverhampton, May 25, 1869, after a severe storm. It is said that the stones were precisely like those that had fallen in Birmingham, the year before, and resembled Rowley ragstone outwardly, but had a different appearance when broken.

16

UPON page 287, Popular Astronomy, Newcomb says that it is beyond all "moral probability" that unknown worlds should exist in such numbers as have been reported, and should be seen crossing the solar disc only by amateur observers and not by skilled astronomers.

Most of our instances are reports by some of the best-known astronomers.

Newcomb says that for fifty years, prior to his time of writing (edition of 1878) the sun had been studied by such men as Schwabe, Carrington, Secchi, and Sporer, and that they had never seen unknown bodies cross the sun--

Aug. 30, 1863--an unknown body that was seen by Sporer to cross the sun (Webb, Celestial Objects, p. 45).

Sept. 1, 1859--two star-like objects that were seen by Carrington to cross the sun (Monthly Notices, 20-13, 15, 88).

Things that crossed the sun, July 31, 1826, and May 26, 1828--

see Comptes Rendus, 83-623, and Webb's Celestial Objects, p. 40. From Sept. 6 to Nov. 1, 1831, an unknown luminous object was seen every cloudless night, at Geneva, by Dr. Wartmann and his assistants (Comptes Rendus, 2-307). It was reported from nowhere else. What all the other astronomers were doing, September- October, 1831, is one of the mysteries that we shall not solve. An unknown, luminous object that was seen, from May 11 to May 14, 1835, by Cacciatore, the Sicilian astronomer (Amer. Jour. Sci., 31-158). Two unknowns that, according to Pastorff, crossed the sun, Nov. 1, 1836, and Feb. 16, 1837 (An. Sci. Disc., 1860- 410)--De Vico's unknown, July 12, 1837 (Observatory, 2-424)-observation by De Cuppis, Oct. 2, 1839 (C. R., 83-314)-by Scott and Wray, last of June, 1847; by Schmidt, Oct. 11, 1847 (C. R., 83-623)-two dark bodies that were seen, Feb. 5, 1849, by Brown, of Deal (Rec. Sci., 1-138)--object watched by Sidebotham, half an hour, March 12, 1849, crossing the sun (C. R., 83-622)--Schmidt's unknown, Oct. 14, 1849 (Observatory, 3-137)--and an object that was watched, four nights in October, 1850, by James Ferguson, of the Washington Observatory. Mr. Hind believed this object to be a Trans-Neptunian planet, and calculated for it a period of 1,600 years. Mr. Hind was a great astronomer, and he miscalculated magnificently: this floating island of space was not seen again (Smithson. Miscell. Cols., 20- 20).

About May 30, 1853--a black point that was seen against the sun, by Jaennicke (Cosmos, 20-64).

A procession--in the Rept. B. A., 1855-94, R. P. Greg says that, upon May 22, 1854, a friend of his saw, near Mercury, an object equal in size to the planet itself, and behind it an elongated object, and behind that something else, smaller and round.

June 11, 1855--a dark body of such size that it was seen, without telescopes, by Ritter and Schmidt, crossing the sun (Observatory, 3-137). Sept. 12, 1857--Ohrt's unknown world; seemed to be about the size of Mercury (C. R., 83-623)--Aug. 1, 1858-unknown world reported by Wilson, of Manchester (Astro. Reg., 9-287).

I am not listing all the unknowns of a period; perhaps the object reported by John
H. Tice, of St. Louis, Mo., Sept. 15, 1859, should not be included; Mr. Tice was said not to be trustworthy--but who has any way of knowing? However, I am listing enough of these observations to make me feel like a translated European of some centuries ago, relatively to a wider existence--lands that may be the San Salvadors, Greenlands, Madagascars, Cubas, Australias of extra-geography, all of them said to have crossed the sun, whereas the sun may have moved behind some of them.

Jan. 29, 1860--unknown object, of planetary size, reported from London, by Russell and three other observers (Nature, 15-505). Summer of 1860--see Sci. Amer., 35- 340, for an account, by Richard Covington, of an object that, without a telescope, he saw crossing the sun. An unknown world, reported by Loomis, of Manchester, March 20, 1862 (Monthly Notices, 22-232)--a newspaper account of an object that was seen crossing the sun, Feb. 12, 1864, by Samuel Beswick, of New York (Astro.Reg., 2-161)--unknown that was seen, March 18, 1865, at Constantinople (L'Ann. Sci., 1865-16)--unknown "cometic objects" that were seen, Nov. 4, 9, and 18, 1865 (Monthly Notices, 26-242).

Most of these unknowns were seen in the daytime. Several reflections arise. How could there be stationary regions over Irkutsk, Comrie, and Birmingham, and never obscure the stars--or never be seen to obscure the stars? A heresy that seems too radical for me is that they may be beyond the nearby stars. A more reasonable idea is that if nightwatchmen and policemen and other persons who do stay awake nights, should be given telescopes, something might be found out. Something else that one thinks of is that, if so many unknowns have been seen crossing the sun, or crossed by the sun, others not so revealed must exist in great numbers, and that instead of being virtually blank, space must be archipelagic.

Something that was seen at night; observer not an astronomer—

Nov. 6, 1866--an account, in the London Times, Jan. 2, 1867, by Senor De Fonblanque, of the British Consulate, at Cartagena, U. S. Colombia, of a luminous object that moved in the sky. "It was of the magnitude, color, and brilliance of a ship's red light, as seen at a distance of 200 yards." The object was visible three minutes, and then disappeared behind buildings. De Fonblanque went to an open space to look for it, but did not see it again.

17

IF we could stop to sing, instead of everlastingly noting vol. this and p. that, we could have the material of sagas--of the bathers in the sun, which may be neither intolerably hot nor too uncomfortably cold; and of the hermit who floats across the moon; of heroes and the hairy monsters of the sky. I should stand in public places and sing our data--sagas of parades and explorations and massacres in the sky--having a busy band of accompanists, who set off fireworks, and send up balloons, and fire off explosives at regular intervals--extra-geographic songs of boiling lakes and floating islands--extra-sociologic meters that express the tramp of space-armies upon inter-planetary paths covered with little black pebbles-- biologic epics of the clouds of mammoths and horses and antelopes that once upon a time fell from the sky upon the northern coast of Siberia--

Song that interprets the perpendicular white streaks in the repeating mirages at Youghal--the rhythmic walruses of space that hang on by their tusks to the edges of space-islands, sometimes making stars variable as they swing in cosmic undulations--so a round space-island with its border of gleaming tusks, and we frighten children with the song of an ogre's head, with a wide-open mouth all around it--fairy lands of the little moon, and the tiny civilizations in rocky cups that are sometimes drained to their slums by the wide-mouthed ogres. The Maelstrom of Everlasting Catastrophe that overhangs Genoa, Italy--and twines its currents around a living island. The ground underneath quakes with the struggle-- then the fall of blood--and the fall of blood--three days the fall of blood from the broken red brooks of a living island whose mutilations are scenery--

But after all, it may be better that we go back to Rept. B. A.--see vol. 1849, p. 46--a stream of black objects, crossing the sun, watched, at Naples, May II, 1845, by Capocci and other astronomers-- things that may have been seeds.

A great number of red points in the sky of Urrugne, July 9, 1853 (An. Soc. Met. de France, 1853-227). Astro. Reg., 5-179--C. L. Prince, of Uckfield, writes that, upon June 11, 1867, he saw objects crossing the field of his telescope. They were seeds, in his opinion.

Birmingham Daily Post, May 31, 1867:

Mr. Bird, the astronomer, writes that, about 11 A.M., May 30, he saw unknown forms in the sky. In his telescope, which was focused upon them and upon the planet Venus, they appeared to be twice the size of Venus. They were far away, according to focus; also, it may be accepted that they were far away because an occasional cloud passed between them and this earth. They did not move like objects carried in the wind: all did not move in the same direction, and they moved at different speeds. "All of them seemed to have hairy appendages, and in many cases a distinct tail followed the object and was highly luminous." Flashes that have been seen in the sky--and they're from a living island that wags his luminous peninsula. Hair-like substances that have fallen to this earth--a meadow has been shorn from a monster's mane. My animation is the notion that it is better to think in tentative hysteria of pairs of vast things, traveling like a North and South America through the sky, perhaps one biting the other with its Gulf of Mexico, than to go on thinking that all things that so move in the sky are seeds, whereas all things that swim in the sea are not sardines.

In the Post, June 3, 1867, Mr. W. H. Wood writes that the objects were probably seeds. Post, June 5--Mr. Bird says that the objects were not seeds. "My intention was simply to describe what was seen, and the appearance was certainly that of meteors." He saves himself, in the annals of extra-geography--"whether they were meteors of the ordinary acceptation, is another matter."

And the planet Venus, and her veil that is dotted with blue-fringed cupids--in the Astronomical Register, 7-138, a correspondent writes, from Northampton, that, upon May 2, 1869, he was looking at Venus, and saw a host of shining objects, not uniform in size. He thinks that it is unlikely that so early in the spring could these objects be seeds. He watched them about an hour and twenty minutes--"many of the larger ones were fringed on one side, the fringe appearing somewhat bluish."

Or that it is better even to sentimentalize than to go on stupidly thinking that all such things in the sky are seeds, whereas all things in the sea are not the economically adjusting little forms without which critics of underground traffic in New York probably could not express themselves--the planet Venus--she approaches this lordly earth--the blue-fringed ecstasies that suffuse her skies.

With the phenomena of Aug. 7, 1869, I suspect that the "phantom soldiers" that have been seen in the sky, may have been reflections from, or mirages of, things or beings that march, in military formations, in space. In Popular Astronomy, 3- 159, Prof. Swift writes that, at Mattoon, Ill., during the eclipse of the sun, of Aug. 7, 1869, he had seen, crossing the moon, objects that he thought were seeds. If they were seeds, also there happened to be seeds in the sky of Ottumwa, Iowa: here, crossing the visible part of the sun, twenty minutes before totality of the eclipse, Prof. Himes and Prof. Zentmayer saw objects that marched, or that moved, in straight, parallel lines (Les Mondes, 21-241). In the Jour. Frank. Inst., 3-58- 214, it is said that some of these objects moved in one direction across the moon, and that others moved in another direction across another part of the moon, each division moving in parallel lines. If these things were seeds, also there happened to be seeds in the sky, at Shelleyville, Kentucky. Here were seen, by Prof. Winlock, Alvan Clark, Jr., and George W. Dean, things that moved across the moon, during the eclipse, in parallel, straight lines (Pop. Astro., 2-332). Whatever these things may have been, I offer another datum indicating that the moon is nearby: that these objects probably were not, by coincidence, things in three widely separated skies, parallelness giving them identity in two of the observations; and, if seen, without parallax, from places so far apart, against the moon, were close to the moon; that observation of such detail would be unlikely if they were near a satellite 240,000 miles away--unless, of course, they were mountain-sized.

It may be that out from two floating islands of space, two processions had marched across the moon. Observatory, 3-137--that, at St. Paul's Junction, Iowa, four persons had seen, without telescopes, a shining object close to the sun and moon, apparently; that, with a telescope, another person had seen another large object, crescentically illuminated, farther from the sun and moon in eclipse. See Nature, 18-663, and Astro. Reg.,. 7-227.

I have many data upon the fall of organic matter from the sky. Because of my familiarity with many records, it seems no more incredible that up in the seemingly unoccupied sky there should be hosts of living things than that the seeming blank of the ocean should swarm with life. I have many notes upon a phosphorescence, or electric condition of things that fall from the sky, for instance the highly luminous stones of Dhurmsulla, which were intensely cold-- Amer. Jour. Sci., 2-28-270:

It is said that, according to investigations by Prof. Shepard, a luminous substance was seen falling slowly, by Sparkman R. Striven, a young man of seventeen, at his home, in Charleston, S. C., Nov. 16, 1857. It is said that the young man saw a fiery, red ball, the size and shape of an orange, strike a fence, breaking, and disappearing. Where this object had struck the fence, was found "a small bristling mass of black fibers." According to Prof. Shepard, it was "a confused aggregate of short clippings of the finest black hair, varying in length from one tenth to one third of an inch." Prof. Shepard says that this substance was not organic. It seems to me that he said this only because of the coercions of his era. My reason for so thinking is that he wrote that when he analyzed these hairs they burned away, leaving grayish skeletons, and that they were "composed in part of carbon," and burned with an odor "most nearly bituminous."

For full details of the following circumstances, see Comptes Rendus, 13-215, and Rept. B. A., 1854-302:

Feb. 17, 1841--the fall, at Genoa, Italy, of a red substance from the sky--another fall upon the 18th-- a slight quake, at 5 P.M., February 18th--another quake, six hours later--fall of more of the red substance, upon the 19th. Some of this substance was collected and analyzed by M. Canobbia, of Genoa. He says it was oily and red.

18

IN a pamphlet entitled Wonderful Phenomena, by Curtis Eli, is the report of an occurrence, or of an alleged occurrence, that was investigated by Mr. Addison A. Sawin, a spiritualist. He interpreted in the only way that I know of, and that is the psychochemic process of combining new data with preconceptions with which they seem to have affinity. It is said that, at Warwick, C. W., Oct. 3, 1843, somebody named Charles Cooper heard a rumbling sound in the sky, and saw a cloud, under which were three human forms, "perfectly white," sailing through the air above him, not higher than the tree-tops. It is said that the beings were angels. They were male angels. That is orthodox. The angels wafted through the air, but without motions of their own, and an interesting observation is that they seemed to have belts around their bodies--as if they had been let down from a vessel above, though this poor notion is not suggested in the pamphlet. They "moaned." Cooper called to some men who were laboring in another field, and they saw the cloud, but did not see the forms of living beings under it. It is said that a boy had seen the beings in the air, "side by side, making a loud and mournful noise." Another person, who lived six miles away, is quoted: "he saw the clouds and the persons and heard the sounds." Mr. Sawin quotes others, who had seen "a remarkable cloud," and had heard the sounds, but had not seen the angels. He ends up: "Yours is the glorious hope of the resurrection of the soul." The gloriousness of it is an inverse function of the dolefulness of it: Sunday Schools will not take kindly to the doctrine--be good and you will moan forever. One supposes that the glorious hope colored the whole investigation.

Someday I shall publish data that lead me to suspect that many appearances upon this earth that were once upon a time interpreted by theologians and demonologists, but are now supposed to be the subject-matter of psychic research, were beings and objects that visited this earth, not from a spiritual existence, but from outer space. That extra-geographic conditions may be spiritual, or of highly attenuated matter, is not my present notion, though that, too, may be some day accepted. Of course all these data suffer, in one way, about as much distortion as they would in other ways, if they had been reported by astronomers or meteorologists. As to all the material in this chapter, I take the position that perhaps there were appearances in the sky, and perhaps they were revelations of, or mirages from, unknown regions and conditions of outer space, and spectacles of relatively nearby inhabited lands, and of space-travelers, but that all reports upon them were products of the assimilating of the unknown with figures and figments of the nearest familiar similarities. Another position of mine that will be found well- taken is that, no matter what my own interpretations or acceptances may be, they will compare favorably, so far as rationality is concerned, with orthodox explanations. There have been many assertions that "phantom soldiers" have been seen in the sky. For the orthodox explanation of the physicists, see Brewster's Natural Magic, p. 125: a review of the phenomenon of June 23, 1744; that, according to 27 witnesses, some of whom gave sworn testimony before a magistrate, whether that should be mentioned or not, troops of aerial soldiers had been seen, in Scotland, on and over a mountain, remaining visible two hours and then disappearing because of darkness. In Clarke's Survey of the Lakes (fol. 1789) is an account in the words of one of the witnesses. See Notes and Queries, 1-7-304.

Brewster says that the scene must have been a mirage of British troops, who, in anticipation of the rebellion of 1745, were secretly maneuvering upon the other side of the mountain. With a talent for clear-seeing, for which we are notable, except when it comes to some of our own explanations, we almost instantly recognize that, to keep a secret from persons living upon one side of a mountain, it is a very sensible idea to go and maneuver upon the other side of the mountain; but then how to keep the secret, in a thickly populated country like Scotland, from persons living upon that other side of the mountain--however, there never has been an explanation that did not itself have to be explained. Or the "phantom soldiers" that were seen at Ujest, Silesia, in 1785--see Parish's Hallucinations and Illusions, p. 309. Parish finds that at the time of this spectacle, there were soldiers, of this earth, marching near Ujest; so he explains that the "phantom soldiers" were mirages of them. They were marching in the funeral procession of General von Cosel. But sometime later they were seen again, at Ujest--and the General had been dead and buried several days, and his funeral procession disbanded--and if a refraction can survive independently of its primary, so may a shadow, and anybody may take a walk where he went a week before, and see some of his shadows still wandering around without him. The great neglect of these explainers is in not accounting for an astonishing preference for, or specialization in, marching soldiers, by mirages. But if often there be, in the sky, things or beings that move in parallel lines, and, if their betrayals be not mirages, but their shadows cast down upon the haze of this earth, or Brocken specters, such frequency, or seeming specialization, might be accounted for.

Sept. 27, 1846--a city in the sky of Liverpool (Rept. B. A., 1847-39) . The apparition is said to have been a mirage of the city of Edinburgh. This "identification" seems to have been the product of suggestion: at the time a panorama of Edinburgh was upon exhibition in Liverpool.

Summer of 1847--see Flammarion's The Atmosphere, p. 160--story told by M. Grellois: that he was traveling between Ghelma and Bone, when he saw, to the east of Bone, upon a gently sloping hill, "a vast and beautiful city, adorned with monuments, domes, and steeples." There was no resemblance to any city known to M. Grellois.

In the Bull. Soc. Astro. de France, 21-180, is an account of a spectacle that, according to 20 witnesses, vas seen for two hours in the sky of Vienne dans le Dauphine, May 3, 1848. A city--and an army, in the sky. One supposes that a Brewster would say that nearby was a terrestrial city, with troops maneuvering near it. But also vast lions were seen in the sky--and that is enough to discourage any Brewster. Four months later, according to the London Times, Sept. 13, 1848, a still more discouraging--or perhaps stimulating--spectacle was, or was not, seen in Scotland. Afternoon of Sept. 9, 1848--Quigley's Point, Lough Foyle, Scotland--the sky turned dark. It seemed to open. The opening looked reddish, and in the reddish area, appeared a regiment of soldiers. Then came appearances that looked like war vessels under full sail, then "a man and a woman and a swan and a peahen." The "opening" closed, and that was the last of this shocking or ridiculous mixture that nobody but myself would record as being worth thinking about.

"Phantom soldiers" that were seen in the sky, near the Banmouth, Dec. 30, 1850 (Rept. B. A., 1852-30).

"Phantom soldiers" that were seen at Buderich, Jan. 22, 1854 (Notes and Queries, 1-9-267).

"Phantom soldiers" that were seen by Lord Roberts (Forty-One Years in India, p. 219) at Mohan, Feb. 25, 1858. It is either that Lord Roberts saw indistinctly, and described in terms of the familiar to him, or that we are set back in our own motions. According to him, the figures wore Hindoo costumes.

Extra-geography--its vistas and openings and fields--and the Thoreaus that are upon this earth, but undeveloped, because they cannot find their ponds. A lonely thing and its pond, afloat in space--they crossed the moon. In Cosmos, n. s., 11-200, it is said that, night of July 7, 1857, two persons of Chambon had seen forms crossing the moon--something like a human being followed by a pond.

"Phantom soldiers" that were seen, about the year 1860, at Paderborn, Westphalia (Crowe, Nightside of Nature, p. 416).

19

WE attempt to co-ordinate various streaks of data, all of which signify to us that, external to this earth, and in relation with, or relatable to, this earth are lands and lives and a generality of conditions that make of the whole, supposed solar system one globule of circumstances like terrestrial circumstances. Our expressions are in physical terms, though in outer space there may be phenomena known as psychic phenomena, because of the solid substances and objects that have fallen from the sky to this earth, similar to, but sometimes not identified with, known objects and substances upon this earth. Opposing us is the more or less well-established conventional doctrine that has spun like a cocoon around mind upon this earth, shutting off research, and stifling even speculation, shelling away all data of relations and relatability with external existences, a doctrine that, in its various explanations and disregards and denials, is unified in one expression of Exclusionism.

An unknown vegetable substance falls from the sky. The datum is buried: it may sprout someday. The earth quakes. A luminous object is seen in the sky. Substance falls from the unknown. But the event is catalogued with subterranean earthquakes.

All conventional explanations and all conventional disregards and denials have Exclusionism in common. The unity is so marked, all writings in the past are so definitely in agreement, that I now think of a general era that is, by Exclusionism, as distinctly characterized as ever was the Carboniferous Era.

A pregnant woman stands near Niagara Falls. There are sounds, and they are vast circumstances; but the cells of an unborn being respond, or vibrate, only as they do to disturbances in their own little environment. Horizons pour into a gulf, and thunder rolls upward: embryonic consciousness is no more than to slight perturbations of maternal indigestion. It is Exclusionism.

Stones fall from the sky. To the same part of this earth, they fall again. They fall again. They fall from some region that, relatively to this part of the earth's surface, is stationary. But to say this leads to the suspicion that it is this earth that is stationary. To think that is to beat against the walls of uterine dogmas--into a partly hairy and somewhat reptilian mass of social undevelopment comes exclusionist explanation suitable for such immaturity.

It does not matter which of our subjects we take up, our experience is unvarying: the standardized explanation will be Exclusionism. As to many appearances in the sky, the way of excluding foreign forces is to say that they are auroras, which are supposed to be mundane phenomena. School children are taught that auroras are electric manifestations encircling the poles of this earth. Respectful urchins are shown an ikon by which an electrified sphere does have the polar encirclements that it should have. But I have taken a disrespectful, or advanced, course through the Monthly Weather Review, and have read hundreds of times of auroras that were not such polar crownings: of auroras in Venezuela, Sandwich Islands, Cuba, India; of an aurora in Pennsylvania, for instance, and not a sign of it north of Pennsylvania. There are lights in the sky for which "auroral" is as good a name as any that can be thought of, but there are others for which some other names will have to be thought of. There have been lights like luminous surfs beating upon the coasts of this earth's atmosphere, and lights like vast reflections from distant fires; steady pencils of light and pulsating clouds and quick flashes and seeming objects with definite outlines, all in one poverty of nomenclature, for which science is, in some respects, not notable, called "auroral." Nobody knows what an aurora is. It does not matter. An unknown light in the sky is said to be auroral. This is standardization, and the essence of this standardization is Exclusionism.

I see one resolute, unified, unscrupulous exclusion from science of the indications of nearby lands in the sky. It may not be unscrupulousness: it may be hypnosis. I see that all seeming hypnotics, or somnambulists of the past, who have most plausibly so explained, or so denied, have prospered and have had renown.

According to my impressions, if a Brewster, or a Swift, or a Newcomb ever had written that there may be nearby lands and living beings in the sky, he would not have prospered, and his renown would be still subject to delay. If an organism flourishes, it is said to be in harmony with environment, or with higher forces. I now conceive of successful and flourishing Exclusionism as an organization that has been in harmony with higher forces. Suppose we accept that all general delusions function sociologically. Then, if Exclusionism be general delusion; if we shall accept that conceivably the isolation of this earth has been a necessary factor in the development of the whole geo-system, we see that exclusionistic science has faithfully, though falsely, functioned. It would be world-wide crime to spread world-wide too soon the idea that there are other existences nearby and that they have been seen and that sounds from them have been heard: the peoples of this earth must organize themselves before conceiving of, and trying to establish, foreign relations. A premature science of such subjects would be like a United States taking part in a Franco-Prussian War, when such foreign relations should be still far in the future of a nation that has still to concentrate upon its own internal development.

So in the development of all things--or that a stickleback may build a nest, and so may vaguely and not usefully and not explicably at all, in terms of Darwinian evolution, foreshadow a character of coming forms of life; but that a fish that should try to climb a tree and sing to its mate before even the pterodactyl had flapped around with wings daubed with clay would be an unnoticed little clown in cosmic drama. But I do conceive that when the Carboniferous Era is dominant, and when not a discordant thing will be permitted to flourish, though it may adumbrate, restrictions will not last forever, and that the rich and bountiful curse upon rooted things will someday be lifted.

20

PATCHED by a blue inundation that had never been seen before--this earth, early in the 60's of the 19th century. Then faintly, from far away, this new appearance is seen to be enveloped with volumes of gray. Flashes like lightning, and faintest of rumbling sounds--then cloud-like envelopments roll away, and a blue formation shines in the sun. Meteorologists upon the moon take notes.

But year after year there are appearances, as seen from the moon, that are so characterized that they may not be meteorologic phenomena upon this earth: changing compositions wrought with elements of blue and of gray; it is like conflict between Synthesis and Dissolution: straight lines that fade into scrawls, but that reform into seeming moving symbols: circles and squares and triangles abound.

Having had no mean experience with interpretations as products of desires, given that upon the moon communication with this earth should be desired, it seems likely to me that the struggles of hosts of Americans, early in the 60's of the 19th century, were thought by some lunarians to be maneuvers directed to them, or attempts to attract their attention. However, having had many impressions upon the resistance that new delusions encounter, so that, at least upon this earth, some benightments have had to wait centuries before finally imposing themselves generally, I'd think of considerable time elapsing before the coming of a general conviction upon the moon that, by means of living symbols, and the firing of explosives, terrestrians were trying to communicate.

Beacon-like lights that have been seen upon the moon. The lights have been desultory. The latest of which I have record was back in the year 1847. But now, if beginning in the early 60's, though not coinciding with the beginning of unusual and tremendous manifestations upon this earth, we have data as if of greatly stimulated attempts to communicate from the moon--why one assimilates one's impressions of such great increase with this or with that, all according to what one's dominant thoughts may be, and calls the product a logical conclusion. Upon the night of May 15, 1864, Herbert Ingall, of Camberwell, saw a little to the west of the lunar crater Picard, in the Mare Crisium, a remarkably bright spot (Astro. Reg., 2-264).

Oct. 24, 1864--period of nearest approach by Mars--red lights upon opposite parts of Mars (C. R., 85-538). Upon Oct. 16, Ingall had again seen the light west of Picard. Jan. I, 1865--a small speck of light, in darkness, under the east foot of the lunar Alps, shining like a small star, watched half an hour by Charles Grover (Astro. Reg., 3-255). Jan. 3, .1865--again the red lights of Mars (C. R., 85-538). A thread of data appears, as an offshoot from a main streak, but it cannot sustain itself. Lights on the moon and lights on Mars, but I have nothing more that seems to signify both signals and responses between these two worlds.

April 10, 1865--west of Picard, according to Ingall--"a most minute point of light, glittering like a star" (Astro. Reg., 3-189).

Sept. 5, 1865--a conspicuous bright spot west of Picard (Astro. Reg., 3-252) It was seen again by Ingall. He saw it again upon the 7th, but upon the 8th it had gone, and there was a cloudlike effect where the light had been.

Nov. 24, 1865--a speck of light that was seen by the Rev. W. O. Williams, shining like a small star in the lunar crater Carlini (Intel. Obs., II-58).

June to, 1866--the star-like light in Aristarchus; reported by Tempel (Denning, Telescopic Work, p. 121).

Astronomically and seleno-meteorologically, nothing that I know of has ever been done with these data. I think well of taking up the subject theologically. We are approaching accounts of a different kind of changes upon the moon. There will be data seeming so to indicate not only persistence but devotedness upon the moon that I incline to think not only of devotedness but of devotions. Upon the 16th of October, 1866, the astronomer Schmidt, of the land of Socrates, announced that the isolated object, in the eastern part of the Mare Serenitatis, known as Linne, had changed Linne stands out in a blank area like the Pyramid of Cheops in its desert. If changes did occur upon Linne, the conspicuous position. seems to indicate selection. Before October, 1866, Linne was well-known as a dark object. Something was whitening an object that had been black.

A hitherto unpublished episode in the history of theologies: The new prophet who had appeared upon the moon--

Faint perceptions of moving formations, often almost rigorously geometric, upon one part of this earth, and perhaps faintest of signal-like sounds that reached the moon--the new prophet--and that he preached the old lunar doctrine that there is no god but the Earth-god, but exhorted his hearers to forsake their altars upon which had burned unheeded lights, and to build a temple upon which might be recited a litany of lights and shades.

We are only now realizing how the Earth-god looks to the beings of the moon--who know that this earth is dominant; who see it frilled with the loops of the major planets; its Elizabethan ruff wrought by the complications of the asteroids; the busy little sun that brushes off the dark.

God of the moon, when mists make it expressionless--a vast, bland, silvery Buddha.

God of the moon, when seeing is clear--when the disguise is off--when, at night, from pointed white peaks drip the fluctuating red lights of a volcano, this earth is the appalling god of carnivorousness. Sometimes the great roundish earth, with the heavens behind it broken by refraction, looks like something thrust into a shell from external existence-- clouds of tornadoes as if in its grasp--and it looks like the fist of God, clutching rags of ultimate fire and confusion.

That a new prophet had appeared upon the moon, and had excited new hope of evoking response from the bland and shining Stupidity that has so often been mistaken for God, or from the Appalling that is so identified with Divinity--from the clutched and menacing fist that has so often been worshiped.

There is no intelligence except era-intelligence. Suppose the whole geo-system be a super-embryonic thing. Then, by the law of the embryo, its parts cannot organize until comes scheduled time. So there are local congeries of development of a chick in an egg, but these local centers cannot more than faintly sketch out relations with one another, until comes the time when they may definitely integrate. Suppose that far back in the 19th century there were attempts to communicate from the moon; but suppose that they were premature: then we suppose the fate of the protoplasmic threads that feel out too soon from one part of an egg to another. In October, 1866, Schmidt, of Athens, saw and reported in terms of the concepts of his era, and described in conventional selenographic language. See Rept. B. A., 1867.

Upon Dec. 14, 16, 25, 27, 1866, Linne was seen as a white spot. But there was something that had the seeming more of a design, or of a pattern, an elaboration upon the mere turning to white of something that had been black--a fine, black spot upon Linne; by Schmidt and Buckingham, in December, 1866 (The Student, 1- 261). The most important consideration of all is reviewed by Schmidt in the Rept. B. A., 1867-22--that sunlight and changes of sunlight had nothing to do with the changing appearances of Linne. Jan. 14, 1867--the white covering, or, at least, seeming of covering, of Linne, had seemingly disappeared--Knott's impression of Linne as a dark spot, but "definition" was poor. January 16--Knott's very strong impression, which, however, he says may have been an illusion, of a small central dark spot upon Linne. Dawes' observation, of March 15, 1867--"an excessively minute black dot in the middle of Linne."

A geometric figure that was white-bordered and centered with black, formed and dissolved and formed again.

I have an impression of spectacles that were common in the United States, during the War: hosts of persons arranging themselves in living patterns: flags, crosses, and in one instance, in which thousands were engaged, in the representation of an enormous Liberty Bell. Astronomers have thought of trying to communicate with Mars or the moon by means of great geometric constructions placed conspicuously, but there is nothing so attractive to attention as change, and a formation that could appear and disappear would enhance the geometric with the dynamic. That the units of the changing compositions that covered Linne were the lunarians themselves-- that Linne was terraced--hosts of the inhabitants of the moon standing upon the ridges of their Cheops of the Serene Sea, some of them dressed in white and standing in a border, and some of them dressed in black, centering upon the apex, or the dark material of the apex left clear for the contrast, all of them unified in a hope of conveying an impression of the geometric, as the product of design, and distinguishable from the topographic, to the shining god that makes the stars of their heavens marginal.

It is a period of great activity--or of conflicting ideas and purposes--upon the moon: new and experimental demonstrations, but also, of course, the persistence of the old. In the Astronomical Register, 5-114, Thomas G. Elger writes that upon the 9th of April, 1867, he was surprised to see, upon the dark part of the moon, a light like a star of the 7th magnitude, at 7:30 P.M. It became fainter, and looked almost extinguished at 9 o'clock. Mr. Elger had seen lights upon the moon before, but never before a light so clear--"too bright to be overlooked by the most careless observer." May 7, 1867--the beacon-like light of Aristarchus--observed by Tempel, of Marseilles, when Aristarchus was upon the dark part of the moon (Astro. Reg., 5-220). Upon the night of June 10, 1867, Dawes saw three distinct, roundish, black spots near Sulpicius Gallus, which is near Linne; when looked for upon the 13th, they had disappeared (The Student, 1-261).

Aug. 6, 1867--

And this earth in the sky of the moon--smooth and bland and featureless earth--or one of the scenes that make it divine and appalling--jaws of this earth, as seem to be rims of more or less parallel mountain ranges, still shining in sunlight, but surrounded by darkness--

And, upon the moon, the assembling of the Chiaroscuroans, or the lunar communicationists who seek to be intelligible to this earth by means of lights and shades, patterned upon Linne by their own forms and costumes. The Great Pyramid of Linne, at night upon the moon--it stands out as a bold black triangularity pointing to this earth. It slowly suffuses white--the upward drift of white-clad forms, upon the slopes of the Pyramid. The jaws of this earth seem to munch, in variable light. There is no other response. Devotions are the food of the gods.

Upon Aug. 6, 1867, Buckingham saw upon Linne, which was in darkness, "a rising oval spot" (Rept. B. A., 1867-7). In October, 1867, Linne was seen as a convex white spot (Rept. B. A., 1867-8) .

Also it may be that the moon is not inhabited, and is not habitable. There are many astronomers who say that the moon has virtually no atmosphere, because when a star is passed over by the moon, the star is not refracted, according to them. See Clerke's History of Astronomy, p. 264--that, basing his calculations upon the fact that a star is never refracted out of place when occulted by the moon, Prof. Comstock, of Washburn Observatory, had determined that this earth's atmosphere is 5,000 times as dense as the moon's.

I did think that in this secondary survey of ours we had pretty well shaken off our old opposition, the astronomers: however, with something of the kindliness that one feels for renewed meeting with the familiar, here we are at home with the same old kind of demonstrations: the basing of laborious calculations upon something that is not so.

See index of Monthly Notices, R. A. S.--many instances of stars that have been refracted out of place when occulted by the moon. See the Observatory, 24-210, 313, 315, 345, 414; English Mechanic, 23-197, 279; 26-229; 52--index,"atmosphere"; 81-60; 84-161; 85-108.

In the year 1821, Gruithuisen announced that he had discovered a city of the moon. He described its main thoroughfare and branching streets. In 1826, he announced that there had been considerable building, and that he had seen new streets. This formation, which is north of the crater Schroeter, has often been examined by disagreeing astronomers: for a sketch of it, in which a central line and radiating lines are shown, see the English Mechanic, 18-638. There is one especial object upon the moon that has been described and photographed and sketched so often that I shall not go into the subject. For many records of observations, see the English Mechanic and L'Astronomie. It is an object shaped like a sword, near the crater Birt. Anyone with an impression of the transept of a cathedral, may see the architectural here. Or it may be a mound similar to the mounds of North America that have so logically been attributed to the Mound Builders. In a letter, published in the Astronomical Register, 20-167, Mr. Birmingham calls attention to a formation that suggests the architectural upon the moon--"a group of three hills in a slightly acute-angled triangle, and connected by three lower embankments." There is a geometric object, or marking, shaped like an "X," in the crater Eratosthenes (Sci. Amer. Sup., 59-24, 469); striking symbolic-looking thing or sign, or attempt by means of something obviously not topographic, to attract attention upon this earth, in the crater Plinius (Eng. Mec., 35-34); reticulations, like those of a city's squares, in Plato (Eng. Mec., 64-253); and there is a structural-looking composition of angular lines in Gassendi (Eng. Mec., 101466). Upon the floor of Littrow are six or seven spots arranged in the form of the Greek letter Gamma (Eng. Mec., 101-47). This arrangement may be of recent origin, having been discovered Jan. 31, 1915. The Greek letter makes difficulty only for those who do not want to think easily upon this subject. For a representation of something that looked like a curved wall upon the moon, see L'Astronomie, 1888-110. As to appearances like viaducts, see L'Astronomie, 1885-213. The lunar craters are not in all instances the simple cirques that they are commonly supposed to be. I have many different impressions of some of them: I remember one sketch that looked like an owl with a napkin tucked under his beak. However, it may be that the general style of architecture upon the moon is Byzantine, very likely, or not so likely, domed with glass, giving the dome-effect that has so often been commented upon.

So then the little nearby moon--and it is populated by Liliputians. However, our experience with agreeing ideas having been what it has been, we suspect that the lunarians are giants. Having reasonably determined that the moon is one hundred miles in diameter, we suppose it is considerably more or less.

A group of astronomers had been observing extraordinary lights in the lunar crater Plato. The lights had definite arrangement. They were so individualized that Birt and Elger, and the other selenographers, who had combined to study them, had charted and numbered them. They were fixed in position, but rose and fell in intensity.

It does seem to me that we have data of one school of communicationists after another coming into control of efforts upon the moon. At first our data related to single lights. They were extraordinary, and they seem to me to have been signals, but there seemed to be nothing of the organization that now does seem to be creeping into the fragmentary material that is the best that we can find. The grouped lights in Plato were so distinctive, so clear and even brilliant, that if such lights had ever shone before, it seems that they must have been seen by the Schroeters, Gruithuisens, Beers and Madlers, who had studied and charted the features of the moon. For several of Gledhill's observations, from which I derive my impressions of these lights, see Rept. B. A., 1871-80--"I can only liken them to the small discs of stars, seen in the transit-instrument"; "just like small stars in the transit instrument, upon a windy night!"

In August and September,. 1860, occurred a notable illumination of the spots in Group I. It was accompanied by a single light upon a distant spot.

February and March, 1870--illumination of another group. April 17, 1870--another illumination in Plato, but back to the first group.

As to his observations of May 10-12, 1870, Birt gives his opinion that the lights of Plato were not effects of sunlight.

Upon the 13th of May, 1870, there was an "extraordinary display," according to Birt: 27 lights were seen by Pratt, and 28 by Elger, but only 4 by Gledhill, in Brighton. Atmospheric conditions may have made this difference, or the lights may have run up or down a scale from 4 to 28. As to independence of sunlight, Pratt says (Rept. B. A., 1871-88) as to this display, that only the fixed, charted points so shone, and that other parts of the crater were not illuminated, as they would have been to an incidence common throughout. In Pratt's opinion, and, I think, in the opinion of the other observers, these lights were volcanic. It seems to me that this opinion arose from a feeling that there should be something of an opinion: the idea that the lights might have been signals was not expressed by any of these astronomers that I know of. I note that, though many observers were, at this time, concentrating upon this one crater, there are no records find-able by me of such disturbance of detail as might be supposed to accompany volcanic action. The clear little lights seem to me to have been anything but volcanic.

The play of these lights of Plato--their modulations and their combinations--like luminous music--or a composition of signals in a code that even in this late day may be deciphered. It was like orchestration--and that something like a baton gave direction to Light 22, upon Aug. 12, 1870, to shine a leading part--"remarkable increase of brightness." No. 22 subsided, and the leading part shone out in No. 14. It, too, subsided, and No. 16 brightened.

Perhaps there were definite messages in a Morse-like code. There is a chance for the electricity in somebody's imagination to start crackling. Up to April, 1871, the selenographers had recorded 1,600 observations upon the fluctuations of the lights of Plato, and had drawn 37 graphs of individual lights. All graphs and other records were deposited by W. R. Birt in the Library of the Royal Astronomical Society, where presumably they are to this day. A Champollion may someday decipher hieroglyphics that may have been flashed from one world to another.

21

OUR data indicate that the planets are circulating adjacencies.

Almost do we now conceive of a difficulty of the future as being not how to reach the planets, but how to dodge them. Especially do we warn aviators away from that rhinoceros of the skies, Mercury. I have a note somewhere upon one of the wickedest-looking horns in existence, sticking out far from Mercury. I think it was Mr. Whitmell who made this observation. I'd like to hear Andrew Barclay's opinion upon that. I'd like to hear Capt. Noble's.

If sometimes does the planet Mars almost graze this earth, as is not told by the great telescopes, which are only millionaires' memorials, or, at least, which reveal but little more than did the little spy glasses used by Burnham and Williams and Beer and Madler--but if periodically the planet Mars comes very close to this earth, and, if Mars, an island with perhaps no more surface-area than has England, but likely enough inhabited, like England--

June 19, 1875--opposition of Mars.

Flashes that were seen in the sky upon the 25th of June, 1875, by Charles Gape, of Scole, Norfolk (Eng. Mec., 21-488). The Editor of Symons' Met. Mag. (see vol. 10- 116) was interested, and sent Mr. Gape some questions, receiving answers that nothing had appeared in the local newspapers upon the subject, and that nothing could be learned of a display of fireworks, at the time. To Mr.

Gape the appearances seemed to be meteoric.

The year 1877--climacteric opposition of Mars. There were some discoveries.

We have at times wondered how astronomers spend their nights. Of course, according to many of his writings upon the subject, Richard Proctor had an excellent knowledge of whist. But in the year 1877, two astronomers looked up at the sky, and one of them discovered the moons of Mars, and the other called attention to lines on Mars--and, if for centuries, the moons of Mars could so remain unknown to all inhabitants of this earth except, as it were, Dean Swift--why, it is no wonder that we so respectfully heed some of the Dean's other intuitions, and think that there may be Liliputians, or Brobdingnagians, and other forms not conventionally supposed to be. As to our own fields of data, I have a striking number of notes upon signal-like appearances upon the moon, in the year 1877, but have notes upon only one occurrence that, in our interests, may relate to Mars. The occurrence is like that of July 31, 1813, and June 19, 1875.

Sept. 5, 1877--opposition of Mars.

Sept. 7, 1877--lights appeared in the sky of Bloomington, Indiana. They were supposed to be meteoric. They appeared and disappeared, at intervals of three or four seconds; darkness for several minutes; then a final flash of light. See Sci. Amer., 37-193.

That all luminous objects that are seen in the sky when the planet Venus is nearest may not be Venus; may not be fire-balloons:

In the Dundee Advertiser, Dec. 22, 1882, it is said that, between 10 and 11 A.M., December 21, at Broughty Ferry, Scotland, a correspondent had seen an unknown luminous body near and a little above the sun. In the Advertiser, December 25, is published a letter from someone who says that this object had been seen at Dundee, also; that quite certainly it was the planet Venus and "no other." In Knowledge,

2-489, this story is told by a writer who says that undoubtedly the object was Venus. But, in Knowledge, 3-13, the astronomer J. E. Gore writes that the object could not have been Venus, which upon this date was 1 h. 33 m., R. A., west of the sun. The observation is reviewed in L'Astronomie, 1883-109. Here it is said that the position of Mercury accorded better. Reasonably this object could not have been Mercury: several objections are comprehended in the statement that superior conjunction of Mercury had occurred upon December 16.

Upon Feb. 3, 1884, M. Staevert, of the Brussels Observatory, saw, upon the disc of Venus, an extremely brilliant point (Ciel et Terre, 5-127). Nine days later, Niesten saw just such a point of light as this, but at a distance from the planet. If no one had ever heard that such things cannot be, one might think that these two observations were upon something that had been seen leaving Venus and had then been seen farther along. Upon the 3rd of July, 1884, a luminous object was seen moving slowly in the sky of Norwood, N. Y. It had features that suggest the structural: a globe the size of the moon, surrounded by a ring; two dark lines crossing the nucleus (Science Monthly, 2-136). Upon the 26th of July, a luminous globe, size of the moon, was seen at Cologne; it seemed to be moving upward from this earth, then was stationary "some minutes," and then continued upward until it disappeared (Nature, 30-360). And in the English Mechanic, 40-130, it is not said that a luminous vessel that had sailed out from Venus, in February, visiting this earth, where it was seen in several places, was seen upon its return to the planet, but it is said that an observer in Rochester, N. Y., had, upon August 17, seen a brilliant point upon Venus.

22

EXPLOSIONS over the towns of Barisal, Bengal, if they were aerial explosions, were continuing. As to some of these detonations that were heard in May, 1874, a writer in Nature, 53-197, says that they did seem to come from overhead. For a report upon the Barisal Guns, heard between April 28, 1888, and March 1, 1889, see Proc. Asiatic Soc. of Bengal, 1889-199.

Phenomena at Comrie were continuing. The latest date in Roper's List of Earthquakes is April 8, 1886, but this list goes on only a few years later. See Knowledge, n. s., 6-145--shock and a rumbling sound at Comrie, July 12, 1894--a repetition upon the corresponding date, the next year. In the English Mechanic, 74-155, David Packer says that, upon Sept. 17, 1901, ribbon-like flashes of lightning, which were not ordinary lightning, were seen in the sky (I think of Birmingham) one hour before a shock in Scotland. According to other accounts, this shock was in Comrie and surrounding regions (London Times, Sept. 19, 1901).

Smithson. Miscell. Cols., 37-Appendix, p. 71:

According to L. Tennyson, Quartermaster's Clerk, at Fort Klamath, Oregon, at daylight, Jan. 8, 1867, the garrison was startled from sleep by what he supposed to be an earthquake and a sound like thunder. Then came darkness, and the sky was covered with black smoke or clouds. Then ashes, of a brownish color, fell--"as fast as I ever saw it snow." Half an hour later there was another shock, described as "frightful." No one was injured, but the sutler's store was thrown a distance of ninety feet, and the vibrations lasted several minutes. Mr. Tennyson thought that somewhere near Fort Klamath, a volcano had broken loose, because, in the direction of the Klamath Marsh, a dark column of smoke was seen. I can find record of no such volcanic eruption. In a list of quakes, in Oregon, from 1846, to 1916, published in the Bull. Seis. Soc. Amer., September, 1919, not one is attributed to volcanic eruptions. Mr. W. D. Smith, compiler of the list, says, as to the occurrence at Fort Klamath--"If there was an eruption, where was it?" He asks whether possibly it could have been in Lassen Peak. But Lassen Peak is in California,, and the explosion upon Jan. 8, 1867, was so close to Fort Klamath that almost immediately ashes fell from the sky.

The following is of the type of phenomena that might be considered evidence of signaling from some unknown world nearby:

La Nature, 17-126--that, upon June 17, 1881, sounds like cannonading were heard at Gabes, Tunis, and that quaking of the earth was felt, at intervals of 32 seconds, lasting about 6 minutes.

July 30, 1883--a somewhat startling experience--steamship Resolute alone in the Arctic Ocean--six reports like gunfire--Nature, 53-295.

In Nature, 30-19, a correspondent writes that, upon the 3rd of January, 1869, a policeman in Harlton, Cambridgeshire, heard six or seven reports, as if of heavy guns far away. There is no findable record of an earthquake in England upon this date. In the London Times, Jan. 12, 15, 16, 1869, several correspondents write that upon the 9th of January a loud report had been heard and a shock felt at places near Colchester, Essex, about 30 miles from Harlton. One of the correspondents writes that he had heard the sound but had felt no shock. In the London Standard, January 12, the Rev. J. F. Bateman, of South Lopham, Norfolk, writes as to the occurrence upon the 9th--"An extraordinary vibration (described variously by my parishioners as being 'like a gunpowder explosion,' 'a big thunder clap,' and 'a little earthquake') was noticed here this morning about 11.20." In the Morning Post, January 14, it is said that at places about twenty miles from Colchester it was thought that an explosion had occurred, upon the 9th, but, inasmuch as no explosion had been heard of, the disturbance was attributed to an earthquake. Night of January 13--an explosion in the sky, at Brighton (Rept. B. A., 1869-307). In the Standard, January 22, a correspondent writes from Swaffham, Norfolk, that, about 8 P.M., January 15, something of an unknown nature had frightened flocks of sheep, which had burst. from their bounds in various places. All these occurrences were in adjoining counties in southeastern England.

Something was seen in the sky upon the 13th, and, according to the Chudleigh Weekly Express, Jan. 13, 1869, something was seen in the sky, night of the 10th, at Weston-super-Mare, near Bristol, in southwestern England. It was seen between 9 and 10 o'clock, and is said to have been an extraordinary meteor. Five hours later were felt three shocks said to have been earthquakes.

Upon the night of March 17, 1871, there was a series of events in France, and a series in England. A "meteor" was seen at Tours, at 8 P.M.--at 10:45, a "meteor" that left a luminous cloud over Saintes (Charante- Inferieure)--another at Paris, 11:15, leaving a mark in the sky, of fifteen minutes' duration--another at Tours, at 11:45 P.M. See Les Mondes, 24-190, and Comptes Rendus, 72-789. There were "earthquakes" this night affecting virtually all England north of the Mersey and the Trent, and also southern parts of Scotland. As has often been the case, the phenomena were thought to have been explosions and were then said to have been earthquakes when no terrestrial explosions could be heard of (Symons' Met. Mag., 6-39). There were six shocks near Manchester, between 6 and 7 P.M., and others about 11 P.M.; and in Lancashire about 11 P.M., and continuing in places as far apart as Liverpool and Newcastle, until 11:30 o'clock. The shocks felt about u o'clock correspond, in time, with the luminous phenomena in the sky of France, but our way of expressing that these so- called earthquakes in England may have been concussions from repeating explosions in the sky, is to record that, according to correspondence in the London Times, there were, upon the 20th, aerial phenomena in the region of Lancashire that had been affected upon the 17th--"sounds that seemed to come from a number of guns at a distance" and "pale flashes of lightning in the sky."

Whether these series of phenomena be relatable to Mars or Martians or not, we note that in 1871 opposition of Mars was upon March 19; and, in 1869, upon February 13; and in 1867 two days after the explosions at Fort Klamath. In our records in this book, similar coincidences can be found up to the year 1879. I have other such records not here published, and others that will be here investigated. There is a triangular region in England, three points of which appear so often in our data that the region should be specially known to us, and I know it myself as the London Triangle. It is pointed in the north by Worcester and Hereford, in the south by Reading, Berkshire, and in the east by Colchester, Essex. The line between Colchester and Reading runs through London.

Upon Feb. 18, 1884, at West Mersea, near Colchester, a loud report was heard (Nature, 53-4). Upon the 22nd of April, 1884, centering around Colchester, occurred the severest earthquake in England in the 19th century. For several columns of description, see the London Times, April 23. There is a long list of towns in which there was great damage: in 24 parishes near Colchester, 1,250 buildings were damaged. One of the places that suffered most was West Mersea (Daily Chronicle, April 28).

There was something in the sky. According to G. P. Yeats (Observations upon the Earthquake of Dec. 17, 1896, p. 6) there was a red appearance in the sky over Colchester, at the time of the shock of April 22, 1884.

The next day, according to a writer in Knowledge, 5-336, a stone fell from the sky, breaking glass in his greenhouse, in Essex. It was a quartz stone, and unlike anything usually known as meteoritic.

The indications, according to my reading of the data, and my impressions of such repeating occurrences as those at Fort Klamath, are that perhaps an explosion occurred in the sky, near Colchester, upon Feb. 18, 1884; that a great explosion did occur over Colchester, upon the 22nd of April, and that a great volume of debris spread over England, in a northwesterly direction, passing over Worcestershire and Shropshire, and continuing on toward Liverpool, nucleating moisture and falling in blackest of rain. From the Stonyhurst Observatory, near Liverpool, was reported, occurring at a 11 A.M., April 26, "the most extraordinary darkness remembered"; forty minutes later fell rain "as black as ink," and then black snow and black hail (Nature, 30-6). Black hail fell at Chaigley, several miles from Liverpool (Stonyhurst Magazine, 1-267). Five hours later, black substance fell at Crowle, near Worcester (Nature, 30-32). Upon the 28th, at Church Stretton and Much Wenlock, Shropshire, fell torrents of liquid like ink and water in equal proportions (The Field, May 3, 1884). In the Jour. Roy. Met. Soc., 11-7, it is said that, upon the 28th, half a mile from Lilleshall, Shropshire, an unknown pink substance was brought down by a storm. Upon the 3rd of May, black substance fell again at Crowle (Nature, 30-32). In Nature, 30-216, a correspondent writes that, upon June 22, 1884, at Fletching, Sussex, southwest of Colchester, there was intense darkness, and that rain then brought down flakes of soot in such abundance that it seemed to be "snowing black." This was several months after the shock at Colchester, but my datum for thinking that another explosion, or disturbance of some kind, had occurred in the same local sky, is that, as reported by the inmates of one house, a slight shock was felt, upon the 24th of June, at Colchester, showing that the phenomena were continuing. See Roper's List of Earthquakes.

Was not the loud report heard upon February 18 probably an explosion in the sky, inasmuch as the sound was great and the quake little? Were not succeeding phenomena sounds and concussions and the fall of debris from explosions in the sky, acceptably upon April 22, and perhaps continuing until the 24th of June? Then what are the circumstances by which one small part of this earth's surface could continue in relation with something somewhere else in space?

Comrie, Irkutsk, and Birmingham.

23

UPON the night of the 13th of July, 1875, at midnight, two officers of H.M.S. Coronation, in the Gulf of Siam, saw a luminous projection from the moon's upper limb (Nature, 12-495). Upon the 14th it was gone, but a smaller projection was seen from another part of the moon's limb. This was in the period of the opposition of Mars.

Upon the night of Feb. 20, 1877, M. Trouvelot, of the Observatory of Meudon, saw, in the lunar crater Eudoxus, which, like almost all other centers of seeming signaling, is in the northwestern quadrant of the moon, a fine line of light (L'Astronomie, 1885-212). It was like a luminous cable drawn across the crater.

March 21, 1877--a brilliant illumination, and not by the light of the sun, according to C. Barrett, in the lunar crater Proclus (Eng. Mec., 25-89).

May 15 and 29, 1877-the bright spot west of Picard (Eng. Mec., 25-335).

The changes upon Linne were first seen by Schmidt, in 1866, near the time of opposition of Mars. In May, 1877, Dr. Klein announced that a new object had appeared upon the moon. It was close to the center of the visible disc of the moon, and was in a region that had been most carefully studied by the selenographers. In the Observatory, 2-238, is Neison's report from his own memoranda. In the years 1874 and 1875, he had studied this part of the moon, but had not seen this newly reported object in the crater Hyginus, or the object, Hyginus N, according to the selenographers' terminology. In the Astronomical Register, 17-204, Neison lists, with details, 20 minute examinations of this region, from July, 1870, to August, 1875, in which this conspicuous object was not recorded.

June 14, 1877--a light on the dark part of the moon, resembling a reflection from a moving mirror; reported by Prof. Henry Harrison (Sidereal Messenger, 3-150).

June 15--the bright spot west of Picard, according to Birt (Jour. B. A. A., 19- 376). Upon the 16th, Prof. Harrison thought that again he saw the moving light of the 14th, but shining faintly. In the English Mechanic, 25-432, Frank Dennett writes, as to an observation of June 17, 1877--"I fancied I could detect a minute point of light shining out of the darkness that filled Bessel."

These are data of extraordinary activity upon the moon preceding the climacteric opposition of Mars, early in September, 1877. Now we have an account of an occurrence during an eclipse of the moon:

On the night of the eclipse (Aug. 27, 1877) a ball of fire, of the apparent size of the moon, was seen, at ten minutes to eleven, dropping apparently from cloud to cloud, and the light flashing across the road (Astro. Reg., 1878-75).

Astro. Reg., 17-251:

Nov. 13, 1877--Hyginus N standing out with such prominence as to be seen at the first glance;

Nov. 14, 1877--not a trace of Hyginus N, though seeing was excellent: [p. 443]

Oct. 3, 1878--the most conspicuous of all appearances of Hyginus N; Oct. 4, 1878--not a trace of Hyginus N.

Upon the night of Nov. 1, 1879, again in the period of opposition of Mars (opposition November 12) again the bright spot west of Picard (Jour B. A. A., 19- 376). But I have several records of observations upon this appearance not in times of opposition of Mars. Whether there be any relation with anything else or not, at five o'clock, morning of Nov. 1, 1879, a "vivid flash" was seen and a shock was felt at West Cumberland (Nature, 21-19).

In the autumn of the year 1883, began extraordinary atmospheric effects in the sky of this earth. For Prof. John Haywood's description of similar appearances upon the moon, Nov. 4, 1883, and March 29, 1884, see the Sidereal Messenger, 3-121.

They were misty light-effects upon the dark part of the moon, not like "earth- shine." Our expression is that so close is the moon to this earth that it, too, may be affected by phenomena in the atmosphere of this earth.

Something like another luminous cable, or like a shining wall, that was seen in Aristarchus, by Trouvelot, Jan. 23, 1880 (L'Astro., 1885-215); a speck of light in Marius, Jan. 13, 1881, by A. S. Williams

(Eng. Mec., 32-494); unexplained light in Eudoxus, by Trouvelot, May 4, 1881 (L'Astro., 1885-213); an illumination in Kepler, by Morales, Feb. 5, 1884 (L'Astro., 9-149).

In Knowledge, 7-224, William Gray writes that, upon Feb. 19, 1885, he saw, in Hercules, a dull, deep, reddish appearance. In L'Astronomie, 1885-227, Lorenzo Kropp, an astronomer of Paysandu, Uruguay, writes that, upon Feb. 21, 1885, he had seen, in Cassini, a formation not far from Hercules, both of them in the northwestern quadrant of the moon, a reddish smoke or mist. He had heard that several other persons had seen, not a misty appearance, but a star-like light here, and upon the 22nd he had seen a definite light, himself, shining like the planet Saturn.

May 11, 1885--two lights upon the moon (L'Astro., 9-73).

May 11, 1886--two lights upon the moon (L'Astro., 6-312).

24

THAT through lenses rimmed with horizons, inhabitants of this earth have seen revelations of other worlds--that atmospheric strata of different densities are lenses--but that the faults of the wide glasses in the observatories are so intensified in atmospheric revelations that all our data are distortions. Our acceptance is that every mirage has a primary; that in human mind all poetry is based upon observation, and that imagery in the sky is similarly uncreative. If a mirage cannot be traced to the known upon this earth, one supposes that it is either a derivation from the unknown upon this earth, or from the unknown somewhere else. We shall have data of a series of mirages in Sweden, or upon the shores of the Baltic, from October, 1881, to December, 1888. I take most of the data from Nature, Knowledge, Cosmos, and L'Astronomie, published in this period. I have no data of such appearances in this region either before or after this period: the suggestion in my own mind is that they were not mirages from terrestrial primaries, or they would not be so confined to one period, but were shadows or mirages from something that was in temporary suspension over the Baltic and Sweden, all details distorted and reported in terms of familiar terrestrial appearances.

Oct. 10, 1881--that at Rugenwalde, Pomerania, the mirage of a village had been seen: snow-covered roofs from which hung icicles; human forms distinctly visible. It was believed that the mirage was a representation of the town of Nexo, on the island of Bornholm. Rugenwalde is on the Baltic, and Nexo is about 100 miles northwest, in the Baltic.

The first definite account of the mirages of Sweden, findable by me, is published in Nature, June 29, 1882, where it is said that preceding instances had attracted attention--that, in May, 1882, over Lake Orsa, Sweden, representations of steamships had been seen, and then "islands covered with vegetation." Night of May 19, [paragraph continues] 1883--beams of light at Lake Ludyika, Sweden--they looked like a representation of a lake in moonshine, with shores covered with trees, showing faint outlines of farms (Monthly Weather Review, May, 1883). May 28, 1883--at Finsbo, Sweden--changing scenes, at short intervals: mountains, lakes, and farms. Oct. 16, 1884--Lindsberg--a large town, with four-storied houses, a castle and a lake. May 22, 1885--Gothland--a town surrounded by high mountains, a large vessel in front of the town. June 15, 1885--near Oxelosund--two wooded islands, a construction upon one of them, and two warships. It is said that at the time two Swedish warships were at sea, but were at considerable distance north of Oxelosund. Sept. 12, 1885--Valla--a representation that is said to have been a "remarkable mirage" but that is described as if the appearances were

cloud-forms-- several monitors, one changing into a spouting whale, and the other into a crocodile--
then forests--dancers--a wooded island with buildings and a park. Sept. 29, 1885--again at Valla--be-
tween 8 and 9 o'clock, P.M.; a lurid glare upon the northwestern horizon; a cloud bank--animals,
groups of dancers, a forest, and then a park with paths. July 15, 1888--Hudikwall--a tempestuous
sea, and a vessel upon it; a small boat leaving the vessel. Upon Oct. 8, 1888, at Merexull, on the
Baltic, but in Russia, was seen a mirage of a city that lasted an hour. It is said that some buildings
were recognized, and that the representation was identified with St. Petersburg, which is about 200
miles from the Baltic.

That a large, substantial mass, presumably of land, can be in at least temporary suspension over a
point upon this earth's surface, and not fall, and be, in ordinary circumstances, invisible--

In L'Astronomie, 1887-426, MM. Codde and Payan, both of them astronomers, well- known for
their conventional observations and writings, publish accounts of an unknown body that appeared
upon the sun's limb, for twenty or thirty seconds, after the eclipse of Aug. 19, 1887. They saw a
round body, apparent diameter about one tenth of the apparent diameter of the sun, according to
the sketch that is published. In L'Astronomie, these two observers write separately, and, in the city
of Marseilles, their observations were made at a distance apart. But the unknown body was seen by
both upon the same part of the sun's limb. So it is supposed that it could not have been a balloon,
nor a circular cloud, nor anything else very near this earth. But many astronomers in other parts
of Europe were watching this eclipse, and it seems acceptable that others, besides two in Marseilles,
continued to look, immediately after the eclipse; but from nowhere else came a report upon this
object, so that all indications are that it was far from the sun and near Marseilles, but farther than
clouds or balloons in this local sky. I can draw no diagram that can satisfy all these circumstances,
except by supposing the sun to be only a few thousand miles away.

If little black stones fall four times, in eleven years, to one part of this earth's surface, and fall
nowhere else, we are, in conceiving of a fixed origin somewhere above a stationary earth, at least conceiving in terms of data, and, whether we are fanatics or not, we are not of the type of other upholders of stationariness of this earth, who care more for Moses than they do for data. I'd not like to have
it thought that we are not great admirers of Moses, sometimes.

The rock that hung in the sky of Servia--

Upon Oct. 13, 1872, a stone fell from the sky, to this earth, near the town of Soko-Banja, Servia. If
it were not a peculiar stone, there is no force to this datum. It is said that it was unknown stone. A
name was invented for it. The stone was called banjite, after the town near which it fell.

Seventeen years later (Dec. 1, 1889) another rock of banjite fell in Servia, near Jelica.

For Meunier's account of these stones, see L'Astronomie, 1890272, and Comptes Rendus, 92-331.
Also, see La Nature, 1881-1-192. According to Meunier these stones did fall from the sky; indigenous to this earth there are no such stones; nowhere else have such stones fallen from the sky; they
are identical in material; they fell seventeen years apart.

At times when we think favorably of this work of ours, we see in it a pointing-out of an evil of modern specialization. A seismologist studies earthquakes, and an astronomer studies meteors;. neither
studies both earthquakes and meteors, and consequently each, ignorant of the data collected by the
other, sees no relation between the two phenomena. The treatment of the event in Servia, Dec. 1,

1889, is an instance of conventional scientific attempts to understand something by separately, or specially, focusing upon different aspects, and not combining into an inclusive concept. Meunier writes only upon the stones that fell from the sky, and does not mention an earthquake at the time. Milne, in his Catalogue of Destructive Earthquakes, lists the occurrence as an earthquake, and does not mention stones that fell from the sky. All combinations greatly affect the character of components: in our combination of the two aspects, we see that the phenomenon was not an earthquake, as earthquakes are commonly understood, though it may have been meteoric; but was not meteoric, in ordinary terms of meteors, because of the unlikelihood that meteors, identical in material, should, seventeen years apart, fall upon the same part of this earth's surface, and nowhere else.

This occurrence was of course an explosion in the sky, and its vibrations were communicated to the earth below, with all the effects of any other kind of earthquakes. Back in our earliest confusion of the data of a century's first quarter, we had awareness of this combination and its conventional misinterpretation: that many concussions that have been communicated from explosions in the sky have been catalogued in lists of subterranean earthquakes. We are farther along now, in our data of the 19th century, and now we come across awareness, in other minds, of this distinguishment. At 8:20 A.M., Nov. 20, 1887, was heard and felt something that was reported from many places in the region that is known to us as the London Triangle, as an earthquake, though in some towns it was thought that a great explosion, perhaps in London, had occurred. It was reported from Reading, and from four towns near Reading, and Reading is said to be one of the places where the concussion was greatest. There were several accounts of slight alarm among sheep, which are sensitive to meteors and earthquakes. But, in Symons' Met. Mag., Mr. H. G. Fordham wrote that the occurrence was not an earthquake; that a meteor had exploded. He had very little to base this opinion upon: out of scores of descriptions, he had record of only two assertions that something had been seen in the sky. Nevertheless, because the sound was so much greater than the concussion, Mr. Fordham came to his conclusion.

In Symons' Met. Mag., 23-154, Dr. R. H. Wake writes that, upon the evening of Nov. 3, 1888, in a region about four miles wide and ten or fifteen miles long, in the Thames Valley (near Reading) flocks of sheep had rushed from their folds in a common alarm. About a year later, in the Chiltern Hills, which extend in a northeasterly direction from the Thames Valley, near Reading, there was another such occurrence. In the London Standard, Nov. 7, 1889, the Rev. J. Ross Barker, of Chesham, a town about 25 miles northeast of Reading, writes that, upon Oct. 25, 1889, many flocks of sheep, in a region of 30 square miles, had, by common impulse, broken from their folds. Mr. Barker asks whether anyone knew of a meteor or of an earthquake at the time. In vol. 24, Symons' Met. Mag., Mr. Symons accepts that all three of these occurrences were effects of meteoric explosions in the sky. The phenomena are insignificant relatively to some that we have considered: the significance is in this definite recognition in orthodoxy, itself, that some supposed earthquakes, or effects of supposed earthquakes, are reactions to explosions in the sky.

25

EXPLODING monasteries that shoot out clouds of monks into cyclonic formations with stormy nuns similarly dispossessed --or collapsing monasteries--sometimes slowly crumbling confines of the cloistered--by which we typify all things: that all developments pass through a process of walling-away within shells that will break.

Once upon a time there was a shell around the United States. The shell broke. Some other things were smashed.

The doctrines of great distances among heavenly bodies, and of a moving earth are the strongest elements of Exclusionism: the mere idea of separations by millions of miles discourages thoughts of communication with other worlds; and only to think that this earth shoots through space at a velocity of 19 miles a second puts an end to speculation upon how to leave it and how to return. But, if these two conventions be features of a walling-away like that of a chick within its shell, or that of the United States within its boundaries, and if some day all such confinements of the embryonic break, our own prophecy, in the vague terms of all successful prophecies, is that a matured view of astronomic phenomena will be from a litter of broken demonstrations.

Our expression now is upon the function of Isolation in Development. Specially it is not ours, because I think we learned it from the biologists, but we are applying it generally. If the general expression be accepted, we conceive that functionally have the astronomers taught that planets are millions of miles away, and that this earth moves at such terrific velocity that it is encysted with speed. Whether isolations function or not, that exclusions that break down are typical of all developments is signified by data upon all growing things, beginning with the aristocratic seeds, which, however, liberalize to intercourse with mean materials or die. All animal-organisms are at first walled away. In human circumstances conditions are the same. The development of every science has been a series of temporary exclusions, and the story of every industry tells of inventions that were resisted, but that were finally admitted. At the beginning of the nineteenth century, Hegel published his demonstration that there could be only seven planets: too late to recall the work, he learned that Ceres had been discovered. It is our expression that the mental state of Hegel partook of a general spirit of his time, and that it was necessary, or that it functioned, because early astronomers could scarcely have systematized their doctrine had they been bewildered by seven or eight hundred planetary bodies; and that, besides the functions of the astronomers, according to our expressions, there was also their usefulness in breaking down the walls of the older, and outlived, orthodoxy. We conceive that it is well that a great deal of experience should be withheld from children, and that, anyway, in their early years, they are sexually isolated, for instance, and our idea is that our data have been held back by no outspoken conspiracy, but by an inhibition similar to that by which a great deal of biology, for instance, is not taught to children. But, if we think of something of this kind, equally acceptable is it that even in the face of orthodox principles, these data have been preserved in orthodox publications, and that, in the face of supposed principles of Darwinism, as applied generally they have survived, though not in harmony with their environment.

Tons of paper have been consumed by calculations upon the remoteness of stars and planets. But I can find nothing that has been calculated, or said, that is sounder than Mr. Shaw's determination that the moon is 37 miles away. It is that the Vogels and the Struves and the Newcombs have been functionally hypnotized and have usefully spread the embryonic delusion that there is a vast, untraversible expanse of space around this earth, or that they have had some basis that it has been my misfortune to be unable to find, or that there is no pleasant and unaccusatory way of explaining them. April 10, 1874--a luminous object that exploded in the sky of Kuttenberg, Bohemia. It is said that the glare was like sunlight, and that the "terrifying flash" was followed by a detonation that rumbled about a minute. April 9, 1876--an explosion that is said to have been violent, over the town of Rosenau, Hungary. See Rept. B. A., 1877-147.

These two objects which appeared in virtually the same local sky of this earth-- points of explosion 250 miles apart--came from virtually the same point in the sky: constellation of Cassiopeia; different by two degrees in right ascension, and with no difference in declination. About the same time in the evening: one at 8:09 P.M., and the other at 8:20 P.M. Same night in the year, according to extra- terrestrial calendars: the year 1876 was a leap year.

If they had been ordinary meteors, by coincidence two ordinary meteors of the same stream might, exactly two years apart, come from almost the same point in the heavens and strike almost the same point over this earth. But they were two of the most extraordinary occurrences in the records of explosions in the sky. Coincidences multiply, or these objects did come from the not far-distant constellation Cassiopeia, and their striking so closely together indicates that this earth is stationary; and something of the purposeful may be thought of. Serially related to these events, or representing some more coincidence, there had been, upon June 9, 1866, a tremendous explosion in the sky of Knysahinya, Hungary, and about a thousand stones had fallen from the sky (Rept. B. A., 1867-430). Rosenau and Knysahinya are about 75 miles apart. Of course one can very much extend our own circumscribed little notions, and think of the firing of projectiles from beyond the stars, just as one can think of our unknown lands as being not in the immediate sky of Servia or Birmingham or Comrie, but as being beyond the nearby stars, reducing everything more than we have reduced--but the firing of stones to this earth seems crude to me. Of course, objects, or fragments of objects made of steel, like the manufactured steel of this earth, have fallen to this earth, and are now in collections of "meteorites." There is a story in a book that is not very accessible to us, because it can't be found along with C. R., or Eng. Mec., or L'Astro., of tablets of stone that were once upon a time fired to this earth. It may be that inhabitants of this earth have been receiving instructions ever since, engravings arriving very badly damaged, however.

I have data upon repeating appearances, said to have been "auroral," in a local sky. If they were auroral, repetitions at regular intervals and so localized are challengers to the most resolute of explainers. If they were of extra-mundane origin, they indicate that this earth is stationary. The regularity is suggestive of signaling. For instance--a light in the sky of Lyons, N. Y., Dec. 9, 1891, Jan. 5, Feb. 2, Feb. 29, March 27, April 23, 1892. In the Scientific American, May 7,

1892, Dr. M. A. Veeder writes that, from Dec. 9, 1891, to April 23, 1892, there had been a bright light that he calls "auroral" in the sky of Lyons, every 27th night. He associates the lights with the sun's synodic period, and says that upon each of the days preceding a nocturnal display, there had been a disturbance in the sun. How a disturbance in the sun could, at night, sun somewhere near the antipodes of Lyons, N. Y., so localize its effects, one can't clear up. In Nature, 46-29, Dr. Veeder associates the phenomena with the synodic period of the sun, but he says that this period is of 27 days, 6 hours, and 47 minutes, noting that this period is inconsistent with the phenomena at Lyons, making more than a day's difference in the time of his records. This precise determination is more of the "exact science" that is driving some of us away from refinements into hoping for caves. Different parts of the sun move at different rates: I have read of sun spots that moved diagonally across the sun.

In Nature, 15-451, a correspondent writes that, at 8:55 P.M., he saw a large red star in Serpens, where he had never seen such an appearance before--Gunnersbury, March 17, 1877. Ten minutes later, the object increased and decreased several times, flashing like the revolving light of a lighthouse, then disappearing. This correspondent writes that, about Jo P.M., he saw a great meteor. He suggests no relation between the two appearances, but there may have been relation, and there may be indication of something that was stationary at least one hour over Gunnersbury, because the object said to have been a "meteor" was first seen at Gunnersbury. In the Observatory, 1-20, Capt. Tupman writes that, at 9:57 o'clock, a great meteor was seen first at Frome, Tetbury, and Gunnersbury. The red object might not have been in the local sky of Gunnersbury; might have been in the constellation Serpens, unseen in all the rest of the world.

There is a great field of records of "meteors" that, with no parallax, or with little parallax, or with little parallax that may be accounted for by supposing that observations were not quite simultaneous, have been seen to come as if from a star or from a planet, and that may have come from such points, indicating that they are not far away. For instance, Rept. B. A., 1879-77--the great meteor of Sept. 5, 1868. It was seen, at Zurich, Switzerland, to come from a point near Jupiter; at Tremont, France, origin was so close to Jupiter that this object and the planet were seen in the same telescopic field; at Bergamo, Italy, it was seen five or six degrees from Jupiter. Zurich is about 140 miles from Bergamo, and Tremont is farther from Zurich and Bergamo than that.

So there are data that indicate that objects have come to this earth from planets or from stars, enforcing our idea that the remotest planet is not so far from this earth as the moon is said, conventionally, to be; and that the stars, all equi- distant from this earth might be reached by traveling from this earth. One notices that I always conclude that, if phenomena repeatedly occur in one local sky of this earth, their origin is traceable to a fixed place over a stationary earth.

The fixed place over this earth is indicated, but that fixed place--island of space, foreign coast, whatever it may be--may be conceived of as accompanying this earth in its rotations and revolutions around the sun. Accepting that nothing much is known of gravitation; that gravitational astronomy is a myth; that attraction may extend but a few miles around this earth, if I can think of something hanging unsupported in space, I always think of an island, say, over Birmingham, or Irkutsk, or Comrie, as soon flying off by the centrifugal force of a rotating earth, or as being soon left behind in a rush around the sun. Nevertheless there is good room for discussion here. But when it comes to other orders of data, I find one convergence toward the explanation that this earth is stationary. But the subject is supposed to be sacred. One must not think that this earth is stationary. One must not investigate. To think upon this subject, except as one is told to think, is, or seems to be considered, impious.

But how can one account for an earth that moves?

By thinking that something started it and that nothing ever stopped it. Earth that doesn't move? That nothing ever started it. Some more sacrilege.

26

IF a grasshopper could hop on a cannon ball, passing overhead, I could conceive, perhaps, how something, from outer space, could flit to a moving earth, explore a while, and then hop off.

But suppose we have to accept that there have been instances of just such enterprise and agility, relatively to the planet Venus. Irrespective of our notion that it may be that sometimes a vessel sails to this earth from Venus and returns, there are striking data indicating that, whether conceivable or not, luminous objects have appeared from somewhere, or presumably from outer space, and have been seen temporarily suspended over the planet Venus. This is in accord with our indications that there are regions in the sky suspended over and near this earth. It looks bad for our inference that this earth is stationary, but it is the supposed rotary motion of this earth more than the supposed orbital motion that seems to us would dislodge such neighboring bodies; and all astronomers, except those who say that Venus rotates in about 24 hours, say that Venus rotates in about 224 days, a velocity that would generate little centrifugal force.

I have a note upon a determined luminosity that was bent upon Saturn, as its objective. In the English Mechanic, 63-496, a correspondent writes that, upon July 13, 1896, he saw, through his telescope, from 10 until after 11:15 P.M., after which the planet was too near the horizon for good seeing, a luminous object moving near Saturn. He saw it pass several small stars. "It was certainly going toward Saturn at a good rate." There may be swifts of the sky that can board planets. If they can swoop on and off an earth moving at a rate of 19 miles a second, disregarding rotation, because entrance at a pole may be thought of, why, then, for all I know smaller things do ride on cannon balls. Of course if our data that indicate that the supposed solar system, or the geo-system, is to an enormous degree smaller than is conventionally taught be accepted, the orbital velocity of Venus is far cut down.

About the last of August, 1873--Brussels; eight o'clock in the evening--rising above the horizon, into a clear sky, was seen a star-like object. It mounted higher and higher, until, about ten minutes later, it disappeared (La Nature, 1873-239). It seems that this conspicuous object did appear in a local sky, and

was therefore not far from this earth. If it were not a fire-balloon, one supposes that it did come from outer space, and then returned.

Perhaps a similar thing that visited the moon, and was then seen sailing away--in the Astronomical Register, 23-205, Prof. Schafarik, of Prague, writes that upon April 24, 1874, he saw "an object of so peculiar a nature that I do not know what to make of it." He saw a dazzling white object slowly traversing the disc of the moon. He had not seen it approaching the moon. He watched it after it left the moon. Sept. 27, 1881--South Africa--an object that was seen near the moon, by Col. Markwick--like a comet but moving rapidly (Jour. Liverpool Astro. Soc., 7-117).

Our chief interest is in objects, like ships, that have "boarded" this moving earth with the agility of a Columbus who could dodge a San Salvador and throw out an anchor to an American coast screeching past him at a rate of 19 miles a second, or in objects that have come as close as atmospheric conditions, or unknown conditions, would permit to the bottom of a kind of stationary sea. We now graduate Capt. Noble to the extra-geographic fold. In Knowledge, 4-173, Capt.

Noble writes that, at 10:35 o'clock, night of Aug. 28, 1883, he saw in the sky something "like a new and most glorious comet." First he saw something like the tail of a comet, or it was like a searchlight, according to Capt. Noble's sketch of it in Knowledge. Then Capt. Noble saw the nucleus from which this light came. It was a brilliant object. Upon page 207, W. K. Bradgate writes that, at 12:40 A.M., August 29, at Liverpool, he saw an object like the planet Jupiter, a ray of light emanating from it. Upon the nights of September 11 and 13, Prof. Swift saw, at Rochester, N. Y., an unknown object like a comet, perhaps in the local sky of Rochester, inasmuch as it was reported from nowhere else (Observatory, 6-345). In Knowledge, 4-219, Mrs. Harbin writes that, upon the night of September 21, at Yeovil, she saw the same brilliant searchlight-like light that had been seen by Capt. Noble, but that it had disappeared before she could turn her telescope upon it. And several months later (November, 1883) a similar object was seen obviously not far away, but in the local sky of Porto Rico and then of Ohio (Amer. Met. Jour., 1-110, and Sci. Amer., 50-40, 97). It may be better not to say at this time that we have data for thinking that a vessel carrying something like a searchlight, visited this earth, and explored for several months over regions as far apart as England and Porto Rico. Just at present it is enough to record that something that was presumably not a fire-balloon appeared in the sky of England, close to this earth, if seen nowhere else, and in two hours traversed the distance of about 200 miles between Sussex and Liverpool.

Aug. 22, 1885--Saigon, Cochin-China--according to Lieut. Reveillere, of the vessel Guiberteau--object like a magnificent red star, but larger than the planet Venus-- it moved no faster than a cloud in a moderate wind; observed 7 or 8 minutes, then disappearing behind clouds (C. R., 101-680).

In this book it is my frustrated desire to subordinate the theme of this earth's stationariness. My subject is New Lands--things, objects, beings that are, or may be, the data of coming expansions But the stationariness of this earth cannot be subordinated. It is crucial.

Again--there is no use discussing possible explorations beyond this earth, if this earth moves at a rate of 19 miles a second, or 19 miles a minute.

As to voyagers who may come to or near this earth from other planets--how could they leave and return to swiftly moving planets? According to our principles of Extra-geography, the planets move part of the time with the revolving stars, the remotest planets remaining in, under, or near one constellation years at a time.

Anything [p. 457]

that could reach, and then travel from, a swiftly revolving constellation in the ecliptic could arrive at a stellar polar region, .where, relatively to a central, stationary body, there is no motion.

27

IT may be that we now add to our sins the horse that swam in the sky. For all I know, we contribute to a wider biology. In the New York Times, July 8, 1878, is published a dispatch from Parkersburg, West Virginia: that, about July 1, 1878, three or four farmers had seen, in a cloudless sky, apparently half a mile high, "an opaque substance." It looked like a white horse, "swimming in the clear atmosphere." It is said to have been a mirage of a horse in some distant field. If so, it is interesting not only because it was opaque, but because of a selection or preference: the field itself was not miraged.

Black bodies and the dark rabbles of the sky--and that rioting thing, from floating anarchies, have often spotted the sun. Then, by all that is compensatory, in the balances of existence, there are disciplined forces in space. In the Scientific American, 44-291, it is said that, according to newspapers of Delaware, Maryland, and Virginia, figures had been seen in the sky in the latter part of September, and the first week in October, 1881, reports that "exhibited a mediaeval condition of intelligence scarcely less than marvelous." The writer suggests that, though probably something had been seen in the sky, it was only an aurora. Our own intelligence and that of astronomers and meteorologists and everybody else with whom we have had experience had better not be discussed, but the accusation of mediaevalism is something that we're sensitive about, and we hasten to the Monthly Weather Review, and if that doesn't give us a modern touch, I mistake the sound of it. Monthly Weather Review, September and October, 1881--an auroral display in Maryland and New York, upon the 23rd of September; all other auroras in September far north of the three states in which it was said phenomena were seen. October--no auroras until the 18th; that one in the north. There was a mirage upon September 23, but at Indianola; two instances in October, but late in the month, and in northern states.

It is said, in the Scientific American, that, according to the Warrentown (Va.) Solid South, a number of persons had seen white-robed figures in the sky, at night. The story in the Richmond Dispatch is that many persons had seen, or had thought they had seen, an alarming sight in the sky, at night: a vast number of armed, uniformed soldiers drilling. Then a dispatch from Wilmington, Delaware--platoons of angels marching and countermarching in the sky, their white robes and helmets gleaming. Similar accounts came from Laurel and Talbot. Several persons said that they had seen, in the sky, the figure of President Garfield, who had died not long before. Our general acceptance is that all reports upon such phenomena are colored in terms of appearances and subjects uppermost in minds.

L'Astronomie, 1888-392:

That, about the first of August, 1888, near Warasdin, Hungary, several divisions of infantry, led by a chief, who waved a flaming sword, had been seen in the sky, three consecutive days, marching several hours a day. The writer in L'Astronomie says that in vain does one try to explain that this appearance was a mirage of terrestrial soldiers marching at a distance from Warasdin, because widespread publicity and investigation had disclosed no such soldiers. Even if there had been terrestrial soldiers near Warasdin repeating mirages localized would call for explanation.

But that there may be space-armies, from which reflections or shadows or Brocken specters are sometimes cast--a procession that crossed the sun: forms that moved, or that marched, sometimes four abreast; observation by M. Bruguiere, at Marseilles, April 15 and 16, 1883 (L'Astro., 5-70). An army that was watched, forty minutes, by M. Jacquot, Aug. 30, 1886 (L'Astro., 1886-71)--things or beings that seemed to march and to counter-march: all that moved in the same direction, moved in parallel lines. In L'Annee Scientifique, 29-8, there is an account of observations by M. Trouvelot, Aug. 29, 1871. He saw objects, some round, some triangular, and some of complex forms. Then occurred something that at least suggests that these things were not moving in the wind, nor sustained in space by the orbital forces of meteors; that each was depending upon its own powers of flight, and that an accident occurred to one of them. All of them, though most of the time moving with great rapidity, occasionally stopped, but then one of them fell toward the earth, and the indications are that it was a heavy body, and had not been sustained by the wind, which would scarcely suddenly desert one of its flotsam and continue to sustain all the others. The thing fell, oscillating from side to side like a disc falling through water.

New York Sun, March 16, 1890--that, at 4 o'clock, in the afternoon of March 12th, in the sky of Ashland, Ohio, was seen a representation of a large, unknown city. By some persons it was supposed to be a mirage of the town of Mansfield, thirty miles away; other observers thought that they recognized Sandusky, sixty miles away. "The more superstitious declared that it was a vision of the New Jerusalem."

May have been a revelation of heaven, and for all I know heaven may resemble Sandusky, and those of us who have no desire to go to Sandusky may ponder that point, but our own expression is that things have been pictured in the sky, and have not been traced to terrestrial origins, but have been interpreted always in local terms. Probably a living thing in the sky--seen by farmers--a horse. Other things, or far-refracted images, or shadows--and they were supposed to be vast lions or soldiers or angels, all according to preconceived ideas. Representations that have been seen in India--Hindoo costumes described upon them. Suppose that, in the afternoon of Jan. 17, 1892, there was a battle in the sky of Montana--we know just about in what terms the description would be published. Brooklyn Eagle, Jan. 18, 1892--a mirage in the sky of Lewiston, Montana--Indians and hunters alternately charging and retreating. The Indians were in superior numbers and captured the hunters. Then details--hunters tied to stakes; the piling of faggots; etc. "So far as could be ascertained last night, the Indians on the reservations are peaceable." I think that we're peaceable enough, but, unless the astronomers can put us on reservations, where we'll work out expressions in beads and wampum instead of data, we'll have to carry on a conflict with the vacant minds to which appear mirages of their own emptiness in the sometimes swarming skies.

Altogether there are many data indicating that vessels and living things of space do come close to this earth, but there is absence of data of beings that have ever landed upon this earth, unless someone will take up the idea that Kaspar Hauser, for instance, came to this earth from some other physical world. Whether spacarians have ever dredged down here or not, or "sniped" down here, pouncing, assailing, either wantonly, or in the interests of their sciences, there are data of seeming seizures and attacks from somewhere, and I have strong objections against lugging in the fourth dimension, because then I am no better off, wondering what the fifth and sixth are like.

In La Nature, 1888-2-66, M. Adrian Arcelin writes that, while excavating near de Solutre, in August, 1878, upon a day, described as superb, sky clear to a degree said to have been parfaitement, several dozen sheets of wrapping paper upon the ground suddenly rose. Nearby were a dozen men, and not one of them had felt a trace of wind. A strong force had seized upon these conspicuous objects, touching nothing else. According to M. Arcelin, the dust on the ground under and around was not disturbed. The sheets of paper continued upward, and disappeared in the sky.

A powerful force that swooped upon a fishing vessel, raising it so far that when it fell back it sank--see London Times, Sept. 24, 1875. A quarter of a mile away were other vessels, from which set out rescuers to the sailors who had been thrown into the sea. There was no wind: the rescuers could not use sails, but had to row their boats.

Upon Oct. 2, 1875, a man was trundling a cart from Schaffhausen, near Beringen, Germany. His right arm was perforated from front to back, as if by a musket ball (Pop. Sci., 15-566). This man had two companions. He had heard a whirring sound, but his companions had heard nothing. At one side of the road there were laborers in a field, but they were not within gunshot distance. Whatever the missile may have been, it was unfindable.

La Nature, 1879-1-166, quotes the Courrier des Ardennes as to an occurrence in the Commune Signy-le-Pettit, Easter Sunday, 1879--a conspicuous, isolated house-- suddenly its slate roof shot into the air, and then fell to the ground. There had not been a trace of wind. The writer of the account says that the force, which he calls a trouble inoui had so singled out this house that nothing in its surroundings beyond a distance of thirty feet had been disturbed.

Scientific American, July 10, 1880--that, according to the Plain-dealer, of East Kent, Ontario, two citizens of East Kent were in a field, and heard a loud report. They saw stones shooting upward from a field. They examined the spot, which was about 16 feet in diameter, finding nothing to suggest an explanation of the occurrence. It is said that there had been neither a whirlwind nor anything else by which to explain.

It may be that witnesses have seen human beings dragged from our own existence either into the objectionable fourth dimension, perhaps then sifting into the fifth, or up to the sky by some exploring thing. I have data, but they are from the records of psychic research. For instance, a man has been seen walking along a road--sudden disappearance. Explanation--that he was not a living human being, but an apparition that had disappeared. I have not been able to develop such data, finding, for instance, that someone in the neighborhood had been reported missing; but it may be that we can find material in our own field.

Upon Dec. 10, 1881, Walter Powell and two companions ascended from Bath in the Government balloon Saladin (Valentine and Tomlinson, Travels in Space, p. 227). The balloon descended at Bridport, coast of the English Channel. Two of the aeronauts got out, but the balloon, with Powell in it, shot upward. There was a report that the balloon had been seen to fall in the English Channel, near Bridport, but according to Capt. Temple, one of Powell's companions, probably something thrown from the balloon had been seen to fall.

A balloon is lost near or over the sea. If it should fall into the sea it would probably float and for considerable time be a conspicuous object; nevertheless the disappearance of a balloon last seen over the English Channel, cannot, without other circumstances, be considered very mysterious. Now one expects to learn of reports from many places of supposed balloons that had been seen. But the extraordinary circumstance is that reports came in upon a luminous object that was seen in the sky at the time that this balloon disappeared. In the London Times, it is said that a luminous object had been seen, evening of the 13th, moving in various directions in the sky near Cherbourg. It is said that upon the night of the 16th three customhouse guards, at Laredo, Spain, had seen something like a balloon in the sky, and had climbed a mountain in order to see it better, but that it had shot out sparks and had disappeared--and had been reported from Bilbao, Spain, the next day. In the Morning Post, it is said that this luminous display was the chief feature; that it was this sparkling that had made the object visible. In the Standard, December 16, is an account of something that was seen in the sky, five o'clock, morning of December 15, by Capt. McBain, of the steamship Countess of Aberdeen, off the coast of Scotland, 25 miles from Montrose. Through glasses, the object seemed to be a light attached to something thought to be the car of a balloon, increasing and decreasing in size--a large light--"as large as the light at Girdleness." It moved in a direction opposite to that of the wind, though possibly with wind of an upper stratum. It was visible half an hour, and when it finally disappeared, was moving toward Bervie, a town on the Scottish coast about 12 miles north of Montrose. In the Morning Post it is said that the explanation is simple: that someone in Monfreith, 8 miles from Dundee, had, late in the evening of the 15th, sent up a fire-balloon, "which had been carried along the coast by a gentle breeze, and, after burning all night, extinguished and collapsed off Montrose, early on Thursday morning (16th)." This story of a balloon that wafted to Montrose, and that was evidently traced until it collapsed near Montrose does not so simply explain an object that was seen 25 miles from Montrose. In the Standard, December 19, it is said that two bright lights were seen over Dartmouth Harbor, upon the 11th.

Walter Powell was Member of Parliament for Malmesbury, and had many friends, some of whom started immediately to search. His relatives offered a reward. A steamboat searched the Channel, and did not give up until the 13th; fishing vessels kept on searching. A "sweeping expedition" was organized, and the coast guard was doubled, searching the shore for wreckage, but not a fragment of the balloon, nor from the balloon, except a thermometer in a bag, was found.

In L'Astronomie, 1886-312, Prof. Paroisse, of the College Bar-sur-Aube, quotes two witnesses of a curieux phenomene that occurred in a garden of the College, May 22, 1886--cloudless sky; wind tres faible. Within a small circle in the garden were some: baskets and ashes and a window frame that weighed sixty kilograms. These things suddenly rose from the ground. At a height of about forty feet, they remained suspended several minutes, then falling back to the place from which they had risen. Not a thing outside this small circle had been touched by the seizure. The witnesses said that they had felt no disturbance in the air.

Scientific American, 56-65--that in June, 1886, according to the London Times, "a well-known official" was entering Pall Mall,, when he felt a violent blow on the shoulder and heard a hissing, sound. There was no one in sight except a distant policeman. At home, he found that the nap of his coat looked as if a hot wire had been pressed against the cloth, in a long, straight line. No. missile was found, but it was thought that something of a meteoritic nature had struck him.

Charleston News and Courier, Nov. 25, 1886--that, at Edina,, Mo., November 23, a man and his three sons were pulling corn on a farm. Nothing is said of meteorologic conditions, and, for all I know, they may have been pulling corn in a violent thunder storm. Something that is said to have been lightning flashed from the sky. The man was slightly injured, one son killed, the other seriously injured--the third had disappeared. "What has become of him is not known, but it is supposed that he was blinded or crazed by the shock, and wandered away."

Brooklyn Eagle, March 17, 1891--that, at Wilkes-Barre, Pa., March 16th, two men were "lifted bodily and carried considerable distance in a whirlwind." It was a powerful force, but nothing else was affected by it. Upon the same day, there was an occurrence in Brooklyn. In the New York Times, March 17, 1891, it is said that two men, Smith Morehouse, of Orange Co., N. Y., and William Owen, of Sussex Co., N. J., were walking in Vanderbilt Avenue, Brooklyn, about 2 o'clock, afternoon of the 16th, when a terrific explosion occurred close to the head of Morehouse, injuring him and stunning Owen, the flash momentarily blinding both. Morehouse's face was covered with marks like powder-marks, and his tongue was pierced. With no one else to accuse, the police arrested Owen, but held him upon the technical charge of intoxication. Morehouse was taken to a hospital, where a splinter of metal, considered either brass or copper, but not a fragment of a cartridge, was removed from his tongue. No other material could be found, though an object of considerable size had exploded.

Morehouse's hat had been perforated in six places by unfindable substances. According to witnesses there had been no one within a hundred feet of the men. One witness had seen the flash before the explosion, but could not say whether it had been from something falling or not. In the Brooklyn Eagle, March 17, 1891, it is said that neither of the men had a weapon of any kind, and that there had been no disagreement between them. According to a witness, they had been under observation at the time of the explosion, her attention having been attracted by their rustic appearance. There is an interesting merging here of the findable and the unfindable. I suppose that no one will suppose that someone threw a bomb at these men. But enough substance was found to exclude the notion of "lightning from a clear sky." Something of a meteoritic nature seems excluded.

28

OUT from a round, red planet, a little white shaft--a fairy's arrow shot into an apple. June 10, 1892--a light like a little searchlight, projecting from the limb of Mars. Upon July 11 and 13, it was seen again, by Campbell and Hussey (Nature, 50-500).

Aug. 3, 1892--climacteric opposition of Mars.

Upon Aug. 12, 1892, flashes were seen by many persons, in the sky of England. See Eng. Mec., vol. 56. At Manchester, so like signals were they, or so unlike anything commonly known as "auroral" were they, that Albert Buss mistook them for flashes from a lighthouse. They were seen at Dewsbury; described by a correspondent to the English Mechanic, who wrote: "I have never seen such an appearance of an aurora." "Rapid flashes" reported from Loughborough.

A shining triangle in a dark circle.

In L'Astronomie, 1888-75, Dr. Klein publishes an account of de Speissen's observation of Nov. 23, 1887--a luminous triangle on the floor of Plato. Dr. Klein says it was an effect of sunlight.

In this period, there were in cities of the United States, some of the most astonishing effects at night, in the history of this earth. If Rigel should run for the Presidency of Orion, and if the stars in the great nebula should start to march, there would be a spectacle like those that Grover Cleveland called forth in the United States, in this period.

So then--at least conceivably--something similar upon the moon. Flakes of light moving toward Plato, this night of Nov. 23, 1887, from all the other craters of the moon; a blizzard of shining points gathering into light-drifts in Plato; then the denizens of Aristarchus and of Kepler, and dwellers from the lunar Alps, each raising his torch, marching upon a triangular path, making the triangle shine in the dark--conceivably. Other formations have been seen in Plato, but, according to my records, this symbol that shone in the dark had never been seen before, and has not been seen since. About two years later--a demonstration of a more exclusive kind--assemblage of all the undertakers of the moon. They stood in a circular formation, surrounded by virgins in their nightgowns--and in nightgowns as nightgowns should be. An appearance in Plinius, Sept. 13, 1889, was reported by Prof. Thury, of Geneva--a black spot with an "intensely white" border.

March 30, 1889--a black spot that was seen for the first time, by Gaudibert, near the center of Copernicus (L'Astro., 1890-235). May 11, 1889--an object as black as ink upon a rampart of Gassendi (L'Astro., 1889-275). It had never been reported before; at the time of the next lunation, it was not seen again. March 30, 1889--a new black spot in Plinius (L'Astro., 1890-187).

The star-like light of Aristarchus--it is a long time since latest preceding appearance (May 7, 1867). Then it cannot be attributed to commonplace lunar circumstances. The light was seen Nov. 7, 1891, by M. d'Adjuda, of the Observatory of Lisbon--"a very distinct, luminous point" (L'Astro., 11-33) Upon April 1, 1893, a shaft of light was seen projecting from the moon, by M. de Moraes, in the Azores. A similar appearance was seen, Sept. 25, 1893, at Paris, by M. Gaboreau (L'Astro., 13-34).

Another association like that of 1884--in the English Mechanic, 55-310, a correspondent writes that, upon May 6, 1892, he saw a shining point (not polar) upon Venus. Upon the 13th of August, 1892, the same object--conceivably--was seen at a short distance from Venus--an unknown, luminous object, like a star of the 7th magnitude that was seen close to Venus, by Prof. Barnard (Ast. Nach., no.4106).

Upon Aug. 24, 1895, in the period of primary maximum brilliance of Venus, a luminous object, it is said, was seen in the sky, in day time, by someone in Donegal, Ireland. Upon this day, according to the Scientific American, 73-374, a boy, Robert Alcorn, saw a large luminous object falling from the sky. It exploded near him. The boy's experience was like Smith Morehouse's. He put his hands over his face: there was a second explosion, shattering his fingers. According to Prof. George M. Minchin no substance of the object that had exploded could be found. Whether there be relation or not, something was seen in the sky of England a week later. In the London Times, Sept. 4, 1895, Dr. J. A. H. Murray writes that, at Oxford, a few minutes before 8 P.M., Aug. 31, 1895, he saw in the sky a luminous object, considerably larger than Venus at greater brilliance, emerge from behind tree tops, and sail slowly eastward. It moved as if driven in a strong wind, and disappeared behind other trees. "The fact that it so perceptibly grew fainter as it receded seems to imply that it was not at a great elevation, and so favors a terrestrial origin, though I am unable to conceive how anything artificial. could have presented the same appearance." In the Times, of the 6th, someone who had read Dr. Murray's letter says that, about the same time, same evening, he, in London, had seen the same object moving eastward so slowly that he had thought it might be a fire-balloon from a neighboring park. Another correspondent, who had not read Dr. Murray's letter, his own dated September 3, writes from a place not stated that about 8:20 P.M., August 31, he had seen a star-like object, moving eastward, remaining in sight four or five minutes. Then someone who, about 8 P.M., same evening, while driving to the Scarborough station, had seen "a large shooting star," astonishing him, because of its leisurely rate, so different from the velocity of the ordinary "shooting star." There are two other accounts of objects that were seen in the sky, at Bath and at Ramsgate, but not about this time, and I have looked them up in local newspapers, finding that they were probably meteors.

In the Oxford Times, September 7, Dr. Murray's letter to the London Times is reprinted, with this comment--"We would suggest to the learned doctor that the supposed meteor was one of the fire-balloons let off with the allotments show."

Let it be that when allotments are shown, balloons are always sent up, and that this Editor did not merely have a notion to this effect. Our data are concerned with an object that was seen, at about the same time, at Oxford, about 50 miles southeast of Oxford, and about 170 miles northeast of Oxford, with a fourth observation that we cannot place.

And, in broader terms, our data are concerned with a general expression that objects like ships have been seen to sail close to this earth at times when the planet Venus is nearest this earth. Sept. 18, 1895--inferior conjunction of Venus.

Still in the same period, there were, in London, two occurrences perhaps like that at Donegal. London Morning Post, Nov. 16, 1895--that, at noon, November 15, an "alarming explosion" occurred somewhere near Fenchurch Street, London. No damage was done; no trace could be found of anything that had exploded. An hour later, near the Mansion House, which is not far from Fenchurch Street, occurred a still more violent explosion. The streets filled with persons who had run from buildings, and there was investigation, but not a trace could be found of anything that had exploded. It is said that somebody saw "something falling." However, the deadly explainers, usually astronomers, but this time policemen, haunt or arrest us. In the Daily News, though it is not said that a trace of anything that had exploded had been found, it is said that the explanation by the police was that somebody had mischievously placed in the streets fog-signals, which had been exploded by passing vehicles.

Observation by Muller, of Nymegen, Holland--an unknown luminous object that, about three weeks later, was seen near Venus (Monthly Notices, R. A. S., 52-276).

Upon the 28th of April, 1897, Venus was in inferior conjunction. In Popular Astronomy, 5-55, it is said that many persons had written to the Editor, telling of "airships" that had been seen, about this time. The Editor writes that some of the observations were probably upon the planet Venus, but that others probably related to toy balloons, "which were provided with various colored lights."

The first group of our data, I take from dispatches to the New York Sun, April 2, 11, 16, 18. First of April--"the mysterious light" in the sky of Kansas City-- something like a powerful searchlight. "It was directed toward the earth, traveling east at a rate of sixty miles an hour." About a week later, something was seen in Chicago. "Chicago's alleged airship is believed to be a myth, in spite of the fact that a great many persons say that they have seen the mysterious night-wanderer. A crowd gazed at strange lights, from the top of a downtown skyscraper, and Evanston students declare they saw the swaying red and green lights." April 16--reported from Benton, Texas, but this time as a dark object that passed across the moon.

Reports from other towns in Texas: Fort Worth, Dallas, Marshall, Ennis, and Beaumont--"It was shaped like a Mexican cigar, large in the middle, and small at both ends, with great wings, resembling those of an enormous butterfly. It was brilliantly illuminated by the rays of two great searchlights, and was sailing in a southeasterly direction, with the velocity of the wind, presenting a magnificent appearance."

New York Herald, April 11--that, at Chicago, night of April 9-10, "until two o'clock in the morning, thousands of amazed spectators declared that the lights seen in the northwest were those of an airship, or some floating object, miles above the earth.... Some declare they saw two cigar-shaped objects and great wings." It is said that a white light, a red light, and a green light had been seen.

There does seem to be an association between this object and the planet Venus, which upon this night was less than three weeks from nearest approach to this earth. Nevertheless this object could not have been Venus, which had set hours earlier. Prof. Hough, of the Northwestern University, is quoted--that the people had mistaken the star Alpha Orionis for an airship. Prof. Hough explains that astronomeric effects may have given a changing red and green appearance to this star. Alpha Orionis as a northern star is some more astronomy by the astronomers who teach astronomy day-times and then relax when night comes. That atmospheric conditions could pick out this one star and not affect other brilliant stars in Orion is more astronomy. At any rate the standardized explanation that the thing was Venus disappears.

There were other explainers--someone who said that he knew of an airship (terrestrial one) that had sailed from San Francisco; and had reached Chicago. Herald, April 12--said that the object had been photographed in Chicago: "a cigar- shaped, silken bag," with a framework--other explanations and identifications, not one of them applying to this object, if it be accepted that it was seen in places as far apart as Illinois and Texas. It is said that, upon March 29th, the thing had been seen in Omaha, as a bright light sailing to the northwest, and that, for a few moments, upon the following night, it had been seen in Denver. It is said that, upon the night of the 9th, despatches had bombarded the newspaper offices of Chicago, from many places in Illinois, Indiana, Missouri, Iowa, and Wisconsin. "Prof. George Hough maintains that the object seen is Alpha Orionis."

April 14--story, veritable observation, yarn, hoax--despatch from Carlensville, Ill.--that upon the afternoon of the 10th, the airship had alighted upon a farm, but had sailed away when approached--"cigar-shaped, with wings, and a canopy on top."

April 15--shower of telegrams--development of jokers and explainers--thing identified as an airship invented by someone in Dodge City, Kansas; identified as an airship invented by someone in Brule, Wisconsin--stories of letters found on farms, purporting to have been dropped by the unknown aeronauts (terrestrial ones)--jokers in various towns, sending up balloons with lights attached--one laborious joker who rigged up something that looked like an airship and put it in a vacant lot and told that it had fallen there--yarn or observation, upon a "queer-looking boat" that had been seen to rise from the water in Lake Erie-- continued reports upon a moving object in the sky, and its red and green lights.

Against such an alliance as this, between the jokers and the astronomers, I see small chance for our data. The chance is in the future. If, in April, 1897, extra- mundane voyagers did visit this earth, likely enough they will visit again, and then the alliance against the data may be guarded against.

New York Herald, April 20--that, upon the 19th, about 9 P.M., at Sistersville, W. Va., a luminous object had approached the town from the northwest, flashing brilliant red, white, and green lights. "An examination with strong glasses left an impression of a huge cone-shaped arrangement 180 feet long, with large fins on either side."

My own general impression:

Night of Oct. 12, 1492--if I have that right. Some night in October, 1492, and savages upon an island-beach are gazing out at lights that they had never seen before. The indications are that voyagers from some other world are nearby. But the wise men explain. One of the most nearly sure expressions in this book is upon how they explain. They explain in terms of the familiar. For instance, after all that is spiritual in a fish passes away, the rest of him begins to shine nights.

So there are three big, old, dead things out in the water--

29

THERE have been published several observations upon a signal-like regularity of the Barisal Guns, which, because unaccompanied by phenomena that could be considered seismic, may have been detonations in the sky, and which, because, according to some hearers, they seemed to come from the sky, may have come from some region stationary in the local sky of Barisal. In Nature, 61-127, appears a report by Henry S. Schurr, who investigated the sounds in the years 1890-91:

"These Guns are always heard in triplets, i.e., three guns are always heard, one after the other, at regular intervals, and, though several guns may be heard, the number is always three or a multiple of three. Then the interval between the three is always constant, i.e., the interval between the first and the second is the same as the interval between the second and the third, and this interval is usually three seconds, though I have heard it up to ten seconds. The interval, however, between the triplets varies, and varies largely, from a few seconds up to hours and days. Sometimes only one series of triplets is heard in a day; at others the triplets follow with great regularity, and I have counted as many as forty-five of them, one after the other, without pause."

In vols. 16 and 17, Ciel et Terre, M. Van den Broeck published a series of papers upon the mysterious sounds that had been heard in Belgium.

July, 1892--heard near Bree, by Dr. Raemaekers, of Antwerp--detonations at regular intervals of about 12 seconds, repeated about 20 times.

Aug. 5, 1892--near Dunkirk, by Prof. Gerard, of Brussels--four reports like sounds of cannons.

Aug. 17, 1893--between Ostend and Ramsgate, by Prof. Gerard--a series of distinct explosions--state of the sky giving no reason to think that they were meteorological manifestations.

Sept. 5, 1893--at Middelkirke--loud sounds of remarkable intensity. Sept. 8, 1893--English Channel near Dover--by Prof. Gerard--an explosive sound.

In Ciel et Terre, 16-485, M. Van den Broeck records an experience of his own. Upon June 25, 1894, at Louvain, he had heard detonations like discharges of artillery: he tabulates the intervals in a series of sounds. If there were signaling from some unknown region over Belgium, and not far from the surface of this earth, or from extra-mundane vessels, and if there were something of the code-like, resembling the Morse alphabet, perhaps, in this series of sounds, there can be small hope of interpreting such limited material, but there may be suggestion to someone to record all sounds and their intervals and modulations, if, with greater duration, such phenomena should ever occur again. The intervals were four minutes and twenty-three minutes; then three minutes, four, three quarters, three and three quarters, three quarters.

Sept. 16, 1895--a triplet of detonations, heard by M. de Schryvere, of Brussels.

There were attempts to explain. Some of M. Van den Broeck's correspondents thought that there had been firing from forts on the coast of England, and somebody thought that the phenomena should be attributed to gravitational effects of the moon. Upon Sept. 13, 1895, four shocks were felt and sounds heard at Southampton: a series of three and then another (Nature, 52-552); but I have no other notes upon sounds that were heard in England at this time, except the two explosions that were explained by the police of London. However, M. Van den Broeck says that Mr. Harmer, of Aldeburgh, Suffolk, had, about the first of November, heard booming sounds that had been attributed to cannonading at Harwich. Mr. Harmer had heard other sounds that had been attributed to cannonading somewhere else. He could not offer a definite opinion upon the first sounds, but had investigated the others, learning that the attribution was a mistake.

It was M. de Schryvere's opinion that the triplet of detonations that he had heard was from vessels in the North Sea. But now, according to developments, the sounds of Belgium cannot very well be attributed to terrestrial cannonading in or near Belgium: in Ciel et Terre, 16-614, are quoted two artillery officers who had heard the sounds, but could not so trace them: one of these officers had heard a series of detonations with intervals of about two minutes. A variety of explanations was attempted, but in conventional terms, and if these localized, repeating sounds did come from the sky, there's nothing to it but a new variety of attempted explanations, and in most unconventional terms. There are recorded definite impressions that the sounds were in the sky: Prof. Peleseneer's positivement aerien. In Ciel et Terre, 17-14, M. Van den Broeck announced that General Hennequin, of Brussels, had co-operated with him, and had sent enquiries to army officers and other persons, receiving thirty replies. Some of these correspondents had heard detonations at regular intervals. It is said that the sounds were like cannonading, but not in one instance were the sounds traced to terrestrial gunfire. Jan. 24, 1896--a triplet of triplets--between 2:30 and 3:30, P.M.--by M. Overloop, of Middelkirke, Belgium--three series of detonations, each of three sounds.

The sounds went on, but, after this occurrence, there seems to me to be little inducement to me to continue upon the subject. This is indication that from somewhere there has been signaling: from extra-mundane vessels to one another, or from some unknown region to this earth, as nearly final as we can hope to find.

There are persons who will see nothing but a susceptibility to the mysticism of numbers in a feeling that there is significance in threes of threes. But, if there be attempt in some other world to attract attention upon this earth, it would have to be addressed to some kind of a state of mind that would feel significances. Let our three threes be as mystic as the eleven horns on Daniel's fourth animal; if throughout nature like human nature there be only superstition as to such serialization, that superstition, for want of something more nearly intelligent, would be a susceptibility to which to appeal, and from which response might be expected. I think that a sense of mystic significance in the number three may be universal, because upon this earth it is general, appearing in theologies, in the balanced compositions of all the arts, in logical demonstrations, and in the indefinite feelings that are supposed to be superstitious.

The sounds went on, as if there were experiments, or attempts to communicate by means of other regularizations and repetitions. Feb. 18, 1896--a series of more than 20 detonations, at intervals of 2 or 3 minutes, heard at Ostend, by M. Pulzeys, an engineer of Brussels. Four or five sounds were heard at Ostend by someone else: repeated upon the 21st of February. Heard by M. Overloop, at Ostend, April 6: detonations at 11:57:30 A.M., and at 12:1:32 P.M. Heard the next day, by M. Overloop, at Blankenberghe, at 2:35 and 2:51 P.M.

The last occurrence recorded by M. Van den Broeck was upon the English Channel, May 23, 1896: detonations at 3:20 and 3:40 P.M. I have no more data, as to this period, myself, but I have notes upon similar sounds, by no means so widely reported and commented upon, in France and Belgium about 15 years later. One notices that the old earthquake-explanation as to these sounds has not appeared.

But there were other phenomena in England, in this period, and to considerable degree they were conventionally explained. They were not of the type of the Belgian phenomena, and, because manifestations were seen and felt, as well as heard, they were explained in terms of meteors and earthquakes. But in this double explanation, we meet a divided opposition, and no longer are we held back by the uncompromising attempt by exclusionist science to attribute all disturbances of this earth's surface to a subterranean origin. The admission by Symons and Fordham that we have recorded, as to occurrences of 1887-89, has survived.

The earliest of the accounts that I have read of the quakes in the general region of Worcester and Hereford (London Triangle) that associated with appearances in the sky, was published by two church wardens in the year 1661, as to occurrences of October, 1661, and is entitled, A True and Perfect Relation of the Terrible Earthquake. It is said that monstrous flaming things were seen in the sky, and that phenomena below were interesting. We are told, "truly and perfectly," that Mrs. Margaret Petmore fell in labor and brought forth three male offsprings all of whom had teeth and spoke at birth. Inasmuch as it is not recorded what the infants said, and whether in plain English or not, it is not so much an extraordinary birth such as, in one way or another, occurs from time to time, that affronts our conventional notions, as it is the idea that there could be relation between the abnormal in obstetrics and the unusual in terrestrics. The conventional scientist has just this reluctance toward considering shocks of this earth and phenomena in the sky at the same time. If he could accept with us that there often has been relation, the seeming discord would turn into a commonplace, but with us he would never again want to hear of extraordinary detonating meteors exploding only by coincidence over a part of this earth where an earthquake was occurring, or of concussions of this earth, time after time, in one small region, from meteors that, only by coincidence, happened to explode in one little local sky, time after time. Give up the idea that this earth moves, however, and coincidences many times repeated do not have to be lugged in.

Our subject now is the supposed earthquake centering around Worcester and Hereford, Dec. 17, 1896; but there may have been related events, leading up to this climax, signifying long duration of something in the sky that occasionally manifested relatively to this corner of the London Triangle. Mrs. Margaret Petmore was too sensational a person for our liking, at least in our colder and more nearly scientific moments, so we shall not date so far back as the time of her performance; but the so-called earthquakes of Oct. 6, 1863, and of Oct. 30, 1868, were in this region, and we had data for thinking that they were said to be earthquakes only because they could not be traced to terrestrial explosions.

At 5:45 P.M., Nov. 2, 1893, a loud sound was heard at a place ten miles northeast of Worcester, and no shock was felt (Nature, 49-245); however at Worcester and in various parts of the west of England and in Wales a shock was felt.

According to James G. Wood, writing in Symons' Met. Mag., 29-8, at 9:30 P.M., Jan. 25, 1894, at Llanthomas and Clifford, towns less than 20 miles west of Hereford, a brilliant light was seen in the sky, an explosion was heard, and a quake was felt. Half an hour later, something else occurred: according to Denning (Nature, 49-325) it was in several places, near Hereford and Worcester, supposed to be an earthquake. But, at Stokesay Vicarage, Shropshire (Symons' Met. Mag., 29-8) was seen the same kind of an appearance as that which had been seen at Llanthomas and Clifford, half an hour before: an illumination so brilliant that for half a minute everything was almost as visible as by daylight.

In the English Mechanic, 74-155, David Packer calls attention to "a strange meteoric light" that was seen in the sky, at Worcester, during the quake of Dec. 17, 1896. I should say that this was the severest shock felt in the British Isles, in the 19th century, with the exception of the shock of April 22, 1884, in the eastern point of the London Triangle. There was something in the sky. In Nature, 55-179, J. Lloyd Bozward writes that, at Worcester, a great light was seen in the sky, at the time of the shock, and that, in another town, "a great blaze" had been seen in the sky. In Symons' Met. Mag., 31-180, are recorded many observations upon lights that were seen in the sky. In an appendix to his book, The Hereford Earthquake of 1896, Dr. Charles Davison says that at the time of the quake (5:30 A.M.) there was a luminous object in the sky, and that it "traversed a large part of the disturbed area." He says that it was a meteor, and an extraordinary meteor that lighted up the ground so that one could have picked up a pin. With the data so far considered, almost anyone would think that of course an object had exploded in the sky, shaking the earth underneath. Dr. Davison does not say this. He says that the meteor only happened to appear over a part of this earth where an earthquake was occurring, "by a strange coincidence."

Suppose that, with ordinary common sense, he had not lugged in his "strange coincidence," and had written that of course the shock was concussion from an explosion in the sky.

Shocks that had been felt before midnight, December 17, and at 1:30 or 1:45, 2, 3, 3:30, 4, 5, and 5:20, and then others at 5:40 or 5:45 and at 6:15 o'clock--and were they, too, concussions, but fainter and from remoter explosions in the sky-- and why not, if of course the great shock of 5:30 o'clock was from a great explosion in the sky--and by what multiplication of strangeness of coincidence could detonating meteors, or explosions of any other kind, so localize in the one little sky of Worcester, if this earth be a moving earth--and how could their origin be otherwise than a fixed region nearby?

In some minds it may be questionable that the earth could be so affected as it was at 5:30 A.M., Dec. 17, 1896, by an explosion in the sky. Upon Feb. 10, 1896, a tremendous explosion occurred in the sky of Madrid: throughout the city windows were smashed; a wall in the building occupied by the American Legation was thrown down. The people of Madrid rushed to the streets, and there was a panic in which many were injured. For five hours and a half a luminous cloud of debris hung over Madrid, and stones fell from the sky.

Suppose, just at present, we disregard all the Worcester-Hereford phenomena except those of Dec. 17, 1896. Draw a diagram, illustrating a stream of meteors pursuing this earth, now supposed to be rotating and revolving, for more than 400,000 miles in its orbit, and curving around gracefully and unerringly after the rotating earth, so as to explode precisely in this one little local sky and nowhere else.

But we can't think very reasonably even of a flock of birds flying after and so precisely pecking one spot on an apple thrown in the air by somebody. Another diagram--stationary earth--bombardment of any kind one chooses to think of--same point hit every time--thinkable.

The phenomena associate with an opposition of Mars. Dec. 10, 1896--opposition of Mars.

But we have gone on rather elaborately with perhaps an insufficiency to base upon. We cannot say, directly, that all the phenomena of the night of Dec. 16-17, 1896, were shocks from explosions in the sky: only during the greatest of the concussions was something seen, or was something near enough to be seen.

We apply the idea of the diagrams to another series of occurrences in this period. Now draw a diagram relatively to the sky of Florida, and see just what the explanation of coincidence demands or exacts. But then consider the diagram as one of an earth that does not move and of something that is fixed over a point upon its surface. Things can be thought of as coming down from somewhere else to one special sky of this earth, as logically as precariously placed objects on one special window sill sometimes come down to a special neighbor.

In the Monthly Weather Review, 23-57, is a report, by the Director of the Florida Weather Service, upon "mysterious sounds" and luminous effects in the sky of Florida. According to investigation, these phenomena did occur in the sky of Florida, about noon, Feb. 7, 1895, again at 5 o'clock in the morning of the 8th, and again between 6 and to o'clock, night of the 8th. The Editor of the Review thinks that three meteors may have exploded so in succession in the sky of Florida, and nowhere else, "by coincidence."

30

CHAR me the trunk of a redwood tree. Give me pages of white chalk cliffs to write upon. Magnify me thousands of times, and replace my trifling immodesties with a titanic megalomania--then might I write largely enough for our subjects. Because of accessibility and abundance of data, our accounts deal very much with the relatively insignificant phenomena of Great Britain. But our subject, if not so restricted, would be the violences that have screamed from the heavens, lapping up villages with tongues of fire. If, because of appearances in the sky, it be accepted that some of the so- called earthquakes of Italy and South America represented relations with regions beyond this earth, then it is accepted that some of this earth's greatest catastrophes have been relations with the unknown and the external. We have data that seem to be indications of signaling, but not unless we can think that foreign giants have hurled explosive mountains at this earth can we see such indications in all the data.

Our data do seem to fall into two orders of phenomena: sounds of Melida, Barisal, and Belgium, and nothing falling from the sky, and nothing seen in the sky, and excellently supported observations for accepting a signal-like intent in intervals and grouping of sounds, at least in Barisal and Belgium; and the unregularized phenomena of Worcester-Hereford, Colchester, Comrie, and Birmingham, in which appearances are seen in the sky, or in which substances fall from the sky, and in which effects upon this earth, not noted at all in Belgium and Bengal, are great, and sometimes tremendous. It seems that extra-geography divides into the extra- sociologic and the extra-physical; and in the second type of phenomena, we suppose the data are of physical relations between this earth and other worlds. We think of a difference of potential. There were tremendous detonations in the sky at the times of the falls of the little black stones of Birmingham and Wolverhampton, and the electric manifestations, according to descriptions in the newspapers, were extraordinary, and great volumes of water fell. Consequently the events were supposed to be thunderstorms. I suppose, myself, that they were electric storms, but electric storms that represented difference of potential between this earth and some region that was fixed, at least eleven years, over Birmingham and Wolverhampton, bringing down stones and volumes of water from some other world, or bringing down stones, and dislodging intervening volumes of water, such as we have many data for thinking exist in outer space, sometimes in bodies of warm or hot water, and sometimes as great masses, or fields; of ice.

Let two objects be generically similar, but specifically different and a relation that may be known as a difference of potential, though that term is usually confined to electric relations, generates between them. Quite as the Gulf Stream-- though there are no reasons to suppose that there is such a Gulf Stream as one reads of--represents a relation between bodies of water heated differently, given any two worlds, alike in general constitution, but differing, say, electrically, and given proximity, we conceive of relations between them other than gravitational.

But this cloistered earth, and its monkish science--shrinking from, denying, or disregarding, all data of external relations, except someone controlling force that was once upon a time known as Jehovah, but that has been re-named Gravitation--

That the electric exchanges that were recognized by the ancients, but that were anthropomorphically explained by them, have poured from the sky and have gushed to the sky, afferently and efferently, between this earth and the nearby planets, or between this mainland and its San Salvadors, and have been recognized by the moderns, or the neo-ancients, but have been meteorologically and seismologically misconstrued by them.

When a village spouts to the sky, it is said to have been caught up in a cyclone: when unknown substances fall from the sky, not much of anything is said upon the subject.

Lost tribes and the nations that have disappeared from the face of this earth-- that the skies have reeked with terrestrial civilizations, spreading out in celestial stagnations, where their remains to this day may be. The Mayans--and what became of them? Bones of the Mayans, picked white as frost by space- scavengers, regioned to this day in a sterile luxuriance somewhere, spread upon existence like the pseudo-breath of Death, crystallized on a sky-pane. Three times gaps wide and dark the history of Egypt--and that these abysses were gulfed by disappearances--that some of the eliminations from this earth may have been upward translations in functional suctions. We conceive of Supervision upon this earth's development, but for it the names of Jehovah and Allah seem old-fashioned-- that the equivalence of wrath, but like the storms of cells that, in an embryonic thing, invade and destroy cartilage-cells, when they have outlived their usefulness, have devastated this earth's undesirables. Likely enough, or not quite likely enough, one of these earlier Egypts was populated by sphinxes, if one can suppose that some of the statuary still extant in Egypt were portraitures. This is good, though also not so good, orthodox Evolutionary doctrine--that between types occur transitionals--

That Elimination and Redistribution swept an earlier Egypt with suctions--because it was written, in symbols of embryonic law, that life upon this earth must form onward--and the crouching sphinx on the sands of Egypt, blinking the mysticism of her morphologic mixtures, would perhaps detain forever the less interesting type that was advancing--

That often has Clarification destroyed transitionals, that they shall not hold back development.

One conceives of their remains, to this day, wafting still in the currents of the sky: floating avenues of frozen sphinxes, solemnly dipping in cosmic undulations, down which circulate processions of Egyptian mummies.

An astronomer upon this earth notes that things in parallel lines have crossed the sun.

We offer this contribution as comparing favorably with the works of any other historian. We think that some of the details may need revision, but that what they typify is somewhere nearly acceptable: Latitudes and longitudes of bones, not in the sky, but upon the surface of this earth. Baron Toll and other explorers have, upon the surface of this earth, kicked their way through networks of ribs and protrusions of skulls and stacks of vertebrae, as numerous as if from dead land they had sprouted there. Anybody who has read of these tracts of bones upon the northern coast of Siberia, and of some of the outlying islands that are virtually composed of bones cemented with icy sand, will agree with me that there have been cataclysms of which conventionality and standardization tell us nothing. Once upon a time, some unknown force translated, from somewhere, a million animals to Colorado, where their remains now form great bone-quarries. Very largely do we express a reaction against dogmatism, and sometimes we are not dogmatic, ourselves. We don't know very positively whether at times the animal life of some other world has been swept away from that world, eventually pouring from the sky of Siberia and of Colorado, in some of the shockingest floods of mammoths from which spattered cats and rabbits, in cosmic scenery, or not. All that we can say is that when we turn to conventionality it is to blankness or suppression. Every now and then, to this day, occurs an alleged fall of blood from the sky, and I have notes upon at least one instance in which the microscopically examined substance was identified as blood. But now we conceive of intenser times, when every now and then a red cataract hung in the heavens like the bridal veil of the goddess of murder. But the science of today is a soporific like the idealism of Europe before the War broke out. Science and idealism--wings of a vampire that lulls consciousness that might otherwise foresee catastrophe. Showers of frogs and showers of fishes that occur to this day--that they are the dwindled representatives to this day of the cataclysms of intenser times when the skies of this earth were darkened by afferent clouds of dinosaurs. We conceive of intenser times, but we conceive of all times as being rhythmic times. We are too busy to take up alarmism, but, if Rome, for instance, never was destroyed by terrestrial barbarians, if we cannot very well think of Apaches seizing Chicago, extramundane vandals may often have swooped down upon this earth, and they may swoop again; and it may be a comfort to us, someday, to mention in our last gasp that we told about this.

History, geology, palaeontology, astronomy, meteorology--that nothing short of cataclysmic thinking can break down these united walls of Exclusionism.

Unknown monsters sometimes appear in the ocean. When, upon the closed system of normal preoccupations, a story of a sea serpent appears, it is inhospitably treated. To us of the wider cordialities, it has recommendations for kinder reception. I think that we shall be noted in recognitions of good works for our bizarre charities. Far back in the topography of the nineteenth century, Richard Proctor was almost submerged in an ocean of smugness, but now and then he was a little island emerging from the gently alternating doubts and satisfactions of his era, and by means of several papers upon the "sea serpent" he so protruded and gave variety to a dreary uniformity. Proctor reviewed some of the stories of "sea serpents." He accepted some of them. This will be news to some conventionalists. But the mystery that he could not solve is their conceivable origin. To be sure this earth may not be round, or top- shaped, and may tower away somewhere, perhaps with the great Antarctic plateau as its foothills, to a gigantic existence commensurate throughout with the sea monsters that sometimes reach regions known to us. Judging by our experience in other fields of research, we suspect that this earth never has been traversed except in conventional trade-routes and standard explorations. One supposes that enormous forms of life that have appeared upon the surface of the ocean, did not come from conditions of great pressure below the surface. If there be no habitat of their own, in unknown seas of this earth, the monsters fell from the sky, surviving for a while. In his day, Charles Lyell never said a more preposterous thing than this--however, we have no idea that mere preposterousness is a criterion.

Then at times the things have fallen upon land, presumably. To scientific minds in their present anaemia of .malnutrition, we offer new nourishment. There are materials for a science of neo-palaeontology--as it were--at least a new view of animal-remains upon this earth. Remains of monsters, supposed to have lived geologic ages ago, are sometimes found, not in ancient deposits, but upon, or near, the surface of the ground, sometimes barely covered. I have notes upon a great pile of bones, supposed to be the remains of a whale, out in open view in a western desert.

In the American Museum of Natural History, New York City, is the mummified body of a monster called a trachodon, found in Converse County, Wyoming. It was not found upon the surface of the ground, which is bad for our attempts to stimulate palaeontology. But the striking datum to me is that the only other huge mummy that I know of is another trachodon, now in the Museum of Frankfort. If only extraordinarily would geologic processes mummify remains of a huge animal, doubly extraordinarily would two animals of the same species be so exclusively affected.

One at least gives some consideration to the idea that these trachodons are not products of geologic circumstances, but were affected, in common, by other circumstances. By inspiration, or progressive deterioration, one then conceives of the things as having wafted and dried in space, finally falling to this earth. Our swooping vandals are relieved with showering mummies. Life is turning out to be interesting.

Organic substances like life-fluids of living things have rained from the sky. However, it is enough for our general purposes to make acceptable simply that unknown substances have, in large quantities, fallen from the sky. That is neo-ism enough, it seems to me. I consider, myself, all such data relatively to this earth's stationariness or possible motions. In Ciel et Terre, 22-198, it is said that, about 2 P.m., June 8, 1901, a glue-like substance fell at Sart. The story is told by an investigator, M. Michael, a meteorologist. He says that he saw this substance falling from the sky, but does not give an estimate of duration: he says that he arrived during the last five minutes of the shower. Editors and extra- geographers can't help trying to explain. The Editor of Ciel et Terre writes that, three days before, there had been, at Antwerp, a great fire, in which, among other substances, a large quantity of sugar had been burned. He asks whether there could be any connection. Antwerp is about 80 miles from Sart.

Sept. 2, 1905--the tragedy of the space-pig:

In the English Mechanic, 86-100, Col. Markwick writes that, according to the Cambrian Natural Observer, something was seen in the sky, at Llangollen, Wales, Sept. 2, 1905. It is described as an intensely black object, about two miles above the earth's surface, moving at the rate of about twenty miles an hour. Col. Markwick writes: "Could it have been a balloon?" We give Col. Markwick good rating as an extra-geographer, but of the early, or differentiating type, a transitional, if not a sphinx: so he was not quite developed enough to publish the details of this object. In the Cambrian Natural Observer, 1905-35--the journal of the Astronomical Society of Wales--it is said that, according to accounts in the newspapers, an object had appeared in the sky, at Llangollen, Wales, Sept. 2, 1905. At the schoolhouse, in Vroncysylite--I think that's it: with all my credulity, some of these Welsh names look incredible to me, in my notes--the thing in the sky had been examined through powerful field glasses. We are told that it had short wings, and flew, or moved, in a way described as "casually inclining sideways." It seemed to have four legs, and looked to be about ten feet long. According to several witnesses it looked like a huge, winged pig, with webbed feet. "Much speculation was rife as to what the mysterious object could be."

Five days later, according to a member of the Astronomical Society of Wales--see Cambrian Observer, 1905-30--a purple-red substance fell from the sky, at Llanelly, Wales.

I don't know that my own attitude toward these data is understood, and I don't know that it matters in the least; also from time to time my own attitude changes: but very largely my feeling is that not much can be, or should be, concluded from our meager accounts, but that so often are these occurrences, in our fields, reported, that several times every year there will be occurrences that one would like to have investigated by someone who believes that we have written nothing but bosh, and by someone who believes in our data almost religiously. It may be that, early in February, 1892, a luminous thing traveled back and forth, exploring for ten hours in the sky of Sweden. The story is copied from a newspaper, and ridiculed, in the English Mechanic, 55-34. Upon March 7, 1893, a luminous object shaped like an elongated pear was seen in the sky of Val-de-la-Haye, by M. Raimond Coulon (L'Astro., 1893-169). M. Coulon's suggestion is that the light may have been a signal suspended from a balloon. The signal-idea is interesting.

In the summer of 1897, several weeks after Prof. Andree and his two companions had sailed in a balloon, from Amsterdam Island, Spitzbergen, it was reported that a balloon had been seen in British Columbia. There was wide publicity: the report was investigated. It may be that had a terrestrial balloon escaped from somewhere in the United States or Canada, or if there had been a balloon-ascension at this time, the circumstances would have been reported: it may well be that the object was not Andree's balloon. President Bell, of the National Geographic Society, heard of this object, and heard that details had been sent to the Swedish Foreign Office, and cabled to the American Minister, at Stockholm, for information. He publishes his account in the National Geographic Magazine, 9-102. He was referred to the Swedish Consul, at San Francisco. In reply to inquiry, the Consul telegraphed the following data, which had been collected by the President of the Geographical Society of the Pacific:

"Statement of a balloon passing over the Horse-Fly Hydraulic Mining Camp, in Caribou, British Columbia, 52 degrees, 20', and Longitude 120 degrees, 30'--

"From letters of J. B. Robson, manager of the Caribou Mining Co., and of Mrs. Wm. Sullivan, the blacksmith's wife, there, and a statement of Mr. John J. Newsome, San Francisco, then at camp. About 2 or 3 o'clock, in the afternoon, between fourth and seventh of August last, weather calm and cloudless, Mrs. Sullivan, while looking over the Hydraulic Bank, noticed a round, grayish-looking object in the sky, to the right of the sun. As she watched, it grew larger and was descending. She saw the larger mass of the balloon above, and a smaller mass apparently suspended from the larger. It continued to descend, until she plainly recognized it as a balloon and a large basket hanging thereto. It finally commenced to swing violently back and forth, and move very fast toward the eastward and northward. Mrs. Sullivan called her daughter, aged 18, and about this time Mrs. Robson and her daughter were observing it."

If someone saw a strange fish in the ocean, we'd like to know--what was it like? Stripes on him--spots--what? It would be unsatisfactory to be told over and over only that a dark body had crossed some waves. In Cosmos, n.s., 39-356, a satisfactory correspondent writes that, at Lille, France, Sept. 4, 1898, he saw a red object in the sky. It was like the planet Mars, but was in the position of no known planet. He looked through his telescope, and saw a rectangular object, with a violent-colored band on one side of it, and the rest of it striped with black and red. He watched it ten minutes, during which time it was stationary; then, like the object that was seen at the time of the Powell-mystery, it cast out sparks and disappeared.

In the English Mechanic, 75-417, Col. Markwick writes that, upon May 10, 1902, a friend of his had seen in the sky, in South Devon, a great number of highly colored objects like little suns or toy-balloons. "Altogether beats me," says Col. Markwick.

Upon March 2, 1899, a luminous object in the sky, from 10 A.M., until 4 P.M., was reported from El Paso, Texas. Mentioned in the Observatory, 22-247--supposed to have been Venus, even though Venus was then two months past secondary maximum brilliance. This seems reasonable enough, in itself, but there are other data for thinking that an unknown, luminous body was at this time in the especial sky of the southwestern states. In the U. S. Weather Bureau Report (Ariz. Sec., March, 1899) it is said that, at Prescott, Arizona, Dr. Warren E. Day had seen a luminous object, upon the 8th of March, "that traveled with the moon" all day, until 2 P.M. It is said that, the day before, this object had been seen close to the moon, by Mr. G. O. Scott, at Tonto, Arizona. Dr. Day and Mr. Scott were voluntary observers for the Weather Review. This association with the moon and this localization of observation are puzzling.

La Nature (Sup.) Nov. 11, 1899--that at Luzarches, France, upon the 28th of October, 1899, M. A. Garrie had seen, at 4:50 P.M., a round, luminous object rising above the horizon. About the size of the moon. He watched it for 15 minutes, as it moved away, diminishing to a point. It may be that something from external regions was for several weeks in the especial sky of France. In La Nature (Sup.) Dec. 16, 1899, someone writes that he had seen, Nov. 15, 1899, 7 P.M., at Dourite (Dordogne) an object like an enormous star, at times white, then red, and sometimes blue, but moving like a kite. It was in the south. He had never seen it before. Someone, in the issue of December 30th, says that without doubt it was the star Formalhaut, and asks for precise position. Issue of Jan. 20, 1900--the first correspondent says that the object was in the southwest, about 35 degrees above the horizon, but moving so that the precise position could not be stated. The kite-like motion may have been merely seeming motion--object may have been Formalhaut, though 35 degrees above the horizon seems to me to be too high for Formalhaut--but, then, like the astronomers, I'm likely at times to expose what I don't know about astronomy. Formalhaut is not an enormous star. Seventeen are larger.

May 1, 1908, between 8 and 9 P.M., at Vittel, France--an object, with a nebulosity around it, diameter equal to the moon's, according to a correspondent to Cosmos, n.s., 58-535. At 9 o'clock a black band appeared upon the object, and moved obliquely across it, then disappearing. The Editor thinks that the object was the planet Venus, under extraordinary meteorologic conditions.

Dark obj., by Prof. Brooks, July 21, 1896 (Eng. Mec., 64-12); dark obj., by Gathmann, Aug. 22, 1896 (Sci. Amer. Sup., 67-363); two luminous objs., by Prof. Swift, evidently in a local sky of California, because unseen elsewhere in California, Sept. 20, and one of them again, Sept. 21, 1896 (Astro. Jour. 17-8, 103); "Waldemath's second moon," Feb. 5, 1898 (Eng. Mec., 67-545); unknown obj., March 30, 1908 (Observatory, 31-215); dark obj., Nov. 10, 1908 (Bull. Soc. Astro. de France, 23-74).

31

COLD HARBOR, Hanover Co., Virginia--two men in a field--"an apparently clear sky." In the Monthly Weather Review, 28-29, it is said that upon Aug. 7, 1900, two men were struck by lightning. The Editor says that the weather map gave no indication of a thunderstorm, nor of rain, in this region at the time.

In July, 1904, a man was killed on the summit of Mt. San Gorgionio, near the Mojave desert. It is said that he was killed by lightning. Two days later, upon the summit of Mt. Whitney, 180 miles away, another man was killed "by lightning" (Ciel et Terre, 29-120).

It is said, in Ciel et Terre, 17-42, that, in the year 1893, nineteen soldiers were marching near Bourges, France, when they were struck by an unknown force. It is said that in known terms there is no explanation. Some of the men were killed, and others were struck insensible. At the inquest it was testified that there had been no storm, and that nothing had been heard.

If there occur upon the surface of this earth pounces from blankness and seizures by nothings, and "sniping" with bullets of unfindable substance, we nevertheless hesitate to bring witchcraft and demonology into our fields. Our general subject now is the existence of a great deal that may be nearby, or temporarily nearby, ordinarily invisible, but occasionally revealed by special circumstances. A background of stars is not to be compared, in our data, with the sun for a background, as a means of revelations. We accept that there are sunspots, but we gather from general experience and special instances that the word "sunspot" is another of the standardizing terms like "auroral" and "meteoric" and "earthquakes." See Webb's Celestial Objects for some observations upon large definite obscurations called "sunspots" but which were as evanescent against the sun as would be islands and jungles of space, if intervening only a few moments between this earth and the swifting moving sun. According to Webb, astronomers have looked at great obscurations upon the sun, have turned away, and then looked again, finding no trace of the phenomena. Eclipses are special circumstances, and rather often have large, unknown bulks been revealed by different light-effects during eclipses. For instance, upon Jan. 22, 1898, Lieut. Blackett, R.N., assisting Sir Norman Lockyer, at Viziadrug, India, during the total eclipse of the sun, saw an unknown body between Venus and Mars (Jour. Leeds Astro. Soc., 1906- 23). We have had other instances, and I have notes upon still more. The photographic plate is a special condition, or sensitiveness. In Knowledge, 16-234, a correspondent writes that, in August, 1893, in Switzerland, moonlighted night, he had exposed a photographic plate for one hour. Upon the photograph, when developed, were seen irregular, bright markings, but there had been no lightning to this correspondent's perceptions.

The details of the sheep-panic of Nov. 3, 1888, are extraordinary. The region affected was much greater than was supposed by the writer whom we quoted in an earlier chapter. It is said in another account in Symons' Meteorological Magazine, that, in a tract of land twenty-five miles long and eight miles wide, thousands of sheep had, by a simultaneous impulse, burst from their bounds; and had been found the next morning, widely scattered, some of them still panting with terror under hedges, and many crowded into corners of fields. See London Times, Nov. 20, 1888.

An idea of the great number of flocks affected is given by one correspondent who says that malicious mischief was out of the question, because a thousand men could not have frightened and released all these sheep. Someone else tries to explain that, given an alarm in one flock, it might spread to the others. But all the sheep so burst from their folds at about eight o'clock in the evening, and one supposes that many folds were far from contiguous, and one thinks of such contagion requiring considerable time to spread over 200 square miles. Something of an alarming nature and of a pronounced degree occurred somewhere near Reading, Berkshire, upon this evening.

Also there seems to be something of special localization: the next year another panic occurred in Berkshire not far from Reading.

I have a datum that looks very much like the revelation of a ghost-moon, though I think of it myself in physical terms of light-effects. In Country Queries and Notes, 1-138, 417, it is said that, in the sky of Gosport, Hampshire, night of Sept. 14, 1908, was seen a light that came as if from an unseen moon. It may be that I can here record that there was a moon-like object in the sky of the Midlands and the south of England, this night, and that, though to human eyesight, this world, island of space, whatever it may have been, was invisible, it was, nevertheless, revealed. Upon this evening of Sept. 14, 1908, David Packer, then in Northfield, Worcestershire, saw a luminous appearance that he supposed was auroral, and photographed it. When the photograph was developed, it was seen that the "auroral" light came from a large, moon-like object. A reproduction of the photograph is published in the English Mechanic, 88-211. It shows an object as bright and as well-defined as the conventionally accepted moon, but only to the camera had it revealed itself, and Mr. Packer had caught upon a film a space- island that had been invisible to his eyes. It seems so, anyway.

In Country Queries and Notes, 1-328, it is said that, upon Aug. 2, 1908, at Ballyconneely, Connemara coast of Ireland, was seen a phantom city of different- sized houses, in different styles of architecture; visible three hours. It is said that no doubt the appearance was a mirage of some city far away--far away, but upon this earth, of course. This apparition is not of the type that we consider so especially of our own data. The so-called mirages that so especially interest us are interesting to us not in themselves, but in that they belong to the one order of phenomena or evidence that unifies so many fields of our data: that is, repetitions in a local sky, signifying the fixed position of something relatively to a small part of this earth's surface. We cannot think that mirages, terrestrial or extraterrestrial, could so repeat. But if in a local sky of this earth there be a fixed region, perhaps not a city, but something of rugged and featureful outlines, with projections that might look architectural, reflections from it, shadows, or Brocken specters repeating always in one special sky are thinkable except by the Chinese-minded who regard all our data as "foreign devils." The writer in Country Queries and Notes says--"Circumstantial accounts have even been published of the city of Bristol being distinctly recognized in a mirage seen occasionally in North America." If we shall accept that anywhere in North America repeated representations of the same city or city-like scene have appeared in the same local sky, I prefer, myself, a foreign devil of a thought, and its significance, whether hellish or not, that this earth is stationary, to such a domestic vagrant of a thought as the idea that mirage could so pick out the city of Bristol, or any other city, over and over, and also invariably pick out for its screen the same local sky, thousands of miles, or five miles, away.

In the English Mechanic, Sept. 10, 1897, a correspondent to the Weekly Times and Echo is quoted. He had just returned from the Yukon. Early in June, 1897, he had seen a city pictured in the sky of Alaska. "Not one of us could form the remotest idea in what part of the world this settlement could be. Some guessed Toronto, others Montreal, and one of us even suggested Pekin. But whether this city exists in some unknown world on the other side of the North Pole, or not, it is a fact that this wonderful mirage occurs from time to time yearly, and we were not the only ones who witnessed the spectacle. Therefore it is evident that it must be the reflection of some place built by the hand of man." According to this correspondent, the "mirage" did not look like one of the cities named, but like "some immense city of the past."

In the New York Tribune, Feb. 17, 1901, it is said that Indians of Alaska had told of an occasional appearance, as if of a city, suspended in the sky, and that a prospector, named Willoughby, having heard the stories, had investigated, in the year 1887, and had seen the spectacle. It is said that, having several times attempted to photograph the scene, Willoughby did finally at least show an alleged photograph of an aerial city. In Alaska, p. 140, Miner Bruce says that Willoughby, one of the early pioneers in Alaska, after whom Willoughby Island is named, had told him of the phenomenon, and that, early in 1899, he had accompanied Willoughby to the place over which the mirage was said to repeat. It seems that he saw nothing himself, but he quotes a member of the Duc d'Abruzzi's expedition to Mt. St. Elias, summer of 1897, Mr. C. W. Thornton, of Seattle, who saw the spectacle, and wrote--"It required no effort of the imagination to liken it to a city, but was so distinct that it required, instead, faith to believe that it was not in reality a city." Bruce publishes a reproduction of Willoughby's photograph, and says that the city was identified as Bristol, England. So definite, or so un-mirage-like, is this reproduction, trees and many buildings shown in detail, that one supposes that the original was a photograph of a good- sized terrestrial city, perhaps Bristol, England.

In Chapter 10, of his book, Wonders of Alaska, Alexander Badlam tries to explain. He publishes a reproduction of Willoughby's photograph: it is the same as Bruce's, except that all buildings are transposed, or are negative in positions. Badlam does not like to accuse Willoughby of fraud: his idea is that some unknown humorist had sold Willoughby a dry plate, picturing part of the city of Bristol.

My own idea is that something of this kind did occur, and that this photograph, greatly involved in accounts of the repeating mirages, had nothing to do with the mirages. Badlam then tells of another photograph. He tells that two men, near the Muir Glacier, had, by means of a pan of quicksilver, seen a reflection of an unknown city somewhere, and that their idea was that it was at the bottom of the sea near the glacier, reflecting in the sky, and reflecting back to and from the quicksilver. That's complicated. A photographer named Taber then announced that he had photographed this scene, as reflected in a pan of quicksilver. Badlam publishes a reproduction of Taber's photograph, or alleged photograph. This time, for anybody who prefers to think that there is, somewhere in the sky of Alaska, a great, unknown city, we have a most agreeable photograph: exotic-looking city; a structure like a coliseum, and another prominent building like a mosque, and many indefinite, mirage-like buildings. I'd like to think this photograph genuine, myself, but I do conceive that Taber could have taken it by photographing a panorama that he had painted. Badlam's explanation is that mirages of glaciers are common, in Alaska, and that they look architectural. Some years ago, I read five or six hundred pounds of literature upon the Arctic, and I should say that far- projected mirages are not common in the Arctic: mere looming is common. Badlam publishes a photograph of a mirage of Muir Glacier. The looming points of ice do look Gothic, but they are obviously only loomings, extending only short distances from primaries, with no detachment from primaries, and not reflecting in the sky.

For the first identification of the Willoughby photograph as a photograph of part of the city of Bristol, see the New York Times, Oct. 20, 1889. That this photograph was somebody's hoax seems to be acceptable. But it was not similar to the frequently reported scene in the sky of Alaska, according to descriptions. In the New York Times, Oct. 31, 1889, is an account, by Mr. L. B. French, of Chicago, of the spectral representation, as he saw it, near Mt. Fairweather. "We could see plainly houses, well-defined streets, and trees. Here and there rose tall spires over huge buildings, which appeared to be ancient mosques or cathedrals.... It did not look like a modern city--more like an ancient European City."

Jour. Roy. Met. Soc., 27-158:

That every year, between June 21 and July 10, a "phantom city" appears in the sky, over a glacier in Alaska; that features of it had been recognized as buildings in the city of Bristol, England, so that the "mirage" was supposed to be a mirage of Bristol. It is said that for generations these repeating representations had been known to the Alaskan Indians, and that, in May, 1901, a scientific expedition from San Francisco would investigate. It is said that, except for slight changes, from year to year, the scene was always the same.

La Nature, 1901-1-303:

That a number of scientists had set out from Victoria, B. C., to Mt. Fairweather, Alaska, to study a repeating mirage of a city in the sky, which had been reported by the Duc d'Abruzzi, who had seen it and had sketched it.

32

NIGHT of Dec 7, 1900--for seventy minutes a fountain of light played upon the planet Mars. Prof. Pickering--"absolutely inexplicable" (Sci. Amer., 84-179).

It may have been a geyser of messages. It may be translated someday. If it were expressed in imagery befitting the salutation by a planet to its dominant, it may be known someday as the most heroic oration in the literature of this geo-system. See Lowell's account in Popular Astronomy, 10-187. Here are published several of the values in a possible code of long flashes and short flashes. Lowell takes a supposed normality for unity, and records variations of two thirds, one and one third, and one and a half. If there be, at Flagstaff, Arizona, records of all the long flashes and short flashes that were seen, for seventy minutes, upon this night of Dec. 7, 1900, it is either that the greetings of an island of space have been hopelessly addressed to a continental stolidity, or there will have to be the descent, upon Flagstaff, Arizona, by all the amateur Champollions of this earth, to concentrate in one deafening buzz of attempted translation.

It was at this time that Tesla announced that he had received, upon his wireless apparatus, vibrations that he attributed to the Martians. They were series of triplets.

It is our expression that, during eclipses and oppositions and other notable celestial events, lunarians try to communicate with this earth, having a notion that at such times the astronomers of this earth may be more nearly alert.

An eclipse of the moon, March 10-11, 1895--not a cloud; no mist--electric flashes like lightning, reported from a ship upon the Atlantic (Eng. Mec., 61-100).

During the eclipse of the sun, July 29, 1897, a strange image was taken on a sensitive plate, by Mr. L. E. Martindale, of St. Mary's, Ohio. It looks like a record of knotted lightning. See Photography, May 26, 1898.

In the Bull. Soc. Astro. de France, 17-205, 315, 447, it is said that upon the first and the third of March, 1903, a light like a little star, flashing intermittently, was seen by M. Rey, in Marseilles, and by Maurice Gheury, in London, in the lunar crater Aristarchus. March 28, 1903--opposition of Mars.

In Cosmos, n.s., 49-259, M. Desmoulins writes, from Argenteuil, that, upon Aug. 9, 1903, at 11 P.M., moving from north to south, he saw a luminous object. The planet Venus was at primary greatest brilliance upon Aug. 13, 1903. In three respects it was like other objects that have been observed upon this earth at times of the nearest approach of Venus: it was a red object; it appeared only in a local sky, and it appeared in the time of the visibility of Venus. With M. Desmoulins were four persons, one of whom had field glasses. The object was watched twenty minutes, during which time it traveled a distance estimated at five or six kilometers. It looked like a light suspended from a balloon, but, through glasses, no outline of a balloon could be seen, and there were no reflections of light as if from the opaque body of a balloon. It was a red body, with greatest luminosity in its nucleus. The Editor of Cosmos writes that, according to other correspondents, this object had been seen, at 11 P.M., July 19th and 26th, at Chatou. Argenteuil and Chatou are 4 or 5 miles apart, and both are about 5 miles from Paris. All three of these dates were Sundays, and even though nothing like a balloon had been seen through glasses, one naturally supposes that somebody near Paris had been amusing himself sending up fire-balloons, Sunday evenings. The one great resistance to all that is known as progress is what one "naturally supposes."

In the English Mechanic, 81-220, Arthur Mee writes that several persons, in the neighborhood of Cardiff, had, upon the night of March 29, 1905, seen in the sky, "an appearance like a vertical beam of light, which was not due (they say) to a searchlight, or any such cause." There were other observations, and they remind us of the observations by Noble and Bradgate, Aug. 28-29, 1883: then upon an object that cast a light like a searchlight; this time an association between a light like a searchlight, and a luminosity of definite form. In the Cambrian Natural Observer, 1905-32, are several accounts of a more definite-looking appearance that was seen, this night, in the sky of Wales--"like a long cluster of stars, obscured by a thin film or mist." It was seen at the time of the visibility of Venus, then an "evening star"--about 10 P.M. It grew brighter, and for about half an hour looked like an incandescent light. It was a conspicuous and definite object, according to another description--"like an iron bar, heated to an orange-colored glow, and suspended vertically."

Three nights later, something appeared in the sky of Cherbourg, France--L'Astre Cherbourg--the thing that appeared, night after night, in the sky of the city of Cherbourg, at a time when the planet Venus was nearest (inferior conjunction April 26, 1905).

Flammarion, in the Bull. Soc. Astro. de France, 19-243, says that this object was the planet Venus. He therefore denies that it had moved in various directions, saying that the supposed observations to this effect were illusions. In L'Illustration, April 22, 1905, he tells the story in his own way, and says some things that we are not disposed to agree with, but also he says that the ignorance of some persons in inenarrable. In Cosmos, n.s., 42-420, months after the occurrence, it is said that many correspondents had written to inquire as to L'Astre Cherbourg. The Editor gives his opinion that the object was either Jupiter or Venus. Throughout our Venus-visitor expression, the most important point is appearance in a local sky. That unifies this expression with other expressions, all of them converging into our general extra-geographic acceptances. The Editor of Cosmos says that this object, which was reported from Cherbourg, was reported from other towns as well. He probably means to say that it was seen simultaneously in different towns. For all guardians of this earth's isolation, this is a convenient thing to say: the conclusion then is that the planet Venus, exceptionally bright, was attracting unusual attention generally, and that there was nothing in the especial sky of Cherbourg. But we have learned that standardizing disguisements often obscure our data in later accounts, and we have formed the habit of going to contemporaneous sources. We shall find that the newspapers of the time reported a luminous object that appeared, night after night, only over the city of Cherbourg, as the name by which it was known indicates. It was a reddish object. The Editor of Cosmos explains that atmospheric conditions could give this coloration to Venus. I suppose this could be so occasionally: not night after night, I should say. We shall find that this object, or a similar object, was reported from other places, but not simultaneously with its appearance over Cherbourg. In the Journal des Debats, the first news is in the issue of April 4, 1905. It is said that a luminous body was appearing, every evening, between 8 and to o'clock, over the city of Cherbourg.

These were about the hours of the visibility of Venus. In this period, Venus set at 9:30 P.M., and Jupiter at 8 P.M. It is enough to make any conventionalist feel most reasonable, though he'd feel that way anyway, in thinking that of course then this object was Venus. In my own earlier speculations upon this subject, this one datum stood out so that had it not been for other data, I'd have abandoned the subject. But then I read, of other occurrences: time after time has something been seen in a local sky of this earth, sometimes so definitely seen to move, not like Venus, but in various directions, that one has to think that it was not Venus, though appearing at the time of visibility of Venus. Between these appearances and visibility of Venus there does seem to be relation.

In the Journal, it is said that L'Astre Cherbourg had an apparent diameter of 15 centimeters, and a less definite margin of 75 centimeters--seemed to be about a yard wide--meaningless of course. In the Bull. Soc. Astro. de France, it is said that, according to reports, its form was oval. In the journal des Debats, we are told that at first the thing was supposed to be a captive balloon but that this idea was given up because it appeared and disappeared.

Journal des Debats, April 12:

That every evening the luminous object was continuing to appear above Cherbourg; that many explanations had been thought of: by some persons that it was the planet Jupiter, and by others that it was a comet but that no one knew what it was. The comet-explanation is of course ruled out. The writer in the journal expresses regret that neither the Meteorological Bureau nor the Observatory of Paris had sent anybody to investigate, but says that the prefet maritime of Cherbourg had commissioned a naval officer to investigate. In Le Temps, of the 12th, is published an interview with Flammarion, who complains some more against general inenarrable-ness, and says that of course the object was Venus. The writer in Le Temps says that soon would the matter be settled, because the commander of a warship had undertaken to decide what the luminous body was. Le Figaro, April 13:

The report of Commander de Kerillis, of the Chasseloup-Laubut--that the position of L'Astre Cherbourg was not the position of Venus, and that the disc did not look like the crescentic disc of Venus, but that the observations had been made from a vessel, under unfavorable conditions, and that the commander and his colleagues did not offer a final opinion.

I think that there was inenarrable-ness all around. Given visibility, I can't think what the unfavorable conditions could have been. Given, however, observations upon something that all the astronomers in the world would say could not be, one does think of the dislike of a naval officer, who, though he probably knew right ascension from declination, was himself no astronomer, to commit himself. In Le Temps, and other newspapers published in Paris, it is said that, according to the naval officers, the object might have been a comet, but that they would not positively commit themselves to this opinion, either.

I think that somebody should be brave; so, though not positively, of course, I incline, myself, to relate these appearances over Cherbourg with the observations in Wales, upon March 29th; also I suggest that there is another report that may relate. In Le Temps, April 12, it is said that, at midnight, April 9-10, a luminous body, like L'Astre Cherbourg, was seen in the sky of Tunis. Though it was visible several minutes, it is said that this object was probably a meteor.

Every night, from the first to the eleventh of April, a luminous body appeared in the sky of Cherbourg. Then it was seen no longer. It may have been seen sailing away, upon its final departure from the sky of Cherbourg. In Le Figaro, April 15, it is said that, upon the night of the eleventh of April, the guards of La Blanche Lighthouse had seen something like a lighted balloon in the sky. Supposing it was a balloon, they had started to signal to it, but it had disappeared. It is said that the lighthouse had been out of communication with the mainland, and that the guards had not heard of L'Astre Cherbourg.

In the London Times, Nov. 23, 1905, a correspondent writes that, at East Liss, Hants, which is about 40 miles from Reading, he and his gamekeeper had, about 3:30 P.M., Nov. 17th, heard a loud, distant rumbling. According to this hearer, the rumbling seemed to be a composition of triplets of sounds. We shall accept that three sounds were heard, but we have no other assertion that each sound was itself so sub-serialized. This correspondent's gamekeeper said that he had heard similar sounds at 11:30 A.M., and at 1:30 P.M. It is said that the sounds were not like gunfire, and that the direction from which they seemed to come, and the time in the afternoon, precluded the explanation of artillery-practice at Aldershot or Portsmouth. Aldershot is about 15 miles from East Liss, and Portsmouth about 20.

Times, November 24--that the "quake" had been distinctly felt in Reading, about 3:30 P.M., November 17th. Times, November 25--heard at Reading, at 11:30, 1:30, and 3:30 o'clock, November 17th.

Reading Standard, November 25:

That consternation had been caused in Reading, upon the 17th, by sounds and vibrations of the earth, about 11:30 A.M., 1:30 P.M., and 3:30 P.M. It is said that nothing had been seen, but that the sounds closely resembled those that had been heard during the meteoric shower of 1866.

Mr. H. G. Fordham appears again. In the Times, December 1, he writes that the phenomena pointed clearly to an explosion in the sky, and not to an earthquake of subterranean origin. "The noise and shock experienced are no doubt attributable to the explosion (or to more than one explosion) of a meteorite, or bolide, high up in the atmosphere, and setting up a wave (or waves) of sound and aerial shock. It is probable, indeed, that a good many phenomena having this source are wrongly ascribed to slight and local earth-shock."

Mr. Fordham wrote this, but he wrote no more, and I think that somewhere else something else was written, and that, in the year 1905, it had to be obeyed; and that it may be interpreted in these words--"Thou shalt not." Mr. Fordham did not inquire into the reasonableness of thinking that, only by coincidence, meteors so successively exploded, in a period of four hours, in one local sky of this earth, and nowhere else; and into the inference, then, as to whether this earth is stationary or not.

We have data of a succession occupying far more than four hours.

In the Times, Mrs. Lane, of Petersfield, 20 miles from Portsmouth, writes that, at 11:30 A.M., and at 3:30 P.M., several days before the 17th, she had heard the detonations, then hearing them again, upon the 17th. Mrs. Lane thinks that there must have been artillery-practice at Portsmouth. It seems clear that there was no cannonading anywhere in England, at this time. It seems clear that there was signaling from some other world.

In the English Mechanic, 82-433, Joseph Clark writes that, a few minutes past 3 P.M., upon the 18th a triplet of detonations was heard at Somerset--"as loud as thunder, but not exactly like thunder."

Reading Observer, November 25--that, according to a correspondent, the sounds had been heard again, at Whitechurch (20 miles from Reading) upon the 21st, at 1:35 P.M., and 3:08 P.M. The sounds had been attributed to artillery-practice at Aldershot, but the correspondent had written to the artillery commandant, at Bulford Camp, and had received word that there had been no heavy firing at the times of his inquiry. The Editor of the Observer says that he, too, had written to the commandant, and had received the same answer.

I have searched widely. I have found record of nobody's supposition that he had traced these detonations to origin upon this earth.

33

IN Coconino County, Arizona, is an extraordinary formation. It is known as Coon Butte and as Crater Mountain. Once upon a time, something gouged this part of Arizona. The cavity in the ground is about 3,800 feet in diameter, and it is approximately 600 feet deeps from the rim of the ramparts to the floor of the interior. Out from this cavity had been hurled blocks of limestone, some of them a mile or so away, some of these masses weighing probably 5,000 tons each. And in the formation, and around it, have been found either extraordinary numbers of meteorites, or fragments of one super-meteorite. Barringer, in his report to the Academy of Natural Science of Philadelphia (Proceedings, A. N. S. P., December, 1905) says that, of the traffickers in this meteoritic material, he knew of two men who had shipped away fifteen tons of it. But Barringer's minimum estimate of a body large enough so to gouge the ground is ten million tons.

It was supposed that a main mass of meteoritic material was buried under the floor of the formation, but this floor was drilled, and nothing was found to support this supposition. One drill went down 1,020 feet, going through too feet of red sandstone, which seems to be the natural, undisturbed substructure. The datum that opposes most strongly the idea that this pit was gouged by one super-meteorite is that in it and around it at least three kinds of meteorites have been found: they are irons, masses of iron-shale, and shale-balls that are so rounded and individualized that they cannot be thought of as fragments of a greater body, and cannot be very well thought of as great drops of molten matter cast from a main, incandescent mass, inasmuch as there is not a trace of igneous rock such as would mark such contact.

There are data for thinking that these three kinds of objects fell at different times, presumably from origin of fixed position relatively to this point in Arizona. Within the formation, shales were found, buried at various distances, as if they had fallen at different times, for instance seven of them in a vertical line, the deepest-buried 27 feet down; also shales outside the formation were found buried. But, quite as if they had fallen more recently, the hundreds of irons were found upon the surface of the ground, or partly covered, or wholly covered, but only with superficial soil.

There is no knowing when this great gouge occurred, but cedars upon the rim are said to be about 700 years old.

In terms of our general expression upon differences of potential, and of electric relations between nearby worlds, I think of a blast between this earth and a land somewhere else, and of something that was more than a cyclone that gouged this pit.

Other meteorites have been found in Arizona: the 85-pound iron that was found at Weaver, near Wickenburg, 130 miles from Crater Mountain, in 1898, and the 960- pound mass, now in the National Museum, said to have been found at Peach Springs, 140 miles from Crater Mountain. These two irons indicate nothing in particular; but, if we accept that somewhere else in Arizona there is another deposit of meteorites, also extraordinarily abundant, such abundance gives something of commonness of nature if not of commonness of origin to two deposits. There are several large irons known as the Tucson meteorites, one weighing 632 pounds and another 1,514 pounds, now in museums. They came from a place known as Iron Valley, in the Santa Rita Mountains, about 30 miles south of Tucson, and about 200 miles from Crater Mountain. Iron Valley was so named because of the great number of meteorites found in it. According to the people of Tucson, this fall occurred about the year 1660. See Amer. Jour. Sci., 2-13-290.

Upon June 24, 1905, Barringer found, upon the plain, about a mile and a half northwest of Crater Mountain, a meteorite of a fourth kind. It was a meteoritic stone, "as different from all the other specimens as one specimen could be from another." Barringer thinks that it fell, about the 15th of January, 1904. Upon a night in the middle of January, 1904, two of his employees were awakened by a loud hissing sound, and saw a meteor falling north of the formation. At the same time, two Arizona physicians, north of the formation, saw the meteor falling south of them. For analysis and description of this object, see Amer. Jour. Sci., 4-21-353. Barringer, who believes that once upon a time one super-meteorite, of which only a very small part has ever been found, gouged this hole in the ground, writes--"That a small stony meteorite should have fallen on almost exactly the same spot on this earth's surface as the great Canon Diablo iron meteorite fell many centuries ago, is certainly a most remarkable coincidence. I have stated the facts as accurately as possible, and I have no opinion to offer, as to whether or not these involve anything more than a coincidence."

Other phenomena in Arizona:

Upon Feb. 24, 1897, a great explosion was heard over the town of Tombstone. It is said that a fragment of a meteor fell at St. David (Monthly Weather Review, 1897- 56). Yarnell, Arizona, Sept. 12, 1898--"a loud, deep, thundering noise" that was heard between noon and 1 P.M. "The noise proceeded from the Granite Range, this side of Prescott. From all accounts, a large meteor struck the earth at this time" (U. S. Weather Bureau Rept., Ariz. Section, September, 1898).Upon July 19, 1912, at Holbrook, Arizona, about 50 miles from Crater Mountain, occurred a loud detonation and one of the most remarkable falls of stones recorded. See Amer. Jour. Sci., 4-34-437. Some of the stones are very small. About 14,000 were collected. Only twice, since the year 1800, have stones in greater numbers fallen from the sky to this earth, according to conventional records.

About a month later (August 18) there was another concussion at Holbrook. This was said to be an earthquake (Bull. Seis. Soc. Amer., 1-209).

34

THE climacteric opposition of Mars, of 1909--the last in our records--the next will be in 1924--Aug. 8, 1909--see Quar. Jour. Met. Soc., n.s., 35-299--flashes in a clear sky that were seen in Epsom, Surrey, and other places in the southeast of England. They could not be attributed to lightning in England. The writer in the Journal finds that there was a storm in France, more than one hundred miles away. For an account of these flashes, tabulated at Epsom--"night fine and starlight"--see Symons' Met. Mag., 44-148. During each period of five minutes, from 10 to 11:15 P.m., the number of flashes-16-14-20-31-15-26-12-20-30-18-27-22-14-12-10-21-8-5-3-1-0-1-0.

With such a time-basis, I can see no possibility of detecting anything of a code- like significance. I do see development. There were similar observations at times in the favorable oppositions of Mars of 1875 and 1877. In 1892, such flashes were noted more particularly. Now we have them noted and tabulated, but upon a basis that could be of interest only to meteorologists. If they shall be seen in 1924, we may have observation, tabulation, and some marvelously different translations of them.

After that there will be some intolerably similar translations, suspiciously delayed in publication. Sept. 23, 1909--opposition of Mars.

Throughout our data, we have noticed successions of appearances in local skies of this earth, that indicate that this earth is stationary, but that also relate to nearest approaches of Mars. Upon the night of Dec. 16-17, 1896, concussion after concussion was felt at Worcester, England; a great "meteor" was seen at the time of the greatest concussion. Mars was seven days past opposition. We thought it likely enough that explosion after explosion had occurred over Worcester, and that something in the sky had been seen only at the time of the greatest, or the nearest, explosion. We did not think well of the conventional explanation that only by coincidence had a great meteor exploded over a region where a series of earthquakes was occurring, and exactly at the moment of the greatest of these shocks.

In November, 1911, Mars was completing its cycle of changing proximities of a duration of fifteen years, and was duplicating the relationship of the year 1896. About to o'clock, night of Nov. 16, 1911, a concussion that is conventionally said to have been an earthquake occurred in Germany and Switzerland. But plainly there was an explosion in the sky. In the Bulletin of the Seismological Society of America, 3-189, Count Montessus de Ballore writes that he had examined 112 reports upon flashes and other luminous appearances in the sky that had preceded the "earthquake" by a few seconds. He concludes that a great meteor had only happened to explode over a region where, a few seconds later, there was going to be an earthquake. "It therefore seems highly probable that the earthquake coincided with a fall of meteors or of shooting stars."

The duplication of the circumstances of December, 1896, continues. If of course this concussion in Germany and Switzerland was the effect of something that exploded in the sky--of what were the concussions that were felt later, the effects? De Ballore does not mention anything that occurred later. But, a few minutes past midnight, and then again, at 3 o'clock, morning of the 17th, there were other, but slighter, shocks. Only at the time of the greatest shock was something seen in the sky. Nature, 88-117--that this succession of phenomena did occur. We relate the phenomena to the planet Mars, but also we ask--how, if most reasonably, all three of these shocks were concussions from explosions in the sky, if of course one of them was, meteors could ever so hound one small region upon a moving earth, or projectiles be fired with such specialization and preciseness?

Nov. 17th, 1911, was seven days before the opposition of Mars. Though the opposition occurred upon the 24th of November, Mars was at minimum distance upon the 17th.

No matter how difficult of acceptance our own notions may be, they are opposed by this barbarism, or puerility, or pill that can't be digested:

Seven days from the opposition of Mars, in 1896, a great meteor [p. 506]

exploded over a region where there had been a succession of earthquakes--by coincidence;

Seven days from the next similar opposition of Mars, a great meteor exploded over a region where there was going to be a succession of earthquakes--by coincidence.

The Advantagerians of the moon--that is the cult of lunar communicationists, who try to take advantage of such celestial events as oppositions and eclipses, thinking that astronomers, or night watchmen, or policemen of this earth might at such times look up at the sky.

A great luminous object, or a meteor, that was seen at the time of the eclipse of June 28, 1908--"as if to make the date of the eclipse more memorable," says W. F. Denning (Observatory, 31-288).

Not long before the opposition of Mars, in 1909, the bright spot west of Picard was seen twice: March 26 and May 23 (Jour. B. A. A., 19-376).

Nov. 16, 1910--an eclipse of the moon, and a "meteor" that appeared, almost at the moment of totality (Eng. Mec., 92-430). It is reported, in Nature, 85-118, as seen by Madame de Robeck, at Naas, Ireland, "from an apparent radiant, just below the eclipsed moon." The thing may have come from the moon. Seemingly with the same origin, it was seen far away in France. In La Nature, Nov. 26, 1910, it is said that, at Besancon, France, during the eclipse, was seen a meteor like a superb rocket, "qui serait partie de la lune." There may have been something occurring upon the moon at the time. In the Jour. B. A. A., 21-100, it is said that Mrs. Albright had seen a luminous point upon the moon throughout the eclipse. Our expression is that there is an association between reported objects, like extra-mundane visitors, and nearest approaches by the planet Venus to this earth. Perhaps unfortunately this is our expression, because it makes for more restriction than we intend. The objects, or the voyagers, have often been seen during the few hours of the visibility of Venus, when the planet is nearest. "Then such an object is Venus," say the astronomers. If anybody wonders why, if these seeming navigators can come close to this earth--as they do approach, if they appear only in a local sky--they do not then come all the way to this earth, let him ask a sea captain why said captain never purposely descends to the bottom of the ocean, though traveling often not far away. However, I conceive of a great variety of extra-mundanians, and I am now collecting data for a future expression--that some kinds of beings from outer space can adapt to our conditions, which may be like the bottom of a sea, and have been seen, but have been supposed to be psychic phenomena.

Upon Oct. 31, 1908, the planet Venus was four months past inferior conjunction, and so had moved far from nearest approach, but there are vague stories of strange objects that had been seen in the skies of this earth--localized in New England-- back to the time of nearest approach. In the New York Sun, Nov. 1, 1908, is published a dispatch, from Boston, dated Oct. 31. It is said that, near Bridgewater, at four o'clock in the morning of October 31, two men had seen a spectacle in the sky. The men were not astronomers. They were undertakers. There may be a disposition to think that these observers were not in their own field of greatest expertness, and to think that we are not very exacting as to the sources of our data. But we have to depend upon undertakers, for instance: early in our investigations, we learned that the prestige of astronomers has been built upon their high moral character, all of them most excellently going to bed soon after sunset, so as to get up early and write all day upon astronomical subjects. But the exemplary in one respect may not lead to much advancement in some other respect. Our undertakers saw, in the sky, something like a searchlight. It played down upon this earth, as if directed by an investigator, and then it flashed upward. "All of the balloons in which ascensions are made, in this State, were accounted for today, and a search through southeastern Massachusetts failed to reveal any further trace of the supposed airship." It is said that "mysterious bright lights," believed to have come from a balloon, had been reported from many places in New England. The week before, persons at Ware had said that they had seen an illuminated balloon passing over the town, early in the morning. During the summer such reports had come from Bristol, Conn., and later from Pittsfield, Mass., and from White River Junction, Vt. "In all these cases, however, no balloon could be found, all the known airships being accounted for." In the New York Sun, Dec. 13, 1909, it is said that, during the autumn of 1908, reports had come from different places in Connecticut, upon a mysterious light that moved rapidly in the sky.

Venus moved on, traveling around the sun, which was revolving around this earth, or traveling any way to suit anybody. In December, 1909, the planet was again approaching this earth. So close was Venus to this earth that, upon the 15th of December, 1909, crowds stood, at noon, in the streets of Rome, watching it, or her (New York Sun, December 16). At 3 o'clock, afternoon of December 24th crowds stood in the streets of New York, watching Venus (New York Tribune, December 25). One supposes that upon these occasions Venus may have been within several thousand miles of this earth. At any rate I have never heard of one fairly good reason for supposing otherwise. If again something appeared in local skies of this earth, or in the skies of New England, and sometimes during the few hours of the visibility of Venus, the object was or was not Venus, all according to the details of various descriptions, and the credibility of the details. The searchlight, for instance; more than one light; directions and motions. Venus, at the time, was for several hours after sunset, slowly descending in the southwest: primary maximum brilliance Jan. 8th, 1910; inferior conjunction February 12th.

There is an amusing befuddlement to clear away first. Upon the night of Sept. 8, 1909, a luminous object had been seen sailing over New England, and sounds from it, like sounds from a motor, had been heard. Then Mr. Wallace Tillinghast, of Worcester, Mass., announced that this light had been a lamp in his "secret aeroplane," and that upon this night he had traveled, in said "secret aeroplane," from Boston to New York, and back to Boston. At this time the longest recorded flight, in an aeroplane, was Farman's, of 111 miles, from Rheims, August, 1909; and, in the United States, according to records, it was not until May 29, 1910, that Curtiss flew from Albany to New York City, making one stop in the 150 miles, however. So this unrecorded flight made some stir in the newspapers. Mr. Tillinghast meant his story humorously of course. I mention it because, if anybody should look the matter up, he will find the yarn involved in the newspaper accounts. If nothing else had been seen, Mr. Tillinghast might still tell his story, and explain why he never did anything with his astonishing "secret aeroplane"; but something else was seen, and upon one of the nights in which it appeared, Tillinghast was known to be in his home.

According to the New York Tribune, Dec. 21, 1909, Immigration Inspector Hoe, of Boston, had reported having seen, at one o'clock in the morning of December 20, "a bright light passing over the harbor" and had concluded that he had seen an airship of some kind.

New York Tribune, December 23--that a "mysterious airship" had appeared over the town of Worcester, Mass., "sweeping the heavens with a searchlight of tremendous power." It had come from the southeast, and traveled northwest, then hovering over the city, disappearing in the direction of Marlboro. Two hours later, it returned. "Thousands thronged the streets, watching the mysterious visitor." Again it hovered, then moving away, heading first to the south and then to the east.

The next night, something was seen, at 6 o'clock, at Boston. "The searchlights shot across the sky line." "As it flew away to the north, queries began to pour into the newspaper offices and the police stations, regarding the remarkable visitation." It is said that an hour and a half later, an object that was supposed to be an airship with a powerful searchlight, appeared in the sky, at Willimantic, Conn., "hovering" over the town about 15 minutes. In the New York Sun, December 24, are more details. It is said that, at Willimantic, had been seen a large searchlight, approaching from the east, and that then dark outlines of something behind the searchlight had been seen. Also, in the Sun, it is said that whatever it may have been that was seen at Boston, it was a dark object, with several red lights and a searchlight, approaching Boston from the west, hovering for 10 minutes, and then moving away westward. From Lynn, Mass., it was described as "a long black object," moving in the direction of Salem, and then returning, "at a high speed." It is said that the object had been seen at Marlboro, Mass., nine times since December 14.

New York Tribune, Jan. 1, 1910--dispatch from Huntington, West Virginia, Dec. 31, 1909--"Three huge lights of almost uniform dimensions appeared in the early morning sky, in this neighborhood, today. Joseph Green, a farmer, declared that they were meteors, which fell on his farm. An extensive search of his land by others who saw the lights was fruitless, and many persons believe that an airship had sped over the country."

In the Tribune, Jan. 13, 1910, it is said that, at 9 o'clock, morning of January 12, an airship had been seen at Chattanooga, Tenn. "Thousands saw the craft, and heard the 'chug' of its engine." Later the object was reported from Huntsville, Alabama. New York Tribune, January 15--dispatch from Chattanooga, January 14--"For the third successive day, a mysterious white aircraft passed over Chattanooga, about noon today. It came from the north, and was traveling southeast, disappearing over Missionary Ridge. On Wednesday, it came south, and on Thursday, it returned north."

In the middle of December, 1909, someone had won a prize for sailing in a dirigible from St. Cyr to the Eiffel Tower and back.

St. Cyr is several miles from Paris.

Huntsville, Ala., and Chattanooga, Tenn., are 75 miles apart.

An association between the planet Venus and "mysterious visitors" either illumines or haunts our data. In the New York Tribune, Jan. 29, 1910, it is said that a luminous object, thought to be Winnecke's comet, had been seen, January 28, near Venus; reported from the Manila Observatory.

I have another datum that perhaps belongs to this series of events. Every night, from the 14th to the 23rd of December, 1909, if we accept the account from Marlboro, a luminous object was seen traveling, or exploring, in the sky of New England. Certainly enough it was no "secret airship" of this earth, unless its navigator went to extremes with the notion that the best way to kept a secret is to announce it with red lights and a searchlight. However, our acceptance depends upon general data as to the development of terrestrial aeronautics. But upon the night of December 24th, the object was not seen in New England, and it may have been traveling or exploring somewhere else. Night of the 24th--Venus in the southwest in the early hours of the evening. In the English Mechanic, 104-71, a correspondent, who signs himself "Rigel," writes that, upon December 24, at 8:30 o'clock in the evening, he saw a luminous object appear above the northeastern horizon and slowly move southward, until 8.50 o'clock, then turning around, retracing, and disappearing whence it came, at two minutes past nine. The correspondent is James Fergusen, Rossbrien, Limerick, Ireland. He writes frequently upon astronomical and meteorological subjects, and is still contributing to the somewhat enlightened columns of the English Mechanic.

Nov. 19, 1912--explosive sounds reported from Sunninghill, Berkshire. No earthquake was recorded at the Kew Observatory, and, in the opinion of W. F. Denning (Nature, 9-363, 417) the explosion was in the sky. It was a terrific explosion, according to the Westminster Gazette (November 19). There was either one great explosion that rumbled and echoed for five minutes, or there were repeated detonations, resembling cannonading--"like a tremendous discharge of big guns" according to reports from Abingdon, Lewes, and Epsom. Sunninghill is about ten miles from Reading, and Abingdon is near Reading, but the sound was heard in London, and down by the English Channel, and even in the island of Alderney. In the Gazette, November 28, Sir George Fordham (H. G. Fordham) writes that, in his opinion, it was an explosion in the sky. He says--"The phenomena of airshock never have, I believe, been very fully investigated." His admissions and his omissions remain the same as they have been since occurrences of the year 1889. He does not mention that, according to Philip T. Kenway, of Hambledon, near Godalming, about thirty miles southeast of Reading, the sounds were heard again the next day, from 1:45 to 2 P.M. Mr. Kenway thinks that there had been big-gun firing at Portsmouth (West. Gaz., November 21). In the London Standard, a correspondent, writing from Dorking, say that the phenomena of the 19th were like concussions from cannonading--"at regular intervals"--"at quick intervals, lasting some seconds each time, for five minutes, by the clock."

It develops that Reading was the center over which the detonations occurred. In the Westminster Gazette, November 30, it is said that the shocks had been felt in Reading, upon the 19th, 20th, and 21st. Only from Reading have I record of phenomena upon the 21st. Mr. H. L. Hawkins, Lecturer in Geology, of the Reading University, writes that according to his investigations there had been no gun-firing in England, to which the detonations could be attributed. He says that Fordham's explanation was in accord with his own investigations, or that the detonations had occurred in the sky. He writes that, inasmuch as the detonations had occurred upon three successive days, a shower of meteors, of long duration, would have to be supposed. How he ever visualized that unerring shower, striking one point over this earth's surface, and nowhere else, day after day, if this earth be a rotating and revolving body, I cannot see. If he should say that by coincidence this repetition could occur, then by what coincidence of coincidences could the same repetitions have occurred in this same local sky, centering around Reading, seven years before? The indications are that this earth is stationary, no matter how unreasonable that may sound.

In the Westminster Gazette, December 9, W. F. Denning writes that without doubt the phenomena were "meteoric explosions." But he alludes to the "airquake and strange noises" that were heard upon the 19th. He does not mention the detonations that were heard upon the following days. Not one of these writers mentions the sounds that were heard in Reading, in November, 1905.

London Standard, Nov. 23, 1912--that, according to Lieut. Col. Trewman, of Reading, the sounds had been heard at Reading, at 9 A.M., upon the 19th; 1:45 P.M., the 10th; 3:30 P.M., the 21st.

35

  "UNKNOWN Aircraft Over Dover."

According to the Dover correspondent to the London Times (Jan. 6, 1913) something had been seen, over Dover, heading from the sea.

In the London Standard, Jan. 24, 1913, it is said that, upon the morning of January 4, an unknown airship had been seen, over Dover, and that, about the same time, the lights of an airship had been seen over the Bristol Channel. These places are several hundred miles apart.

London Times, January 21--report by Capt. Lindsay, Chief Constable of Glamorganshire: that, about five o'clock, in the afternoon of January 17, he saw an object in the sky of Cardiff, Wales. He says that he called the attention of a bystander, who agreed with him that it was a large object. "It was much larger than the Willows airship, and left in its trail a dense smoke. It disappeared quickly." The next day, according to the Times, there were other reports: people in Cardiff saw something that was lighted or that carried lights, moving rapidly in the sky. In the Times, of the 28th, it is said that an airship that carried a brilliant light had been seen in Liverpool. "It is stated at the Liverpool Aviation School that none of the airmen had been out on Saturday night." Dispatches from town after town--a traveling thing in the sky, carrying a light, and also a searchlight that swept the ground. It is said that a vessel, of which the outlines had been clearly seen, had appeared in the sky of Cardiff, Newport, Neath, and other places in Wales. In the Standard, January 31, is published a list of cities where the object had been seen. Here a writer tries to conclude that some foreign airship had made half a dozen visits to England and Wales, or had come once, remaining three weeks; but he gives up the attempt, thinking that nothing could have reached England and have sailed away half a dozen times without being seen to cross the coast; thinking that the idea of anything having made one journey, and remaining three weeks in the air deserved no consideration.

If the unknown object did carry something like a searchlight, an idea of its powers is given in an account in the Cardiff Evening Express, Jan. 25, 1913--"Last evening brilliant lights were seen, sweeping skyward, and now, this evening, the lights grow bolder. Streets and houses in the locality of Totterdown were suddenly illuminated by a brilliant, piercing light, which, sweeping upward, gave many spectators a fine view of the hills beyond." In the Express, February 6, is a report upon this light like a searchlight, and the object that flashed it, by the police of Dulais Valley. Also there is an account, by a police sergeant, of a luminous thing that was for a while stationary in the sky, and then moved away. Still does the conventional explanation, or suggestion, survive. It is said that members of the staff of the Evening Express had gone to the roof of the newspaper building, but had seen only the planet Venus, which was brilliant at this time.

Then writes a correspondent, to the Express, that the object could not have been Venus, because he had seen it traveling at a rate of 20 or 30 miles an hour, and had heard sounds from it. Someone else writes that not possibly could the thing be Venus: he had seen it as "a bright red light, going very fast." Still someone else says that he had seen the seeming vessel upon the 5th of February, and that it had suddenly disappeared.

There is a hiatus. Between the 5th and the 21st of February, nothing like an airship was seen in the sky of England and Wales. If we can find that somewhere else something similar was seen in the sky, in this period, one supposes that it was the same object, exploring or maneuvering somewhere else. It seems however that there were several of these objects, because of simultaneous observations at places far apart. If we can find that, during the absence from England and Wales, similar objects were seen somewhere else, a great deal of what we try to think upon the subject will depend upon how far from Great Britain they were seen. It seems incredible that the planet Venus should deceive thousands of Britons, up to the 5th of February, and stop her deceptions abruptly upon that date, and then abruptly resume deceptions upon the 21st, in places at a distance apart. These circumstances oppose the idea of collective hallucinations, by which some writers in the newspapers tried to explain. If they were hallucinations, the hallucinations renewed collectively, upon the 21st, in towns one hundred miles apart. One extraordinary association is that all appearances, except the first, were in the hours of visibility of Venus, then an "evening star."

Upon the night of the 21st, a luminous object was reported from towns in Yorkshire and from towns in Warwickshire, two regions about one hundred miles apart; about 10 P.M. All former attempts to explain had been abandoned, and the general supposition was that airships were maneuvering over England. But not a thing had been seen to cross the coast of England, though guards were patroling the coasts, especially commissioned to watch for foreign airships. Sailors in the North Sea, and people in Holland and Belgium had seen nothing that could be thought a German airship sailing to or from England. A writer in Flight takes up as especially mysterious the appearance far inland, in Warwickshire. Then came reports from Portsmouth, Ipswich, Hornsea, and Hull, but, one notes, no more, at this time, from Wales.

Also in Ipswich, which is more than a hundred miles from the towns in Warwickshire, and more than a hundred miles from the Yorkshire towns, a luminous object was seen upon the night of the 21st. Ipswich Evening Star, February 25-- something that carried a searchlight that had been seen upon the nights of the 21st and 24th, moving in various directions, and then "dashing off at lightning speed"--that, at Hunstanton, had been seen three bright lights traveling from the eastern sky, remaining in sight 30 minutes, stationary, or hovering over the town, and then disappearing in the northwest. Portsmouth Evening News, February 25--that soon after 8 P.M., evening of the 24th, had been seen a very bright light, appearing and disappearing, remaining over Portsmouth about one hour, and then moving away. Portsmouth and Ipswich are about 120 miles apart. In the London newspapers, it is said that, upon the evening of the 25th, crowds stood in the streets of Hull, watching something in the sky, "the lights of which were easily distinguishable." Hull is about 190 miles northeast of Portsmouth. Hull Daily Mail, February 26--that a crowd had watched a light high in the air. It is said that the light had been stationary for almost half an hour and had then shot away northward. In the Times, February 28, are published reports upon "the clear outlines of an airship, which was carrying a dazzling searchlight," from Portland, Burcleaves, St. Alban's Head, Papplewich, and the Orkneys. The last account, after a long interval, that I know of, is another report from Capt. Lindsay: that, about 9 o'clock, evening of April 8th, he and many other persons had seen, over Cardiff, something that carried a brilliant light and traveled at a rate of sixty or seventy miles an hour.

Upon April 24, 1913, the planet Venus was at inferior conjunction.

In the Times, February 28, it is said that a fire-balloon had been found in Yorkshire, and it is suggested that someone had been sending up fire-balloons.

England were as credulous as the people of Cherbourg, and had permitted themselves to be deceived by the planet Venus.

If German airships were maneuvering over England, without being seen either approaching or departing, appearing sometimes far inland in England without being seen to cross the well-guarded coasts, it was secret maneuvering, inasmuch as the accusation was denied in Germany (Times, February 26 and 27). It was then one of the most brilliantly proclaimed of secrets, or it was concealment under one of the most powerful searchlights ever seen. Possibly an airship from Germany could appear over such a city as Hull, upon the east coast of England, without being seen to arrive or to depart, but so far from Germany is Portsmouth, for instance, that one does feel that something else will have to be thought of. The appearances over Liverpool and over towns in Wales might be attributed to German airships by someone who has not seen a map since he left school. There were more observations upon sudden appearances and disappearances than I have recorded: stationariness often occurred.

The objects were absent from the sky of Great Britain, from February 5 to February 21.

According to data published by Prof. Chant, in the Journal of the Royal Astronomical Society of Canada, 7-148, the most extraordinary procession in our records was seen, in the sky of Canada, upon the night of Feb. 9, 1913. Either groups of meteors, in one straight line, passed over the city of Toronto, or there was a procession of unknown objects, carrying lights. According to Prof. Chant, the spectacle was seen from the Saskatchewan to Bermuda, but if this long route was traversed, data do not so . The supposed route was diagonally across New York State, from Buffalo, to a point near New York City, but from New York State are recorded no observations other than might have been upon ordinary meteors, this night. A succession of luminous objects passed over Toronto, night of Feb. 9, 1913, occupying from three to five minutes in passing, according to different estimates. If one will think that they were meteors, at least one will have to think that no such meteors had ever been seen before. In the Journal, 7-405, W. F. Denning writes that, though he had been watching the heavens since the year 1865, he had never seen anything like this. In most of the observations, the procession is described as a whole--"like an express train, lighted at night"--"the lights were at different points, one in front, and a rear light, then a succession of lights in the tail." Almost all of the observations relate to the sky of Toronto and not far from Toronto. It is questionable that the same spectacle was seen in Bermuda, this night. The supposed long flight from the Saskatchewan to Bermuda might indicate something of a meteoric nature, but the meteor-explanation must take into consideration that these objects were so close to this earth that sounds from them were heard, and that, without succumbing to gravitation, they followed the curvature of this earth at a relatively low velocity that cannot compare with the velocity of ordinary meteors.

If now be accepted that again, the next day, objects were seen in the sky of Toronto, but objects unlighted, in the daytime--I suppose that to some minds will come the thought that this is extra-ordinary, and that almost immediately the whole subject will then be forgotten. Prof. Chant says that, according to the Toronto Daily Star, unknown objects, but dark objects this time, were seen at Toronto, in the afternoon of the next day--"not seen clearly enough to determine their nature, but they did not seem to be clouds or birds or smoke, and it was suggested that they were airships cruising over the city." Toronto Daily Star, February 10--"They passed from west to east, in three groups, and then returned west in more scattered formation, about seven or eight in all."

36

AUGUST, 1914--this arena-like earth, with its horizon banking high into a Coliseum, when seen from not too far above--faint, rattling sounds of the opening of boundaries--tawny formations slinking into the arena--their crouchings and seizures and crunchings. Aug. 13, 1914--things that were gathering in the sky.

They were seen by G. W. Atkins, of Elstree, Herts, and were seen again upon the 16th and the 17th (Observatory, 37-358). Sept. 9, 1914--a host in the sky; watched several hours by W. H. Steavenson (Jour. B. A. A., 25-27). There were round appearances, but some of them were shaped like dumb-bells. They were not seeds, snowflakes, insects, nor anything else that they "should" have been, according to Mr. Steavenson. He says that they were large bodies.

Oct. 10, 1914--a ship that was seen in the sky--or "an absolutely black, spindle- shaped object" crossing the sun. It was seen, at Manchester, by Albert Buss (Eng. Mec., 100-236). "Its extraordinarily clear-cut outline was surrounded by a kind of halo, giving the impression of a ship, plowing her way through the sea, throwing up white-foamed waves with her prow."

Mikkelsen (Lost in the Arctic, p. 345):

"During the last few days (October, 1914) we have been much tumbled up and down in our minds, owing to a remarkable occurrence, somewhat in the nature of Robinson Crusoe's encounter with the footprints in the sand. Our advance load has been attacked--an empty petroleum cask is found, riddled with tiny holes, such as would be made by a charge of shot! Now a charge of shot is scarcely likely to materialize out of nowhere; one is accustomed to associate the phenomenon with the presence of human beings. It is none of our doing--then whose doing is it? We hit upon the wildest theories to account for it, as we sit in the tent, turning the mysterious object over and over. No beast of our acquaintance could make all those little round holes: what animal could even open its jaws so wide? And why should anybody take the trouble to make a target of our gear? Are there Eskimos about--Eskimos with guns? There are no footprints to be seen: it could scarcely have been an animal--the whole thing is highly mysterious."

Jan. 31, 1915--a symbolic-looking formation upon the moon--six or seven white spots, in Littrow, arranged like the Greek letter Gamma (Eng. Mec., 101-47).

Feb. 13, 1915--Steep Island, Chusan Archipelago--a lighthouse-keeper complained to Capt. W. F. Tyler, R.N., that a British warship had fired a projectile at the lighthouse. But no vessel had fired a shot, and it is said that the object must have been a meteor (Nature, 97-17).

In the middle of February, 1915, the planet Venus was about two months and a half past inferior conjunction. If objects like navigating constructions were seen in the sky, at this time, there may be an association, but I am turning against that association, feeling that it is harmful to our wider expression that extra-mundane vessels have been seen in the sky of this earth, and that they come from regions at present unknown. New York Tribune, Feb. 15, 1915--that, at 10 P.M., February 14, three aeroplanes had been seen to cross the St. Lawrence river, near Morristown, N. Y., according to reports, but that, in the opinion of the Dominion police, nothing but fire-balloons had been seen. It is said that two "responsible residents" had seen two of the objects cross the river, between 8 and 8:30 P.M., and then return five hours later. In the Canadian Parliament, Sir Wilfred Laurier had said that, at 9 P.M., he had been called up by the Mayor of Brockwell, telling him that three aeroplanes with "powerful searchlights" had crossed the St. Lawrence. The story is told in the New York Herald. Here it is said that, according to the Chief of Police, of Ogdensburg, N.Y., a farmer, living five miles from Ogdensburg, had reported having seen an aeroplane, upon the 12th. Then it is said that the mystery had been solved: that, while celebrating the one hundredth anniversary of peace between the United States and Canada, some young men of Morristown had sent up paper balloons, which had exploded in the sky, after 9 P.M., night of the 14th. New York Times--that the objects had been seen first at Guananoque, Ontario. Here it is said that the balloon-story is absurd. According to the Dominion Observatory, the wind was, at the time, blowing from the east, and the objects had traveled toward the northeast. It is said that one of the objects had, for several minutes, turned a powerful searchlight upon the town of Brockwell.

Upon Dec. 11, 1915, Bernard Thomas, of Glenorchy, Tasmania, saw a "particularly bright spot upon the moon" (Eng. Mec., 103-10). It was on the north shore of the Mare Crisium, and "looked almost like a star." In Dr. Thomas' opinion, it was sunlight reflected from the rim of a small crater. The crater Picard is near the north shore of the Mare Crisium, and most of the illuminations near Picard have occurred several months from an opposition of Mars.

In December, 1915, another new formation upon the moon--reported from the Observatory of Paris--something like a black wall from the center to the ramparts of Aristillus (Bull. Soc. Astro. de France, 30-383).

Jan. 12, 1916--a shock in Cincinnati, Ohio. Buildings were shaken. The quake was from an explosion in the sky. Flashes were seen in the sky. (New York Herald, Jan. 13, 1916.)

Feb. 9, 1916--opposition of Mars.

In the English Mechanic, 104-71, James Ferguson writes that someone had seen, at

11 o'clock, night of July 31, 1916, at Ballinasloe, Ireland, just such a moving thing, or just such a sailing, exploring thing as is now familiar in our records. For fifteen minutes it moved in a northwesterly direction. For three quarters of an hour it was stationary. Then it moved back to the point where first it had been seen, remaining visible until four o'clock in the morning. Whatever this object may have been, it left the sky at about the time that Venus appeared, as a "morning star," in the sky at Ballinasloe, and resembles the occurrence of Sept. 11, 1852, reported by Lord Wrottesley. Inferior conjunction of Venus was upon July 3, 1916. We have noticed that all occurrences that we somewhat reluctantly associate with nearness of Venus associate more with times of greatest brilliance, five weeks before and after inferior conjunction, than with dates of conjunction. Somebody may demonstrate that at these times Venus comes closest to this earth.

Oct. 10, 1916--a reddish shadow that spread over part of the lunar crater Plato; reported from the Observatory of Florence, Italy (Sci. Amer., 121-181).

Nov. 25, 1916--about twenty-five bright flashes, in rapid succession, in the sky of Cardiff, Wales, according to Arthur Mee (Eng. Mec., 104-239).

Col. Markwick writes, in the Jour. B. A. A., 27-188, that, at 6:10 P.M., April 15, 1917, he had seen, upon the sun, a solitary spot, different from all sunspots that he had seen in an experience of forty-three years. Col. Markwick had written to Mr. Maunder, of the Greenwich Observatory, and had been told that, in photographs taken of the sun upon this day, one at 11:17 and another at 11:20 o'clock, there was no sign of a sunspot.

July 4, 1917--an eclipse of the sun, and an extraordinary luminous object said to have been a meteor, in France (Bull. Soc. Astro. de France, 31-299). About 6:20 P.M., this day, there was an explosion over the town of Colby, Wisconsin, and a stone fell from the sky (Science, Sept. 14, 1917).

Aug. 29, 1917--a luminous object that was seen moving upon the moon (Bull. Soc. Astro. de France,

31-439).

Feb. 21, 1919--an intensely black line extending out from the lunar crater Lexall (Eng. Mec., 109-517).

Upon May 19, 1919, while Harry Hawker was at sea, untraceable messages, meaningless in the languages of this earth, were picked up by wireless, according to dispatches to the newspapers. They were interpreted as the letters K U J and V K A J.

In October, 1913, occurred something that may not be so very mysterious because of nearness to the sea. One supposes that if extra-mundane vessels have sometimes come close to this earth, then sailing away, terrestrial aeronauts may have occasionally left this earth, or may have been seized and carried away from this earth. Upon the morning of Oct. 13, 1913, Albert Jewel started to fly in his aeroplane from Hempstead Plains, Long Island, to Staten Island. The route that he expected to take was over Jamaica, Brooklyn, Coney Island, and the Narrows. New York Times, Oct. 14, 1913--"That was the last seen or heard of him ... he has been as completely lost as if he had evaporated into air." But as to the disappearance of Capt. James there are circumstances that do call for especial attention. New York Times, June 2, 1919--that Capt. Mansell R. James was lost somewhere in the Berkshire Hills, upon his flight from Boston to Atlantic City, or, rather, upon the part of his route between Lee, Mass., and Mitchel Field, Long Island. He had left Lee upon May 29th. Over the Berkshires, or in the Berkshires, he had disappeared. According to later dispatches, searching parties had "scoured" the Berkshires, without finding a trace of him. Upon June 4th, army planes arrived and searched systematically. There was general excitement, in this mystery of Capt. James. Rewards were offered; all subscribers of the Southern New England Telephone Company were enlisted in a quest for news of any kind; boy scouts turned out. Up to this date of writing there has been nothing but a confusion of newspaper dispatches: that two children had seen a plane, about thirteen miles north of Long Island Sound; that two men had seen a plane fall into the Hudson River, near Poughkeepsie; that, in a gully of Mount Riga, near Millerton, N. Y., had been found the remains of a plane; that part of a plane had been washed ashore from Long Island Sound, near Branford. The latest interest in the subject that I know of was in the summer of 1921. A heavy object was known to be at the bottom of the Hudson River, near Poughkeepsie, and was thought to be Capt. James' plane. It was dredged up and found to be a log.

For an extraordinary story of windows, in Newark, N. J., that were perforated by unfindable bullets, see New York Evening Telegram, Sept. 19, 1919, and the Newark Evening News. The occurrence is a counterpart of Mikkelsen's experience.

The detonations at Reading were heard seven years apart. Here it is not quite seven years later. London Times, Sept. 26, 1919--that upon September 25, a shock had been felt at Reading; that inquiries had led to information of no known explosion near Reading. In the Times, October 14, Mr. H. L. Hawkins writes that the shock was "quite definitely an earthquake, but its origin was superficial" and that the shock "was transmitted through the earth more than through the air." In the London Daily Chronicle, September 27, Mr. Hawkins, having considered all suggestions that the shock was a subterranean earthquake, had written: "However, as the whole thing terminated in a bump and a big bang, without subsequent shaking of the ground, it points more to an explosion of a natural type up in the air than to a real earthquake." And, in the London Daily Mail, Mr. Hawkins is quoted:

that if the detonation were local, he would believe that it was an aerial explosion ("meteoric"); but, if it were widespread, it would be considered an earthquake. And in the whole series of the Reading phenomena, this violent detonation was most distinctly local to Reading.

Reading Observer, Sept. 27, 1919--"The most probable explanation of the occurrence is that there was an explosion somewhere near enough to affect the town....

Officials at the Greenwich Observatory were unable to throw any light on the matter, and said that their instruments showed no signs of earth-disturbance."

It is said that the sound and shock were violent, and that, in the residential parts of Reading, the streets were crowded with persons discussing the occurrence.

There was a similar shock in Michigan, Nov. 27, 1919. In many cities, persons rushed from their homes, thinking that there had been an earthquake (New York Times, November 28). But, in Indiana, Illinois, and Michigan, a "blinding glare" was seen in the sky. Our acceptance is that this occurrence is, upon a small scale, of the type of many catastrophes in Italy and South America, for instance, when just such "blinding glares" have been seen in the sky, data of which have been suppressed by conventional scientists, or data of which have not Impressed conventional scientists.

English Mechanic, 110-257--J. W. Scholes, of Huddersfield, writes that, upon Dec. 19, 1919, he saw, near the lunar crater Littrow, "a, very conspicuous black-ink mark." Upon page 282, W. J. West, of Gosport, writes that he had seen the mark upon the 7th of December.

March 22, 1920--a light in the sky of this earth, and an illumination upon the moon (Eng. Mec., III-142). That so close to this earth is the moon that illuminations known as "auroral" often affect both this earth and the moon.

July 20 and 21, and Sept. 13, 1920--dull rumbling sounds and quakes at Comrie, Perthshire (London Times, July 23 and Sept. 14, 1920).

According to a dispatch to the Los Angeles Times--clipping sent to me by Mr. L. A. Hopkins, of Chicago--thunder and lightning and heavy rain, at Portland, Oregon, July 21, 1920: objects falling from the sky; glistening, white fragments that looked like "bits of polished china." "The explanation of the local Weather Bureau is that they may have been picked up by a whirlwind and carried to the district where they were found." The objection to this standardized explanation is the homogeneousness of the falling objects. How can one conceive of winds raging over some region covered with the usual great diversity of loose objects and substances, having a liking for little white stones, sorting over maybe a million black ones, green ones, white ones, and red ones, to make the desired selection?

One supposes that a storm brought to this earth fragments of a manufactured object, made of something like china, from some other world.

In the Literary Digest, Sept. 2, 1921, is published a letter from Carl G. Gowman, of Detroit, Michigan, upon the fall from the sky, in southwest China, Nov. 17 (1920?) of a substance that resembled blood. It fell upon three villages close together, and was said to have fallen somewhere else forty miles away. The quantity was great: in one of the villages, the substance "covered the ground completely." Mr. Gowman accepts that this substance did fall from the sky, because it was found upon roofs as well as upon the ground. He rejects the conventional red-dust explanation, because the spots did not dissolve in several subsequent rains. He says that anything like pollen is out of the question, because at the time nothing was in bloom.

Nov. 23, 1920--a correspondent writes, to the English Mechanic; 112-214, that he saw a shaft of light projecting from the moon, or a spot so bright that it appeared to project, from the limb of the moon, in the region of Funerius.

About Jan. 1, 1921--several irregular, black objects that crossed the sun. To the Rev. William Ellison (Eng. Mec., 112-276) they looked like pieces of burnt paper.

July 25, 1921--a loud report, followed by a sharp tremor, and a rumbling sound, at Comrie (London Times, July 27, 1921).

July 31, 1921--a common indication of other lands from which come objects and substances to this earth--but our reluctance to bother with anything so ordinarily marvelous.

Because we have conceived of intenser times and furies of differences of potential between this earth and other worlds: torrents of dinosaurs, in broad volumes that were streaked with lesser animals, pouring from the sky, with a foam of tusks and fangs, enveloped in a bloody vapor that was falsely dramatized by the sun, with rainbow-mockery. Or, in terms of planetary emotions, such an out-pouring was the serenade of some other world to this earth. If poetry is imagery, and if a flow of images be solid poetry, such a recitation was in three-dimensional hyperbole that was probably seen, or overheard, and criticized in Mars, and condemned for its extravagance in Jupiter. Some other world, meeting this earth, ransacking his solid imagination and uttering her living metaphors: singing a flood of mastodons, purring her butterflies, bellowing an ardor of buffaloes. Sailing away--sneaking up close to the planet Venus, murmuring her antelopes, or arching his periphery and spitting horses at her-- Poor, degenerate times--nowadays something comes close to this earth and lisps little commonplaces to her--

July 31, 1921--a shower of little frogs that fell upon Anton Wagner's farm, near Stirling, Conn. (New York Evening World, Aug. I, 1921).

At sunset, Aug. 7, 1921, an unknown luminous object was seen, near the sun, at Mt. Hamilton, by an astronomer, Prof. Campbell, and by one of those who may someday go out and set foot upon regions that are supposed not to be: by an aviator, Capt. Rickenbacker. In the English Mechanic, 114-211, another character in these fluttering vistas of the opening of the coming drama of Extra-geography, Col. Markwick, a conventional astronomer and also a recorder of strange things, lists other observations upon this object, the earliest upon the 6th, by Dr. Emmert, of Detroit. In the English Mechanic, 114-241, H. P. Hollis, once upon a time deliciously "exact" and positive, says something, in commenting upon these observations, that looks like a little weakness in Exclusionism, because the old sureness is turning slightly shaky--"that there are more wonderful things in the sky than we suspect, or that it is easy to be self-deceived."

It is funny to read of an "earthquake," described in technical lingo, and to have a datum that indicates that it was no earthquake at all, in the usual seismologic sense, but a concussion from an explosion in the sky. Aug. 7, 1921--a severe shock at New Canton, Virginia. See Bull. Seis. Soc. Amer., 11-197--Prof. Stephen Taber's explanation that the shock had probably originated in the slate belt of Buckingham County, intensity about V on the R.-F. scale. But then it is said that, according to the "authorities" of the McCormick Observatory, the concussion was from an explosion in the sky. The time is coming when nothing funny will be seen in this subject, if someday be accepted at least parts of the masses of data that I am now holding back, until I can more fully develop them--that some of the greatest catastrophes that have devastated the face of this earth have been concussions from explosions in the sky, so repeating in a local sky weeks at a time, months sometimes, or intermittently for centuries, that fixed origins above the ravaged areas are indicated.

New York Tribune, Sept. 2, 1921:

"J. C. H. Macbeth, London Manager of the Marconi Wireless Telegraph Company, Ltd., told several hundred men, at a luncheon of the Rotary Club, of New York, yesterday, that Signor Marconi believed he had intercepted messages from Mars, during recent atmospheric experiments with wireless on board his yacht Electra, in the Mediterranean. Mr. Macbeth said that Signor Marconi had been unable to conceive of any other explanation of the fact that, during his experiments, he had picked up magnetic wavelengths of 150,000 meters, whereas the maximum length of wave-production in the world today is 14,000 meters. The regularity of the signals, Mr. Macbeth declared, disposed of any assumption that the waves might have been caused by electrical disturbance. The signals were unintelligible, consisting apparently of a code, the speaker said, and the only signal recognized was one resembling the letter V in the Marconi code." See datum of May 19, 1919. But, in the summer of 1921, the planet Mars was far from opposition. The magnetic vibrations may have come from some other world.

They may have had the origin of the sounds that have been heard at regular intervals--

The San Salvadors of the sky--

And we return to the principle that has been our re-enforcement throughout: that existence is infinite serialization, and that, except in particulars, it repeats--

That the dot that spread upon the western horizon of Lisbon, March 4, 1493, cannot be the only ship that comes back from the unknown, cargoed with news--

And it may be September this, nineteen hundred and twenty or thirty something, or February that, nineteen hundred and twenty or thirty something else--and, later, see record of it in Eng. Mec., or Sci. Amer., vol. and p. something or another--a speck in the sky of this earth--the return of somebody from a San Salvador of the sky--and the denial by the heavens themselves, which may answer with explosions the vociferations below them, of false calculations upon their remotenesses. If the heavens do not participate with snow, the skyscrapers will precipitate torn up papers and shirts and skirts, too, when the papers give out.

There will be a procession. Somebody will throw little black pebbles to the crowds. Over his procession will fly blue-fringed cupids. Later he will be insulted and abused and finally hounded to his death. But, in that procession, he will lead by the nose an outrageous thing that should not be: about ten feet long, short-winged, waddling on webbed feet. Insult and abuse and death--he will snap his fingers under the nose of the outrageous thing. It will be worth a great deal to lead that by the nose and demonstrate that such things had been seen in the sky, though they had been supposed to be angels. It will be a great moment for somebody. He will come back to New York, and march up Broadway with his angel.

Some now unheard-of De Soto, of this earth, will see for himself the Father of Cloudbursts.

A Balboa of greatness now known only to himself will stand on a ridge in the sky between two auroral seas.

Fountains of Everlasting Challenge.

Argosies in parallel lines and rabbles of individual adventurers. Well enough may it be said that they are seeds in the sky. Of such are the germs of colonies.

37

THAT the Geo-system is an incubating organism, of which this earth is the nucleus--but an organism that is so strongly characterized by conditions and features of its own that likening it to any object internal to it is the interpreting of a thing in terms of a constituent--so that we think of an organism that is incompletely, or absurdly inadequately, expressible in terms of the egg- like and the larval and other forms of the immature--a geo-nucleated system that is dependent upon its externality as, in one way or another, is every similar, but lesser and included, thing--stimulated by flows of force that are now said to be meteoric, though many so-called "meteoric" streams seem more likely to be electric, that radiate from the umbilical channels of its constellations-- vitalized by its sun, which is itself replenished by the comets, which, coming from external reservoirs of force, impart to the sun their freightages, and, unaffected by gravitation, return to an external existence, some of them even touching the sun, but showing no indication of supposed solar attraction.

In a technical sense we give up the doctrine of Evolution. Ours is an expression upon Super-embryonic Development, in one enclosed system. Ours is an expression upon Design underlying and manifesting in all things within this one system, with a Final Designer left out, because we know of no designing force that is not itself the product of remoter design. In terms of our own experience we cannot think of an ultimate designer, any more than we can think of ultimacy in any other respect. But we are discussing a system that, in our conception, is not a final entity; so then no metaphysical expression upon it is required.

I point out that this expression of ours is not meant for aid and comfort to the reactionaries of the type of Col. W. J. Bryan, for instance: it is not altogether anti-Darwinian: the concept of Development replaces the concept of Evolution, but we accept the process of Selection, not to anything loosely known as Environment, but relatively to underlying Schedule and Design, predetermined and supervised, as it were, but by nothing that we conceive of in anthropomorphic terms.

I define what I mean by dynamic design, in the development of any embryonic thing: a pre-determined, or not accidental, or not irresponsible, passage along a schedule of phases to a climax of unification of many parts. Some of the aspects of this process are the simultaneous varying of parts, with destiny, and not with independence, for their rule, or with future co-ordinations and functions for their goal; and their survival while still incipient, not because they are fittest relatively to contemporaneous environment, so not because of usefulness or advantage in the present, inasmuch as at first they are not only functionless but also discordant with established relations, but surviving because they are in harmony with the dynamic plan of a whole being: and the presence of forces of suppression, or repression, as well as forces of stimulation and protection, so that parts are held back, or are not permitted to develop before their time.

If we accept that these circumstances of embryonic development are the circumstances of all wider development, within one enclosed system, the doctrine of Darwinian Evolution, as applied generally, will, in our minds, have to be replaced by an expression upon Super-embryonic Development, and Darwinism, unmodified, will become to us one more of the insufficiencies of the past. Darwinism concerns itself with the adaptations of the present, and does heed the part that the past has played, but, in Darwinism, there is no place for the influence of the future upon the present. Consider any part of an embryonic thing--the heart of an embryo--and at first it is only a loop. It will survive, and it will be nourished in its functionless incipiency; also it will not be permitted to become a fully developed heart before its scheduled time arrives; its circumstances are dominated by what it will be in the future. The eye of an embryo is a better instance.

Consider anything of a sociologic nature that ever has grown: that there never has been an art, science, religion, invention that was not at first out of accord with established environment, visionary, preposterous in the light of later standards, useless in its incipiency, and resisted by established forces so that, seemingly animating it and protectively underlying it, there may have been something that in spite of its unfitness made it survive for future usefulness. Also there are data for the acceptance that all things, in wider being, are held back as well as protected and prepared for, and not permitted to develop before comes scheduled time. Langley's flying machine makes me think of something of the kind--that this machine was premature; that it appeared a little before the era of aviation upon this earth, and that therefore Langley could not fly. But this machine was capable of flying, because, some years later, Curtis did fly in it. Then one thinks that the Wright Brothers were successful, because they did synchronize with a scheduled time. I have heard that it is questionable that Curtis made no alterations in Langley's machine. There is no lack of instances. One of the greatest of secrets that have eventually been found out was for ages blabbed by all the pots and kettles in the world--but that the secret of the steam engine could not, to the lowliest of intellects, or to supposititiously highest of intellects, more than adumbratively reveal itself until came the time for its co-ordination with the other phenomena and the requirements of the Industrial Age. And coal that was stored in abundance near the surface of the ground--and the needs of dwellers over coal mines, veins of which were often exposed upon the surface of the ground, for fuel--but that this secret, too, obvious, too, could not be revealed until the coming of the Industrial Age. Then the building of factories, the inventing of machines, the digging of coal, and the use of steam, all appearing by simultaneous variation, and co-ordinating, Shores of North America--nowadays, with less hero-worship than formerly, historians tell us that, to English and French fishermen, the coast of Newfoundland was well-known, long before the year 1492; nevertheless, to the world in general, it was not, or, according to our acceptances, could not be, known. About the year 1500, a Portuguese fleet was driven by storms to the coast of Brazil, and returned to Europe. Then one thinks that likely enough, before the year 1492, other vessels had been so swept to the coasts of the western hemisphere, and had returned--but that data of westward lands could not emerge from the suppressions of that era--but that the data did survive, or were preserved for future usefulness--that there are "Thou shalt nots" engraved upon something underlying all things, and then effacing, when phases pass away.

We conceive now of all buildings--within one enclosed system--in terms of embryonic building, and of all histories as local aspects of Super-embryonic Development. Cells of an embryo build falsely and futilely, in the sense that what they construct will be only temporary and will be out of adjustment later. If, however, there are conditions by which successive stages must be traversed before the arrival of maturity, ours is an expression upon the functioning of the false and the futile, in which case these terms, as derogations, should not be applied. We see that the cells that build have no basis of their own; that for their formations there is nothing of reason and necessity of their own, because they flourish in other formations quite as well. We see that they need nothing of basis, nor of guidance of their own, because basis and guidance are of the essence of the whole. All are responses, or correlates, to a succession of commandments, as it were, or of dominant, directing, supervising spirits of different eras: that they take on appearances that are concordant with the general gastrula era, changing when comes the stimulus to agree with the reptilian era, and again responding harmoniously when comes the time of the mammalian era. It is in accordance with our experience that never has human mind, scientific, religious, philosophic, formulated one basic thought, one finally true law, principle, or major premise from which guidance could be deduced. If any thought were true and final it would include the deduced. We conceive that there has been guidance, just the same, if human beings be conceived of as cellular units in one developing organism; and that human minds no more need foundations of their own than need the sub-embryonic cells that build so preposterously, according to standards of later growth, but build as they are guided to build. In this view, human reason is tropism, or response to stimuli, and reasoning is the trial-and-error process of the most primitive unicellular organisms, a susceptibility to underlying mandates, then a groping in perhaps all possible distortions until adjustment with underlying requirements is reached. In this view, then, though there are, for instance, no atoms in the Daltonian sense, if in the service of a building science, the false doctrine of the atoms be needed, the mind that responds, perhaps not to stimulus, but to requirement, which seems to be a negative stimulus, and so conceives, is in adjustment and reaches the state known as success. I accept, myself, that there may be Final Truth, and that it may be attainable, but never in a service that is local or special in any one science or nation or world.

It is our expression that temporary isolations characterize embryonic growth and super-embryonic growth quite as distinctly as do expansions and co-ordinations. Local centers of development in an egg--and they are isolated before they sketch out attempting relations. Or in wider being--hemisphere isolated from hemisphere, and nation from nation--then the breaking down of barriers--the appearance of Japan out of obscurity--threads of a military plasm are cast across an ocean by the United States.

Shafts of light that have pierced the obscurity surrounding planets--and something like a star shines in Aristarchus of the moon. Embryonic heavens that have dreamed--and that their mirages will be realized someday. Sounds and an interval; sounds and the same interval; sounds again--that there is one integrating organism and that we have heard its pulse.

38

FEB. 7, 1922--an explosion "of startling intensity" in the sky of the northwestern point of the London Triangle (Nature, Feb. 23, 1922).

Repeating phenomena in a local sky--in L'Astronomie, 36-201, it is said that, at Orsay (Seine-et-Oise), Feb. 15, 1922, a detonation was heard in the sky, and that 9 hours later a similar sound was heard, and that an illumination was seen in the sky. It is said that, 10 nights later, at Verneuil, in the adjoining province, Oise, a great, fiery mass was seen falling from the sky.

March 12, 1922--rocks that had been falling "from the clouds," for three weeks, at Chico, a town in an "earthquake region" in California (New York Times, March 12, 1922). Large, smooth rocks that "seemed to come straight from the clouds."

In the San Francisco Chronicle, in issues dating from the 12th to the 18th of March--clippings sent to me by Mr. Maynard Shipley, writer and lecturer upon scientific subjects, if there be such subjects--the accounts are of stones that, for four months, had been falling intermittently from the sky, almost always upon the roofs of two adjoining warehouses, in Chico, but, upon one occasion, falling three blocks away: "a downpour of oval-shaped stones"; "a heavy shower of warm rocks." San Francisco Call, March 16--"warm rocks." It is said that crowds gathered, and that upon the 17th of March a "deluge" of rocks fell upon a crowd, injuring one person. The police "combed" all surroundings: the only explanation that they could think of was that somebody was firing stones from a catapult. One person was suspected by them, but, upon the 14th of March, a rock fell when he was known not to be in the neighborhood.

The circumstances point to one origin of these stones, stationary in the sky, above the town of Chico.

Upon the first of January, 1922, the attention of Marshal J. A, Peck, of Chico, had been called to the phenomena. After investigating more than two months, he said (San Francisco Examiner, March 14) "I could find no one through my investigations who could explain the matter. At various times I have heard and seen the stones. I think someone with a machine is to blame."

Prof. C. K. Studley, vice-president of the Teachers' College, Chico, is quoted in the Examiner: "Some of the rocks are so large that they could not be thrown by any ordinary means. One of the rocks weighs 16 ounces. They are not of meteoric origin, as seems to have been hinted, because two of them show signs of cementation, either natural or artificial, and no meteoric factor was ever connected with a cement factory."

Once upon a time, dogmatists supposed, asserted, angrily declared sometimes, that all stones that fall from the sky must be of "true meteoric material." That time is now of the past. See Nature, 105-759--a description of two dissimilar stones, cemented together, seen to fall from the sky, at Cumberland Falls, Ky., April 9, 1919.

Miriam Allen de Ford (P. O. Box 573, San Francisco, Cal.--or see the Readers' Guide) has sent me an account of her own observations. About the middle of March, 1922, she was in Chico, and investigated. Went to the scene of the falling rocks; discussed the subject with persons in the crowd. "While I was discussing it with some bystanders, I looked up at the cloudless sky, and suddenly saw a rock falling straight down, as if becoming visible when it came near enough. This rock struck the roof with a thud, and bounced off on the track beside the warehouse, and I could not find it." "I learned that the rocks had been falling since July, 1921, though no publicity arose until November."

There have been other phenomena at Chico. In the New York Times, Sept. 2, 1878, it is said that, upon the 20th of August, 1878, according to the Chico Record, a great number of small fishes fell from the sky, at Chico, covering the roof of a store, and falling in the streets, upon an area of several acres. Perhaps the most important observation is that they fell from a cloudless sky. Several occurrences are listed as earthquakes, by Dr. Holden, in his Catalogue; but the detonations that were heard at Oroville, a town near Chico, Jan. 2, 1887, are said, in the Monthly Weather Review, 1887-24, to have been in the sky. Upon the night of March 5-6, 1885, according to the Chico

Chronicle, a large object, of very hard material, weighing several tons, fell from the sky, near Chico (Monthly Weather Review, March, 1885). In the year 1893, an iron object, said to be meteoritic, was found at Oroville (Mems. Nat. Acad. Sci., 13- 345).

My own idea is either that there is land over the town of Chico, and not far away, inasmuch as objects from it fall with a very narrow distribution, or that far away, and therefore invisible, there may be land from which objects have been carried in a special current to one very small part of this earth's surface. If anyone would like to read an account of stones that fell intermittently for several days, clearly enough as if in a current, or in a field of special force, of some kind, at Livet, near Clavaux, France, December, 1842, see the London Times, Jan. 13, 1843. There have been other such occurrences. Absurdly, when they were noticed at all, they were supposed to be psychic phenomena. I conceive that there is no more of the psychic to these occurrences than there is to the arrival of seeds from the West Indies upon the coast of England. Stones that fell upon a house, near the Pantheon, Paris, for three weeks, January, 1849--see Dr. Wallace's Miracles and Modern Spiritualism, p. 284. Several times, in the course of this book, I have tried to be reasonable. I have asked what such repeating phenomena in one local sky do indicate, if they do not indicate fixed origins in the sky. And if such occurrences, supported by many data in other fields, do not indicate the stationariness of this earth, with new lands not far away--tell me what it is all about. The falling stones of Chico--new lands in the sky--or what?

Boston Transcript, March 21, 1922--clipping sent to me by Mr. J. David Stern, Editor and Publisher of the Camden (N. J.) Daily Courier--

"Geneva, March 21--During a heavy snow storm in the Alps recently thousands of exotic insects resembling spiders, caterpillars, and huge ants fell on the slopes and quickly died. Local naturalists are unable to explain the phenomenon, but one theory is that the insects were blown in on the wind from a warmer climate."

The fall of unknown insects in a snow storm is not the circumstance that I call most attention to. It is worth noting that I have records of half a dozen similar occurrences in the Alps, usually about the last of January, but the striking circumstance is that insects of different species and of different specific gravities fell together. The conventional explanation is that a wind, far away, raised a great variety of small objects, and segregated them according to specific gravity, so that twigs and grasses fell in one place, dust some other place, pebbles somewhere else, and insects farther along somewhere. This would be very fine segregation. There was no very fine segregation in this occurrence. Something of a seasonal, or migratory, nature, from some other world, localized in the sky, relatively to the Alps, is suggested.

May 4, 1922--discovery, by F. Burnerd, of three long mounds in the lunar crater Archimedes. See the English Mechanic, 115-194, 218, 268, 278. It seems likely that these constructions had been recently built.

St. Thomas, Virgin Islands, May 18, 1922 (Associated Press)--particles of matter falling continuously for several days. "The phenomenon is supposed here to be of volcanic origin, but all the volcanoes of the West Indies are reported as quiet."

New York Tribune, July 3, 1922--that, for the fourth time in one month, a great volume of water, or a "cloudburst," had poured from one local sky, near Carbondale, Pa.

Oct. 15, 1922--a large quantity of white substance that fell upon the shores of Lake Michigan, near Chicago. It fell upon the clothes of hundreds of persons, fell upon the campus of Northwestern University, likely enough fell upon the astronomical observatory of the University. It occurred to one of these hundreds, or thousands, of persons to collect some of this substance. He is Mr. L. A. Hopkins, 111 West Jackson Blvd., Chicago. He sent me a sample. I think that it is spider web, because it is viscous: when burned it chars with the crinkled effect of burned hair and feathers, and the odor is similar. But it is strong, tough substance, of a cottony texture, when rolled up. The interesting circumstance to me is that similar substance has fallen frequently upon this earth, in October, but that, in terrestrial terms, seasonal migrations of aeronautical spiders cannot be thought of, because in the tropics and in Australia, as well as in the United States and in England, such showers have occurred in October. Then something seasonal, but seasonal in an extra-mundane sense, is suggested. See the Scientific Australian, September, 1916--that, from October 5 to 29, 1915, an enormous fall of similar substance occurred upon a region of thousands of square miles, in Australia.

Time after time, in data that I have only partly investigated, occur declarations that, during devastations commonly known as "earthquakes," 'in Chile, the sky has flamed, or that "strange illuminations" in the sky have been seen. In the Bull.

Seis. Soc. Amer., for instance, some of these descriptions have been noted, and have been hushed up with the explanation that they were the reports of unscientific persons.

Latest of the great quakes in Chile--1,500 dead "recovered" in one of the cities of the Province of Atacama. New York Tribune, Nov. 15, 1922--"Again, today, severe earthquakes shook the Province of Coquimbo and other places, and strange illuminations were observed over the sea, off La Serena and Copiapo."

Back to Crater Mountain, Arizona, for an impression--but far more impressive are similar data as to these places of Atacama and Copiapo, in Chile. In the year 1845, M. Darlu, of Valparaiso, read, before the French Academy, a paper, in which he asserted that, in the desert of Atacama, which begins at Copiapo, meteorites are strewn upon the ground in such numbers that they are met at every step. If these objects fell all at one time in this earthquake region, we have another instance conceivably of mere coincidence between the aerial and the seismic. If they fell at different times, the indications are of a fixed relationship between this part of Chile and a center somewhere in the sky of falling objects commonly called "meteorites" and of cataclysms that devastate this part of Chile with concussions commonly called "earthquakes." There is a paper upon this subject in Science, 14-434. Here the extreme abundance asserted by M. Darlu is questioned: it is said that only thirteen of these objects were known to science.

But, according to descriptions, four of them are stones, or stone-irons, differing so that, in the opinion of the writer, and not merely so interpreted by me, these four objects fell at different times. Then the nine others are considered. They are nickel-irons. They, too, are different, one from another. So then it is said that these thirteen objects, all from one place, were, with reasonable certainty, the products of different falls.

Behind concepts that sometimes seem delirious, I offer--a reasonable certainty--

That, existing somewhere beyond this earth, perhaps beyond a revolving shell in which the nearby stars are openings, there are stationary regions, from which, upon many 'occasions, have emanated "meteors," sometimes exploding catastrophically over Atacama, Chile, for instance. Coasts of South America have reeled, and the heavens have been afire. Reverberations in the sky--the ocean has responded with islands. Between sky and earth of Chile there have been flaming intimacies of destruction and slaughter and woe--

Silence that is conspiracy to hide past ignorance; that is imbecility, or that is the unawareness of profoundest hypnosis.

Hypnosis--

That the seismologists, too, have functioned in preserving the illusion of this earth's isolation, and by super-embryonic processes have been hypnotized into oblivion of a secret that has been proclaimed with avalanches of fire from the heavens, and that has babbled from books of the blood of crushed populations, and that is monumentalized in ruins.

L^{o!}

Part I

I

A NAKED man in a city street-the track of a horse in volcanic mud-the mystery of reindeer's ears-a huge, black form, like a whale, in the sky, and it drips red drops as if attacked by celestial swordfishes-an appalling cherub appears in the sea-

Confusions.

Showers of frogs and blizzards of snails-gushes of peri winkles down from the sky-

The preposterous, the grotesque, the incredible-and why, if I am going to tell of hundreds of these, is the quite ordinary so regarded?

An unclothed man shocks a crowd-a moment later, if nobody is generous with an overcoat, somebody is collecting handkerchiefs to knot around him.

A naked fact startles a meeting of a scientific society and whatever it has for loins is soon diapered with conventional explanations.

Chaos and muck and filth-the indeterminable and the unrecordable and the unknowable-and all men are liars- and yet-.

Wigwams on an island-sparks in their columns of smoke. Centuries later-the uncertain columns are towers. What once were fluttering sparks are the motionless lights of windows. According to critics of Tammany Hall, there has been monstrous corruption upon this island: nevertheless, in the midst of it, this regularization has occurred. A woodland sprawl has sprung to stony attention.

The Princess Caraboo tells, of herself, a story, in an unknown language, and persons who were themselves liars ' have said that she lied, through nobody has ever known what she told. The story of Dorothy Arnold had been told thousands of times, but the story of Dorothy Arnold and the swan has not been told before. A city turns to a crater, and casts out eruptions, as lurid as fire, of living things and where Cagliostro came from, and where he went, are so mysterious that only historians say they know-venomous snakes crawl on the sidewalks of London-and a star twinkles-

But the underlying oneness in all confusion.

An onion and a lump of ice-and what have they in common?

Traceries of ice, millions of years ago, forming on the surface of a pond-later, with different materials, these same forms will express botanically. If something had examined primordial frost, it could have predicted jungles. ' Times when there was not a living thing on the face of this earth-and, upon pyrolusite, there were etchings of forms that, after the appearance of cellulose, would be trees. Dendritic sketches, in silver and copper, prefigured ferns and vines.

Mineral specimens now in museums-calcites that are piles of petals-or that long ago were the rough notes of a rose. Scales, horns, quills, thorns, teeth, arrows, spears, bayonets-long before they were the implements and weapons of living things they were mineral forms. I know of an ancient sketch that is today a specimen in a museum-a colorful, little massacre that was composed of calcites ages before religion was dramatized-pink forms impaled upon mauve spears, sprinkled with drops of magenta. I know of a composition of barytes that appeared ages before the Israelites. made what is said to be history-blue waves heaped high on each side of a drab streak of forms like the horns of cattle, heads of asses, humps of camels, turbans, and up held hands.

Underlying oneness-

A new star appears-and just how remote is it from drops of water, of unknown origin, falling on a cottonwood tree, in Oklahoma? Just what have the tree and the star to do with the girl of Swanton Novers, upon whom gushed streams of oils? And why was a clergyman equally greasy? Earthquakes and droughts and the sky turns black with spiders, and, near Trenton, N. J., something pegged stones at farmers. If lights that have been seen in the sky were upon the vessels of explorers from other worlds-then living in New York City, perhaps, or in Washington, D. C., perhaps, there are inhabitants of Mars, who are secretly sending reports upon the ways of this world to their governments?

A theory feels its way through surrounding ignorance the tendrils of a vine feel their way along a trellis-a wagon train feels its way across a prairie-

Underlying oneness

Projections of limonite, in a suffusion of smoky quartz it will be ages before this little mineral sketch can develop into the chimneys and the smoke of Pittsburgh. But it re produces when a volcano blasts the vegetation on a mountain, .and smoke-forms hang around the stumps of trees. Broken shafts of ·an ancient city in a desert-they are projections in the tattered gusts of a sandstorm. It's Napoleon Bonaparte's retreat from Moscow-ragged bands, in the grimy snow, stumbling amidst abandoned cannon.

Maybe it was only coincidence-or what may there be to Napoleon's own belief that something was supervising him? Suppose it is that, in November, 1812, Napoleon's work, as a factor in European readjustments, was done. There was no military power upon this earth that could re move this one, whose work was done. There came coldness so intense that it destroyed the Grand Army.

Human knowledge-and its fakes and freaks. An astronomer, insulated by his vanity, seemingly remote from the flops and frailties of everybody else, may not be so far away as he thinks he is. He calculates where an undiscovered planet will be seen. "Lo!"-as the astronomers like to say it is seen. But, for some very distressing, if no delightful, particulars, see, later: an account of Lowell's planet. Stars are said to be trillions of miles away, but there are many alleged remotenesses that are not so far away as they are said to be.

The Johnstown flood, and the smash of Peru, and the little nigger that was dragged to a police station-

Terrified horses, up on their hind legs, hoofing a storm of frogs.

Frenzied springboks, capering their exasperations against frogs that were tickling them.

Storekeepers, in London, gaping at frogs that were tapping on their window panes.

We shall pick up an existence by its frogs.

Wise men have tried other ways. They have tried to understand our state of being, by grasping at its stars, or its arts, or its economics. But, if there is an underlying oneness of all things, it does not matter where we begin, whether with stars, or laws of supply and demand, or frogs, or Napoleon Bonaparte. One measures a circle, beginning anywhere.

I have collected 294 records of showers of living things. Have I?

Well, there's no accounting for the freaks of industry.

It is the profound conviction of most of us that there· never has been a shower of living things. But some of us have, at least in an elementary way, been educated by surprises out of much that we were "absolutely sure" of, and are suspicious of a thought, simply because it is a profound conviction.

I got the story of the terrified horses in the storm of frogs from Mr. George C. Stoker, of Lovelock, Nev. Mr. John Reid, of Lovelock, who is known to me as a writer upon geological subjects, vouches for Mr. Stoker, and I vouch for 1 Mr. Reid. Mr. Stoker vouches for me. I have never heard of anything-any pronouncement, dogma, enunciation, or pontification-that was better substantiated.

What is a straight line? A straight line is the shortest distance between two points. Well, then, what is a shortest distance between two points? That is a straight line. According to the test of ages, the definition that a straight line is a straight line cannot be improved upon. I start with a logic as exacting as Euclid's.

Mr. Stoker was driving along the Newark Valley, one of the most extensive of the desert regions of Nevada. Thunderstorm. Down came frogs. Up on their hind legs went the horses.

The exasperated springboks. They were told of, in the Northern News (Vryburg, Transvaal) March 21, 1925, by Mr. C. J. Grewar, of Uitenhage. Also I have a letter from Mr. Grewar.

The Flats-about 50 miles from Uitenhage-springboks leaping and shaking themselves unaccountably. At a distance, Mr. Grewar could conceive of no explanation of such eccentricities. He investigated, and saw that a rain of little frogs and fishes had pelted the springboks. Mr. Grewar heard that some time before, at the same place, there had been a similar shower.

Coffins have come down from the sky: also, as everybody knows, silk hats and horse collars and pajamas. But these things have come down at the time of a whirlwind. The two statements that I start with are that no shower exclusively of coffins, nor of marriage certificates, nor of alarm clocks has been recorded: but that showers exclusively of living things are common. And yet the explanation by orthodox scientists who accept that showers of living things have occurred is that the creatures were the products of whirlwinds. The explanation is that little frogs, for instance, fall from the sky, unmixed with anything else, because, in a whirlwind, the creatures were segregated, by differences in specific gravity. But when a whirlwind strikes a town, away go detachables in a monstrous mixture, and there's no findable record of washtubs coming down in one place, all the town's cats in one falling battle that lumps it in felicities in one place, and all the kittens coming down together somewhere else, in a distant bunch that miaows for its lump of mothers.

See London newspapers, Aug. 18 and 19, 1921-innumerable little frogs that appeared, during a thunderstorm, upon the 17th, in streets of the northern part of London. I have searched in almost all London newspapers, and in many provincial newspapers, and in scientific publications. There is, findable by me, no mention of a whirlwind upon the 17th of August, and no mention of a fall from the sky of anything else that might be considered another segregated discharge from a whirlwind, if there had been a whirlwind.

A whirlwind runs amok, and is filled with confusions:

and yet to the incoherences of such a thing have attributed the neatest of classifications. I do not say that no wind ever scientifically classifies objects. I have seen orderly, or logical, segregations by wind-action. I ask for records of whirlwinds that do this. There is no perceptible science by a whirl-wind, in the delivery of its images. It rants trees, doors, frogs, and parts of cows. But living things have fallen from the sky, or in some unknown way have appeared, and have arrived homogeneously. If '"they have not been segregated by winds, something has selected them. There have been repetitions of these arrivals. The phenomenon of repetition, too, is irreconcilable with the known ways of whirlwinds. There is an account, in the London Daily News, Sept. 5; 1922, of little toads, which for two days had been dropping from the sky, at Chalon-sur-Saône, France. Lies, yarns, hoaxes, mistakes-what's the specific gravity of a lie, and how am I to segregate?

That could be done only relatively to a standard, and I have never heard of any standard, in any religion, philosophy, science, or complication of household affairs that could not be made to fit any requirement. We fit standards to judgments, or break any law that it pleases us to break, and fit to the fracture some other alleged law that we say is higher and nobler. We have conclusions, which are the products of senility or incompetence or credulity, and then argue from them to premises. We forget this process, and then argue from the premises, thinking we began there.

There are accounts of showering things that came from so far away that they were unknown in places where they arrived.

If only horses and springboks express emotions in these matters, we'll be calm thinking that even living things may have been transported to this earth from other worlds.

Philadelphia Public Ledger, Aug. 8, 1891-a great shower of fishes, at Seymour, Ind. They were unknown fishes. Public Ledger, Feb. 6, 1890-a shower of fishes, in Montgomery County, California. "The fishes belong to a species altogether unknown here." New York Sun, May 29, 1892-a shower, at Coalburg, Alabama, of an enormous number of eels that were unknown in Alabama Somebody said that he knew of such eels, in the Pacific Ocean. Piles of them in the streets -people alarmed-farmers coming with carts, and taking them away for fertilizing material.

Our subject has been treated scientifically, or too scientifically. There have been experiments. I have. no more of an ill opinion of experimental science than I have of everything else, but I have been an experimenter, myself, and have impressions of the servile politeness of experiments. They have such an obliging, or ingratiating, way that there's no trusting the flatterers. In the Redruth (Cornwall, England) Independent, Aug. 13, and following issues, 1886, correspondents tell of a shower of snails near Redruth. There were experiments. One correspondent, who believed that the creatures were sea snails, put some in salt water. They lived. Another correspondent, who believed that they were not sea snails, put some in salt water. They died.

I do not know how to find out anything new without being offensive. To the ignorant, all things are pure: all knowledge is, or implies, the degradation of something. One who learns of metabolism, looks at a Venus, and realizes she's partly rotten. However, she smiles at him, and he renews his ignorance. All things in the sky are pure to those who have no telescopes. But spots on the sun, and lumps on the planets-and, being a person of learning, or, rather, erudition, myself, I've got to besmirch something, or nobody will believe I am-and I replace the pure, blue sky with the wormy heavens-

London Evening Standard, Jan. 3, 1924-red objects falling with snow at Halmstead, Sweden.

They were red worms, from one to four inches in length. Thousands of them streaking down with the. snowflakes red ribbons in a shower of confetti-a carnival scene that boosts my discovery that meteorology is a more picturesque science than most persons, including meteorologists, have suspected-and I fear me that my attempt to besmirch has not been successful, because the worms of heaven seem to be a jolly lot. However, I cheer up at thought of chances to come, because largely I shall treat of human nature.

But how am I to know whether these things fell from the sky in Sweden, or were imagined in Sweden?

I shall be scientific about it. Said Sir Isaac Newton-or virtually said he-"If there is no change in the direction of a moving body, the direction of a moving body is not changed." "But," continued he, "if something be changed, it is changed as much as it is changed." So red worms fell from the sky, in Sweden, because from the sky, in Sweden, red worms fell. How do geologists determine the age of rocks? By the fossils in them. And how do they determine the age of the fossils? By the rocks they're in. Having started with the logic of Euclid, I go on with the wisdom of a Newton.

New Orleans Daily Piayune, Feb. 4, 1892-enormous numbers of unknown brown worms that had fallen from the sky, near Clifton, Indiana. San Francisco Chronicle, Feb. 14, 1892-myriads of unknown scarlet worms-somewhere in Massachusetts-not seen to fall from the sky, but found, covering several acres, after a snowstorm.

It is as if with intelligence, or with the equivalence of intelligence, something has specialized upon transporting, or distributing, immature and larval forms of life. If the gods send worms, that would be kind if we were robins. Insect Life, 1892, p. 335, the Editor, Prof. C. V. Riley, tells of four other mysterious appearances of worms, early in the year 1892. Some of the specimens he could not definitely identify. It is said that at Lancaster, Pa., people in a snowstorm caught falling worms on their umbrellas.

The wise men of our tribes have tried to find God in a poem, or in whatever they think they mean by a moral sense in people, or in inscriptions in a book of stone, which by one of the strangest freaks of omission is not now upon exhibition in from fifteen to twenty synagogues in Asia Minor, and all up and down Italy-

Crabs and periwinkles-

Ordinary theologians have overlooked crabs and periwinkles-

Or mystery versus the fishmonger.

Upon May 28, 1881, near the city of Worcester, England, a fishmonger, with a procession of carts, loaded with several kinds of crabs and periwinkles, and with a dozen energetic assistants, appeared at ·a time when nobody on a busy road was looking. The fishmonger and his assistants grabbed sacks of periwinkles, and ran in a frenzy, slinging the things into fields on both sides of the road. They raced to gardens, and some assistants, standing on the shoulders of other assistants, had sacks lifted to them, and dumped sacks over the high walls. Meanwhile other assistants, in a dozen carts, were furiously shoveling out periwinkles, about a mile along the road. Also, meanwhile, several boys were busily mixing in crabs. They were not advertising anything. Above all there was secrecy. The cost must have been hundreds. of dollars. They appeared without having been seen on the way, and they melted away equally mysteriously. There were houses all around, but nobody saw them.

Would I be so kind as to tell what, in the name of some slight approximation to sanity, I mean by telling such a story?

But it is not my story. The details are mine, but I have put them in, strictly in accordance with the circumstances. There was, upon May 28, 1881, an occurrence near Worcester, and the conventional explanation was that a fishmonger did it. Inasmuch as he did it unobserved, if he did it, and inasmuch as he did it with tons upon acres, if he did it, he did it as I have described, if he did it.

In Land and Water, June 4, 1881, a correspondent writes that, in a violent thunderstorm, near Worcester, tons of periwinkles had come down from the sky, covering fields and a road, for about a mile. In the issue of June 11th, the Editor of Land and Water writes that specimens had been sent to him. He notes the mysterious circumstances, or the indication of a selection of living things, that appears in virtually all the accounts. He comments upon an enormous fall of sea creatures, unaccompanied by sand, pebbles, other shells and seaweed.

In the Worcester Daily Times, May 30, it is said that, upon the 28th, news had reached Worcester of a wonderful fall from, the sky of periwinkles on Cromer Gardens Road, and spread far around in fields and gardens. Mostly, people of Worcester were incredulous, but some had gone to the place. Those who had faith returned with periwinkles.

Two correspondents then wrote that they had seen the periwinkles upon the road before the storm, where probably a fishmonger had got rid of them. So the occurrence conventionalized, and out of these surmises arose the story of "the fishmonger, though it has never been told before, as I have fold it.

Mr. J. Lloyd Bozward, a writer whose notes on meteorological subjects are familiar to readers of scientific periodicals of this time, was investigating, and his findings were published in the Worcester Evening Post, June 9th. As to the story of the fishmonger, note his statement that the value of periwinkles was 16 shillings a bushel. He says that a wide area on both sides of the road was strewn with periwinkles, hermit crabs, and small crabs of an unascertained species. Worcester is about 30 miles from the mouth of the River Severn, or, say, about 50 miles from the sea. Probably no fishmonger in the world ever had, at one time, so many periwinkles, but as to anybody having got rid of a stock, because of a glutted market, for instance, Mr. Boz ward says: "Neither upon Saturday, the 28th, nor Friday, the 27th, was there such a thing procurable in Worcester as a live periwinkle." Gardens as well as fields were strewn. There were high walls around these gardens. Mr. Bozward tells of about 10 sacks of periwinkles, of a value of about £20, in the markets of Worcester, that, to his knowledge, had been picked up. Crowds had filled pots and pans and bags and trunks before he got to the place. "In Mr. Maund's garden, two sacks were filled with them." It is his conclusion that the things fell from the sky during the thunderstorm. So his is the whirlwind-explanation.

There are extraordinary occurrences, and conventionalization cloaks them, and the more commonplace the cloakery, the more satisfactory. Periwinkles appear upon a tract of land, through which there is a road. A fishmonger did it.

But the crabs and the fishmonger-and if the fishmonger did the periwinkles, did he do the crabs, if he did it?

Or the crabs and the whirlwind-and, if the periwinkles were segregated from pebbles and seaweed, why not from the crabs, if segregation did it?

The strongest point for the segregationists is in their own mental processes, which illustrate that segregations, whether by wind action, or not, do occur. If they have periwinkles and crabs to explain, and, say, that with a story of a fish monger, or of a whirlwind, they can explain the periwinkles, though so they cannot explain the crabs, a separation of data occurs in their mentalities They. forget the crabs and tell of the periwinkles.

II

FROGS and fishes and worms-and these are the materials of our expression upon all things. Hops and Hops and squirms-and these are the motions. But we have been considering more than matter and motion, to start with: we have been considering attempts by scientists to explain them. By explanation, I mean organization. There is more than matter and motion in our existence: there is organization of matter and motion.

Nobody takes· a little clot that is central in a disease germ, as Absolute Truth; and the latest scientific discovery is only something for ideas to systematize around. But there is this systematization, or organization, and we shall have to consider it.

There is no more meaning-though that may be utmost meaning-to arrangements of observations, than there is to arrangements of protoplasm in a microbe, but it must be noted that scientific explanations do often work out rath er well-but say in medical treatments, if ailments are mostly fancied; or in stock-market transactions, except in a crisis; or in expert testimony· in the courts, except when set aside by other expert testimony-

But they are based upon definitions-

And in phenomenal existence there is nothing that is in dependent of everything else. Given that there is Continuity, everything is a degree or aspect of whatever everything else is. Consequently there is no way of defining anything, except in terms of itself. Try any alleged definition. What is an island? An island is a body of land completely sur rounded by water. And what is a body of land that is completely surrounded by water?

Among savage tribesmen, there is a special care for, or even respectfulness for, the mentally afflicted. They are regarded as in some obscure way representing God's chosen. We recognize the defining of a thing in terms of itself, as a sign of feeble-mindedness. All scientists begin their works with just such definitions, implied, if not stated. And among our tribes there is a special care for, or even respectfulness for, scientists.

It will be an expression of mine that there is a godness in this idiocy. But, no matter what sometimes my opinion may be, I am not now writing that God is an Idiot. Maybe he, or it, drools comets and gibbers earthquakes, but the scale would have to be considered at least super-idiocy.

I conceive, or tell myself that I conceive, that if we could have a concept of our existence as a whole, we could have a kind of understanding of it, rather akin to what, say, cells in an animal organism could have of what is a whole to them, if they should not be mere scientists, trying to find out what a bone is, or the :flow of blood in a vein is, in itself; but if they could comprehend: what the structures and functions of the Organism are, in terms of itself.

The attempted idea of Existence as Organism is one of the oldest of the pseudo-thoughts of philosophy. But the idea in this book is not metaphysical. Metaphysical speculations are · attempts to think unthinkably, and it is quite hard enough to think thinkably. There can be nothing but bafflement for anybody who tries to think of Existence as Organism: our attempt will be to think of an existence as an organism. Having a childish liking for a little rhetoric, now and then, I shall call it God.

Our expressions are in terms of Continuity. li all things merge away into orle another, or transmute into one another, so that nothing can be defined, they are of a oneness, which may be the oneness of one existence. I state that, though I accept that there is continuity, I accept that also there is discontinuity. But there is no need, in this book, to go into the subject of continuity-discontinuity, because no statement that I shall make, as a monist, will be set aside by my pluralism. There is a Oneness that both submerges and individualizes.

By the continuity of all things we have, with a hop and a flop and a squirm, jumped from frogs toward finality. We have rejected whirlwinds and the fishmonger, and have incipient notions upon a selectiveness and an intelligent. or purposeful, distribution of living things.

What is selecting and what is distributing?

The old-fashioned theologian thinks of a being, with the looks of himself, standing aside somewhere and directing operations.

What, in any organism, is selecting and distributing say, oxygen in lungs, and materials in stomachs? The organism itself.

If we can think of our existence as a conceivable-sized formation-perhaps one of countless things, beings, or formations in the cosmos-we have graspableness, or we have the outlines and the limits within which to think.

We look up at the stars. The look is of a revolving shell that is not far away. And against such a view there is no opposition except by an authoritative feeble-mindedness, which most of us treat respectfully, because· such is the custom in all more or less savage tribes.

Mostly in this book I shall specialize upon indications that there exists a transportory force that I shall call Teleportation. I shall be accused of having assembled lies, yarns, hoaxes, and superstitions. To some degree I think so, myself. To some degree I do not. I offer the data.

III

THE subject of reported falls from the sky, of an edible substance, in Asia Minor, is confused, because reports have been upon two kinds of substances. It seems that the sugar like kind cannot be accepted. In July, 1927, the Hebrew University of Jerusalem sent an expedition to the Sinai Pen. insula to investigate reported showers of "manna.,. See the New York Times, Dec. 4, 1927. Members of the expedition found what they called "manna" upon leaves of tamarisk trees, and on the ground underneath, and explained that it was secreted by insects. But the observations of this expedition have nothing to do with data, or stories, of falls from the sky of fibrous, convoluted lumps of a substance that can be ground into an edible Hour. A dozen times, since early in the 19th century-and I have no definitely dated data upon still earlier occurrences-have been re ported showers of "manna" in Asia Minor.

An early stage within the shell of an egg-and a protoplasmic line of growth feels out through surrounding sub stance-and of itself it has no means of subsistence, or of itself it is lost. Nourishment and protection and guidance come to it from the whole.

Or, in wider existence-several thousand years ago-a line of fugitives feels out in a desert. It will be of use to coming social organizations. But in the desert, it is un provided for and is withering. Food falls from the sky.

It is one of the most commonplace of miracles. Within any womb an embryonic thing is unable to provide for itself, but "manna" is sent to it. Given an organic view of an existence, we think of the supervision of a whole upon its parts.

Or that once upon a time, a whole responded to the need ·of a part, and then kept on occasionally showering "manna" thousands of years after a special need for it had ceased. This looks like stupidity. It is in one of my moments of piety that I say this, because, though in our neo-theology there is no worship, I note that in this conception of what we may call godness, I supply grounds for devotions. Let a god change anything, and there will be reactions of evil as much as of good. Only stupidity can be divine.

Or occasional falls of "manna," to this day, in Asia Minor, may be only one factor in a wider continuance. It may be that an Organism, having once showered a merely edible substance upon its chosen phenomena, has been keeping this up, as a symbol of favoritism, by which said chosen phenomena have been receiving abundances of "manna" in many forms, ever since.

The substance that occasionally falls from the sky, in Asia Minor, comes from far away. The occurrences are far apart, in time, and always the substance is unknown where it falls, and its edibleness is sometimes found out by the sight of sheep eating it. Then it is gathered and sold .in the markets. We are told that it had been identified as a terrestrial product. We are told that these showers are aggregations of Lecanora esculenta, a lichen that grows plentifully in Algeria. We are told that whirlwinds catch up these lichens; lying loose, or easily detachable, on the ground. But note this:

There have been no such reported showers in Algeria. There have been no such reported showers in places between Algeria and Asia Minor.

The nearest similarity that I can think of is of tumble weeds, in the Western States, though tumble weeds are much larger. Well, then, new growths of them, when they're not much larger. But I have never heard of a shower of tumble weeds. Probably the things are often carried far by whirlwinds, but only scoot along the ground. A story that would be similar to stories of lichens, from Algeria, falling in Asia Minor, would be of tumble weeds, never falling in showers, in Western States, but repeatedly showering in Ontario, Canada, having been carried there by whirlwinds.

Out of a dozen records, I mention that, in Nature, 43- 255, and in La Nature, 36-82, are accounts of one of the showers, in Asia Minor. The Director of the Central Dispensary of Bagdad had sent to France specimens of an edible substance that had fallen from the sky, at Meridin, and at Diarbekis (Turkey in Asia) in a heavy rain, the last of May, 1890. They were convoluted lumps, yellow outside and white inside. They were ground into flour from which excellent bread was made. According to the ready made convention, botanists said that the objects were specimens of Lecanora esculenta, lichens that had been carried in a whirlwind.

London Daily Mail, Aug. 13, 1913-that streets in the town of Kirkmanshaws, Persia, had been covered with seeds, which the people thought were the manna of biblical times. The Royal Botanical Society had been communicated with, and had explained that the objects had been carried from some other part of this earth's surface by a whirlwind. "They were white in substance, and of a consistency of Indian com."

I believe nothing. I have shut myself away from the rocks and wisdoms of ages, and from the so-called great teachers of all time, and perhaps because of that isolation I am given to bizarre hospitalities. I shut the front door upon Christ and Einstein, and at the back door hold out a welcoming hand to little frogs and periwinkles: I believe nothing of my own that I ha.ve ever written. I cannot accept that the products of minds are subject-matter for beliefs. But I accept, with reservations that give me freedom to ridicule the statement at any other time, that showers of an edible substance that has not been traced to an origin upon this earth, have fallen from the sky, in Asia Minor.

There have been suggestions that unknown creatures and unknown substances have been transported to this earth from other fertile worlds, or from other parts of one system, or organism, a composition of distances that are small relatively to the unthinkable spans that astronomers think they can think of. There have been suggestions of a purposeful distribution in this existence. Purpose in Nature is thinkable, without conventional theological interpretations, if we can conceive of our existence, or the so-called solar system, and -the stars around, as one organic static formation, or being. I can make no demarcation between the organic, or the functional, and the purposeful. Then, in an animal-organ ism, osteoblasts appear and mend a broken bone, they rep resent purpose, whether they know what they're doing or not. Any adaptation may be considered an expression of purpose, if by purpose we mean nothing but intent upon adaptation. If we can think of our whole existence, per haps one of countless organisms in the cosmos, as one organism, we can call its functions and distributions either organic or purposeful, or mechanically purposeful.

IV

OVER the town of Noirfontaine, France, one day in April, 1842, there was a cloudless sky, but drops of water were falling. See back to data upon repetitions. The water was falling, as if from a fixed appearing-point, somewhere above the ground, to a definite area beneath. The next day water was still falling upon this one small area, as mysteriously as if a ghost aloft were holding the nozzle of an invisible hose.

I take this account from the journal of the French Aca demy of Sciences (Comptes Rendus), vol 14, p. 664.

What do I mean by that?

I don't mean anything by that. At the same time, I do mean something by the meaninglessness of that. I mean that we are in the helpless state of a standardless existence, and that the appeal to authority is as much of a wobble as any other of our insecurities.

Nevertheless, though I know of no standards by which to judge anything, I conceive-or accept the idea-of some thing that is The Standard, if I can think of our existence as an Organism. If human thought is a growth, like all other growths, its logic is without foundation of its own, and is only the adjusting constructiveness of all other growing things. A tree cannot find out, as it were, how to blossom, until comes blossom-time. A social growth cannot find out the use of steam engines, until comes steam-engine-time. For whatever is supposed to be meant by progress, there is no need in human minds for standards of their own: · this is in the sense that no part of a growing plant needs guidance of its own devising, nor special knowledge of its own as to how to b come a leaf or a root. It needs no base of its own, because the relative wholeness of the plant is relative ·baseness to its parts. At the same time, in the midst of this theory of submergence, I do not accept that human minds are absolute nonentities, just as I do not accept that a leaf, or a root, of a plant, though so dependent upon a main body, and so clearly only a part, is absolutely without something of an individualizing touch of its own.

It is the problem of continuity-discontinuity, which perhaps I shall have to take up sometime.

However-

London Times, April 26, 1821-that the inhabitants of Truro, Cornwall, were amused, astonished, or alarmed, "according to nerve and judgment," by arrivals of stones, from an unfindable source, upon a house in Carlow Street. The mayor of the town visited the place, and was made so nervous by the rattling stones that he called out a military guard. He investigated, and the soldiers investigated, and the clatter of theorists increased the noise. Times, May I stones still rattling, theorists still clattering, but nothing found out.

Flows of frogs-flows of worms-flows of water-flows of stones-just where'd we expect to draw a line? Why not go on to thinking that there have been mysterious transportations of human beings?

Well go on.

A great deal of the opposition to our data is connotative. Most likely when Dr. Gilbert rubbed a rod and made bits of paper jump on a table, the opposition to his magic was directed not so much against what he was doing as against what it might lead to. Witchcraft always has a hard time, until it becomes established and changes its name.

We hear much of the conflict between science and religion, but our conflict is with both of these. Science and religion always have agreed in opposing and suppressing the various witchcrafts. Now that religion is inglorious, one of the most fantastic of transferences of worships is that of glorifying science, as a beneficent being. It is the attributing of all that is of development, or of possible betterment to science. But no scientist had ever upheld a new idea, without bringing upon himself abuse from other scientists. Science has done its utmost to prevent whatever Science has done.

There are cynics who deny the existence of human gratitude. But it seems that I am no cynic. So convinced am I of the existence of gratitude that I see in it .one of our strongest oppositions. There are millions of persons who receive favors that they forget: but gratitude does exist, and they've got to express it somewhere. They take it out by being grateful to science for all that science has done for them, a gratitude, which, according to their dull perceptions won't cost them anything. So there is economic indignation against anybody who is disagreeable to science. He is trying to rob the people of a cheap gratitude.

I like a bargain as well as does anybody else, but I can't save expenses by being grateful to Science, if for every scientist who had perhaps been of benefit to me, there have been many other scientists who have tried to strangle that possible benefit.. Also, if I'm dead broke, I don't get benefits to be grateful for·

Resistance to notions in this book will come from per sons who identify industrial science, and the good of it, with the pure, or academic, or aristocratic sciences that are living on the repute of industrial science. In my own mind there is distinguishment between a good watchdog and the fleas on him. If the fleas, too, could be taught to bark, there'd be a little chorus that would be of some tiny value. But fleas are aristocrats.

London Times, Jan. 13, 1843-that, according to the Courrier de l'Isere, two little girls, last of December, 1842, were picking leaves from the ground, near Clavaux (Livet), France, when they saw stones falling around them. The stones fell with uncanny slowness. The children ran to their homes, and told of the phenomenon, and returned with their parents. Again stones fell, and with the same uncanny slowness. It is said that relatively to these falls the children were attractive agents. There was another phenomenon, an upward current, into which the children were dragged, as if into a vortex. We might have had data of mysterious disappearances of children, but the parents, who were un affected by the current, pulled them back.

In the Toronto Globe, Sept. 9, 1880, a· correspondent writes that he had heard reports of most improbable occurrences upon a farm, near the township of Wellesley, Ontario. He went to the place, to interview the farmer, Mr. Manser. As he approached the farmhouse, he saw that all the windows were boarded up. He learned that, about the end of July, windows had begun to break, through no missiles had been seen. The explanation by the incredulous was that the .old house was settling. It was a good explanation, except for what it overlooked. To have any opinion, one must overlook something. The disregard was that, quite as authentic as the stories of breaking windows, were stories of falls of water in the rooms, having passed through walls, showing no trace of such passage. It is said that water had fallen in such volumes, from appearing-points in rooms, that the furniture of the house had been moved to a shed. In all our records openness of phenomena is notable. The story is that showers fell in rooms, when the farmhouse was crowded with people. For more details see the Halifax Citizen, September 13.

I omit about sixty instances of seeming teleportations of stones and water, of which I have records. Numerousness hasn't any meaning, as a standard to judge by.

The simplest cases of seeming teleportations are flows of stones, into open fields, doing no damage, not especially annoying anybody, and in places where there were no means of concealment for mischievous or malicious persons. There is a story of this kind, in the New York Sun, June 22, 1884. June 16th-a farm near Trenton, N. J.-two young men, George and Albert Sanford, hoeing in a field-stones falling. There was no building anywhere near, and there was not even a fence behind which anybody could hide. The next day stones fell again. The young men dropped their hoes and ran to Trenton; where they told of their experiences. They returned with forty or fifty amateur detectives, who spread out and tried to observe something, or more philosophically sat down and arrived at conclusions without observing anything. Crowds came to the cornfield. In the presence of crowds, stones continued to fall from a point overhead. Nothing more was found out.

A pig and his swill

Or Science and data-

Or that the way of a brain is only the way of a belly We can call the process that occurs in them either assimilative or digestive. The mind-worshiper might as well take guts for his god.

For many strange occurrences there are conventional explanations. In the mind of a conventionalist, reported phenomena assimilate with conventional explanations. There must be disregards. The mind must reject some data. This process, too, is both alimentary and mental.

The conventional explanation of mysterious flows of stones is that they are peggings· by neighbors. I have; given' data as I have found them. Maybe they are indigestible. The conventional explanation of mysterious flows of water is that they are exudations from insects. If so there must some times be torrential bugs.

New York Sun, Oct. 30, 1892-that, day after day, in Oklahoma, where for weeks there had been a drought, water was falling upon a large cottonwood tree, near Stillwater. A conventionalist visited this tree. He found insects. In In sect Life, 5-204, it is said that the Stillwater mystery had been solved. Dr. Neel, Director of the Agricultural Experimental Station, at Stillwater, had gone to the tree, and had captured some of the insects that were causing the precipitation. They were Proconia undata Fab.

And how am l going to prove that this was a senseless, or brutal, or anyway mechanical, assimilation? We don't have proofs. We have expressions.

Our expression is that this precipitation in Oklahoma was only one of perhaps many. We find three other recorded instances, at this time, and if they be not attributable to exudations from insects-but well not prove anything. There is a theorem that Euclid never attempted. That is to take Q.E.D. as a proposition.

In Science-, 21-94, Mr. H. Chaplin, of Ohio University, writes that, in the town of Akron,. Ohio- about while water was falling upon a tree in Oklahoma-there had· been a continuous fall of water, during a succession of clear days. Members of the faculty of Ohio University had investigated, but had been unable to solve the problem. There was a definite and- persisting appearing-point from which fu a small area near a brickyard, Wilter was falling. Mr. Chaplin, who had probably never heard of similar occurrences far from damp places, thought that vapors from this brickyard were rising, and condensing, and falling back. If so there would often be such precipitations over ponds and other bodies of water.

About the same time, water was mysteriously appearing at Martinsville, Ohio, according to the Philadelphia Public Ledger, Oct. 19, 1892. Behind a house, a mist was falling upon an area not more than a dozen feet square. St. Louis Globe-Democrat, November 19-that, in Water Street, Brownsville, Pa., there was a garden, in which was a peach tree, upon which water was falling. As to the insect-explanation, we note the statement that the water "seemed to fall from some height above the tree, and covered an area about 14 feet square."

For all I know, some trees may have occult powers. Perhaps some especially gifted trees have 'power to transport water, from far away, in times of need. I noted the drought in Oklahoma, and then I looked up conditions in Ohio and Pennsylvania. Rainfall was below normal. In Ohio, according to the Monthly Weather Review, of November, there was a drought. A watery manna came to chosen trees.

There is no sense in trying to prove anything, if all things are continuous, so that there isn't anything, except the inclusive of all, which may be Something. But aesthetically, if not scientifically, there may be value in expressions, and we'll have variations of our theme. There were, in places far apart, simultaneous flows of water from stationary appearing-points, in and around Charleston, S. C., in the period of the long series of earthquake shocks there. Later I shall touch more upon an idea that will be an organic interpretation of falls of water in places that have been desolated by catastrophes. About the middle of September, 1886, falling water from "a cloudless sky," never falling outside a spot 25 feet wide, was reported from Dawson, Ga. This shower was not intermittent. Of course the frequently mentioned circumstance of the "cloudless sky" has no significance. Water falling all the way from the sky, even at times of the slightest breezes, cannot be thought of as localizing strictly upon an area a few yards in diameter. We think of appearing-points a short distance above the ground. Then showers upon a space 10 feet square were reported from Aiken, S; C. There were similar falls of water at Cheraw, S. C. For particulars, see the Charleston News and Courier, October 8, 21, 25, 26. For an account of falls of water, "from a cloudless sky," strictly to one point, in Charlotte, N. C., according to investigations by a meteorologist, see the Monthly Weather Review, October, 1886. In the New York Sun, October 24, it is said that, for 14 days, water had been falling from "a cloudless sky," to a point in Chesterfield County, S. C., falling so heavily that streams of it had gushed from roof pipes.

Then came news that water was falling from a point in Charleston.

Several days before, in the News and Courier, had been published. the insect-explanation of falls of water. In the News and Courier, November 5, a reporter tells that he had visited the place in Charleston, where it was said that water was falling, and that he had seen· a fall of water. He had climbed a tree to investigate. He had seen insects.

But there are limits to what can be attributed, except by the most desperate explainers, to insects.

In the Monthly Weather Review, August, 1886, it is said that, in Charleston, September 4th, three showers of hot stones had been reported.

"An examination of some of these stones, shortly after they had fallen, forced the conviction that the public was being made the victim of a practical joke."

How an examination of stones could demonstrate whether they had been slung humorously or not, is more than whatever brains I have can make out. Upon September 4th, Charleston was desolated. The great earthquake had occurred upon August 31st, and continuing shocks were terrorizing the people. Still, I'd go far from my impressions of what we call" existence, if I'd think that terror, or anything else, was ever homogeneous at Charleston, or anywhere else. Battles and shipwrecks, and especially diseases, are materials for humorists, and the fun of funerals never will be exhausted. ·1 don't argue that in the midst of desolation and woe, at Charleston, there were no jokers. I tell a story as I found it recorded in the Charleston News and Courier, September 6, and mention of my own conclusion, which is that wherever jocular survivors of the catastrophe may have been cutting up capers, they were not concerned in this series of occurrences.

At 2:30 o'clock, morning of September 4th, stones, which were found to be "warm," fell near the News and Courier building, some of them bounding into the press room. Five hours later, when there was no darkness to hide mischievous survivors, more stones fell. It was a strictly localized repetition, as if one persisting current of force. At 1:30 o'clock in the afternoon again stones fell, and these were seen, coming straight down from a point overhead. If any conviction was forced, it was forced in the same old way as that in which for ages convictions have been forced, and that is by forcing agreements with prior convictions. Other details were published in the Richmond Whig: it was told that the stones which were flint pebbles, ranging from the size of a grape to the size of a hen's egg, had fallen upon an area of 75 square feet, and that about gal on of them had been picked up. In A Descriptive Narrative of the Earthquake of August 31, 1886, Carl McKinley, an editor of the News and Courier, tells of the two of these showers of stones, which, according to him, "undoubtedly fell."

The localized repetitions of showers of stones are so much like the localized repetitions of showers of water, that one, inclusive explanation, or expression, is called for. Insects did them? Or the fishmonger of Worcester had moved to South Carolina?

A complication has been developing. Little frogs fell upon Mr. Stoker and his horses, but we had no reason to think that either Mr. Stoker or his horses had anything to do with bringing about the precipitation. But the children of Clavaux did seem to have something to do with showers of stones, and trees did seem to have something to do with the precipitations of water.

Rand Daily Mail, May 29, 1922-that Mr. D. Neaves, living near Roodeport, employed as a chemist in Johannesburg, having for several months endured showers of stones, had finally reported to the police. Five constables, having been sent to the place, after dark, had hardly taken positions around the house, when a stone crashed on the roof. Phenomena were thought to associate with the housemaid, a Hottentot girl. She was sent into the garden, and stones fell vertically around her. This is said to have been one of the most mysterious of the circumstances: stones fell vertically, so that there was no tracing of them to .an origin. Mr. Neaves' home was an isolated building, except for outhouses. These outhouses were searched, but nothing to suspect was found. The stones continued to fall from an unknown source.

Police Inspector Cummings took charge. He ordered all members of the family, servants, and newspaper men to remain in the house for a while: so everybody was under inspection. Outside were constables, and all around were open fields, with no means of concealment. Stones fell on the roof. Watched by the police, the Hottentot girl went to the well. A large stone fell near her. She ran back to the house, and a stone fell on the roof. It is said that everything that could be done was done, and that the cordon of police was complete. More stones fell. Convinced that in some way the girl was implicated, the Inspector tied her hands. A stone fell on the roof.

Then everything was explained. A "civilian," concealed in one of the outhouses, had been caught throwing a stone.

If so, whoever wrote this account did not mention the name of the culprit, and it is not said that the police made any trouble for him for having made them work.

Then everything was explained again. It was said that the girl, Sara, had been taken to the police station, where she had confessed. "It is understood that Sara admits being a party to all the stone-throwing, and has implicated two other children and a grown native. So ends the Roodeport ghost story, shorn of all its alleged supernatural trappings. Though usually we do not think piously of the police, their stations are confessionals. But they're confessionals more in a scientific than in a religious sense. When a confessor holds a club over a conscience, he can bully statements with the success of any scientist who slugs data with a theory. There is much brutality in police stations and in laboratories, but I can't say that we're trying to reform anything; and if there never has been a Newton, or a Darwin, or an Einstein-or a Moses, or a Christ, or a St. Augustine-who has practiced other than the third degree upon circumstances, I fear me that sometimes we are not innocent of one or two degrees, ourselves.

However, the story reads more as if the girl had been taken to a barber shop. Her story was shorn, we read. It was clipped bald of all details, such as the cordon of police, search of the outhouses, and the taking of precautions, such as will not fit in with this yarn of the tricky. kids. In this book we shall note much shearing.

The writer, in the Monthly Weather Review, is not the only clipper who forces a conviction, when he can. There was a case, in another part of South· Africa, not long before the bombardments at Roodeport began. In the Klerksdorp Record, Nov. 18, 1921, it is said that for several weeks there had been "mysterious stone-throwing by invisible agencies" at the houses of Mr. Gibbon Joseph and Mr. H. J. Minnaar, in North Street. A detective was put upon the case. He was a logician. It was a ghost story, or it was a case of malicious mischief. He could not pinch a ghost. So he accused two Negroes, and arrested them. The Negroes were tried upon testimony given by two boys of their race. But the boys contradicted each other, and it was brought out that they were lying. They admitted that the logical detective had promised them five shillings to substantiate his syllogisms.

In the Journal of the Society for Psychical· Research, 12-260, is published a letter from Mr. W. G. Grottendieck, telling that, about one o'clock, one morning in September, 1903, at Dortrecht, Sumatra, he was awakened by hearing something fall on the floor of his room. Sounds of falling objects went on. He found that little, black stones were falling, with uncanny slowness, from the ceiling, or the roof, which was made of large, overlapping, dried leaves. Mr. Grottendieck writes that these stones, were appearing near the inside of the roof, not puncturing the material, if through this material they were passing. He tried to catch them at the appearing-point, but, though they moved with extraordinary slowness, they evaded him. There was a coolie boy, asleep in the house, at the time. "The boy certainly did not do it, because at the time that I bent over him, while he was sleeping on the floor, there fell a couple of stones." There was no police station handy, and this story was not finished off with a neat and fashionable cut.

I point out that these stories of flows of stones are not conventional stories, and are not well known. Their details are not ·standardized, like "clanking chains,. in ghost stories, and "eyes the size of saucers," in sea serpent yarns. Some body in France, in the year 1842, told of slow-moving stones, and somebody in Sumatra, in the year 1903, told of slow-moving stones. It would be strange, if two liars should invent this circumstance-

And that is where I get, when I reason.

If strangeness be a standard for unfavorable judgment, I damn at a swipe most of this book.

But damnation is nothing to me. I offer the data. Suit yourself.

Nobody can investigate the reported phenomena that we're taking up, without noticing the number of cases in which boys and girls, but a great preponderance of girls, appear. An explanation by those who disregard a great deal-or dis regard normally-is that youngsters are concerned so much, because it is their own mischief. Poltergeist-phenomena, or teleportations of objects, in the home of Mr. Frost, 8 Ferrostone-road, Hornsey, London, for several months, early in the year 1921, cannot be so explained. There were three children. Phenomena so frightened one of, them that, in a nervous breakdown, she died (London Daily Express, April 2, 1921). Another, in a similar condition, was taken to the Lewisham (London) Hospital (London Daily News, April 30, 1921).,

In attempting to rationalize various details that we have come upon, or to assimilate them, or to digest them, the toughest meal is swallowing statements upon mysterious appearances in closed rooms, or passages of objects and substances through walls of houses, without disturbing the material of the walls. Oh, yes, I have heard of "the fourth dimension," But I am going to do myself some credit by not lugging in that particular way of showing that I don't know what I'm writing about. There's a story in the St. Louis Globe-Democrat, Jan. 27, 1888-large stones that were appearing and "falling slowly" in closed rooms .in the home of Mr. P. C. Martin, Caldwell County, N. C. Aladras (India) Maa, March 5, 1888-pieces of brick that, in the presence of many investigators, were falling in a school room, in Pondicherry.

I can understand this phenomenon, or alleged phenomenon, of appearances in closed rooms, no more than I can understand the passage of a magnetic field of force through the wall of a house, without disturbing the material. But lines of this force do not transport objects through a dense material. Then I think of X-rays, which do something like this, if it be accepted that X-rays are aggregations of very small objects, or particles. X-rays do, or sometimes do, disturb materials penetrated by them, but this disturbance is not evident until after long continuance.

If there is Teleportation, it is in two orders, or fields: electric and non-electric-or phenomena that occur during thunderstorms, and phenomena that occur under "a cloudless sky," and in houses. In the hosts of stories that I have gathered-but with which I have not swamped this book -showers of living things, the rarest of all statements is of injury to the falling creatures. Then, from impressions that have arisen from other data, we think that the creatures may not have fallen all the way from the sky, but may have fallen from appearing-points not high above the ground or may have fallen a considerable distance under a counter- gravitational influence. ·

I think that there may be a counter-gravitational influence upon transported objects, because of the many agreeing accounts-more than I have told of-of slow-falling stones, by persons who had probably never heard of other stories of slow-falling stones, and because I have come upon records of similar magic, or witchcraft, in what will be accepted as sane and sober meteorological observations. See the Annual Register 1859-70-an account by Mr. E. J: Lowe, meteorologist and an astronomer, of a fall of hail stones, at Nottingham, England, May 29, 1859. Though the objects were more than an inch across, they fell slowly. In September, 1873, near Clermont-Ferrand, France, according to La Nature, 7-289. hailstones, measuring from an inch to an inch and a half across, fell. They were under an unknown influence. Notwithstanding their size, they fell so slowly that they did no damage. Some fell upon roofs, and rebounded, and it was as if these shook off the influence. Those that • rebounded then fell faster than fell those that came down in an unbroken fall. For other records of this phenomenon, see Nature, 36-445; Illustrated London News, 34-546; Bull. Soc. Astro. de France, June 19, 1900.

If in the general electric conditions of a thunderstorm there be sometimes a counter-gravitational effect upon objects, somebody might find out how counter-gravitationally to electrify aircraft and aviators. if all work is opposition to gravitation, somebody may make a big discovery of bene fit to general laziness. Elevators in skyscrapers might be run with half the power now needed. Here is an idea that may revolutionize industry, but just now I am too busy revolutionizing everything else, and I give this idea to the world, with the generosity of somebody who bestows some thing that isn't any good to him.

But mysterious disappearances?

Our data have been upon mysterious appearances.

If I could appeal to what used to be supposed to be known as common sense, to ask whether something that mysteriously appears somewhere had not mysteriously dis appeared somewhere else.

Annals of Electricity, 6-499-Liverpool, May 11, 1842- "not a breath of air." Suddenly clothes on lines on a common shot upward. They moved away slowly. Smoke from chimneys indicated that above ground there was a south ward wind, but the clothes moved away northward.

There was another instance, a few weeks later. London Times, July 5, 1842-a bright, clear day, at Cupar, Scot land, June 30th-women hanging out clothes on a common. There was a sharp detonation, and clothes on lines shot upward. Some fell to the ground but others went on and vanished. There was a seeming of selection, which, because of possible bearing upon various observations of ours interests me. Though this was a powerful force nothing but the clothes it seized was affected. I wonder about the detonation, largely because it is in agreement with a detail of still another story.

The closeness in time of these two occurrences attracts my attention. They were a few weeks apart, and I have no other such record, until seventy-seven years later. A sensible suggestion is that somebody, in Cupar, having read the Liverpool story, had faked a similar story from his town. A suggestion that is not so sensible is that, in this year 1842, somebody had learned the secrets of teleportation, and to avoid attracting much attention in any one place was experimenting in places far apart. It seems likely enough to me that, if there be teleportation, human beings may have come upon knowledge of it, and may have used it. "Likely enough?" a spiritualist would say. "Has he never heard of apports?".

But whether it's narrowness and bigotry, upon my part, or not, I do not go to seances for data. I have collected notes upon "mysterious robberies," wondering whether a teleportative power has ever been used criminally. As to apports, if a medium could transport sea shells from the sea to his cabinet, he could abstract funds from a bank to his pocket. If he could, but would not, how account for his being a medium? Looking through newspapers, I have had a searching eye for something like an account of a medium, who had become mysteriously rich, in a town where there had been shortages of funds: clerks accused of embezzlement, and convicted, but upon evidence that was not altogether satisfactory. Although usually I can find data to "prove" anything that I want to "prove," I have come upon no such account, and I am skeptical as to apports, and think that mediums are like most of the rest of us, who are not criminals, having no exceptional abilities. How ever, there may be criminal adepts who are not known mediums.

There was, in June, 1919, at Islip, Northampton, England, an occurrence like the occurrences at Liverpool and Cupar. London Daily Express, June 12, 1919-a loud detonation-basketful of clothes shooting into the air. Then the clothes came down. There may be ineffective teleportative seizures.

London Dally Mall, May 6, 1910--phenomena near Cantillana, Spain. From ten o'clock in the morning until noon, May 4th, stones shot up from a spot in the ground. Loud detonations were heard. Traces of an extinct volcano are visible at the spot, and it is believed that a new crater is being formed." But there is no findable record of volcanic activity in Spain, at this time-nor at any other time. I am reminded of the loud noises that often accompany poltergeist disturbances.

In Niles' Weekly Register, Nov. 4, 1815, there is an account of stones that had been watched rising in a Held, near Marbleton, Ulster County, New York-that these stones he'd been seen to rise three or four feet from the ground, then moving horizontally, from thirty to sixty feet.

Out in open fields, there have been showers of water, strictly localized, and of unknown origin. A Dr. Neel will be heard from. He has captured, not indefinitely alluded to insects, but Proconia undata Fab. Every mystery has its fishmonger. Considered figuratively, he need not be a seller of fish. His name may be Smith, or O'Brien, or it may be Proconia Undata Fab.

But presumably in the wintertime, in England, members of the Proconia family are not busy and available for explanations. In the Chorley (Lancashire) Standard, Feb. 15, 1873, is a story of excitement in the town of Eccleston. At Bank House, occupied by two elderly women and their niece, streams of water started falling, about the first of February, seemingly from ceilings. Furniture· was soaked, and the occupants of the house were alarmed. The falls seemed to come from the ceiling, but probably the most singular feature of the affair is that ceilings were apparently quite dry." See back to Mr. Grottendieck's story of objects that were appearing near a ceiling, or roof, with no signs of penetrating the material. Workmen bad been called to the house, and had investigated, but were unable to explain. Openness again. House packed with neighbors, watching the showers. These data would make trouble for spiritualistic mediums and their requirements for special, or closed, conditions, and at least semi-darkness, if mediums were bothered by more than unquestioning or, occasionally politely questioning, faith. If some of them have been knocked about a bit, they were relatively few. Nobody in this house sat in a cabinet. Nobody was a logician. Nobody reasonably argued that chemists, for instance, must have special conditions, or their reactions will not. work out. "For instance," said nobody, "how could you develop a photograph, except in the special conditions of darkness, or semi-darkness?"

The look to me is that, throughout what is loosely called Nature, teleportation exists, as a means of distribution of things and materials, and that sometimes human beings have command, mostly unconsciously, though perhaps some times as a development from research and experiment, of this force. It is said that in savage tribes there are "rain makers," and it may be that among savages there are teleportationists. Some years ago, I'd have looked superior, if anybody had said this to me but a good many of us are not so given to the "tut-tut!" as we used to be. It may be that in civilized communities, because of their storages, a power to attract flows of water, being no longer needed, has virtually died out, still appearing occasionally, however. It could be that, in reading what most persons think are foolish little yarns of falling stones, we are, visionarily, in the presence of cosmic constructiveness-or that once upon a time this whole earth was built up by streams of rocks, teleported from other parts of an existence. The crash of falling islands-the humps of piling continents-and then the cosmic humor of it all-or utmost spectacularity functioning, then declining, and surviving only as a vestige or that the force that once heaped the peaks of the Rocky Mountains now slings pebbles at a couple of farmers, near Trenton, N. J.

So I'd conceive of the existence of a force, and the use of it, unconsciously mostly, by human beings. It may be that, if somebody, gifted with what we think we mean by "agency," fiercely hates somebody else, he can, out of intense visualizations, direct, by teleportation, bombardments of stones upon his enemy.

Water falls on a tree, in Oklahoma. It is told of in an entomological magazine. Water falls in a house in Eccleston. I read that in a spiritualists' periodical, though I went to a newspaper for the data. These are the isolations, or the specializations, of conventional treatments. I tell of water falling upon a tree, in Oklahoma, and of water falling in a house, in Eccleston, and. think that both phenomena are manifestations of one force. It is my attempt to smash false demarcations: to take data away from narrow and exclusive treatments by spiritualists, astronomers, meteorologists, entomologists: also denying the validity of usurpations of words and ideas by metaphysicians and theologians. But my interest is not only that of a unifier: it is in bringing together seeming incongruities and finding that they have affinity. I am very much aware of the invigoration of products of ideas that are foreign to each other, if they mate. This is exogamy, practiced with thoughts-to fertilize a volcanic eruption with a storm of frogs-or to mingle the fall of an edible substance from the sky with the unexplained appearance of Cagliostro. But I am a pioneer and no purist, and some of these stud-stunts of introducing vagabond ideas to each other may have about the eugenic value of some of the romances in houses of ill fame. I cannot expect to be both promiscuous and respectable. Later, most likely, some of these unions will be properly licensed.

Sometimes, in what I call "teleportations," there seems to be "agency" and sometimes not. That the "agency• is not exclusively human, and has nothing to do with "spirits of the departed" is indicated, I suppose, if we accept that sometimes there are "occult powers" of trees. Some other time I may be able more clearly to think out an expression upon flows of pigeons to their homes, and flows of migratory birds, as teleportative, or quasi-teleportative. My suggestion as to the frequently reported "agency" of children, is that "occult forces" were, in earlier times of human affairs, far more prevalent, and far more necessary to the help and maintenance of human communities than they are now, with political and economic mechanisms somewhat well established, or working, after a fashion; and that, wherein children are atavistic, they may be in rapport with forces that mostly human beings have outgrown.

Though just at present I am no darling of the popes, I expect to end up holy, some other time, with a general expression that all stories of miracles are not lies, or are not altogether lies; and that in the primitive conditions of the Middle Ages there were hosts of occurrences that now, considerably, though not altogether, have been outgrown. Anybody who broadly accepts the doctrine of relativity should accept that there are phenomena that exist relatively to one age, that do not, or do not so pronouncedly, exist in another age. I more or less accept a great deal that religionists piously believe. As I see myself, I represent a modernization of the old fashioned atheist, who so sweepingly denied everything that seemed to interfere with his disbeliefs.

There of course other explanations of the "occult powers" of children. One is that children, instead of being atavistic, may occasionally be far in advance of adults, foreshadowing coming human powers, because their minds are not stifled by conventions. After that, they go to school and lose their superiority. Few boy-prodigies have survived an education.

The outstanding suggestion, which, however, like many other suggestions, I cannot now develop, is that, if Teleportation exists, it may be used. It may be criminally used, or it may be used commercially. Cargoes, without ships, and freights, without trains, may be of the traffics of the future. There may be teleportative voyages from planet to planet. Altogether, so many of our data are bound up with jokes, hoaxes, and flippant treatments that I think of the toy and play genesis of many practical inventions. Billions of dollars are today seriously drawing dividends from toys and games that were put to work. Billions of laughs and jeers have preceded solemn expressions of satisfaction with fat bank accounts. But this is only reasoning, and is nothing hut logic and argument, and there have been billions of laughs that never turned into anything more satisfactory -though where do I get the idea that there is anything more satisfactory than a laugh?

If, in other worlds, or in other parts of one relatively little existence, there be people who are far ahead of terrestians, perhaps, teleportatively, beings from other places have come to this earth. And have seen nothing to detain them. Or perhaps some of the more degraded ones have felt at home here, and have hung around, or-have stayed here. I'd think of these fellows as throw-hacks concealing their origin, of course, having perhaps only a slightly foreign appearance; having affinity with our barbarisms, which their own races had cast off. I'd think of a feeling for this earth, in other worlds, as corresponding to the desire of most of us, now and then, to go to South Sea Island and be degraded. Throw-hacks, translated to this earth, would not, unless intensely atavistic, take to what we regard as vices·, but to what their own far-advanced people regard as perhaps unmentionable, or anyway, un printable, degradations. They would join our churches, and wallow in pews. They'd lose all sense of decency and be come college professors. Let a fall start, and the decline is swift. They'd end up as members of Congress.

There is another view, for which I am now gathering material.

New York Times, Dec. 6, 1930-"Scores die; 300 stricken by poison fog in Belgium; panic grips countryside. Origin complete mystery. War scenes recalled." It may be that it was war.

Mostly, explanations by the scientists were just about what one would expect, but, in the New York Telegram, December 6, Prof. H. H. Sheldon was quoted-"If there is a wide spread, lethal fog in the Meuse Valley, the conclusion of science would be that it is being deliberately caused by men or women."

It may be that inhabitants of other worlds, or other parts of one, organic existence, have declared war upon this earth, and have discharged down here, sometimes under cover of fogs, volumes of poisonous gases. I have other records that may indicate something of this kind, but, reluctantly, I give up this interesting •notion, as applied to the occurrence of Dec. 5, 1930, because it associates with another phenomenon, of which I shall tell later.

Only two weeks after the tragedy in Belgium, appeared the fishmonger. The writer of an editorial, in the New York Herald Tribune, Dec. 19, 1930, started the conventionalizing and the minimizing and the obscurizing that always cloak events that are inconsistent with a main norm of supposed knowledge. "One may suspect that a sensational newspaper man, counting up the deaths, some dark day, in the smoky steel towns on the Allegheny River, could pro duce a story not far behind that from Belgium."

Seventy-seven men and women were struck dead in Belgium. Oh, there's always some commonplace explanation for these occurrences, if we only use our common sense.

V

UPON the 9th of January, 1907, Mr. McLaughlin, of the town of Magilligan, County Derry, Ireland, hadn't a red light. Neither had his sister, nor his niece, nor his maid servant. They hadn't a cabinet. But a show was staged at their house, as if they knew altogether too much about phosphorescent paint, and as if Mr. McLaughlin bought false whiskers. There were phenomena in sunlight, and there was an atmosphere as unmystical -as pigs and neighbors. If any spiritualistic medium can do stunts, there is no more need for special conditions than there is for a chemist to turn down lights, start operations with a hymn, and ask whether there's any chemical present that has affinity with something named Hydrogen.

Mr. McLaughlin had cleaned soot from the chimney. I wonder what relation there may be. It is said that immediately afterward, phenomena began. There were flows of soot from undetectable sources, in rooms, and from room to room, independent of drafts, sometimes moving against drafts. Also there were flows of stones, or bombardments. About thirty panes of glass were broken by stones, in the daytime, some of them in the presence of neighbors. This is the story, as it was told by reporters of the Derry Journal and the Coleraine Constitution, who had been sent to investigate. Probably there was a girl, aged 14 or 15, in this house, but as to the ages of Mr. McLaughlin's niece and maidservant, I could not learn particulars.

The conventionally scientific, or fishmongerish, thing to do would be to think of some commonplace explanation of the soot, and forget the stones. There would not be so much science, if people had good memories. The flows of stones cart be explained peggings by neighbors, if the soot be forgotten.

Our data have been bullied by two tyrannies. On one side, the spiritualists have arbitrarily taken over strange occurrences, as manifestations of "the departed." On the other side, conventional science has pronounced against everything that does not harmonize with its systematizations. The scientist goes investigating, about as, to match ribbons, a woman goes shopping. The spiritualist stuffs the maws of his emotions. One is too dainty, and the other is gross. Perhaps, between these two, we shall someday be considered models of well-bred behavior.

Showers of frogs and worms and periwinkles-and now it's showers of nails. St. Louis Globe-Democrat, Oct. 18, 1888-dispatch from Brownsville, Texas-that, on the night of the 12th, the lighthouse at Point Isabel, occupied by Mrs. Schreiber, widow of the keeper, ·who had departed not long before, had been struck by a rain of nails. The next night, about dark, came another shower of nails. More variety-also down pelted clods of earth and oyster shells. Bombardments continued. People gathered and saw showers, mostly of nails but could not find out where they were coming from.

In Human Nature, March, 1871, is a story of flows of com that were passing from a locked crib, in Buchanan, Virginia. But, in this case, it was said that apparitions were seen, and mostly, at least so far as apparitions are concerned, our accounts are not ghost stories.

There have been mysterious showers of money, in public places. I have gravely copied accounts from newspapers, but there must have been something the matter with my gravity; because I put the notes away, without indexing them, and just pow can't find them, among about 60,000. One of the stories was of coins that, for several days, a few years ago, fell intermittently into Trafalgar Square, London. Traffic was so interfered with by scramblers that the police investigated, but could trace nothing to the buildings around the Square. Every now and then there was a jingle of coins, and a scramble, and the annoyance of the police was in creased. They investigated.

Maybe there are experimenters who have learned to do such things, teleportatively. I'd see some sport in it, my self, if it wouldn't cost too much.

There was a piker with pennies, in London, several years ago. New York Evening World, Jan. 18, 1928-flows of copper coins and chunks of coal, in a house in Battersea, London, occupied by a family named Robinson. "The Robinsons are educated people, and scout the idea of a supernatural agency. However they are completely baffled, and declare the phenomena take place in closed rooms, thus precluding the possibility of objects being thrown from out side."

There's small chance of such phenomena being understood, just at present, because everybody's a logician. At most everybody reasons: "There are not supernatural occurrences: therefore these alleged phenomena did not occur." However through some closed skulls, mostly independently of eyes and ears and noses, which tell mostly only what they should tell, is penetrating the idea that flows of coins and chunks of coal may be as natural as the flows of rivers. Those of us who have taken this degree of our initiation may now go on to a more. advanced stage of whatever may be the matter with us.

Aug. 30, 1919-Swanton Novers Rectory, near Melton Constable, Norfolk, England-oil "spurting" from walls and ceilings. It was thought that the house was over an oil well, the liquid percolating and precipitating, but it was not crude oil that was falling: the liquids were paraffin and petrol. Then came showers of water. Oil was falling from one of the appearing-points, at a rate of a. quart in ten minutes. Methylated spirits and sandalwood oil were falling. In an account, dated September 2nd, it is said that receptacles had been placed under appearing-points, and that about 50 gallons of oil had been caught. Of thirteen showers, upon September 1st, two were of water.

The circumstance that is of most importance in this story is that such quantities of oils and water appeared here that the Rector, the Rev. Hugh Guy, and been driven out, and had moved his furniture to another house.

London Times, September 9-"Norfolk Mystery Solved." We are told that Mr. Oswald Williams, the "illusionist," or the stage magician, and his wife, who were investigating, had seen the housemaid, aged 15, enter the house, which for several days had been unoccupied, and throw a glass of water, which they had salted, to a ceiling, then crying that another shower had occurred. They had shut off the water supply, in the house, and had placed around glasses and pails of water, salted so that it could be identified.

As Mr. and Mrs. 'Williams told it, they, in hiding, saw the girl throw the salted water, and rushed "out of their hiding place and accused her. Conceivably all for the sake of science, and conceivably with not a thought of publicity values, Mr. Williams told newspaper reporters of his successful stratagem, and put completeness into his triumph, by telling that the girl had confessed. "She admitted that she had done it, and finally she broke down and made a clean breast of it."

Times, September 12-girl interviewed by a. representative of a Norwich newspaper-denied that she had con fessed-denied that she had played tricks of any kind denied that the Williamses had been in hiding-told that she had gone to the house, with Mr. and Mrs. Williams, and that a wet spot had appeared upon a ceiling, and that she had been wrongfully accused of having thrown water. "According to the little girl's statement, she was at no time alone in the kitchen" (London Daily News, September 10). "She insists that she was the victim of a trick, and that great pressure was put upon her to admit that she had thrown salted water to the ceiling. 'I was told,' she said that I would be given one minute to say I had done it, or go to prison. I said that I didn't do it.

Having an interest in ways in which data are suppress ed, I have picked up some information upon how little girls are "pressed." No details of the "pressure" were published in the London newspapers. Norfolk News, November 8-that, in the Holt Petty Sessions had come up the case of the girl Mabel Louisa Philippo-spelled Phillips, in the other ac counts-complainant against Mrs. Oswald Williams, who was charged with having assaulted her: The girl said that Mrs. Williams had time after time struck her in the face, and had called attention to her face, reddened by blows, as evidence of her guilt. Mrs. Philippo testified that, when she arrived at the Rectory, her daughter's first words were that she had been beaten. The Rev. Hugh Guy testified, but he did not testify that he was in the house, at the time. According to details picked up from other accounts, he was not in the house, at the time.

It is said that legal procedure in Great Britain is superior to whatever goes under that name in the United States. I can't accept that legal procedure anywhere is superior to anything. Mr. Guy, who had not been present, testified that he had not seen the girl struck, and I found no record of any objection by the girl's attorney to such testimony. The case was dismissed.

And then a document closed investigation. It was a letter from Mr. Guy, published in the Times, September 13. Mr. Guy wrote that he had tasted the water, upon the ceiling, and had tasted, salt in it: so he gave his opinion that the girl had thrown the water. Most likely there is considerable salt, reminders of long successions of hams and bacons, on every kitchen ceiling.

According to Mr. and Mrs. Williams, the girl had confessed. But see Mr. Guy's letter to the Times-that the girl had not confessed.

So, because of Mr. Guy's letter, the Williamses cannot be depended upon. But we're going to find that Mr. Guy cannot be depended upon. To be sure, I am going to end up with something about photographs, but photographs cannot be depended upon. I can't see that out of our own reasoning, we can get anywhere, if there isn't anything phenomenal that can be depended upon. It is my expression that, if we are entering upon an era of a revised view of many formerly despised and ridiculed data, there will be a simultaneous variation of many minds, more favorably to them, and that what is called reasoning in those minds will be only supplementary to a general mental tropism.

The investigation was stopped by Mr. Guy. The inquiry shearer, or the mystery-bobber, was this statement, in his letter-"It would have taken only a small quantity to create the mess."

The meaning of this statement is that, whereas gallons, or barrels, of oils, at a cost of hundreds of dollars, could not be attributed to a mischievous girl, "only a small quantity" could be.

Flows of frogs-flows of worms-flows of lies-read this: London Daily Express, August 30-"The rector, in response to a request from the Daily Express, for the latest news, reported as follows:

" 'To the Editor of the Daily Express:

" 'Expert engineer arriving Monday. Drippings ascribed to exudations, on August 8, of petrol, methylated spirits, and paraffin. House evacuated; vapor dangerous; Every room affected; downpour rather than dripping-Guy.' ".

In the Daily Express, September 2, is published Mr. Guy's statement that he had been compelled to move his furniture from the house.

According to other accounts, the quantities were · great. In the London Daily News were published reports by an architect, a geologist, and a chemist, telling of observations upon profuse flows. In the Norwich newspapers, the accounts are similar. For instance, the foreman of an oil company, having been asked to give an opinion, had visited the house, and had caught in a tub, two gallons of oil, which had dripped, in four hours, from one of the appearing-points. Just how, as a matter of tricks, a girl could have been concerned in these occurrences is not picturable to me. The house was crowded, while the oil-expert, for instance, was investigating. But it does seem that unconsciously she was concerned. The first of the showers occurred in her room. Ceilings were bored and ripped off, but nothing by which to. explain was found. Then another stage magician, Mr. N. Maskelyne, went to Swanton Novers, with the idea of exposing trickery. Possibly this competition made the Williamses hasty. But Mr. Maskelyne could find nothing by which to explain the mystery. According to him (Daily Mail, September 10) "barrels of it" had appeared, during the time of his observations. Just how effective, as an inquiry-stopper, was the story of the girl and the "small quantity," is shown by the way the Society for Psychical Research was influenced by it. See the Journal S.P.R, 'October, 1919. Mr. Guy's letter to the Times is taken as final No knowledge of conflicting statements by him is shown. The Society did not investigate. "A small quantity" can be explained, as it should be explained, but "barrels of it" must be forgotten. Case dis missed.

If the Rev. Hugh Guy described at one time a "downpour," which had driven out him and his tables, chairs, beds, rugs, all those things that I think of seriously, because I have recently done some moving, myself, and then told of "a small quantity," why have I not an explanation of this contradiction? I wrote to Mr. Guy, asking him to explain, having the letter registered for the sake of a record. I have received no answer.

In the London Dally Maa, Sept. 3, 1919, are reproduced two photographs of oil drippings from different ceilings. Large drops of oil are clearly visible.

VI

Flows of blood from "holy images-

I take for a proposition that, though nothing can be proved-because, if all phenomenal things are continuous, there is, in a final sense, nothing phenomenal-anything can be said to be proved-because, if all phenomenal things are continuous, the most preposterous nonsense must some where be linked with well-established beliefs. If I had the time for an extra job, I'd ask readers to think up loony theories, and send them to me, and I'd pick out the looniest of all, and engage to find abundant data to make it rea sonable to anybody who wanted to think it reasonable.

Once upon a time I thought that stories of Flows of blood from "holy images" were as ridiculous as anything that I had ever read in any astronomical, or geological text book, or in any treatise upon economics or mechanics.

Well, then, what happened?

It occurred to me that stories of flows of blood from "holy images" are assimilable with our general expressions upon teleportations. Whereupon, automatically, the formerly despised became the somewhat reasonable. Though now and then I am ill-natured with scientific methods, it is no pose of mine that I am other than scientific, myself, in our expressions. l am tied down like any' college professor or Zulu wise -

As a start-off, I suggest that if we accept that flows of water ever have appeared at points in objects, called "houses," a jolt is softened, and we pass easily into thinking that other fluids may have appeared at· points in other objects, called "holy images." The jolt is softened still more, if we argue that other fluids did appear at points in the object, called a "house," at Swanton-Novers.

There may be Teleportation, and maybe for ages the secret of it has been known by esoteric ones. It may be that priests, especially in the past, when, sociologically, they were of some possible use, have known how to teleport a red fluid, or blood, to points upon images. They may have been "agents," able to do this, without knowing how they got their effects. If I can accept that our whole existence is an organism, I can accept that, if by so-called miracles, its masses of social growths can best be organized and kept coordinated, then appear so-called miracles. The only flaw that I note in this argument is that it overlooks that there is no need for miracles. If there is a need for belief in miracles, miracles can be said to have occurred. ·

We shall have an expression in terms of some of the other of our expressions. If we arrange the ideas of it neatly, if not nattily, no more will be required to impress anybody who would like to be impressed.

Out in open fields there have been mysterious, or miraculous, showers of water. Then has appeared the seeming "agency" of human beings, and similar showers have occurred in houses-

Out in open places, there are electrical manifestations, and they are known as "lightning." The general specializes, and human beings use electricity, in their houses, or· in images that are called "machines." Or we'd say that electricians are trained "agents" in the uses of lightning.

Out in open places there have been flows of a red liquid. In La Nature, Sept. 25, 1880, Prof. J. Brun, of the University of Geneva, writes that, near Djebel-Sekra, Morocco, he had heard rumors of a fall of blood from the sky. He visited the place of the reported phenomenon. He says that, to his stupefaction, he found rocks and vegetation covered with scales of a red, shining material. Examining specimens under a microscope, he found them composed of minute organisms, which he tells us were Protococcus fluvialis.

The identification may be doubted. I don't like it. The ease with which any writer can pick to pieces any statement made by anybody who is not present to bandy delusions with him is becoming tiresome, but if I will write a book, I will write it triumphantly.

So this identification may be doubted. First we note that Prof. Brun says that, instead of having the features of the Algae that he had named, these organisms were simple, or undifferentiated. To explain this appearance, the Professor, who had perhaps recovered from his stupefactions, says that the things. were young ones. But an aggregation exclusively of young Protococci is as extraordinary as would be a vast assemblage, say filling Central Park, New York, of human infants, without a sign of a parent.

The explanation sublimates segregationism. It attributes to a grab, an exquisite discrimination. Somewhere in a swamp, said Prof. Brun, there were hosts of Protococci-venerable -ones, middle-aged ones, and their brats-or "all sizes," as he worded it. Along came a whirlwind. Carrying away all the minute organisms, this big, rough disturbance removed, with microscopic fastidiousness, old Protococci from young Protococci, according to differences in specific gravity. It cast down at one place all the bereft parents, and precipitated, at Djebel-Sekra, a rain of little, red orphans.

When we recover from the sadness into which this tragedy cast us, we reflect that of all organisms, red blood cells are of the simplest, or least differentiated. Anyway, here is an orthodox scientist who accepted that a red fluid did fall from the sky. I have about a dozen other records of showers of red fluids that were not rains colored by dusts. Upon several of these occasions the substance was identified as blood.

Or that once upon a time, or once upon an archaic time, there came to this earth, along arterial paths in space, red Flows of a primitive plasm that deluged continents, and out of which, by the plan, purpose, guidance, or design that governs developments in all organisms, higher forms of life developed-And that maybe this mechanism has not altogether ceased, so that to this day, but in a vestigial sense, or in a very much dwindled representation, such flows are continuing And that, if human beings ever have had "agency" in directing such flows, that is only a specialization of the general.

Once upon a time, it was the fashion with those of us who say that they are of the enlightened, to reject all stories of the "Miracles at Lourdes.,. The doctors had much to do 'with this rejection. Somewhere behind everything that everybody believes, or disbelieves, is somebody's pocket. But now, as to those "miracles," the explanation of auto-suggestion is popular. Some of us who were not interested are beginning to think. The. tendency that I point out is that of so often rejecting both data and an explanation, simply because one rejects an explanation. Many of our data are in this position of phenomena at Lourdes. Explanations have been taken over by theologians, or by spiritualists, and scientists, instead of opposing this usurpation, have denied the data. Whether it is only because I now want so to accept, or not, I now accept that the phenomenon of the stigmata, or flows of blood from points upon living images, has occurred:

Most likely those who deny the phenomenon of the stigmata are those who have not read, or have not recently read, the story of Louise Lateau, for instance. One would have to be of a very old-fashioned resistfulness not to accept this story, half an hour after reading it. For the latest instance, that of Theresa Neumann, of the village of Konnersreuth, near Munich, Germany, see the New York Times, April 18, ·1928. In recent years, several cases have been re ported, in the United States. Flows of blood from points in living images lead us to flows of blood from points in graven images. If one accepts the phenomenon of the stigmata, I don't know that acceptance is monstrously stretched by transferring the idea from bodies to statues.

"On Saturday (Aug. 21, 1920) all statues and holy pictures, in the home of Thomas Dwan, of Templemore, Tiperary, Ireland, began to bleed." See newspapers of August 24th.

A boy, James Walsh, a devout youngster, aged sixteen, was the center of the reported phenomena, at Templemore. Perhaps the bleeding statues and pictures were trickeries of his.

All boys and girls are little rascals. This is a generalization that one can feel somewhat nearly sure of, until it is examined. Then, because of continuity, we find that we cannot define boys and girls, because no definite line can be drawn between youngsters and adults. Also rascality and virtue merge. Well, without arguing, I say that if all the boys and girls who appear in our records were rascals, they were most expert little rascals.

"Towns in ruins-terrible bloodshed-bombs and burntality and terrorism-hangings, ambushes, raids."

Whatever the association may be, I note conditions in Ireland, at this time.

Here is one newspaper heading, telling of occurrences of one day-"Reign of terror in Ireland-terrible massacre-appalling loss of life-holocaust-bloodshed and horror.

Five days before the phenomena at Templemore were first reported, this town· was raided. The Town Hall was burned down, and other buildings were destroyed. Temple more was terrorized. All shops were closed. Few persons dared to be seen in the streets. On the road to Templemore there was not a cart. The town was partly in ruins. It was god-forsaken and shilling-and-pence-deserted.

I take from the Tipperary Star:

"In Dwan's house, and in the house of his sister-in-law, Mrs. Maher, where lived and worked the young man, James Walsh, statues started to bleed simultaneously."

This news sneaked up and down the roads. Its carriers were stealths amidst desolation and ruin. Then they scurried from farm to farm, and people were coming out from their homes. They went to Templemore to see. Then they went in droves. The roads began to hum. Sounds of tramping and the creaking of wheels-men and horses and primitive old carts and slickest of new cars from cities-it was medievalism honked with horns-or one of the crusades, with chariots slinging out beer bottles-and anachronism is just one more of the preposterous errors of Life, Nature, or an Organism, or whomever, or whatever may be the artist that does these things. The roads began to roar. Strings of people became ropes of marching thousands. Hope and curiosity, piety, and hilarity, and the incentive to make it a holiday: out for the fun of it, out to write letters to the newspapers, exposing the fakery of it, out to conform religious teachings-but maybe all this cannot be explained in terms only of known human feelings: it was as enormous as some of the other movements of living things that I shall tell of. Then the news that was exciting Ireland was going out to the world.

The terror that chanting processions were threading may have had relation with these rhythms of marchers. · They were singing their songs of the long, long way, and the arriving shiploads took up the song. Messrs. Cook, the tourist agents, sent inquiries as to whether the inns of Templemore could provide for 2,000 pilgrims from England. Scotchmen and Englishmen and Frenchmen-tourist agencies in the Uni ted States, European countries, and Japan sent inquiries. Waves that billowed from this excitement beat upon Table Mountain, South Africa, and in the surf that fell upon Cape Town, people bobbed into a Committee that. was sent to investigate. Drops of blood from a statue in Ireland-and a trickle of turbans down a gangway at Bombay-a band of pilgrims set out from Bombay. I am far from making a religion of it, but whatever was directing all this would make hats come off at Hollywood. Also, whether somebody was monkeying with red ink, or not, is getting lost in this story. Because of a town that the world had never before heard of, Paris and London were losing Americans.

Other phenomena, which may have been teleportations were reported. In the earthen floor of the

Walsh boy's room, a hollow, about the size of a teacup, filled with water. No matter how it was drained-and thousands of persons took away quantities-water, from an unknown source, always returned to this appearing-point. The subject of '"holy wells• occurs to me, as a field of neglected data. Everything that I can think of occurs to me as a field of disregard and neglect. Statues in Walsh's room bled-that's the story-and, as in poltergeist doings-or as in other poltergeist doings objects moved about in an invisible force.

I take notice of these stories of objects that moved in the presence of a boy, because scarcely can it be said that they were of value to priestcraft, and it can be said that· they are common in accounts of occult phenomena of adolescence. I now offer as satisfactory an expression upon phenomena at Templemore, as has ever been conceived of by human mind. A Darwiri writes a book about species. By what constitutes a species? He does not know. A Newton explains all things in terms· of gravitation. But what is gravitation? But he has stopped. .I explain the occurrences at Templemore in terms of poltergeist phenomena. Any questions? But I claim scientific license for myself, too.

Marvelous cures were reported at Templemore. What teleportations have to do with cures, I don't see: but I do see that if people believe that any marvel, such as a new arrival at a Zoo, has curative powers, there will be a pile of crutches outside the cage of that thing.

Walkers, bicyclists, motor cars, donkey carts, lorries, charabancs, wheelbarrows with cripples in them: jaunting cars, special trains rushing from Dublin. Some. of the quietest old towns were in up-roars. Towns all around and towns far away were reported streets resounding with tramping thousands. There were not rooms enough in the towns. From storms of people, drifts slept on door steps. Temple more, partly in ruins, stood black in the center of a wide growth of tents. This new city, mostly of tents, was named Pilgrimsville.

I have not· taken up' definite accounts of the bleeding statues. See statements published in various issues of the Tipperary Star. They are positively convincing, or they are fairy stories for grown-up brats. I could fill pages, if I wanted to, but that would imply that I think there is any meaning in solemn assertions, or in sworn testimony, with hands on Bibles. For instance, I have notes upon an account by· Daniel Egan, a harness-maker of Templemore, of blood that he had seen oozing from a statue-but this statement may be attributed to a sense of civic responsibility. He would be a bad citizen who would testify otherwise, considering the profit that was flowing into Templemore. The town's druggist, a man of what is said to be education, stated that he had seen the phenomena. He was piling up a fortune from people who had caught bad colds sleeping in the field,. I suppose that some of them had come devoutly from far away, but had begun to sneeze, and had back-from piety to pills. However, something that I cannot find a hint of is that either Dwan or the Mahers charged. admission. At first, people were admitted in batches of fifty, somebody, holding a watch, saying, every five minutes: "Time, please!" Soon Dwan and the Mahers placed the statues in windows, for all to see. There were crowds all day, and torchlight processions moved past these windows all night. The blood that was shed in Ireland continued to pour from human beings: but the bleeding statues stopped, or statements that statues were bleeding stopped. However, wherever the water was coming from, it continued to flow from the appearing-point in the Walsh boy's room. In the Tipperary Star, September 25, the estimate is that, in about one month, one million persons had visited Pil-grimsville. To some degree the excitement kept up the rest. of the year.

They were threading terror with their peaceful processions. They marched through "a terrible toll of bloodshed-wild scenes at Nenagh-the Banshaw Horror". Past burned and blackened fields in which corpses were lying, streamed these hundreds of thousands: chanting their song of the long, long way; damning the farmers, who were charging them two shillings apiece for hard-boiled eggs; praying, raiding chicken houses, telling their beads, stealing bi cycles. "Mr. John McDonnell gave a pilgrim a lift, and was robbed of £250."

But one of these detachments enters a town. In another street, a man runs from a house-"My God! I'm shot!" Not far away-the steady sound of tramping pilgrims. These flows of beings are as mysterious as the teleportations of substances. They may mean organic control, or maintenance of balance, even in a part that is diseased with bombs and ambuscades and arson.

But it is impossible, except to the hopelessly pious, to consider, with anything like veneration, any such maintenance of a balance, because, if a god of order be conceived of, also is he, or it, a god of murder.

But, regarded aesthetically, sometimes there are effects that are magnificent.

"Bloody Sunday in the County Cork!" But, upon this day, somewhere upon every road in Ireland is maintained a rhythm.

Somewhere, a lorry of soldiers is moving down, a road. Out of bushes come bullets, and the sides of the car are draped with a droop of dead men. Not far away, men and women and children are marching. Along the roads of distracted Ireland-steady pulsations of people and people and people.

VII

NOSE in the mud, and the bend of a thing to the ground. There are postures from which life is acting to escape: one of them, the embryonic crouch; another, whether in the de gradation of worship, or as a convenience in eating grass, the bend of a neck to the ground. The all-day gnaw of the fields. But the eater of meat is released from the munch. One way to broaden horizons is to climb a tree, but another way is to stand on one's own hind legs, away from the grass. A Bernard Shaw dines on hay, and still looks behind for a world that's far ahead.

These are the disgusts for vegetarians, felt by the planters of Ceylon, in July, 1910. Very likely, I am prejudiced, myself. Perhaps I think that it is gross and brutal to eat any thing at all. Why stop at vegetarianism? Vegetarianism is only a semi-ideal. The only heavenly thing to do is to do nothing. It is gross and brutal and animal-like to breathe.

We contribute to the records of strange alarms. There was one in Ceylon. Gigantic vegetarians were eating trees. Millions of foreigners, big African snails (Achatena fulica), had suddenly appeared, massed in the one small district of Kalutara, near Colombo. Shells of the largest were six inches long. One of them that weighed three quarters of a pound was exhibited at the Colombo Museum. They were crowded, or massed, in one area of four square miles. One of the most important of the data is that this was in one of the mostly thickly populated parts of Ceylon. But nothing had been seen of these "gigantic snails," until suddenly trees turned knobby with the monsters. It was as surprising as it would be, in New York, going out one morning, finding everything covered with huge warts. In Colombo was shown a photograph of a tree trunk, upon the visible part of which 227 snails were counted. The ground was· as thick with them as were the trees.

They were explained.

So were the periwinkles of Worcester: but we had rea sons for omitting from our credulities the story of the mad fishmonger of Worcester and his frenzied assistants.

In the Zoologist, February, 1911, Mr. E. Ernest Green, the Government Entomologist, of Ceylon, explained. Ten years before, Mr. Oliver Collet, in a place about fifty miles from Kalutara, had received "some of these snails" from Africa, had had turned them loose in his garden. Then, be cause of the damage by the monsters, he had destroyed all. he thought: but he was mistaken, some of them having survived. In Kalutara lived a native, who was related to other natives, in this other place (Watawella). In a parcel of vegetables that he had brought from Watawella two of these snails had been found, and had been turned loose in Kalutara, and the millions had descended from them. No names: no date.

All the accounts, in the Ceylon Observer, in issues from July 27 to September 23, are of a sudden and monstrous appearance of huge snails, packed thick, and not an observation upon them until all at once appeared millions. It takes one of these snails two years to reach full size. All sizes were in this invasion. "Never known in Ceylon before." "How they came here continues to be a mystery." According to Mr. Green's report, published in a supplement of the Ceylon Observer, September 2, stories of the multitudes were not exaggerations: he described "giant snails in enormous numbers," "a horde in a comparatively small space," "a foreign pest." This was in a region of many plantations, and even if the hordes could have been hidden from sight in a jungle, the sounds of their gnawing and of the snapping of branches of trees under the weight of them would have been heard far.

Plantations-and the ceaseless sound of the munch. The vegetarian bend-the sagging of trees, with their tops to the ground, heavy with snails. Natives, too, and the vegetarian bend-they bowed before the invasion. They would destroy no snails: it would be a sin. A bubonic crawl lumps fall off and leave skeletons. There would be a sight like this, if a plague could hypnotize a nation, and eat, to their bones, rigid crowds. Tumors that crawl and devour -clothing and flesh disappearing-congregations of bones. There was a hope for infidels. When a lost soul was found, there was rejoicing in Kalutara, and double pay was hand ed out, satanically. The planters raked up infidels, who sinfully gathered snails into mounds and burned them.

One of our reasons for being persuaded into accepting what we wanted to accept, in the matter of the phenomenon at Worcester, was that not only periwinkles appeared: also appeared crabs, which could not fit in with the conventional explanation. Simultaneously with the invasion of snails, there was another mysterious appearance. It was of unusually large scale-insects, which, according to Mr. Green (Ceylon Observer, August 9), had never before been recorded in Ceylon.

Maybe, in September, 1929, somebody lost an alligator. According to some of our data upon the insecurities of human mentality, there isn't anything that can't be lost by somebody. A look at Losts and Founds-but especially at Losts - confirms this notion. New York American, Sept. 19, 1929-an alligator, 31 inches long, killed in the Hacken sack Meadows, N. J., by Carl Weise, 14 Peerless Place, North Bergen N. J. But my attention is attracted by another "mysterious appearance" of an alligator, about the same time.

New York Sun, September 23-an alligator, 28 inches long, found by Ralph Miles, in a small creek, near Wolcott, N. Y. In the Gentleman's Magazine, August, 1866, somebody tells of a young crocodile, which, about ten years before, had been killed on a farm, at Over-Norton, Oxfordshire, England. In the November issue of this magazine, C. Parr, a well-known writer upon antiquarian subjects, says that, thirty years before, near Over-Norton, another young crocodile had been killed. According to Mr. Parr, still another young crocodile had been seen, at Over-Norton. In the Field, Aug. 23, 1862, is an account of a fourth· young crocodile that had been seen, near Over-Norton.

It looks as if, for about thirty years, there had been a translatory current, especially selective of young crocodiles, between somewhere, say in Egypt, and an appearing-point near Over-Norton. If, by design and functioning, in the distribution of life in an organism, or in one organic existence, we mean anything so misdirected as a teleportation of young crocodiles to a point in a land where they would be out of adaptation, we evidently mean not so very intelligent design and functioning. Possibly, or most likely. It seems to me that an existence that is capable of sending young butchers to medical schools, and young boilermakers to studios, would be capable of sending young crocodiles to Over-Norton, Oxfordshire, England. When I go on to think of what gets into the Houses of Congress, I expect to come upon data of mysterious distributions of cocoanuts in Green land.

There have often been sudden, astonishing appearances of mice, in great numbers. In the autumn of 1927, millions of mice appeared in the fields of Kem County, California. Kem County, California, is continuous with all the rest of a continent: so a sudden appearance of mice there is not very mysterious.

In May, 1832, mice appeared in the fields of Invemessshire, Scotland. They were in numbers so great that foxes turned from their ordinary ways of making a living and caught mice. It is my expression that these mice may have arrived in Scotland, by way of neither land nor sea. If they were little known in Great Britain, the occurrence of such multitudes is mysterious. If they were unknown in Great Britain, this datum becomes more interesting. They were brown; white rings around necks; tails tipped with white. In the Magazine of Natural History, 7-182, a correspondent writes that he had examined specimens, and had not been able to find them mentioned in any book.

I have four records of snakes that were said to have fallen from the sky, in thunderstorms. Miss Margaret Mc Donald, of Hawthorne, Mass, has sent me an account of many speckled snakes that appeared in the streets of Hawthorne, one time, after a thunderstorm.

Because of our expressions upon teleportative currents, I am most interested in repetitions in one place. Upon May 26, 1920, began a series of tremendous thunderstorms, in England, culminating upon the 29th, in a flood that destroyed 50 houses, in Louth, Lincolnshire. Upon the 26th, in a central part of London-Gower Street-near the British Museum, a crowd gathered outside Dr. Michie's house. Gower Street is in Bloomsbury. To the Bloomsbury boarding houses go the American schoolmarms who visit London, and beyond the standards of Bloomsbury-primly pronounced Bloomsbry-respectability does not exist. Dr. Michie went out and asked the crowd what it, or anything else, could mean by being conspicuous in Bloomsbry. He was told that in an enclosure behind his house had been seen a snake·

In a positive sense, he did not investigate. He simply went to a part of the enclosure that was pointed out to him. Though, in his general practice, Dr. Michie was probably as scientific as anybody else, I must insist that this was no scientific investigation. He caught the snake.

The creature was explained. It was said to be a nafa hafa, a venomous snake from Egypt. Many oriental students live in Gower Street, to be near the British Museum and University College: in all probability the oriental snake had escaped from an oriental student.

You know, I don't see that oriental students having oriental snakes is any more likely than that American students should have American snakes: but there is an association here that will impress some persons. According to my experience, and according to data to come, l think that some body "identified" an English adder, as an oriental snake, to fit in. with the oriental students, and then fitted in the oriental students with the oriental snake, arguing reasonably that if an oriental snake was found where there were oriental students, the· oriental snake had probably escaped from the oriental students. As I have pointed out, often enough; I know of no reasoning process that is not parthenogenetic, and if this is the way the identification and the explanation came about, the author of them has companionship with Plato and Darwin and Einstein, and earth worms.

The next day, there was another crowd: this one in a part of London far from Gower Street (Sydenham). A snake had been seen in a garden. Then a postman killed it. Oriental students do not live in Sydenham. This snake was an adder (London Daily Express, May 28).

Upon the 29th, in Store Street, near Gower Street, a butcher, Mr. G. H. Hill, looked out from his shop, and saw a snake wriggling along the sidewalk. He caught the snake which was probably an adder-picture of it in the Weekly Dispatch, of the 30th.

So there were some excitements, but they were mild, compared with what occurred in a crowded part of Lon don, June 2nd. See the Daily Express, June 3. Outside the Roman Catholic Cathedral (Westminster) an adder appeared. This one stopped traffic, and, had a wide audience that approached and retreated, and reacted with a surge to every wriggle, in such a disproportion that there's no seeing how action and reaction can always he equal. Three men jumped on it. This one is told ·of, in the Westminster and Pimlico News, June 4 and 11, and here it is said that another adder had appeared in Westminster, having been caught under a mat at Morpethmansions. About this time, far away in North London "(Willesden), an adder was killed in a field (Times, June 21).

Common sense tells me that probably some especially vicious joker had been scattering venomous snakes around. But some more common sense tells me that I cannot de pend upon common sense. I have received letters upon strange appearances of living things in tanks of rain water that seemed inaccessible except to falls from the sky. Mr. Edward Foster, of Montego Bay, Jamaica, B. W. I., has told me of crayfishes that were found in a cistern of rain water at Port Antonio, Jamaica. Still, such occurrences may be explained, convention ally. But, in the London Daily Mail, Oct. 6, 1921, Major Harding Cox, of Newick, Sussex, tells of an appearance of fishes that is more mysterious. A pond near his house had been drained, and the mud had been scraped out. It was dry from July to November, when it was re-filled. In the following May, this pond teemed with tench. One day, 37 of them were caught. Almost anybody, interested, will try to explain in terms of spawn carried by winds, or in mud on the feet of water birds, but I am going right ahead with ideas different from Darwinian principles of biologic distributions. Major Cox, who is a well-known writer, probably reviewed all conventional explanations, but still he was mystified. There would not be so much of the interesting in this story, were ·in not for his statement that never before had a tench been caught in this pond. Eels are mysterious beings. It may be that what are called their "breeding habits" are teleportations. According to what is supposed to be known of eels, appearances of eels any where cannot be attributed to transportations of spawn. In the New York Times, Nov. 30, 1930, a correspondent tells of mysterious appearances of eels in old moats and in mountain tarns, which had no connection with rivers. Eels can travel over land, but just how' they rate as mountain climbers, I don't know. ·

In the Amer. Jour. Sci., 16-41, a correspondent tells of a ditch that had been dug on his farm, near Cambridge, Maryland. It was in ground that was a mile from any body of water. The work was interrupted by rain, which fell for more than a week. Then, in the rain water that filled the ditch, were found hundreds of perch, of two species. The fishes could not have developed from spawn, in so short a time: they were from four to seven inches long. But there was, here, a marksmanship that strikes my attention. Nothing is said of dead fishes living upon the ground, at sides of the ditch: hundreds of perch arrived from some where, exactly in this narrow streak of water. There could have been nothing so scattering as a "shower." Accept this story, and it looks as if to a new body of water, vibrating perhaps with the needs of vacancy, there was response somewhere else, and that, with accuracy, hundreds of fishes were teleported. If somebody should have faith in us, and dig a ditch and wait for fish, and get no fish, and then say that we're just like all other theorists, we explain that, with life now well established upon this earth, we regard many teleportations as mere atavisms, of no functional value. This idea of need and response, or of the actively function al, is taking us into a more advanced stage of a conception of an organic existence. For a while, we shall make no progress with this expression, having much work to do, to make acceptable that there is teleportation, whether organic, or not.

Perhaps some sudden and widespread appearances of exotic plants were teleportations. Such appearances in Australia and new Zealand seem to be satisfactorily explained, as ordinary importations: but, in the London Daily News, April 1, 1924, Dr. F. E. Weiss, Professor Botany, University of Manchester., tells of the Canadian pond weed that suddenly infested the canals and slow-moving rivers of England, about the year 1850, and says that the phenomenon never had been satisfactorily explained.

Cardiff (Wales) Evening Express, July 1, 1919-"The countryside is set by the ears!" That's a queer way for a countryside or anything else to be set. There may have been a queer occurrence. It is said that, upon land, belonging to Mr. William Calvert, between the villages of Sturton and Stowe, ten miles from Lincoln, wheat had appeared. It was ten years since wheat had grown here. There had been bar ley, but this year the field had been left fallow. "It was a fine crop of wheat, apparently of more robust growth than some in the cultivated fields around. Farmers from far and near were going to see this phenomenon, but nobody could explain it."

Perhaps, at the same· time, another "mystery crop" appeared somewhere else. Sunday Express (London), Aug. 24, 1919-that, in a field, near Ormskirk, West Lancashire, where, the year before, because of a drought, wheat had died off so that there was nothing worth harvesting, a crop of wheat had appeared. That some of the seeds that had been considered worthless should sprout would not have been considered extraordinary, but this was "one of the best crops of vigorous, young wheat in West Lancashire, for the season.

Though I am not a very pious theologian, I take respectful notice here. A Providence that gives one snails, or covers one's property with worms, has to be called "inscrutable": but we can understand a good crop of wheat better, and that is enough to make anybody grateful, until come following seasons, with no more benefactions.

We take up again the phenomenon of localized repetitions, which suggest the existence of persisting translatory currents. If again we come to the seemingly preposterous, we reflect that we have only preposterous pseudo-standards to judge by. In this instance, the sending of salt water fishes to a fresh water lake is no more out of place than, for instance, is the sending of chaplains to battleships: and, of course, in our view, it is what is loosely called Nature that is doing all things. Perhaps what is called Nature amuses itself by occasionally sending ·somewhat intelligent fellows to theological seminaries, and salt water fishes to fresh water. Whether we theologians believe in God, or accept that there is an Organism, wherein we agree is in having often to apologize for him or it.

In Science, Dec. 12, 1902, Dr. John M. Clarke writes that a strange-looking fish had been caught in Lake Onondaga, Western New York, and had been taken to Syracuse. Here is was identified as a squid. Then a second specimen was caught. ·

Whatever thoughts we're trying to develop did not belong away back in the Dark Age, or the other Dark Age, of the year 1902. Just where they do belong has not been decided yet. Said Dr. Clarke, with whatever reasoning abilities people had in the year 1902: "There are salt springs near Lake Onondaga: so perhaps there is, in the lake, a sub stratum of salt water." The idea is that, for millions of years, there had been, in Lake Onondaga, ocean life down below, and fresh water things swimming around, overhead, and never mixing. Perhaps, by way of experiment, Dr. Clarke put salt water and a herring in an aquarium, and then fresh water and a goldfish on top, and saw each fish keeping strictly to his own floor, which is the only way to get along as neighbors.

Another scientist turned on his reasoning abilities. Prof. Ortman, of Princeton University, examined one of the specimens, which, according to him, was "a short-finned squid, of the North Atlantic, about 13 inches long." Prof. Ortman reasoned that Atlantic fishermen used squid for bait. Very well: then other fishermen may use squid for bait. So somebody may have sent for squid, to go fishing in Lake Onondaga, and may have lost a couple of live ones.

This is the science that is opposing our own notions. But for all I know, it may be pretty good science. An existence that would produce such explainers, might very well pro duce such fishermen. So perhaps fishermen of Lake Onondaga, with millions of worms around, send several hundred miles for squid, for bait, and perhaps Atlantic fishermen, with millions of squid available, send all the way to Lake Onondaga for worms. I've done foolisher, myself.

It seems to me that there is something suggestive in the presence of large deposits of salt near this lake, but I have heard nothing of salt water in it. There's no telling about a story that was published, in the New York Times, May 2, 1882, but if it could be accepted, here would be some thing worth thinking about-that a seal had been shot, in Lake Onondaga. Some· years before the appearance of the squid, another sea creature, a sargassum fish, had been caught in Lake Onondaga. It had been exhibited in Syracuse, according to Prof. Hargitt, of Syracuse University (Science, n.s., 17-114). It has to be thought that these things were strays. If they were indigenous and propagated, they'd be common.

For various reasons, I do not think much of an idea of an underground passage, all the way from the ocean to Lake Onondaga: but, in the London Daily Mail, July 1, 1920, a correspondent expresses an idea, like this, as to mysterious appearances and disappearances of the Barbary apes of Gibraltar, conceiving of a submarine tunnel from Gibraltar to Africa. "All these creatures were well-known to the staff of the signal station on the Rock, may of the apes being named. The numbers sometimes change in the most unaccountable way. Well-known monkeys are absent for months, and then reappear with new, strange, adult monkeys of a similar breed. Those who know Gibraltar will agree that there is not a square yard on the Rock where they could have hidden."

Chicago Citizen, Feb. 27, 1892-an alligator, 51' feet long, found frozen to death on a bank of the Rock River, near Janesville, Wisconsin. In the Field, Sept. 21, 1895, it is said that a parrakeet had appeared in a farm yard, where it was caught, at Gledhill, Ardgay, Scotland, and that, about two years later, another parrakeet appeared in this farm yard, and was caught. Both birds were males. "No one living anywhere near had missed a bird, upon either occasion."

Later, we shall have expressions upon psychological, and also physiological, effects of teleportative seizures. It may be that a living thing, in California, was, upon the first of August, 1869, shot from point to point, and was tom to pieces, in the passage.

Flesh and blood that fell "from the sky," upon. Mr. J.

Hudson's farm, in Los Nietos Township, California-a show er that lasted three minutes and covered an area of two acres. The conventional explanation is that these substances had been disgorged by Hying buzzards. "The day was perfectly clear, arid the sun was shining, and there was no perceptible breeze,,. and if anybody saw buzzards, buzzards were not mentioned.

The story is told, in the San Francisco Evening Bulletin, Aug. 9, 1869. The flesh was in fine particles, and also in strips, from one to six inches long. There were short, fine hairs. One of the witnesses• took specimens to Los Angeles, and showed them to the Editor of the Los Angeles News, as told.in the News, August 3rd. The Editor wrote that he had. seen, but had not kept the disagreeable objects, to the regret of many persons who had besieged him for more information. "'That the meat fell, we cannot doubt. Even the parsons of the neighborhood are willing to vouch for that. Where it came from, we cannot even conjecture: In the Bulletin, it is said that, about two months before, flesh and blood had fallen from the sky, in Santa Clara County, California.

London Dally Express, March 24, 1927-a butterfly, a Red Admiral, that had appeared in a comer of the Girls' National School, at Whittlesey. It is said that every year, for sixteen years, a butterfly had so appeared in this comer of this room, about the end of February, or the first of March. I wrote to Miss Clarke, one of the teachers, and she replied, verifying the story, in general, though not vouching for an appearance every one of the sixteen years. I kept track, and wrote again, early in 1928. I copy the letter that I received from Miss E. Clarke, 95 Station Road, Whittlesey.

As to the idea of jokes by· little girls, I do not think that little girls could get Red Admiral butterflies, in the winter time, in England.

"On the 9th of February, a few days before I received your letter, a lovely Red Admiral again appeared at the same window. The girls were all quietly at work, when suddenly a voice exclaimed: 'Oh Miss Clarke-the butterfly!' This child was with me, last year, and remembered the sudden appearance then, which I may add, was later, March 2nd, in fact.

"As I am writing, the visitor is fluttering about the window, and seems quite lively. Last year's visitor lived about a month after its appearance, and then we found it dead.

"There is nothing else that I can tell you about our annual visitor, but really it does seem remarkable."

Early in the year 1929, I again wrote to Miss Clarke, but this time she-did not answer me. Maybe a third letter was considered too much of a correspondence with some body who had not been properly introduced. Anyway, people do not like to go upon record, in such matters.

There are circumstances in the story of the children of Clavaux that linger in my mind. It was a story of a double, or reciprocating, current. I have searched for accounts of a mysterious disappearance and an equally mysterious appearance, or something in the nature of an exchange, in the same place. Upon Dec. 12, 1910, a handsome, healthy girl disappeared somewhere in New York City. The only known man in her affairs lived in Italy. It looks as if she had no intention of disappearing: she was arranging for a party, a tea, whatever those things are, for about sixty of her former schoolmates, to be held upon the 17th of the month. When last seen, in Fifth Avenue, she said that she intended to walk through Central Park. on her way to her home, near the 79th Street entrance of the park. It may be that some where in the eastern part of the park. between the 59th Street and the 79th Street entrances, she disappeared. No more is known of Dorothy Arnold.

This day something appeared in Central Park. There was no record of any such occurrence before. As told, in the New York Sun, December 13th, scientists were puzzled. Upon the lake, near the 79th Street entrance, appeared a swan.,

Mountainous districts of Inverness-shire, Scotland-mysterious footprints in the bogs-sheep and goats slaughtered. "A large, fierce, yellow animal of unknown species" was seen by a farmer, who killed it. More mysterious tracks in the bogs, and ·continued slaughter-another large, fierce, yellow animal was shot. Soon a third specimen was caught in a trap. "The body was sent to the London Zoo, where it was identified as that of a lynx." See the London Daily Express, Jan. 14, 1927. There is no record of the lynx, as indigenous to Great Britain. "It is found, in Europe, in the Alps, and the Carpathians, and more often in the Caucasus. The last specimen, in France, was killed 100 years ago."

I have a feeling of impiety, in recording this datum. So many of our data are upon a godness that so much resembles idiocy that to attribute intelligence to it may be even blasphemous. Early in this theological treatise we noted a widespread feeling that there is something of the divine, imbecility. But, if these three lynxes were teleported, say from somewhere in the Carpathians, there was good sense to this teleportation, and there was a good shot this time, because they landed in a lynx's paradise. There is no part of Great Britain that is richer in game than is Inverness-shire, and the country abounds with deer and sheep. However, if into this Eden were shot an Adam and two Eves, and these two Eves cats, we may think of this occurrence with a restored piety.

·In the London evening newspapers, Aug. 26, 1926, it was told that a mystery had been solved. People in Hampstead (London) had reported that, in the pond, in Hampstead Heath, there was a mysterious creature. Sometimes it was said that the unknown inhabitant was a phantom, and there were stories of dogs that had been taken to the pond, and had sniffed, and had sneaked away, "with their tails between their legs." All this in a London park. There was a story of "a huge, black creature, with the head of a gorilla, and a bark like that of a dog with a sore throat." Mostly these were fishermen's tales. Anglers sit around this pond, and sometimes they catch something.

Upon the night of August 25th, the line of one of these anglers, named Trevor, was grabbed. He landed something.

This is Mr. Trevor's story. For all I know, he may have been out on an iceberg somewhere, hunting for materials for his wife's winter coat, catching something that was insufficient, if he had a large wife. All that can be said is that Trevor appeared at a hotel, near the pond, carrying a small animal that he said he had caught in the pond.

Mr. F. G. Gray, proprietor of the hotel, had an iron tank, and in this the creature was lodged: and the next day the newspapers told that a young seal had been caught.

. Reporters went to the place, and one of them, the Evening News representative, took along Mr. Shelley, of the London Zoo. Mr. Shelley identified the animal as a young seal and no tame specimen, but a wild one that snapped at fingers that were poked anywhere near him.

So it was said that a mystery had been solved.

But there were stories of other seals that had been seen or had been heard barking, before the time of the birth of this seal, in this London pond. One would think that the place was somewhere in Greenland. It was Mr. Gray's statement that for several years, there had been, intermittently, these sounds and appearances. The pond is connected with the River Fleet, which runs into the Thames, and conceivably a seal could make its way, without being reported, from the ocean to this park, far inland in London: but the idea of seals coming ·and going, without being seen on the way, in a period of several years, whereas in centuries be fore nothing of the kind had been heard of, was enough to put this story where most of our other stories, or data, have been put. Mostly the opinion is that they should stay there.

London Dally Mail, Nov. 2, 1926-"Tale that taxes credulity!" "A story of two seals, within three months, in a local pond, is taxing the credulity of residents of Hamp stead." But there is a story of another seal that had been caught, after a struggle, dying soon after capture. In the Daily Chronicle, it is said that the "first mystery-catch" was still in the tank, in a thriving condition.

I have come upon more, though to no degree enlighteningly more, about the apes-of Gibraltar. In the New York Sun, Feb. 6, 1929, Dr. Raymond L. Ditmars tells of an "old legend" of a tunnel, by which apes travel back and forth, between Africa and Spain. No special instances, or alleged instances, are told of. In Gilbard's History of Gibraltar, published in 1881, is mention of the "wild and impossible theory of communication, under sea, between Gibraltar and the Barbary coast." Here it is said that the apes were kept track of, so that additions to families were announced in the Signal Station newspaper. The notion of apes in any way passing across the Mediterranean is ridiculous to Gilbard, but he notes that there are so many apes upon the mountain on the African side of the Strait of Gibraltar that it is known as the Hill of Apes.

In November, 1852, a much talked about subject, in England, was reindeer's ears. There were letters to the news papers. Reindeer's ears came up for discussion in Parliament. Persons who had never seen a reindeer were dogmatizing upon reindeer's ears. It has been reported that among reindeer's skins that had arrived at Tromso, Norway, from Spitzbergen were some with the ears clipped.

Many Englishmen believed that Sir John Franklin had sailed through the Northwest Passage, and that survivors of his expedition were trying to communicate with occasional hunters in Spitzbergen, by marking reindeer. Spitzbergen was uninhabited, and no other explanation could be thought of. Spitzbergen is about 450 miles north of North Cape, Norway, and possibly an exceptional reindeer could swim this distance, but this is a story of many reindeer. All data upon drifting ice are upon southward drats.

Branded reindeer, presumably from Norway or Finland, continued to be reported in Spitzbergen, but by what means they made the journey never has been found out. Lamont, in Yachting In Arctic Seas, p. 110, says that he had heard of these marked animals, and that, in August, 1869, he had shot two stags, each having the left ear "back half-cropped." "I showed them to Hans, a half-bred Lapp, accustomed to deal with reindeer since infancy, and had no doubt whatever of these animals having been mark ed by the hands of men." Upon page 357, Lamont tells of having shot two more reindeer, similarly marked. Nordenskiold (Voyage of the Vega, vol 1, p. 135) tells of these marked reindeer, some of them marked also upon antlers, and traces reports back to the year 1785. Upon one of these antlers was tied a bird's leg.

. Wherever they are coming from, and however they are doing it, or however it is being done to them, the marked reindeer are still appearing in Spitzbergen. Some of them that were shot, in the summer of 1921, are told of in the Field, Dec. 24, 1921. It must be that hundreds, or thousands, of these animals have appeared in Spitzbergen. There is no findable record of one reindeer having ever been seen drifting on ice in that direction. As to the possibility of swimming, I ·note that Nova Zembla is much nearer the mainland than is Spitzbergen, but the Nordenskiold says that the marked reindeer do not appear in Nova Zembla.

VIII

THERE is no way of judging these stories. Every canon, or device, of inductive logic, conceived of by Francis Bacon and John Stuart Mill has been employed in investigating some of them, but logic is ruled by the fishmonger. Some of us will think as we're told to think, and be smug and superior, in rejecting the yarns: others will like to flout the highest authority, and think that there may be some thing in them, feeling that they're the ones who know better, and be just as smug and superior. Smug, we're going to be, anyway, just so long as we're engaged in any profession, art, or business, and have to make balance some where against a consciousness of daily stupidities. I should think that somebody in a dungeon, where it is difficult to make bad mistakes, would be of the least smug. Still, I don't know: I have noted serene and self-satisfied looks of mummies. The look of an egg is of complacency.

There is no way of judging our data. There are no ways, except arbitrary ways, of judging anything. Courts of Ap peals are of the busiest of human institutions. The pragmatist realizes all this, and says that there is no way of judging anything except upon the basis of the work-out. I am a pragmatist, myself, in practice, but I see no meaning in pragmatism, as a philosophy. Nobody wants a philosophy of description, but does want a philosophy of guidance. But pragmatists are about the same as guides on the top of a mountain, telling climbers, who have reached the top, that they are on the summit. "Take me to my destination," says a traveler. "Well, I can't do that," says a guide, "but I can tell you when you get there."

My own acceptance is that ours is an organic existence, and that our thoughts are the phenomena of its eras, quite as its rocks and trees and forms of life are; and that I think as I think, mostly, though not absolutely, because of the era I am living in. This is very much the philosophy of the Zeitgeist, but that philosophy, as ordinarily outlined, is Absolutism, and I am trying to conceive of a schedule of predetermined-though not absolutely predetermined-developments in one comprehensibly-sized existence, which may be only one of hosts of other existences, in which the scheduled eras correspond to the series of stages in the growth, say, of an embryo. There is, in our expressions, considerable. of the philosophy of Spinoza, but Spinoza conceived of .no outlines within which to think.

In anything like a satisfactory sense there is no way of judging our data, nor of judging anything else: but of course we have ways of forming opinions that are often somewhat serviceable. By means of litmus, a chemist can decide whether a substance is an acid, or an alkali. So nearly is this a standard to judge by that he can do business upon this basis. Nevertheless there are some substances that illustrate continuity, or represent the merging-point be tween acids and alkalis; and there are some substances that under some conditions are acids, and under other conditions are alkalis. If there is any mind of any scientist that can absolutely pronounce either for or against our data, it must be more intelligent than litmus paper.

A barrier to rational thinking, in anything like a final sense, is continuity, because of which only fictitiously can anything be picked out of a ·nexus of all things phenomenal, to think about. So it· is not mysterious that philosophy, with its false, or fictitious, differences, and therefore false, or fictitious, problems, is as much baffled as it was several thousand years ago.

But if, for instance, no two leaves of any tree are exactly alike, so that all appearances are set apart from ·all other appearances, though at the same time all interrelated, there is discontinuity, as well as continuity. So then the frustrations of thought are double. Discontinuity is a barrier to anything like a finally sane. understanding, because the process of understanding is a process of alleged assimilation of something with something else: but the discontinuous, or the individualized, or the unique, is the unassimilable·

One explanation of our survival is that there is underlying guidance, or control, or organic governments which to high degree regularizes the movements of the 'planets, but is less efficient in its newer phenomena. Another explanation is that we survive, because everybody with whom we are in competition, is equally badly off, mentally.

Also, in other ways, how there can be survivals of persons and prestiges, or highest and noblest of reputations, was illustrated recently. About April Fool's Day, 1930, the astronomers announced that, years before, the astronomer Lowell, by mathematical calculations of the utmost complexity, or bewilderingly beyond the comprehension of any body except an astronomer, had calculated the position of a ninth major planet in this solar system: and that it had been discovered almost exactly in the assigned position. Then columns, and pages of special articles, upon this tri umph of astronomical science. But then a doubt appeared there were a few stray paragraphs telling that, after all, the body might not be the planet of Lowell's calculations the subject was dropped for a while. But, in the public mind, the impressions worked up by spreadheads enormously outweighed whatever impressions came from obscure para graphs, and the general idea was that, whatever it was, there had been another big, astronomical triumph. It is probable that the prestige of the astronomers, instead of suffering, was boomed by this overwhelming of obscure paragraphs by spreadheads.

I do not think that it is vanity, in itself, that is so necessary to human beings: it is compensatory vanity that one must have. Ordinarily, one pays little if any attention to astronomers, but now and then come consoling reflections upon their supposed powers. Somewhere in everything that one does there is error. Somebody is not an astronomer, but he classes himself with astronomers, as differentiated from other and "lower" forms of life and mind. Consciousness of the irrationality, or stupidity, pervading his own daily affairs, is relieved by a pride in himself and astronomers, as contrasted with dogs and cats.

According to the Lowell calculations, the new planet was at a mean distance of about 45 astronomical units from the sun. But, several weeks after April Fool's Day, the object was calculated to be at a mean, or very mean, distance of 217 units. I do not say that an educated cat or dog could do as well, if not better: I do say that there is a great deal of delusion in the gratification that one feels when thinking of himself and astronomers, and then looking at a cat or a dog.

The next time anybody thinks of astronomers, and looks at a cat, and feels superior, and would like to keep on feeling superior, let him not think of a cat and a mouse. The cat lies down and watches a mouse. The mouse moves away. The cat knows it. The mouse wobbles nearer. The cat knows whether it's coming or going.

In April, 1930, the astronomers told that Lowell's planet was receding so fast from the sun that soon it would be come dimmer and dimmer.

New York Times, June 1, 1930-Lowelfs planet approaching the sun-for fifty years it would become brighter and brighter.

A planet is rapidly approaching the sun. The astronomers publish highly technical "determinations" upon its rate of recession. Nobody that I know of wrote one letter to any newspaper. One reason is that one fears to bring upon one self the bullies of science. In July, 1930, the artist, Walter Russell, sent some views that were hostile to conventional science to the New York Times. Times, August 3rd-a letter from Dr. Thomas Jackson-a quotation from it, by which we have something of an idea of the self-apotheosis of these pundits, who do not know, of a thing in the sky, whether it is coming or going:

"For nearly three hundred years no one, not even a scientist, has had the temerity to question Newton's laws of gravitation. Such an act on the part of a scientist would be akin to blasphemy, and for an artist to commit such an absurdity is, to treat it kindly, an evidence of either mis guidance or crass ignorance of the enormity of his act."

If we're going to be kind' about this, I simply wonder, without commenting, what such a statement as that for nearly three hundred years nobody had ever questioned Newton's law of gravitation, is evidence of.

But in the matter of Lowell's planet, I neglected to point out how the astronomers corrected their errors, and that is a consideration of importance to us. Everything that was determined by their mathematics turned out wrong-planet coming instead of going-period of revolution 265 years, instead of 3,000 years-eccentricity of orbit three tenths instead of nine tenths. They corrected, according to photo graphs.

It is mathematical astronomy that is opposing our own notions.

Photographic astronomy can be construed any way one pleases: say that the stars are in a revolving shell, about a week's journey away from this earth.

Everything mathematical cited by me, in this Lowell-planet-controversy, was authoritatively said by somebody one lime, and equally authoritatively denied by somebody else, some other time. Anybody who dreams of a mathematician's heaven had better reconsider, if of its angels there be more than one mathematician.

IX

I HAVE come upon a story of somebody, in Philadelphia, who, having heard that a strange wild animal was prowling in New Jersey, announced that he had caught it.. He exhibited something, as the "Jersey Devil." I have to accept that this person was the press agent of a dime museum, . and that the creature that he exhibited was a kangaroo, to which he had attached tin wings and green whiskers. But, if better-established branches of biology are subject to Nature-fakery, what can be expected in our newer biology, with all the insecurities of newness?

"Jersey Devils" have been reported other times, but, though I should not like to be so dogmatic as to say that there are no "Jersey Devils," I have had no encouragement investigating them. One of the stories, according to a clipping that was sent to me by Miss F. G. Talman, of Woodbury, N. J., appeared in the Woodbury Daily Times, Dec. 15, 1925. William Hyman, upon his farm, near Woodbury, had been aroused by a disturbance in his chicken coop. He shot and killed a never-before-heard-of-animal. I have writ ten to Mr. Hyman, and have no reason to. think that there is a Mr. Hyman. I have had an extensive, though one sided, correspondence, with people who may not be, about things that probably aren't. For the latest accounts of the. "Jersey Devil," see the New York Times, Aug. 6, 1930. Remains of a strange animal, teleported to this earth from Mars or the moon-very likely, or not so likely-found on a bank of a stream in Australia. See the Adelaide Observer, Sept. 115, 1883-that Mr. Hoad, of Adelaide, had found on a bank of Brungle Creek, a headless trunk of a pig-like animal, with an appendage that curved inward, like the tail of a lobster. New Zealand Times, May 9, 1883 -excitement near Masterton-unknown creature at large curly hair, short legs, and broad muzzle. Dogs sent after it-one of the dogs flayed by it-rest of the dogs running away-probably "with their tails between their legs," but the reporter overlooking this convention.

There have been stories of strange animals that have appeared at times of earthquakes and volcanic eruptions. See Sea Serpent stories, about the time of the Charleston earthquake. About the same time, following a volcanic eruption in New Zealand, there were stories in New Zealand. The volcano Rotomahana was a harsh, black cup that had spilled scenery. Or the somber thing was a Puritan in finery. It had belied its dourness with two broad decorations of siliceous deposits, shelving down to its base, one of them the White Terrace and the other the Pink Terrace. These gay formations sloped from the bare, black crater to another inconsistency, which was a grove of acacias. All a round, the famous flowering bushes of this district made more sinful contrast with a gaunt, towering thing. Upon the 10th of June, 1886, this Black Fanatic slung a constitutional amendment. It was reformation, in the sense that virtue is uniformity that smothers variation. It drabbed its gay Terraces: the grove of acacias was a mound of mud: it covered over the flowering bushes with smooth, clean mud. It was a virtuously dismal scene, but, as in all other reformations, a hankering survived in it. A left-over living thing made tracks in the smoothness of mud. In the New Zealand Herald, Oct. 13,

1886, a correspondent writes of having traversed this dull, dead expanse, having seen it marked with the footprints of a living creature. He thought that the marks were a horse's. But there was another story that was attracting attention at this time, and his letter was in allusion to it. Maoris were telling of a wandering animal, unknown to them, that had appeared in this desert of mud. It was a creature with antlers, or a stag, according to descriptions, an animal that had never been seen, or had never before been seen, by Maoris.

Just what relation I think I can think of, between volcanic eruptions and mysterious appearances of living things may seem obscure. But I have been impressed with several accounts of astonishing revivifications in regions that were volcanically desolated. Quick growths of plants have been attributed to the fertilizing properties of volcanic dust: nevertheless writers have expressed astonishment. If we can have an organic view of our existence, we can think of restorative teleportions to a place of desolation, quite as we think of restorations occurring in places of injury in an animal-organism. ·

There are phenomena upon the. borderline between the organic and the inorganic that we can think of: such as restorations of the forms of broken crystals in a solution. It is by automatic purpose, or design or providence, or guidance by which lost parts of a starfish are regenerated. In higher animal-organisms, distinct structures, if lost, mostly are not restored, but. injured tissues are. Still even in the higher organisms there are some restorations of mutilated parts, such as renewals of forms of a bird's clipped wing feathers. The tails of some lizards, if broken off, renew.

For a conventional explanation of reviving plants in a fem forest that had been destroyed by Flows of liquid lava, from the volcano Kilauea, Hawaii, see an account, by Dr. G. R. Wieland, in Science-, April 11, 1930. Dr. Wieland considers his own explanation "amazing." I'd not say that ours is more than that.

Strange animals have appeared and they may have been teleported to this earth from other parts of an existence, but the easiest way of accounting for strange animals is to say that they are hybrids. Of course I could handle, or man handle, this subject any way to suit me, and be about as reasonable one way as another. I could quote many authorities against the occurrence of bizarre hybrids, leaving hard to explain, in terms of terrestrial origin, strange creatures that have appeared upon this earth. There are biologists who will not admit fertility between creatures as much alike as hares and rabbits. Nevertheless, I think that there have been strange hybrids.

The cow that gave birth to two lambs and a calf.

I don't. know how that will strike all minds, but to the mind of a standardized biologist, I'd not be much more preposterous, if I should tell of an elephant that had produced two bicycles and a baby elephant.

The story is told in the Toronto Globe, May 25, 1889. It is said that a member of the staff of the Globe had been sent to investigate this outrage upon conventional obstetrics. The. reporter went to the farm of Mr.: John H. Carter at South Simcoe, and then wrote that be bad seen the lambs, which were larger and coarser than ordinary, or less romantically derived, lambs, having upon their breasts tufts of hair like calves' hair. Other newspapers Quebec Dally Mercury, for instance-published other details, such as statements by well-known stockbreeders that had examined the lambs, and were compelled to accept the story of their origin.

So I am harming our idea that creatures, unlike anything known upon this earth, but that have appeared upon this earth, may have been teleported from Mars or the moon: but I am supporting our general principle that, whether in biology, astronomy, obstetrics, or any other field of research, everything that is, also isn't; and that everywhere there are data, partly sense and partly nonsense, that oppose established nonsense that has partly some sense to it.

It does not matter what scientific dictum may be brought against us. I will engage to find that it is only an approximation, or that it is a work-out only in imaginary conditions. The most rigorous science is frosted childishness. Every severe, or chaste, treatise upon mechanics is only a fairy story of frictionless and non-extensible characters that interact up to the "happy ending." Nowadays, a scenario-writer will sometimes tone down the absolute happiness of a conclusion, with just a sug-gestion that there is a little trouble in the offing: but the tellers of theorems represent about the qual-ity of intellect in the most primitive times of Holly wood. For everything that is supposed to be so well-known that it is proverbial, there are Perception. A mule is a symbol of sterility. For instances of fertility in mules, Jook over indexes of the Field. As to anything else that we're talking as absolute truth-look it up.

One afternoon, in October, 1878, Mr. Davy, a naturalist, who was employed at the London Aquar-ium, took a stroll with a new animal. I think of a prayer that is said to have been uttered by King Louis XIV. He was tired of lamb chops and beef and bacon-"Oh, God! send me a new animal." Mr. Davy took a stroll with one People far away were attracted by such screeches as are seldom heard in London. Some ex-slaves, who were playing in Uncle Tom's Cabin, were following the new animal, and were letting loose their excitability. The creature was about two feet long, and two feet high, and was formed like nothing known to anatomists-anyway to anatomists of this earth. It was cov-ered with wiry hair: head like a boar's, and curly tail like a boar's. It was described as "a living cube." As if with abdomen missing, its hind legs were closed to its forelegs. If Mr. Davy's intention had been to attract attention, he was succeeding. Almost anybody with the modem view of things will think what a pity he wasn't advertising something. The crowd jammed around so that he ran into an Underground Railway Station. Here there was an uproar. He was compelled to ride in the brake, because of a fear that there would be a panic among the passengers. At the Aquarium, Davy told that an acquaintance of his, named Leman, had seen this creature with some peasants, in the South of France, and had bought it, but, unable to speak the patois of the district, had been unable to learn anything of its origin. At the Aquarium the only explanation that could be thought of was that it was a dog-boar hybrid. ·

Davy's publicity continued. He took the new animal to his home, and a crowd went with him. His landlord looked at the animal. When the animal looked at the landlord, the landlord ran to his room, and from behind closed doors, ordered Davy to take away the monster. There was another hold-up of traffic all the way to the home of Frank Buckland.

In Land and Water, of which he was the editor, issue of October 5, Buckland wrote an account of this "demon," as he called it, saying that it looked like a gargoyle, or like one of Fuseli's satanic an-imals. He did not try to explain, but mentioned what was thought at the Aquarium. In the next issue of Land and Water, Thomas Worthington, the naturalist, wrote that the idea of the hybrid was "utterly untenable": but his own idea that the creature was "a tame hyena of some abnormal kind" leaves mysterious how the "demon" ever got into the possession of peasants in the South of France.

It would be strange if they had a tame hyena of a normal kind.

In January, 1864 (Tasmanian Journal of Science, 3-147), a skull was found on a bank of the river Murrumbridgee, Australia. It was examined by Dr. James Grant, who said that the general form and arrangement of the teeth were different from those of any animal known to him. He noted some-body's suggestion that it might be the skull of one of the camels that had been sent to Australia, in the year 1839. He accounted for its having characters that were unknown to him, by thinking that it might be foetal. So then, whether in accordance with a theory or not, he found that some of the bones were imperfectly ossified, and that the teeth were covered with a. membrane. It was not a fossil. It was a skull of a large, herbivorous ' animal, and had not been exposed long.

Melbourne Argus, Feb. 28, and March 1, 1890-a wandering monster. A list of names and addresses of persons who said that they had seen it, was published. It was a creature about thirty feet long, and was terrorizing the people of Euroa. "The existence of some altogether un heard-of monster is vouched for by a cloud of credible wit nesses."

I am tired of the sensible explanations that are holding back new delusions. So I suggest that this thing, thirty feet long, was not a creature, but was a construction, in which explorers from some-where else, were traveling back and forth, near one of this earth's cities, having their own rea sons for not wanting to investigate too closely.

I don't know what will be thought of zoologists of Melbourne, but whatever will be thought of me can't be altogether focused. upon me because there were scientists in Melbourne who were as enlightened as I am, or as preposterous and sensational as I am. Officials of the Melbourne Zoological Gardens thought that, whether this story was nonsense or not, it should be looked into. They got a big net, and sent a man with the net to Euroa. Forty men, with the man with the net, set out. They hunted all day, but no huge bulk, more or less in the distance, was seen, and a statement that enormous tracks were found may be only a sop to us enlightened, or preposterous, ones.

But the man with the net is a significant character. He had not the remotest of ideas of using it, but, just the same, he went along with it. There are other evidences of occasional open-mindedness among biologists, and touches of indifference, now and then, to whatever may be the fascinations of smugness. Why biologists should be some what less dogmatic than astronomers, or why association with the other animals should be rather more liberalizing than is communion with the stars is not mysterious. One can look at a rhinoceros and at the same time be able to think. But the stupefying, little stars shine with a hypnotic effect, like other glittering points. The little things are taken too seriously. They twinkle humorously enough, themselves. A reported monster is told of, in the Scientific American, July, 1922. Dr. Clement Onelli, Director of the. Zoology Gardens, of Buenos Aires, had published a letter that had been sent to him by an American prospector named Sheffield, who said that, in the Argentine Territory of Chebut, he had seen huge tracks, which he had followed to a lake. "There I saw in the middle of the lake an animal with a huge neck, like that of a swan, and the movement of the water made me suppose the beast to have a body like. that of a crocodile." I wrote to Dr. Onelli, and received a reply, dated Aug. 15, 1924, telling that again he had heard of the monster. Maybe this same huge-necked creature was seen somewhere else, however we explain its getting there. The trouble in trying to understand all reported monsters is their mysterious appearances and disappearances. In the London Daily Mall, Feb. 8, 1921, a huge, unknown animal, near the Orange

River, South Africa, is told of by Mr. F. C. Cornell, F.R.G.S. It was something with a neck like a bending tree trunk, "something huge, black, and sinuous." It devoured cattle. "The object may have been a python, but if it was it was of incredible size." It is only preposterously unreasonable to think that the same thing could have appeared in South Africa and then in South America.

The "blonde beast of Patagonia," which was supposed to be a huge ground sloth, parts of which are now in various museums, attracted attention, in the year 1899. See the Zoologist, August, 1899. Specimens of the blonde's hide were brought to England, by Dr. F. P. Moreno, who believed that the remains had been preserved for ages. We prefer to think otherwise: so we note that Dr. Ameghino, who got specimens of the hide from the natives, said that it was their story that they had killed it.

There was a volley of monsters from some other world, 'about the time of the Charleston earthquake, or some one thing skipped around with marvelous agility, or it is that, just. before the quake, there were dull times for the news papers. So many observations in places far apart can be reconciled by thinking that not a creature but explorers in a construction, had visited this earth. They may have settled down in various places. However, it is pretty hard to be reconciled to our reconciliations.

New York Sun, Aug. 19; 1886-a homed monster, in Sandy Lake, Minnesota. More details, in the London (Ontario) Advertiser-Chris. Engstein fired a shot at it, but missed. Then came dispatches from the sea coast. According to them, Mr. G. P. Putnam, Principal of a Boston grammar school, had seen a monster, in the sea, at Gloucester. In Science, 8-258, Mr. B. A. Colona, of the U. S. Coast Sur vey, writes that, upon the 29th of August, he had seen an unknown creature in the sea off Cape Cod. In the New York newspapers, early in September, a monster was reported as having been seen at sea, off Southport, and off Norwalk, Conn.: in Michigan, in the Connecticut River, and in the Hudson River. The conventional explanation is that this was simply an epidemic of fancied observations. Most likely some of them were only contagions.

There's a yarn, or a veritable account, in the New York Times, June 10, 1880-monstrous, dead thing, floating on the sea, bottom up. Sailors rowed to it, and climbed up its sides. they danced on its belly. That's a merry, little story, but I know a more romantic one. It seems that a monster was seen from a steamship. Then the lonely thing mistook the vessel for a female of his species. He overwhelmed her with catastrophic endearments.

But I am avoiding stories of traditional serpentine monsters of the sea. One reason is that collections of these stories are easily available. The astronomer has not lived, who had ever collected and written a book upon data not sanctioned by the dogmas of his cult, but my slightly favorable opinion of biologists continues, and I note that a big book of Se Serpent stories was written by Dr. Oudemans, Director of the Zoo, at The Hague, Holland. When that book came out, a review of it, in Nature, was not far from abusive. Away back in the year 1848, conventionalists were outraged, because of the source of one of these stories. For the account, by Capt. M'Quhae, of H. M. S. Daedalus, of a huge, unknown creature, said by him to have been seen by him, in the ocean, Aug. 6, 1848, see the Zoologist, vol. 6. Someone else who bothered the conventionalists was the Captain of the Royal Yacht, the Osborne, who, in an official report to the Admiralty, told of having seen a monster-not serpent-like-off the coast of Sicily, May 2, 1877. See the London Times, June 14, 1877, and Land and Water, Sept. 8, 1877. The creature was turtle-like, visible part of the body about fifty feet long.

There was an at tempt to correlate this appearance with a submarine eruption, but I have found that this eruption-in the Gulf of Tunis-had occurred in February.

The suggestion was that in the depths of· the ocean may live monsters, which are occasionally cast to the surface by submarine disturbances.

It is a convenience. Accept that unknown sea monsters exist, and how account for the relatively few observations upon things so conspicuous? That they live in ocean depths, and come only occasionally to the surface.

I have gone into the subject of deep-sea dredging, and, in museums, have looked at models of deep-sea creatures, but I have never heard of a living thing of considerable size that has been brought up from profound ocean depths. William Beebe has never brought up anything of the kind. On his Arcturus Adventure, anything that got away from him, and his hooks and his nets and his dredges, must have been small and slippery. It seems that anything with an exposure of wide surfaces could not withstand great pressure. However, this is only reasoning., Before the days of deep-sea dredging, scientists reasoned that nothing at all could live far down in the sea. Also, now most of them would argue that, because of the great difference between pressures, any living thing coming up from ocean depths would burst. Not necessarily so, according to Beebe. Some of the deep-sea creatures that he brought up were so un conventional as to live several hours, and to show no sign of disruption. So-, like everybody else, I don't know what to think, but, rather uncommonly, I know that.

In October, 1883, there was a story in the newspapers-I take from the Quebec Daily Mercury, Oct. 7, 1883-of an unknown animal, which was seen by Capt. Seymour, of the bark Hope On, off the Pearl Islands, about 50 miles from Panama. In Knowledge, Nov. 30, 1883, Richard Proctor tells of this animal, and says that also it had been reported by officers of a steamship. This one was handsome. Anyway, it had a head like that of a "handsome horse." It had either four legs or four "jointed fins." Covered with a brownish hide, upon which were large, black spots. Circus-horseish. About twenty feet long. There was another story told, about the same time. New Zealand Times, Dec. 12, 1883- report by a sea captain, who had seen something like a turtle, 60 feet long, and 40 feet wide.

Perhaps stories of turtle-backed objects of large size re late to submersible' vessels. If there were no submersible vessels of this earth, in the year 1883, we think of submersibles from somewhere else. Why they should be so secretive, we can't much inquire into now, because we are so much concerned with other concealments and suppressions. I suspect that, in other worlds, or in other parts of one existence, there is esoteric knowledge of the human beings of this earth, kept back from common knowledge. This is easily thinkable, because even upon this earth there is little knowledge of human beings.

There have been suggestions of an occult control upon the minds of the inhabitants of this earth. Let anybody who does not like the idea that his mind may be most subtly controlled, without his knowledge of it, think back to what propagandists did with his beliefs in· the years 1914-18. Also he need not think so far back as that.

The standardized explanations by which conventional scientists have checked inquiry into alleged appearances of strange living things, in the ocean, are mentioned in the following record:

Something was seen, off the west coast of Africa, Oct. 17, 1912. Passengers on a vessel said that they had seen the head and neck of a monster. They appointed a committee to see to it that record should be. made of their observations. in the Cape Times (Cape Town) Oct. 29, 1912, Mr. Wilmot, former member of the Cape Legislative Council, records this experience, saying that there is no use trying to think that four independent witnesses had seen nothing but a string of dolphins or a gigantic strand of sea weed, or anything else, except an unknown monster. It's the fishmonger of Worcester in his marine appearance.

In this field of reported observations, so successful has been a seeming control of minds upon this earth, and guidance into picturing nothing but a string of dolphins or a gigantic strand of sea weed, that, now that the ghost has been considerably rehabilitated-though in my own records of hundreds of unexplained occurrences, the ghost-like scarcely ever appears-the Sea Serpent is foremost in representing what is supposed to be the mythical. I don't know how many books I have read, in each of which is pictured a long strand of sea weed, with the root-end bulbed and gnarled grotesquely like a head. I suppose that hosts of readers have been convinced by these pictures.

But, if a monster from somewhere else should arrive upon the land of this earth, and, perhaps being out of adaptation, should die upon land, probably it would not be seen. I have noted several letters to newspapers, by big-game hunters who had never heard of anybody coming upon a dead elephant. Sir Emersoll Tennent has written that, though he had often inquired of Europeans and Cingalese, he had never heard of anybody who had seen the remains of an elephant in the forests of Ceylon. A jungle soon vegetates euphemisms around its obscenities, but the frank ocean has not the pruderies of a jungle.

Strange bones have often been found on land. They have soon been conventionalized. When bones of a monster are found, the pattern-makers of a museum arrange whatever they can into conventional structures, and then fill in with plaster, colored differently, so that there shall be no deception. After a few years, these differences become undetectable. There is considerable dissatisfaction with the paleontologists. I notice in museums that, even when plaster casts are conspicuously labeled as nothing but plaster casts, some honest fellow has dug off chips to expose that there isn't a bone in them.

What we're looking for is an account of something satisfactorily monstrous, and not more or less in the distance: something that is not of paleontologic memory that has been jogged so plasterfully. The sea is the best field for data.

In the Mems. Wernerian Nat. Hist. Soc., 1-418, is published a paper by Dr. Barclay, who tells of the remains of an unknown· monster that had been cast up by the sea, in September, 1808, at Stronsa, one of the Orkneys. We've got ahold of something now that was well observed. As fast as they could, observers got rid of this hunk, which for weeks, under a summer sun, had been making itself evidential. But the evidence came back. So again the observers got a rope and towed it out to sea. Sultry day soon-a flop on the beach-more observations. According to different descriptions, in affidavits by inhabitants of Stronsa, the remains of this creature had six "arms," or "paws," or "wings." There is a suggestion of stumps of fins here, but it is said that the bulk was "without the least resemblance or affinity to fish." Dr. Barclay told that in his possession was part of the "mane" of the monster.

A perhaps similar bulk· was, upon the 1st of December, 1896, cast upon the coast of Florida, twelve miles south of St. Augustine. There were appendages, or ridges, upon it, and at first these formations were said to be stumps of tentacles. But, in the American Naturalist, 31-304, Prof. A. E. Verrill says that this suggestion that the mass of flesh was the remains of an octopus, is baseless. The mass was 21 feet long, 7 feet wide and 4" feet high: estimated weight 7 tons. Reproductions of several photographs are published in the American Naturalist. Prof. Verrill says that, despite the great size of this mass, it was only part of an animal. He argues that it was part of the head of a creature like a sperm whale, but he says that it was decidedly unlike the head of any ordinary sperm whale, having no features of a whale's head. Also, according to a description in the New York Herald, Dec. 2, 1896, the bulk seems not to have been whale-like. "The hide is of a light pink color, nearly white, and in the sunshine has a distinct silvery appearance. It is very tough and cannot be penetrated even with a sharp knife." A pink monster, or an appalling thing with the look of a cherub, is another of our improvements upon conventional biology.

For a yarn, or an important record, of a reptile of "prehistoric size and appearance," said to have been found on the beach of the Gulf of Fonseca, Salvador, see the New York Herald Tribune, June 16, 1928. It was about ninety feet long, marked with black and white stripes, and was "exceedingly corpulent." Good-natured, fat monsters, too, are new to me.

I have searched especially for sea stories of hairy, or fur covered monsters. Such creatures would not be sea animals, in the exclusive sense that something covered with scales might be. If unknown, they would have to be considered inhabitants of lands. Then up comes the question-what lands?

English Mechanic, April 7, 1899-that, according to Australian newspapers, the captain of a trading vessel had arrived in Sydney, with parts of an unknown monster. "The hide, or skin, of the monster was covered with hair."

The arrival of these remains is reported in the Sydney Morning Herald, in issues from Feb. 23 to March 2, 1899. It is said that, according to Capt. Oliver, of the trading ship Emu, he had found, upon the beach of Suarro Is land, the carcass of a two-headed monster.

That is just a little too interesting.

We find that the reporter who told this story dropped the most interesting part of it, in his subsequent accounts, which were upon two skulls, a vertebra, and a rib bone: but he was determined to discredit the find, and told that the bones were obviously fossils, implying that the Captain had invented a story of bodies of two animals that had recently been alive.

When we come upon assurances that a mystery has been solved, we go on investigating.

In the Sydney Daily Telegraph, February 28, it is said that an attempt to identify the bones as fossils had been refuted. Professional and amateur scientists had accepted an invitation to examine the bones, and, according to the testimony of their noses, these things decidedly were not fossils. Each skull was more than two feet long, and was shaped somewhat like a horse's, but upon it was a beak. There are beaked whales, but these remains were not remains of beaked whales, if be accepted Captain Oliver's unsupported statements as to hairiness and great size. It is said that no specimens of the hairy hide had been taken, because all parts, except the scraped bones, of these bulks that had been lying under a tropical sun weren't just what one would want to take along in a small ship. According to Capt. Oliver, one of the bodies was sixty feet long. The largest beaked whales are not known to exceed thirty feet in length.

Mr. Waite, of the Australian Museum, examined the bones.

He said that they were of beaked whales.

Mr. F. A. Mitchell-Hedges, in Battles with Giant Fish, tells of remains of a tremendous, unknown mammal, which was washed ashore, at Cape May, N. J., November, 1921.

"This mammal whose weight was estimated at over 15 tons, which-to give a comparison of size-is almost as large as five fully grown elephants, was visited by many scientists, who were unable to place it, and positively stated that nothing yet known to science could in any way compare with it."

I investigated the story of the Cape May monster, wherever I got the idea that I could find out anything in particular.

Somebody in Cape May wrote to me that the thing was a highly undesirable · carcass of a whale, which had been towed out to sea. Somebody else wrote to me that it was a monster with a tusk twelve feet long, which he had seen. He said that, if I'd like to have it, he'd send me a photo graph of the monster. After writing of having seen some thing with a tusk twelve feet long, he sent me a photograph of something with two tusks, each six feet long. But only one of the seeming tusks is clear in the picture, and it could be, not a tusk, but part of the jaw bone of a whale, propped up tuskwise.

In the London- Daily Mail, Dec. 27, 1924, appeared a story

of an extraordinary carcass that was washed up, on the coast of Natal, Oct. 25, 1924. It was 47 feet long, and was covered with white hair, like a polar bear's-

1 won't go into this, because I consider it a worthless yam. In accordance with my methods, considering this a foolish and worthless yarn, I sent out letters to South African newspapers, calling upon readers, who could, to investigate this story. Nobody answered.

In the New Zealand Times, March 19, 1883, it is said that bones of an unknown monster, about 40 feet long, had been found upon the coast of Queensland, .and had been taken to Rockhampton, Queensland, "There are. the remains of what must have been an enormous snout, 8 feet long, in which the respiratory passage are yet traceable." These could not have been the remains of a beaked whale. Whatever hip bones a cetacean has are only vestigial structures. In a sperm whale, 55 feet long, the hip bones are detached and atrophied relics of former uses, each about one foot long. A hip bone of the Queensland monster is described as enormous.

In looking over the London Daily News, I came upon an item. Trawlers of the steamship Balmedic had brought to Grimsby the skull of an unknown monster, dredged up in the Atlantic, north of Scotland (Daily News, June 26, 1908). The size of the skull indicated an animal the size of an elephant, and it was in ""a wonderful state of preservation. It was unlike the skull of any cetacean, having eye sockets a foot across. From the jaws hung a leathery tongue, three feet long. I found, in the Grimsby Telegraph, June 29th, a reproduction of a photograph of this skull, with the long tongue hanging from the beak-like jaws. I made a sketch of the skull, as pictured, and sent it with a description to the British Museum (Natural History). I received an answer from Mr. W. P. Pycraft, who wrote that he had never seen any animal with such a skull-•and I have seen a good many!" It is just possible that nobody else has ever seen anything much resembling a sketch that I'd make of anything, but that has nothing to do with -descriptions of the tongue. According to Mr. Pycraft no known cetacean has such a tongue.

I went on searching, trying to come upon something about a hairy monster: furred, anything except scaled, or with a hide like a whale's.

London newspapers, July 6, 1913-a lengthy telegram that had been sent by Mr. Hartwell Conder,

Tasmanian State Mining Engineer, to Mr. Wallace, the Secretary of Mines; of Tasmania-that, upon April 20, 1913, two of Mr. Conder's companions, named Davies and Harris, had seen a huge, unknown animal, near Macquarie Harbor, Tasmania. "The animal was about fifteen feet long. It had ·a very small head, only the size of the head of a kangaroo dog. It had a thick, arched neck, passing gradually into the barrel of the body. It had no definite tail and no fins. It was furred, the coat in appearance resembling that of a horse of chestnut color, well-groomed and shining. It had four dis tinct legs. It traveled by bounding-i.e., by arching its back and gathering up its body, so that the footprints of the fore feet were level with those of the hind feet. It made definite footprints. These showed circular impressions, with a diameter (measured) of 9 inches, and the marks of claws, about 7 inches long, extending outward from the body. There was no evidence for or against webbing."

In reply to my inquiries, Mrs. Conder-North Terrace, Burnie, Tasmania, wrote to me, as asked to by Mr. Con der, saying that the published description is accurate, and that, unless there be a seal with jointed flippers, upon which the creature could raise itself and run, Mr. Conder "could not be altogether convinced that the animal was a seal."

I have not looked for record of any such known seal. I take for granted that the· seal type had conventionalized so that there is no such seal.

It may be that there have been several finds of remains of a large, long-snouted animal that is unknown to the paleontologists, because, though it has occasionally appeared here, it has never been indigenous to this earth. New York Sun, Nov. 28, 1930-"Monster in ice has long snout." Skeleton and considerable flesh, of an unknown animal found in the ice, upon Glacier Island, Alaska. The animal was 24 feet long; head 59 inches long; snout 39 inches long. In some of the reports it was said that the animal was covered with hair, or fur. Conventionally one thinks of mammoths of Siberia, preserved for ages in ice. But, if nothing proves any thing, simply that something is found in ice may not mean that for ages it was preserved in ice.

X

UNKNOWN, luminous things, or beings, have often been seen, sometimes close to this earth, and. sometimes high in the sky. It may be that some of them were living things that occasionally come from somewhere else in our existence, but that others were lights on the vessels of explorers, or voyagers, from somewhere else.

From time to time, luminous objects, or beings, have been reported from Brown Mountain, North Carolina. They appear, and then for a long time are not seen, and then they appear again. See the Literary Digest, Nov. 7, 1925. I have other records. The luminosities travel, as if with motions of their own. They are brilliant, globular forms, and move in the sky, with a leisureliness and duration that exclude any explanation in meteoric terms. For many years, there had been talk upon this subject, and then, in the year 1922, people of North Carolina, asking for a scientific investigation, were referred to the United States Geological Survey. A geologist was sent from Washington to investigate these things in the sky.

One imagines, but most likely only faintly, the superiority of this geologist from Washington. He heard stories from the natives. He contrasted his own sound principles with the irresponsible gab of denizens, and went right to the investigation, scientifically. He went out on a road, and saw lights, and made his report. 47% of the lights that he saw were automobile headlights; 33% of them were locomotive headlights; 10% were lights in houses, and 10% were bush fires. Tot that up, and see that efficiency can't go further. The geologist from Washington, having investigated nothing that he had been sent to investigate, returned to Washing ton, which also, by the way, is a place where· there's plenty to investigate, and I suppose that the people of North Carolina will be no wiser, as to these things in the sky, if some other time they appeal to a United States Fish Commission, or the Department of Labor.

I don't know to just what degree my accusation, in these matters, is of the laziness and feeble-mindedness of scientists. Or, instead of accusing, I am simply pointing out every body's inability seriously to spend time upon something, which, according to his preconceptions, is nonsense. Scientists, in matters of our data, have been like somebody in Europe, before the-year 1492, hearing stories of lands to the west, going out on the ocean for an hour or so, in a row-boat, and then saying, whether exactly in these words, or not: "Oh, hell there ain't no America."

In Knowledge, September, 1913, Count de Sibour enjoyed his laziness, or incompetence, which a merciful providence, bent upon keeping us human beings reconciled to being human beings, made him think was his own superiority. He told a story of foolish, credulous people, in North Norfolk, England, who, in the winter of 1907.:8, believed that a pair 'Of shining things, moving about the fields, could not be explained as he explained them. We are told of a commonplace ending of this alleged mystery: that finally a game keeper shot one of these objects, and found that it was a common barn owl, phosphorescent with decayed wood from its nesting place, or with a fungous disease of its feathers. According to other accounts, these thin s were as brilliant as electric lights. But a phosphorescent owl could not shine with a light like an electric light. So De Sibour described the light as "a pale, yellow glow," such as a phosphorescent owl could shine with.

Science concerns itself with adaptions, and science itself is adaptation. We are reminded of the Rev. Hugh Guy. He could not explain downpours: so he turned downpours into "a small quantity," which he could explain.

De Sibour knew nothing about this subject, from his own experiences. We go to the same records to which he went. Like him, we find just about what we want to find. In the London Times, Dec. 10, 1907, and in following issues, are accounts of these luminous objects, which were flying about the fields of North Norfolk, having been reported by Mr. R. W. Purdy, a well-known writer upon biologic subjects.

Among · other attempts to assimilate with the known, or among other expressions of a world-wide antipathy to the finding out of anything new, was the idea that owls are sometimes luminous. 'The idea came first, or the solution of the problem was published first; and then the problem was-. fitted to the solution. This is said to be a favorite method of ratiocination with inmates of a home for the mentally deficient, but I should think that one of these inmates should feel at home anywhere. De Sibour and others fitted in a story that a luminous owl had been shot. I think that at times there may be faintly luminous owls, because I accept that, under some circumstances, almost anything may be luminous. English Mechanic, 10-15-case of a man with a luminous toe.

Shining things, flying like birds, in the fields of North Norfolk continued to be reported. The brilliant things look ed electric. When they rested on trees, everything around them was illuminated.

Purdy's descriptions are very' different from "a pale, yellow glow." Upon the night of December 1st, he saw something that he thought was the lamp of a motorcycle, moving rapidly toward him, in a field, stop ping, then rising several yards, moving higher, and then retreating. It moved in various directions. See the Field, Jan. 11, 1908.

De Sibour was uncareful, in his mystery-squelching story, his bobbed story, a story that forced a mystery to a commonplace ending. No gamekeeper shot a luminous owl, at this time, in North Norfolk. But somebody did say that he had conventionally solved the mystery. Eastern Daily Press (Norwich), Feb. 7, 1908 -that, early in the morning of the 5th, Mr. E. S. Cannell, of Lower Hellesdon, saw something shining on a grass bank. According to him, it fluttered up to him, and he found that it was the explanation of a mystery. It was a luminous owl, he said; and, as told by him, he carried it to his home, where it died, "still luminous."

But see the Press of the 8th-that Mr. Cannell's dead owl had been taken to a taxidermist, who had been interview ed. Of course a phosphorescence of a bird, whether from decayed wood, or feather fungi, would be independent of life or. death of the bird. Questioned as to whether the body of the owl was luminous or not, the taxidermist said: "I have seen nothing luminous about it."

In zoological journals, one frequently comes upon allusions to these things, or beings, of North Norfolk. No gamekeeper killed one of them, but the story of the game keeper who had killed a luminous owl is told in these records that are said to be scientific. It 'is not necessary that a gamekeeper should kill a luminous owl, and so put an end to a mystery. A story that he did will serve just as well. The finding, or the procuring in some way or another, of the body of an owl, did not put an end to the mystery, except in most of the records, that are said to be scientific. There were at first two lights, and there continued to be two lights. The brilliant things continued to be seen in the fields, flitting about, appearing and disappearing. The last observation findable by me (May 3, 1908) is record ed in the Trans. Norfolk and Norwich Naturalists' Society, 8-550. Purdy records an observation upon the two lights, seen together, more than a month after the date upon which Mr. Cannell said that his owl had fluttered right up to him. Something else was reported, in this region. In the Eastern Daily Press, Jan. 28, 1908, it is said that, at night moon bright-"a dark, globular object, with a structure of some kind upon the side of it, traveling at a great pace," had been seen in the sky, by employees of the Norwich Transportation Company, at Mousehead. "It · seemed too large for a kite, and, besides, its movements seemed under control, for it was traveling against the wind."

I am here noting only a few of the many records of un known, seemingly living, luminous things that used to be called will-o'-the-wisps. They come and they go, and their reappearances in a small region makes me think of other localized repetitions that we have noted.

London Daily Express, February 15, and following issues, 1923-brilliant, luminous things moving across fields, some times high in the air, at Fenny Compton, Warwickshire. They were "intense lights," like automobile headlights. Some times these luminous things, or beings, hovered over a farmhouse. It was a deserted farmhouse, according to the London Daily News, February 13. About a year later, one of these objects, or whatever they were, returned, and was reported as "a swiftly moving light," by several persons, one of them Miss Olive Knight, a school teacher, of Fenny Compton (London Sunday News, Jan. 27, 1924).

The Earl of Erne tells, in the London Daily Mall, Dec. 24, 1912, of brilliant luminosities that, from time to time; in a period of seven or eight years, had been appearing near Lough Erne, Londonderry, Ireland, "in size and shape very much like a motor car lamp." In later issues of the Daily Mail, the ·Countess of Erne tells of these things, or creatures, "like motor car lamps, large and round."

XI

SCIENCE is very much like the Civil War, in the U.S.A. No matter which side won, it would have been an. American victory. By Science, I mean conventionalization of alleged knowledge. It, or maybe she, acts to maintain itself, or what ever, against further enlightenment, or alleged enlightenment but when giving in, there is not surrender, but partnership, and something that had been bitterly fought then becomes another factor in its, or her, prestige. So, seventy years ago, no matter whether evolutionists or anti-evolutionists had won, it would have been a big, scientific victory anyway. No wonder so many of us are humbled by a reputation that can't lose any way.

Science is a maw, or a headless and limbless stomach, an amoeba-like gut that maintains itself by incorporating the assimilable and rejecting the indigestible. There are whirl winds and waterspouts, and it seems acceptable that there · have been rare occurrences of faintly luminous owls. Then by a process of sorting over data, rejecting the objection able, and taking in the desirable, Science saves itself great pains, because a bellyache is something that is only a gut in torment. So, with alimentary treatments, a shower of living things can always be made to assimilate with the whirl wind-explanation, and a brilliant, electric thing can be toned down digestibly-:In extreme cases there is a secretion of fishmongers or gamekeepers.

In some cases of obstinate unassimilableness, it is sup posed to be necessary to swat a little girl. But I doubt the necessity, because there is in human beings such a fondness, or sometimes such a passion, for confessing, that sooner or later somebody will come forward with almost any desired confession. Sometimes, from time to time, half a dozen per sons confess to having committed the same murder. The police pay scarcely any attention any more to a new confession, in the matter of the Hall murder, in New Jersey. There was a case, in an English police court, of a man who had given. himself up, as a deserter from the army.

But a policeman testified that this was his fifth or sixth confession, and that he had never been in the army. The man admitted the charge. "But," said he, "I have something else that I wish to confess." "I'll hear no more of your confessions. Six months!" said the magistrate. In some cases the incentive for false confessions is not obvious, but in others it is obviously to come out of one's pale, yellow glow, and be brilliant in limelight. There have been cases not quite of confessions, but of somebody attributing to himself unexplained occurrences, or taking advantage of them for various kinds of profit. I accept that, if explorers from some where else should visit this earth, and if their vessels, or the lights of their vessels, should be seen by millions of the inhabitants of this earth, the data would soon be conventionalized. If beings, like human beings, from somewhere else, should land upon this earth, near New York, and parade up Broadway, and then sail away, somebody, a year or so later, would "confess" that it had been a hoax by him and some companions, who had dressed up for their parts, and had jabbered, as they thought extra-mundanians should jabber. New Yorkers would say that from the first they had suspected something wrong. Who ever heard of distinguished foreigners coming to New York, and not trying to borrow something? Or not coinciding with propaganda in newspapers?

Probably, in August, 1929, an aeroplane from somewhere in Europe passed over a jungle, and was reported, in a village, by a native. Savages are highly scientific. They rea son upon generalizations that are so exclusive as to be nothing short of academic. They are as keen as any Newton, or Einstein, in understanding that, in order to arrive at what is said to be the known, they must start with something that they don't know about. We'd have a pretty good idea of what the wisemen said, when they heard the story of a vessel, probably occupied by passengers, that had been seen in the sky, if we could accept that. they could be so undignified 'as to say anything-

New York Herald Tribune, Aug. 29, 1929-a traveling light in the sky-about 400 miles off the coast of Virginia. It was reported by Thomas Stuart. third mate of the steam ship Coldwater, of the South Atlantic Steamship Line. "There was something that gave the impression that it was a large, passenger craft." It was traveling at an estimated speed of 100 miles an hour, in the direction of Bermuda. There was an investigation that "failed to reveal any trans-Atlantic or Bermuda flight."

We shall have an expression upon luminous appearances that may have been of a. kind with the seeming creatures of the preceding chapter, or that may belong in another category.

Before I could find out the date, and look the matter up, I came upon several humorous allusions, in English newspapers, to a time when there was a scare in England, because of moving lights in the sky. And all the excitement was about the advertising scheme of an automobile manufacturer, who had sent up an imitation-airship, with lanterns tied to it. There was a lesson in this: presumably other alleged mysteries could be explained in similar, commonplace terms. I was doing one of my relatively minor jobs, which was going through the London Daily Mail, for a period of about twenty-five years, when I came upon this-

March 25, 1909-that, upon the 23rd of March, at 5:10 o'clock in the morning two constables, in different parts of the city of Peterborough, had reported having seen an object, carrying a light, moving over the city, with sounds of a motor. In the Peterborough Advertiser, March 27th, is published an interview with one of these constables, who described "an object, somewhat oblong and narrow in shape, carrying a powerful light." To suit whatever anybody should prefer, I could give data to show that only lights, and no object were seen, and that no ·sound was heard; or that a vessel, carrying lights, was seen, and that sounds, like sounds of a motor, were heard.

It is said, in the Daily Mail, May 17th, that many other stories of unaccountable objects and lights in the sky had reached the office of the Mail. If so, these stories were not published. The newspapers are supposed to be avid for sensational news, but they have their conventions, and un accountable lights and objects in the sky are not supposed to have sex, and it is likely that hosts of strange, but sexless, occurrences have been reported, but have not been told of in the newspapers. In the Dally Mail, it is said that no attention had been paid to the letters, because every thing that was mentioned in them, as evidence, was unsatisfactory. It is said that the object reported at Peterborough was probably a kite. with a lantern tied to it. On the 15th of May, a Constable at Northampton had sent to headquarters a written report, upon lights that he, had seen in the sky at 9 P.M.; but Chief Constable Madlin had learned that a practical joker had sent up a fire balloon.

The practical joker of Northampton, amusing himself at

9 o'clock in the evening, is an understandable representative of his species; but some other representative of his species, flying · a kite and lanterns, at Peterborough, at 5 o'clock in the morning, limiting

his audience mostly to milkmen, though maybe a joker, could not have been a very practical joker. He must have been fond of travel. There were other reports from various places in England and in Wales. There were reports from places far apart.

Dally Mail, May 20-that a man, named Lithbridge, of 4 Roland Street, Cardiff, Wales, had, in the office of the Cardiff Evening Express, told a marvelous story. This story was that, upon the 18th of May, about 11 P.M., while walking along a road, near the Caerphilly Mountains, Wales, he had seen, on the grass, at a side of the road, a large tube-shaped construction. In it were two men, in heavy fur-overcoats. When they saw Mr. Lithbridge, they spoke excitedly to each other, in a foreign language, and sailed away. Newspaper men visited the place, and found the grass trampled, and found a scattering of tom newspapers and other debris.

If anybody else wants to think that these foreigners were explorers from Mars or the moon, here is a story that of course can be reasoned out quite, or almost, satisfactorily. At any rate, still more satisfactorily it may be said that no foreigners of this earth were sailing in the sky of Great Britain. In the Western Mail (Cardiff), May 21st, is published an interview with Mr. C. S. Rolls, the motorist, and the founder of the Aero Club, who gave his opinion that some of the stories of a strange object in the sky were hoaxes, but that not all of them could be explained so. Chiefly for the reason that there was no known airship of this earth, with such powers of flight, the reported observations were discredited, at least sometimes, in all news papers that I have looked at. In the London Weekly Dispatch, May 23rd, the stories are so discredited, and it is argued at to be seen so often, without having been seen to cross the Channel, an airship would have to have a base in England, to which, in view of the general excitement, it would certainly have been traced: a base where it would be seen, especially during the tedious preliminaries of ascent." Then, in the Weekly Dispatch, are listed reports from 22 places, in the week preceding the 23rd of May, and 19 reports earlier in May and in March.

Mr. Lithbridge was a Punch and Judy man. Perhaps his story was of some profit to him. Not much attention was paid to it. But then came another explanation.

Upon May 26th, it was told in the newspapers that the mystery of the lights in the sky had been solved. A large imitation-airship had "come down at Dunstable, and the lights had been upon that. It was an advertising scheme. An automobile manufacturer had been dragging the thing around in England and Wales. There had been reports from Ireland, but Ireland was omitted in this explanation. We are told that this object, roped to an automobile, had been dragged along the roads, amusingly exciting persons who were not very far advanced mentally. With whatever degree of advancement mine may be, I suppose that such a thing could be dragged slowly, and for a short time, perhaps only a few minutes, because it was of hot-air-inflation, along a road, and conceivably through a city or two, with a policeman, who reported lights in the sky, not seeing a rope going up from an automobile: but, with whatever degree of advancement that of mine may be, I do not think. of any such successful imposition in about forty large cities, some of them several hundred miles apart. No one at Dunstable saw or heard the imitation-airship come down from the sky. An object, to which was tied a card, upon which was a request to communicate with a London automobile manufacturer, "in case of accident," was found in a field, morning of May 26th.

The explanation, as I want to see it, is that probably the automobile manufacturer took advantage of the interest in lights in the sky, and at night dumped a contrivance into a field, having tied his card to it. If so this was only one of many occurrences that have been exploited by persons who had a liking, or a use, for publicity. Probably Mr. Cannell and his dead owl can be so explained: and, though I should very much like to accept Mr. Lithbridge's story, I fear me that we shall have to consider him one of these exploiters. Also, there was the case of the press agent, who, taking advantage of stories of a prowling animal, tied in wings and green whiskers to a kangaroo.

The range of the reported observations • was from Ipswich, on the east coast of England, to Belfast, Ireland, a distance of 350 miles and in Great Britain, from Hull to Swansea, a distance of 200 miles. Perhaps a gas bag could be dragged around a little, but the imitation-airship that was found at Dunstable, was a Oiinsy contrivance, consisting of two hot air balloons, and a frame about 20 feet long, connecting them.

The lights in the sky were frequently reported, upon the same night, from places far apart. Upon the night of May 9th, reports came from Northampton, Wisbech, Stamford, and Southend. In the Weekly Dispatch, May 23rd, it is pointed out that to be seen at Southend about 11 P.M., and then twenty minutes later, at Stamford, seventy miles away, the object in the sky must have traveled at a rate of 210 miles an hour.

The question that comes up is whether, after the finding of the object at Dunstable, or after a commonplace ending of a mystery, lights continued to be seen traveling in the sky.

The stoppage was abrupt. Or the stoppage of publication of reports was abrupt.

XII

IT may be that upon new principles we can account for the ·mystery of the Marie Celeste. If there is a selective force, which transports stones exclusively, or larvae, and nothing but larvae, or transports living things of various sizes, but nothing but living things, such a selective force might affect a number of human beings, leaving no trace, because unaffective to everything else.

I take from the report by the Queen's Proctor, in the Admiralty Court, published in the London Times, Feb. 14, 1873. Upon the 5th of December, 1872, between the Azores and Lisbon, the crew of the British ship Dei Gratia saw a vessel, made out to be the American brigantine Marie Celeste. Her sails were set, and she was tacking, but so erratically that attention was attracted. Whether ships are really females, or not, this one looked so helpless, or woe begone, that all absence of male protection was suspected. The Britons shoved out and boarded the vessel. There was nobody aboard. There was findable nothing by which to account for the abandonment. "Every part of the vessel. inside and outside, was· in good order and condition." In the log book, the latest entry, having in it no suggestion of impending trouble of any kind, was dated November 25th. There was no sign of any such trouble as mutiny. A phial of oil, used by the captain's wife, upon a sewing machine, stood upright, indicating that there had been no rough weather. Investigation of this mystery was worldwide. The State Department of the United States communicated with all representatives abroad, and every custom house in the world was more or less alert for information of any kind: but fourteen persons, in a time of calm weather, and under circumstances that gave no indication of any kind of violence, disappeared, and either nothing, or altogether too much, was found out. I have a collection of yarns, by highly individualized liars, or artists who scorned, in any particular, to imitate one another; who told, thirty, forty, or fifty years later, of having been members of this crew.

London Times, Nov. 6, 1840-the Rosalie, a large French ship, bound from Hamburg to Havana-abandoned ship no clew to an explanation. Most of the sails set-no leak valuable cargo. There was a half-starved canary in a cage. But I suggest that, with our hints of Teleportation, we are on the wrong track. Crews of vessels have disappeared, and vessels have disappeared. It may be that something of which the inhabitants of this earth know nothing, is concerned in these disappearances, or seizures. In the New York Sun, April 24, 1930, the French astronomer and meteorologist, Gen. Frederic Chapel. is quoted, saying that aircraft, missing at sea, and seacraft, may have been struck by meteors. That there is something of the unexplained in these disappearances, many writers have felt. But there is no recorded instance of aircraft, flying over land, having been struck by a meteor, and I know of few instances of reported falls of meteoric matter upon vessels, and no instance of a vessel that has been much damaged by a meteor.

The disappearance of the Cyclops, a fuel ship of the U. S. N., even though in war time, is considered mysterious-sometime after March 4, 1918, after leaving Barbados, B. W. I., for Hampton Roads, Va. When the Titanic went down, April 15, 1912, flotsam was ported months afterward, and there were many survivors: but, after the disappearance of the White Star steam ship, the Naronic, in February, 1893, two empty lifeboats, supposed to be hers, were reported by a sea captain, and nothing more was seen. In the report by the London Board of Trade, it was considered highly improbable that the Naronic had struck an iceberg. It was said that this vessel was "almost perfect," in construction and equipment. She was a freighter, with 75 men aboard. There were life belts for all New York Times, June 21, 1921-disappearance of three American ships-difficult to think of piracy-seemed to be no other explanation-five departments of the Washington Government investigating. In February, the Carol Deering, a five-masted schooner, of Portland, Me., had gone ashore, near Diamond Shoals, North Carolina. The mystery is similar to that of the Marie Celeste. Nobody aboard. Everything was in good condition. The circumstance that attracted most attention was that the crew had disappeared about the time a meal was to be served. A little later, a bottle was picked up on shore, and in it was a message purporting to have been written by the mate. of the vessel "An oil-burning tanker has boarded us, and placed our crew in irons. Get word to headquarters of the Company at once." Just how somebody in irons could get a container for a message makes me wonder: still, if it's a bottle, they say that that could be got by anybody in double irons.

In the London Dally Maa, June 22, the finding of another bottle, with a message in it, is told of-from the Captain of the Deering, this time-that he had been taken prisoner by the crew, and had been put upon another vessel. After the Waratah "mysteriously disappeared," off the coast of South Africa, July, 1909, five bottles, all said to be hoaxes, were found. There is as much complication and bafflement Jn this subject, as in anything that Science is said absolutely to have proved. If some of us tire of our existence, and would like to try some other existence, they had better think it over, because anything merrier than ours is hard to conceive of. Every shipwreck, or any other catastrophe, brings out merrymakers. The tragedy of the Waratah was enjoyed a long time. More than thirteen years later (Nov. 21, 1922) another bottle, said to be a hoax, was found near Cape Town. Still, I am affected just the other way, and am taking on a new pessimism. Heretofore I have thought cheerfully of bottles. But there's a depression from anything, once the humorists get ahold of it. I wonder how comes it that nobody has reported Boding an old bottle, and in it a sea captain's account of an impending mutiny, signed "Christopher Columbus.

New York Times, June 22, 1921-"More ships added to the mystery-list-almost simultaneous disappearances, with out a trace, regarded as significant." Times, June 24-about a dozen vessels in the list. And yet such a swipe by an unknown force, of the vessels of a nation, along its own coast, was soon thought of no more. Anything could occur, and if not openly visible, or if observed by millions, would soon be gulfed in forgetful ness. Or soon it would be conventionalized. In the year 1921, it was customary to accuse the Russians. I think that the climax was reached, in the year 192'1, when unruliness of natives in the jungles of Peru was attributed to Russian agents. Still, I suppose that, for years, whenever there is re volt against misrule and oppression, propagandists will tell us the same old yarn of otherwise contented natives, misled by those Russians. In June, 1921, the way of explaining the disappearance of a dozen vessels was by saying that it was thought that the Soviet Government was stealing them.

It may be that constructions from somewhere else have appeared upon this earth, and have seized crews of this earth's ships.

In their book, The Cruise of the Bacchante, the two young princes, sons of the Prince of Wales, one of them now the King of England, tell of "a strange light, as if of a phan tom vessel all aglow" that was, at four o'clock, morning of the 11th of June, 1881, between Melbourne and Sydney, reported by the lookout of the Bacchante. The unknown appearance was seen by twelve other members of the crew. Whether there be relation, or not, five hours later, the lookout fell from a crosstree and was killed.

Brooklyn Eagle, Sept. 10, 1891-something that was seen at Crawfordsville, Indiana, 2 A.M., Sept. 5th. Two icemen saw it. It was a seemingly headless monster, or it was a construction, about 20 feet long, and 8 feet wide, moving in the sky, seemingly propelled by fin-like attachments. It mov ed toward the icemen. The icemen moved. It sailed away and made such a noise that the Rev. G. W. Switzer, pastor of the Methodist church was awakened, and, looking from his window, saw the object circling in the sky.

I suppose that there was no such person as the Rev. G. W. Switzer. Being convinced that there had probably never been Rev. G. W. Switzer, of Crawfordsville-and taking for a pseudo-standard that if I'm convinced of something that is something to suspect-I looked him up. I learned that the Rev. G. W. Switzer had lived in Crawfordsville, in September, 1891. Then I found out his present address in Michigan. I wrote to him, and received a reply that he was traveling in California, and would send me an account of what he had seen in the sky, immediately after returning home. But I have been unable to get him to send that ac count. li anybody sees a "headless monster" in the sky, it is just as well to think that over, before getting into print. Altogether, I think that I make here as creditable and scientific a demonstration as any by any orthodix scientist, so far encountered by us. The problem is: Did a "headless monster" appear in Crawfordsville, in September, 1891? And I publish the results of my researches: "Yes, a Rev. G. W. Switzer did live in Crawfordsville, at the time."

I'd like to know what Mr. W. H. Smith saw, Sept. 18, 1877, in the sky, moving over the city of Brooklyn. It looked like a winged human form (New York Sun, Sept. 21, 1877).

Zoologist, July, 1868-something that was seen in the sky, near Copiapo, Chile-a construction that carried lights, and was propelled by a noisy motor-or "a gigantic bird; eyes wide open and shining like burning coals; covered with immense scales, which clashed together with a metallic sound."

I don't know what will be thought generally of our data, but in the New York Times, July 6, 1873, the writer of General Notes tells of something that he considered "the very worst case of delirium tremens on record." This was before my time. He copied from the Bonham (Texas) Enterprise-that a few days before the time of writing, a man living 5 or 6 miles from Bonham, had told of having seen something like an enormous serpent, floating over his farm; and that other men working in the fields had seen the thing and had been frightened. I suppose that, equally deliriously, inhabitants of the backwoods of China, would similarly describe one of this earth's airships floating over their farms. I don't know that this one account, considered alone, amounts to anything, but, in the Times, of the 7th of July, I found something else noted. A similar object had been reported from Fort Scott, Kansas. "About halfway above the horizon, the form of a huge serpent, apparently perfect in form, was plainly seen.

New York Times, May 30, 1888-reports from several places, in Darlington County, South Carolina-huge serpent in the sky, moving with a hissing sound, but without visible means of propulsion.

In the London Daily Ex-press, Sept. 11, 1922, it is said that, upon September 9th, John Morris, coxswain of the Barmouth (Wales) Life Boat, and William James, looking out at sea, from the shore, at Barmouth, saw what they thought was an aeroplane falling into the ocean. They rushed out in a motor boat, but found nothing. In the Barmouth Advertiser, of the 14th, it is said that this object had fallen so slowly that features described as features of an aeroplane had been seen. In newspapers and aeronautical journals of the time, there is no findable record of an aeroplane of this earth reported missing.

There was a series of occurrences, in the summer of 1910. Early in July, the crew of the French fishing smack. Jeune Frédéric, reported having seen, in the sky, off the coast of Normandy, a large, black. bird-like object. Suddenly it fell into the sea, bounded back, fell again, and disappeared, leaving no findable traces. Nothing was known of the flight of any terrestrial aircraft, by which to explain (London Weekly Dispatch, July 10). Upon August 17th (London Times, August 19) laborers at work in the forest east of Dessau, Germany, saw in the sky an object that they thought was a balloon. It suddenly flamed, and something that was thought to be its car, fell into the forest. The Chief Forester was notified, and a hunt, on a large scale, was made, but nothing was found. Aeronautical societies report that no known balloon had been sent up. It was thought that the object must have been somebody's large toy balloon. About this time, the fall from the sky of a white cylinder of marble was reported. One of us pioneers, or whatever we are, Mr. F. T. Mayer, looked up this matter, and learned that the reported occurrence was upon the farm of Mr. Daniel Lawyer, Rural Route 4, Westerville. Ohio. I wrote to Mr. Lawyer, asking whether the object could be considered artificial. l had an idea that it might, or might not be a container of a message that had been fired to this earth from Mars or the moon or somewhere else. Mr. Lawyer did not like the suggestion of artificiality, which he interpreted as meaning that he had picked up something that had been made in Ohio. He said that it was not an artificial object, but a meteorite. For a reproduction of a photograph of this symmetric, seemingly carved cylinder, 12 inches long, weight about 3 pounds, see Popular Mechanics, 14-801. About 9 P.M., August 30th-lights as if upon an airship, moving over New York City (New York World, August 31). Aviators were interviewed, but all known aircraft were accounted for. World, September 2-that two men had sent up a large kite. Upon the 21st of September

(New York Tribune, September 22) a great number of round objects were seen passing from west to east over the lower part of New York City. Crowds stood in the streets, watching them. They were thought to be little balloons. I have records of similar objects, in large numbers, that could not be considered little balloons. For several hours this procession continued. If somebody in Jersey City was advertising, he kept quiet in his bid for publicity. The next day, at Dunkirk, N.Y., an object, described as an unknown cigar-shaped balloon, was seen in the sky, O\'.er Lake Erie, seeming to be unmanageable, gradually disappearing, late in the evening. There was so much excitement in Dunkirk that tugboats went out and searched all night. Toronto Daily Maa and Empire, September 24-that someone on a tugboat had found a large box-kite, which had been sent up by a party of campers, and was undoubtedly the reported object.

Mr. A. H. Savage-Landor, in Across Unknown South America, vol. II, p. 425, tells a· story that was told to him, by the people of Porto Principal, Peru, in January, 1912- that, some years before, a ship had been seen in the sky passing over the town, not far above the tree tops. According to his interpretations, it was a "square globe," flying a flag of Stars and Stripes. Mr. Savage-Landor thinks that the object might have been the airship, which, upon Oct. 17, 1910, Wellman abandoned about 400 miles east of Hatteras. In newspaper accounts of this unsuccessful attempt to cross the Atlantic, it is said that, when abandoned, this airship was leaking gas rapidly. li a vessel from somewhere else, flying the Stars and Stripes, is pretty hard to think of, except by thinking that there are Americans everywhere, also the "square globe" is not easy, at least for the more conventional of us. Probably these details· are faults of interpretation; Whatever this thing in the sky may have been, if we will think that it may have been, it returned at night, and this time it showed lights.

In the New York newspapers, September, 1880, are allusions to an unknown object that had been seen traveling in the sky, in several places, especially in St. Louis and Louisville. I have not been able to get a St. Louis newspaper of this time, but I found accounts in the Louisville Courier-Journal July 29, Aug. 6, 1880. Unless an inventor of this earth was more self-effacing than biographies of inventors indicate, no inhabitant of this earth succeeded in making a dirigible aerial contrivance, in the year 1880, then keeping quiet about it. The story is that, between 6 and 7 o'clock, evening of July 28th, people in Louisville saw in the sky "an object like a man, surrounded by machinery, which he seemed to be working with his hands and feet. The object moved in various directions, ascending and de scending, seemingly under control. When darkness came, it disappeared. Then came dispatches, telling of something that had been seen in the sky, at Madisonville, Ky. "It was something with a ball at each end." "It sometimes appeared in a circular form, and then changed to an oval. It passed out of sight, moving south.

These are stories of at least harmless things that were, or were not, seen over lands of this earth. It may be that if beings from somewhere else would seize inhabitants of this earth, wantonly, or out of curiosity, or as a matter of scientific research, the preference would be for an operation at sea, remote from observations by other humans of this earth. If such beings exist, they may in some respects be very wise, but-supposing secrecy to be desirable-they must have neglected psychology in their studies, or un concernedly they'd drop right into Central Park, New York, and pick up all the specimens they wanted, and leave it to the wisemen of our tribes to explain that there had been a whirlwind, and that the Weather Bureau, with its usual efficiency, had published warnings of it.

Now and then admirers of my good works write to me, and try to convert me into believing things that I say. He would have to be an eloquent admirer, who could persuade me into thinking that our present expression is not at least a little fanciful; but just the same I have labored to support it. I labor, like workers in a beehive, to support a lot of vagabond notions. But how am I to know? How am I to know but that sometime a queen-idea may soar to the sky, and from a nuptial flight of data, come back fertile from one of these drones?

In the matter of the disappearance of the Danish training ship Kobenhoven, which, upon Dec. 14, 1928, sailed, with fifty cadets and sailors aboard, from Montevideo, I note that another training ship, the Atalanta (British) set sail, early in the year 1880, with 250 cadets and sailors aboard, from Bermuda, and was not heard of again.

Upon Oct. 3, 1902, the German bark, Freya, cleared from Manzanilla for Punta Arenas, on the west coast of Mexico. I take from Nature, April 25, 1907. Upon the 20th of October, the ship was found at sea, partly dismasted, lying on her side, nobody aboard. The anchor was still hanging free at her bow, indicating that calamity had occurred soon after the ship had left port. The date on a calendar, on a wall of the Captain's cabin, was October 4th. Weather re ports showed that there had been only light winds in this region. But upon the 5th, there had been an earthquake in Mexico·

Several weeks after the disappearance of the crew of the Freya, another strange sea-occurrence was reported.

Zoologist, 4-7-38-that, according to the log of the steam ship Fort Salisbury, the second officer, Mr. A. H. Raymer, had, Oct. 28, 1902, in Lat. 5°, 31' S., and Long. 4°, 42' W., been called, at 3:05 A.M., by the lookout, who reported that there was a huge, dark object, bearing lights in the sea ahead. Two lights were seen. The steamship passed a slowly sinking bulk, of an estimated length of five or six hundred feet. Mechanism of some kind-fins, the observers thought-was making a commotion in the water. "A scaled back" was slowly submerging.

One thinks that seeing for such details as "a scaled back" could not have been very good, at three o'clock in the morning. So doubly damned is this datum that the attempt to explain it was in terms of the accursed Sea Serpent.

Phosphorescence of the water is mentioned several times, but that seems to have nothing to do with two definite lights, like those of a vessel. The Captain of the Fort Salisbury was interviewed. "I can only say that he (Mr. Raymer) is very earnest on the subject, and has, together with the lookout and helmsman, seen something in the water, of a huge nature, as specified."

One thinks that this object may have been a large, terrestrial vessel that had been abandoned, and was sinking.

I have looked over Lloyd's List, for the period, finding no record, by which to explain.

XIII

As to data that we shall now take up, I say to myself: "You are a benign ghoul, digging up dead, old legends and superstitions, trying to breathe life into them. Well, then, why have you neglected Santa Claus?"

But I am particular in the matter of data, or alleged data. And I have come upon no record, or alleged record, of mysterious footprints in snow, on roofs of houses, leading to chimneys, Christmas Eves.

There is a great deal, in the most acceptable of the science of today, that represents a rehabilitation of supposed legends, superstitions, and folk lore. Recall Voltaire's incredulity as to fossils, which according to him only a peasant would believe in. And note that his antagonism to fossils was probably because they had been taken over by theologians, in their way of explaining. Here is one of the keenest of minds: but it could not accept data, because it rejected explanations of the data. And so one thinks of, say, the transmutation of metals, which is now rehabilitated. And so on. There are some backward ones, today, who do not believe in witches: but every married man knows better. In the month of May, 1810, something appeared at Ennerdale, near the border of England and Scotland, and kill ed sheep, not devouring them, sometimes seven or eight of them in a night, but biting into the jugular vein and sucking the blood. That's the story. The only mammal that I know of that does something like this is the vampire bat. It has to be accepted that stories of the vampire bat are not myths. Something was ravaging near Ennerdale, and the losses by sheep farmers were so serious that the whole region was aroused. It became a religious duty to hunt this marauder. Once, when hunters rode past a church, out rushed the whole congregation to join them, the vicar throwing off his surplice, on his way to a horse. Milking, cutting of hay, feeding of stock were neglected. For more details, see Chambers' Journal, 81-470. Upon the 12th of September, someone saw a dog in a cornfield, and shot it. It is said that this dog was the marauder, and that with its death the killing of sheep stopped.

For about four months, in the year 1874, beginning upon January 8th, a killer was abroad, in Ireland. In land and Water, March 7, 1874, a correspondent writes that he had· heard of depredations by a wolf, in Ireland, where the last native wolf had been killed in the year 1712. According to him, a killer was running wild, in Cavan, slaying as many· as 30 sheep in one night. There is another account, in Land and Water, March 28. Here, a correspondent writes that, in Cavan sheep had been killed in a way that led to the belief that the marauder was not a dog. This correspondent knew of 42 instances, in three townlands, in which sheep had been similarly killed -throats cut and blood suck ed, but no flesh eaten. The footprints were like a dog's, but were long and narrow, and showed traces of strong claws. Then, in the issue of April 11th, of Land and Water, came the news that we have been expecting. The killer had been shot. It had been shot by Archdeacon Magenniss, at Lismoreville, and was only a large dog.

This announcement ends the subject, in Land and Water. Almost anybody, anyway in the past, before suspiciousness against conventions had the development that it has today, reading these accounts down to the final one, word say

"Why, of course! It's the way these stories always end up. Nothing to them." But it is just the way these stories always end up that has kept me busy. Because of our experience with pseudo-endings of mysteries, or the mysterious shearing and bobbing and clipping of mysteries, I went more into this story that was said to be no longer mysterious. The large dog that was shot by the Archdeacon was sacrificed not in vain, if its story shut up the minds of readers of Land and Water, and if it be desirable somewhere to shut up minds upon this earth.

See the Clare Journal, issues up to April 27th-the shooting of the large dog, and no effect upon the depredations -another dog shot, and the relief of the farmers, who believed that this one was the killer-still another dog shot, and supposed to be the killer-the killing of sheep continuing. The depredations were so great as to be described as "terrible losses for poor people." It is not definitely said that something was killing sheep vampirishly, but that "only a piece was bitten off, and no flesh sufficient for a dog ever eaten. "The scene of the killings shifted.

Cavan Weekly News, April 17-that, near Limerick, more than-100 miles from Cavan, "a wolf or something like it" was killing sheep. The writer says that several persons, alleged to have been bitten by this animal, had been taken to the Ennis Insane Asylum, "laboring under strange symptoms of insanity."

It seems that some of the killings were simultaneous near Cavan and near Limerick. At both places, it was not said that finally any animal, known to be the killer, was shot or identified. If these things that may not be dogs be, their disappearances are as mysterious as their appearances.

There was a marauding animal in England, toward the end of the year 1905. London Daily Mail, Nov. 1, 1905- "the sheep-slaying mystery of Badminton." It is said that, in the neighborhood of Badminton, on the border between Gloucestershire and Wiltshire, sheep had been killed. Sergeant Carter, of the Gloucestershire Police, is quoted-'1 have seen two of the carcasses, myself, and can say definitely that it is impossible for it to be the work of a dog. Dogs are not vampires, and do not suck the blood of a sheep, and leave the flesh almost untouched."

And, going over the newspapers, just as we're wondering what's delaying it, here it is-

London Daily Mail, December 19- "Marauder shot near Hinton." It was a large, black dog.

So then, if in London any interest had been aroused, this announcement stopped it.

We go to newspapers published nearer the scene of the sheep-slaughtering. Bristol Mercury, November 25-that the killer was a jackal, which had escaped from a menagerie in Gloucester. And that stopped mystification and inquiry, in the minds of readers of the Bristol Mercury.

Suspecting that there had been no such escape of a jackal, we go to Gloucester newspapers. In the Gloucester Journal, November 4, in a long account of the depredations, there is no mention of the escape of any animal in Gloucester, nor anywhere else. In following issues, nothing is said of the escape of a jackal, nor of any other animal So many reports were sent to the editor of this newspaper that he doubted that only one slaughtering thing was abroad. "Some even go so far as to call up the traditions of the werewolf, and superstitious people are inclined to this theory."

We learn that the large, black dog had been shot upon December 16th, but that in its region there had been no reported killing of sheep, from about November 25th. The look of data is of another scene-shifting. Near Gravesend, an unknown animal had up to December 16th, killed about 30 sheep (London Daily Mail, December 19). "Small armies" of men went hunting, but the killing stopped, and the unknown animal remained unknown.

I go on with my yams. I no more believe them than I believe that twice two are four.

If there is continuity, only fictitiously can anything be picked out of the nexus of all phenomena; or, if there is only oneness, we cannot, except arbitrarily, find any two units with which even to start the sequence that twice two are four. And, if there is also discontinuity, all things are so individualized that, except arbitrarily and fictitiously, nothing can be classed with, or added to, anything else.

London Daily Express, Oct. 14, 1925-the district of Edale, Derbyshire, terrorized, quite as, centuries ago, were regions by stories of werewolves. Something, "black in color and of enormous size," was slaughtering sheep, at night, "leaving the carcasses· strewn about, with legs, shoulders, and heads tom off; broken backs, and pieces of flesh ripped off." Many hunting parties had gone out, but had been unable to track the animal. "People in many places are so frightened that they refuse to leave their homes after dark, and keep their children in the house." If something had mysteriously appeared, it then quite as mysteriously disappeared. There are stories of wanton killings, or of bodies that were not fed upon. London Daily Express, Aug. 12, 1919 -something that, at Llanelly, Wales, was killing rabbits, for the sake of killing-entering hutches at night, never taking rabbits, killing them by breaking their backbones.

Early in the morning of March 3, 1906, the sentry at Windsor Castle saw something, and fired a shot at it (London Daily Mail, March 6). The man's account of what he thought he saw was not published. It was said that he had shot at one of the ornamental, stone elephants, which had looked ghostly in moonlight. He was sentenced to three days' confinement in barracks, for firing without proper cause. It would be interesting to know what he thought he saw, with such conviction that he fired and risked punishment and whether it had anything to do with-

Daily Mail, March 22-that about a dozen of the King's sheep, in a field near Windsor Castle, had been bitten by something, presumably a dog, so severely that they had to be killed. In the Daily Mail, March 19, is an account of extraordinary killing of sheep, "by dogs," near Guildford, about 17 miles from Windsor. 51 sheep were killed in one night.

A woman in a field-something grabbed her. At first the story as of a marauding panther that must have escaped from a menagerie, See the Field, Aug. 12, 19, 1893-an animal, supposed to be an escaped panther, that was preying upon human beings, in Russia. Look up records of werewolves, or supposed werewolves. and note instances of attacks almost exclusively upon women. For a more particularized account, by General R. G. Burton, who was in Russia, at the time, see the Field, Dec. 9, 1893. General Burton had no opportunity to visit the place "haunted by this mysterious animal," but he tells the story, as he got it from Prince Sherincki, who was active in the hunt. An unknown beast was terrorizing a small district in the Orel Government, south of Moscow. The first attack was upon the evening of July 6th. Three days later, another woman was grabbed by an undescribed animal, which she beat off, until help arrived. That day, a boy, aged 10, was killed and devoured. July 11th-a woman killed, near Trosna. "At four o'clock, on the 14th, the beast severely wounded another woman and at five o'clock, made' another attack upon a peasant girl, but was beaten off by a companion, who pulled the animal off by the tail. These details are taken from the official accounts of the events."

There was a panic, and the military authorities were appealed to. 3 officers and 40 men were sent from Mos cow. They organized beats that were composed of from 500 to 1,000 peasants, but all hunts were unsuccessful. On the 24th of July, four women were attacked, and one of them was killed.

Something was outwitting 3 officers and 40 men, and armies of 1,000 peasants. War was declared. Prince Sherincki, with 10 officers and 130 men, arrived from St. Petersburg. We notice that in wicanny occurrences, when there is wide publicity, or intense excitement, phenomena stop-or are stopped. War was declared upon something, but it dis appeared. "According to general descriptions, the animal was long, with a black muzzle, and round, standing-up ears, with a long, smooth, hanging tail."

We know what to expect.

In the Field, Dec. 23, 1893, it is said that, after a study of sketches of the spoor of the animal, the naturalist Alferachi gave his opinion that the animal was a large dog. He so concluded because of the marks of protruding nails in the sketches.

But also it is said that plaster casts of the footprints show ed no such marks. It is said that the nail marks had been added to the sketches, because of assertions by hunters that nail marks had been seen. Writing 30 years later (Chambers' Journal, ser. 7, vol 14, p. 308) General Bur ton tells of the animal as something that had never been identified.

This is fringing upon an enormous subject that leads away from the slaughtering of sheep to attacks, some of them mischievous, some ordinarily deadly, and some of the Jack the Ripper kind, upon human beings. Though I have hundreds of notes upon mysterious attacks upon human beings, I cannot develop occult criminology now.

XIV

IN October, 1904, a wolf, belonging to Captain Bains of Shotley Bridge, twelve miles· from Newcastle, England, es caped, and soon afterward, killing of sheep were reported from the region of Hexham, about twenty miles from Newcastle.

There seems to be an obvious conclusion.

We have had some experience with conclusions that were said to be obvious.

A story of a wolf in England is worth space, and the London newspapers rejoiced in this wolf story. Most of them did, but there are several that would not pay much attention to a dinosaur-hunt in Hyde Park. Special correspondents were sent to Hexham, Northumberland. Some of them, because of circumstances that we shall note, wrote that there was no wolf, but probably a large dog that had turned evil. Most of them wrote that undoubtedly a wolf was ravaging, and was known to have escaped from Shotley Bridge. Something was slaughtering sheep, killing for food. and killing wantonly, sometimes mutilating four or five sheep, and devouring one. An appetite was ravaging, in Northumberland. We have impressions of the capacity of a large and hungry dog, but, upon reading these accounts, one has to think that they were exaggerations, or that the killer must have been more than a wolf. But, according to developments, I'd not say that there was much exaggeration. The killings were so serious that the farmers organized into the Hexham Wolf Committee, offering a reward, and hunting systematically. Every hunt was fruitless, except as material for the special correspondents, who told of continuing depredations, and reveled in special announcements. It was especially announced that, upon December 15th, the Haydon foxhounds, one of the most especial packs in England, would be sent out. These English dogs, of degree so high as to be incredible in all other parts of the world, went forth. It is better for something of high degree not to go forth. Mostly in times of peace arise great military reputations. So long as something is not tested it may be of high renown. But the Haydon foxhounds went forth. They returned with their renown damaged.

This takes us to another of our problems:

Who can blame a celebrity for not smelling an absence?

There are not only wisemen: there are wisedogs, we learn. The Wolf Committee heard of Monarch, "the celebrated bloodhound." This celebrity was sent for, and when he arrived, it was with such a look of sagacity that the sheepfarmer's troubles were supposed to be over. The wisedog was put on what was supposed to be the trail of the wolf. But, if there weren't any wolf, who can blame a celebrated bloodhound for not smelling something that wasn't? The wisedog sniffed. Then he sat down. It was impossible to set this dog on the trail of a wolf, though each morning he was taken to a place of fresh slaughter.

Well, then, what else is there in all this? If, locally, one of the most celebrated intellects in England could not solve the problem, it may be that the fault was in taking it up locally.

Throughout my time of gathering material for this book, it was my way to note something, and not to regard it as isolated; and to search widely for other occurrences that might associate with it. So, then, I noted this wolf story, and I settled upon this period, of the winter of 1904-5, with the idea of collecting records of seemingly most in congruous occurrences, which, however, might be germs of correlations.

Such as this, for instance-but what could one of these occurrences have to do with the other?-

That in this winter of 1904-5, there were two excitements in Northumberland. One was the wolfhunt, and the other was a revival-craze, which had spread from Wales to England. At the time of the wolf-hunt, there was religious mania in Northumberland. Men and women staggered, as they wept and shouted, bearing reeling lights, in delirious torchlight processions.

If Monarch, the celebrated bloodhound sniffed and then sat down, I feel, myself, that the trail cannot be picked up in Hexham.

It was a time of widespread, uncanny occurrences in Great Britain. But in no account of any one uncanny occurrence have I read of any writer's awareness that there were other uncanny occurrences, or more than one or two other uncanny occurrences. There were many, special scares, at this time, in Great Britain. There was no general scare. The contagions of popular delusions cannot be lugged in, as a general explanation.

Strange, luminous things, or beings, were appearing in Wales.

In Wales had started one of the most widely hysterical religious revivals of modem times.

A light in the sky-and a pious screech-I sniff, but I don't sit down.

A wolf and a light and a screech.

There are elaborate accounts of the luminous things, or beings, in the Proc. S. P. R., vol. 19, and in the first volume of the Occult Review. We are told that, over the piously palpitating principality of Wales, shining things traveled, stopping and descending when they came to a revival meeting, associating in some unknown way with these centers of excitation, especially where Mrs. Mary Jones, was the leader. There is a story of one shining· thing that persistently followed Mrs. Jones' car, and was not shaken off, when the car turned abruptly from one road to another. So far as acceptability is concerned, I prefer the accounts by newspaper men. It took considerable to convince them. Writers, sent to Wales by London newspapers, set out with blithe incredulity. Almost everybody had a hankering for mysteries. but it is likely to be an abstract hankering, and when a mystery comes up in one's own experience, one is likely to treat it in a way that warns everybody else that one is not easily imposed upon. The reports that were sent back by the Londoners were flippant: but, in the London Daily Mail, Feb. 13, 1905, one of these correspondents describes something like a ball of fire, which he saw in the sky, a brilliant object that was motionless for a while, then disappearing. Later he came upon such an appearance, near the ground, not 500 feet away. He ran toward it, but the thing disappeared. Then Bernard Redwood was sent, by the Daily Mail, to investigate. In the Mail, February 13, he writes that there were probably will-o'-the-wisps, helped out by practical jokers. As we very well know, there are no more helpful creatures than practical jokers, but, as inquiry stoppers, will-o'-the-wisps have played out. A conventionalist, telling the story, today, would say that they were luminous bats from a chapel belfry, and that a sexton had shot one. Almost every writer who accepted that these things were, thought that in some unknown, or unknowable, way. they were associated with the revival. It is said that they were seen hovering over chapels.

According to my methods, I have often settled upon special periods, gathering data, with the idea of correlating, but I have never come upon any other time in which were reported so many uncanny occurrences.

There were teleportations in a butcher shop, or things were mysteriously Hying about, in a butcher shop, in Portmadoc, Wales. The police were called in, and they accused a girl who was employed .in a butcher shop. "She made a full confession" (News of the World, February 26). A ghost in Barmouth: no details (Barmouth Advertising, January 12). Most of the records are mere paragraphs, but the newspapers gave considerable space to reported phenomena in the home of Mr. Howell, at Lampeter, Wales. As told in the London Daily News, February 11 and 13, "mysterious knockings" were heard in this house, and crowds gathered outside. The Bishop of Swansea and Prof. Harris investigated, but could not explain. Crowds in the street became so great that extra police had to be called out to regulate traffic, but nothing was learned. There were youngsters in this house, but they did not confess. Mr. Howell had what is known as "standing," in his community. It's the housemaid, or the girl in a butcher shop, with parents who probably haven't much "standing" who is mocked about, or more gently slugged, or perhaps only· slapped on the face, who confesses, or is said to have confessed. Also, as told in the Liverpool Echo, February 15, there was excitement at Rhymney, Wales, and investigations that came to nothing. Tapping sounds had been heard, and strange lights had been seen, in one of the revival-centers, the Salvation Army Barracks. Whether these lights were like the other lights that were appearing in Wales. I cannot say. It was the assertion of the Rev. J. Evans and other investigators, who had spent a night in the Barracks, that they had seen "very bright lights...

In the Southern Daily Echo, February 23, is an account of "mysterious rappings on a door of a house in Crewe, and of a young woman, in the house who was said to have dropped dead. A physician "pronounced" her dead. But there was an inquest, and the coroner said: "There is not a single sign of death... Nevertheless she was officially dead, and she was buried, anyway. I am too dim in my notions of possible correlations, go into details, but, along with my supposition that ordinarily catalepsy is of rare occurrence, I note that I have records of three persons, who, in this period, were aroused from trances, in time to save them from being buried alive. There are data of "strange suicides; that I shall pass over. I have several dozen records of "mysterious fires," in this period.

Slaughter in Northumberland-farmers, who could, housing their sheep, at night-others setting up lanterns in their fields. Monarch, the celebrated bloodhound, who could not smell something that perhaps was not, got no more space in the newspapers, and, to a woman, the inhabitants of Hexham stopped sending him chrysanthemums. But faith in celebrities kept up, as it always will keep up, and when the Hungarian Wolf Hunter appeared, the only reason that a brass band. did not escort him, in showers of torn-up telephone books, is that, away back in this winter, Hexham, like most of the other parts of England, was not yet Americanized. It was before the English were educated. The moving pictures were not of much influence then. The Hungarian Wolf Hunter, mounted on a Shaggy Hungarian pony, galloped over hills and dales, and, with strange Hungarian hunting cries, made what I think is called the welkin ring. He might as well have sat down and sniffed. He might as well have been a distinguished General. or Admiral, at the outbreak of a war.

Four sheep were killed at Low Eschelles, and one at Sed ham, in one night. Then came the big hunt, of December 20th, which, according to expectations, would be final. The big hunt set out from Hexham: gamekeepers, woodmen, farmers, local sportsmen and sportsmen from far away. There were men on horseback, and two men in "traps," a man on a bicycle, and a mounted policeman: -two women with guns, one of them in a blue walking dress, if that detail's any good to us.

They came wandering back, at the end of the day, not having seen anything to shoot at. Some said that it was be cause there wasn't anything. Everybody else had some thing to say about Capt. Bains. The most unpopular person, in the north of England, at this time, was Capt. Bains, of Shotley Bridge. Almost every night. something, presumably Capt. Bains' wolf, even though there was no findable statement that a wolf had been seen, was killing and devouring sheep.

In Brighton, an unknown force, or thing, struck notes on a musical instrument (Daily Mail, December 24). Later, there were stories of "a phantom bicyclist" near Brighton (London Daily Mirror, February 6). In the Jour. S. P. R., 13-259, is published somebody's statement that. near the village of Hoe Benham, he had seen something that looked like a large dog turn into a seeming donkey. Strange sounds heard near Bolton, Lancashire-"nothing but the beating of a rope against a flagstaff." Then it was said that a figure had been seen (Lloyd's Weekly News, January 15). A door bell was mysteriously ringing at Blackheath, London: police watching the house, but unable to find out any-thing (Daily Mirror, February 13). But in not one of these accounts is shown knowledge that. about the same time, other accounts were being published. Look in the publications of the S. P. R., and wonder what that Society was doing. It did investigate two of the cases told of in this chapter, but no awareness is shown of a period of widespread occurrences. Other phenomena, or alleged phenom-ena-a ghost, at Exeter Deanery: no details (Daily Mail, December -24). Strange sounds and lights, in a house in Epworth (Liver pool Echo, January 25). People in Bradford thought that they saw a figure enter a club house-police notified-fruit less search (Weekly Dispatch, January 15). At Edinburgh, Mr. J. E. Newlands, who held the Fulton chair, at the Uni ted Free College, saw a "figure" moving beside him (Weekly Dispatch, April 16).

But the outstanding phenomenon of this period was the revival-
Liverpool Echo, January 18--Wales in the Grip of Supernatural Forces!"

This was in allusion to the developing frenzies of the revival, and the accompanying luminous things, or beings, that had been reported. "Supernatural" is a word that has no place in my vocabu-lary. In my view, it has no meaning, or distinguishment. If there never had been, finally, ·a natural explanation of anything, everything is, naturally enough. the supernatural.

The P was a grab by a craze. The excitement was combustion, or psycho-electricity, or almost any-thing except what it was supposed to be, and perhaps when flowing from human batteries there was a force that was of use to the luminous things that hung around. Maybe they fed upon it, and grew, and glowed, brilliant with nourishing ecstasies. See data upon astonishing growths of plants, when receiving other kinds of radio-active nourishment, or stimulation. If a man can go drunk on God, he may usefully pass along his exhilarations to other manifestations of godness.

There were flares where they'd least be expected. In the big stores, in the midst of waiting on customers, shop girls would suddenly, or electrically, start clapping hands and singing. Very likely some of them cut up such capers for the sport of it, and enjoyed keeping hard customers waiting. I notice that, though playing upon widely different motives, popular excitements are recruited and kept going, as if they were homogeneous. There's no understanding huge emotional revolts against sin, without considering all the fun there is in them. They are monotony-breakers. Drab, little personalities have a chance to scream themselves vivid. There were confession-addicts who, past possibility of being believed, proclaimed their own wickedness, and then turned to public confessions for their neighbors, until sinful neighbors appealed to the law for protection. In one town, a man went from store to store, •returning" things that he had not stolen. Bands of girls roved the streets, rushing earnestly and mischievously into the more sedate churches, where the excitation was not encouraged, singing and clap ping their hands, all of them shouting, and some of them blubbering, and then some of the most sportive ones blubbering, compelled into a temporary uniformity. This clapping of hands, in time with the singing, was almost irresistible: some vibrational reason: power of the rhythm to harmonize diverse units; primitive power of the drum beat. Special trains set out from Liverpool to Welsh meeting places, with sightseers, who hadn't a concern for the good of their souls; vendors of things that might have a sale; some earnest ones. Handclapping started up, and emotion al furies shot through Wales.

There were ghost-scares in the towns of Blyth and Dover. Blyth News, March 14-crowds gathered around a school house-something of a ghostly nature inside-nothing but the creaking of a partition. I pick up something else. We wonder how far om neo mediaevalism is going to take us. Perhaps-though our interpretations will not be the same-only mediaevalism will be the limit. Blyth News, February 28-smoke that was seen coming from the windows of a house, in Blyth. Neighbors broke in, and found the body of the occupant, Barbara Bell, aged 77, on a sofa. Her body was burned, as if for a long time it had been in the midst of intense flames. It was thought that the victim had fallen into the fireplace. "The body was fearfully charred.

Something was slaughtering sheep-and things In the sky of Wales-and it may be that there were things, or beings, that acted like fire, consuming the bodies of wo men. London Daily News, Dec. 17, 1904-"Yesterday morning, Mrs. Thomas Cochrane, of Rosehall, Falkirk, widow of a well-known, local gentlemen, was found burned to death in her bedroom. No fire in the grate-burned almost beyond recognition'"-no outcry-little, if anything else burned-body found, "sitting in a chair, surrounded by pillows and cushions." London Daily Mail, December 24-inquest on the body of a woman, who had died of the effects of "mysterious burns." "She could give no account of her In juries." An almshouse, late at night-and something burned a woman. Trinity Almshouse, Hull-story told, in the Hull Daily Mail, January 6. Body covered with bums-woman still living, when found in the morning-strange that there had been no outcry-bed unscorched. The woman, Elizabeth Clark, could tell nothing of her injuries, and she died without giving a clew to the mystery. "There was no fire nor light in the room.

On both sides of the River Tyne, something kept on slaughtering. It crossed the Tyne, having killed on one side, then killing on the other side. At East Dipton, two sheep were devoured, all but the fleece and the bones, and the same night two sheep were killed on the other side of the river.

"The Big Game Hunter from India!"

Another celebrity came forth. The Wolf Committee met him at the station. There was a plaid shawl strapped to his back, and the flaps of his hunting cap were considered unprecedented. Almost everybody had confidence in the shawl, or felt that the flaps were authoritative. The devices by which he covered his ears made holders feel that they were in the presence of Science.

Hexham Herald-"The right man, at last!"

So finally the wolf hunt was taken up scientifically. The ordinary hunts were going on, but the wiseman from India would have nothing to do with them. In his cap, with flaps such as had never before been seen in Northumberland, and with his plaid shawl strapped to his back, he was going from farm to farm, sifting and dating and classifying observations: drawing maps, card-indexing his data. For some situations, this is the best of methods: but something that the methodist-wiseman cannot learn is that a still better method is that of not being so tied to any particular method. It was a serious matter in Hexham. The ravaging thing was an alarming pest. There were some common hunters who were unmannerly over all this delay, but the Hexham Herald came out strong for Science-"The right man in the right place, at last!"

There was, in this period, another series of killings. Upon a farm, near Newcastle, late .in this year 1904, something was killing poultry. The depredations were so persistent, and the marauder was so evasive that persons who are said to be superstitious began to talk in a way that is said to be unenlightened.

Then the body of an otter was found.

The killing of poultry stopped.

For a discussion of the conclusion that to any normal logician looks obvious, see the Field, Dec. 3, 1904. Here we learn that otters, though ordinarily living upon fish, do some times vary their diet. But no data upon persistent killing of poultry, by otters, came out.

This body of an otter was found, lying on a railroad line. France in the grip of military forces. August 1914-France was invaded, and the people of France knew that France was invaded. It is my expression that so they knew, only because it was a conventional recognition. There were no wisemen to say that reported bodies of men moving along roads had nothing to do with mutilated persons appearing in hospitals, and that only by coincidence was there devastation. The wiseman of France did not give only a local explanation to every local occurrence, but of course correlated all, as the manifestations of one invasion. Human eyes have been made to see human invaders.

Wales in the grip of "supernatural forces." People in England paid little attention, at first, but then hysterias mobbed across the border. To those of us who have some failings, and now and then give a thought to correcting them, if possible, but are mostly too busy to bother much, cyclones of emotions relating to states that are vaguely known as good and evil, are most mysterious. In the Barmouth Advertiser (April· 20) it is said that, in the first three months of this year 1905, there had been admitted to the Denbigh Insane Asylum, 16 patients, whose dementias were attributed to the revival. It is probable that many cases were not reported. In the Liverpool Echo, November 25 are accounts of four insane revivalists, who were under restraint in their own homes. Three cases in one town are told of in this newspaper, of January 10th. The craze spread in England. and in some parts of England it was as in tense as anywhere in Wales. At Bromley, a woman wrote a confession of sins, some of which, it was said, she could not have committed, and threw herself under a railroad train. In town after town, police stations were invaded by exhorters. In both England and Wales, bands stood outside theaters, calling upon people not to enter. In the same way they tried to prevent attendance at football games.

December 29-"Wolf killed on a railroad line!"

It was at Cumwinton, which is near Carlisle, about thirty miles from Hexham. The body was found on a railroad line-"Magnificent specimen of male gray wolf-total length five feet-measurement from foot to top of shoulder; thirty inches." -

Captain Bains, of Shotley Bridge, went immediately to Cumwinton. He looked at the body of the wolf. He said that it was not his wolf.

There was doubt in the newspapers. ·Everybody is supposed to know his own wolf, but when one's wolf has made material for a host of damage suits, one's recognitions may be dimmed.

This body of a wolf was found, and killings of sheep stopped.

But Capt. Bains' denial that the wolf was his wolf was accepted by the Hexham Wolf Committee. Data were with him. He had reported the escape of his wolf, and the description was on record in the Shotley Bridge police station. Capt. Bains' wolf was, in October, no "magnificent" full. grown specimen, but a cub, four and a half months old. Though nobody had paid any attention to this circumstance, it had been pointed out, in the Hexham Herald, October 15. The wolf of Cumwinton was not identified, according to my reading of the data. Nobody told of an escape of a grown wolf, though the news of this wolfs death was published throughout England. The animal may have come from somewhere far from England. Photographs of the wolf were sold, as picture postal cards. People flocked to Cumwinton. Men in the show business offered to buy the body, but the decision of the railroad company was that the body had not been identified, and belonged to the company. The head was preserved, and was sent to the central office, in Derby.

But what became of the Shotley Bridge wolf The mystery begins with this statement:

That, in October, 1904, a wolf, belonging to Capt. Bains, of Shotley Bridge, escaped, and that about the same time began a slaughtering of sheep, but that Capt. Bains' wolf bad nothing to do with the slaughter.

Or the statement is that there was killing of sheep, in Northumberland, and that then came news of the escape of a wolf, by which the killing of a few sheep might be explained-

But that then there were devourings, which could not be attributed to a wolf-cub.

The wolf-cub disappeared, and there appeared another wolf, this one of a size and strength to which the devourings could be attributed.

Somewhere there was science.

If it had not been for Capt. Bains' prompt investigation, the reported differences between these two animals would have been overlooked, or disbelieved, and the story would be simply that a wolf had escaped from Shotley Bridge, had ravaged, and had been killed at Cumwinton. But Capt. Bains did investigate, and. his statement that the wolf of Cumwinton was not his wolf was accepted. So then, instead of a satisfactory explanation, there was a new mystery. Where did the wolf of Cumwinton come from?

There is something that is acting to kill off mysteries. Perhaps always, and perhaps not always, it can be under stood in common place terms. If luminous things that move like dying birds are attracting attention, a Mr. Cannell appears, and says that he has found a luminous owl. In the newspapers, about the middle of February, appeared a story that Capt. Alexander Thompson, of Tacoma, Washington-and I have looked this up, learning that a Capt Alexander Thompson did live in Tacoma, about the year 1905-was walking along a street in Derby, when he glanced in a taxidermist's window, and there saw the supposed wolfs head. He recognized it, not as the head of a wolf, but the head of a malmoot, an Alaskan sleigh dog, half wolf and half dog. This animal, with other malmoots, had been taken to Liverpool, for exhibition, and had escaped in a street in Liverpool. Though I have not been able to find out the date, I have learned that there was such an exhibition, in Liverpool. No date was mentioned by Capt. Thompson. The owners of the malmoot had said nothing, and rather than to advertise, had put up with the loss, because of their fear that there would be damages for sheep-killing. Not in the streets of Liverpool, presumably. No support for this subject is findable in Liverpool newspapers.

Liverpool is 120 miles from Hexham. It is a story of an animal that escaped in Liverpool. and, leaving no trail of slaughtering behind it, went to a distant part of England, exactly to a place where a young wolf was at large, and there slaughtered like a wolf.

I prefer to think that the animal of Cumwinton was not a malmoot.

Derby Mercury, February 22-that the animal had been identified as a wolf, by Mr. A. S. Hutchinson, taxidermist to the Manchester Museum of Natural History. Liverpool Echo, December 31-that the animal had been identified as a wolf, by a representative of Bostock and Wombell's Circus, who had traveled from Edinburgh to see the body.

Killing of poultry, and the body of an otter on the railroad line-and the killing of poultry stopped. Or that there may be occult things, beings and events, and that also there may be something of the nature of an occult police force, which operates to divert human suspicions, and to supply explanations that are good enough for whatever, somewhat of the nature of minds, human beings have-or that, if there be occult mischiefmakers and occult ravagers, they may be of a world also of other beings that are acting to check them, and to explain them, not benevolently, but to divert suspicion from themselves, because they, too, may be exploiting life upon this earth, but in ways more subtle, and in orderly, or organized fashion. We have noticed, in investigating obscure, or occult phenomena, or alleged phenomena, that sometimes in matters that are now widely supposed to be rank superstitions, orthodox scientists are not so uncompromising in their op positions, as are those who have not investigated. In the New Orleans Medical and Surgical Journal, April, 1894, is an account of a case of "spontaneous combustion of human bodies." The account is by Dr. Adrian Hava, not as observed by him, but as reported by his father. In Science, 10-100, is quoted a paper that was read by Dr. B. H. Hartwell, of Ayer, Mass., before the Massachusetts Medico-Legal Society. It was Dr. Hartwell's statement that, upon May 12, 1890, while driving through a forest, near Ayer, be had been called, and, going into the wood, saw in a clearing, the crouched form of a woman. Fire which was not from clothing, was consuming the shoulders, both sides of the abdomen, and both legs. See Dr. Dixon Mann's Forensk Medicine (edition of 1922), p. 216. Here, cases are told of and are accepted as veritable-such as the case of a woman, consumed so by fire that on the floor of her room there was only a pile of calcined bones left of her. The fire, if in an ordinary sense it was fire, must have been of the intensity of a furnace: but a table cloth, only three feet from the pile of cinders, was unscorched. There are other such records.

I think that our data relate, not to "spontaneous combustion of human bodies," but to things, or beings, that, with a flaming process, consume men and women, but, like were wolves, or alleged werewolves, mostly pick out women. Occurrences of this winter of 1904-5 again. Early in February, in London, a woman, who was sitting asleep, before a fire in a grate, awoke, finding herself flaming. A commonplace explanation would seem to be sufficient: nevertheless it is a story of "mysterious burns," as worded in Lloyd's Weekly News, February 5. A coroner had expressed an inability to understand. In commenting upon the case, the coroner l said that a cinder might have shot from the grate, ignite this woman's clothes, but that she had been sitting, facing the fire, and that the burns were on her back.

Upon the morning of February 26th (Hampshire Advertiser, March 4) at Butlock Heath, near Southampton, neighbors of an old couple, named Kiley, heard a scratching sound. They entered the house, and found it in Hames, inside. Kiley was found, burned to death, on the Boor. M Kiley, burned to death, was sitting in a chair, in the sit room, "badly charred, but recognizable."

A table was overturned, and a broken lamp was on the floor.

So there seems to be an obvious explanation. But, at the inquest, it was said that an examination of this lamp showed that it could not have caused the fire. The verdict was: "Accidental death, but by what means, they (the jury) were unable to determine.

Both bodies had been fully dressed, "judging by fragments of clothes." This indicates that the Kileys had been burned before their time for going to bed. Hours later, the house was in flames. At the inquest, the mystery was tbs two persons, neither of whom had cried for help, presumably not asleep in an ordinary sense, should have been burned to death in a fire that did not manifest as a general fire, until hours later.

Something had overturned a table. A lamp was broken Again the phenomenon of scene-shifting-Soon after the killing of poultry ceased, near Newcastle there were uncanny occurrences upon Binbrook Farm, near Great Grimsby. There is an account, in the Jour. S. P. R. 12-138, by the Rev. A. C. Custance, of Binbrook Rectory. There was no confession, this time, but this time the girl in the case-the young housemaid again-was in no condition to be dragged to a police station. It will not be easy to think that it was trickery by the girl in this case. The story is that objects were thrown about rooms: that three times, near "a not very good, or big, fire," things burst into flames, and that finally a servant girl was burned, or was attacked by something that burned her. In the Liver pool Echo, January 25, is published a letter from a school teacher of Binbrook, in which it is said that a blanket had been found burning in a room in which there was no fire place. According to the report by Col. Taylor, to the S. P.R., the first manifestations occurred upon the 31st of December. Something was killing chickens, in the farm yard, and in the henhouse. All were killed in the same way. A vampirish v-ay? Their throats were tom.

I go to a newspaper for an account of phenomena, at Binbrook: The writer was so far from prejudice in favor of cult phenomena, that he began by saying: "Superstition lies hard." In the Louth and North Lincolnshire News, January 28, he tells of objects that unaccountably fell from shelves in the farmhouse and of mysterious transportations of objects, "according to allegations." "A story that greatly dismays the unsophisticated is that of the servant girl, who, while sweeping the floor, was badly burned on the back.

This is how the farmer relates it: 'Our servant girl, whom we had taken from the workhouse, and who had neither kin nor friend in the world that she knows of, was sweeping the kitchen. There was a very small fire in the grate: there was a guard there, so that no one can come within two feet or more of the fire; and she was at the other end of the room, and had not been near. I suddenly came into the kitchen, and there she was, sweeping away, while the back of her dress was afire. She looked around, as I shouted, and, seeing the flames, rushed through the door. She tripped, and I smothered the fire out with wet sacks. But she was terribly burned, and she is at the Louth Hospital, now, in terrible pain.'·

"This last sentence is very true. Yesterday our representative called at the hospital, and was informed that the girl was burnt extensively on the back, and lies in a critical condition. She adheres to the belief that she was in the middle of the room, when her clothes ignited."

A great deal, in trying to understand this occurrence, depends upon what will be thought of the unseen killing of chickens-

"Out of 250 fowls, Mr. White says that he has only 24 left. They have all been killed in the same weird way. The skin around the neck, from the head to the breast, has been pulled off, and the windpipe drawn from its place and snapped. The fowl house has been watched night and day, and, whenever examined, four or five birds would be found dead."

In London, a woman sat asleep, near a grate, and something, as if taking advantage of this means of commonplace explanation, burned her, behind her. Perhaps a being, of incendiary appetite, had crept up behind her, but I had no data upon which so to speculate. But, if we accept that, at Binbrook Farm, something was savagely killing chickens, we accept that whatever we mean by a being was there. It seems that, in the little time taken by the farmer to put out the fire of the burning girl, she. could not have been badly scorched. Then the suggestion is that, unknown to her, something behind her was burning her, and that she was un conscious of her own scorching flesh. All the stories are no table for absence of outcry, or seeming unconsciousness of victims that something was consuming them.

The town of Market Rasen is near ·Binbrook Farm. The address of the clergyman who reported, to the S. P. R., the fires and the slaughterings of chickens, upon the farm, is "Binbrook Rectory, Market Rasen." Upon January 16th, as told in the Louth and North Lincolnshire News, January 21, there was, in a chicken house, at Market Rasen, a fire in which 57 chickens were consumed. Perhaps a fire in a chicken house is not much of a circumstance to record, but I note that it is said that how this fire started could not be found out.

The girl of Binbrook Farm was taken to the Louth Hospital. In Lloyd's Weekly News, February 5, there is an account of "mysterious bums." It is the case of Ashton Clodd, a man aged 75, who, the week before, had died in the Louth Hospital. It is said that he had fallen into a grate, while putting coals in it, and that, for some reason probably because of his rheumatism, had been unable to rise, and had been fatally burned. But a witness at the inquest is quoted: "If there was a fire in the fire-place, it was very little."

All around every place that we have noted, the revival was simmering, seething, or raging. In Leeds, women, who said that they were directed by visions, stood in the streets, stopping cars, trying to compel passengers to join them. A man in Tunbridge Wells, taking an exhortation literally, chopped his right hand off. "Holy dancers'.' appeared in London. At Driffield, someone led a procession every night, trundling his coffin ahead of him. And all this in England. And, in England, it is very much the custom to call attention to freaks and extravagances in other parts of the world, or more partic-ularly in one other part of the world as if only there occurred all the freaks and extravagances.

Riots broke out in Liverpool, where the revivalists, with a mediaeval enthusiasm, attacked Catholics. The Liverpool City Council censured "certain so-called religious meetings, which create danger to life and property." Also at South end, there were processions of shouters, from which rushed mis-sionaries to slug Catholics, and to· sling bricks at houses in which lived Catholics. In the Liverpool Echo, February 6, is quoted a magistrate, who said to a complainant, who, because of differences in a general doctrine of loving one's neighbors, had been assaulted: "When you see one of these proces-sions, you should run away, as you would from a mad-bull."

Upon all the occurrences that we have noted was the one enveloping phenomenon of the revival. There is scarcely a place that I have mentioned, in any of the accounts, that was unagitated.

Why is it that youngsters have so much to do with psychic phenomena? I have gone into that sub-ject, according to my notions. Well, then, when a whole nation, -or hosts of its people, goes primi-tive, or gives in to atavism, or reverts religiously, it may be that conditions arise that are susceptible to phenomena that are repelled by matured mentality. A hard-headed materialist says, dogmatically: "There are no occult phenomena." Perhaps he is right about this, relatively to himself. But what he says may not apply to child ren. When, at least to considerable degree, a nation goes childish with mediaevalism, it may bring upon itself an invasion of phenomena that in the middle ages were com-mon, but that were discouraged, or alarmed, and were driven more to concealment, when minds grew up somewhat.

If we accept that there is Teleportation, and that there are occult beings, that is going so far that we may as well consider the notion that, to stop inquiry, a marauding thing, to divert suspicion, tele-ported from somewhere in Central Europe, a wolf to England: or that there may be some thing of the nature of an occult police force, which checks mischief and slaughter by the criminals of its kind, and takes teleportative means to remove suspicion-often solving one problem, only by making an-other, but relying upon conventionalizations of human thought to supply cloakery.

The killing of poultry-the body on the railroad line stoppage-scene-shifting.

The killing. of sheep-the body on the railroad linestoppage-

Farm and Homes, March 16-that hardly. had the wolf been killed, at Cumwinton, in the north of England, when farmers, in the south of England, especially in the districts between Tunbridge and Seven Oaks, Kent, began to tell of mysterious attacks upon their flocks. "Sometimes three or four sheep would be found dying in one flock, having in nearly every case been bitten in the shoulder and disemboweled. Many persons had caught sight of the animal, and one man had shot at it. The inhabitants were living in a state of terror, and so, on the first of March, a search party of 60 guns beat the woods, in an endeavor to put an end to the depredations.

A big dog? Another malmoot? Nothing?

"This resulted in its being found and dispatched by one of Mr. R. K. Hodgson's gamekeepers, the animal being pronounced, on examination, to be a jackal."-

The story of the shooting of a jackal, in Kent, is told in the London newspapers. See the' Times, March 2. There is no findable explanation, nor attempted explanation, of how the animal got there. Beyond the mere statement of the shooting, there is not another line upon this extraordinary ·appearance of an exotic animal in England, findable in any London newspaper. It was in provincial newspapers that I came upon more of this story.

Blyth News, March 14-"The Indian jackal. which was killed recently, near Seven Oaks, Kent, after destroying sheep and game to the value of £ 100, is attracting attention in the shop windows of a Derby taxidermist."

Derby Mercury, March 15-that the body of this jackal was upon exhibition in the studio of Mr. A. S. Hutchinson. London Road, Derby.

XV

IN every organism, there are, in its governance as a whole. mysterious transportations of substances and forces, some times in definite, circulatory paths, and sometimes specially, for special needs. In the organic view, Teleportation is a distributive force that is acting to maintain the balances of a whole; with the seeming wastefulness sometimes, and niggardliness sometimes, of other forces: providing, or some times providing, new islands with vegetation, and new ponds with fishes: Edens with Adams, and Adams with Eves; al ways dwindling when other mechanisms become established, but surviving sporadically.

Our expression is that once upon a time, showers of little frogs were manifestations of organic intelligence, in the choice of creatures that could survive, in the greatest variety of circumstances, if indefinitely translated from place to place. They'd survive in water, or on land; in warmth, or in coldness. But, if organic intelligence is like other intelligence, there is no understanding it, except as largely stupid; and, if it keeps on sending little frogs to places where they're not wanted, we human phenomena cheer up, thinking of the follies of Existence, itself. I have never done foolisher, myself, than did Nature when it, or she-probably she fatally loaded the tusks of mammoths, and planted a tree on the head of the Irish elk, losing species for the sake of displays. By intelligence I mean nothing that can be thought of as exclusively residing in, or operating in, brain substance: I mean equilibration, or adaptation, which pervades all phenomena. The scientific intelligence in human grains, and the physiologic intelligence that pervades the bodies of living things, wisely-foolishly acts to solve problems, and somewhere in the beauty of a theorem, or ·of a peacock, lurks the grotesque. When Nature satisfies us critics with such a graceful stroke as a swimming seal, she fumbles her seal on land.

But there is another view. We apologizing theologians always have another view. Cleverness and stupidity are relative, and what is said to be stupidity has functional value. To keep on sending little frogs, where, so far as can be seen, there is no need for little frogs, is like persistently, if not brutally, keeping right on teaching Latin and Greek, for instance. What's that for? Most of the somewhat good writers know little of either. According to my experience, both of· these studies, if at all extended, are of no active value, except to somebody who wants to write up to the highest and noblest standards of the past, and considers himself literary. But this is an expression upon the functions of stupidity. It is likely that showers of little frogs, and the vermiform appendix, and classical studies are necessary for the preservation of continuity between the past and the present. Some persons, who know nothing about it. must for ages go on piously believing in Sir Isaac Newton's doctrine. People who go to fortune tellers and people who go to church are functioning conservatives. If the last platypus or the last churchgoer should die off, there would be broken continuity. It would be a crack in existence. Perhaps today, a chink is stuffed with iguanas, which are keeping alive the dinosaur-strain. Why is it that. when one's mind is not specializing upon anything, it is given to recalling past experiences? It is preserving continuity with the past, or is preserving· whatever one can be thought of, as having, of identity. We shall have instances of the interruption of this process, in human minds. Perhaps if Existence should stop sending little frogs, and stop teaching Latin and Greek. a whole would be in a state of amnesia. Our expression is that Teleportation is enormously useful to · life upon this earth, but our data have been, and for a while will continue to be, mostly of its vagaries, or its conservations.

If our existence is an organism, it would seem that it must be one of the most notorious old rascals in the cosmos. It is a fabric of lies. Everywhere it conjures up appearances of realness and finality and trueness-words that I use as synonyms for one state-and then, when examined, everything is found not to be real, or final, or true, but to be depending upon something else, or some other chimera, merging away, and losing its appearance of individuality, into everything else, or every other fraud. That this pseudo-individualizing may in some cases realize itself is a view that I am not taking up in this book. Here it is our concern to and out, if we think we can, whether we be the phenomena of an organism or not. Whether that organism be producing something, or' be graduating realness out of the phenomenal, is a question that I shall take up some other time.

Imposture pervades all things phenomenal. Everything is a mirage. Nevertheless, accepting that there is continuity, I cannot accept that anybody ever has been an absolute impostor. If he's a Tichborne Claimant, after a· while he thinks that there may be some grounds for his claims. If good and evil are' continuous, any crime can be linked with any virtue. Imposture merges away into self-deception so that only relatively has there ever been an impostor.

Every scientist who has played a part in any developing science has, as can be shown, if he's been dead long enough, by comparing his views with more modern views, deceived himself. But there have been cases that look more flagrant. To what degree did Haeckel doctor illustrations in his book, to make a theory work out right? What must one think of Prof. Kammerer? In August, 1926, he was accused of faking what he called acquired characters on the feet of toads. In September, he shot himself. The only polite way of explaining Prof ..Smyth, Astronomer Royal of Scot land, who founded a cult upon his measurements of the Great Pyramid, is to say that his measuring rod must have slipped. If in his calculations, Prof. Einstein made the error that two distinguished mathematicians say he made, but, if eclipses came out, as they should come out, as reported by astronomers who did not know of the error, there is very good encouragement for anybody to keep on deceiving himself. I can draw no line between imposture and self-deception. I can draw no line between anything phenomenal and anything else phenomenal, even though I accept that also there are lies. But there are scientists who have deceived others so rankly that it seems an excess of good manners to say that also they deceived themselves. If among scientists there have been instances of rank imposture, we shall expect to come upon much imposture in our data of irresponsible persons. The story told by Prof. Ma o-Fusco, of Naples, when, in August, 1924, he announced that he had discovered the 109 missing volumes of Livy's History of Rome, is not commonly regarded as imposture, because when the Professor could not produce the missing volumes, his explanation that he had been indiscreet was published and accepted. This scientist's indiscretion was glossed over, as in the time of full power of the preceding orthodoxy, the indiscretion of any priest was hushed up. The impression went abroad that all that was wrong was that the Professor had been too ardent, or so hopeful of finding. the books that prematurely he had announced having found them. But there are other impressions. They are of credulous American millionaires, and of the unexpected interest that the Italian Government showed in the matter.

What about the other Professors, who told that they had seen the volumes? See Current Literature, 77-594. Here is published a facsimile of four lines, which Dr. Max Funcke said that he had copied from one of the manuscripts, which according to Prof. Fun's explanation, he had only hoped to find. I can find no explanation by Dr. Funcke.

One explanation is that perhaps there was not forgery, and that perhaps the volumes were found, and by evasion of representatives of the Italian Government. are in the col lection of a silent American millionaire, today. But I do not think that collectors care much for treasures that they can't tell about.

The tale of an itch-Dr. Grimme and the inscribed stone- and the irritation it was to a pious Professor, until he was able to translate it, as it should be translated. In the year 1923, Dr. Grimme, Professor of Semitic Languages, at the University of Munich, sent out good cheer to the faithful. God. who had been doing poorly, got a boost. Dr. Grimme announced that. from an inscribed stone, which had been discovered in a temple, at Sinai, he had deciphered the story of the rescue of the infant Moses, from the Nile, by an Egyptian Princess.

London Observer, Oct. 25, 1925-a letter from Sir Flinders Petrie-that Dr. Grimme had made his translation by adding cracks in the stone, and some of its weather marks, to the hieroglyphics-that. in one division of the inscription, he had "translated" as many scratches as he had veritable characters, to make the thing come out right.

If Dr. Grimme alleviated an itch with scratches, that fs the temporary way by which problems always have been said to be solved.

Only to be phenomenal is to .be a least questionable. Any scientist who claims more is trying to register divinity. If Life cannot be positively differentiated from anything else. the appearance of Life itself is deception. li, in mentality, there is no absolute dividing-line between intellectuality and imbecility, all wisdom is partly idiocy. The seeker of wisdom departs more and more from the state of the idiot, only to find that he is returning. Belief after belief fades from his mind: so his goal is the juncture of two obliterations. One is of knowing nothing, and the other is of knowing that there is nothing to know.

But here are we, at present not so wise as no longer to have ideas. Suppose we accept that anything phenomenal ever has developed, though only relatively, into considerable genuineness, or a good deal of a look of genuineness so long as it is not examined. But it began in what we call fraudulency. Everybody who can exceptionally do anything, began with a pose, with false claims, and with extreme self-deception. Our expression is that, in human affairs, rank imposture is often a sign of incipiency, or that astrologers, alchemists, and spiritualistic mediums are fore runners of what we shall have to call values, if we can no longer believe in truths. It could be that, with our data, we tell of nothing but lies, and at the same time be upon the track of future values.

Snails, little frogs, seals, reindeer have mysteriously appeared.

The standardized explanation of mysterious human strangers, who have appeared at points upon this earth, acting as one supposes inhabitants of some other world· act, if arriving here, or acting as inhabitants of other parts of this earth, transported in a state of profound hypnosis, would probably act, is that of imposture. Having begun with a pretty liberal view of the prevalence of impostors, I am not going much to say that the characters of our data were not impostors, but am going to examine the reasons for saying that they were. If, except fraudulently, some of them never have been explained conventionally, we are just where we are in everything else that we take up, and that is in the position of having-to pretend to think for ourselves.

The earliest of the alleged impostors in my records-for which, though not absolutely, I draw a deadline at the year 1800-is the Princess Caraboo, if not Mary Willcocks, though possibly Mrs. Mary Baker, but perhaps Mrs. Mary Burgess, who, the evening of April 3, 1817, appeared at the door of a cottage, near Bristol, England, and in an un known language asked for food.

But I am not so much interested in whether the Princess, or Mary, was a rascal- as I am in the reasons for saying that she was. It does not matter whether we take up a theorem in celestial mechanics, or the case of a girl who jabbered, we come upon the bamboozlements by which conventional thought upon this earth is made and preserved. The case of the angles in a triangle that equal two right angles has never been made out: no matter what refinements of measurement would indicate, ultra-refinement would show that there had been errors. Because of continuity, and be cause of discontinuity, nothing has ever been proved. If only by making a very bad error to start with, Prof. Einstein's prediction of the curvature of lights worked out as it should work out, we suspect, before taking up the case of the Princess Caraboo that the conventional conclusion in her case was a product of mistakes.

That the Princess Caraboo was an impostor-first we shall take up the case, as it has been made out: London Observer, June 10, 1923-that the girl, who spoke unintelligibly, was taken before a magistrate, Samual Worrall, of Knowle Park, Bristol, who instead of committing her as a vagrant, took her to his home. It is not recorded just what Mrs. Worrall thought of that. It is recorded that the girl was at least what is said to be "not unprepossessing." When questioned the "mysterious stranger" wrote in unknown characters, many of which looked like representations of combs. Newspaper correspondents interviewed her. She responded with a fluency of "combs," and a smattering of "birdcages" and "frying pans." The news spread, and linguists traveled far to try their knowledge, and finally one of them was successful. He was "a gentleman from the East Indies," and, speaking in the Malay language to the girl, he was answered. To him she told her story. Her name was Caraboo, and one day while walking in her garden in Java, she was seized by pirates, who carried her aboard a vessel, from which, after a long imprisonment, she escaped to the coast of England. The story was colorful with details of Javanese life. But then Mrs. Willcocks, not of Java, but of a small town in Devonshire, appeared and identified her daughter Mary. Mary broke down and confessed. She was not prosecuted for her imposture: instead, Mrs. Worrall was so kind as to pay her passage to America.

Mostly our concern is in making out that this case was not made out-or, more widely, that neither this nor any other case ever has been made out-but I notice a little touch of human interest entering here. I notice that we feel a disappointment, because Mary broke down and confessed. We much prefer to hear of impostors who stick to their impostures. If no absolute line can be drawn be tween morality and immorality, I can show, if I want to, that this touch of rascality in all of us-or at any rate in me-is a virtuous view, instead. So when an impostor sticks to his imposture, and we are pleased, it is that we approve a resolutely attempted consistency, even when applied to a fabric of lies.

Provided I can find material enough, I can have no trouble in making it appear "reasonable," as we call it, to accept that Mary, or the Princess, confessed, or did not confess, or questionably confessed. Chambers' Journal, .66-753-that Caraboo, the imposter, had told her story of alleged adventures, in the Malay language.

Farther along, in this account-that the girl had spoken in an unknown language, .

This is an inconsistency worth noting. We're on the trail of bamboozlement, though we don't have to go away back to the year 1817 to get there. We hunt around. We come upon a pamphlet, entitled Caraboo, published by J. M. Cutch, of Bristol, in the year 1817. We learn in this account, which is an attempt to show that Caraboo was unquestionably an impostor, that it was not the girl. but the "gentle man from the East Indies," whose name was Manual Eyenesso, who was the impostor, so far as went the whole Javanese story. To pose as a solver of mysteries,. he had pre tended that to his questions, the girl was answering him in the Malay language, and pretending to translate her gibberish, he had made up a fanciful story of his own.

Caraboo had not told any story, in any known language, about herself. Her writings were not in Malay characters. They were examined by scientists, who could not identify them. Specimens were sent to Oxford, where they were not recognized. Consequently, the "gentleman from the East In dies" disappeared. We are told in the pamphlet that every Oxford scholar who examined the writings, "very properly and without a moment's hesitation, pronounced them to be humbug." That is swift propriety.

li the elaborate story of the Javanese Princess had been attributed to a girl who had told no under standable story of any kind, it seems to us to be worthwhile to look over the equally elaborate confession, which has been attributed to her. It may be that regretfully we shall have to give up a notion that a girl had been occultly transported from the planet Mars, or from somewhere up in Orion or Leo, but we are seeing more of the ways of suppressing mysteries. The mad fishmonger of Worcester shovels his periwinkles everywhere.

According to what is said to be the confession, the girl was Mary Willcocks, born in the village of Witheridge, Devonshire, in the year 1791, from which at the age of 16 she had gone to London, where she had married twice. It is a long, detailed story. Apparently the whole story of Mary's, adventures, from the time of her departure from Witheridge, to the time of her arrival in Bristol, is told in what is said to be the confession. Everything is explained-and then too much is explained. We come to a question that would be an astonisher, if we weren't just a little sophisticated, by this time-By what freak of accomplishment did a Devonshire girl learn to speak Javanese?

The author of the confession explains that she had pick ed up with an East Indian, who bad taught her the language.

If we cannot think that a girl, who bad not even pretended to speak Javanese, would explain bow she bad picked up Javanese, it is clear enough that this part of the alleged confession is forgery. I explain it by thinking that somebody bad been hired to write a confession, and with too much of a yarn for whatever skill be bad, bad overlooked the exposed imposture of the "gentleman from the East Indies." All that I can make of the story is that a girl mysteriously appeared. It cannot be said that her story was imposture, because she told no intelligible story. It may be doubted that she confessed, if it be accepted that at least part of the alleged confession was forgery. Her mother did not go to Bristol and identify her, as, for the sake of a neat and convincing finish, the oonventionalized story goes. Mrs. Worrall told that she had gone to Witheridge, where she bad found the girl's mother, who bad verified whatever she was required to verify. Caraboo was shipped away on the first vessel that sailed to America; or, as told in the pamphlet, Mrs. Worrall, with forbearance and charity, paid her passage far away. In Philadelphia, somebody took charge of her affairs, and, as if having never beard that she was supposed to have confessed, she gave exhibitions, writing in an unknown language. And I wouldn't give half this to the story of the Princess Caraboo, were it not for the epitomization, in her story, of all history. If there be God, and if It be ubiquitous, there must be a jostle of ubiquities because the Fishmonger of Worcester, too, is everywhere. I should like to think that inhabitants of other worlds, or other parts of one existence, have been teleported to this earth. How I'd like it, if I were teleported the other way has nothing to do with what I'd like to think had befall somebody else. But I can't say that our own stories, any way so far, have the neat and convincing finish of the conventional· stories. Toward the end of the year 1850, a stranger, or I should say a "mysterious stranger," was found wandering in a village near Frankfort-on-the-Oder. How he got there, nobody knew. See the Athenaeum, April 15, 1851. We are told that his knowledge of German was imperfect. If the imperfections were filled out by another Manuel Eyenesso, I fear me that suggestions of some new geographical, or cosmographical, knowledge can't develop. The man was taken to Frankfort where he told his story, or where, to pose as a linguist, somebody told one for him. It was told that his name was Joseph Vorin, and that he had come from Laxaria. Laxaria is in Salena. and Salcria is far from Europe-"beyond vast oceans."

In the London Daily Mail, Sept. 18, 1905, and following issues, are accounts of a young man who had been arrested in Paris, charged with vagrancy. It was impossible to understand him. In vain had he been tried with European and Asiatic languages, but by means of signs, he had made known that he had come from Lisblan. Eisar was the young man's word for a chair: a table was a lotoba, and his sonar was his nose. Mr. George R. Sims, well-known criminologist, as well as a story writer, took the matter up scientifically. As announced by him, the mystery had been solved by him. The young man, an impostor, had trans posed letters, in fashioning his words. So the word raise, transposed, becomes eisar. But what has a raise to do with a chair? It is said that true science is always simple. A chair raises one, said Mr. Sims, simply. Now take the word sonar. As we see, when Mr. Sims points it out to us, that word is a transposition of the word snore, or is almost. That's noses, or relation to noses.

The criminologists are not banded like some scientists. In Paris, the unhanded wisemen said that Mr. Sims' trans positions were farfetched. With a freedom that would seem reckless to more canny scientists, or without waiting three or four months to find out what each was going to say, they expressed opinions. The savants at Glozel, in the year 1927, were cannier, but one can't say that their delays boosted the glories of science. One of the wisemen of Paris, who accused Mr. Sims of fetching too far, was the eminent scientist, M. Haag. "Take the young man's word Odir, for God," said M. Haag: "transpose that, and we have Dio, or very nearly. Dio is Spanish for God. The young man is Spanish." Another distinguished wiseman was M. Roty. He rushed into print, while M. Haag was still explaining. "Consider the word sacar, for house said M. Roty. "Unquestionably we have a transposition of the word casa, with a difference of only one letter, and casa is Italian for house. The young man is Italian." Le Temps, September 18-an other wiseman, a distinguished geographer, this time, identified the young man as one of the Russian Doukhobors.

Where would we be, and who would send the young ones to school, if all the other wisemen of our tribes had such independence? If it were not for a conspiracy that can be regarded as nothing short of providential, so that about what is taught in one school is taught in the other schools, one would spend one's lifetime, learning and unlearning, in school after school. As it is, the unlearning can be done, after leaving one school.

The young man was identified by the police, as Rinaldo Agostini, an Austrian, whose fingerprints had been taken several times before, somewhere else, when he had been arrested for vagrancy.-

Whether the police forced this mystery to a pseudo-conclusion, or not, a suggestive instance is told of, in the London Daily Express, Oct. 16, 1906. A young woman had been arrested in Paris, charged with picking pockets, and to all inquiries she answered in an unknown language. Interpreters tried her with European and Asiatic languages, without success, and the magistrate ordered her to be kept under surveillance, in a prison infirmary. Almost immediately, watchers reported that she had done exactly what they wanted to report that she had done-that she had talked in her sleep, not mumbling in any way that might be questionable, but speaking up "in fluent French, with the true Parisian accent." If anybody thinks that this book is an attack upon scientists, as a distinct order of beings, he has a more special idea of it than I have. As I'm seeing things, everybody's a scientist.

If there ever have been instances of teleportations of human beings from somewhere else to this earth, an examination of inmates of infirmaries and workhouses and asylums might lead to some marvelous astronomical disclosures. I suppose I shall be blamed for a new nuisance, if after the publication of these notions, mysterious strangers start cropping up, and when asked about themselves, point up to Orion or Andromeda. Suppose any human being ever should be translated from somewhere else to this earth, and should tell about it. Just about what chance would he have for some publicity? I neglected to note the date, but early in the year 1928, a man did appear in a town in New Jersey, and did tell that he had come from the planet Mars. Wherever he came from, everybody knows where he went, after telling that.

But, if human beings ever have been teleported to this earth from somewhere else, I should think that their clothes, different in cut and texture, would attract attention. Clothes were thought of by Manuel Eyenesso. He pretended that Caraboo had told him that, before arriving in Bristol, she had exchanged her gold-embroidered, Javanese dress for English clothes. Whatever the significance may be, I have noted a number of "mysterious strangers," or "wild men," who were naked.

A case that is mysterious, and that may associate with other mysteries, was reported in the London newspapers (Daily Mail, April 2; Daily News, April 3, 1923). It was at the time that Lord Camarvon was dying, in Cairo, Egypt, of a disease that physicians said, was septic pneumonia, but that, in some minds, was associated with the opening of Tut-Ankh-Amen's tomb. Upon Lord Carnarvon's estate, near Newbury, Hampshire, a naked man was running wild, often seen, but never caught. He was first seen, upon March 17th. Upon March 17th, Lord Camarvon fell ill, and' he died upon April 5th. About April 5th, the wild man of New bury ceased to be reported.

If human beings from somewhere else ever have been translated to this earth-

There are mysteries at each end, and in between, in the story of Cagliostro.

He appeared in London, and then in Paris, and spoke with an accent that never has been identified with any known language of this earth. If, according to most ac counts of him, he was Joseph Balsamo, a Sicilian criminal, who, after a period of extraordinarily successful imposture, was imprisoned in Rome, until he died, that is his full life-story.

The vagueness of everything-and the merging of all things into everything else, so that stories that we, or some of us, have been taking, as "absolutely proved," turned out to be only history, or merely science. Hosts of persons suppose that the exposure of Cagliostro, as an imposter, is as firmly, or rationally, established, as are the principles of geology, or astronomy. And it is my expression that they are right about this.

Wanted-well, of course if we could find data to support our own notions-but, anyway, wanted-data for at least not accepting the conventionalized story of Cagliostro:

See Trowbridge's story of Cagliostro. According to Trowbridge, the identification of Cagliostro was fraudulent. At the time of the Necklace Affair, the police of Paris, needing a scapegoat, so "identified" him, in order to discredit him, according to Trowbridge. No witness appeared; to identify him. There was no evidence, except that handwritings were similar. There was suggestion, in the circumstance that Balsamo had an uncle, whose name was Giuseppe Cagliostro. One supposes that a police official, whose labors were made worthwhile by contributions from the doctors of Paris, searched records until he came upon an occurrence of the name of Cagliostro in the family of a criminal. and then went on from that finding. Then it was testified that the handwritings of Balsamo and Cagliostro were similar. For almost everybody's belief that of course Cagliostro was identified as Joseph Balsamo, there is no more than this for a base. In February, 1928, the New York newspapers told of a graphologist, who had refused to identify handwriting, according to the wishes of the side that employed him. Ac cording to all other cases that I have ever read of, any body can get, for any handwriting, any identification that he pays for. li in any court, in any land, any scientific pronouncement should be embarrassing to anybody, that is because he has been too stingy to buy two expert opinions.

Cagliostro appeared, and nothing more definite can be said of his origin. He rose and dominated, as somebody from Europe, if transported to a South Sea Island, might be expected to capitalize his superiority. He was bounded by the medical wisemen, as Mesmer was hounded by them, and as anybody who, today, would interfere with flows of fees, would be hounded by them. Whether in their behalf, or because commonplace endings of all mysteries must be published, we are told, in all conventional accounts, that Cagliostro was an impostor, whose full life-story is known, and is without mystery.

It is said that, except where women were concerned, where not much can be expected, anyway, Cagliostro bad pretty good brains. Yet we are told that, having been identified as an Italian criminal, he went to Italy.

There are two accounts of the disappearance of Cagliostro. One is a matter of mere rumors: that he had been seen in Aix-les-Bains; that he had been seen in Turin. The other is a definite story that he went to Rome, where, as Joseph Balsamo, he was sent to prison. A few years later, when Napoleon's forces were in Rome, somebody went to the prison and investigated. Cagliostro was not there. Perhaps he had died.

XVI

HERE is the shortest story that I know of:
St. Louis Globe-Democrat, :Nov. 2. 1886-a girl stepped from her home, to go to a spring.
Still, though we shall have details and comments, I know of many occurrences of which, so far as definitely finding out anything is concerned, no more than that can be told.
After all, I can tell a shorter story:
He walked around the horses.
Upon Nov. 25, 1809 Benjamin Bathurst, returning from Vienna, where, at the Court of the Emperor Francis, he had been representing the British Government, was in the small town of Perleberg, Germany. In the presence of his valet and his secretary, he was examining horses, which were to carry his coach over more of his journey back to England. Under observation, he walked around to the other side of the horses. He vanished. For details, see the Cornhill Magazine; 55-279.
I have not told much of the disappearance of Benjamin Bathurst, because so many accounts are easily available: but the Rev. Sabine Baring-Gould, in Historic Oddities, tells of a circumstance that is not findable in all other accounts that I have read. It is that, upon Jan. 23, 1810, in a Hamburg newspaper, appeared a paragraph, telling that Bathurst was safe and well, his friends having received a letter from him. But his friends had received no such letter. Wondering as to the origin of this paragraph, and the reason for it, Baring-Gould, asks: "Was it inserted to make the authorities abandon the search. Was it an inquiry-stopper? is the way I word this. Some writers have thought that, for political reasons, at the instigation of Napoleon Bonaparte, Bathurst was abducted. Bonaparte went to the trouble to deny that this was so.
In the Literary Digest, 46-922, it is said that the police records of London show that 170,472 persons mysteriously disappeared, in the years 1907-13, and that nothing had been found out, in 3,260 of the cases. Anybody who has an impression of 167,212 cases all explained ordinarily, may not think much of 3,260 cases left over. But some of us, now educated somewhat, or at least temporarily, by experience with pseudo-endings of mysteries, will question that · the 167,212 cases were so satisfactorily explained, except relatively to not very exacting satisfactions. If it's a matter of re-marriage and collection of insurance, half a dozen bereft ones may "identify" a body found in a river, or cast up by the sea. They settle among themselves which shall marry again and collect. Naturally enough, wher-

ever Cupid is, cupidity is not far away, and both haunt morgues. Whether our astronomical and geological and biological know ledge is almost final, or not, we know very little about ourselves. Some of us can't, or apparently can't, tell a husband or a wife from someone else's husband or wife. About the year 1920, in New York City, a woman, whose husband was in an insane asylum, was visited by a man, who greeted her fondly, telling her that he was her husband. She made everything cheerful and homelike for him. Sometime later, she learned that her husband was still in the asylum. She seemed resentful about this, and got the other man arrested. Cynical persons will think of. various explanations.

I have notes upon another case. A man appeared and argued with a woman, whose husband was a sailor that· he was her husband. "Go away!" said she: "you are darker than my husband." "Ahl" said be: "I have bad yellow fever." So she listened to reason, but something went wrong, and the case got into a police court.

Because of the flux and the variation of all supposed things, I typify all judgments in all matters-in trifles and in scientific questions that are thought to be of utmost importance-with this story of the woman and her uncertain ties. If a husband, or a datum, would stay put, a mind, if that could be kept from varying, might be said to know him, or it, after a fashion.

There have been many mysterious disappearances of human beings. Here the situation is what it is in every other subject, or so-called subject, ff there is no subject that has independent existence. Only those who know little of a matter can have a clear and definite opinion upon it. Whole civilizations have vanished. There are statistical reasons for doubting that five sixths of the Tribes of Israel once upon a time disappeared, but that is tradition, anyway. Historians tell us what became of the Jamestown Colonists, but what becomes of historians? Persons as well-known as Bathurst have disappeared. As to the disappearance of Conant, one of the editors of Harpers Weekly, see the New York newspapers beginning with Jan. 29, 1885. Nothing was found out. For other instances of well-known persons who have disappeared, see the New York Tribune, March 29, 1903, and Harper's Magazine, 38-504.

Chicago Tribune, Jan. 5, 1900-"Sherman Church, a young man employed in the Augusta Mills' (Battle Creek, Mich.) has disappeared. He was seated in the Company's office, when be arose and ran into the mill. He has not been seen since. The mill has been almost taken to pieces by the searchers, and the river, woods, and country have been scoured, but to no avail. Nobody saw Church leave town, nor is there any known reason for his doing so.

Because of the merging of everything-without entity, identity, or soul of its own-into everything else, anything, or what is called anything, can somewhat reasonably be argued anyway. Anybody who feels so inclined will be as well justified- as anybody can be, in arguing about all mysterious disappearances, in terms of Mrs. Christie's mystery. In December, 1926, Mrs. Agatha Christie, a writer of detective stories, disappeared from her home in England. The newspapers, noting her occupation, commented good naturedly, until it was reported that, in searching moors and forests and villages and towns, the police bad spent £10,000. Then the frugal Englishmen became aware of the moral aspect of the affair, and they were severe. Mrs. Christie was found. But, according to a final estimate, the police had spent only £ 25. Then everybody forgot the moral aspect and was good-natured again. It was told that Mrs. Christie, in a hotel, somewhere else in England, having been keen about getting newspapers every morning, bad appeared at the hotel, telling fictions about her identity. She was taken home by her husband. She remembered nobody, her friends said, but, thinking this over, they then said that she remembered nobody but her husband. Several weeks later, a new book by Mrs. Christie was published. It seems to have been a somewhat readable book, and was pleasantly reviewed by frugal Englishmen, who are very good-humored and tolerant, unless put to such expense as to make them severe and moral.

Late in the year 1913, Ambrose Bierce disappeared. It was explained. He had gone to Mexico, to join Villa, and had been killed at the. Battle of Torreon. New York Time, April 3, 1915-mystery of Bierce's disappearance solved he was upon Lord Kitchener's staff, in the recruiting ser vice, in London. New York Times, April 7, 1915-no know ledge of Bierce, at the War Office, London. In March, 1920, newspapers published a dispatch from San Francisco, telling that Bierce had gone to Mexico, to fight against Villa, and had been shot. It would be a fitting climax to the life of this broad-minded writer to be widely at work in London, while in Mexico, and to be killed while fighting for and against Villa. But that is pretty active for one, who, as Joseph Lewis French points out, in Pearsons Magazine, 39- 245, was incurably an invalid and was more than seventy years old. For the latest; at this writing, see the New York Times, Jan. 1, 1928. Here there is an understandable explanation of the disappearance. It is that Bierce had criticized Villa.

London Daily Chronicle, Sept. 29, 1920-a young man, evening of September 27th, walking in a street, in South London-

Magic-houses melting-meadows appearing Or there was a gap between perceptions.

However he got there, he was upon a road, with fields around. The young man was frightened. He might be far away, and unable to return. It was upon a road, near Dunstable, 30 miles from London, and a policeman finding him exclaiming, pacing back and forth, took him to the station house. Here he recovered sufficiently to tell that he was Leonard Wadham, of Walworth, South London, where he was employed by the Ministry of Health. As to how he got to this point near Dunstable, he could tell nothing Of a swish, nobody could tell much.

Early in the year 1905, there were many mysterious disappearances in England. See back to the chapter upon the extraordinary phenomena of this period. Here we have an account of one of them, which was equally a mysterious appearance. I take it from the Liverpool Echo, February 8. Upon the 4th of February, a woman was found, lying unconscious, upon the shore, near Douglas, We of Man. No one had see her before, but it was supposed that she had arrived by the boat from England, upon the 3rd of February. She died, without regaining consciousness. There were many residents of the island, who had, in their various callings, awaited the arrival of this boat, and had, in their various interests, looked more than casually at the passengers: but 200 Manxmen visited the mortuary, and not one of them could say that he had seen this woman arrive. The news was published, and then came an inquiry from Wigan, Lancashire. A woman had "mysteriously dis appeared" in Wigan, and by her description of the body found near Douglas was identified as that of Mrs. Alice Hilton, aged 66, of Wigan. & told, in the Wigan Observer, somebody said that Mrs. Hilton had been last seen, upon February 2nd, on her way to Ince, near Wigan, to visit a cousin. But nobody saw her leave Wigan, and she had no known troubles. According to the verdict, at the inquest, Mrs. Hilton had not been drowned, but had. died of the effects of cold and exposure upon her heart.

I wonder whether Ambrose Bierce ever experimented with self-teleportation. Three of his short stories are of "mysterious disappearances." He must have been uncommonly interested to repeat so. Upon ·Sept. 4, 1905, London newspapers reported the disappearance, at Ballycastle, Co. Antrim, Ireland, of Prof. George A. Simcox, Senior Fellow of Queen's College, Ox ford; Upon August 28th, Prof. Simcox had gone for a walk, and had not returned. There was a search, but nothing was learned.

Several times before, Prof. Simcox had attracted attention by disappearing. The disappearance at Ballycastle was final.

XVII

As interpreters of dreams, I can't say that we have ambitions, but I think of one dream that many persons have had, repeatedly, and it may have relation to our present subject. One is snoring along, amidst the ordinary marvels of dreamland-and there one is, naked, in a public place, with no impression of how one got there. I'd like to know what underlies the prevalence of this dream, and its dis agreeableness, which varies, I suppose, according to one's opinion of oneself. I think that it is· subconscious aware ness of something that has often befallen human beings, and that in former times was commoner. It may be that occult transportations of human beings do occur, and that, because of their selectiveness, clothes are sometimes not included.

"Naked in-the street-strange conduct by a strange man. See the Chatham (Kent, England) News, Jan: 10, -1914. Early in the evening of January 6th-"weather bitterly cold -a naked man appeared, from nowhere that could be found out, in High Street, Chatham.

The man ran up and down the street, until a policeman caught him. He could tell nothing about himself. "Insanity," said the doctors, with their customary appearance of really saying something.

I accept that, relatively, there is insanity, though no definite lines can be drawn as to persons in asylums, persons not in asylums, and persons not yet in asylums. U by insanity is meant processes of thought that may be logical enough, but that are built upon false premises, what am I showing but the insanity of all of us? I accept that as extremes of the state that is common to all, some persons may be considered insane: but, according to my experience with false classifications, or the impossibility of making anything but false classifications, I suspect that many per sons have been put away, as insane, simply because they were gifted with uncommon insights, or had been through uncommon experiences. It may be that, hidden under this cloakery, are the subject-matters of astonishing, new inquiries. There may be stories that have been told by alleged lunatics that someday will be listened to, and investigated, leading to extraordinary disclosures. In this matter of insanity, the helplessness of science is notorious, though it is only of the helplessness of all science. Very likely the high-priced opinions of alienists are sometimes somewhat nearly honest: but, as in every other Geld of so-called human knowledge, there is no real standard to judge by: there is no such phenomenon as insanity, with the noumenal quality of being distinct and real in itself. If it should ever be somewhat difficult to arrange with professional wisemen to testify either for or against any person's sanity, I should have to think that inorganic science, in this Held, may not be so indefinite.

This naked man of Chatham appeared suddenly. Nobody had seen him on his way to his appearingpoint. His clothes were searched for, but could not be found. Nowhere near Chatham was anybody reported missing.

Little frogs, showers of stones, and falls of water-and they have repeated, indicating durations of transportory currents to persisting appearing-points, suggesting the existence of persisting disappearing-points somewhere else. There is an account, in the London Times, Jan. 30, 1874, of repeating disappearances of young. men, in Paris. Very likely, as a development of feminism, there will be female Bluebeards, but I don't think of them away back in the year 1874. "In every case, their relatives and friends declare that they were unaware of any. reason for evasion, and the missing persons seem to have left their homes for their usual avocations."

A field, somewhere near Salem, Va., in the year 1885-and that in this Held there was a suction. In the New York Sun, April 25, 1885, it is said that Isaac Martin, a young farmer, living near Salem, Va., had gone into a Held, to work. and· that he had disappeared. It is said that in this region there had been other mysterious disappearances. In Montreal, in July and August, 1892, there were so many unaccountable disappearances that, in the newspapers, the headline "Another Missing Man" became common. In July, 1883, there was a similar series, in Montreal. London Evening Star. Nov. 2, 1926-"mysterious series of disappearances-eight persons missing, in a few days." It was in and near Southend. First went Mrs. Kathleen Munn, .and her two small children. Then a girl aged 15-girl aged 16, girl aged 17, another girl aged 16. Another girl, Alice Stevens, disappeared. "She was found in a state of collapse. and was taken to hospital."

New York Sun, Aug. 14, 1902-disappearances, in about a week, of five men, in Buffalo, N. Y.

Early in August, 1895, in the city of Belfast, Ireland, a little girl named Rooney disappeared. Detectives investigated. While they were investigating, a little boy, named Webb, disappeared. Another child disappeared. September 10--disappearance of a boy, aged seven, named Watson. Two days later, a boy, named Brown, disappeared. See the Irish New, (Belfast), September 20. Io following issues of this newspaper, no more information is findable.

London Daily Mirror, Aug. 5, 1920-"Belfast police are in possession of the sensational news that eight girls, all under twelve years of age, are missing since last Monday, week, from the Newtownards-road, East Belfast."

In August, 1869, English newspapers reported disappearances of 13 children, in Cork, Ireland. I take from the Tiverton Times, August 31. It may be that the phenomenon cannot be explained in terms of local kidnapers, because somewhere else, at the same time, children were disappearing. London Daily News, August 31-excitement in Brussels, where children were disappearing.

Five "wild men" and a "wild girl appeared in Connecticut, about the first of January, 1888. See the St. Louis Globe Democrat, January 5, and the New York Times, Jan. 9, 1888.

I have records of six persons, who, between Jan. 14, 19.20, and Dec. 9, 1923, were found wandering in or near the small town of Romford, Essex, England, unable to tell how they got there, or anything else about themselves. I have satisfactorily come upon no case in which somebody has stated that he was walking, say, in a street in New York, and was suddenly seized upon and set down somewhere, say in Siberia, or Romford. I have come upon many cases like that of a man who told that he was walking along Euston Road, London, and-but nine months later-when next he was aware of where he was, found himself working on a farm, in Australia. If human beings ever have been tele ported, and, if some mysterious appearances of human beings be considered otherwise unaccountable·, an effect of the experience is effacement of memory.

There have been mysterious appearances of children in every land. In India, the explanation of appearances of children of an unknown past is that they had been brought up by she-wolves.

There have been strange fosterings: young rabbits adopted by cats, and young pigs welcomed to strangely foreign founts. But these cases are of maternal necessity, and of unlikely benevolence, and we're asked to believe in bene volent she-wolves. I don't deny that there is, to some degree, benevolence in wolves, cats, human beings, ants: but benevolence is erratic, and not long to be depended upon. Sometimes I am benevolent, myself, but pretty soon get over it. The helplessness of a human infant outlasts the suckling period of a wolf. How do she-wolves, any of the rest of us, keep on being unselfish, after nothing's made by unselfishness?

For an account of one of the later of the "wolf child ren" of India (year 1914) see Nature, 93-566. In the Zoologist, 3-12-87, is an account of a number of them, up to the year 1852. In the Field, Nov. 9, 1895, the story of the "wolf child" of Oude is told by an Assistant Commissioner, who had seen it. It was a speechless, little animal, about four years old. Policemen said that, in a wolf's den, they had found this child, almost devoid of human intelligence. The child grew up and became a policeman. In Human Nature, 7-302, is a story of two "wolf children" that were found at different times, near Agra, Northern India. Each was seven or eight years old. For a recent case, see the London Observer, Dec. 1926. Hindus had brought two "wolf children, one aged two, and the other about eight years old, to the Midnapore Orphanage. The idea of abandonment of young idiots does not look so plausible, in cases of more than one child. Also, in a case of several children. a she-wolf would seem very graspingly unselfish. The child ren crawled about on all fours, ate only raw meat, growled, and avoided other inmates of the Orphanage. I suppose that they ate only raw meat, because to confirm a theory that was all they got.

London Daily Mail, April 6, 1927-another wolf child- boy aged seven-found in a cave, near Allahabad. For an instance that is the latest, at this writing, see the New York Times, July 16, 1927. Elephant youngsters and rhinoceros brats have still to be heard of, but, in the London Morning Post, Dec. 31, 1926, is a story of a "tiger child." A "leopard boy" and a "monkey girl" are told of, in the London Observer, April 10, 1927.

Our data are upon events that have astonished horses and tickled springboks. They have shocked policemen. I have notes upon an outbreak of ten "wild men," who appeared in different parts of England, in that period of extraordinary phenomena, the winter of 1904-5. One of them, of origin that could not be found out, appeared in a street in Cheadle. He was naked. An indignant policeman, trying to hang his overcoat about the man, tried to reason with him, but had the same old trouble that Euclid and New ton and Darwin had, and that everybody else has, when trying to be rational, or when trying, in the inorganic, or scientific, way, to find a base to argue upon. I suppose the argument was something like this-

Wasn't he ashamed of himself?

Not at all. Some persons might have reasons for being ashamed of themselves, but he had no reason for being ashamed of himself. What's wrong with nakedness? Don't cats and horses and dogs go around without clothes on?

But they are clothed with natural, furry protections. Well, Mexican dogs, then.

Let somebody else try-somebody who thinks that, as products of logic, the teachings of astronomy, biology, geo logy, or anything else are ·pretty nearly final, though with debatable minor points, to be sure. Try this simple, little problem to start with. Why shouldn't the man walk around naked? One is driven to argue upon the basis of conventionality. But we are living in an existence, which itself may be base, but in which there are not bases. Argue upon the basis of conventionality and one is open to well-known counterarguments. What is all progress but defiance of conventionality?

The policeman, in Euclid's state of desperation, took it as self-evident disgracefulness. Euclid put theorems in bags. He solved problems by encasing some circumstances in an exclusion of whatever interfered with a solution. The policeman of Cheadle adopted the classical method. He dumped the "wild man" into a sack, which he dragged to the station house.

Another of these ten "wild men" spoke in a language that nobody had ever heard of before, and carried a book, in which were writings that could not be identified at Scot land Yard. Like a traveler from far away, he had made sketches of things that he had seen along the roads. At Scotland Yard, it was said of the writings: "They are not French. German, Dutch, Italian, Spanish, Hungarian, Turkish. Neither are they Bohemian, Greek, Portuguese Arabic, Persian, Hebrew, nor Russian." See London newspapers, and the East Anglian Daily Times, Jan.. 12, 1905.

I have come upon fragments of a case, which I reconstruct:

Perhaps in the year 1910, and perhaps not in this year, a Hindu magician teleported a boy from somewhere in England, perhaps from Wimbledon, London, perhaps not. The effect of this treatment was of mental obliteration; of profound hypnosis, or amnesia. The boy could learn, as if starting life anew, but mostly his memory was a void. Later the magician was dying. He repented, and his problem was to restore the boy, perhaps not to his home, but to his native land. He could not tell of the occult transportation, but at first it seemed to him that nobody would believe a story of ordinary kidnaping. It would be a most improbable story: that, in London, a Hindu had kidnaped a boy, and on the way to India had spent weeks aboard a vessel with this boy, without exciting inquiry, and with ability to keep the boy from appealing to other passengers. Still, a story of kidnaping is a story in commonplace terms. No story of ordinary kidnaping could account for the boy's lapsed memory, but at the most some person would think that some of the circumstances were queer, and would then forget-the matter.

For fragments of this story, see Lloyd's Sunday News (London), Oct. 17, 1920. Sometime in the year 1917, the Society for the Propagation of the Gospel, in Nepal, India, received a message from a native priest, who was dying, and wanted to tell something. With the priest was a well grown boy. The priest told that, about the year 1910, in a street in Wimbledon (South London) he had kidnaped this boy. Details of a voyage to India not given. The boy was taken to Gorakapur, and was given employment in a railway workshop. He could speak a. little English, but had no recollection of ever having been in England.

This is the account that the Society sent to its London representative, Mrs. Sanderson, Earl's Court, London. A confirmation of the story, by Judge Muir, of Gorakapur, was sent. Mrs. Sanderson communicated with Scotland Yard.

Lloyd's Sunday News, October 24-"boy not yet identified by Scotland Yard. Ari even more extraordinary development of the story is that quite a number of boys disappeared in Wimbledon, ten years ago." It is said that the police had no way of tracing the boy, because, in Scotland Yard, all records of missing children were destroyed, after a few years. I have gone through the Wimbledon New., for the year 1910, without finding mention of any missing child. Someone else may take a fancy to the job, for 1909, or 1911. In Thomson's Weekly News, Oct. 23, 1920, there are additional details. It is said that without doubt the boy was an English boy: as told by the priest, his Christian name was Albert Hant, and Sussex News, Feb. 25, 1920-"one of the most sensational discoveries and most mysterious cases of tragedy that we have ever been called upon to record" - a naked body of a man, found in a plowed field, near Petersfield, Hampshire, England.

The mystery is in that there had not been a murder. A body had not been thrown from a car into this field. Here had appeared a naked man, not in possession of his. senses. He had wandered, and he had died. It was not far from a road, and was about a mile from the nearest house. Prints of the man's bare feet were traced to the road, and across the road into another field. Police and many other persons searched for clothes, but nothing was found. A photograph of the man was published through out England, but nobody had seen him, clothed or un clothed, before the finding of the body. At the inquest, the examining physician testified that the body was that of a man, between 35 and 40; well-nourished, and not a manual worker; well-cared-for, judging from such particular as carefully trimmed finger nails. There were scratches upon the body, such as would be made by bushes and hedges, but there was no wound attributable to a weapon, and in the stomach there was no poison, nor drug. Death had been from syncope, due to exposure. "The case remains one of the most amazing tragedies that could be conceived of. This mystery did not immediately subside. From time to time there. were comments in the newspapers. London Daily News, April 16-"Although his photograph has been circulated north, east, south, and west, throughout the United Kingdom, the police are still without a clew, and there is no record of any missing person, bearing the slightest re semblance to this man, presumably of education and good standing.

XVIII

THERE was the case of Mrs. Guppy, June 3, 1871, for in stance. As the spiritualists tell it, she shot from her home, in London. Several miles away, she flopped down through a ceiling. Mrs. Guppy weighed 200 pounds. But Mrs. Guppy was a medium. She was a prominent medium, and was well-investigated, and was, or therefore was, caught playing tricks, several times. I prefer to look elsewhere for yarns, or veritable accounts.

In the New York World, March 25, 1883, is told a story of a girl, the daughter of Jesse Miller, of Greenville Town ship, Somerset Co., Pa., who was transported several times, out of the house, into the front yard. But it was her belief that apparitions were around, and most of our data are not concerned with ghostly appearances.

told in the Cambrian Dally Leader (Swansea, Wales), July 7, 1887, poltergeist phenomena were occurring in the home of the Rev. David Phillips, of Swansea. Some time I am going to try to find out why so many of these disturbances have occurred in the homes of clergymen. Why have so many supposed spirits of the departed tormented clergymen? Perhaps going to heaven makes people atheists. However, I do not know that poltergeists can be considered spirits. It may be that many of our records-see phenomena of the winter of 1904-5-relate not to occult beings, as independent creatures, but to projected mentalities of living human beings. A woman of Mr. Phillips' household had been transported over a wall, and toward a brook, where she arrived in a "semi-conscious condition." I note that, not in agreement with our notions upon teleportation, it was this woman's belief that an apparition had carried her. Mr. Phillips and his son, a Cambridge graduate, who had probably been brought up to believe in nothing of the kind. asserted that this transportation had occurred.

A great deal has been written upon the phenomena, or the alleged phenomena, of the Pansini boys. Their story is told in the Occult Review, 4-17. These boys, one aged seven, and the other aged eight, were sons of Mauro Pansini, an architect, of Bari, Italy. Their experiences, or their alleged experiences, began in the year 1901. "One day Alfredo and his brother were at Ruvo, at 9 A.M., and at 9:30 A.M., they were found in the Capuchin Convent, at Malfatti, thirty miles away... In the Annals of Psychic Science, it is said that, about the last of January, 1901, the Pansini boys were transported from. Ruvo to a relative's house, in Trani, arriving in a state of profound hypnosis. In volumes 2 and 3, of the Annals, a discussion of these boys continues:

"But I haven't told the damnedest. Oh, well. we'll have the damnedest A Mediterranean harbor-a man in a boat and, like Mrs. Guppy, down the Pansini boys flop into his boat.

Into many minds flops this idea-"It isn't so much the preposterousness of this story alone: but, if we'd accept this, what else that would threaten all conventional teachings. would we be led into?"

I can't help arguing I have cut down smoking some, and our home brew goes Sat so often that at times I have gone without much of that, but I can't stop arguing. It has no meaning, but I argue that much that is commonplace today was once upon a time denounced from pulpits as the way to hell For all I know, a couple of kids Sopped into a boat I don't feel hellish about it The one thought that I do so little to develop JS that if there be something that did switch the Pansini boys from place to place, it may be put to work, and instead of wharves and railroad stations, there may be built departing and receiving points for commodities, which may be "wished, as it were, from California to London Let stockholders of transportation companies get ahold of this idea, and, if I'm not satisfied with having merely science and religion against me, I'll have opposition enough to suit anybody who can get along without popularity. Just at present, however, I am not selling short on New York Central.

Has anybody, walking along a street, casually looking at someone ahead of him, ever seen a human being vanish? It is a common experience to think that one has seen something like this occur. Another common experience, which has been theorized upon by James and other psychologists, is to be somewhere and have an uncanny feeling that, though so far as one knows, one was never there before, one, nevertheless, was at some time there. It may be that many persons have been teleported back and forth. with out knowing it, or without having more than the dimmest impression of the experience.

But about walking along a street, and having a feeling that somebody has vanished-there have been definitely reported observations upon disappearances. In these instances, the explanation has been that someone had seen a ghost, and that the ghost had vanished. We shall have accounts that look as if observers have seen, not ghosts, but beings like themselves, vanish.

In the Jour. Soc. Psychical Research, 11-189, is published a story by a painter, named John Osborne, living at 5 Hurst Street, Oxford, England. He said that, about the last of March, 1895, he was walking along a road to Wolverton, when he heard sounds of a horse's hoofs behind him, and, turning, saw a man on horseback, having difficulty in controlling his horse. He scurried out of the way, and, when safe, looked again. Horse and man had vanished. Then came the conventionalization, even though it would be widely regarded as an unorthodox conventionalization. It was said that, the week before, a man on horseback had been killed in this part of the road, and that the horse, badly injured, had been shot. Usually there is no use searching for anything further In a publication in which a conventionalization has appeared, but this instance is an exception. In the June number of the Journal, there is a correction: it is said that the accident with which this disappearance had been associated, had not occurred a week before, but years before, and was altogether different, having been an accident to a farmer in a hayfield. Several per sons investigated, among them a magistrate, who wrote that he was convinced at least that Osborne thought that he had seen the "figures" disappear.

Well, then, why didn't I get a Wolverton newspaper, and even though it would be called "a mere coincidence" and noted the disappearance of somebody who had been last seen on horseback? I forget now why I did not, but I think it was because no Wolverton newspaper was obtainable. I haven't the item, but with all our experience with explanations, I should have the knack, myself, by this time. I think of a man on horseback, who was suddenly transported, but only a few miles. If, when he got back, he was a wise man on horseback, he got off the back of his horse and said nothing about this. Our general notion is that he would have been unconscious of the experience. Perhaps, if Osborne had lingered, he would have seen this man and his horse reappear.

In the Jour. S. P. R., 4-50, is a story of a young woman, who was more than casually looked at, near the foot of Milton Hill, Massachusetts. She vanished. She was seen several times. So this is a story of a place that was "haunted," and the "figure" was supposed to be a ghost. For a wonder there was no story of a murder that was committed, years before, near this hill. For all I know, some young wo man, living in Boston, New York, some distant place, may have had teleportative affinity with an appearing-point, or terminal of an occult current, at this hill, having been translated back and forth several times, without knowing it, or without being able to remember, or remembering dimly, thinking that it was a dream. Perhaps, sometime happening to pass this hill, by more commonplace means of transportation, she would have a sense of uncanny familiarity, but would be unable to explain, having no active consciousness of having ever been there before. Psychologists have noted the phenomenon of a repeating scene in different dreams, or supposed dreams. The phenomenon may not .be of fancifulness, but of dim impressions of teleportations to one persisting appearing point. A naive, little idea of mine is that so many ghosts in white garments have been reported, because persons, while asleep, have been tele ported in their nightclothes.

In Real Ghost Stories, published by the Review of Re views (English), a correspondent tells of having seen a woman in a field, vanish. Like others who have had this experience, he does not say that he saw a woman vanish. but he saw "a figure of a woman" vanish. He inquired for some occurrence by which to explain, and learned that somewhere in the neighborhood a woman had been murdered, and that her "figure" had haunted the place. In the Proc. S. P. R., 10-98, someone tells of having walked, with her father, upon a sandy place, near Aldershot, hearing footsteps, turning, seeing a soldier. The footsteps suddenly ceased to be heard. She turned again to look. The soldier had vanished. This correspondent writes that her father never would believe anything except that it was "a real soldier, who somehow got away." In the Occult Review, 23-168, a. correspondent writes that, while walking in a street in Twickenham, he saw, walking toward him, "a figure of a man." The "figure" turned and vanished, or "disappeared through a garden wall." This correspondent failed to learn of a murder that had been committed in the neighborhood, but influenced by' the familiar convention, mentions that there was an old dueling ground nearby.

The most circumstantial of the stories appears in the]our. S. P. R., November, 1893. Miss M. Scott writes that, upon the afternoon of the 7th of May, 1893, between five and six o'clock, she was walking upon a road, near St. Boswells (Roxburghshire) when she saw ahead of her a tall man, who, dressed in black, looked like a clergyman. There is no assertion that this "figure" looked ghostly, and there is a little circumstance that indicates that the "figure," or the living being, was looked at more than casually. Having considerable distance to go, Miss Scott started to run: but it occurred to her that it would not be dignified to run past this stranger: so she stood still, to let the distance increase. She saw the clerical-looking man turn a corner of the road, the upper part of his body visible above a low hedge "be was gone in an instant." Not far beyond this vanishing point, Miss Scott met her sister, who was standing in the road, looking about her in bewilderment, exclaiming that she had seen a man disappear, while she was looking at him.

One of our present thoughts is that teleportations, back and forth, often occur. There are many records, some of yarns, of persons who have been seen far from where, so far as those persons, themselves, knew, they were, at the time. See instances in Gurney's Phantasms of the Living. The idea is that human beings have been switched away some. where, and soon switched back, and have been seen, away somewhere, and have been explained to the perceivers, as their own hallucinations.

It may be that I can record a case of a man who was about to disappear, but was dragged back, in time, from a disappearing point. I think of the children of Clavaux, who were about to be taken into a vortex, but were dragged back by their parents, who were not susceptible. Data look as if there may have been a transporting current through so-called solid substance, which "opened" and then "closed, with no sign of a yawning. It may be that what we call substance is as much open as closed. I accept, myself, that there is only relative substance, so far as the phenomenal is concerned: so I can't take much interest in what the physicists are doing, trying to find out what mere phenomenal substance really, or finally, is. It isn't, or it is intermediate to existence and non-existence. If there is an organic existence that is more than relative, though not absolute, it may be The Substantial, but its iron and lead and gold are only phenomenal The greatest seeming security is only a temporary disguise of the abysmal All of us are skating over thin existence.

Early in the morning of Dec. 9, 1873, Thomas B. Cumpston and his wife, "who occupied good positions in Leeds, were arrested in a railroad station, in Bristol, England, charged with disorderly conduct, both of them in their nightclothes, Cumpston having fired a pistol See the London Times, Dec. 11, 1873. Cumpston excitedly told that he and his wife had arrived the day before, from Leeds, and had. taken a room in a Bristol hotel, and that, early in the morning. the floor had "opened," and that, as he was about to be dragged into the "opening," his wife had saved him, both of them so terrified that they had jumped out the window, running to the railroad station. looking for a policeman. In the Bristol Daily Pod, December 10, is an account of proceedings in the police court. Cumpston's excitement was still so intense that he could not clearly express himself. Mrs. Cumpston testified that, early in the evening, both of them had been alarmed by loud sounds, but that they had been re-assured by the landlady. At three or four in the morning the sounds were heard again. They jumped out on the floor, which was felt giving away under them. Voices repeating their exclamations were heard, or their own voices echoed strangely. Then, according to what she saw, or thought she saw, the floor opened wide. Her husband was falling into this "opening" when she dragged him back.

The landlady was called, and she testified that sounds had been heard, but she was unable clearly to describe them. Policemen said that they had gone to the place, the Victoria Hotel, and had examined the room, Boding nothing to justify the extraordinary conduct of the Cumpstons. They suggested that the matter was a case of collective hallucination. I note that there was no suggestion of intox-ication. The Cumpstons, an elderly couple, were discharged in the custody of somebody who had come from Leeds. Collective hallucination is another of the dismissal-labels by which conventional-ists shirk thinking. Here is another illustration of the lack of standards, in phenomenal existence, by which to judge anything. One man's story, if not to the liking of conventionalists, is not accepted, because it is not supported: and then testimony by more than one is not accepted, if undesirable, because that is collective hallucination. In this kind of jurisprudence, there is no hope for any kind of testimony against the beliefs in which conventional scientists agree. Among their amusing disre-gards is that of overlooking that, quite as truly may their own agreements be collective delusions.

The loud sounds in the Cumpston case suggest something of correlation with poltergeist phenom-ena. Spiritualists have persistently called poltergeist-sounds "raps." Sometimes they are raps, but of-ten they are detonations that shake buildings. People up and down a street have been kept awake by them. Maybe existences open and shut noisily. From my own experience I don't know that there ever has been a poltergeist. At least, I have had only one experience, and that is explainable several ways. But what would be the use of writing a book about things that we think we're sure of? -Un-less, like a good deal in this book, to show the dooce we are.

In the Sunday Express (London), Dec. 5, 1926, Lieut. Colonel Foley tells of an occurrence that resembles the Cumpstons' experience. A room in Corpus Christi College (Cambridge University) was, in October, 1904, said to be haunted. Four students, of whom Shane Leslie, the writer, was one, investigated. Largely the story is of an invisible, but tangible, thing; or being, which. sometimes became dimly visible, inhabiting, or visiting, this room. The four students went into the room, and one of them was dragged away from the others. His companions grabbed him. "Like some power-ful magnet" something was drawing him out of their grasp. They pulled against it, and fought in ·a frenzy, and they won the tug. Other students, out side the room, were shouting. Undergraduates came running down the stairs, and, crowding into the room, wrecked it, even tearing out the oak paneling. Appended to the story, in the Sunday Express, is a statement by Mr. Leslie-"Colonel Foley has given an accurate account of the occurrence.

XIX

IN the Encyclopedia Britannica, the story of Kaspar Hauser is said to be one of the most baffling mysteries in history. This is an unusual statement. Mostly we meet denials that there are mysteries. In everything that I have read upon this case, it is treated as if it were· unique. A writer like Andrew Lang, who had a liking for mysteries, takes up such a case, with not an indication of a thought in his mind that it should not be studied as a thing in itself, but should be correlated with similars. That, inductively, any thing of an ultimate nature could be found out, is no delusion of mine: I think not of a widening of truth, but of a lessening of error. I am naive enough in my own ways, but I have not the youthful hopes of John Stuart Mill and Francis Bacon.

As to one of the most mysterious of the circumstances in the story of Kaspar Hauser, I have many records of attacks upon human beings, by means of an unknown, missile-less weapons. See the newspapers for several dozen accounts of somebody, or something, that was terrorizing people in New Jersey, in and around Camden. in the winter of 1927- 28·. People were fired upon, and in automobiles there were bullet holes, but bullets were unfindable. I know of two other instances, in the State of New Jersey. In France, about the year 1910, there was a long series of such attacks, attributed to "phantom bandits."

It may be that telepathically, human beings have been induced to commit suicide. Look up the drowning of Frank Podmore. It may be that the mystery of Kaspar Hauser was attracting too much attention. There is a strange similarity in the taking off of Frank Podmore, Houdini, Washington Irving Bishop, and perhaps Dr. Crawford. The list is long, of the deaths that followed the opening of the tomb of Tut-Ankh-Amen.

Psychologically and physiologically the case of the Rev. Thomas Hanna is so much like the case of Kaspar Hauser that the suggestion is that if Hanna were not an impostor, Hauser was not. For particulars of the Hanna case, see Sidis, Multiple Personality. In both cases there was said to be obliteration of memory, or reduction to the mental state of the newborn, with, however, uncommon, or marvelous, ability to learn. Phenomena common to both cases were no idea of time; no idea of sex; appearance of all things, as if at the same distance, or no idea of distance; and in ability, or difficulty, in walking. Kaspar Hauser was no impostor, who played a stunt of his own invention, as tellers of his story have thought. If he were an imposter, some where, back in times when little was known of amnesia, he had gotten ahold of detailed knowledge of profoundest amnesia. And he was about seventeen years old. Perhaps he was in a state of profound hypnosis. If the boy of Nepal, India, had wandered from the priest-who may have kidnaped him ordinarily, and may not have kidnaped him ordinarily-and had appeared in an English community, he would have been unable to account for himself, and there would have been a mystery similar to Hauser's.

If a "wolf child," when found, was "almost devoid of human intelligence," and when grown up became a police man, ours is not quite the cynicism of a scenario-writer, or a writer of · detective stories. If we do not think that this child had been the associate of wolves from an age of a few months, we think of an obliterative process that rendered it "a speechless, little animal," but that did not so impair its mentality that the child could not start anew. Our expression upon Kaspar Hauser will be that he was a "wolf child, and that, if he had appeared somewhere in India, he would, according to local conventions, probably be called a "wolf child," and that if he had found any place of refuge, it would be called a "wolf's den": but. in our expression, the lupine explanation is not accepted in his case and the cases of all so-called "wolf children." "Wolf children" have appeared, and the conventional story of their origin is not satisfactory. If "wolf children" have ·had something the matter with their legs, or have crawled on all fours, it is not satisfactory to say that this was because they had been brought up with wolves, any more than it would be to say that a young bird, even if not taught by its parents, would not be able to fly, if brought up with mammals.

If we accept that the Pansini boys ever were teleported, we note the mental effects of the experience, in that they were in a state of profound hypnosis.

Little frogs bombard horses-and, though there have been many attempts to explain Kaspar Hauser, it has never before occurred to anybody to bring little frogs into an explanation-

Or seals in a pond in a park-and the banded reindeer of Spitzbergen-see back to everything else in this book. Later, especially see back to lights in the sky, and the dis appearances of them, when a story was told, and. so long as the story was not examined, seemed to account for them. The luminous owl-the malmoot-and if anybody can't be explained conventionally, he's an impostor-or, if we're all to some degree, impostors, he's an exceptional impostor.

Upon Whit Monday afternoon, May, 1828, a youth, aged sixteen or ·seventeen, staggered, with a jaunty stride, into the town of Nuremberg, Germany. Or, while painfully dragging himself along the ground, he capered into the town. The story has been told by. theorists. The tellers have fitted descriptions around their theories. The young. man was un able fully to govern the motions of his legs, according to Andrew Lang, for instance. He walked with firm, quick steps, according to the Duchess of Cleveland. The Duchess' theory required that nothing should be the matter with his legs. By way of the New Gate, he entered the town, and there was something the matter with his legs, according to all writers, except the one who preferred that there should be nothing the matter with his legs.

To Nurembergers who gathered around, the boy held out two letters, one of which was addressed to a cavalry captain. He was taken to the captain's house, but, because the captain was not at home, and because he could give no account of himself, he was then taken to a police station. Here it was recorded that he could speak only two sentences in the German language, and that when given paper and pencil he wrote the name Kaspar Hauser. But he was not put away and forgotten. He had astonished and mystified Nurembergers, in the captain's house, and these townsmen had told others, so that a crowd had gone with him to the station house, remaining outside, discussing the strange arrival. It was told in the crowd, as recorded by von Feuerbach, that near the New Gate of the town had appeared a boy who seemed unacquainted with the commonest objects and experiences of everyday affairs of human beings. The astonishment with which he had looked at the captain's saber had attracted attention. He had been given a pot of beer. The luster of the pot and the color of the beer affected him, as if he had never seen anything of the kind before. Later, seeing a burn-ing candle, he cried out in delight with it, and before anybody could stop him, tried to pick up the Same. Here his education began.

This is the story that has been considered imposture by everybody who wanted to consider it impos-ture. I cannot say whether all alleged cases of amnesia are fakes, or not. I say that, if there be amnesia, the phenomena of Kaspar Hauser are aligned with phenomena of many cases that are said to be well-known. The safest and easiest and laziest of explanations is that of imposture.

Of the two letters, one purported to be from the boy's mother, dated sixteen years before, telling that she was abandoning her infant, asking the finder to send him to Nuremberg, when he became seventeen years old, to enlist in the Sixth Cavalry Regiment, of which his father had been a member. The other letter purported to be from the finder of the infant, telling that he had ten children of his own, and could no longer support the boy.

Someone soon found that these letters had not been writ ten by different ·persons, sixteen years apart. One of them was in Latin characters, but both were written with the same ink, upon the same kind of paper. In the "later.. letter, it was said: "I have taught him to read and write, and he writes my handwriting exactly as I do." Whereupon the name that Kaspar had written, in the police station, was examined. and it was said that the writings were similar. Largely with this circumstance for a basis, it has been said that Kaspar Hauser was an impostor-or that he had written the letters himself. With what expectation of profit to himself is not made clear. If I must argue, I argue that an impostor, aware that handwritings might be compared, would, if he were a good impostor, pretend to be unable to write, as well as unable to speak. And those who consider Kaspar Hauser an impostor, say that he was a very good impostor. The explanation in the letter, of the similarity of handwritings, seems to be acceptable enough.

People living along the road leading to the New Gate were questioned. Not an observation upon the boy, before he appeared near the Gate, could be heard of. But we see, if we accept that someone else wrote his letters, that this Gate could not have been his "appearing-point," in the sense we' thinking of. He must have been with, or in the custody of someone else, at least for a while. Streets near the jail, where for a time he was lodged, were filled with crowds, clamoring for more information. Excitement and investigation spread far around Nuremberg. A reward was offered, and, throughout Germany, the likeness of Kaspar Hauser was posted in public places. People in Hungary took up the investigation. Writers in France made much of the mystery, and the story was published in England. People from all parts of Europe went to see the boy. The mystery was so stimulated by pamphleteers that, though "feverish" seems an extreme word, writers described the excitement over this boy, "who had appeared as if from the clouds," as a "fever."'. Because of this international interest, Kaspar Hauser was known as "The Child of Europe."

The city of Nuremberg adopted Kaspar. He was sent to live with Prof. Daumer, a well-known scientist, and the Mayor of Nuremberg notified the public to "keep away from his present residence, and thereby avoid collision with the police." The seeming paralysis of his legs wore' off. He quickly learned the German language, but spoke always with. a foreign accent. I have been unable to learn anything of the peculiarities of this accent. Except to students of revivals of obliterated memories, his quickness in learning would seem incredible. Writers have said th.at so marvelous was his supposed ability to learn that he must have been an impostor, having a fair education, to start with. Though the impostor theory is safest and easiest, some writers have held th.at the boy was an idiot, who had been turned adrift. This explanation can be held simply and honestly by anybody who refuses to believe all records after the first week or so of observations. Whether impostor or idiot, the outstanding mystery is the origin of this continentally advertised boy.

The look of all the circumstances to me is that somebody got rid of Kaspar, considering him an imbecile, having been able to teach him only two German sentences. Then the look is that he had not for years known Kaspar, but had known him only a few weeks, while his disabilities were new to him. Where this custodian found the boy is the mystery.

Kaspar Hauser, in the year 1829, wrote his own story, telling that, until the age of sixteen or seventeen, he had lived upon bread and water, in a small, dark cell He had known only one person, alluded to by as "the man" who, toward the end of his confinement had taught him two sentences, one of them signifying that he wished to join a ,;tie cavalry regiment, and the other, "I don't know." He had not been treated kindly, except once, when he had been struck for being noisy. Almost anybody, reading this account, will, perhaps regretfully, perhaps not, say farewell to our idea of a teleported boy. "That settles it." But nothing ever has settled anything, except relatively to a d ire for settlement, and if ours is a desire for unsettlement, we have assurance that we, or any other theorist, can find in the uncertainties of any human document, whether supposed to have been dictated from on high, or written by a boy, material for thinking as our theories require.

We note in Kaspar's story a statement that he had no idea of time. That is refreshing to our wilting theory. We may think that he had lived in a small, dark room all his life of which he had remembrance, and that that may have been a period of only a few weeks. We pick upon his statement that once he had been struck for being noisy. To us that means that he had been confined, not in a cell, or a dungeon, but in a room in a house, with neighbors around, and that there was somebody's fear that sounds from him would attract attention-or that there were neighbors so close to this place that the imprisonment of a boy could not have been kept a secret more than a few weeks.

We're not satisfied. We hunt for direct data for thinking It· that, if Kaspar Hauser had been confined in a dark room, it had not been for more than a few weeks.

"He had a healthy color" (Hiltel). "He had a very healthy color: he did not appear pale or delicate, like one who had been some time in confinement" (Policeman Wüst).

According to all that can be learned of another case, a, man, naked, almost helpless, perhaps in a state of hypnosis so profound that also it was physical, so that he could scarcely walk, and in whom memory was obliterated so that he did not know enough to make his way along a road, which he crossed, appeared near Petersfield, Hampshire, Feb. 21, 1920. If we can think that a peasant, near Nuremberg, found on his farm a boy in a similar condition, and took him in, then considering him an imbecile, and wanting to get rid of him, keeping him in confinement, fearing he might be held responsible for him, then writing two letters that would lain an abandonment in commonplace terms that would not excite inquiry, but not being skillful in such matters, that looks as if we're explaining somewhat.

Because of the continuation of Kaspar's story, we think that this place was near Nuremberg. Whit Monday was a holiday, and the farmers, or the neighbors, were probably not laboring in the fields: so this was the day for the shifting of the supposed imbecile. Upon this day, as told by Kaspar, "the man" carried the boy from the dark room, and carried, or led, him, compelling him to keep his eyes downward, toward Nuremberg. Kaspar's clothes were changed for the abandonment.

Perhaps he had been found naked, and had been given makeshift garments. Perhaps he had been found in clothes, of cut and texture that were remarkable and that would have caused inquiry. The clothes that were given to him were a peasant's. It was noted in Nuremberg that they seemed not to belong to him, because Kaspar was not a peasant boy, judging by the softness of his hands (von Feuerbach) .

The story has resemblances to the story of the English boy of Nepal. In each case somebody got rid of a boy, and in each case it is probable that a false story was told. If "the man" in Kaspar's case had the ten children that, to excuse an abandonment, he told of, there'd have been small chance for him to keep his secret. There are differences in these two stories. It will be my expression that they came about be cause of the wide difference in attention that was attracted.

Oct. 17, 1829-Kaspar was found in the cellar of Prof. Daumer's house, bleeding from a cut in the forehead. He said that a man in a black mask had appeared suddenly, and had stabbed him.

It has been explained that this was attempted suicide. But stabbing oneself in the forehead is a queer way to attempt suicide, and in Nuremberg arose a belief that Kaspar's life was in danger from unknown enemies, and two policemen were assigned to guard him.

Upon an afternoon in May, 1831, one of these policemen, while in one room, heard a pistol shot, in another room. He ran there, and found Kaspar again wounded in the fore head. Kaspar said that it was an accident: that he had climbed upon the back of a chair, and, reaching for a book, had slipped, and, catching out wildly, had grasped a pistol that was hanging on the wall, discharging it.

Dec. 14, 1833-Kaspar Hauser ran from a park, crying that he had been stabbed. Deeply wounded in his side, he was taken to his home. The park, which was covered with new-fallen mow, was searched, but no weapon was found, and only Kaspar's footprints were seen in the snow. Two of the attending physicians gave their opinion that Kaspar could not so have injured himself. The opinion of the third physician was an indirect accusation of suicide: that the blow had been struck by f left-handed person. Kaspar was not. left-handed, but was ambidextrous.

Kaspar lay on his bed, with his usual publicity. He was surrounded by tormentors, who urged him/ to gasp plugs in his story. He was the only human being who had been in the park, according to the testimony of the snow tracks. It was not only Kaspar who was wounded. There was a wound in circumstances. Tormentors urged him' to confess, so that in terms of the known they could fill out his story. Faith in confessions and the desire to end a mystery with a confession are so intense that some writers have said that Kaspar did confess. As a confession, they have interpreted his protest against his accusers-"My Codi that I should so die in shame and disgrace!"

Kaspar Hauser died. The point of his heart had been pierced by something that had cut through the diaphragm, penetrating stomach and liver. In the opinion of two of the doctors and of many of the people of Nuremberg, this wound could not have been self-inflicted. Rewards for the capture of an assassin were offered. Again, throughout Germany, posters appeared in public places, and in Germany and other countries there were renewed outbursts of pamphlets. The boy appeared "as if from the clouds," and nothing more was learned.

It was Kaspar's story that a man in the park had stabbed him. If anybody prefers to think that it cannot be maintained that there was only one track of footprints in the snow, let him look up various accounts, and he will find assurances any way he wants to find them. For almost every statement that I have made, just as good ·authority for denying it, as for stating it, can be found, provided any two conflicting theories depend upon it. One can read that Kaspar Hauser was highly intelligent or brilliant. One can read that the autopsy showed that his brain was atrophied to the size of a small animal's, accounting for his idiocy. One comes upon just about what one comes upon in looking up any other matter of history. It is said that history is a science. I think that it must be.

A great deal, such as Kaspar's alleged ability to see in the dark, and his aversion to eating meat, and his inability to walk would be understandable, if could be accepted the popular theory that Kaspar Hauser was the rightful Crown Prince of Bavaria, who for political reasons had been kept for sixteen or seventeen years in a dungeon. There would be an explanation for two alleged attacks upon him. But see back to his own story of confinement in a house, or a peasant's hut, near Nuremberg, where probably his imprisonment could not have been kept secret more than a few weeks. See testimony by Hiltel and Wüst.

See back to a great deal more in this book-

The wolf of Shotley Bridge, and the wolf of Cumwinton or that something removed one wolf and procured another wolf to end a mystery that was attracting too much attention. It was said that Kaspar Hauser was murdered to suppress political disclosures. If it be thinkable that Kaspar was murdered to suppress a mystery, whether political, or not so easily defined, there are statements that support the idea that also some of the inhabitants of Nuremberg, who were prominent in Kaspar's affairs, were murdered. One can read that von Feuerbach was murdered, or one can read that von Feuerbach died ..of a paralytic stroke. See Evans (Kaspar Hauser, p. 150)-that, soon after the death of Kaspar Hauser, several persons, who had shown much interest in his case, died, and that it was told in Nuremberg that they had been poisoned. They were Mayor Binder, Dr. Osterhauser, Dr. Preu, and Dr. Albert.

"Kaspar Hauser showed such an utter deficiency of words and ideas, such perfect ignorance of the commonest things and appearances of Nature, and such horror of all customs, conveniences, and necessities of civilized life, and, withal, such extraordinary peculiarities in his social, mental, and physical disposition, that one might feel oneself driven to the alternative of· believing him to be a citizen of another planet, transferred by some miracle to our own" (von Feuerbach).

Part II

XX

According to appearances, this earth is a central body, within a revolving, starry globe.

But am I going to judge by appearances?

But everything of the opposing doctrine is judgment by other appearances. Everybody who argues against judging by appearances bases his argument upon other appearances. Monistically, it can be shown that everybody who argues against anything bases his argument upon some degree or aspect of whatever he opposes. Everybody who is attacking something is sailing on a windmill, while denouncing merry go-rounds.

"You can't judge by appearances," say the astronomers. "Sun and stars seem to go around this earth, but they are like a field that seems to go past a train, whereas it is the train that is. passing the field." Judging by this appearance, they say that we cannot judge by appearances.

Our judgments must depend upon evidence, the scientists tell us.

Let somebody smell, hear, taste, see, and feel something that is unknown to me, and then tell me about it. Like every body else, I listen politely, if he's not too long about it, and then instinctively consult my preconceptions, before deciding whether all this is evidence. An opinion is a matter of evidence, but evidence is a matter of opinion.

We can depend upon intuition, says Bergson.

I could give some woebegone accounts of what has befallen me, by depending upon intuitions, whether called "hunches," or "transcendental consciousness"; but similar experiences have befallen everybody else. There would have been what I call good sport, if Bergson had appeared upon the floor of the Stock Exchange, and preached his doctrine, in October, 1929.

We have only faith to guide us, say the theologians. Which faith?

It is my acceptance that what we call evidence, and whatever we think we mean by intuition and faith are the phenomena of eras, and that the best of minds, or minds in rapport with the dominant motif of an era, have intuition and faith and belief that depend upon what is called evidence, relatively to pagan gods, then to the god of the Christians, and then to godlessness-and then to whatever is coming next.

We shall have data for thinking that our existence, as a whole, is an organism. First we shall argue that it is a thinkable-sized formation, whether organic, or not. If now, affairs upon this earth be fluttering upon the edge of a new era, and I give expression to coming thoughts of that era, thousands of other minds are changing, and all of us will take on new thoughts concordantly, and see, as important evidence, piffle of the past.

Even in orthodox speculations there are more or less satisfactory grounds for thinking that ours is an existence, perhaps one of countless other existences, that is an egg-like formation, shelled away from the rest of the cosmos. Many astronomers have noted that the Milky Way is a broad band in the sky, with the look of a streak around a globular object. For conventional reasons for thinking that the "solar system" is central in "a mighty cluster of stars," see Dolmage, Astronomy, p. 327. Dolmage even speculates upon a limiting demarcation which is akin to the notion of a shell, shutting off this existence from everything else.

Back in the pessimistic times of Sir Isaac Newton was formulated the explanation of existence in general that is our opposition. It was the melancholy doctrine of universal fall. It was in agreement with the theology of the time: fallen angels, the fall of mankind: so falling planets, falling moons, everything falling. The germ of this despair was the supposed fall of the moon, not to, but around, this earth. But if the moon is falling away from observers upon one part of this earth's surface, it is rising in the sky, relatively to other observers. If something is quite as truly rising as it is falling, only minds that belong away back in times when everything was supposed to be falling, can be satisfied with this yam of the rising moon that is falling. Sir Isaac Newton looked at the falling moon, and ·explained· all things in terms of attraction. It would be just as logical to look at the rising moon, and explain all things in terms of repulsion. It would be more widely logical to cancel falls with rises, and explain that there is nothing.

I think of this earth as central, and as almost stationary, and with the stars in a shell, revolving around. By so thinking, I have the concept of an object, and the visualization of an existence as a whole. But the trouble with this idea is that it is reasonable. Not absolutely can it be said that human minds reason according to reasonableness. There is the love of the paradox to consider. We are in agreement with observations, but peasants, or clodhoppers, think as we think. We offer no paradox to make one feel superior to somebody who hops from clod to clod.

What is the test? Of course, if there are no standards, all tests must be fakes. But if we have an appearance of reasonableness, and if the other side says that it is reason able, how choose?

We read over and over that prediction is the test of science.

The astronomers can predict the movements of some of the parts of what they call the solar system.

But so far are they from a comprehensive grasp upon the system as a whole that, if, for a basis of their calculations, be taken that this earth is stationary, and that the sun and the planets, and the stars in a shell, move around this earth, the same motions of heavenly bodies can be foretold. Take for a base that the earth moves around the sun, or take that the sun moves around the earth: upon either base the astronomers can predict an eclipse, and enjoy renown and prestige, as if they knew what they were telling about. Either way there are inaccuracies.

Our opposition is ancient and at least uppish.

Prof. Todd, in his book, Stars and Telescopes, says: "Astronomy may be styled a very aristocrat among the sciences."

For similar descriptions, by implication, of themselves, by themselves, see all other books by astronomers.

There are aristocratic human beings. I'll not contend otherwise. There are aristocratic dogs, and all cats, except for relapses, are aristocrats. There are aristocratic goldfish. In whatever is bred, is the tendency to aristocraticize. Porcupines, as the untouchable and the stupid, are verier aristocrats than the merely very. The aristocratic state is supposed to be the serene; the safe, and the established. It is unintelligent, because intelligence is only a means of making adaptations, and the aristocratic is the made. If this state of the relatively established and stupid were the really, or finally, established and stupid, we'd see good reason for the strivings and admirations and imitations of strugglers, climbers, or newcomers to stabilize themselves into stupors. But, in phenomenal being, the aristocratic, or the academic, is, though thought of as they arrived, only a poise between the arriving and the departing. When far advanced it is the dying. Wherein it is a goal, our existence is, though only locally, suicidal. The literature of the academic ends with the obituary. Prof. Todd's self-congratulation is my accusation.

But there is only relative aristocracy. If I can show that, relatively to a viewpoint, other than the astronomers' own way of adoring themselves, the supposed science of astronomy is only· a composition of yarns, evasions, myths, errors, disagreements, boasts, superstitions, guesses, and bamboozlements, I am spreading the good cheer that it is still very faulty and intellectual and still alive, and may be able to adjust, and keep on exciting its exponents with admiration for themselves.

We shall see what mathematical astronomy. is said to start with. If we can't accept that it ever fairly started, we'll not delay much with any notion that it could get anywhere.

The early mathematical astronomers, in their calculations upon moving bodies, could not treat of weights, because these inconstancies are relative; nor of sizes, because sizes are relative and variable. But they were able to say that they had solved their problem of how to begin, because nobody else interfered and asked whether they had or not. They gave up weight and size and said that their treatment was of mass.

If there were ultimate particles of matter, one could think of mass as meaning a certain number of those things. When atoms were believed in, as finals, an astronomer could pretend that he knew what he meant by a quantity of matter, or mass. Then, with electrons, he could more or less seriously keep on pretending. But now the sub-electron is talked of.

And, in tum, what is that composed of? Perhaps the pretensions can stretch, but there is too much strain to the serious ness. If nobody knows what constitutes a quantity of matter, the astronomer has no idea of what he means by mass. His is a science of masses.

But it may be said that, even though he has not the remotest idea of what he is calculating about, the astronomers' calculations work out, just the same.

There was the mass of Mars, once upon a time, for instance: or the "known" unknowables constituting the planet. Once upon a time, the mass of Man was said to be known. Why shouldn't it be said to be known? The equations were said to work out, as they should work out.

In the year 1877, two satellites of the planet Mars were discovered. But their distances and their periods were not what they should be, theoretically. So then everything that had worked out so satisfactorily as it should work out, turned out to have worked out as it shouldn't have worked out. A new mass had to be assigned to the planet Mars.

Now that works out as it should work out.

But I think that it is cannier not to have things so marvelously work out, as they should work out, and to have an eye for something that may come along and show that they had worked out as they shouldn't have worked out. For data upon these work-outs, see Todd, Astronomy, p. 78.

It would seem that the mistake by the astronomers is in thinking that, in a relative existence, there could be more than relative mass, is the idea of mass could be considered as meaning anything. But it is more of a dodge than a mistake. It is just relativity that the astronomers have tried to dodge, with a pseudo-concept of a constant, or a final. Instead of science, this is metaphysics. It is the childish attempt to find the absolutely dependable in a flux, or an intellectually not very far-advanced attempt to find the absolute in the relative. The concept of mass is borrowing from the theologians, who are in no position to lend anything. The theologians could not confidently treat of human characters, personalities, dispositions, temperaments, nor intellects, all of which are shifts: so they said that they conceived of finals, or unchangeables, which they called "souls." If economists and psychologists and sociologists should disregard all that is of hopes and fears and wants and other changes of human nature, and take "souls" for their units, they would have sciences as aristocratic and sterile as the science of astronomy, which is concerned with souls, under the name of masses. A final, or unchangeable, must be thought of as a state of unrelated ness. Anything that is reacting with something else must be thought of as being in a state of change. So when an astronomer formulates, or says he formulates, the effects of one mass, or one planet, as a mass, upon another, his meaningless statement might as well be that the subject of his equations is the relations of unrelatedness.

Starting with nothing thinkable to think about, if constants, or finals, are unknown in human experience, and are unrepresentable in human thought, the first and the simplest of the astronomers' triumphs, as they tell it, is the Problem of the Two Masses.

This simplest of the problems of celestial mechanics is simply a fiction. When Biela's comet split, the two masses did not revolve around a common center of gravity. Other comets have broken into parts that did not so revolve. They have been no more subject to other attractions than have been this earth and its moon. The theorem is Sunday School Science. It is a mathematician's story of what bodies in space ought to do. In the textbooks, it is said that the star Sirius and a companion star exemplify the theorem, but this is another yarn. If this star has moved, it has not moved as it was calculated to move. It exemplifies nothing but the inaccuracies of the textbooks. It is by means of their inaccuracies that they have worked up a reputation for exactness.

Often in his book, an astronomer will sketchily take up a subject, and then drop it, saying that it is too complex, but that it can be mathematically demonstrated. The reader, who is a good deal of a dodger, himself, relieved at not having to go into complexities, takes this lazily and faithfully. It is bamboozlement. There are many of us, nowadays, who have impressions of what mathematicians can do to, or with. statistics. To say that something can be mathematically demonstrated has no more meaning than to say of something else that it can be politically demonstrated. During any campaign, read newspapers on both sides, and see that anything can be politically demonstrated. Just so can be mathematically shown that twice two are four, and it can be mathematically shown that two can never become four. Let somebody have two of arithmetic's favorite fruit, or two apples, and under take to add two more to them. Although he will have no trouble in doing this, it can be mathematically shown to be impossible. Or that, according to Zeno's paradoxes, nothing can be carried over intervening space and added to something else. Instead of ending up skeptically about mathematics, here am I upholding that it can prove anything.

We are told in the textbooks, or the tracts, as I regard these propagandist writings of Sunday School Science, that by parallax, or annual displacement of stars, relatively to other stars, the motion of this earth around the sun has been instrumentally determined. Mostly, these displacements are about the apparent size of a fifty cent piece, held up by someone in New York City, as seen by somebody in Saratoga. This is much refinement. We ask these ethereal ones-where is their excuse, if they get an eclipse wrong by a millionth of an inch, or a millionth of a second?

We look up this boast.

We find that the disagreements are so great that some astronomers have reported what is called negative parallax, or supposed displacement of stars, the wrong way, according to theory. See Newcomb, The Stars, p. 152. See the English Mechanic, 114-100, 112. We are out to show that astronomers themselves do not believe parallactic determinations, but believe those that they want to believe. Newcomb says that he does not believe these determinations that are against what he wants to believe.

Spectroscopic determinations are determined by whatever the spectroscopists want to determine if one thinks not, let one look up the "determinations" by astronomers who were for and against Einstein. Grebe and Bachem, at Bonn, found
shifts of spectral lines in Einstein's favor. They were for Einstein. St. John, at the Mt. Wilson Observatory, found the testimony of the spectroscope not in Einstein's favor. He was against Einstein. The spectroscope is said to be against us. But, if we had a spectroscope of our own, it would be for us.

In The Earth and the Stars, Abbot says that the spectroscope "seems to indicate" that variable stars, known as the Cepheid Variables, are double stars. But he says: "The distance between the supposed pairs turns out to be impossibly small." When a spectroscopic determination is not what it should be, it only "seems to indicate."

The camera is another of the images in astronomical idolatry. I note that bamboozlements that have played out everywhere else, still good in astronomy. Spirit photographs fall flat. At the movies, if we see somebody capering seemingly near an edge a roof, we do not think that he had been photographed anywhere near an edge of a roof. Nevertheless, even in such a religious matter as photography in astronomy. A camera tells what it should tell, or the astronomers will not believe it.

If the astronomers would fight more among themselves, more would come out. How can I be a· pacifist, just so long as I am trying to educate myself? Much comes out, war times. Considerable came out, in astronomical matters, during the Mars controversy. Everything that was determined by Lowell, with his spectroscope, and his camera, and his telescope, to be not so. The question is now what an instrument deter mines. The question is-whose instrument All the astronomers in the world may be against our notions, but most of their superiority is in their more expensive ways of deceiving themselves.

Foucault's experiment, or the supposed demonstration with the pendulum is supposed to show that this earth rotates daily. If a pendulum does-at least for a while-swing somewhat nearly in a constant line, though changing relatively to environment, and if we think that neither religiously, nor accidentally, has it received some helpful little pushes, we accept that here there may be indication of an annual. and not daily, rotation of this earth. That would account for the annual shift, and not the daily shift, of the stars. I don't know that I accept this, but I have no opposing prejudice. When I write of this earth as "almost stationary," as I have to regard it, if I think of it as surrounded by a starry shell that is not vastly far away, I mean that relatively to the tremendous velocities of conventionality. But this alleged experiment has never been more than part of an experiment. I quote from one of the latest textbooks, Astronomy, by Prof. John C. Duncan, published in 1926. We are told that a pendulum, if undisturbed, swings for "several hours," in very nearly" the same plane. Farther along we read that, in the latitude of Paris, where Foucault made his experiment, the time for a complete demonstration is 32 hours. Prof. Duncan makes no comment, but it is the reader's own fault if he reads in these statements that the swing of a pendulum, through more than part of an experiment, and in more than "'very nearly" the same plane, ever has demonstrated the daily rotation of this earth.

In the textbooks, which are pretty good reading for contrary persons like ourselves, it is said that the circumstance that this earth is approximately all oblate spheroid indicates the rapid rotation of this earth. But our negative principle is that nothing exclusively indicates anything. It does not matter what an astronomer, or anybody else says to support any statement, the support must be a myth. Even if I could accept that the astronomers are right, I could not accept that they can demonstrate that they're right. So we hunt around for opposing data, knowing that they must be findable somewhere. We come upon the shape of the sun. The sun rotates rapidly, but the sun is not an oblate spheroid: if there by any departure from sphericity, the sun is a prolate spheroid. Or we argue that oblateness may be an indication that in early, formative times this earth rotated rapidly, but that now this earth could be oblate and almost stationary. It may be another instance of my many credulities, but here I am accepting that this roundish, or perhaps pear-shaped, earth is flattened at the poles, as it is said to be.

Astronomers cite relative numerousness of meteors, as indication of this earth's motion in an orbit. Prof. Duncan (Astronomy, p. 262) says that meteors seen after midnight are about twice as numerous as are those that are seen before midnight. "This is because, in the latter half of the night, we are riding on the front side of the Earth, as it moves along its orbit, and receives meteors from all directions, whereas in the earlier half we see none of those which the Earth meets 'head on.' "

There is no use comparing little sparks of meteors, seen at different times of night, because of course soon after midnight more of these little things are likely to be seen than earlier in the evening, in lingering twilight. Here, Prof. Dun can's statement is that when meteors can be seen morely, more meteors can be seen. That is wisdom that we shall not defile.

In the records of great meteors that were seen in England, in the year 1926-see Nature, Observatory, English Mechanic-eighteen were seen before midnight, and not one was seen after midnight. All other records that I know of are against this alleged indication that this earth moves in an orbit. For instance, see the catalogue of meteors and meteorites published in the Rept. Brit .Assoc. Ad. Sci., 1860. See page 18. 51 after midnight (from midnight to noon); 146 before midnight (noon to midnight). I have records of my own, for 125 years, in which the preponderance of early meteors is so great that; if there were any sense to this alleged indication, it would mean that this earth.is. running backward, or going around the way it shouldn't. Of course I note that great meteors are more likely to be reported before midnight, because, though many persons are out after midnight, mostly they're not out reporting meteors. But Prof. Duncan has made a statement, which depends upon records, and I am checking it up, according to records. Year 1925, for instance-meteors of France and England-14 before midnight: 3 after midnight. This record, as I have it, is not complete, but I will hold out for the proportionality. Most of the great meteors of 1930 were seen before midnight. Whatever becomes of Prof. Duncan's statement, I'll make one, myself, and that is that, if nobody looks up, or checks up, what the astronomers tell us, they are free to tell us anything that they want to tell us. Their system is a slippery imposition of evasions that cannot be checked up, or that, for various reasons, mostly are not checked up. But at least once there was a big checkup.

The 24th of January, 1925-excitement in New York City. It was such as, in all foreign countries, is supposed to arise in America only when somebody finds out a new way of making dollars.

It was the morning of the eclipse of the sun, total over a part of New York City.

Open spaces in Central Park were crowded down to a line, as exactly as possible at 83rd Street. Up in the air were planes full of observers. Coogan's Bluff was lively with scientific gab. Hospitals were arranging that patients should see the eclipse. There was scarcely a dollar in it, and this account will be believed, in England and France, no more than will most of our other accounts. At the Fifth Avenue Police Court, Magistrate Dale adjourned court, and went, with lawyers and cops and persons out on bail to the roof. In Brooklyn, the Chamber of Commerce dropped all matters of exports and imports and went to the roof. I don't suppose everybody was looking. I can't accept homogeneity. There were probably some contrary ones who went down into cellars, simply because most of their neighbors were up on roofs. But the New York Telephone Company reported that when the eclipse came, not one call came into one of its offices, for ten minutes. When there are uproars in New York, they are such uproar as have never been heard anywhere else: but I think that most striking in the records of silences is this hush that came for ten minutes upon New York City.

Along the line of 83rd Street, which had been exactly predicted by the astronomers, as the southern limit of the path of totality, and in places north and south, were stationed 149 observers, sent by the New York lighting companies, to report upon light effects. With them were photographers.

At Petropaulovsk, Kamchatka, and at Cachapoyas, Peru, an eclipse is all that it should be, and books by astronomers tell of the minute exactness of the astronomers. But this was in New York City.

Coogan's Bluff got into this. There were cops and judges and gunmen on roofs, and the telephones were silent. There were 149 expert observers, who were not astronomers. They had photographers with them.

In time, the astronomers did pretty well. But, hereafter when they tell of their refinements, as with discs several hundred miles away, I shall think, not of fifty cent pieces, but of Ferris Wheels. Their prediction was wrong by four seconds.

The 149 observers for the lighting companies reported that the astronomers were wrong, in space, by three quarters of a mile.

It was the day of the big checkup.

If the sun and the planets compose a system that is enormously remote from everything else in existence, what is it that regularizes the motions, and why does not the mechanism run down? The astronomers say that the planets keep moving, and that a w ole system does not run down, because space is empty, and there is "absolutely" nothing to tend to stop the moving bodies. See Abbot, The Earth and the Starr, p. 71. Astronomers say this early in their books. Later, they forget. Later, when something else requires ex planation, they tell a different story. They explain the zodiacal light in terms of enormous quantities of matter in space. In their chapters upon meteors, they tell of millions of tons of meteoric dusts that fall from space to this earth, every year. Abbot says that space is "absolutely" empty. Ball, for instance, explains the shortening of the orbit of Funcke's comet as a result of friction with enormous quantities of matter in space. I don't know how satisfactory, except to ourselves, our own expression will be, but compare it with a story of an absolute vacancy that is enormously occupied.

There is a tendency to regularize. Crystals, flowers, and butterflies' wings proportionately as they become civilized, people regularize, or move in orbits. People regularize in eating and sleeping. There are clockwork Romeo's·and Juliet's.

Everywhere, where the tendency is not toward irregularization, the tendency is toward regularization. Here's a specimen of my wisdom. Something is so, except when it isn't so.

Not in terms of gravitation, but in terms of this tendency to regularize, celestial periodicities may be explained.

Why does not the mechanism of what the astronomers think is a solar system run down?

The astronomer say that this is because. it is unresisted by a resisting medium.

Why does not a heart run down? Anyway for a long time? It is only a part, and, as a part, is sustained by what may be considered a whole. If we think of the so-called solar system, not as a virtually isolated, independent thing, with stars trillions of miles away, but as part 'of what may be considered an organic whole, within a starry shell, our expression is that it is kept going organically, as the heart of a lesser organism is kept going.

Why does not the astronomers' own system, or systematized doctrine, run down, or why so slow about it? It is only a part of wider organization, from which it is receiving maintenance in the form of bequests. donations, and funds of various kinds.

Our opposition is a system of antiquated thought, concerned primarily with the unthinkable. It is supported by instruments that are believed when they tell what they should tell. The germ of the system is the fall of the rising moon. Its simplest problem is a fairy-theorem, fit for top-heavy infants, but too fanciful for grownup realists. Its prestige is built upon its predictions. We have noted one of them that was three-quarters of a mile wrong.

Newtonism is no longer satisfactory. There is too much that it cannot explain.

Einsteinism has arisen.

If Einsteinism is not satisfactory, there is room for other notions.

For records of eclipses during which the stars were not displaced, as, according to Einstein, they should be, see indexes of Nature. See vols. 104 and 105. Displacement of spectral lines-see records of astronomers who have disagreed. Perihelion-motion of Mercury's orbit-Einstein calculated without knowing what he was calculating about. Nobody knows what this eccentricity is. See records of the transits of Mercury. Neither Newtonians nor Einsteinites have pre-dieted them right. See the London Times, April 17 and 24, 1923. Here Sir J. Larmor shows that, if Einstein's predictions of light-effects during eclipses were verified, they disproved his theory-that, though Prof. Einstein would be a great mathematician, if in our existence anything could really be anything, relativity is so against him that he is only a relatively great mathematician, and made a bad error calculations, having mistakenly doubled certain effects.

Defeat has been unconsciously the quest of all religions, all philosophies, and all sciences. If they were consciously trying to lose, they would be successes. Their search has been for the Absolute, in terms of which to explain the phenomenal, or for the Absolute to relate to. Supposed to have been found, it has been named Jehovah, or Gravitation, or the Persistence of Force. Prof. Einstein has taken the Velocity of Light, as the Absolute to relate to.

We cannot divorce the idea of reciprocity from the idea of relations, and relating something to the Absolute would be relating the Absolute to something. This is defeating an alleged concept of the Absolute, with the pseudo-idea of the Relative Absolute. The doctrine of Prof. Einstein's is based not upon an absolute finding, but upon a question: Which is the more graspable interpretation of the Michelson-Morley experiments:

That no motion of this earth in an orbit is indicated, because the velocity of light is absolute;

Or that no motion of this earth in an orbit is indicated, because this earth is stationary?

Unfortunately for my own expressions, I have to ask a third question:

Who, except someone who was out to boost a theory ever has demonstrated that light has any velocity?

Prof. Einstein is a Girondist of the Scientific Revolution. His revolt is against classical mechanics, but his methods and his delusions are as antiquated as what he attacks. But it is my expression that he has functioned. Though his strokes were wobbles, he has shown with his palsies the insecurities of that in Science which has been worshipfully regarded as the Most High.

It is my expression that the dissolution of phenomenal things is as much a matter of internal disorders as the effect of any external force, and that the slump of so many astronomers in favor of Einstein, who has made good in nothing, indicates a state of dissatisfaction that may precede a revolution-or that. if a revolt starts in the Observatories, hosts of irreconcilable observations will be published by the astronomers themselves, cutting down distances of planets and stars enormously. I shall note an observation by an astronomer, such as probably no astronomer, in the past. would have published. It seems to have been recorded reluctantly, and a conventional explanation was attempted-but it was published.

I take from a clipping, from the Loa Angela, Evening Herald, April 28, 1930, which was sent to me by Mr. L. E. Stein, of Los Angeles. In an account of the eclipse of the sun, April 28, 1930, Dr. H. M. Jeffers, staff astronomer of Lick Observatory, says: "We expected the shadow to be but half a mile in width. Instead of that. I think that it was nearer five miles broad." He says: "It may be suggested by others that the broad shadow was cast by astronomical errors due to the moon being closer to the earth than we have placed it in theory. But I don't believe that this broad belt was caused by anything but refraction."

The difference between half a mile and five miles is great. If the prophets of Lick Observatory did not take refraction into consideration, all the rest of their supposed knowledge may be attributable to incompetence. This difference may mean that the moon is not more than a day's journey away from this earth.

In The Earth and the Stars, p. 211, Abbot tells of the spectroscopic determinations, by which the new star in Perseus (Feb. 22, 1901) was "found" to be at a distance of 300 light years from this earth. The news was published in the newspapers. A new star had appeared, about the year 1600, and its light was not seen upon this earth, until Feb. 22, 1901. And the astronomers were able to tell this-that away back, at a moment when Queen Elizabeth-well, whatever she was doing-maybe it wouldn't be any too discreet to inquire into just what she was doing-but the astronomers told that just when Queen Elizabeth was doing whatever she was doing, the heavens were doing a new star. And where am I comparatively? Where are my poor, little yarns of flows and methylated spirits from ceilings, and "mysterious strangers," and bodies on railroad lines, compared with a yarn of the new star and Queen Elizabeth?

But the good, little star restores my conceit. In the face of all spectroscopes .in, all Observatories, it shot out nebulous rings that moved at a rate of 2 or 3 seconds of arc a day. If they were 300 light years away, this was a velocity far greater than that of light is said to be. If they were 300 light years away, it was motion at the rate of 220,000 miles a second. There were dogmas that could not stand this, and the spectroscopic determinations, which were in agreement, were· another case of agreements working out, as they shouldn't have worked out. The astronomers bad to cut· down one of their beloved immensities. Whether as a matter of gallantry, or not, they spread a denial for Queen Elizabeth's reputation to tread upon, saving that from the mud of an inquiry into just what Her Majesty was doing, and substituting unromantic speculations upon what, say, Andrew Jackson was up to.

Abbot's way of explaining the mistake· is by attributing the first "pronouncements" to "the roughness of the observations."

All over this earth, astronomers were agreeing in these "determinations." They were refinements until something else appeared and roughened them.

It would seem that, after this fiasco of the readjusted interest in what historical personages were doing, astronomers should have learned something. But, if Prof. Todd is right, in his characterization of them, that is impossible. About twenty years later, this situation, essentially the same in all particulars, repeated. Upon May 27, 1925, a new star was discovered in the southern constellation Pictor. By spectroscopic determination, its distance was "determined" to be 540 light years. See this stated in a bulletin of the Harvard Observatory, November, 1927.

March 27, 1928-the new star split. When the split was seen, astronomers of the South African Observatory repudiated the gospel of their spectroscopes of three years before. There must have been much roughness, even though there had been three years in which to plane down the splinters. They cut the distance from 540 to 40 light years. If there should be any more reductions like this, there may start a slump of immensities down toward a conception of a thinkable-sized formation of stars. A distance cut down 60 x 60 x 24 x 365 x 500 x 186,000 miles is a pretty good start.

Prof. Einstein, having no means of doing anything of the kind, predicts a displacement of the stars. Astronomers go out upon an expedition to observe an eclipse, and, not knowing that Einstein has no special means of predicting anything, they report, presumably because they want so to report, that he is right.

Then eclipse after eclipse-and Einstein is wrong.

But he has cast an ancient system into internal dissensions, and has cast doubts upon antiquities of thought almost as if his pedantic guesses had had better luck.

Whether the time has come, or not, here is something that looks as if it is coming:

An editorial in the New York Sun, Sept. 3, 1930: views of somebody else quoted:

· "The public is being played upon and utterly misled by the dreamery of the rival mathematical astronomers and physicists-not to mention the clerics-who are raising the game of notoriety to a fine art; in rivalry to religious mysticis a scientific pornography is being developed, and attracts the more because it is mysterious."

These are the views of Professor Henry E. Armstrong, emeritus head of the department of chemistry, at City and Guilds College, South Kensington, London.

This is revolt inside. That is what develops into revolution. Prof. Armstrong's accusation of pornography may seem unduly stimulating: but, judging by their lecheries in other respects, one sees that all that the astronomers have to do is to discover that stars have sex, and they'll have us sneaking to bookstores, for salacious "pronouncements" and "determinations" upon the latest celestial scandals. This would popularize them. And after anything becomes popular-then what?

That the time has come-or is coming-or more of the revolt within-

Or that, if they cannot continue upon their present pretenses of progress, the astronomers must return from their motionless excursions. A generation ago, they told of inconceivable distances of stars. Then they said that they had, a thousand times, multiplied any number some of these distances: but, if the inconceivable be multiplied any number of times, it is still the same old inconceivability. If, at the unthinkable, thought stops, but if thought must move somewhere, the astronomers, who cannot go on expansively, will, if they do think, have to think in reductions. If the time has come, there will be a crash in the Observatories, with astronomers in a panic selling short on inconceivabilities.

Upon Sept. 2, 1930, began a meeting of the American Astronomical Society, in Chicago. A paper that was read by Dr. P. Van de Kamp may be a signal for a panic. Said he: "Some of the stars may actually be thousands of light years nearer than astronomy believes them to be."

That-with some extensions-is about what I am saying.

Says the astronomer Leverrier-back in times when an astronomical system is growing up, and is of use in combating an older and, decaying orthodoxy, and needs support and prestige-says he-"Look in the sky, and at the point of my calculations, you will find the planet that is perturbing Uranus."

"Lo!" as some of the astronomers say in their books. At a point in the sky that can be said-to anybody who does not inquire into the statements-to be almost exactly the point of Levenier's calculations, is found the planet Uranus, to which-for all the public knows-can be attributed the perturbations of Uranus.

Up goes the useful renown of the astronomers. Supported by this triumph, they function.

But, if they're only the figments of one of the dream-like developments of our pseudo-existence, they, too, must pass away, and they must go by way of slaughter, or by way of laughter. Considering all their doings, I think that through hilarity would be the fitter exit.

Later:

"Look at the sky," we are told that the astronomer Lowell said, "and at the point of the calculations. The exultations, you will find the planet that is perturbing Neptune."

But this is in the year 1930.

Nevertheless we are told that a planet is found almost exactly at the point of the calculations. The exultations of the astronomers are spreadheaded.

But this is later. The damned thing takes a tack that shows that it could no more have been perturbing Neptune than I, anyway just at present, could cast a meeting of the National Academy of Sciences into disorder by walking past it.

They must be murdered, or we shall laugh them away. There is always something that can be said in favor of murder, but in the case of the astronomers that would be willful waste of the stuff for laughter. Orthodox astronomers have said that Levenier used no mathematical method by which he could have determined the position of Neptune. See Lowell's The Evolution of World's, p. 124. By way of stuff for the laugh, I mention that one of these disbelieving astronomers was Lowell.

One time, in a mood of depression, I went to the New York Public Library, and feeling a want for a little, light reading, I put in a slip for Lowell's Memoir on a Trans-Neptunian Planet. I got even more amusement than I had expected.

Just where was this point, determined by Lowell, almost exactly in which his planet was found? The spreadbeads the special articles-over and over in the newspapers of the world-"almost exactly."

Says Lowell, page 105: "Precise determination of its place does not seem possible. A general direction alone is predicable."

The stuff for a laugh that is as satisfactory as murder is in the solemn announcements by the astronomers about April Fool's Day, 1930, that they had found Lowell's planet almost exactly in the place, precise determination of which does not seem possible –

Their chatter over Lowell's magnificent accuracy in pointing in a general direction-

Then the tack of a thing that showed that it could not have been all this indefiniteness, anyway-

265 years, instead of 3,000 years-

And instead of going the thing was coming.

If they can't tell whether something is coming or going, their solemn announcements upon nearness or farness may be equally laughable.

If by mathematical means Adams and Leverrier did not determine the position of the planet Neptune, or if it was, in an opinion that Lowell quotes, "a happy accident," bow account for such happiness, or for this timely and sensational boost to a prestige, if we suspect that it was not altogether an accident?

My expression is that herein I'd typify my idea of organic control which, concealed under human vanity, makes us think that we are doing all things ourselves, gives support to human institutions, when they are timely and are functioning, and then casts its favorites into rout and fiasco, when they have outlived their functioning-period.

If Leverrier ·really had bad powers by which he could have pointed to an unseen planet, that would have been a finality of knowledge that would be support to a prestige that could never be overthrown. Suppose a church had ever been established upon foundations not composed of the stuff of lies and frauds and latent laughter. Let the churchman stand upon other than gibberish and mummery, and there'd be nothing by which to laugh away his despotisms.

Say that. whether it be a notion of organic control, or not, we accept any theory of Growth, or Development, or Evolution-

Then we accept that the solemnest of our existence's phenomena are of a wobbling tissue-rocks of ages that are only hardened muds-or that a lie is the heart of everything sacred.

Because otherwise there could not be Growth, or Development, or Evolution.

XXI

A TREK of circumstances that kicks up a dust of details-a vast and dirty movement that is powdered with particulars- the gossip of men and women, and the yells of brats whether dinner is ever going to be ready, or not-young couples in their nightly sneaks-and what the hell has be
come of the grease for the wheels? -who's got a match?

It's a wagon train that feels out across a prairie.

A drink of water-a chaw of tobacco-just where to borrow a cupful of flour-and yet, even though at its time any of these wants comes first. there is something behind all

The hope for Californian gold.

The wagon train feels out across the prairie. It traces a path that other wagon trains make more distinct-and then so rolls a movement that to this day can be seen the ruts of its wheels.

But behind the visions of gold, and the imagined feel of nuggets, there is something else-

The gold plays out. A dominant motive turns to something else. Now a social growth feels out. Its material of people, who otherwise would have been stationary, has been moved to the west.

The first, faint structures in an embryonic organism are of cartilage. They are replaced by bone.

The paths across prairies turn to lines of steel.

Or that once upon a time, purposefully, to stimulate future developments, gold was strewn in California-and that there had been control upon the depositions, so that only enough to stimulate a development, and not enough to destroy a financial system had been strewn-

That in other parts of this earth, in far back times, there had been purposeful plantings of the little, yellow slugs that would-when their time should come-bring about other extensions of social growths.

But the word purposeful, and the word providential, are usurped words. They are of the language

of theologians, and are meant to express an idea of a presiding being, ruling existence, superior to it, and not of it, or not implicit to it. I'd rather go on using these words, denying their ownership by any special cult, than to coin new words. With no necessity for thinking of an external designer and controller, I can think of design and control and providence and purpose and preparation for future uses, if I can think not loosely of Nature, but of a Nature, as an organic whole. Every being, except for its dependence upon environment, is God to its parts.

It is upon the northern parts of this earth that the civilizations that have persisted have grown up, then extending themselves colonially southward. History, like South America and Africa, tapers outwards. There are no ruins of temples, pyramids, obelisks, in Australia, Argentina, South Africa. Preponderantly peninsulas are southward droops. As if by design, or as if. concordantly with an ac-centuation of lands and peoples in the north, the sun shines about a week longer in the north, each year, than in the south. The coldness in the less important Antarctic regions is more intense than in the Arctic, and here there is no vegetation like the grasses and flowers of the Arctic, in the sum-mertime. Life withers southward. Musk oxen, bears, wolves, foxes,- lemmings in the Far North-but there are only amphibious mammals in the Antarctic. Fields of Arctic poppies in the Arctic summer time-but summer in the Antarctic is gray with straggling lichens. If this earth be top-shaped as some of the geodesists think, it is a bloom that is stemmed with desolation.

There are no deposits of coal in the southern parts that compare with deposits in the northern parts. The greatest abundance of oil supplies is north of the equator. It looks like organic preparation, in formative times, before human life appeared upon this earth, for civilizations that would grow up in the north. For ages, peoples of this earth were ignorant of the uses of coal and oil; upon which their later developments would depend.

But so conventionalized are the thoughts of most persons, upon this subject, that if, for instance, my expression is that gold was strewn in California in preparation for future uses, there must be either a visualization of an aggrandized man, who walked about, slinging nuggets, or a denial that, except in the mind of a man, there can be purpose, or control, or design, or providence.

But the making of a lung in an embryonic being that cannot breathe-but it will breathe. This mak-ing of a lung is a preparation for future uses. Or the depositions of tissues that are muscles that are not, but that will be used. Mechanical foresight, or preparation for future uses, pervades every em-bryonic being. There is a fortune teller in every womb.

Still, not altogether only theological have been speculations upon the existence of purpose, or de-sign, control, or guidance in "Nature." There are philosophical doctrines known as orthogenesis and entelechy. Again we are in a situation that we have noted. If there be orthogenesis, or guidance from within-within what? Heretofore, this doctrine has provided no outlines within which to think. All that is required for thinkableness, instead of bafflement, is to give up attempted notions upon Na-ture, as Universality, and conceive of one thinkable-sized existence, of shape that is representable in thought, and conceive of an organic orthogenesis within that. In the organic sense, there is, in the Arctic regions, no great need for water. Though the coldness is not so intense here as it is commonly supposed to be, the climate nevertheless prevents much colonization. I have never read of a deluge in the Arctic. Thunderstorms are very uncommon. Some explorers have never seen a thunderstorm in the Artie regions. And at the same time there are oppressively warm, or almost tropical, summer days in the Arctic. Instead of the enormous falls of snow, of common suppositions, the fall of snow, in the Far North, is "very light" (Stefansson). It looks like organically economic neglect of a part that cannot be used. Where, as reliefs, thunderstorms are not needed, there are, except as vagaries, no thunderstorms, though the summertime conditions in places of need and no need are much alike. See Heilprin's account of his experiences in Greenland-summer days so nearly tropical that pitch melted from the seams of his ship.

The alternations that are known as the seasons are beneficial. They have come about accidentally, or they have been worked out by Automatic Design, or by all-pervasive intelligence, or by equilibration, if that word be preferred to the word "intelligence." It looks as if more complexly a problem was solved. It is commonly thought that only brains solve problems, or, rather, approximate to solutions: but every living thing that carries a weapon, or a tool, has, presumably not with its brains, but with the intelligence that pervades all substances-so then with the intelligence of its body solved a problem. It looks as if more complexly a problem was solved, as I say, though in anything like a real, or final, sense, no problem ever has been solved. By the varying incidence of the sun, alternations of fruitfulness and rest could be brought about in the north and the south, but that left rhythms small in the tropics. It · looks as if here, intelligently, were brought about the changes that are known. as the dry season and the rainy season. l have never read a satisfactory explanation of this alternation, in conventional, meteorological terms.

In the April rains there is evidence, or might be, if we could have a rational idea as to what we mean by evidence, of design, and an automatically intelligent provision and control. Something is controlling the motions of the planets, according to all appearances that we take as appearances of control. Accepting this, I am only amplifying. Rains, of a gentle and frequent kind that is most beneficial to young plants, or best adapted to them, fall in April. Conventional biology is too one-sided. It treats of adaptation of plants to rain. We see also the adaptation of rains to plants. But there must be either the conventionally theological, or the organic, view, to see this reciprocity. If one prefers to think of a kind and loving deity, who is sending the April rains, he will have to consider-or, rather, will be faced by-records of other rains, which are of the loving kindness of slaughter and desolation and woe.

There is some, unknown condition that ameliorates the climate of Great Britain, as if this center of colony-sporing were prepared for by an automatic purposefulness, and protected from the rigors of the same latitudes in the west. Once upon a time, one of the wisemen's most definite concepts was the Gulf Stream. They wrote about its "absolute demarcation" from surrounding waters. They were as sure of the Gulf Stream as they are today that the stars are trillions of miles away. Lately so much has been written upon the in conceivability of the Gulf Stream having effect upon climate farther from its source than somewhere around Cape Hat teras, that I shall not go into that subject. Something is especially warming Great Britain,. and it cannot be thought to be the Gull Stream. It may be an organically providential amelioration. It may play out, when the functioning period of Great Britain passes away. I am not much given to prophecy, but I'll take this chance-that if England loses India, we may expect hard winters in England.

Our acceptance is that nations work together, or operate against one another functionally, or as guided by the murderous supervisions of a whole Organism. Or, apologizing again, I call such organized slaughter, super-metabolism. So enormous is the subject of human history, as affected by its partness in a whole, that I shall reserve it for treatment some other time. Monistically-,though some other time I shall pluralistically take another view, as well-the acceptance is that human beings have not existed as individuals any more than have cells in an animal organism existences of their own. Still, one must consider that there is something of individuality, or contrariness in- every cell. This view of submergence is now so widespread that it is expressed by writers in many fields of thought. But they lack the concept of a whole, trying to think of a social organism as a whole, though clearly every social quasi-organism has relations with other social quasi-organisms, and is dependent enormously, or vitally., upon environment. Other thinkers, or more than doubtful thinkers, say that they think of the unthinkable Absolute as the whole.

I have a notion that, for ages, as a factor in an automatic plan. the Australian part of our existence's nucleus, this earth, was reserved. If this be not easy to think it is equally hard to think why Australia, in its fertile parts, was not colonized by Asiatics. There was relative isolation. But it was not geographical isolation: the distance between Cape York, Australia, and New Guinea is only 100 miles. There was an approximation to isolation so extreme that one type of animal life grew up and prevailed. This gap was jumped by the marsupials of Australia. Then the question is-why, if not obediently to an inhibition, was it not jumped the other way? Of course we can have no absolute expressions, but just when the dingoes and the wild cattle of Queensland first arrived in Australia is still considered debatable.

There were civilizations in the Americas, but they were civilizations that could not resist the relatively late-appearing Europeans. Long before, there had been other civilizations in Central America, but they had disappeared, or they had been removed. The extinction of them is, by archaeologists, considered as mysterious, as is the extinction of the dinosaurs, by the paleontologists--or as, by cells of a later period, ·might be considered the designed and scheduled, or purposeful, extinction of cartilage cells in an embryo.

The expression is that Australia and the Americas were reserved, as relative blanks, in which human life upon this earth could shake off, after a fashion, many. conventions and traditional hamperings, and start somewhat anew.

Drones appear in a beehive. They are reserved. At first they contribute nothing to the welfare of the hive, but there is a providence that looks after them just so long as they will be of future use. This is automatic foresight and purpose, according to automatic ·plan, in a beehive, regarded as a whole. The God of the bees is the Hive. There is no necessity to think of an external control, nor of any being, presiding over the bees and directing their affairs.

Reservations besides those in the affairs of bees and men are common. Some trees have buds that are not permitted to develop; These are known as dormants, and are held in reserve, against the possibility of a destruction of the tree's developed leaves. In one way or another, there are reservations in every organism.

We think of inter-mundane isolations that have been maintained, as once the Americas were kept separated from Europe, not by vast and untraversable distances, but by belief in vast and untraversable distances. I have no sense of loneliness in thinking that the inorganic sciences that are, by inertia, holding out for the isolation of this earth, have lost much power over minds. There are dissatisfactions and contempts everywhere.

There may be civilizations in the lands of the stars, or it may be that, in the concavity of a starry shell, vast, habitable regions have been held in reserve for colonization from this earth. Though there is considerable opposition to wars, they are, as at any moving picture place, one can see, still popular: but other elimination of human beings have waned, and it is likely that for a long time birth control will have no more than its present control upon births. The pestilences that used to remove millions are no longer so much heard of. It may be that an organic existence is, by lessening eliminations, preparing a pressure of populations upon this earth that can have relief only in enormous colonizing outlets somewhere else. It is as if concordantly, the United States has shut down, as a relief, to superabundances of people in Europe, and-as if representing the same purpose or plan, Australia and Canada, as well as the United States, are shutting out Asiatics. It is as if co-operatively with the simultaneous variations of need, aviation is developing, as the means of migratory reliefs-

If there be a nearby land that is a revolving shell of stars

And if, according to data that I have collected, there be not increasing coldness and attenuation of air past a zone not far from this earth.

Nineteen hundred and thirty something or another-may be nineteen hundred and forty or fifty-

There's a flash in the sky. It is said to be a meteor. There's a glow. That is said to be an aurora borealis

The time has come.

The slogan comes-

Skyward ho!

The treks to the stars. Flows of adventurers-and the movietone news-press agents and interviews-and some body about to sail to Lyra reduces expenses by letting it be known what brand of cigarettes he'll take along-

Caravels with wings-and the covered planes of the sky and writers of complaints to the newspapers: this dumping of milk bottles and worse from the expeditions is an outrage. New comets are watched from this earth-long trains of voyagers to the stars, when at night they turn on their lights. New constellations appear-the cities of the lands of the stars.

And then the commonplaceness of it all.

Personally conducted tours to Taurus and Orion. Summer vacations on the brink of Vega. "My father tells of times, when people, before going to the moon, made their wills." "Just the same the e was something peaceful about those old skies. It's getting on my nerves, looking up at all those lip stick and soap and bathing suit signs."

Or my own acceptance that there can be no understanding of our existence, if be overlooked the irony of it all-

The aristocratic astronomers-their alleged rapport with infinitude-their reputed familiarity with the ultra-remote the academic-the classical-

One looks up and sees, instead, an illuminated representation of a can of spaghetti in tomato sauce, in the sky.

The commonplaceness of it all. Of course the stars are near. Who, but a few old fossils, ever thought otherwise? Does the writer of this book think that he found out anything new? All these notions of his were matters of common knowledge, away back in the times of ancient Greece.

XXII

THAT, in the summer of 1880, some other world, or whatever we'll call it, after a period of hard luck, cheered up-and cast off its despairs-which came. to this earth, where there is always room for still more melancholy-in long, black, funereal processions.

Aug. 18, 1880-people, near the waterfront of Havre, France, saw the arrival of a gloom. Sails, in the harbor of Havre, suddenly turned black. But, like every other gloom, this one alternated with alleviations. The sails flapped white. There was a flutter of black and white. Then enormous numbers of the units of these emotions were falling into the streets of Havre. They were long, black flies.

In an editorial, in the London Daily Telegraph, August 21st, it is said that this appearance of flies, at Havre, was a "puzzle of the most mysterious kind." These flies had come down from a point over the English Channel. They had not come from England. I have searched widely in Continental publications, and there is no findable record of any observation upon this vast swarm of flies, until it came down from the sky, over the English Channel. Pilot boats, returning to Havre, came in black with them. See the Journal des Debates (Paris) August 20-that they were exhausted flies, which fell, when touched, and could not move, when picked up. Or they may have been chilled into torpidity. Presumably there were survivors, but most of these helpless flies fell into the water, and the swarm, as a swarm, perished. If this is a puzzle of the "most mysterious kind," I am going to be baffled for a description, as we go along. I don't know what comes after the superlative.

Three days later, another vast swarm of long, black flies appeared somewhere else. Just how much we're going to be puzzled by more than the most mysterious depends upon how far this other place was from Havre. See the New York Times, September 8-that, upon August 21st, a cloud of long, black flies, occupying twenty minutes in passing, had ap eared at East Pictou, Nova Scotia. Halifax Citizen, August 21-that they had passed Lismore, flying low, some of them appearing to fall into the water.

Upon the 2nd of-September, another swarm came down from the sky. It appeared suddenly, at one place, and there is no findable record that it was seen anywhere else, over land or water of this earth. It is told of, in the Entomologists' Monthly Magazine, November, 1880-off the coast of Norfolk, England-an avalanche that overwhelmed a schooner. - "millions and millions of flies." The sailors were forced to take shelter, and it was five hours before they could return to the decks. "The air became clear, about 4 P.M., when the flies were thrown overboard by shovelfuls, and the remainder were washed off the decks by buckets of water and broo s." It was another appearance of exhausted, or torpid, flies.

Scientific American, 43-193: "0n the afternoon of Saturday, September 4th, the steamboat Martin encountered, on the Hudson River, between New Hamburg and Newburgh, a vast cloud of flies. It reached southward, from shore to shore, as far as the eye could reach, and resembled a drift of black snow. The insects were flying northward, as thick as snowflakes driven by a strong wind." They were long, black flies. Halifax Citizen, September 7-that, upon the 5th of September, a compact cloud of flies, occupying half an hour in passing, had appeared at Guysboro, Nova Scotia, hosts of them falling into the water.

I think that this crowd of flies was not the same as the Hudson River crowd, even though that was flying northward. So I think, because the flies of Guysboro, like the flies of Havre, came as if from a point over the ocean. "They came from the east" (Brooklyn Eagle, September 7).

The look of the data is that, with an ocean between appearing points, a bulk of flies, of the size of a minor planet, dividing into swarms, somewhere in outer space, came to this earth from somewhere else. It is simply a matter of thinking of one origin. and then thinking that that origin could not have been in either North America or Europe.

If we can think that these flies came to this earth from the moon, or Mars, or from a fertile region in the conclave land of the stars, that is interesting; but by this time we have passed out of the kindergarten of our notions, and are ready to take up not merely mysterious appearances, but mysterious appearances that will be data for our organic expressions. In data upon insect-swarms of the summer of 1921, there is suggestion not only of conventionally unaccountable appearances of insect-swarms, but of appearances in response to need. If one has no very active awareness for any need for insects, that is because one is not thinking far enough back into interrelations of bugs and all other things.

In the summer of 1921, England was bereft of insects. The destruction of insects, in England, by the ·drought of 1921, was, very likely unequaled at any other time, anyway for a century or more. The story of dwindling and disappearing is told, in Garden Life, for instance-aphides becoming fewer and fewer-absence of mosquitoes, because of the drying of the ponds-not one dragon Hy all summer-scarcity of ants-midges almost entirely absent-stricken fields in which not a butterfly was seen ordinary - flies uncommon, and blue bottles exterminated. See the Field and the Entomologists' Record for similar accounts.

Then came clouds of insects and plagues of insects: foreign insects, and unknown insects. Anybody can find the data in various English publications. I note here that one of the swarms of exotic insects was of large fireflies that appeared in Wales (Cardiff Western Maa, July 12). Locusts appeared (London Weekly Dispatch, July 6). I suppose that almost any conventional entomologist will question my statement that vast swarms of unknown insects appeared at this time, in England: nevertheless, in the London Daily Express, September 24, Prof. Le Froy is quoted as saying, of a species of stinging insects, that it was unknown to him.

Destructions that were approaching extermination-and then multitudinous replenishments. I have searched without finding one datum for thinking that one of these replenishment was seen crossing the Channel. Three of them were of foreign insects.

Once upon a time, according to ancient history; Somebody so loved this world that be gave to it his only begotten son. In this year 1921, according to more recent records, Some thing gave to the streets of London its many forgotten women. To starving humans it gave a dole. But, when it's insects dwindled away, it bestowed profusions of bugs.

All our expressions are in terms of relative importance.

In the summer of 1869, in many parts of England, there was a scarcity of insects that was in some ways more remark able than that of 1921. This scarcity was discussed in all entomological magazines of the time, and was mentioned in newspapers and other publications. For one of the discussions, see the Field, July 31 and Aug. 14, 1869. Most widely noticed was the absence of one of the commonest of insects, the small, white butterfly, Pieris rapae. Some of the other ordinarily plentiful species were scarcely findable.

In the London Times, July 17, a correspondent, in Ashford, Kent, writes that a tropical, or sub-tropical insect, a firefly (Lampyris Italia) had been caught in his garden. In the Times, of the 20th, the presence of this insect in England is seemingly explained. Someone else writes that, upon June 29th, at Dover, only fifteen miles from Ashford, he had released twelve fireflies, which he had brought in a bottle from Coblenz. But in the same issue of this newspaper, a third correspondent writes that, at Catherham, Surrey, had appeared many fireflies. Weekly Dispatch (London)-"They were so numerous that people called them a nuisance." Even a firefly can't By its fire, without a man with a bottle appearing and saying that he had let it go. There will ·be accounts of other swarms. Only Titans, who bad uncorked Mammoth Caves, in mountains of glass, could put in claims for letting them go from bottles.

The coast of Lincolnshire-and a riddance long and wide. The coast of Norfolk-several miles of tragedy. In the Zoologist, 1869-1839, someone reports belts of water, some a few yards wide, and some hundreds of yards wide, "of a thick, pea-soup appearance," so colored by drowned aphides, off the coast of Lincolnshire; and, off the coast of Norfolk, a band of drowned ladybirds, about ten feet wide, and two or three miles long. Wherever this little dead comet came from, there is no findable record that it had been seen alive anywhere in Europe.

Upon the 26th of July, columns of aphides came down from the sky, at Bury St. Edmunds, about 60 miles south of the coast of Lincolnshire massed so that they gave off a rank odor, and so dense that, for anybody surrounded by them, it was difficult to breathe. Upon the same day, at Chelmsford, about 40 miles south of Bury St. Edmunds, appeared masses of these insects equally vast. See Gardener Chronicle, July 31, August 7.

Aphides had streaked the ocean. Columns of others had come down, like vast, green stems, from their fem-like clouds. Less decoratively, others darkened the sky. A new enormity appeared upon the coast of Essex, about the first of August. According to correspondence, in the Maidstone Journal, August 23, fogs of aphides had shut off sunlight. They appeared in other parts of southeastern England. "They swarmed to such an extent as to darken the air for days together, and to render it almost dangerous to the sight of men and animals to be out of doors."

The 9th of August-the first of the ladybirds that reached England alive were reported at Ramsgate. Three days later, between Margate and Nore Light, near the mouth of the Thames, thousands of ladybirds speckled a vessel. This dis eased appearance took on a more serious look, with blotches of small, yellow, black-marked flies. Then spread a cosmetic of butterflies. '

These were van-swarms. Upon the 13th, an invasion was on. I quote chiefly from the London Times.

A cloud was seen over the Channel, not far from land, moving as if from Calais, reaching Ramsgate, discharging ladybirds upon the town. They drifted into piles in the streets. The town turned yellow. These were not red lady birds.. There would be less mystery, if they were. People in the town were alarmed by the drifting piles in the streets, and a new job, worth the attention of anybody who collects notes

' upon odd employments, appeared. Ladybird shovelers were hired to throw the drifts into sewers. Clouds streaked counties. They moved northward, reaching London, upon the 14th, pelting into

the streets, and filling gutters. Children scooped them up, filling bags and pails with them, and
"played store" with them. Multitudes went on as far as Worcester.

Upon the 14th, "a countless multitude" of other ladybirds arrived upon the coasts of Kent and Surrey, and these clouds too, seemed to have come from France. They rattled, like colored hail, against
windows. They were "yellow perils," and the inhabitants were alarmed, fearing a pestilence from ac-
cumulations of bodies. Fires were built, to burn millions of them, and people who bad never shoveled ladybirds be fore took up the new employment.

The next day, "an enormous multitude" of new arrivals appeared at Dover, coming as if from France.
The people who were out in this storm carried umbrellas, which soon looked like huge sunflow-
ers. People, stopping to discuss the phenomenon, gathered into bouquets. The storm abated, and
umbrellas were closed. All blossomed again. Another cloud rolled in from no place or origin that
has ever been found out. Th living gushes from the unknown moved on toward London, and in
accounts of them, in Land and Water, are amusing descriptions of the astonishment they caused.
There is a story of five hypnotized cats. A multitude alighted upon a lawn. Five cats sat around, motionless, gazing at the insects. A woman tells of her bewilderment, when, looking out at her lines of
wash, which had been spotless, she saw garments hanging blotched and heavy. At Shoeburyness, the
ladybirds pelted so that men in brickyards were driven from their work. Unless from celestial nozzles
living fountains were playing down upon this earth, I cannot conceive of the origin of these deluges.

Some entomologists tried to explain that the insects must have gathered in other parts of England,
having flown toward France, having been home back by winds to the southeastern coast of England.
If anybody accepts, with me, that these insects were not English ladybirds, and that they did not
come from France, and did not keep on coming, day after day, to one point, from Holland, Sweden,
Spain, Africa-and here consider the feeble flight of ladybirds-but if anybody accepts with me that
these ladybirds did not fly from any part of this earth to their appearing-point, I suppose that he will
go on thinking that they must so have flown, just the same.

That there are data for thinking that these insects were not English ladybirds:

In the London Standard, August 20, there is a description of them. "They all seemed to be much
larger than the common ladybirds, of a paler color, with more spots." In the Field, August 28, someone writes that all the insects, except a few, were yellow.. So far as he knew, he had never before seen
specimens of this species. The Editor of the Field writes: "The red is paler, and there are divers slight
differences that rather indicate a foreign origin." He says that, in the opinion of Mr. Jenner Weir, the
naturalist, these ladybirds were different from ordinary English specimens.

But these millions must have been very ordinary somewhere.

That there are data for thinking that these insects came from neither France nor Belgium:

Such as hosts of observations upon the swarms, within a mile or two of the English coast, and no
findable record of an observation farther away, or nearer France. There is, in news papers of Paris, no
mention of appearances of ladybirds any where upon the continent of Europe. There is no mention
in publications of entomological societies of France and Belgium. But any of these enormous clouds
leaving a coast of France or Belgium would have attracted as much attention as did an arrival in England. Other scientific publications in which I have searched, without finding mention of observations upon ladybirds in France, or any other part of the Continent, are Comptes Rendus, Cosmos,

Petites Nouvelles Entomologiques, Rev. et Mag. de Zoologie, La Science Pour Tous, L'Abeille, Bib. Unioerselle, and Rev. Cours. Sci. In Galignani's Messenger (Paris) considerable space is given to accounts of the invasions of England by ladybirds, but there is no mention of observations anywhere, except in England, or within a mile or so of the English coast.

This is the way an invasion began. A great deal was written about conditions in the invaded land. Probably the scarcity of insects in England was unprecedented. There was no drought. It is simply that the insects had died out. And billions were coming from somewhere else.

"Margate Overwhelmed!"

In the Field, August 28, a correspondent writes: "On Wednesday (25th) I went to Ramsgate by steamboat, and, as we approached within five or six miles of Margate, com plaints of wasps began to be heard. I soon ascertained that they were not wasps, but a bee-like fly. As we neared Margate, they increased to millions, and at Margate they were almost, unendurable." Some specimens were sent to the Editor of the Field, and he identified them as Syrphi. There had been a similar multitude at Walton, on the coast, about 30 miles north of Margate, the day before.

The little band of scouts, at Ashford-they carried lanterns. Then green processions-yellow multitudes-the military looking Syrphi, costumed like hussars-

A pilgrimage was on.

"Thunder bugs" appeared between Wingham and Adisham. The tormented people of the region said that they had never seen anything of the kind before (Field, August 21). "Wasps and flies -in overwhelming numbers" besieging Southampton (Gardeners' Chronicle, September 18). London an arriving point-the descent of crane flies upon London doorsteps and pavements looking muddy with them-people turning out with buckets of boiling water, destroying multitudes of them (Illustrated London News, September 18), This is one of the ways of treating tourists.

I think that there is a crowd-psychology of insects, as well as of men, or an enjoyment of communicated importance from a crowd of millions to one of the bugs. They were humming to England, not merely with bands playing, but each of them blowing some kind of a horn of his own. There are persons who would be good, if they thought that they could go to heaven, or so swarm in the sky, with millions of others, all tooting saxophones.

Pilgrims, or expeditionaries, or crusaders-it was more like a crusade, with nation after nation, or species after species, pouring into England, to restore something that had been lost.

In Sci. Op., 3-261, is an account of a new insect that appeared in England, in July of this year, 1869. For accounts of other unknown insects that appeared in England, in this summer, see the Naturalists' Note Book, 1869-318; Sci. Gos., 1870-141; Ent. Mo. Mag., 1869-86, and February, 1870; Sci. Op., 2-359. It was a time of "mysterious strangers."

In the Times, August 21, someone noted the absence of small, white butterflies, and wondered how to account for it. In the Entomologist, Newman wrote that, up to July 12th, he had seen, of this ordinarily abundant insect, only three specimens. Upon pages 313-315, half a dozen correspondents discussed this remarkable scarcity. In the Field, September 4, someone told of the astonishing scarcity of house flies: in more than six weeks, at Axminster, he had seen only four flies. London Standard, August 20-that, at St. Leonard's on-Sea, all insects, except ladybirds and black ants, were "few and far between." In Symons' Met. Mag., August, 1869, it is said that, at Shiffnal, scarcely a white butterfly had been seen, and that, up to July 21st, only one wasps' nest had been found. Correspondents, in the Entomologist, September and October, mentioned the scarcity of three species of white butterflies, and noted the unprecedented fewness of beetles, bees, wasps, and moths. Absence of hornets is commented upon, in the Fie/4, July 24.

They were pouring into England.

An army of beetles appeared in the sky. At Ullswater, this appearance was a military display. Regiment after regiment, for half an hour, passed over the town (Land and Water, September 4).

The spiders were coming.

Countless spiders came down from the sky into the city of Carlisle, and, at Kendal, thirty-five miles away, webs fell enormously (Carlisle Journal, October 5). About the 12th of October, "a vast number" of streamers of spiders' web and spiders came down from the sky, at Tiverton, Devon shire, 280 miles south of Carlisle. See the English Mechanic, November 19, and the Tiverton Times, October 12. As if in one persisting current, there was a repetition. Upon the morning of the 15th, webs, "like pieces of cotton," fell from the sky, at South Molton, near Tiverton. Then fell "wondrous quantities," and all afternoon the fall continued, "covering fields, houses, and persons." It was no place for files, but to this webby place files did come.

Species after species-it was like the internationalism of the better-known crusades- The locusts were coming.

Upon the 4th of September, a locust was caught in York shire (Entomologist, 1870-58). There are no locusts indigenous to England. At least up to May, 1895, no finding of a locust in its immature state had ever been recorded in Great Britain (Sci. Gos., 1895-83). Upon the 8th and 9th of October, locusts appeared in large numbers, in some places, in Pembrokeshire, Derbyshire, Gloucestershire, and Cornwall. They had the mystery of the ladybirds. They were of a species that, according to records, had never before appeared in England. An entomologist, writing in the Journal of the Plymouth Institute, 4-15, says that he had never heard of a previous visit to England by-this insect (Acridium perigrinum). It seems that in all Europe this species had not been seen before. In the Ent. Mo. Mag., 7-1, it is said that these locusts were new to European fauna, and were mentioned in no work upon European Orthoptera.

At the meeting of the Entomological Society of London, Nov. 15, 1869, it was decided, after a discussion, that the ladybirds had not come from France, but had flown from places in England, and had been carried back, by winds to other parts of England. There was no recorded observation to this effect. It was the commonplace ending of a mystery. I add several descriptions that indicate that. in spite of London's most-eminent bugmen, the ladybirds were not English ladybirds. Inverness Courier, September 2- "That they are foreigners, nobody doubts. They are nearly twice the size of the common English ladybirds, and are of a paler color." See the Student, 4-160-"the majority were of a larger size, and of a dull. yellow hue." In the London Standard, August 23, it is said that some of the insects were almost half an inch long.

That the locusts were foreigners was, by the Entomological Society of London, not discussed. Nothing else was discussed. Crane flies and Syrphi and spiders and all the rest of them not a mention. I know of no scientist who tried to explain the ladybirds, and mentioned locusts. I know of no scientist who tried to explain the locusts and mentioned ladybirds no scientist who wrote upon a scarcity of insects, and mentioned the swarms-no scientist who told of swarms, and mentioned scarcity.

The spiders, in a localized fall that lasted for hours, arrived as if from a persisting appearing-point over a town, and the ladybirds repeatedly arrived, as if from an appearing-point a few miles from a coast. The locusts came, not in one migration, but as if successively along a persisting path, or current. because several had been caught more than a month before large numbers appeared (Field, October 23).

A mob from the sky, at Burntisland, Scotland-"spinning jennys" that were making streets fuzzy with their gatherings on cornices and window sills (Inverness Courier, September 9). An invasion at Beccles was "an experience without pre cedent." A war correspondent tells of it, in the Gardeners' Chronicle, September 18. The invaders were gnats-correspondent trying to write about them, from an ink pot filled with drowned gnats-people breathing and eating gnats. Near Reading, "clouded yellow butterflies," insects that had never before been recorded in Berkshire, appeared (Sci. Gos., 1869-210). At Hardwicke, many bees of a species that was unknown to the observer, were seen (Nature, 2-98).

Field, August 21 and November 20-swarms of hummingbird hawkmoths. As described in Science Gossip, 1869-273, there was, at Conway, "a wonderful sight"-a flock of hummingbird hawkmoths and several species of butterflies. Clouds of insects appeared in Battersea Park, London, hovering over trees, in volumes so thick that people thought the trees had been set afire (Field, June 4, 1870). An invasion at Tiverton, seemingly coming with the spiders, "a marvelous swarm of black flies" made its headquarters upon the Town Hall, covering the building, turning it dark inside, by settling upon the window glass (Tiverton Times, October 12). At Maidstone, as if having arrived with the lady birds, a large flight of winged ants was seen (Maidstone Journal, August 23). Midges were arriving at Inverness, August 18th. "At some points the cloud was so dense that people had to hold their breath and run through" (Inverness Courier, August 19). Thrip& suddenly appeared at Scarborough, August 25th (Sci. Op., 2-292). At Long Benton, clouds of Thrips descended upon the town, wafting into houses, where they were dusted from walls, and swept from Boors (Ent. Mo. Mag., 1869- 171). Also, at Long Benton appeared an immense flight of the white butterflies that were so scarce everywhere else, gardeners killing thousands of them (Ent. Mo. Mag., Dec ember, 1869). At Stonefield, Lincolnshire, appeared beetles of a species that had never been seen there fore (Field, October 16).

It was more than a deluge of bugs. It was a pour of species. It was more than that. It was a pour on a want.

Entomologists' Record, 1870-that, in this- summer of 1869, in England, there had been such an "insect famine" that swallows had starved to death.

XXIII

Melbourne Age, Jan. 21, 1869-there was a carter. He was driving a five-horse truck along the bed of a dry creek. Down the gulley shot a watery fist that was knuckled. with boulders. A dead man, a truck, and five horses were punched into trees.

New Orleans Daily Picayune, Aug. 6, 1893-a woman in a carriage, crossing a dried-up stream .in Rawlings County, Kansas. It was a quiet, summery scene.

There was a rush of water. The carriage crumbled. There was a spill of crumbs that were a woman's hat and the heads of horses. Philadelphia Public Ledger, Sept. 16, 1893-people asleep, in the town of Villa-canas, Toledo, Spain. The town was raided by trees. Trees smashed through the walls of houses. People in bed were grabbed by roots. A deluge had fallen into a forest.

Bright, clear day, near Pittsburgh, Pa. From the sky swooped a wrath that incited a river. It was one bulk of water: two miles away, no rain fell (New Orleans Daily Picayune, July 11, 1893). A raging river jeered against former confinements. Some of its gibes were freight can. It scoffed with bridges. Having made a high-water mark of rebellion, it subsided into a petulance of jostling row-boats. Monistically, I have to accept that no line of demarcation can be drawn between emotions of minds and motions of rivers.

These sudden, astonishing leaks from the heavens are not understood. Meteorologists study them meteorologically. This seems logical, and is therefore under suspicion. This is the fallacy of all the sciences: scientists are scientific. They are inorganically scientific. Someday there may be organic science, or the interpretation of all phenomenal things in terms of an organism that comprises all.

If om existence is an organism, in which all phenomena are continuous, dreams cannot be utterly different, in the view of continuity, from occurrences that" are said to be real.

Sometimes, in a nightmare, a kitten turns into a dragon. Louth, Lincolnshire, England, May 29, 1920-the River Lud, which is only a brook, and is known as "Tennyson's Brook," was babbling, or maybe it was purling-

Out of its play, this little thing humped itself twenty feet high. A ferocious transformation of a brook sprang upon the houses of Louth, and mangled fifty of them. Later in the day, between banks upon which were piled the remains of houses, in which were lying twenty-two bodies, and from which hundreds of the inhabitants had been driven homeless, the little brook was babbling, or purling.

In scientific publications, early in the year 1880, an event was told of, in the usual, scientific way: that is, as if it were a· thing in itself. It was said that a "water-spout" had burst upon the island of St. Kitts, B. W. I. A bulk of water had struck this island, splitting it into cracks, carrying away houses and. people, drowning 250 of the inhabitants. A paw of water, clawed with chasms, had grabbed these people.

In accordance with our general treatments, we think that there are waterspouts and cloudbursts, but that the water spout and cloudburst conveniences arise, when nothing else can, or, rather, should, be thought of, and as labels are stuck on events that cannot be so classified except as a matter of scientific decorum and laziness. Some of the sleek, plump sciences are models of good behavior and inactivity, because, with little else to do, they sit all day on the backs of patient fishmongers.

As a monist, I think that there is something meteorological about us. Out of the Libraries will come wraths of data, and we, too, shall jeer against former confinements; Our gibes will be events, and we shall scoff with catastrophes.

The "waterspout" at St. Kitts-as if it were··a single

thing, unrelated to anything else. The West Indian, Feb. 3, 1880-that, while the bulk that was called a waterspout was overwhelming St. Kitts, water was falling upon the island of Grenada, "as it had never rained before, in the history of the island." Grenada is 300 miles from St. Kitts.

I take data of another occurrence, from the Dominican, and The People, published at Roseau, Dominica, B. W. I. About 11 o'clock, morning of January 4th, the town of · Roseau was bumped by midnight. People in the streets were attacked by darkness. People in houses heard the smash of their window panes. Night fell so heavily that it broke roofs. It was a daytime night of falling mud. With the mud came a deluge.

The River Roseau rose, and there was a conflict: The river, armed with the detachables of an island, held up shields of mules, and pierced the savage darkness with spears of goats. Long lines of these things it flung through the black streets of Roseau. In the Boiling Lakes District of Dominica, there had been an eruption of mud, at the time of the deluge, which was like the fall of water upon St. Kitts, eight: days later. There had, in recorded time, never been an eruption here before. Three months before, there had been, in another part of the West Indies, a catastrophe like that of St. Kitts. Upon Oct. 10, 1879, a deluge fell upon the islands of Jamaica, and drowned one hundred of the inhabitants {London Times, Nov. 8, 1879). A flood that slid out from this island was surfaced with jungles-tangles of mahogany logs, trees, and bushes; brambled with the horns of goats and cattle; hung with a moss of the fleece of sheep. Incoming vessels plowed furrows, as if in a passing cultivation of one of the rankest-luxuriances that ever vegetated upon an ocean. Passengers looked at tangles of trees and bodies, as if at picture puzzles.

·In foliage, they saw faces.

For months, there had been, in the Provinces of Murcia and Alicante, Spain, a drought so severe that inhabitants had been driven into emigration to Algeria. Whether we think of this drought and the prayers of the people as having relation or not. there came a downpour that was as intense as the necessities. See London Times, Oct. 20, 1879. Upon October 14th, floods poured upon these parched provinces. Perhaps it was response to the prayers of the people. Five villages were destroyed. Fifteen hundred persons perished. Virtually in the sable zone with Spain and the West Indies (U. S. Colombia) a deluge fell, in December. The River Cauca rose beyond all former high water marks, so suddenly that people were trapped in their houses. This was upon December 19th.

The next day, the earth started quaking in Salvador, near Lake Ilopanga. This lake was the crater of what was supposed to be an extinct volcano. I take data from the Panama Daily Star and Herald, Feb. 10, 1880.

Upon the 31st of December-four days before the occurrences in the island of Dominica-the quaked in Salvador, and from the middle of Lake Ilopanga emerged a rocky formation. Water fell from the sky, in bulks that gouged gullies. Gullies writhed in the quaking· ground. The inhabitants who cried to the heavens prayed to Epilepsy. Mud was falling upon the convulsions. A volcanic island was rising in Lake Ilopanga, displacing the water, in streams that writhed from it violently. Rise of a form that filled the lake-it shook out black torrents-head of a Gorgon, shaggy with snakes.

Beginning upon October 10th, and continuing until the occurrence at St. Kitts, deluge after deluge came down to one zone around this earth-or a flight of, lakes was cast from a constellational reservoir, which was revolving and discharging around a zone of this earth. In the minds of most of us, this could not be. We have been taught to look up at the revolving stars, and to see and to think that they do not revolve.

Our data are of the slaughters of people, who, by fish mongerish explanations, have been held back from an under standing of an irrigational system: of their emotions, and of· the elementary emotions of lands. There's a hope in a mind, and it turns to despair-or there's a fertile region in the materials of a South American country-and an unsuspected volcano chars it to a woe of leafless trees. Plains and the promise of crops that are shining in sunlight-plains crack into disappointments, into which fall expectations. An island appears in the ocean, and after a while young palms feel upward. There's a convulsive relapse, and subliminal filth, from the bottom of the ocean, plasters the little aspirations. Quaking lands have clasped their fields, and have wrung their forests.

Each catastrophe has been explained by the metaphysical scientists, as a thing in itself. Scientists are contractions of metaphysicians, in their local searches for completeness, and in their statements that, except for infinitesimal errors, plus or minus, completenesses have been found. I can accept that there may be Superphenomenal Completeness, but not that there can be phenomenal completenesses. It may be that the widespread thought that there is God, or Allness, is only an extension of the deceiving process by which to an ex planation of a swarm of lady birds, or to a fall of water at St. Kitts, is given a guise of completeness-or it may be the other way around-or that there is a Wholeness-perhaps one of countless Wholenesses, in the cosmos-and that at tempting completenesses and attempting concepts of completenesses are localizing consciousness of an all-inclusive state, or being-so far as its own phenomena are concerned-that is Complete.

There have been showers of ponds. From blue skies there have been shafts of water, golden in sunshine. Reflections from stars have fluted sudden, dark, watery columns. There have been violent temples of water-colonnades of shafts, revealed against darkness by lightning-foaming fa des as white as marble. Nights have been caves, roofed with vast, fluent stalactites.

These are sprinkles. March, 1913.

The meteorologists study meteorologically. The meteor- ologists were surprised.

March 23, 1913-250,000 persons driven from their homes-torrents falling, rivers rising, in Ohio. The Hoods at Dayton, Ohio, were especially disastrous.

Traffics of bodies, in the watery streets of Dayton. The wind whistles, arid holds up a cab. They stop. Night-and the running streets are hustling bodies-but, coming, is worse than the sights of former beings, who never got anywhere in life, and are still hurrying. The wreck of a trolley car speeds down an avenue-down a side street rushes a dead man. Let him catch the car, and he'll get about where all bis lifetime he got catching other cars. A final dispatch from Dayton-"Dayton in total darkness."

March 23, 24, 25-a watery sky sat on the Adirondack Mountains. It began to slide. It ripped its slants on a peak, and the tops of lamp posts disappeared in the streets of Troy and Albany. Literary event, at Paterson, N. J.-something that was called "a great cloudburst" grabbed a fac tory chimney, and on a ruled page of streets scrawled a messy message. With the guts of horses and other obscenities, it put in popularizing touches. The list of dead, in Columbus, Ohio, would probably reach a thousand. Connecticut River rising rapidly. Delaware River, at Trenton, N. J., 14 feet above normal.

. March 26th-in Parkersburg, West Virginia, people who called on their neighbors, rowed boats to second story windows. If they had in their cellars what they have nowadays, there was much demand for divers. New lakes in Vermont, and the State of Indiana was an inland sea. "Farmers caught napping." Surprises everywhere: napping everywhere. Wherever Science was, there was a swipe at a sleep. Floods in Wisconsin, floods and destruction in Illinois and Missouri.

March 27th-see the New York Tribune, of the 28th that the Weather Bureau was issuing storm warnings.

The professional wisemen were not heard from, before this deluge. Some of us would like to know what they had to say, afterward. They said it, in the Monthly Weather Review, April, 1913.

The story, is told "completely." The story is told, as if there had been exceptional rains, only in Ohio and four neighboring States. Reading this account, one thinks-as one should think-of considerable, or of extraordinary, rain, in one smallish region, and of its derivation from other parts of this earth, where unusual sunshine had brought about unusual evaporations.

Canada-and it was not here that the sun was shining. Waters falling and freezing. in Canada, loading trees and telegraph wires with ice-power houses flooded, and towns in darkness-crashes of trees, heavy with ice. California was drenched. Torrents falling in Washington and Oregon. Un-precedented snow in Texas, New Mexico, and Oklahoma Alabama deluged-floods in Florida.

"Ohio and four neighboring States."

Downpours in France and in other parts of Europe. Spain-seems that, near Valencia, one of these nights, there was a rotten theatrical performance. Such a fall of big hailstones that a train was stalled-vast tragedian, in a black cloak, posing on the funnel of an engine-car windows that were foot lights-and disapproval was expressing with the looks of millions of pigeons' eggs. Anyway, near Valencia, a fall of hail, three feet deep, stopped trains. Just where was all that sunshine?

South Africa-moving pictures of the low degree of the now old-fashioned "serials." Something staged Clutching Hands. There were watery grabs from the sky, at Colesburg, Murraysburg, and Prieska. The volume of one of these bulks equaled one-tenth of the total rainfall in South Africa, in one year.

Snow, two months before its season, was falling in the Andes-floods in Paraguay, and people spreading in panics Government vessels carrying supplies to homeless, starving people-River Uruguay rising rapidly.

Heavy rains in the Fiji Islands.

The rains in Tasmania, during the month of March, were 26 points above the average.

Upon the first day of the floods in "Ohio and four neigh boring States" (March 22nd) began a series of terrific thunderstorms in Australia. There was a "rain blizzard" in New South Wales. In Queensland, all mails were delayed by floods.

New Zealand.

Wellington Evening Post, March 31-"The greatest disaster in the history of the Colony!"

Where there had been sluggish rivers, bodies of countless sheep tossed in woolly furies. Maybe there is a vast, old being named God, and reported str ds of tossing sheep were glimpses of his whiskers, in one of those wraths of his. In the towns, there were fantastic savageries. Wherever the floods had been before, it looks as if they had been to college. One of them rioted through the streets of Gore, having broken down store windows. It roystered with the bodies of animals, wrapped in lace curtains, silks, and rib bons. Down the Matura River sounded a torrent of "terrible cries." It was a rush of drowning cattle. It was a delirium of brandishing horns, upon which invisible collegians were blowing a fanfare.

"Ohio and four neighboring States."

The clip of Paraguay, and the bob of New Zealand: the snip of South Africa, and the shearing of everything else that did not fit in with a theory. Whoever said that the pen is mightier than some-thing else, overlooked the mightiest of all, and that's the scissors.

Wherever all this water was coming from, the full account is of North America and four neighbor-ing Continents.

Peaches flying from orchards, in the winds of New Zealand -icicles clattering in the streets of Mon-treal. The dripping palms of Paraguay and the pine trees of Oregon were mounds of snow. At night this earth was a black constellation; sounding with panics. I can think of the origin of the ocean that fell upon it in not less than constellational terms. Perhaps Orion or Taurus went dry.

U a place, say in China, greatly needs water, and if there be stores of water, somewhere else, in one organism, I can think of relations of requital, as I think of need and response in any lesser organism, or suborganism.

Need of a camel-and storages-and reliefs. Hibernating bear-and supplies from his storages.

At a meeting of the Royal Geographical Society, Dec. 11, 1922, Sir Francis Younghusband told of a drought, in August, 1906, in Western China. The chief magistrate of Chung king prayed for rain. He put more fervor into it. Then he prayed prodigiously for rain. It began to rain. Then some thing that was called "a waterspout" fell from the sky. Many of the inhabitants were drowned.

In the organic sense, I conceive of people and forests and dwindling lakes all expressing a need, and finally compelling an answer. By "prayers" I mean utterances by parched mouths, and also the rustlings of dried leaves and grasses. It seems that there have been responses. There are two explana-tions. One is that it is the mercy of God. For an opinion here, see the data. The other is that it is an Organism that is maintaining itself.

The British Government has engineered magnificently for water supply in Egypt. It might have n better to plant persuasive trees and clergymen in Egypt. But clergymen are notoriously eloquent, and I think that preferable would be Jess excitable tipsters to God, who could convey the idea of moderation.

In one year the fall of rain, at Norfolk. England, is about 29 inches. In Symons' Meteorological Mag-azine, 1889, p. 101, Mr. Symons told of this fall of water of 29 inches in a year, and then told of vol-umes of water to depths of from 20 to 24 inches that had fallen, from May 25th to the 28th, 1889, in New South Wales, and of a greater deluge-34 inches that, from the 29th to the 30th, had devastated Hongkong. Mr. Symons called attention to these two bursts from the heavens, thousands of miles apart, saying that they might, or might not, be a coincidence, but that he left it to others to theorize. I point out that a professional meteorologist thought the ·occurrence of only two deluges, about the same time, but far apart, remarkable, or difficult to explain in terms of terrestrial meteorology.

It was left a Jong time to others.

However, when I was due to appear, I appeared, perhaps right on scheduled time; and I got Aus-tralian newspapers. The Sydney newspapers told of the soak in New South Wales. I learned that all the rest of Australia was left to others-or was left, waiting for me to appear, right on scheduled time, most likely. Not rain, but columns of water fell near the town of Avoca, Victoria, and, in the Mel-bourne Argus, the way of accounting for them was to say that "a waterspout" had burst here. There were wide floods in Tas mania. Fields turned to blanks that were then lumpy with rabbits.

There had been drought in Australia, and floods were a relief to a necessity, but the greater downpour in China interests us more in conditions in China.

It was a time of dirt drought and extremest famine in China. Homeward Mail, June 4-that, in some of the more cannibalistic regions, sales of women and children were common. It is said to be almost impossible for anybody to devour his own child. Parents exchanged children.

Down upon monstrous need came relief that was enormous. At Hongkong, houses collapsed under a smash of alleviation. A fury of mercy tore up almost every street in the Colony. The people had prayed for rain. They got it. Godness so loved Hongkong that in the town's morgue it stretched out sixteen of the inhabitants. At Canton, every pietist proclaimed the efficacy of prayer, and I think he was right about that: but the problem is to tone down all this efficacy. If we will personify what I consider an organism, what he, or more likely she, has not, is any conception of moderation. The rise of the river, at Canton, indicated that up country there had been catastrophic efficacy, At Canton efficacy was so extreme that for months the people were rebuilding.

Show me a starving man-I pay no attention. Show me the starving man-I can't be bothered. Show me the starving man; on the point of dying-I grab up groceries and I jump on him. I cram bread down his mouth, and stuff his eyes and his ears with potatoes. I rip open his lips to hammer down more food, and bung in his teeth, the better to stuff him. The explanation-it is the god-like in me.

Now, in a Library, we put in calls for the world's news papers. Not a hint have we had that there is anything else nothing in scientific publications of the period-not another word from Mr. Symons-but there is an implement that is mightier than the pen-and we are led on to one of our attempted correlations, by our experiences with it.

Germany-there was a drought so severe. that there were public prayers for rain. Something that was called " a waterspout" fell from the sky, and people who did not get all the details went to church about it. Liverpool Echo, May 20- one hundred persons perished.

At the same time, there were public rejoicings in Smyrna, where was staged another assuasive tragedy.

Drought in Russia. Straits Times, June 6-droughts ended by downpours in Bengal and Java. In Kashmir and in the Punjab, violent thunderstorms and earthquakes occurred together (Calcutta Statesman, June I and 3). In Turkey, there would have been extremist distress, but about the first of June, amidst woe and thanksgiving, destructive salvations demonstrated efficacy, and for a week kept on spreading. joy and misery. Levant Herald, June 4-earthquakes preceded deluges, and then continued with them.

In conventional meteorology, no relation between droughts and exceptional rains is admitted. Our data are of widespread droughts and enormous flows of water. There are two little, narrow strips of views on the margins of our moving pictures. On one side-that there is a beneficent Cod. On the other side-that there isn't anything. And every one of us who has paid any attention to the annals of controversies knows that such oppositions usually give in to an intermediacy. May, 1889-widely this earth was in need-widely waters were coming from somewhere. Now-in Organic terms-I am telling of what seems to me to be Functional Teleportation, or enormous manifestations of that which is sometimes, say in Oklahoma, a little drip over a tree.

Volcanic eruptions upon this earth, at times of deluges-. and maybe, in a land of the stars, there was an eruption, in May, 1889. In France, May 31st, there was one of the singularly lurid sunsets that are known as "afterglows" and that appear after volcanic eruptions. There was no known volcanic eruption upon this earth, from which a discharge could have gone to the sky of France. It may have come from a volcanic eruption somewhere else. All suggestiveness is that it came to this earth over no such distance as millions of miles.

Other discharges, maybe-red rain coming down from the sky, at Cardiff, Wales (Cardiff Western Mail, May 26). Red dust falling upon the island of Hyeres, off the coast of France, in the Mediter-ranean-see the Levant Herald, May 29. St. Louis Gobe-Democrat, May 30-an unknown sub stance that for several hours had fallen from the sky crystalline particles, some pink, and some white. Que-bec Daily Mercury, May 25-a fine dust that had the appearance of a snowstorm, falling in Dakota. Monstrous festivities in Greece-a land that was bedecked with assassinations. Its rivers were garlands-vast twist of vine, budded with the bodies of cattle.

The Malay States gulped. The mines of Kamunting were suctions, into which flowed floods (Penang Gazette, May 24). The Bahama Islands were thirsts-droughts and loss of crops-then huge swigs from the sky. Other West Indian islands went on Gargantuan sprees-and I'll end up a Prohibitionist. Or-gies in Greece, and more or less everywhere else-this earth went drunk of water. I've experimented-try autosuggestion-you can get a pretty fair little souse from any faucet. Tangier-"great suffering from the drought"- abundant rains, about June 1st. Drought in British Honduras, and heavy rains upon the 1st and 2nd of June. Tremendous downpours described in the newspaper published upon the island of St. Helena. Earthquake at Jackson, California the next day a gush from the sky broke down a dam. I'm on a spree, myself-Library attendants wheeling me stacks of this earth's newspa-pers. Island of Cyprus-a flop from the sky, and the river Pedias went up with a rush from which people at Nicosia narrowly escaped. Torrents in Ceylon, June 4th-a drought of many weeks broken by rains in Cuba. Drought in Mexico-and out of the heavens came a Jack the Ripper. Tom planta-tions and mutilated cities-rise of the river, at Huezutla-when it subsided the streets were strewn with corpses.

In England, Mr. Symons expressed astonishment, because there had been two deluges.

Deluge and falls of lumps of ice, throughout England. France deluged. Water dropped from the sky, at Lausanne, Switzerland, flooding some of the streets five feet deep. It was not rain. There were falling columns of water from what was thought to be a waterspout. The most striking of the state-ments is that bulks dropped. One of them was watched. Or some kind of a vast, vaporous cow sailed over a town, and people looked up her bag of water. Something that was described as "a large body of water" was seen at Coburg. Ontario. It crossed the town, holding its bag-like formation. Two miles away, it dropped. ·It splashed rivers that broke down all dams between Coburg and Lake Ontario. In the Toronto Globe, June 3, this falling bulk is called "a waterspout. Fall of a similar bulk, in Switzer-land-crops and houses and bridges mixing down a valley, at Sargans. Fall of a bulk, at Reichenbach, Saxony. "It was a water spout" (London Times, June 6).

This time the fishmonger is a waterspout.

Spain pounded by falling waters: Madrid flooded: many buildings damaged by a violent hailstorm. Deluges in China continuing. Deluges in Australia continuing. Floods in Argentina: people of Ay-acuchio driven from their homes: sudden rise of the river, at Buenos Aires. In the South Ameri-can Journal, of this period are accounts of tremendous down pours and devastations in Brazil and

Uruguay.

One of these bodies of water that were not rain fell at Chetnole, Dorsetshire, England. The people, hearing crashes, looked up at a hill, and saw it frilled with billows. Watery ruffs, from eight to ten feet high, heaved on the hill. The village was tossed in a surf. "The cause of this remarkable occurrence was for some time unknown but it has now been ascertained that a waterspout burst on Batcombe Hill." So wrote Mr. Symons, in whose brains there was no more consciousness of all that was going on in the world about him than there was in any other pair of scissors.

It was not ascertained that a waterspout had burst on Batcombe Hill. No waterspout was seen. What was ascertained was that columns of water of unknown origin had fallen high on the Hill, gouging holes, some of them eight or nine feet deep. Though Mr. Symons gave the waterspout explanation, it did occur to him to note that-there was no statement that the water was salty-

These bulks of water, and their pendent columns-that they were waterspouts-

Or that Slaughter had lain with Life, and that murderous mothers had slung off their udders, from which this earth drank through teats that were cataracts.

Wherever the deluges were coming from, I note that, as with phenomena of March, 1913, unseasonable snow fell. Here it was about the first of June, and snow was falling in Michigan. The suggestion is that this was not a crystallization in the summer sky of Michigan, but an effect of the intenser coldness of outer regions, upon water that had· come to this earth from storages on a planet, or from a reservoir in Star land. Note back to mention of falls of lumps of ice in England. Wherever the deluges were coming from, meteors, too, were coming. If we can think that falls of water and falls of meteors were related, we have reinforcement to our ex pression that water was coming-to this needful earth from somewhere else. Five remarkable meteors are told of, in the Monthly Weather Review. In the New York Sun, May 30,- is an account of a meteor that exploded in the sky of Putnam County, Florida, and was heard 15 miles around. In Madras, India, where the drought was "very grave," an extraordinary meteor was seen, night of June 4th (Madras Mail, June 26). In South Africa, where the drought was so extreme that a herd of buffaloes had been driven to a pool within five miles of the town of Uitenhage, a meteor exploded, with detonations that were heard in a line 40 miles long (Cape Argus May 28). May 22nd-great, detonating meteor, at Otranto, Italy. The meteor that was seen in England and Ireland, May 29th, is told of in Nature, 40-174. For records of three other great meteors, see Nature and Cosmo,. There was a spectacular occurrence at Dunedin, New Zealand, early in the morning of May 27th (Otago Witness, June 6). Rumbling sounds-a shock-illumination of the sky-exploding meteor.

In some parts of the United States, there had been extreme need for water. In the New Orleans Daily Picayune are accounts of the "gloomy outlook for crops" in six of the Southern States. About twenty reports upon this drought were published in the Monthly Weather Review.

Rushes of violent mercies-they flooded the south and smashed the north-crash of a dam, at Littleton, New Hamp shir&-busted dam near Laurel, Pa.-

May, 1889-and Science and Religion-

It is my expression that the two outstanding blessings, benefits, or "gifts of God" to humanity, are Science and Religion. I deduce this-or that the annals of both are such trails of slaughter, deception, exploitation, and hypocrisy that they must be of enormous good to balance with their appalling evils-

Or the craze of medical science for the vermiform appendix. That played out. Now everybody who can pay for it is. losing his tonsils. Newspaper headings-"Family of eight relieved of their tonsils"- "Save your pets-dogs and cats endangered by their tonsils."

Concentrate in one place this bloody fad, or scientific "racket," and there would be a fury like that at Andover, N. Y., in May, 1889-

A bulk of water, foaming as white as a surgeon-it jabbed a bolt of lightning into Andover. It operated on farms, and cut off their inhabitants. Trained clouds stood around, and handed out more bolts of. lightning. A dam broke, and a township writhed upon its field of operations. Another dam broke-but the operations were successes, and, if there was much-destruction, that was because of a complication of other causes.-

May 31st-Johnstown, Pa.-

If I can't think of massacre apart from devotions, I think that a lake ran mad with religious mania. It rushed down a valley, and, if I'm right about this, it bore on its crest, the most appalling of all symbols-the mast of a ship that was crossed by a telegraph pole. In a pogrom against houses, it clubbed out their occupants, with bridges. It impaled homes upon the steeples of churches. Its watery Cossacks, mounted on billows, flogged factories. And then, along the slopes of the Conemaugh Valley, it told its beads with strings of corpses.

Earthwide droughts-prayers to many gods-something vouchsafed catastrophes-

That from somewhere else in existence, vast volumes of water were sent to this arid earth, or were organically tele ported-

Or that, by coincidence, and unseen, waterspout after waterspout rose from the Atlantic, and rose from the Pacific: &om the Indian Ocean, the Southern Ocean, and from the Mediterranean; from the Gulf of Mexico, the English Channel Lake Ontario-or that such an extension of such fish mongering is a brutalization of conveniences-

Or that from somewhere in a starry· shell that is not enormously far away from this earth, more than a Mississippi streamed to this needful earth. and forked the disasters of its beneficence from Australia to Canada.

15,000 persons were drowned at Johnstown. Chicago Tribune, June 10, 1889-"The people of Johnstown have lost all faith in Providence. Many have thrown away their Bibles, and since the disaster have openly burned them."

By the providential, I mean the organically provided for. By God, I mean an automatic Jehovah.

Part III

XXIV

EMISSIONS of arms-the bubbling of faces, at crevices-fire and smoke and a lava of naked beings. Out from a crater, discharges of bare bodies boiled into fantastic formations-

At five o'clock, morning of Dec. 28, 1908-violent shocks in Sicily. The city of Messina fell in a heap, which caught fire. It is the custom of Sicilians to sleep without nightclothes, and from this crater of blazing wreckage came an eruption of naked beings. Thick clouds of them scudded into thin vapors.

The earth quaked, at Messina, and torrential rains fell. According to Nature, Dec. 31, 1908, a fall of meteorites had been reported in Spain, a few days before the quake. According to the wisemen of our more or less savage tribes, the deluge at Messina, at the time of this quake, fell only by coincidence. No wiseman would mention the fall of meteorites, as having any relation.

There were, at the time, world-wide disturbances, or rather, disturbances, along a zone of this earth-Asia Minor, Greece, Sicily, Spain, Canary Islands, Mexico. But all wise men who wrote upon this subject clipped off everything else, and wrote that there had been a subsidence of land in Sicily. It is the same old local explanation. Scientists and priests are unlike in some respects, but they are about equally parochial.

Dec. 3, 1887-from a plinth of ruins, an obelisk of woe sounded to the sky.

It was at Roggiano, Italy. 900 houses were thrown down by an earthquake. The wail that went up from the nuns continued long before individual cries could be distinguished. Then the column of woe shattered into screams and prayers. The survivors said that they had seen fires in the heavens. In Cosmos, n. s., 69-422, we are told that Prof. Agamennone had investigated these reported celestial blazes. But they were new lights upon old explanations. A blazing sky could have nothing to do with a local, geological disturbance. The orthodox explanation was that a stratum of rocks had slip ped. What could the slip of rocks have to do with sky-fires? We are told that the Professor had reduced all alleged witnesses of the blaze in the heavens to one, who had told about it, "with little seriousness." What bad suggested levity to him, as to scenes of ruination and slaughter, was not enquired into: but the story is recorded as a jest, and it may be all the more subtle, because the fun of it is not obvious.

About 6 A. M., Feb. 23, 1887; at Genoa, Italy, burst a dam of conventional securities. There was a flood of human beings. An earthquake cast thousands of people into the streets. The sky was afire. There was a pour to get out of town. It was a rush in a glare. If, at the time of a forest fire, a dam should burst, thousands of logs, leaping red in the glare, would be like this torrent of human forms under a fiery sky. In other places along the Riviera, the quake was severe. At other places was made this statement that orthodox science will not admit-that the sky was afire. See Pop. Sci. News, 21-58. It will not be admitted, or it is said to be merely a coincidence. See L'Astronomie, 1887, p. 137-that at Apt (Vaucluse) a fiery appearance bad been seen. and that then had come a great light, like a Bengal fire without doubt coincidences."

16th of August, 1906-and suddenly people, living along the road to Valparaiso, Chile, lost sight of the city. There had come •a terrible darkness." With it came an earthquake. The splitting of ground, and the roar of falling houses-intensest darkness-and then a voice in this chaos. It was a scream. People along the road heard it approaching. Chile lit up. Under a flaming sky, the people of Valparaiso were running from the smashing city-people as red as the flames, under the glare in the heavens: screaming and falling, and leaping over the bodies of the fallen-an eruption of spurting rms that leaped and were extinguished. This reddened gush from Valparaiso-rising, falling shapes-brief faces and momentary arms-it was like looking at vast flames and imagining that spurts of them were really living beings.

In Nature, 90-550, it is said that 136 reports upon illuminations 'in the sky, at Valparaiso, had been examined by Count de Ballore, the seismologist. At one stroke, he bobbed off 98 of them, saying that they were indefinite. He said that the remaining 38 reports were more or less explicit, but came from a region where at the time, a deluge was falling. He clipped these, too. For a wonder there was an objection: a writer in the Scientific American, 107-67, pointed out that De Ballore so dismissed the subject, without enquiring into the possibility that the quake and the deluge were related.

Had he admitted the possibility of relationship, dogma would have slipped upon dogma, and upon the face of this earth there would have been a subsidence of some ignorance. "The lights that were seen in the sky; said De Ballore, "were very likely only searchlights from warships."

"The whole sky seemed afire" (Scientific American, 106- 464). In Symons' Met. Mag., 41-226, William Gaw, of Santiago, describing the blazing heavens, writes that it seemed as ff the sober laws of physics had revolted.

"Or," said De Ballore, "the people may have seen lights from tram cars."

It does not matter how preposterous some of my own notions are going to seem. They cannot be more out of accordance with events this earth than is such an attribution -of the blazing sky of a nation to searchlights or to lamps in tram cars. If I should write that the stars are probably between forty and fifty miles away, I'd be not much more of a trimmer of circumstances than is such a barber, whose clips are said to be scientific. Maybe they are scientific. Though, mostly, barbers are artists, some of them do consider themselves scientific.

Upon July 11th, 1856, the sun rose red in the Caucasus. See Ll,oyd's Weekly Newspaper (London), Sept. 21, 1856. At five o'clock in the afternoon, at places where the sun was still shining red, there was an earthquake that destroyed 300 houses. There was another quake, upon the 23rd of July. Two days later, black water fell from the sky, in Ireland (News of the World, Aug. 10, 1856).

And what has any part of that to do with any other part of that? If a red-haired girl, or a red shirt on a clothesline, had been noted here, there would be, according to orthodox science, no more relation with earthquakes than there could be between a red sun and an earthquake. Black water falling in Ireland-somebody spilling in Kansas.

The moon turned green.

For two observations upon a green moon that was seen at a place where an earthquake was going to occur, see the Englishman (Calcutta), July 14 and 21, 1897. One of the observations was six days before, and the other one day before the quake in Assam, June 12, 1897. It was a time of drought and famine, in India.

The seismologist knows of no relation between a green moon, or a red sun, and an earthquake, but the vulcanologist knows of many instances in which the moon and the sun have been so colored by the volcanic dusts and smokes that are known as "dry fogs." The look is that "dry fogs," from a vol-canic eruption, came to the sky of India, one of them six days before, and the other one day before, a catastrophe.

The mystery is this:

If there had been a volcanic eruption somewhere else, why not volcanic appearances in Italy, or Patagonia, or California-why at this place where an earthquake was going to occur?

Coincidence.

Upon the 11th of June, in Upper Assam, where, upon the 12th, the center of the earthquake was going to be, torrents fell suddenly from the sky. A correspondent to the Englishman, July 14, writes that this deluge was of a monstrousness that exceeded that of any other downpour that he had ever seen in Assam, or anywhere else.

At 5:15 P. M., 12th of June, there was a sight at Shillong that would be a marvel to the more innocent of the text-book writers. I tell so much of clipping and bobbing and shearing, but also there may be considerable innocence. Not a cloud in the sky-out of clear, blue vacancy, dumped a lake. This drop of a bulk of water, or transportation, or teleportation, of it, was at the time of one of the most catastrophic of earthquakes, centering farther north in Assam.

This earthquake was an earthstorm. Hills were waves, and houses cast adrift were wrecked on them. Out into fields stormed people from villages, and long strings of them, in white summer garments, were lines of surf on the earthwaves. Breakers of them spumed with infants. In a human storm billows of people crashed against islands of cattle. It is not only in meteorology that there are meteorological occurrences. The convulsions were so violent that there was scene-shifting. When the people recovered and looked around, it was at landscapes, changed as if a curtain had gone down and then up, between acts of this drama. They saw fields, lakes, and roads that, in the lay of the land, before the quake, had been bidden. It is not only in playhouses that there are theatrical performances. It is not exclusively anywhere where anything is,. if ours is one organic existence, in which all things are continuous.

There were more deluges that will not fit into conventional explanations. Allahabad Pioneer, June 23, 1897- extremist droughts-the quakes -enormous falls of water.

There are data for thinking that somewhere there was a volcanic eruption. Another datum is that, at Calcutta, after the earthquake, there was an "afterglow." "Afterglows" are exceptional sunsets, sometimes of an auroral appearance, which are reflections of sunlight &om volcanic dust high in the sky, continuing to be seen an hour or so later than ordinary sunsets. Friend of India, June 15- °"The entire west was a glory of deepest purple, and the colors did not fade out, until an hour after darkness is usually complete."

Something else that I note is that in many places in Assam. the ground was incipiently volcanic, during the earthquake. Countless small craters appeared and threw out ashes.

Considering the volcanic and the incipiently volcanic, I think of a relation between the catastrophe in Assam and a volcanic eruption somewhere else.'

But there is findable no record of a volcanic eruption upon
this earth to which could be attributed effects that we have noted.

I point out again that, if there were a volcanic eruption in some part of our existence, external to this earth, or upon this earth, it would, unless a special relation be thought of, be as likely to cause an "afterglow" in England or South Africa, as in India. The suggestion is that somewhere, external to this earth, if in terrestrial terms there is no explanation, there was a volcanic eruption, and that the earthquake in India was a response to it, and that bulks of water and other discharges came from somewhere else exclusively to a part of this earth that was responsively, or functionally, quaking, because a teleportative current of some kind, very likely electric, existed between the two centers of disturbances.

Upon the 25th of June, dust fell from the sky, near Calcutta (The Englishman, July 3). In the issue of this newspaper, of July 14th, a meteorologist, employed in the Calcutta Observatory, described "a most peculiar mist," like volcanic smoke, which bad been seen in the earthquake regions. In his opinion it was "cosmic dust; or dust that had fallen to this earth from outer space. He said nothing of possible relationship with the earthquakes. He would probably have called it "mere coincidence." Then he told of a fall of mud, upon the 27th of June, at Thurgrain (Midnapur). There was a fall of mud, in the Jessore District of Bengal, night of June 29th. "It fell horn a cloudless sky, while the stars were shining" (Madras Mall, July 8).

Suppose it were "cosmic dust." Suppose with the conventionalists that this earth is a swiftly moving planet that had overtaken a cloud of "cosmic dust." in outer space. In one minute, this earth would be more than one thousand miles away from this point of contact, by orbital motion, and would tum away axially.

But other falls of dust came upon India, while the shocks were continuing, as if settling down from an eruption some where else, to a world that was not speeding away orbitally, and to a point that was not turning away by daily rotation.

Five days after the first fall of dust, "a substance resembling mud" fell at Ghattal (Friend of India, July 14). For descriptions of just such a "dry fog," as has often been seen in Italy, after an eruption of Vesuvius, see the Madras Mail, July 5, and the Friend of India, July 14- "a perpetual haze on the horizon, all around," "sky covered with thick layers of dust, resembling a foggy atmosphere." About the first of July, mud fell at Hetamphore (Beerbhoom) according to the Friend of India, July 14.

I list these falls of dust and mud, but to them I do not give the importance that I give to the phenomena that preceded this earthquake. I have come upon nobody's statement that they were of volcanic material. But it may be that there were other precipitations, and that they were of a substance that is unknown upon this earth.

In the Englishman (Calcutta), July 7, a correspondent wrote that, several days before, at Khurdah, there had been a shower at night, and that the air became filled with the perfume of sandalwood. The next morning everything was found covered with "a colored matter, which emitted the scent of sandalwood. About the same time, somebody else wrote to the Madras Mall (July 8) that, at Nadia, there had been a fall from the sky, of a substance "more or less resembling the sandal used by natives in worshiping their gods.

The moon turns green before an earthquake. Torrential rains precede an earthquake.

We have only begun listing phenomena' that appear before catastrophes. They are interpretable as warnings. Clipped from events, by barber-shop science.

There was an investigation of phenomena in Assam. It was scientific, in the sense that the tonsorial may be the scientific. Dr. Oldham enormously reduced a catastrophe to manageable dimensions. He lathered it with the soap of his explanations, and shaved it clean of all unconventional de tails. This treatment of "Next!" to catastrophes is as satisfactorily beautifying, to neat, little minds, as are some of the marcel waves that astronomers have ironed into tousled circumstances. For a review of Dr. Oldham's report, see Nature, 62-305. There is no mention of anything that was seen in the sky, nor of anything that fell from the sky, nor of occurrences anywhere else. Dr. Charles Davison, in A Study of Recent Earthquakes, gives 57 pages ·to his account of this catastrophe, and be, too, mentions nothing that was seen in the sky, or that fell from the sky. He mentions no simultaneous phenomena anywhere else. It is a neat and well-trimmed account, but there's a smell that I identify as too much bay rum.

Simultaneous phenomena that always are left out of a conventionalist's account of an earthquake-one of the most violent convulsions ever known in Mexico, while the ground in India was quaking. There was a glare in the sky, and the Mexicans thought that the glare was volcanic. If so, no active volcano in Mexico could be found (New Orleans Daily Picayune, June 22). Deluges fell upon this quaking land. One of the falls of water, upon a Mexican town, drowning some of the inhabitants, is told of, in the San Francisco Chronicle, June 17.

In all this part of our job, our opposition is not so much denial of data, as assertions that the occurrences in which we see relationship were only coincidences. If I ever accept any such explanation, I shall be driven into extending it to everything. We'll have a theory that in our existence there is nothing but coincidence: and, according to my experience with theorists, well develop this theory somewhat reason-ably. Chemical reactions, supposed to be well-known and accounted for, do not invariably work out, as, according to formula, they should work out. Failures are attributed to impurities in chemicals, but perhaps it is only by persistent coincidence, like that of glares so often occurring at times of earthquakes, that water appears when oxygen and hydro gen unite. Meteors frequently fall to this earth during earthquakes, but that may be only by coincidence; just as off-springs so often appear after marriage-indicating nothing exclusively of relationships, inasmuch as we have heard of cases of alleged independent reproduction. Let the feminists become only a little more fanatical, and they will probably publish lists of instances of female independence. It is either that our data are not of coincidences, or that everything's a coincidence.

As to some deluges, at times of earthquakes, there is no assertion of coincidence, and there is no mystery. There's an earthquake, and water falls from the sky. Theo it is learned that a volcano-one of this earth's volcanoes-had been in eruption, and that, responsively to it, the earth had quaked, and that volumes of water, some of them black, and some of them not discolored, had been discharged by this volcano, falling in bulks, or falling in torrential rains upon the quaking ground. Sometimes the sky darkens during earthquakes, and there is no assertion of coincidence, and there is no mystery. Upon March 11, 1875, for instance, a vast, black cloud appeared at Guadalajara, Mexico. There was an earthquake. See L'Annee Scientifique, 1876-322. In this instance, the darkened sky at the time of an earthquake was explained, because it was learned that both phenomena were effects of an eruption of the volcano Caborucuco. There have been unmysterious showers of meteors, or of fireballs that looked like meteors, at times of earthquakes. There were eruptions upon this earth, and the fireballs, or meteors, came from them. There were especially spectacular showers of volcanic bombs that looked like meteors, or that were meteors, during the eruptions in Java, August, 1883; New Zealand, June, 1886; West Indies, May 1902.

But our data are of such phenomena in the sky, during earthquakes,, at times when no terrestrial volcano that could have had such effects was active.

So far we have not correlated with anything that could be considered a volcanic eruption anywhere in regions ex ternal to this earth. Now we are called upon, not only for data seemingly of volcanic eruptions in a nearby starry shell around this earth, but for data that may be regarded as observations upon celestial volcanoes in action.

XXV

WITH a surf and a g]are, this earth quaked a picture--

Or, in the monistic sense, there was, in Peru, a catastrophe that was a hideous and magnificent emotion. It is likely that there's a wound in a brain, at a time of intensest excitement Red of the writhing earth, and red of the heaving ocean and, in between, a crimson gash of surf, slashed from Ecuador to Chile –

Or so was visualized a rage, by super-introspection.

According to the midgets of orthodoxy, such a picture cannot be accepted. See the little De Ballore school of criticism. But quakes that were pictures by a very independent artistry –

Snow that was white on the peaks of mountains-cataclysm-peaks struck off-avalanches of snow, glaring red, gushing in jugular spouts from the decapitations. Glints from the fiery sky-upon land and sea, tossing houses and ships were spangles. Forests lashed with whips of fire, from which shot out sparks that were birds and running animals.

Aug. 13, 1868-people in Peru, rushing from their falling houses, stumbling in violations of streets, seeing the heavens afire, crying: "El Vulcan!"

Away back in the year 1868, scientific impudence had not let loose, and there was no scientific clown to laugh off a blazing sky, with a story of lights in horse cars. The mystery of this occurrence is in the belief in Peru that there was, somewhere, at this time, a volcanic eruption.

Cities were flung in the sea. The sea rushed back upon ruins. It doubled all ordinary catastrophes by piling the wrecks of ships upon the ruins of houses. Fields poured over cliffs into the Bay of Arica. 'it was a cataract of meadows. We have gone far in our demonstration of continuity, which has led from showers of frogs to storms of meadows.

Vast volumes of water fell from the sky. It was appalling providence: this water was needed. The waters soaked into the needful earth, and surplus beneficences made new rivers. In the streams, there was a ghastly frou-frou of torrents of corpses, and the coast of Peru was frilled with fluttering bodies. Almost Ultimate Evil could be stimulated by such a lingerie. These furbelows of dead men, flounced in the waves, were the drapery of Providence.

Upon August 19th, there was another violent quake, and again there was a glare in the sky. Both times there was no accounting for such a spectacle except by thinking that there had been an eruption in Peru. According to the New York Herald, September 29, the volcano Moquequa was suspected. London Times, October 21-letter from someone who had seen the flaming sky, and had heard that Canderave was the volcano. It was said that the eruption had been at Aqualonga, and then that it had not been Aqualonga, but Cayambe. An illumination in the sky, lasting several hours, is described in Comptes Rendus, 67-1066, and here a writer gives his opinion that the volcano was Saajama. Other observers of the glare said that it came from Cotopaxi. Cosmos, n. s., 3-3-367-it was supposed that Cotocachi was the volcano. But it is not possible to find anything of this disagreement in any textbook: all agree upon attributing to one volcano-it was Mt. Misti.

New York Herald, Oct. 30, 1868-that Mt. Misti had not been active.

See Comptes Rendus, 69-262-the results of M. Gay's investigations-that, in this period, not one of the suspected volcanoes had been active. See the Student, 4-147.

Sometimes volcanic eruptions upon this earth shine, at a distance, like stars. It will be my acceptance that new stars are new volcanic eruption in Starland. For a description of a terrestrial eruption that shone like a star, see the Amer. Jour. Sci., 2-21-144. See a description, in the New York Times, Sept. 23, 1872, of an eruption of Mauna Loa, which far away looked star-like.

At 12:30 P. M., September 4th, appeared something that has often been seen at Naples, when Vesuvius has discharged. It was like the volcanic discharge that we have noted, at Guadalajara, Mexico. A dense, mountain-like cloud appeared, in the western sky, at Callao, Peru. The earth heaved with violence equal to that of August 13th.

New York Tribune, October 7-that in the southwestern sky was seen a star.

It is my expression that this was the star that broke Peru.

Night of Feb. 4, 1872-another glare in the sky-that the constellation Orion was afire-that a tragedy upon this earth began in the sky, with a spectacle that excited peoples of this earth, from Norway to South Africa-but that, under lying tragedies written by human beings, or wrought in sky-and lands, are the same conventions, and that Organic Drama is no more likely to let catastrophe come, without preceding phenomena that may be interpreted as warnings, than would stagecraft of this earth permit final calamity, without indications of its approach-

That a surprise was preceded by a warning that was perhaps of the magnitude of a burning of all the forests of North America-testimony of the sun and the moon to coming destruction-announcements that were issued in blazes-showers of gleaming proclamations-brilliant and long-enduring advertisement- .

But that mind upon this earth was brutalized with dogmas-and that scientific wisemen, stupefied by a creed, presided over a slaughter, or were surprised when came the long and brilliantly advertised.

This night of Feb. 4, 1872-a blaze in the constellation Orion. From centers of alarm upon this earth there was telegraphing. City called upon city. People thought that a neighboring community was burning. In the West Indies, island called upon island. In each island, the glare in the sky was thought to come from a volcanic eruption in some other island. At Moncalieri, Italy, an earthquake, or a response in this earth to cataclysm somewhere else, was recorded by seismographs. There may have been special relation with the ground in Italy.

With this glare, which was considered auroral, because there was no other way of conventionally explaining it, though auroras never have been satisfactorily explained, came meteors. Denza recorded them, as seen in Italy, and noting the seeming relation to the glare, explained that the seeming relation was only a coincidence. That's got to be thought by everybody who opposes all that this book stands for. If it was not a coincidence, the meteors came to this earth from wherever the glare was. U the glare was in the constellation Orion, Orion may be no farther from Italy than is San Francisco.

Upon the night of February 22nd, another glare was seen in the sky, and "by coincidence," it was identical in all respects, except magnitude, with the glare of the 4th. "'By coincidence" again meteors appeared. See Comptes Rendus, 74-641.

Five days after this second seeming eruption in Orion, dust fell from the sky, at Cosenza, Italy (C. R., 74-826).

The meteors that were seen at the time of the first glare were extraordinary. They appeared only in the zone of Italy. As seen with the glare, in India. they are told of in the Allahabad Pioneer Mail,

February 12, and the Bombay Gazette, February 19. See other records of ours of zoo phenomena. Sixteen days after the second glare in Orion, reddish yellow dust fell in Sicily, and continued to fall the second day, and fell in Italy.

Trembling billions-or a panic of immensities-and the twinkles of the stars are the winks of proximities-and our data are squeezing supposed remoteness in o familiarities because, if from a constellational eruption, dust drifted to this earth in a few weeks, it did not drift billions of miles-

But was this dust a discharge from a volcano?

It was volcanic dust, according to Prof. Silvestri. See the

Jour. Chem. Soc. London, 25-1083. Prof. Silvestri thought that it must have come from an eruption somewhere in South

· America. But my notes upon phenomena of this year 1872 are especially numerous, and I have no record of any eruption in South America, or anywhere else-upon this earth to which could be attributed this discharge.

For records of a stream of events that then started flowing, see Comptes Rendus, vols. 74, 75, and Les Mondes, vol. 28. Io Italy, upon the first of April began successions of "auroral" lights and volleys of meteors. Night of April 7-8-many meteors, at Mondovi, Italy. Solar and lunar haloes, which may, or may not, be attributed to the presence of volcanic dusts, were seen in Italy, April 6th, 7th, and 8th. Two days later, Vesuvius became active, but there were only minor eruptions.

There was uneasiness in Italy. But it was told, in Naples. that the wisemen were watching Vesuvius. Because of the slight eruptions, some of the peasants on the slopes began to move. These were a few of the untrustful ones: the others believed, when the wisemen said that there was no reason for alarm. Night after night, while this volcano in Italy was rumbling, meteors c e   to the skies of Italy. There is no findable record that they; so came anywhere else. They. came down to this one part of this earth, as if this earth were stationary.

April 19th-the third arrival of dust-volumes of dust of unknown origin, fell from the sky, in Italy. There was alarm. The sounds of Vesuvius were louder, but a quiet fall of dust, if from the unknown, spreads an alarm of its own.

The wisemen continued to study Vesuvius. They paid no more attention to arrivals of dusts and meteors in the sky of a land where a volcano was rumbling, than to arrivals of song birds or of tourists, in Italy. Their assurances that there was no reason for alarm, founded only upon their local observations, held back upon the slopes of the volcano all but a few disbelievers-

The 20th of April Eruption of Vesuvius.

Convolutions of clouds'-scrimmages of brains that had broken out of an underground academy of giants-trying to think for themselves-struggling to free themselves from subterranean repressions. But clouds and brains are· of an underlying oneness: struggles soon relapsed into a general fogginess. Volcanic or cerebral-the products are obscurities. Naples was in darkness.

The people of Naples groped in the streets, each in a hellish geometry of his own, each seeing in a circle, a few yards in diameter, and hearing in one dominant roar, no minor sounds more than a few yar<k away. Streams of refugees were stumbling into the streets of Naples. People groped in circles, into which were thrust hands, holding up images, or clutching loot. Fragments of sounds in the one dominant roar-geometricity in bewilderment-or circles in a fog, and something dominant, and

everything else crippled. The flit ting of feet, shoulders, bandaged heads-cries to the saints -profanity 'of somebody who didn't give a damn for Vesuvius-legs of a corpse, carried by invisibles-prayers to God, and jokers screeching false alarms that the lava was coming.

A blast from the volcano cleared away smoke and fog. High on Vesuvius-a zigzag streak of fire. It was a stream of lava that looked fixed in the sky: With ceaseless thunder, it shone like lightning-a bolt that was pinned to a mountain. Glares that were followed by darkness-in an avalanche of bounding rocks and stumbling people, no fugitive knew one passing bulk from another, crashing rocks and screaming women going by in silence, in the one dominant roar of the volcano. When it was dark, there were showers of fire, and then in the glares, came down dark falls of burning cinders. In brilliant illuminations, black rains burned the running peasants. Give me the sting of such an ink, and there'd be running.

Somewhere, in the smoke and flames, on the mountain side, fell a sparrow. According to conventional theologians, this was noted.

The next day there was another flow down the slopes of Vesuvius. It was of carts that were laden with bodies.

· Possibly this was overcooked, if attention was upon the sparrow. See data to come, for a more matured opinion.

In at least one mind, or quasi mind, or whatever we think are minds, upon this earth, there was awareness of more than coincidence between flows of meteors in Italy and a volcanic eruption in Italy. In Comptes Rendus, 74-1183, M. Silberman tells of the meteors in Italy, and the eruption of Vesuvius, and gives his opinion that there was relation. It was a past generation's momentary suspicion. The record is brief. There was no discussion. To this day, no conventional scientist will admit that there is relation. But, if there is, there is also another relation. That is between his dogmas and the slaughters of people.

In orthodox terms of a moving earth crossing orbits of meteor streams, to which any one part of this earth, such as the Italian part, could have no especial exposure to meteors so moving, there is no explanation of the repeated arrivals o' meteors, especially, or exclusively, in Italy, except this

Night after night after night-

Coincidence after coincidence after coincidence.

Our unorthodox expression is that it was because this earth is stationary.

According to data that have been disregarded about sixty years, it may be that there was a teleportative, or electrolytic, current between a volcano of this earth and a stellar volcano. If we think that a volcano in a land that we call the Con stellation Orion interacted with a volcano in Italy-as Vesuvius and Etna often interact-there must be new thoughts upon the distance of Orion.

The one point that every orthodox astronomer would con test, or deride-because its acceptance would be followed by acceptance of this book as a whole-is that the glare that seemed to be in Orion was in Orion.

These are the data for thinking that the glare that seemed to be in Orion, was in Orion, which cannot be vastly far away: The glare in the sky, early in the evening of Feb. 4, 1872, was west of Orion, as if cast by reflection from. an eruption below the horizon. But, when Orion appeared in the east, the glare was in Orion, and it remained in Orion. At Paris, all beams of light came from Orion, after ·8 P.M. (Comptes Rendus 74-385). In England-in Orion (Symons' Met. Mag., 7-1). In South Africa, the point from which all beams di verged was in Orion (Cape Argus, February 10). An account by Prof. A. C. Twining, of observations in the United States, is published in the Amer. Jour. Sci., 3-3-273. This "remark able fact," as Prof. Twining calls it, but without attempting to explain, is noted-that, from quarter past seven o'clock, in the evening, · until quarter · past ten, though Orion had moved one eighth of its whole revolution, the light remained in Orion.

There is no conventional explanation to oppose us. My expression is that the glare so remained in Orion, because it was in Orion. Anybody who thinks that the glare was some where between this earth and the constellation will have to account not only for the fixedness of it in a moving constellation, but for its absence of parallax, as seen in places as far apart as South Africa and the United States.

XXVI

HORSES erect in a blizzard of frogs-and the patter of worms on umbrellas. The hum of lady birds in England the twang of a swarm of Americans, at Templemore, Ireland. The appearance of Cagliostro-the appearance of Prof. Einstein's theories. A policeman dumps a wild man into a sack and there is alarm upon all continents of this earth, because of a blaze in a constellation-

That all are related, because all are phenomena of one, organic existence-just as, upon Aug. 26, 1883, diverse occurrences were related, because all were reasons to something in-common.

Aug. 26, 1883-people in Texas, excitedly discussing a supposed war in Mexico-young men in Victoria, Australia, watching a snowstorm, the first time in their lives--crowds of Chinamen hammering on gongs-staggering sailors in a vessel, off the Cape of Good Hope.

I have data for thinking that, somewhere beyond this earth, and not enormously far away, there was, before these occurrences, an eruption. About August 10th, of this year 1883, at various places, appeared "afterglows" that cannot be traced to terrestrial eruptions. Upon the 13th of August, an "afterglow" was reported from Indiana, and, ten days later, from California (Monthly Weather Review, 1883-289).

Upon the 21st and 22nd, "afterglows" appeared in South Africa (Knowledge, 5-418).

There was no known eruption upon this earth, by which to explain these atmospheric effects, but there was a disturbance upon this earth, and the circumstances were similar to those in Italy, in April, 1872. The volcano Krakatoa, in Java, was in a state of minor activity. It was not considered alarming.

Upon the 25th of August, a correspondent to the Perth (Western Australia) Inquirer-see Nature, 29-388-'was traveling far inland, in Western Australia. He was astonished to see ashes falling from the sky, continuing to fall, all afternoon. ff this material came from regions external. to this earth, it came down, hour after hour, as if to a point upon a stationary earth. An attempt to explain was that there may have been an eruption in some little-known part of Australia.

In Australian newspapers, there is no mention of an eruption in Australia, at this time; and in my own records there is only one instance, and that one doubtful, of an eruption in Australia, at any time. I am not here including New Zealand. There was, at this time, no eruption in New Zealand. Krakatoa was in a state of minor activity. Wisemen from Batavia, localizing, like the wisemen at

Vesuvius, in April, 1872, were studying only Krakatoa. Considered as a thing in itself, out of relation with anything else, conditions here were not alarming. The natives were told that there was no danger-and natives-Columbia University, or east side, west side, New York City-or Java-believe what the men tell them.

April 19, 1872-the dust of unknown origin that fell from the sky-it preceded, by one day, the eruption of Vesuvius.

Aug. 25, 1883-the ashes of unknown origin that fell from the sky-

August 26-Krakatoa exploded. It was one of this world's biggest noises, and surrounding mountains doffed their summits. Or, like a graduating class at Annapolis, they fired off their peaks, which came down. as new reefs in the ocean. The bombs that shot out were like meteors. The mountain was a stationary meteor-radiant, and shot out Krakatoatids. Had the winds gone upward, the new meteor stream would have been also of houses and cattle and people. The ex plosion shattered shores so that all charts were useless.

Krakatoa paused.

Early in the morning of the 27th, the Straits of Sunda went up. The units of its slaughters were villages. 95 villages went up, in waves that were 90 feet high, 100 feet high, 120 feet high. From the 95 villages, tens of thousands of humans were recruited, and they went dead into warring confusions. In the gigantic waves, armies of the dead were flung upon each other. There was no more knowledge of. what it was all about than in many other battles. Charges

of dead men rushed down the waves, and were knocked into rabbles by regiments spouting up from the bottom. Onslaughts of corpses-and in the midst of them appeared green fields that were the tops of palm trees. Raiders in thousands dashed over momentary meadows, and there were stampedes that were as senseless as the charges.

Stone clouds were rolling over the conflict. Furies of dead men were calming under bulks of pumice, and slimy palm trees were protruding. The waves were going down under pressures of pumice, twenty feet deep, in places. Battlefields on land have, after a while, turned quiet with graveyards: but change in the Straits of Sunda was of the quickest of mortuary transitions. The waves were pressed flat by pumice. There was a gray slab of stone over the remains of the people of 95 villages. Black palms, heavy with slime, drooped on each side of this long; gray slab.

By volcanic dust, the sun was dimmed so that unseasonable coldness followed. In places in Victoria, Australia, where for twenty-five years snow had not been seen, mow fell. Td like to have this especially noted-that at places far away, the volcanic bombs were mistaken for meteors-or they were meteors. An account of volcanic bombs from Krakatoa, which looked like "shooting stars," as seen from a vessel, about 90 miles from the volcano, is published in the Cape Times (Weekly Edition), Oct. 3, 1883. At Foo chow, China, the glare in the sky was like an aurora borealis.

For this record, which is important to us, see the Rept. Krakatoa Committee, Roy. Soc., p. 269. People in Texas heard sounds, as if of a battle. Off the Cape of Good Hope, vessels lurched in waves from the catastrophe.

It is my expression that the explosion of Krakatoa was stimulated by, or was a reaction to, an eruption in a land of stars that is not enormously far away. Afterglows that were seen after August 26th were attributed to Krakatoa-

That the preceding afterglows and the fall of ashes were of materials that drifted to this earth, from an eruption somewhere else, passing over a distance that cannot be considered vast, in a few weeks, or a few days-

And that the light of a volcanic eruption somewhere in the sky was seen by people of this earth.

See the Perth Inquirer, October 3-a correspondent tells of several observations, early in September, upon a brilliant light that had been seen in the sky, near the sun. There was a beam of light from it, and the observers. thought that it was a comet. This appearance is described as conspicuous. li so, it was seen at no Observatory in Australia.

The circumstance that no professional astronomer in Australia saw this brilliant light brings up, in any normally respectful mind, doubt that there was such a light. But this appearance in the sky is the central datum of our ex pression, and I am going to make acceptable that, even though it was reported only by amateurs, there was at this time a conspicuous new appearance in the heavens. New Zealand. Silence in the Observatories-but reports from amateurs, upon a "very large" light in the sky. See the New Zealand Times, Sept. 20, 1883. That a yam in Australia could quickly spread to New Zealand? Ceylon-an unknown light that was seen in the sky of Ceylon, about ·a week after the eruption of Krakatoa (Madras Athenaeum, September 22). Straits Times, October 13-an appearance like a comet, in the sky, at Samarung, where the natives and the Chinese were terrified, and burned incense for protection from it. England-observation, upon the night of August 28th, by Captain Noble, a well-known amateur astronomer. Whatever the professionals of Australia and New Zealand were doing, the professionals of England were doing likewise, if doing nothing is about the same wherever it's done. In Knowledge, 4-173 Capt. Noble tells of a sight in the heavens that he describes as "like a new and most glorious comet: An amateur in Liverpool saw it. Knowledge, 4-207-an object that looked like the planet Jupiter, with a beam from it. However, one professional astronomer did report something. Prof. Swift, at Rochester, N. Y., saw, nights of September 11 and 13, an object that was supposed by him to be a comet, but if so, it was not seen again (Observatory, 6-343).

There was a beam of light from this object: so it was thought to be a comet. See coming data upon beams of light that have associated with new stars.

The first observation upon a new light in the sky was two nights after the explosion of Krakatoa. It may have been shining, conspicuous but unseen, at the-time of the eruption.

The matter of the supposed velocity of light, or the hustle of visibility, comes up in the mind of a conventionalist. But, if in the past, scientists have determined the velocity of light to be whatever suited their theories, I feel free for any view that I consider suitable. Look it up, and find that once upon a time the alleged velocity of light agreed with supposed distances in this supposed solar system, and that when changed theories required changes in these distances, the velocity of light was cut down to agree with the new sup posed distances. In the kindergarten of science, the more or less intellectual, infants who have made these experiments have prattled anything that the child-like astronomers have wanted them to prattle. A conventionalist would say-that, even if there were a new star, at a time of terrestrial catastrophe, light of it would not be seen upon this earth, until years later. My expression is that so close to this earth are the stars that when a new star appears, or erupts, effects of it are observable upon this earth, and that, whether because of closeness, or because there is no velocity of light, it is seen immediately-or is seeable immediately-if amateurs happen to be looking.

Upon the night of Aug. 6, 1885, while all the professional astronomers of this earth were attending to whatever may be professional astronomical duties, a clergyman made an astronomical discovery. The Rev. S. If. Saxby looked up at a new star, in the nebula of Andromeda, and he saw it.

There is much of uppishness to anybody who says, or announces, that of all cults his cult is the aristocrat. But most of his upward looking is likely to be at the supposed altitude of himself. All ever this earth, professional astronomers were looking up at themselves. In England and Ireland, three amateurs, besides the Rev. S. H. Saxby, being probably of only ordinary conceit, looked up beyond them selves, and saw the new star. For the records, see the English Mechanic, of this period. Whatever the professional astronomers of the United States were looking up at, they saw nothing new. But somebody, in the U.S. A., did see the new star. Sidereal Messenger, 4-246-an amateur in Red Wing, Minnesota. It was not until August 31st that upness in an Observatory related to the stars. Finally a professional astronomer either looked up, or woke up; or waking up, looked up; or looking up, woke up.

The whole nebula of Andromeda shone with the light of the new star. Several observers thought that the newly illuminated nebula was a comet (Observatory, 8-330). From the light of the new star the whole formation lighted up, like a little West Indian island, at a time of an eruption in it. According to conventionalists, this nebula is 60 x 60 x 24 x 365 x 8 x 186,000 miles in diameter, and light from a new star, central hi it, would occupy four years in traversing the whole. But as if because this nebula may not be so much as 60 x 186,000 inches hi diameter, no appreciable time was occupied, and the whole formation lighted up at once.

Other indications-whatever we think we mean by "indications"-that the nebula of Andromeda is close to this earth:

Sweden-and it was reported that wild fowl began to migrate, at the earliest date (August 16th) ever recorded in Sweden-

Flap of a wild duck's wings-and the twinkles of a star the star and the bird stammered a little story that may someday be vibrated by motors, oscillating back and forth from Vega to Canopus.

So the birds began to fly.

It was because of unseasonable coldness in Sweden. Un seasonable coldness is a phenomenon of this earth's volcanic eruptions. It is attributed to the shutting off of sunlight by volcanic dusts. The temperature was' lower, in Sweden, than ever before recorded, in the middle of August (Nature, 32-427). Upon August 31st, the new star reached maxi mum brilliance, and upon this date the temperature was the lowest that it had ever been known to be, in the last of August, in Scotland (Nature, 32-495).

All very well-except that unusual coldness may be explained in various ways, having nothing to do with volcanoes-

See Nature, 32-466; 625-that nine days after the observation upon the new star in Andromeda, an afterglow was seen in Sweden. There is no findable record of a volcanic eruption, upon this earth, to which this phenomenon could be attributed. September 3rd, 5th, 6th-afterglows appeared in England. These effects continued to be seen in Sweden, until the middle of September.

I don't know whether these data are enough to jolt our whole existence into a new epoch, or not.

From what I know of the velocity of thought, I should say not.

If a volcanic discharge did drift from Andromeda to Sweden, it came from a northern constellation to a northern part of this earth, as if to a part of this earth that is nearest to a northern constellation. But, if Andromeda were trillions of miles away, no part of this earth could be appreciably nearer than any other part, to the constellation. If repeating afterglows, in Sweden, were phenomena of repeating arrivals of dust, from outer space, they so repeated in the sky of one part of this earth. because this earth is stationary.

I note that I have overlooked the new star in Cygnus, late in the year 1876. Perhaps it is because this star was discovered by a professional astronomer that I neglected it. However, I shall have material for some malicious comments. Upon the night of Nov. 24, 1876, Dr. Schmidt, of the Athens Observatory, saw, in the constellation Cygnus, a new star. It was about third magnitude, and increased to about second. Over all the Observatories of this earth. This new star was shining magnificently, but it was not until December 9th that any other astronomer saw it. It was seen, in England, upon the 9th, because, upon the 9th, news reached England of Dr. Schmidt's discovery. I note the matter of possible overclouding of the sky in all other parts of this earth, at this time. I note that,. between the dates of November 24th and December 9th, there were eight favorable nights, in England.

It so happens that I have record of what one English astronomer was doing, in this period. Upon the night of November 25, he was looking up at the sky.

Meteoric observations are conventionalities. New stars are unconventionalities.

See Nature. Dec. 21, 1876-this night. this astronomer observed meteors.

There was a volcanic eruption in the Philippines, upon the 26th of November. About two weeks later, a red rain fell in Italy.

There is considerable in this book that is in line with the teachings of the most primitive theology. We have noted how agreeable I am sometime$ to the most southern Methodists. It is that scientific orthodoxy of today has brutally. or mechanically. or unintelligently. reacted sheerly against all beliefs of the preceding. or theological, orthodoxy. and has reacted too far. All reactions react too far. Then a reaction against this reaction must of course favor, or restore, some of the beliefs of the earlier orthodoxy. ·

Often before disasters upon this earth there have been appearances that were interpretable as warnings.

But if a godness places kindly lights in the sky, also is it spreading upon the minds of this earth a darkness of scientists. This is about the beneficence of issuing warnings, and also seeing to it that the warnings shall not be heeded. This may not be idiocy. It may be "divine plan" that surplus populations shall be murdered. In less pious terms we may call this maintenance of equilibrium.

If surplusages of people upon this earth should reduce, and if then it should, in the organic sense. become desirable that people in disaster-zones should live longer, or die more lingeringly. provided for them are phenomena of a study of warnings, a study that is now, or that has been. subject to inhibitions.

Aug. 31, 1886-"Just before the sun dropped behind the horizon, it was eclipsed by a mass of inky, black clouds. People noted this appearance. It was like the "dense, mountain-like cloud" that appeared, at Callao, Peru. before the earthquake of Sept. 4, 1868. But these people were in a North

American city. Meteors were seen. They were like the fire balls that have shot from this earth's volcanoes.

Luminous clouds, such ·as have been seen at times of this earth's eruptions, appeared, and people watched them. There was no thought of danger. There was a glare. More meteors. The city of Charleston. South Carolina, was smashed.

People running from their houses - telegraph poles falling around them - they were meshed in roils . of wires. Street lamps lights in houses waved above like lights of a fishing fleet that had cast out nets. It was a catch of bodies, but that was because minds had been meshed in the net of a cult, woven out of the impudence of the De Ballorea and the silences of the Davison's, spread to this day upon every school of this earth.

The ground went on quaking. Down from the unknown came, perhaps, a volcanic discharge upon this quaking ground. Whether it were volcanic dust, or not, it is said, in the New York World, September 4, that "volcanic dust" was falling, at Wilmington, North Carolina.

September 5th-a severe shock, at Charleston. and a few minutes later came a brilliant meteor, which left a long train of fire. At the same time, two brilliant meteors were seen, at Columbia, S. C. See almost any newspaper, of the time. I happen to take from the London Times, September 7.

There was another discharge from the unknown-or a "strange cloud" appeared, upon the 8th of September, off the coast of South Carolina. The cloud hung, heavy, in the sky, and was thought to be from burning grass on one of the islands. Charleston News and Courier, September 10-that such was the explanation, but that no grass was known to be burning.

li a procession starts at Washington Square, New York City, and, if soldiers arrive in Harlem, and then keep on arriving in Harlem, I explain that, in spite of all the eccentricities of Harlem, Harlem is neither flying away from the procession, nor turning on 125th Street, for an axis. Meteors kept on coming to Charleston. They kept on arriving at this quaking part of this earth's surface, as if at a point on a stationary body. The most extraordinary display was upon the night of October 22nd. There was a severe quake, at Charleston, while these meteors were falling. About fifty appeared (New York Sun, November 1). About midnight, October 23-24, a meteor exploded over Atlanta, Georgia, casting a light so intense that small objects on the ground were visible (New York Herald; October 25). A large meteor, at Charleston, night of October 24th (Monthly Weather Review, 1886-296). An extraordinary meteor, at Charleston, night of the 28th, is described, in the News and Courier, of the 29th, as "a strange, celestial visitor."

"It was only coincidence."

There is no conventional seismologist, and there is no orthodox astronomer who-says otherwise.

In the Friend of India, June 22, 1897, is another record of some of the meteors that were seen in Charleston: that, at the time of the great quake, Prof. Oswald saw meteor after meteor shoot from an apparent radiant near Leo. Carl McKinley, in his Descriptive Narrative of the Earthquake of August 31, 1886, records a report from Cape Romain Light Station, upon "an unusual fall of meteors during the night."

Again a volcanic discharge came to this point-or a fall of ashes was reported. In the News and Courier, November 20, it is said that, about ten days before, ashes had fallen from the sky, at Summerville, S. C. It is said that the material was undoubtedly ashes. Then it is said that it had been discovered that, upon the day of this occurrence, there had been "an extensive forest fire near Summerville."

Monthly Weather Review, October and November, 1886- under "Forest and Prairie Fires," there is no mention of a forest fire, either small or extensive, in either North or South Carolina.

Summerville, and··not Charleston, was the center of the disturbances. Ashes came from somewhere exactly to this central point.

In A Study of Recent Earthquakes, Dr. Charles Davison gives 36 pages to an account of phenomena at Charleston. He studies neither meteors nor anything else that was seen in the sky. He studiously avoids all other occurrences upon this earth, at this time. Refine such a. study to a finality of omissions, and the vacancy of the imbecile is the ideal of the student. I approve this, as harmless.

The other occurrences were enormous. Destruction 'in South Carolina was small compared with a catastrophe in Greece. Upon the day of the first slight shock, at Charleston (August 27th), there was a violent quake in Greece, and at the same time, torrents poured from the sky, in Turkey, carrying away houses and cattle and bridges (Levant Herald, September 8). Thousands of houses collapsed, and hundreds of persons perished. This day, there was a shock, at Srinagar, Kashmir: shocks in Italy and Malta; and increased activity of Vesuvius. Just such an inky cloud as was seen at Charleston, was seen in the eastern Mediterranean, at the time of the catastrophe in Greece-reported by the captain of the steamship LA Valette-see Malta Standard September 2- "a mass of thick, black smoke, changing into a reddish color." "The sea was perfectly calm, at the time." In the sky of Greece, , there was a glare, like the light of a volcanic eruption (Comptes Rendus, 103-565). I confess to a childish liking for making little designs, or arrangements of data myself. And every formal design de pends upon blanks, as much as upon occupied spaces. But my objection to such a patternmaker as Dr. Davison is to the preponderance of what he leaves out. In Dr. Davison's 36 pages upon the lesser catastrophe, at Charleston, he spins thin lines of argument, in a pretty pattern of agreements, around omissions. It is his convention that all earthquakes are of local, subterranean origin-so he leaves out all appearances in the sky and mentions none of the other violences that disturbed a zone around this earth. It is a monstrous disproportion, when a mind that should be designing embroidery takes· catastrophes for the lines and blanks of its compositions.

It is my expression that if a clipper of data should mislay his scissors, or should accidentally let in an account of one of the many localized repetitions of meteors, he would tell of an indication that this earth is, or is almost, stationary. Night of October 20th-meteors falling at Srinagar, Kashmir. There was an earthquake. Shocks and falls of meteors continued together. According to my searches in Indian newspapers, these repeating meteors were seen nowhere else. As to zone-phenomena I point out th.at there is a difference of only one degree of latitude between Charleston and Srinagar. For the data, see the Times of India, November 5.

If a string of meteors should be flying toward this earth, and if the first of them should fall to this earth, at Srinagar, how is anybody going to think of the rest of them falling exactly here, if this earth is speeding away from them? Sometimes· I am almost inclined to have a little faith, of course not in general reasoning, but anyway in my own reasoning, and I go on to observe that a long string of meteors can be thought of, as coming down to the one point where the first fell, if that point is not moving away from them. But I begin to suspect that the trouble with me is that I am simple-minded, and that mine enemies, whom I call "conventionalists," are more subtle than I am, and prefer their views, because mine are so obvious: Of course this earth is stationary, in a surrounding of revolving stars so far from far away that an expedition could sail to them. But no dialectician, 'of any fastidiousness, would be attracted by a subject so easy to maintain.

Back to data-geysers spouted from the ground, at Charleston, and there were sulphurous emissions. The ground was incipiently volcanic and charged electrically.

Meteors and smoky discharges and glares and falls of ashes and enormous pours of water, as if from a volcano that was moving around a zone of this earth-

And there is no knowing when, in the year 1886, disturbances began in the co stellation Andromeda.

In the Observatory, 9-402, it is said that, upon September 26th, a new star in the Andromeda nebula had been reported by one astronomer, but that, according to another astronomer, there was no such new star. Astronomical Register, 1886-269 -that, upon October 3rd, a new star in this nebula had been photographed.

I think of our existence as a battery-an enormous battery, or, in the cosmic sense, a little battery. So I think of volcanic regions, or incipiently volcanic regions, in a land of the stars and in a land of this earth, as electrodes, which are mutually perturbative, and between which flow quantities of water and other volcanic discharges, in electrolytic, or electrically teleportative, currents. According to the data, I think that some teleportations are instantaneous, and that some are slow drifts. To illustrate what I mean by stimulation, most likely electric, by interacting volcanoes, and transportation, or electric teleportation, of matter, between mutually affective volcanoes, I shall report a conversation, which, unlike mere human dialogue, was seen, as well as heard.

Upon the evening of Sept. 3, 1902, at Martinique, where the volcano Mont Pelée humps high, Prof. Angelo Heilprin, as he tells in his book, Mont Pelée, saw southward, at sea, electric flashes. They were in the direction of La Soufrière, the volcano upon the island of St. Vincent, 90 miles away. La Soufrière was flashing. Then Pelée answered. Pelée hugely answered, in tones befitting greater magnitude. A dozen flickers in the southern sky-and then Pelée speaking up. with a blinding, electric opinion. The little female volcano, or anyway the volcano with a feminine name,. nagged and nagged, and was then answered with a roar. This bickering kept up a long time.

About 5 o'clock, morning of the 4th, there was another appearance, upon the southern horizon. It was a dense bulk of smoke from La Soufrière. It drifted slowly. It went directly to Pelée and massed about Pelée.

There's no use arguing with a little, female volcano: she casts out obscurations. But there may be enormous use for this occurrence regarded as data.

XXVII

ONCE UPON A TIME, one of this earth's earlier scientists pronounced, or enunciated, or he told a story, which was somewhat reasonable, of a flood, and of all the animals of this earth saved, as species, in a big boat. Perhaps the story was not meant seriously by its author, but was a satire upon the ambitious boat-builders of bis day. It is probable that all religions are founded upon ancient jokes and hoaxes. But. considering the relative fewness of the animals that were known to the scientists, or the satirists, of that early time, this story was as plausible as the science. or as the best satire, of any time. However data of such a host of animals piled up that the story of the big boat lost its plausibility.

Note that our data are upon events of which the founders of the present so-called science of astronomy knew little. or knew nothing. Orthodox astronomy has been systematized. without considering new stars, their phenomena and indications. It is a big-boat story. Once upon a time it was plausible. It is in the position of the orthodox geology of former times, when a doctrine was formulated without con sideration for fossils and sedimentary rocks. But, when fossils and sedimentary rocks were incorporated, they forced a radical readjustment. New stars were not taken into the so-called science of astronomy, by the builders of that system, because no astronomer ever saw, or reported, a new star, between the years 1670 and 1848. Presumably new stars have not started appearing all at once in modem times.

Presumably, in this period of 178 years, many new stars appeared, and were not seen, though we shall have data for thinking that some of them shone night after night with the brilliance of first magnitude. One would like to know what. when time after time, the sky was probably spectacular with a new light; the astronomers were doing, in these 178 years. We may be able to answer that question, if we can mid out what the astronomers are doing now.

There is not agreement among the wisemen. · Virtually there is, by the wisemen of our tribes, no ·exploration of new stars. The collision-theory is heard of most.

Always-provided there have been little boys and other amateurs to inform them-the wise ones tell of stars that have collided. They have never told of stars that are going to collide.

Why is a story always of stars that have collided? Assuming now that, instead of being points in a revolving shell, stars are swiftly moving bodies there must be instances of stars that are going to collide, some days, weeks, or years from any given time. It fs too much to assume. that only dark stars collide, or the preponderance of dark stars would be so great that the sky would be black with Inky Ways. So far. we have not fair impression of how frequently new stars appear. It will be said that stars that are so close to each other that, in a year or so, they will collide, have, of their enormous distance from this earth, the appearance of one point of light.

This takes us to one of the solemnest and laughablest of the wisemen's extravagances. It is their statement that, after two stars have collided. they can, by means of the spectroscope, pick out in what is to the telescope only one point of light, the fragments of an alleged collision, the velocities and the directions of these parts.

If any spectroscopist can do this thing that the reading public is told that he can do, never mind about parts where he says there has been a collision, but let him pick out a point in the sky, which is of parts that are going to collide. Let him tell where a new star is going to be: otherwise let him go on being told, by amateurs, where a new star is.

New stars appear. There are disturbances upon this earth -there are. volcanic appearances in the sky-volumes of smoke and dust roll down upon this earth.

And the meaning of it all may someday be-"Skyward ho!" Storms, upon a constellation's vacant areas, of Poles and Russians. A black cloud appears in the sky of Lyra. And down pours a deluge of Italians. Drifting sands of Scandinavians sift down to a star.

Jan. 5, 1892-just such a fiery blast as has often tom down the slopes of Vesuvius, shot across the State ·of Georgia. It was "a black tornado filled with lire" (Chicago Tribune. January 7). About this time, there were shocks in Italy and, in the evening, ·people in many parts of New York State were looking up and wondering at a glare in the sky. The next day they had something else to wonder about. There were shocks in New York State. Upon the 8th, dust that was perhaps volcanic, but that had probably been dis charged from no volcano of this earth, fell from the sky, in Northern Indiana. 14th-"tidal wave" in the Atlantic, and a shock, at Memphis, Tennessee. Snow fell in Mobile, Alabama, where there had been only four falls of snow in seventy years. Floods in New England. Quakes in Japan, 15th, 16th, 17th. At this time began an eruption of Tongariro, New Zealand. "Tidal wave," or seismic wave, in Lake Michigan, upon the 18th. For references, see the New York newspapers. The Philadelphia Public Ledger, January 27, reported a fall, from the sky, of a mass of fire into a town in Massachusetts, upon the 20th. At this time, Rome, Italy, was quaking. Shocks in France, two days later. Shocks in Italy and Sicily. January 24th-a great meteor, with thunderous detonations, shot over Cape Colony, South Africa (Cope Argus, February 2 and 4). A drought, at Durango, Medco, was broken by rain, the first to fall in four years. Upon the night of the 26th, there was a glare in the sky that alarmed people throughout Germany. Severest shock ever known in Tasmania, upon the 27th, and shocks in many places in Victoria, Australia. In the night sky of England, people watched a luminous cloud (Nature, 45-365; 46-127).

There was a new star.

In all the Observatories of this earth, not a professional astronomer had observed anything out of the ordinary; but, in Edinburgh, a man who knew nothing of astronomical technicalities (Nature, 45-365) looked up at the sky, and saw the new star, night of February 1st. Throughout this period of the glares and the shocks and the seeming volcanic discharges, a new star, or a new celestial volcano, had been shining in the constellation Auriga. The amateur, Dr. Ander son, told the professionals. They examined photographs, and learned that they had been photographing the new star since December 1st.

The look of data is that volcanic dust drifted from a new star to the sky of this earth, in Indiana, in not more than 39 days.

For four hours, upon the 8th of January, dust came down from the sky, in Northern Indiana, and if it did come from regions external to this earth, it came settling down, hour after hour, as if to a point upon a stationary earth. I have searched in many scientific periodicals, and in newspapers of all continents, finding record of no volcanic eruption upon this earth, by which to explain.

La Nature, 41-206-that this dust had been analyzed, and had been identified as of volcanic origin. Science, 21-303 -that this dust had been analyzed, and had-been identified as not of volcanic origin. Monthly Weather Review, January, 1892-"It was in all probability of volcanic origin."

I have records of five other new stars, which, from Dec. 21, 1896, to Aug. 10, 1899, appeared at times of disturbances upon this earth; times of deluges and of volcanic discharges that cannot be attributed to terrestrial volcanoes. Two of the discoveries were made by amateurs. The other discoveries were made by professionals, who, with nothing at all resembling celerity, learned, by examining photographic plates, that new stars that had been looked at by astronomers had been recorded by cameras. The period of one of these incelerities was eleven years. See Nature, 85-248.

Star after star has appeared, as a minute point, or as a magnificent sight in the heavens, and the professional astronomers have been unobservatory. They have been notified by amateurs. We shall have records of youngsters who have seen what they were not observing. The first of the bright infants, of whom I have record, is Seth Chandler, of Boston. I have it that anybody who is only 19 years old, or, for that matter, 29, is a youngster. Seth was 19 years old. Upon May 12, 1866, an amateur astronomer, named Birmingham, at Tuam, Ireland, notified the professional astronomers, who were looking somewhere else, that there was a new star in the constellation Corona 'Borealis. In the United States, the professional astronomers were busily engaged-looking in other directions. Upon the night of the 14th, Seth Chandler interrupted their observations, telling them that there was something to look at. For any pessimist, who is interested in what becomes of exceptionally bright boys, and the disappointing records of many of them, I note that when this bright youngster grew up, he became a professional astronomer.

What on earth-or pretty nearly assuredly unrelated to the skies-were the professionals doing, February, 1901? Night of February 22nd-and Dr. Anderson, the amateur who had discovered Nova Aurigae, nine years before, looked up at the constellation Perseus, and, even though he had probably been befoozling himself with astronomical technicalities ever since, saw something new, and knew the new, when he saw it. It was a magnificent new star. It was a splendor that scintillated over stupidity: not a professional, at any of this earth's Observatories, knew of this spectacle, until informed by Dr. Anderson. Usually it is said that Dr. Anderson discovered this star, but his claim has been con tested. In Russia, it was recorded that, nine hours earlier, at ·a time when the sleepiest of the Observatories had not yet closed down, or had not yet quit not observing, the new star had been discovered by Andreas Borisiak, of Kieff. Andreas was a schoolboy.

Before the discovery of this new star in Perseus, or Nova Persei, there had been appearances like volcanic phenomena, unattributable, however, to any volcano of this earth. Upon the morning of February 13th, deep-greenish-yellow clouds, spreading intensest darkness, appeared in France (Bull. Soc. Astro. de France, March, 1901). Upon the 16th, a black substance fell from the sky, in Michigan (Monthly Weather Review, 29-465). There was the extreme coldness that often results from interferences with sunlight, by volcanic dust. At Naples, three persons were found to have frozen to death, night of the 13th (London Daily Mail, February 15). A red substance fell with snow, near Mildenhall (Lon don Daily Mail, February 22). It may have been functionally transmitted organic matter. "Pigeons seemed to feed upon it."

I have data for thinking that at least four nights before Dr. Anderson's observation, this new star, or a beam from it, though unseen at all Observatories, was magnificently visible. I think so, because, upon the night of the 18th, somebody in Finchley (London) and somebody in Tooting (London) saw something that they thought was a comet. Upon the night of the 20th, somebody in Tottenham saw it. A story of three somebodies who had seen something that was missed by all this earth's professional astronomers, would not be worth much, if told after an announcement of a discovery, but these observations were told of in the London Daily Mail, published upon the morning of February 22nd, before Dr, Anderson was heard from.

Sixteen days after the Anderson observation, dust arrived upon this earth-or it fell from the sky: n volumes that• were proportional to this outburst of first magnitude in the heavens. The new star at its brightest was of the magnitude of Vega. Dust, of the redness of no African deserts that one hears anything of-and if African deserts ever are red, moving picture directors, who are strong for realism, or, rather, sometimes are, should hear about this-fell from the sky. It came down, upon the 9th and 10th of March, in Sicily, Tunis, Italy, Germany, and Russia. A thick orange-red stain was reported from Ongar Essex, England, upon the 12th (London Daily Mail, March 19).

The standardized explanation was published. I shall pose it with heresy. Throughout this book, I shall say that all expressions of mine are only mental phenomena, and sometimes may be rather awful specimens, even at that. But, if we examine our opposition, and find it wanting, and if my own expression includes much that it left out, my own expression is not wanting, whether it's wanted, or not. Two wisemen wrote the standardized explanation. The red dust had come from an African desert. See Nature, 66-41.

They wrote that they had traced this dust to a hurricane In an African desert, pointing out that, upon the first day, it bad fallen in Tunis. That looked like a first fall near an African desert.

But the meteorologists are not banded like the astronomers. For a record of a fall, not so near an alleged African desert, see Symons' Met. Mag., 1902-25-that while this dust was falling in Tunis, also it was falling in Russia. That this dust did not come to this earth from outer space-see the Chemical News, 83-159-Dr. Phipson's opinion that it was meteoric. That may be accepted as the same as volcanic.

My own expression-

That a hurricane that could have strewn Europe with dust, from the Mediterranean to Denmark, and from England to Russia, could have been no breeze fluttering obscurely in some African desert, but a devastating force that would have fanned all Northern Africa into taking notice-

Lagos (Gold Coast) Record, March-April, 1901-no mention of a whirlwind of any kind in Africa. In the Egyptian Gazette- (Alexandria) there is nothing relating to atmospheric disturbances. There is nothing upon any such subject noted in the Sierra Leone News. Al-Moghreb (Tangier) reports the falls of dust in Europe, but mentions no raising of dust anywhere in Africa.

But there was a new star.

The standardized explanation is perforated with omissions. It seems unthinkable that mind upon this earth could be so bound down to this earth by this thing of gaps, until we reflect that so are all nets fabricated. In Austria, while this dust was falling, the earth quaked. What could such an occurrence have to do with dust from an African desert? Omitted. But at the time of this quake, something else was seen, and it may have been a volcanic bomb that had been shot from a star. London Daily Mail, March 13-a great meteor was seen. Dust falling in Tunis-and that was told. More of the omitted-see the Levant Herald, March 11-, that while the dust was falling in Tunis, there were violent earthquakes in Algeria. Something else that was left out see the English Mechanic, 73-96, and the Bull. Soc. Astro. de France, April, 1901-that, upon March 12th, ashes fell from the sky, at Avellino, Italy.

Vesuvius, April, 1872-

Krakatoa, August, 1883-

Charleston, August, 1886- Time after time after time-

And now May, 1902 - "in the hollow of their ignorance, two of these conventionalists held 30,000 lives.

May, 1902-there was another surprise. It, too, was preceded by announcements that were published by mountains, and were advertised in columns of fire upon pages of clouds. In April and May, 1902, across a zone of this earth, also outside the zone, there were disturbances. More than earth wide relations are indicated, to start with. Eruptions of Mt. Pelée, Martinique, began upon April 20th, the date of the Lyrids. However, in this book. I am omitting many data upon a seeming · relation between dates of meteor streams and catastrophes. Then the volcano La Soufrière, island of SL Vincent B. W. I., broke out. Upon the day of an earth quake in Siberia (April 12th) mud fell from. the sky, in widely separated places, in Pennsylvania, New York. New Jersey, and Connecticut. See Science, n.s., 15-872; New York Herald, April 14, 1902; Monthly Weather Review, May, 1902.

There may have been an eruption in some other part of one relatively small existence, or organism. There may have been a new star. In the English Mechanic, 75-291, a correspondent in South Africa told that, in the constellation Gemini, night of April 16th, he had seen an appearance like a new red star. He thought that it might have been not a star, but a mirage of the red light of the Cape Town.

The white houses of St. Pierre; Martinique-a white city, spread up on the slopes of Mt. Pelée. Early in May, there were panics in St. Pierre. Pelée was convulsed, and the quavering city of St. Pierre shook out inhabitants. Desertion of the city was objected to by its rulers, and the Governor of the island called upon two wisemen, Prof. Landes and Prof. Doze, for an authoritative opinion.

They had studied the works of Dr. Davison.

There were shocks in Spain, d in France. La Soufrière was of continuing violence. A volcano broke out in Mexico. Quakes in the Fiji Islands. Violent quake in Iceland. Volcanic eruption at Cook's Inlet, Alaska.

Prof. Landes and Prof. Doze were studying Mt. Pelée.

An eruption of cattle and houses and human beings, in Rangoon, Burma-or "the most terrible storm remembered." A remarkable meteor was seen at Calcutta. In Java, the Rooang volcano broke out. There were rumblings of an extinct volcano in France. In Guatemala, with terrific electrical displays, enormous volumes of water fell upon earthquakes.

Profs. Doze and Landes "announced" that there was, in Pelée's activity, nothing to warrant the flight of the people. See Heilprin, Mont Pelée, p. 71. Governor Mouttet ordered a cordon of soldiers to form around the city, and to permit nobody to leave.

Upon the 7th of May, a sky in France turned black with warning. See back to other such "coincidences" with catastrophes. Soot and water, like ink, fell from the sky, at Pare Saint Maur (Comptes Rendus, 134-1108).

Upon this day, the people of St. Pierre were terrified by the blasts from Pelée. No inhabitant of the city was per mitted to leave, but, as recorded by Heilprin, the captain of the steamship Orsolina did break away. They tried to hold him. The "pronouncement" was read to him, and officialdom threatened him, but, with half a cargo, he broke away. His arrival at Havre is told of, in the Daily Messenger (Paris) June 22. The authorities of the Port of St. Pierre had refused to give him clearance papers, but, terrified by the blasts from the volcano, he had sailed without them.

The people of St. Pierre were trying to escape, but they were bound to the town by chains of soldiers. Even in discreetly worded accounts, we read of ·these people, running in droves in the streets. In storms of ashes, turned back at all outskirts, by the soldiers, they were running in whirlpools. Not one word of retreat, nor of any modification, came from the two Professors. They had spoken in accordance with the dogmas of their deadly cult. Considered locally, an effusion of ashes was not considered planning, and no relationship with wider disturbances could be admitted. The Professors had spoken, and the people of St. Pierre were held to the town. The people were hammered and stabbed back, or they were reasoned with, and persuaded to stay.

Just how it was done, one has to visualize for oneself. It was done.

At night there was a lull. Then blasts came from the volcano. Screaming people ran from the houses. The narrow streets of this white city were dark with people, massing one way, only to gather against the military confines, sweeping some other way, only to be turned back by soldiers. Had there been darkness some of them might have escaped, but even at sea the glare from the volcano was so intense that the crew of a passing steamship, Lord Antrim, were almost blinded.

As seen from the sea, the streets, hung up on the mountain side, distorted by smoke and glare, Buttered. Long narrow crowds darkening these fluttering streets - folds of white garments of a writhing being, chained, awaiting burning.

Upon the morning of the 8th, the city of St. Pierre was bound to the stake of Mt. Pelée.

· There was a rush of flames. In the volcanic fires that burned the city, 30,000 persons perished.

XXVIII

EARLY IN OCTOBER, 1902, vast volumes of smoke; of un known origin, obscured all things at sea, and made navigation difficult and dangerous, from the Philippines to Hongkong, and from the Philippines to Australia. I do not know of anything of terrestrial origin that, with equal density, ever has had such widespread effects. Vesuvius has never been known to smoke up the whole Mediterranean. Compared with this obscuration, smoke at a distance from Krakatoa, in August, 1883, was only a haze. For an account, see the Jour. Roy. Met. Soc., 30-285. Hongkong Telegraph, October 25-that a volcanic eruption in Sumatra bad been reported. Science, n.s., 23-193-that there had been no known eruption in Sumatra-that perhaps there had been enormous forest fires in Borneo and Sumatra. Sarawak (Borneo) Gautte, October - November, 1902-no record of any such fires. There came something that was perhaps not vaster, but that was more substantial. If a story of a sand storm in a desert is dramatic, here is a story of a continent that went melodramatic. Upon the 12th of November, upon all Australia, except Queensland, dust and mud fell from ·the sky. Then densest darkness lit up with glares. Fires were falling from the sky.

Sometimes there are abortive embryos that are mixtures an eye looking out from ribs-other features scattered. Fires and dust and darkness-mud that was falling from the sky Australia was a womb that was misconceiving:

A fire ball burst over a mound, which Bickered; and frightened sheep ran from it-or, reflecting glares in the. sky a breast leered, and stuck out a long, red mob of animals. A furrowed field-or ribs in a haze-and a stare from the embers of a bush fire. An avenue of trees, heavy with mud, sagging upon a road that was pulsing with carts-or black lips, far from jaws, closing soggily upon an umbilical cord, in vainly attempted suicide.

Fire balls started up fires in every district in Victoria. They fell into cities, and set fire to houses. At Wycheproof, the whole air seemed on fire." All day of the 12th, and the next day, dust, mostly red dust, sifted down upon Australia, falling, upon the 13th, in Queensland, too. Smoke rolled in. upon Northern Australia, upon the 14th. A substance that fell from it was said to be ashes. One of the descriptions is of "a light, Buffy, grey material" (Sydney Daily Telegraph, November 18).   .

How many of those who have a notion that they're pretty well-read, have ever heard of this discharge upon Australia? And what are the pretty well-read but the pretty well-led? Little of this tremendous occurrence has been told in publications that are said to be scientific, and I take from Australian newspapers: but accounts of some of the fire balls that fell from the sky were published in Nature. vol. 67. There are reports from about 50 darkened, stifled towns. in the Sydney Herald, of the 14th- "business suspended -nothing like it before, in the history of the colony.- "people stumbling around with lanterns." Traffics were gropes. Mail coaches reached Sydney, nine hours late. Ashes with a sulphurous odor fell in New Zealand, upon the 13th (Otago Witness, November 19). The cities into which fell balls of fire that burned houses were Boort, Allendale, Deniliquin, Langdale, and Chiltern.

Smoke appeared in Java, and the earth quaked. A meteorite fell, at Kamsagar (Mysore), India. Upon this day of the 12th, a disastrous deluge started falling, in the Malay States: along one river, seven bridges were carried away.

There was no investigation. However, passing awareness did glimmer in one mind, in England. In a dispatch to the newspapers, Dec. 7, 1902, Sir Norman Lockyer called attention to the similarity between the dust and the me balls in Australia, and the dust and the fire balls from volcanoes in the West Indies, in May, of this year.

Our own expression depends upon whether all this can be attributed to any eruption upon this earth, or not. The smoke, in October, cannot be so explained. But was there any particular volcanic activity upon this earth, about Nov. 12, 1902?

The most violent eruption of Kilauea, Hawaii, in 20 years, was occurring, having started upon the 10th of November. There was a geyser of fire from the volcano Santa Maria, in Guatemala, having started upon October 26th. About the 6th of November, Colima, in Mexico, began to discharge dense volumes of smoke. The volcano Savii, in Samoa, broke out, upon the 13th, having been active, though to no great degree, upon October 30th. According to a dispatch, dated November 14th, there was an eruption in the Windward Islands, West Indies. Stromboli burst into eruption upon the 13th. About this time, Mt. Chullapata, in Peru, broke out.

But the smoke that appeared, with an earthquake, in Java, was the spans of an ocean and a continent far away. New Zealand is nearer all these volcanoes, except Stromboli, than is Australia, but dust and ashes fell a day later in New Zealand. Fire balls fell enormously in Australia. Not one was reported in New Zealand. Nothing appeared be tween New Zealand and these volcanoes, but dense clouds of smoke, between Australia and the Philippines, delayed vessels until at least November 20th (Hobart Mercury, November 21). o we regard the unexplained smoke of October and November, as continuous with the discharges of November 12th, and relate both, as emanations from one source, which is undiscoverable upon this earth.

There was a new star.

popular Astronomy, 30-60-that, in October, 1902, a new star appeared in the southern constellation Puppis.

Upon the 19th of November, a seismic wave, six feet high, crashed upon the coast of South Australia (Sydney Morning Herald, November 20). Upon this day the new star shone at its maximum of 7th magnitude.

See our expression upon phenomena of August, 1885. If, in November, 1902, a volcanic discharge came to Australia from a southern constellation, it came as if from a star-region to the nearest earth-region. But, •if constellations be trillions of miles away, no part of this earth could be appreciably nearer than any other part, to any star.

So extraordinary was terrestrial, volcanic activity at this time, that it will have to be considered. Like other expressions, om expression here is that mutually affective out-. bursts spread from the land of the stars to and through the land of this earth, firing off volcanoes in the disturbances of one organic and relatively little whole.

It was a time of extremest drought in Australia. Thunder storms that came, after November 12th, were described .as terrific.

As a glimmer of awareness, Lockyer told of the fire balls that came with the dust to Australia, and the suggestion to him was that there had been a volcanic discharge. But there was something that he did not tell. He did not know. It was told of in no scientific publication, and it reached no newspaper published outside Australia After the first volley of fire balls, other fire balls came to Australia. I have searched in newspapers of all continents, and it .is my statement that no such fire balls were reported anywhere else. All were so characterized that it will be accepted that all were of one stream. Perhaps they came from an eruption in tire constellation Puppis, but my especial expression is that, if all were of one origin, and if, days apart, they came to this earth, only to Australia, they so localized, because this earth is stationary. For references, see the Sydney Hera/4, and the Melbourne Leader. There was a meteoric explosion, at Parramatta, November 13th. A fire ball fell and exploded terrifically, at Carcoar. At Murrumburrah, N. S. W., dust and a large fire ball fell; upon the 18th. A fire ball passed, over the town of Nyngan, night· of the 22nd, intensely illuminating sky and ground. Upon the night of the 20th, as reported by Sir Charles Todd, of the Adelaide Observatory, a large fire ball was seen, moving so slowly that it was watched four minutes. At 11 o: clock, night of the· 21st, a fire ball of the apparent size of the sun was seen at Towitta. An hour later, several towns were illuminated by a great fire ball. Upon the 23rd, a fire ball exploded at Ipswich, Queensland. It is of especial importance to note the record of one of these bombs, or meteors, that moved so slowly that it was watched for four minutes.

From Feb. 12 to March 1, 1903, dusts and discolored rains fell along the western coast of Africa, upon many parts of the European Continent, and in England. The conventional explanation was published: there had been a whirlwind in Africa.

I have plodded for more than twenty years in the Libraries of New York and London. There are millions of persons who would think this a dreary existence.

But the challenges-the excitements-the finds.

Any pronouncement by any orthodoxy is to me the same as handcuffs. It's brain cuffs. There are times when I don't give a damn whether the stars are billions of miles away or ten miles away-but, at any time, let anybody say to me, authoritatively, or with an air of finality, that the stars are billions of miles away, or ten miles away, and my contrariness stirs, or inflames, and if I can't pick the lock of his pronouncements, I'll have to squirm out some way to save my egotism.

So then the dusts of February and March, 1903-and the whirlwind-explanation-and other egotists will under stand how I suffer. Simply say to me, "Mere dusts from an African desert," arid I begin to squirm like an Houdini.

Feb. 12 to March 1, 1903-"dusts from an African desert."

I get busy.

Nature, 75-589-that some of this dust, which had fallen at Cardiff, Wales, had been analyzed, and that it was

probably volcanic.

But the word "analyzed" is an affront to my bigotries conventional chemist-orthodox procedures-scientific delusions-more coercions.

I am pleased with a find, in the London Standard, Feb. 26, 1903. It is of no service to me, especially, now, but, in general, it is agreeable to my malices - a letter -from Prof. T. G. Bonney, in which the Professor says that the dust was not volcanic, because there was no glassy material in it and a letter from someone else, stating that in specimens of the dust that were examined by him, all the particles were glassy.

"It was dust from an African desert." But I have resources.

One of them is Al-Moghreb. How many persons, besides myself, have ever heard of Al-Moghreb? Al-Moghreb is mine own discovery.

The dust came down in England, Austria, Switzerland, Belgium, Germany, and along the west coast of Africa. Here's the question:

If there had been an African hurricane so violent as to strew a good part of Europe, is it not likely that there would have been awareness of it in Africa?

Al-Moghreb (Tangier)-no mention of any atmospheric disturbance that would bear out the conventional explanation. Lagos Weekly Record-Sierra Leone Weekly News- Egyptian Gazette-no mention. ·

And then one of those finds that make plodding in Libraries as exciting as prospecting for nuggets-February 14th, of this year-one of the most extraordinary phenomena in the history of Australia. In magnitude it was next to the occurrences of the preceding November. In the blackest of darkness, dust and mud fell from the sky. Melbourne Age, February 16-three columns of reports, upon darkness and dust and falling mud, in about 40, widespread towns in New South Wales and Victoria.

The material that fell in Australia fell about as enormously as fell the dusts in Europe. There is no mention of it in any of the dozens of articles by conventionalists, upon the phenomena in Europe. It started falling two days after the first fall of dust, west of Africa. It was coincidence, or here is an instance of two enormous volumes of dust that had one origin.

There was an unnoticed hurricane in Africa, which strewed Europe, and daubed Australia, precipitating nowhere be tween these two continents; or two vast volumes of dust were discharged from a disturbance somewhere beyond this earth, drifting here, arriving so nearly simultaneously that the indications are that the space between the source and this stationary earth is not of enormous extent, but was traversed in a few days, or a few weeks.

There was. no known eruption upon this earth, at this time. If here, but unknown, it would have to be an eruption more tremendous than any of the known eruptions of this earth.

There was a new star.

·It was found, by a professional astronomer, upon photo graphs of the constellation Gemini, taken upon March 8th (Observatory, 22-245). It may have existed a few weeks before somebody happened to photograph this part of the sky.

"Cold-blooded scientists," as we hear them called-and their "ideal of accuracy"-they're more like a lot of spoiled brats, willfully determined to have their own way. In Cosmos, n.s., 69-422 were reported meteoric phenomena, before the destructive earthquake in Calabria, Italy, Sept. 8, 1905. It was said-or it was "announced"-that Prof. Agamennone was investigating. It would have been a smash to conventional science if Prof. Agamennone had confirmed these reports. We know what to expect. According to the account in. Cosmos, first came a fall of meteors, and then, three quarters of ail hour later, to this same place upon this probably stationary earth, came a great meteor. It exploded, and in the ground was a shock by which 4,600 buildings were destroyed, and 4,000 persons were killed.

A volume of sound from crashing walls, in billows of roars from falling roofs, sailed like a ship in a storm. When it sank, lamentations leaped from it.

Because of underlying oneness, the sounds of a catastrophe are renderable in the terms of any other field of phenomena. Structural principles are the same, either in phonetic or biologic anatomy. A woe, or an insect, or a centipede is a series of segments.

Or the wreck of a city was a cemetery. Convulsed into animation, it was Resurrection Day, as not conceived of by religionists. Concatenations of sounds arose from burials. Spinal columns of groans were exhuming from ruination. Articulations of sobs clung to them. A shout that was jointed with oaths reached out from a hole. A church, which for centuries had been the den of a parasite, sank to a heap. It was a maw that engulfed a. congregation. From it came the chant of a litany that was a tapeworm emerging from a ruptured stomach.

Choruses broke into ·moans that were rows of weeping willows. A prayer crossed a field of mur-murs, and was gored by a blasphemy. Tellers of beads told ladders, up which ran profanities. Then came submergence again, in a chorus.

In earthquake lands, it is the belief of the people that there is a godness that, at times of catastrophes, directs them to Hock to churches. My own theology is in agreement It is by such concentrations that the elimination of surplusages is facilitated. But, if Virgin Marys were replaced by images of Mrs. Sanger, there would be no such useful, murders.

It is said, in the Bull. Soc. Astro. de France, October,

1905, that luminous phenomena had preceded this catastrophe, in Calabria. Observations upon ap-pearances in the sky were gathered by Prof. Alfani, and were recorded by him, in the Revista di Fisca. But it was Prof. Agamennone's decision upon reported phenomena in the sky that was awaited by the scientific world. From time to time, in scientific periodicals, there was something to the effect that he was investigating.

Not only meteors were told of. There was a fall of dust. from the sky, at Tiriolo.

Explained.

There had been an eruption of Stromboli.

Comptes Rendus, 141-576-report by M. Lacroix, who had been living near Stromboli, at the time-that, at this time of the fall of dust. there had been no more than normal activity of Stromboli.

Long afterward, the result of Prof. Agamennone's investigation was published. He could find only one witness.

It is not easy to think of an organic control that would beguile its human supernumeraries into manageable concentrations, for eliminative purposes, and also permit a Prof. Agamennone to conceive of warnings for them. But, if in super-metabolism, there is, as in sub-organisms, the katabolism of destructions, also there are restorations. Anabolic vibrations, known to the people of this earth, as "sympathy" and "charity," shook from pockets as far away as California, money that rebuilt the mucilated tissues of Italy.

Something else that every conventionalist will explain as "mere coincidence" is that down from the sky came deluges upon the quaking land of Calabria. There was widespread need for water, at this time.

India-"pitiable," as described in accounts of one of the severest of droughts. The wilt of a province-the ebb of its life is at the rate of 2,000 of its starving inhabitants, a day, into the town of Sind. Its people shrivel. and its fields, burned brown, wrinkle with trains of dark-skinned refugees. A band of natives in a desolate corpse-trunks and limbs of leafless, little trees, and shrunken arms and legs merge in one jungle of emaciation. A starving native, ii.at in a field-he has crawled away from the long. white cloud of dust of a trampled road. It might be hell anywhere, but there are glimpses of the especial hell that is India. Breech clout of the starving natives-pinned to it, a string of jewels, which, though dying, he had stolen. Long, wide cloud of dust that is a landscape-girdle-and it is emblazoned with a rajah's elephants.

There was intense suffering. at Lahore. All the gods were prayed to for rain.

Upon the 9th of September, there was an earthquake, at Lahore. All the gods answered at once, combining their deliveries, with an efficiency that smashed houses. Allahabad Pioneer, Mail, September 15- "houses collapsing in great numbers, .and the occupants wandering homeless." "Such an. occurrence at this season is most unprecedented, and has taken everybody by surprise" (Times of India, September 16).

Main Street-any good-sized American city-a dull after noon-the barber shop and the cigar store on the comer much dullness-

A sudden frenzy-Main Street rushes out of the town. Or a human mind in a monotonous state of smugness there's a temptation, or the smash of a conviction, and something that it has taken for a principle rushes out of it, in a torrent of broken beliefs-

That delirium, or frenzy-or anything else mental or hu man-is not exclusively mental or human-but just what are my data for thinking that the principal street of an American city ever did rush out of the town?

Well, something similar. It was at the time of the deluges in India. There was a monstrous fall of water in Kashmir.

Many of the inhabitants of the city of Srinagar, Kashmir, lived in rows of houseboats, upon the river Jhelum, a sluggish, muddy stream, with so little visible motion that, between the rows, it looked like a smooth pavement. It suddenly went up 17 feet. The two long rows of houseboats rushed away. ·

Another river in Kashmir smashed a village. On its banks it left parallel confusions.

Notch a butterfly's wings-this is mutilation. But correspondingly notch the other wing, and there is balance. Two mutilations may be harmony. The doubly hideous may be beautiful. If, on both sides of the river that was a subsiding axis, mothers simultaneously screamed over the bodies of children, this correspondence was the soul of design. Two anguishes, neatly balanced in parallel lines of wreckage, satisfy the requirements of those who worship godness as only harmony.

A quaking zone of Europe and Asia was deluged. Drought in Turkey-earthquakes-plentiful rain (Levant Herald, September 11, 18). Tremendous falls of water and shocks were continuing in Calabria. Spain was flooded.

September 27th-another severe shock, at Lahore; and, this day, again dust, of unknown origin, fell from the sky, in Calabria. A current of hot air came with it. According to the Levant Herald, October 9, many persons were asphyxiated. According to description. it was a volcanic blast that cannot be traced to origin upon this earth. If it came from somewhere beyond this earth, such a repetition in Calabria is a coincidence, or is an indication that this earth is stationary. It is easier to call it a coincidence.

There was a new star.

Upon the night of August 18th, an "auroral" beam, such as has often been seen in the sky, at times of volcanic emptions upon this earth, and at times of new stars, was seen in England (English Mechanic, 82-88). Upon the 31st of August Mrs. Fleming, at Harvard University Observatory, looking over photographic plates, saw that a new star had been recorded, on and after the 18th. The new star, diminishing, continued to shine during September.

Our expression upon "auroral" beams is that vast beams of light have often been seen in the sky, at times when terrestrial volcanoes were active; that similar appearances have been se n at times of new stars, and may be considered light-effects of volcanoes, not terrestrial. For records of several of these beams, one while Stromboli was violent. Sept. 1, 1891, and one, July 16, 1892, while one of the greatest eruptions of Etna was occurring, see Nature, vols. 44 and 45, and Popular Astronomy, 10 449. For one of the latest instances, see newspapers, April 16, 1926: while Mauna Loa, Hawaii, was in eruption, a beam of light was seen in Nebraska.

There's a new light in the heavens, and there's a disturbance upon this earth, as if an interaction that could not occur, if trillions of miles intervene.

"Mere coincidence."

There's a quake in Formosa, and there's a quake in California.

"Only coincidence; say the conventionalists, who are committed to local explanations.

April 18, 1906-the destruction of San Francisco. The Governor of California appointed a commission of eight Professors to investigate the catastrophe. The eight Professors ignored, as coincidence, everything else that had occurred at the time, and explained in the usual, local, geological terms. In Nature, 73-608, is published Dr. Charles Davison's explanation, which is in terms of a local subsidence. Dr. Davison mentions nothing else that occurred at the time.

At the time-a disastrous quake in Formosa, and the most violent eruption of Vesuvius since April, 1872; activity in a long-dormant volcano in the Canary Islands; quake in Alberta, Canada; sudden rise and fall of Lake Geneva, Switzerland; eruption of Mt. Asama, Japan.

See back to the occurrence of St. Pierre, Martinique. May, 1902-30,000 persons, who perished properly-blackened into cinders, with academic sanction. They turned into ashes, but the principles of an orthodoxy were upheld.

January, 1907-and the ignoramuses of Jamaica. They saved their own lives, because they did not know better.

About 3 o'clock, afternoon of Jan. 14, 1907, there was sudden darkness, at Kingston, Jamaica. People cried to one another that an earthquake was coming, and many of them ran to parks and other open spaces. The earthquake came. The people who ran to the open spaces lived, but a thousand of the others were killed by falling houses.

A web that was spun of dogmas caught a thousand victims. After the quake, the ruins of Kingston sprawled like a spider, stretching out long, black lines that were trains of hearses. But all who ran to the parks, believing that appearances in the sky did mean that catastrophe upon earth was corning, lived. I have given data for thinking that a De Ballore, or any other conventionalist, would ridicule these people for so interpreting "a mere coincidence."

October, 1907, and March, 1908-- falls, from the sky, of substances like soot and ashes-catastrophes upon this earth and new stars that were discovered by amateurs. See the English Mechanic, 86-237, 260, and the Observatory, 31-215.

Dec. 30, 1910-new star-disastrous earthquakes-an enormous fall, from the sky, of a substance like ashes. The new star, Nova Lacertae, was discovered by Dr. Espin, a professional astronomer. Photographic records were looked up. Almost six weeks this star had been shining, unobserved at this earth's Observatories. It was visible without a tele scope (5th magnitude).

For almost six weeks, a new star had shone over the Observatories of this earth, and no milkman had reported it. However, without chagrin, we note this remissness, because, it is no purpose of this book to spout eulogies to the amateur in science. It is only in astronomy that the humiliation of professionals by amateurs is common. I have no records of little boys running into laboratories, startling professors of chemistry, or physics, with important discoveries. The achievements of amateurs in astronomy rank about with a giving of information, upon current events, to a Rip Van Winkle. I'll not apologize, because no night watchman hammered for several hours upon· the front door of an Observatory, rudely disturbing the spiders.

XXIX

WHY DON'T THEY SEE, when sometimes magnificently there is something to see?

The answer is the same as the answer to another question:

Why, sometimes, do they see when there is nothing to see?

In the year 1899, Campbell, the astronomer, "announced that the star Capella is a spectroscopic binary, or has a companion-star, as determined by the spectroscope. Astronomers of Greenwich Observatory investigated. One of them looked, through the telescope, and he said, or rather, announced, that he saw it. Another of the astronomers looked for the companion. He announced that he saw it. Eight other astronomers followed. Each announced that he saw it. But now the astronomers say that this alleged companion cannot be seen in any telescope. See Duncan, Astronomy. 335.

The Andromeda nebula is said to be so far away that, though a description of a nearby view of its parts would read like divorce statistics in the United States, no dissolving motion can be seen by observers on this earth. In astronomical books, published in the past, appeared reproductions of photographs of this nebula, which were as artfully touched up, I should say, as any life of a saint ever was by any theologian. It was given a most definitely spiral appearance, to convey an impression of whirlpool-motion. But the nebula theory of existence has passed away. In astronomical books of recent date we see no such definite look of a whirlpool in pictures of the Andromeda nebula, but more of a stratified appearance. The astronomers see whatever they want to see-when they do see-and then see to it that we see as they see. So, though according to our records, one would think otherwise, there is considerable seeing by astronomers. If I look at a distant house, and see faces appearing at the windows, something seems to tell me that the stoop of the house is not Hying in one direction, the roof of it scooting some other way, and every brick upon a jamboree of its own. Of course, minutely, there are motions. But, if the house is not so far away as to prevent the seeing of new faces looking out; I argue that other changes, such as the roof in a frenzy to get away from the stoop, would, if there were such incompatibilities, be visible. Of course this is only argument. If we can have neat little expressions, that's mentality's profoundest.

The Andromeda nebula is said to be so far away that tremendous motions of its parts cannot be seen.

But more than fifty new stars in it have been seen, looking out.

So we are realizing bow numerous new stars are, if in one little celestial formation more than fifty have been seen. If amateur astronomers were as numerous as amateur golf players, for instance, we'd realize much more.

Pronounced changes, such as appearances and disappearances of stars, have been watched, but no change in relative positions of stars has ever been watched occurring. There are parts of the sky that are dusty with little stars. If they were not such good-looking little things, the heavens would be filthy with them. But no grain of these shining sands has ever been watched changing its position relatively to other grains. All recorded changes of positions are so slight that some of them may be attributed to inaccuracies in charting. in earlier times, and some to various stresses that have nothing to do with independent motions. Just here we are not discussing the alleged phenomena of "companion stars." But our own expressions require that there be small changes in positions of stars, just as terrestrial volcanoes change slightly. Not a star has ever been seen to cross another star, but observations upon other changes in the stars are frequent.

For records of five new stars in five months, see Popular Astronomy, March, 1920.

Many of the so-called new stars have been sudden flares of faint old stars. Upon this earth there have been sudden flares in volcanic craters that were dormant, or that were supposed to be extinct. And it was not by collision. Nothing came along and knocked against them.

Apart· from our expressions upon organic suppressions, it is easy enough to understand one aspect of the origin of the present astronomical doctrine. It was in a time of mathematicians, to whom astronomical observations were secondary. The only one of these earlier ones who was an industrious observer (Tycho Brahe) gave his opinion that this earth is stationary. The rest of them did little observing, and spent their time calculating. Nowadays new stars are seen often, but, for 178 years, the calculators saw not one of them. In their time it was considered crude, or vulgar, to see. Mentality always has been bullied by snobbery. Io the times of the founders of astronomical dogmas, observations were sneered at, and were called empiricism. Any way that is not the easiest way always is held in contempt, until com petition forces harder methods. The easiest of all affectations is the aristocratic pose, if by aristocracy we mean minimization of doing anything. There's a coarseness to anybody who works. Give this a thought-he might sweat. Amateurs, out in their back yards, see, with their little spyglasses, much that the professionals miss, but they catch colds. When a back yard amateur, like W. F. Denning, reports something, that represents patience and snuffling. Denning blew his nose, and kept his eyes open. The inmates of Observatories, when not asleep, are calculating. It's easier on brains, and it's easier on noses. Back in times when little boys were playing hop-scotch and marbles, and had not yet taken up the new sport of giving astronomers astronomical information, or in those times when only astronomers were attending somewhat to astronomical matters, and when therefore changes in stars were unheard of, arose the explanation of vast distances, to account for unobserved changes.

The look is that stars do not change positions relatively to one another for the same reason that Vesuvius and Etna do not. Or there are very slight changes of position, just as relatively to each other Vesuvius and Etna change: but no star has ever been seen to pass over any' other star, any more than has Vesuvius ever been seen sailing in the sky over Etna. Other changes of stars that are said to be so far away that changes of position cannot be seen, have been noted. For a discussion of stars that have disappeared, see Nature, 99-159. For a list of about 40 missing stars, see Monthly Notices, R. A. S., 77-56. This list is only supplementary to other lists.

Upon March 14, 1912, the newspapers told that the discovery of a new star had been "announced" by the-Kiel Observatory, Germany. No reader of newspapers, of that time, would suppose otherwise than that vigilant astronomers, at least worth their keep, knew when a new star appeared.

Early in the morning of March 11th, earthquakes of unusual intensity were recorded at many places in the United States. At Harvard University, the calculators announced that the center of the quake was in the West Indies, or Mexico. Newspaper readers, if they paid any attention, were properly impressed with this ability of intellectuals in Massachusetts to know what was going on in the West Indies, or Mexico. But newspapers the next day told of a quake, upon the 11th, of Triangle Island, off Vancouver, B. C., and of nothing in either the West Indies or Mexico. At Victoria, B. C., it was calculated that the center of the quake was in the Pacific Ocean, 400 miles westward. The same readers, forgetting just where the calculators of Harvard had placed the quake, thought it marvelous how the scientists can know these things.

Sometimes distant skies turn black with the shadows of disasters. Had this quake centered in a densely populated region, we'd have another datum of a distant fall of probably volcanic material, about the time of a catastrophe. Upon the day of the quake, black water fell from the sky, near Colmer, about 30 miles from London (Jour. Roy. Met. Soc., 38-275). The rain was not muddy, but was like diluted ink. Somebody else thought not, pointing out that, if this were so, ink. not much diluted, would often fall in London.

The night of the 12th, the astronomers of this earth's Observatories were calculating. Wherever the town of Dom bass, Norway, is, the astronomers of Kiel, Germany, were disturbed by a telegram. It was from an amateur, named Enebo, telling that there was a new star in the constellation Gemini, and that it was visible without a telescope. Astronomers in other parts of this earth were notified. They looked up from what they considered astronomical matters, and saw what the amateur had discovered.

In November, 1913, an astronomer photographed a part of the constellation Sagittarius. I don't know what his idea was. Perhaps simply and somnambulically he photographed, and had no ideas. Six years later, somebody found out that he had photographed a new star. Then other photographs were examined. Astronomers are pretty keen at detecting something that has been pointed out to them, and they learned that they had photographed this star, rising from 10th to 7th magnitude, between the 21st and the 22nd of November.

Wanted-something by way of data for us. Like every other theorist, we find just that - Nature, 94-372-that, seven days after the maximum of this new star, an afterglow, which can be attributed to no known volcanic eruption upon this earth, was seen in the sky, in Italy, France, Belgium, and England.

April 25, 1917-a professional astronomer photographed a new star (magnitude 6.5) in the constellation Hercules. The next day there was a disastrous earthquake in Italy. Upon May 1st, a great quake-perhaps in the constellation Hercules-whereabouts unknown to this earth's scientists-was registered by this earth's seismographs (Nature, 99-472).

Domes of Observatories look like big snail shells. Architectural symbolism. It took the astronomers about three years to learn that they had photographed the new star in Hercules (Pop. Astro., March, 1920). If newspaper editors were like astronomers, they'd send out photographers, rather busily, and, perhaps years later, if they could condescend from journalism into doing some newspaper work, they'd examine plates. They'd tell of a fire that had occurred long before. They'd write up some fashion notes upon the modes in their readers' childhood. Like dealers in stale stars, they'd wonder at a lack of public interest.

Upon March 6, 1918, black rain fell from the sky, in Ireland (Symons' Met. Mag., 53-29.) If our preconceptions so direct, we relate this occurrence with smoky discharges from factory chimneys of South Wales, or somewhere else in Great Britain-and it is better that we do not ask why black rains are not common near Pittsburgh. Or we note that the next night there was in the sky a crimson appearance that worried many communities in Europe and North America. For a week there were, in the newspapers of New York and London, descriptions of this glare, and comments upon it. People thought that there was a great fire somewhere. I give data for thinking that there may have been a volcanic eruption somewhere.

March 6th-the fall of black rain. March 7th-the glare in the sky. March 9th-down upon this earth fell dusts in volumes that were proportional to the glare. See Amer. Jour. Sci., Monthly Weather Review, and Sci. Amer., of this period. There was a fall of dust in Wisconsin, and in Michigan; and there was a fall of dust in Vermont. These falls, so far apart-in Ireland, in Western States, and in Vermont-look like what is called indication of an origin somewhere beyond this earth. There is no findable record of any disturbance upon this earth, by which to explain. No new star was reported, but there may have been a stellar eruption in a part of the daytime sky, reflecting in a glare, at night. There may have been relation with an occurrence in June. In the meantime, there were several remarkable glares in the sky.

Early in the evening of June 8th, of this year 1918, two men, one of them in Madras, India, and the other in South Africa. looked up at the sky, and saw a brilliant new star in the constellation Aquila. Each of them notified an Observatory, which had not been observing. Evening of the 8th Harvard University Observatory notified .by an amateur. The astronomers of Harvard had seen nothing new, but telegram after telegram came· to them from other amateurs. Whatever else the astronomers of Lick Observatory were doing, I don't know. They were probably calculating. But they, too, ·were receiving telegrams, and when told, by amateurs, to look up at the sky, and see a new star, they looked up at the sky, and saw the new star. See Pubs. Astra. Soc. Pacific, August, 1918. Besides the amateur in Madras, an amateur in Northern India notified the Observatories (Nature, 102-105). English Mechanic, August 9-professionals of New Zealand notified by an amateur. In Nature, 101-285, is published a list of amateurs, who, in England, had reported this new star to official centers of unobservation. There is only one record of a professional astronomer, who, without information from amateurs, saw this new star. One of the astronomers of Greenwich Observatory had looked up at the sky, and had seen this new star. Nature, 101-285- that he had seen it, but had not recognized that it was a new star. One of the amateurs who saw it, and recognized it, was a schoolboy named Wragge (London Times, June 21). The Lisbon Observatory was notified by a boy, aged 14 (Observatory, 41-292).

XXX

I AM thinking of an abstraction that was noted by Aristotle, and that was taken by Hegel, for the basis of his philosophy: That wherever there is a conflict of extremes, there is an outcome that is not absolute victory on either side, but is a compromise, or what Hegel called "the union of complimentaries."·

Our own controversy is an opposition of extremes: That this earth moves swiftly;

That this earth is stationary.

In terms of controversies and their outcomes, I cannot think that either of these sides can be altogether right, or will absolutely defeat the other, when comes some way of finding out, and settling this issue.

The idea of stationariness came first. Then, as sheer mechanical reaction-inasmuch as Copernicus had not one datum that a conventionalist of today would accept as meaning anything-came the idea of a swiftly moving earth. An intermediate view will probably appear and prevail.

My own notion of equilibrium between these extremes, backed up. with our chapters of data, is that, within a revolving, starry shell that, relatively to the extravagances of the astronomical extremists, is not far away, this roundish earth is almost central, but is not absolutely stationary having various slight movements. Perhaps it does rotate, but within a period of a year. Like everybody else, I have my own notions upon what constitutes reasonableness, and this is my idea of a compromise.

The primary view had for its support the highest mathematical authoritativeness of its era. Now, so has the secondary view. Mathematics has been as subservient to one view as the other.

Mostly our data have been suggestive, or correlative, but it may be that there are visual indications of a concave land in the sky, or of a substantial shell around this earth. There are dark places in the sky, and some of them have the look of land. They are called "dark nebulae." Some astronomers have speculated upon them, as glimpses of a limiting outline of a system as a whole. See back to Dolmage, quoted upon this subject. My own notion is of a limiting, outlining substance. that I call a "shell." "Dark nebulae" have the look of bare, or starless, patches of a shell. Some of them may be formations that are projections from a shell. They hang like superstalactites in a vast and globular cave. At least one of these appearances has the look of a mountain peak. In several books by astronomers plates of this object have appeared. See Duncan's Astronomy. It is known as the Horse-head nebula. It stands out, as a vast, sullen refusal to mix into a frenzy of phosphorescent confetti. It is solid looking gloom, such as, some election night, the Woolworth Building world he, if Republican, and all the rest of Broad way hysterical with a Democratic celebration. Over its summit comes light, like the fringe of dawn topping a mountain. Something is shining behind this formation, but penetrates no more than it would shine through a mountain.

It may be that relatively there are few stars-that hosts of tiny lights in the sky are reactions, upon irregularities of the shell-land, from large stars.

Among expressions that I have not developed is one that is suggested by a circumstance that astronomers consider strange. This is that some variable stars have a period of about a year. Just what variations of stars that are said to be trillions of miles away could have to do with a period upon this ultra-remote earth cannot be conceived of in ortho dox terms. The suggestion is that these lights, with variations corresponding with advances and recessions of the sun, moving spirally around this almost stationary earth, are reflections of sunlight from points of land, or from lakes in extinct, or dormant. craters. It may be that many variations of light that have been attributed to companions. are tidal phenomena in celestial lakes that shine as reflections from the sun, or from other stars, which may be lakes of molten Java.

There is a formation in the constellation Cygnus that has often been noted. It is faintly luminescent, but this light, according to Prof. Hubble, is a reflection from the star Deneb. It is shaped like North America, and it is known as the America nebula. Out from its Gulf of Mexico are islands of light. One of these may be a San Salvador someday.

Like Alaska to birds from the north, the Horse-head nebula stands out from its background like something to fly to.

Star after star after star has blazed a story, sometimes publishing tragedy on earth, illustrated with spectacles in the heavens. But, when transcribed into human language, these communications are depopularized with "determinations" and "pronouncements." So our tribes have left these narratives of fires and smokes and catastrophes to the wise men, who have made titanic tales unintelligible with their little technical jargons. The professionals will not unprofessionalize; they will not give up their system. Where have the wisemen ever done so among the Eskimos, or the hairy Ainus, the Zulus, or the Kaffirs? Whatever we are, they are acting to keep us whatever we are, as the Zulus are kept whatever they are. We are beguiled by snoozers, who have been beaten time after time by schoolboys.

There's a fire in the sky, and ashes and smoke and dust reach this earth, as sometimes after an eruption of Vesuvius, discharges reach Paris. There may be volcanoes in a land of the sky, so close to this earth that; if intervening space be not airless and most intensely cold, an expedition could sail away in a dash to the stars that Would be a bold and magnificent trifle.

XXXI

BESIDES THE NEW STAR, which was an object so conspicuous that it was discovered widely, except by astronomers, there was another astronomical occurrence in the month of June, 1918-an eclipse of the sun. It was observed in Oregon. We can't expect such a check up as when Coogan's Bluff and the Consolidated Gas Company get into astronomy, hut Oregonians set their alarm clocks, and looked up .at the sky. See Utchell's Eclipses of the Sun, p. 67-the astronomers admitted an error of 14 seconds in their prediction.

Measurements of ordinary refinement are in hair-breadths, but a hair is coarse material to the ethereal astronomers, who use filaments spun by spiders. And just where do the astronomers get their cobwebs? This book of ours is full of mysteries, but here is something that is not one of them.

My own opinion is that an error of only 14 seconds is a very creditable approximation. But it is a huge and grotesque blunder, when compared with the fairy-like refinements that the astronomers dream are theirs, in matters that cannot be so easily checked up.

To readers who are not clear upon this point, I repeat that predictions of eclipses cannot be cited in support of conventionality against our own expressions, because, whether upon the basis that this earth moves, or is stationary, eclipses can be predicted-and Lo! come to pass. But Lo! if, looking on, there be an intelligent representative of the Consolidated Gas Company, or an Oregonian with an alarm clock, predictions aren't just exactly what they should be.

We have divided astronomers into professionals and amateurs: but, wherever there are differences there is some where the merging-point that demonstrates continuity. W. F. Denning represents the amateur-professional merging-point. He has never had a job-though it does not look to me that job is the right word-in an Observatory, but he has written a great deal upon astronomical matters. He is an accountant, in the city of Bristol, England. Has nothing to do with Observatories, but has a celebrated back yard. Upon the night of Aug. 20, 1920, Denning sat in his back yard, and, in surroundings that were touched up most unacademically by cats on the fences, though Observatory-like enough, with mores from back windows, he discovered Nova Cygni III. This is another instance of a new star appearing close to where there had been preceding new stars, as if all were eruptions in one region of especially active volcanic land. There were earthquakes in this period. In the United States, there were the sudden deluges that are called "cloudbursts." Upon the night of August 28th, a seismic wave drowned 200 persons, on the island of Saghalien, off the east coast of Siberia.

For four nights, astronomers of the so-called Observatories had been photographing this star. Students of phenomena of somnambulism will be interested in our data. When Denning woke up the astronomers, they looked at what they had been unconsciously doing, and learned that from the 16th of the month, this star had risen from 7th to about 3rd magnitude. A star of 3rd magnitude is a conspicuous star. In the whole sky there are (photographic magnitudes) only 111 of this size. At any one time not more than 40 of them are visible. The limit of visibility, without a telescope, is somewhere between the 6th and 7th magnitudes. So it is said. According to our data, the limit of seeing depends upon who's looking.

I wonder what ironic fellow first called these snug, little centers of inattention Observatories. He had a wit of his own, whoever he was.

Discovery of the new star, if not a comet, of Aug. 7, 1921. has been attributed to a professional· wiseman (Director Campbell, of Lick Donrutory). But it was a brilliant and conspicuous appearance. Most of the new stars that professionals have discovered, or have had discovered for them, by the not very eagle-eyed females of Harvard, have been small points on photographic plates. English Mechanic, 114-211- records of observations by four amateurs, before the time of Director Campbell's "discovery." One of these observations was twenty-four hours earlier.

Sometime ago, I read an astronomer's complaint against heavy traffic near an "Observatory." Though now I have different ideas as to an astronomer's dislike for disturbance, night times, I was not so experienced then, and innocently supposed that he meant that delicate instruments were jarred.

A convention of Methodist ministers-and how agreeable it would be to note, in the midst of preciseness and purity, one of these parsons standing on his head-

Or see the Monthly Notices of the Royal Astronomical Society, 1922-upon page 400 there is a diagram. Mistake that I make-and errors of yours-

Contrasting with the much-advertised divinity of the astronomers.

Page 400-in the midst of a learned treatise upon "adiabatic expansions" and "convective equilibrium "is printed a diagram. It is upside down.

I attended this convention of pedantries, of course inspired by a religious faith that is mine that I'd not have to look far for a crook in its bombast, or somewhere a funny little touch of "·waywardness" in its irreproachability, but especially I attended to pick out something of which, in this year 1922, the astronomers were boasting, to contrast with something they were doing. I picked out a Jong Jaudation upon an astronomer who had received a gold medal for predicting the motions of a star-cluster for a term of 100,000,000,000 year to contrast with-

Sept. 20, 1922-an eclipse of the sun-see Mitchell's Eclipses of the Sun, p. 67-and the predictions by the astronomers. They made one error of 16 seconds, and another error of 20 seconds.

There are persons who' do not believe in ordinary fortune tellers Yet, without a quiver in their credulities, they read of an astronomical gypsy who tells the fortunes of a star for 100,000,000,000 years, though, according to conventionality, that star is 60 x 60 x 24 x 365 x 100,000 times frier away than is the moon, motions of which cannot be exactly foretold, unless the observations are going to be, say at Paranagua, or somewhere in Jungaria.

The eclipse of Sept. 20, 1922, was checked up by police constables. in Australia. But the eclipse of Oct. 21, 1930, was observed at Niuafou. This time the dispatches sent by the astronomers told of a complete success." "The eclipse began exactly as predicted."

There are records of seeming new stars that have blazed up, like spasmodic eruptions, then dying out. For Dr. Ander son's report upon one of these appearances, May 8, 1923, see Popular Astronomy, 31-422. Upon the 7th, Etna was active; earthquake in Anatolia; extraordinary rise and fa)) oi the Mediterranean, at Gibraltar. The "Observatories" miss ed Dr. Anderson's observation, but at one of them a small, new star was photographed, night of May 5th. I neglected to note whether, on a photographic plate, this was immediately detected. See Pop. Amo., 31-420.

Upon Feb. 13, 1923, an increase of the star Beta Ceti was reported. There was interest in the newspapers. Maps of the sky were published. If newspapers start first-paging astronomical occurrences, putting down X-marks for stars, as well as for positions of bodies of the murdered, there will be more interest. This is dangerous to the astronomers, but so long as their technicalities hold out, they have good protection.

Even so there might be inquiries into what the "Observatories" are doing, when, time after time, only amateurs are observing. The "Observatories" had of course missed this rise of Beta Ceti, but, when· told by an amateur where to look, professionals at Yerkes and Juvisy confirmed the report. For the fullest account, see the Bull. Soc. Astro. de France, of this period. Upon February 22nd, a yellow dust, perhaps a discharge from an increased volcanic activity in Cetus, fell from the sky, in Westphalia (London Evening Standard, February 27). The amateur, this time, was a schoolboy, aged 16.

Night of May 27, 1925-the Rip Van Winkles of the South African "Observatory" were disturbed by an amateur. He told them that there was a new star in the southern constellation Pictor. When they were aroused, the Rips looked up and saw the new star, and then stayed awake long enough to learn that somnambulically they had, for months, been photographing it. For months it had been gleaming over the "Observatories" of four continents. There are slits in the domes of Observatories.

The fixed grin of a clown-the slit in the dome of an Observatory.

Sept. 21, 1930-that the astronomers had ascertained the heat received from a thirteenth magnitude star to be 631 times that of the heat from the faintest star visible to the unaided eye-that this faintest visible star radiates upon the whole United States no more heat than the sun radiates upon one square yard of said U. S.

A grin in the dark-or the sardonic slit in any Observatory, night times. Most likely the inmates haven't a notion what is symbolized. But we contrast an alleged perception of 631 times the inconceivable with this item, in Popular Astronomy, 1925, p. 540:

That 44 nights before the amateur's discovery, Nova Pictoris had shone as a star of third magnitude, and had been perceived by no astronomer.

The Building That Laughs-as a modern Victor Hugo would call an Observatory with a slit in it.

Forced grin of the clown-and, according to theatrical conventions, his head is full of seriousness.

Sept. 24, 1930-this is what came from a Building That Laughed, though its dome was full of astronomers:

That, according to spectroscopic determinations at Mt. Wilson, a distant nebula is moving away from this earth at a rate of 6,800 miles a second; that, upon the day of the calculations, the distance of this nebula was 75,000,000 x 60 x 60 x 24 x 365 x 186,00 miles.

To appreciate the clownishness of this, see our data for accepting that the spectroscope tells about what is told by tea leaves in a teacup-which is considerable, if one wants to be told considerable. To realize the pathos of this, think of the grinning old clown, whose gags have played out; who is driven to most extravagant antics to hold a little attention. Our general expression is that the inmates of this earth's "Observatories" are not astronomers, but are mathematicians. Since medieval times there has never been a shake-up in this system of ancient lore, comparable with Lyellisms in geology, or Darwinism in biology, or the reconstructions of thought brought about by radio-activity, in physics and chemistry. Einsteinism was a slight shock, but it is concerned with differences of minute quantities. Mathematicians are incurable. They are inert to the new, because the new is a surprise, and mathematics concerns itself with the expected. It does occur to me that there might be good results, if the next millionaire who contemplates donating a big telescope, should, instead, send around to the "Observatories" big quantities of black coffee: but such is the concordance be tween the twinkles of the stars and the nods of drowsy

heads that I'd not much like to disturb such harmony.

Nova Pictoris, like many other so-called new stars, was an increase of an old star. For twenty-five years it had from time to time been photographed as a speck of the, 12th magnitude.

There is nothing on any photographic plate to indicate that another star was going to collide with it. It went up, just as dimly shining, or only slightly active, volcanoes of this earth sometimes become violent.

No star has ever been seen to cross another star, but just such changes as have been seen in volcanoes of this earth have been seen in stars. Mostly, in their books, astronomers, telling of what they call "proper motion," do all that they can to give an impression of the stars as moving with tremendous velocities, but here is Newcomb (Astronomy for Everybody, p. 327) quoted upon the subject: "If Ptolemy should come to life, after his sleep of nearly eighteen hundred years, and be asked to compare the heavens as they are now with those of his time, he would not be able to see the slightest difference in the configuration of a single constellation."

And, if Ptolemy should come back, and be asked to com pare the Mediterranean lands as they are now with those of his time, he would not be able to see the slightest difference in the configuration of any land-even though erosions of various kinds have been constant.

What Orion was, Orion is, in the sense that what the configuration of Italy was, it now is-in the sense that in all recorded time Italy has been booting Sicily, but has never scored a goal.

There is no consistency, and there is no inconsistency in our hyphenated state of phenomenal being: there is consistency-inconsistency. Everything that is inconsistent with some thing is consistent with something else. In the oneness of allness, I am, in some degree or aspect, guilty of, or infected with, or suffering from, everything that I attack. Now I too, am aristocratic. Let anybody else who is as patrician as I now am read this book, and contrast the principles of orthodox astronomy with the expressions in this book, and ask himself:

Which is the easier and lazier way, with the lesser necessity for effort, and with the lesser need for the use of brains, and therefore the more aristocratic view:

That for, say eighteen hundred years, stars have scarcely moved, because, though changes in them have often been seen, they are too far away for changes in them to be observed;

Or that the stars have scarcely moved, because they are points in a shell-like formation that holds them in place?

However, the orthodox visualization of stars rushing at terrific velocities, in various directions, and never getting anywhere, is so in accordance with the unachievements of all other phenomenal things that I'd feel my heresies falter were it not for other data-

But what of other data-or of other circumstances?

In this day of everybody's suspicion against "circumstantial evidence," just what is not generally realized is that orthodox astronomy is founded upon nothing but circumstantial evidence. Also all our data, and repetitions and agreements of data, are nothing but circumstantial evidence. Simply mention "circumstantial evidence" relatively, say to a murder trial, an most of us look doubtful. Consequently I have only expressions and acceptances.

Other data-or other circumstances-

Last of March, 1928-that Nova Pictoris has split into two parts.

Part was seen to have moved from part, as divisions occur in this earth's volcanoes.

So then, when changes of positions of stars do occur, the stars are not so far away that changes of position cannot be seen.

Ten little astronomers squinting through a tube-or more characteristically employed-or looking at a mirror. They had been told, upon the highest authority, that the star Capella had a companion. Said they-or announced they saw it-or perceived it. Having calmed down, in the matter of "dust from an African desert," but seeming to have a need for something to be furious about, I now tom my indignations upon "companion stars." Most persons have, in their everyday affairs, plenty to annoy them: but it seems that I must have something of exclusiveness to my annoyances. If stars be· volcanoes in a concave land, surrounding this earth, the notion of "companion stars" perhaps enrages me, because I do not visualize one volcano revolving around another volcano. If some stars do revolve around other stars, I may as well give up this book, as a whole-or I shall have to do some explaining.

Which won't be much trouble. Explaining is equilibrating. That is what all things phenomenal are doing. I now have a theory that once upon a time our existence was committed as a bad error, and that everything in it has been excusing itself, or has in one way or another been equilibrating, ever since. It is as natural to a human being to explain as it is to a lodestone to adjust to a magnet. ·

Let anybody look up "determinations" upon the "dark companion" of Algol, for instance. He will find record not of a theory, but of theory after theory replacing one another. In the matter of the light ones, let him look up data upon the "right companion" of Sirius. He will read in the textbooks that around Sirius a light star revolves, with a most accurately known period, which. demonstrated the soundness of mathematical astronomy. But, in scientific journals, which are not so uncompromisingly committed to propaganda, he will read that this is not so. A faint light has, at various times, been reported near Sirius, in positions that do not accord with the calculated orbit. For no mention of this discrepancy, read the books that reach the general public.

March, 1928-the split of Nova Pictoris. There was cataclysm in a southern constellation. At the same time there were catastrophes in southern parts of this earth. See back to other expressions upon seeming relations between parts of this earth and parts of the sky that would be nearest to each other, if the stars be points in a shell of land that is not enormously far away, but could not be appreciably nearer, if the stars were trillions of miles away. I take all data from New York newspapers. Quakes in Italy, and a glare in the sky at the time (March 31st) of a quake in Smyrna-"sky aflame." The heaviest rain in 50 years poured in Honduras, April 9th-Peru shaking, this day-such a fall of snow in Chile that 200 persons and thousands of farm animals were reported to be buried in drifts-quakes and panics in Mexico-

Orthodoxy-all this by mere coincidence-

Our expression-that nebulous rings were going out from Nova Pictoris, just as rings of smoke and dust go out from Vesuvius, during an eruption.

The 14th of April was the day of the Bulgarian devastations. Quakes continuing in Bulgaria - quakes in Mexico towns rocking in southern Mexico-quakes continuing in Peru. Quakes in Greece, on the 19th-a violent snowstorm in Poland, this day. Torrents were pouring upon the quaking land of Bulgaria. A De Ballore, or a Davison, or a Milne. would not mention these torrents, in an account of this quake. The severest shock ever recorded in Johannesburg, South Africa, occurred upon the 21st. The next day, Corinth, Greece, was wrecked, and torrents fell from the sky, at the time of this quake.

Nova Pictoris broke into four parts-and the cities of Greece wailed rushes of people. Seeming discharges moving out from the new star-and "A five-hour rain of mud filled streets ankle high, causing terror at Lemburg and Cemowitz, today" (New York Sun, April 27th).

Wails of the cities of Greece-and they subsided into sodden despairs that were processions of stretchers. Some where in a building that collapsed, fell a sparrow.

The road from Corinth-refugees and their belongings Terrified mules, up on their hind legs, hoofing storms of bundles-yells and prayers and the laughter of jokers-a screaming woman, shaking bloody hands-her fingers had been hacked off, for the rings on them. Crying kids, whose parents were pulps-prayers to God, or to the blessed some-thing or another-the screams of the woman, with stumps of fingers-

Sudden consciousness of a pulsation.

A rhythm of gleams appears in distant sunlight.

Stars that are watched through the windows of prisons or through openings in any of the other hells of this earth and it may be that if all the stars should start to twinkle in unison, the hells of this earth would vibrate out of existence.

There's a rhythm of gleams on distant bayonets. Along the road is marching a column of soldiers.

The swing of these gleams-and it tranquilizes panic. It glistens into new formations. There are long lines of sparkles in sunlight-tin cups are undulating toward soup kettles.

Somewhere else there is an injured sparrow. Storages in its body are giving to its needs from their substances-the tranquilizing of its heart beats, and the reduction of its fever -the rebuilding of its tissues.

A British squadron appears in the Bay of Corinth-an Italian warship-an American cruiser. From centers of the American Near East Relief are streaming 6,000 blankets- 10,500 tents-5,000 cases of condensed milk-carloads of flour.

If we can think that around this earth, and not too vastly far away, there is a starry shell, here are the outlines within which to think of our existence as an organism.

Nov. 28, 1930-an enormous fall, from the sky, of dust and mud, in France. I shall not get perhaps all worked up again about this, but I mention that it was attributed to a hurricane in the Sahara Desert. Dec. 5, 1930-the poisonous gas, in Belgium. See back to the account, in this book.

Accept that these two phenomena were probably volcanic discharges, from regions external to this earth-if for them there be no terrestrial explanation-one in France and one in Belgium-arriving relatively near each other, but a week apart-and here is another of our data of this earth's stationariness. This earth broke out, as if responsively to disturbances somewhere else-volcanic eruptions and disastrous quakes.

December 24-26-violent quakes in Argentina and in Alaska-and, between these far-distant places, there was a spectacular arrival of something that may have been a volcanic bomb from a stellar volcano. New York Times, Dec. 26, 1930-the great meteor that was seen and heard in Idaho. "The crash, heard for miles, was described as 'like an earthquake.' "

The deluge that was "only a coincidence," poured upon the quaking land of Argentina. "Rain fell in such torrents that the water was three feet deep in several parts of Mendoza City." A "strange glow" was seen in the sky. "Great spears of colored lights flashed across the sky.

Into the month of January, 1931, disturbances upon this earth continued. There may have been a new star. I have the authority of amateurs for thinking so. New York Times, Jan. 7, 1931-that, at San Juan, Porto Rico, morning of January 6th, from ten o'clock until noon, a strange star had been seen in the western sky. According to an opinion from the Weather Bureau, it may have been, not a star, but the planet Venus. This Venus-explanation of lights that have been seen in the sky, in the daytime, is a standard explanation; but according to records it has often not applied.

Catastrophes and deluges-and, if we can accept that around this earth there is only a thin zone of extreme coldness, which, by the stresses of storms and other variations, may often be penetrated by terrestrial evaporations, so that, unless replenished from reservoirs somewhere else, this earth would go dry, we can understand a mechanism of necessary transportations of floods from the stars to this earth.

Flows of insects and the patter of frogs, and the Pilgrims cross the Atlantic Ocean. Metabolism in the foot of a frog and in the United States a similar readjustment is known as the Civil War. The consciousness of philosophers and theologians and scientists, and to some degree of everybody else, of a state of Oneness-and my expression is that the mis interpretation has been in trying to think of Universality, or the Absolute. Give me more data for thinking that around this earth there is a starry shell that is not vastly far away, and here is the base for a correlation of all things phenomenal.

XXXII

STAB AFTER STAB after star-and the signs that there were. at the times of them. Quake after quake after quake-and the sights in the sky, at the times of them. Star after quake after deluge-the sky boils with significances - there are tempests of indications.

There's a beam of light in the sky, and it dips into a star. Spattering ponds of ink, it scribbles information. The story is that a vast and habitable land surrounds this earth. It is fertile, if showers of organic substances that have fallen from the sky, came from there. The variable stars are intermittent signs that are advertising enormous real estate opportunities. The story is declaimed by meteors, but most of us stolid ones aren't going to be persuaded by any such sensational appeal to the emotions. The story is more obscurely told with clouds of dust that strew Europe. Most of us can't take a hint the size of a continent.

The searchlights of the sun play upon a celebration in the sky. It has been waiting ages to mean something. Just at present known as the Milky Way, it's the Broadway of the Sky, and some day explorers from this earth may parade it-

If this earth is stationary.

According to a great deal in this book, that may be a matter of no importance, nor bearing. If we accept that Teleportation, as a "natural force," exists, and suspect that some human beings have··known this and have used it; and, if we think that the culmination of a series of tele-operations will be the commercial and recreational teleportation of objects and beings, we are concerned little with other considerations, and conceive of inhabitants of this earth willing themselves-if that's the way it's done-to Mars, or the moon, or Polaris. But I take for a proposition that there is an underlying irony, if not sadism, in our existence, which rejoices in driving the most easily driven beings of this earth into doing, at enormous pains and expenses, the unnecessary-the building of complicated telegraph-systems, with the use of two wires-then reducing to one wire-then the discovery that the desired effects could be achieved wirelessly. Labors and sufferings of early Arctic explorers to push north ward over piles of ice, at a rate of three or four miles a day then Byrd does it with a whir.

Consequently, I concern myself with data for what may be a new field of enormous labors and sufferings, costs of lives and fortunes, misery and bereavements, until finally will come awareness that all this is unnecessary.

Upon this basis of mechanical and probably unnecessary voyagings-unless to something disasters to the beings of this earth be necessary-the most important consideration is whether this earth is stationary. There can be no mechanical, or suffering exploration from something that is somewhere one day, and the next day 60 x 60 x 24 x 19 miles away from there.

Then comes the subject of conditions surrounding this earth. If common suppositions be right, or if this earth be surrounded by a void that is intensely cold, penetration to anywhere beyond would probably be, anyway at present, impossible.

I compare ideas upon outer space with former ideas upon spaces in the Arctic regions. Resistances to the idea of exploration are similar. But in the wintertime, Arctic regions are not colder than are some of the inhabited parts of Canada. Stefansson, the Arctic explorer, has written that the worst blizzards ever seen by him were in North Dakota. Prevailing ideas as to the intensity of cold surrounding this earth; and preventing exploration, may be as far astray as are prevailing ideas as to Arctic coldness.

Outer space may not be homogeneously cold, and may be zoned, or pathed, with warm areas. Everything of which one knows little has the guise of homogeneousness. If anybody has a homogeneous impression of anything, that is something that he is going to be surprised about.

In the London Daily Mail, Jan. 29, 1924, Alan Cobham tells of one of his Bights in India. "The air was quite warm, at 17,000 feet, but, as we descended to lower altitudes, it become gradually cooler, and, at 12.000 feet it was icy cold. "The higher the colder• is a fixed idea, just as formerly was the supposition that the farther north the colder the atmosphere. Many reports by aviators and mountain climbers agree. Everybody who does anything out of the ordinary has to think that he suffered. It is one of his compensations. But fixed ideas have a way of not staying fixed.

I'd like to know how astronomers get around their idea that comets are mostly of a gaseous composition, if gases would solidify at the temperature in which they suppose those comets to be moving. But stationariness-and what's the good of any of· these speculations and collections of data, if by no conceivable agility could a returning explorer board a world scooting away from him at a rate of 19 miles a second?

In early times, upholders of the idea of stationariness of this earth argued that a swiftly moving planet would leave its atmosphere behind. But it was said that the air partakes of the planet's motions. Nevertheless, it was agreed that, far from this earth's surface, air, if existing, would not partake of the motions. No motions of this earth away from them have been detected by aviators but it is said that they have not gone p high enough. But will an aviator, starting northward, from somewhere near the equator, partaking we'll say of an axial swing of 1,000 miles an hour, making for a place where the swing is, we'll say, 800 miles an hour, be opposed by the westward motion that he started with, amounting to 200 miles an hour, at his destination? How would he ever get there, without consciously opposing this transverse force, from the beginning of his flight? In the winter of 1927-28, flying south, and then north, Col. Lindbergh re ported no indication of different axial velocities. Whether this earth is stationary, or not, his experience was the same as it would be if this earth were stationary. Or Admiral Byrd over the south pole of this earth. From a point of this earth, theoretically of no axial motion, he flew northward. He flew over land, which, relatively to his progress, spun with in-. creasing velocity, according to the conventionalists. It cannot be said that the air around him was strictly partaking of this alleged motion, because gusts were blowing in various directions. Admiral Byrd started northward, from a point of no axial swing, partaking, himself, of no axial swing, and, as he traveled northward, the land underneath him did not swing away from him. The air was moving in various directions.

There is another field of data. There have been occurrences in the sky which, according to conventionalists, destroy the idea of the stationariness of this earth, and prove its motion. Trying to prove anything is no attempt of mine. We shall have an expression upon luminous night clouds of meteor trains.

Rather often have been observed luminous night clouds, or night clouds, that shine, presumably by reflected sunlight, but with the sun so far below the horizon of observers upon this earth that so to reflect its light the clouds would have to be 50 or 60 miles high, according to calculations. At this height, it is conceded, whatever air there may be does not partake of this earth's motions. If this earth be rotating from west to east, these distant clouds, not partaking of terrestrial motion, would seem to move, as left behind, from east to west. For an article upon this subject. see the New York, Times, April 8, 1928. The statement that such clouds do not partake, and do seem to move from east to west, has been published by conventionalists. To an observer in Central Europe they should, as left behind, seem to move from east to west, at a rate of about 500 miles an hour by terrestrial rotation. The statement has been made that one of these clouds was seen to move, from east to west, the way it should "move," at exactly the rate that should be.

I make the statement that luminous night clouds have moved north, south, east, and west, sometimes rapidly, and sometimes slowly. If somebody can, with data that will have to be accepted, show that, more than once, luminous night clouds have moved from east to west, at a rate of 500 miles an hour in a latitude where they "should.. move at a rate of 500 miles an hour I shall be glad to regret that I have backed the wrong theory-except that you can't down any theorist so easily or at all-and up I'll bob, pointing out that this is another of the should that shouldn't, and that the conventionalists forgot about compounding their 500 miles ·hour with this earth's supposed orbital motion of 19 miles a second.

All data upon this subject that I know anything of are interpretable as indications that this earth is stationary. For instance, look up, in Nature, and other English, and French, scientific journals, observations upon the great meteor train' of Feb. 22, 1909. This appearance was thought to be as high as any luminous night cloud has been thought to be. It was so high that it was watched in France and England. Here was something, which, because it came from externality, was not partaking of any of this earth's supposed motions. Then it should have shot away from observers, by the compounding of two velocities. Whether it came to a stationary earth or not, it hung in the sky, as if it had come to a stationary, earth, drifting considerably, but remaining in sight. about two hours.

According to this datum-and it is only one of many-an explorer could go up from this earth 50 or 60 miles, and though, according to orthodox pronouncements, -the earth would spin away from him, the earth would not spin away from him.

There are data for thinking that aviators, who have gone up from the surface of this earth, as far as they supposed they could go, have missed entering conditions that, instead of being cold, may be even warmish, and may exist all the way to a not so very remote shell of stan. Somebody may want to know how it is that, if there be such data, they are not commonly known. But somebody else, who has read this book at all carefully will not ask that question.

An expression of mine is that all human achievements are compounded with objectives. Let someone go without food for a week, and that is a record of human endurance. Some one else makes his objective a week and a day, and achieves, in a dying condition. The extension goes on, and someone lives a month without food, and reaches the limit of human endurance. Aviators have set their minds upon surpassing the records of other aviators. It is possible that, with its objective a star, an expedition from this earth could, by merely reaching the limit of human endurance, arrive there.

Current Literature, September, 1924-that, 50 miles up, the air is ten times as dense as used to be supposed, and that it is considerably warmer than at lower levels.

See Nature, Feb. 27, 1908, and following issues-experiments with balloons that carried temperature-recording instruments. According to Mr. W. H. Dines, about 30 balloons, which had been sent up, in Great Britain, in June, 1907, had moved through increasing coldness, then coming to somewhat warmer regions. This change was recorded at a height of about 40,000 feet.

Monthly Weather R ,1923, page 316-that, away from this earth, the temperature falls only to a height of about 7 miles, where it is from 00 to 70 degrees below zero (Fahrenheit) . "But from this altitude to as high as balloons have gone, which is about 15 miles, the temperature has remained about the same."

It is said that, according to observations upon light-effects of meteor trains, there are reasons for thinking that, in their zone of from 30 to 50 miles above this earth's surface, conditions are mild, or not even freezing.

For· data that may indicate, in another field of observations, that, not enormously far away, there is a shell around this earth, see the newspapers of Aug. 20, 1925. According to data collected by the Naval Research Laboratory there is something, somewhere in the sky, that is deflecting electro magnetic waves of wireless communications, in a way that is similar to the way in which sound waves are sent back by the dome of the Capitol, at Washington. The published ex planation is that there is an "ionized zone" around this earth. Those waves are rebounding from something. More was published in the newspapers, May 21, 1921. The existence of "a ceiling in the sky" had been verified by experiments at Carnegie Institution. Sept. 5, 193 a paper read by Prof. E. V. Appleton, at a meeting of the British Association for the Advancement of Science. The "ionized zone" is not satisfactory.

"The subject is as puzzling as it is fascinating, and no decisive answer to the problem can be given at present." From Norway had been reported experiments upon short wave transmissions, which had been reflected back to this earth. They had come back, as if from a shell-like formation, around this earth, not unthinkably far away.

Wild Talents

Wild Talents

1

YOU know, I can only surmise about this--but John Henry Sanders, of 75 Colville Street, Derby, England, was the proprietor of a fish store, and I think that it was a small business. His wife helped. When I read of helpful wives, I take it that that means that husbands haven't large businesses. If Mrs. Sanders went about, shedding scales in her intercourses, I deduce that theirs wasn't much of a fish business.

Upon the evening of March 4, 1905, in the Sanders' home, in the bedroom of their housemaid, there was a fire. Nobody was at home, and the firemen had to break in. There was no fireplace in the bedroom. Not a trace of anything by which to explain was found, and the firemen reported: "Origin unknown." They returned to their station, and were immediately called back to this house. There was another fire. It was in another bedroom. Again--"Origin unknown."

The Sanders', in their fish store, were notified, and they hastened home. Money was missed. Many things were missed. The housemaid, Emma Piggott, was suspected. In her parents' home was found a box, from which the Sanders' took, and identified as theirs, L5, and a loot of such things as a carving set, sugar tongs, tablecloths, several dozen handkerchiefs, salt spoons, bottles of scent, curtain hooks, a hair brush, Turkish towels, gloves, a sponge, two watches, a puff box.

The girl was arrested, and in the Derby Borough Police Court, she was charged with arson and larceny. She admitted the thefts, but asserted her innocence of the fires. There was clearly such an appearance of relation between the thefts and the fires, which, if they had burned down the house, would have covered the thefts, that both charges were pressed.

It is not only that there had been thefts, and then fires: so many things had been stolen that--unless the home of the Sanders' was a large household--some of these things would have been missed--unless all had been stolen at once. I have no datum for thinking that the Sanders lived upon any such scale as one in which valuables could have been stolen, from time to time, unknown to them. The indications were of one wide grab, and the girl's intention to set the house afire, to cover it.

Emma Piggott's lawyer showed that she had been nowhere near the house, at the time of the first fire; and that, when the second fire broke out, she, in the street, this off-evening of hers, returning, had called the attention of neighbors to smoke coming from a window. The case was too complicated for a police court, and was put off for the summer assizes.

Derby Mercury, July 19--trial of the girl resumed. The prosecution maintained that the fires could be explained only as of incendiary origin, and that the girl's motive for setting the house afire was plain, and that she had plundered so recklessly, because she had planned a general destruction, by which anything missing would be accounted for.

Again counsel for the defense showed that the girl could not have started the fires. The charge of arson was dropped. Emma Piggott was sentenced to six months' hard labor, for the thefts.

Upon Dec. 2, 1919, Ambrose Small, of Toronto, Canada, disappeared. He was known to have been in his office, in the Toronto Grand Opera House, of which he was the owner, between five and six o'clock, the evening of December 2nd. Nobody saw him leave his office. Nobody--at least nobody whose testimony can be accepted--saw him, this evening, outside the building. There were stories of a woman in the case. But Ambrose Small disappeared, and left more than a million dollars behind. Then John Doughty, Small's secretary, vanished.

Small's safe deposit boxes were opened by Mrs. Small and other trustees of the estate. In the boxes were securities, valued at $1,125,000. An inventory was found. According to it, the sum of $105,000 was missing. There was an investigation, and bonds of the value of $105,000 were found, hidden in the home of Doughty's sister.

All over the world, the disappearance of Ambrose Small was advertised, with offers of reward, in acres of newspaper space. He was in his office. He vanished.

Doughty, too, was sought. He had not only vanished: he had done all that he could to be unfindable. But he was traced to a town in Oregon, where he was living under the name of Cooper. He was taken back to Toronto, where he was indicted, charged with having stolen the bonds, and with having abducted Small, to cover the thefts.

It was the contention of the prosecution that Ambrose Small, wealthy, in good health, and with no known troubles of any importance, had no motive to vanish, and to leave $1,125,000 behind: but that his secretary, the embezzler, did have a motive for abducting him. The prosecution did not charge that Small had been soundlessly and invisibly picked out of his office, where he was surrounded by assistants. The attempt was to show that he had left his office, even though nobody had seen him go: thinkably he could have been abducted, unwitnessed, in a street. A newsboy testified that he had seen Small, in a nearby street, between 5 and 6 o'clock, evening of December 2nd, but the boy's father contradicted this story. Another newsboy told that, upon this evening, after 6 o'clock, Small had bought a newspaper from him: but, under examination, this boy admitted he was not sure of the date.

It seemed clear that there was relation between the embezzlement and the disappearance, which, were it not for the inventory, would have covered the thefts: but the accusation of abduction failed. Doughty was found guilty of embezzlement, and was sentenced to six years' imprisonment in the Kingston Penitentiary.

In the News of the World (London) June 6, 1926, there is an account of "strangely intertwined circumstances." In a public place, in the daytime, a man had died. On the footway, outside the Gaiety Theatre, London, Henry Arthur Chappell, the manager of the refreshment department of the Theatre, had been found dead. There was a post-mortem examination by a well-known pathologist, Prof. Piney. The man's skull was fractured. Prof. Piney gave his opinion that, if, because of heart failure, Chappell had fallen backward, the fractured skull might be accounted for: but he added that, though he had found indications of a slight affection of the heart, it was not such as would be likely to cause fainting.

The indications were that a murder had been committed. The police inquired into the matter, and learned that not long before there had been trouble. A girl, Rose Smith, employed at one of the refreshment counters, had been discharged by Chappell. One night she had placed on his doorstep a note telling that she intended to kill herself. Several nights later, she was arrested in Chappell's back garden. She was dressed in a man's clothes, and had a knife. Also she carried matches and a bottle of paraffin. Presumably she was bent upon murder and arson, but she was charged with trespassing, and was sentenced to two months' hard labor. It was learned that Chappell had died upon the day of this girl's release from prison.

Rose Smith was arrested. Chappell had no other known enemy. Upon the day of this girl's release, he had died.

But the accusation failed. A police inspector testified that, at the time of Chappell's death, Rose Smith had been in the Prisoners' Aid Home.

2

I AM a collector of notes upon subjects that have diversity--such as deviations from concentricity in the lunar crater Copernicus, and a sudden appearance of purple Englishmen--stationary meteor-radiants, and a reported growth of hair on the bald head of a mummy--and "Did the girl swallow the octopus?"

But my liveliest interest is not so much in things, as in relations of things. I have spent much time thinking about the alleged pseudo-relations that are called coincidences. What if some of them should not be coincidences?

Ambrose Small disappeared, and to only one person could be attributed a motive for his disappearance.

Only to one person's motives could the fires in the house in Derby be attributed. Only to one person's motives could be attributed the probable murder of Henry Chappell. But, according to the verdicts in all these cases, the meaning of all is of nothing but coincidence between motives and events.

Before I looked into the case of Ambrose Small, I was attracted to it by another seeming coincidence. That there could be any meaning in it seemed so preposterous that, as influenced by much experience, I gave it serious thought. About six years before the disappearance of Ambrose Small, Ambrose Bierce had disappeared. Newspapers all over the world had made much of the mystery of Ambrose Bierce. But what could the disappearance of one Ambrose, in Texas, have to do with the disappearance of another Ambrose, in Canada? Was somebody collecting Ambroses? There was in these questions an appearance of childishness that attracted my respectful attention.

Lloyd's Sunday News (London) June 20, 1920--that, near the town of Stretton, Leicestershire, had been found the body of a cyclist, Annie Bella Wright. She had been killed by a wound in her head. The correspondent who wrote this story was an illogical fellow, who loaded his story with an unrelated circumstance: or, with a dim suspicion of an unexplained relationship, he noted that in a field, not far from where the body of the girl lay, was found the body of a crow.

In the explanation of coincidence there is much of laziness, and helplessness, and response to an instinctive fear that a scientific dogma will be endangered. It is a tag, or a label: but of course every tag, or label, fits well enough at times. A while ago, I noted a case of detectives who were searching for a glass-eyed man named Jackson. A Jackson, with a glass eye, was arrested in Boston. But he was not the Jackson they wanted, and pretty soon they got their glass-eyed Jackson, in Philadelphia. I never developed anything out of this item--such as that, if there's a Murphy with a hare lip, in Chicago, there must be another hare-lipped Murphy somewhere else. It would be a comforting idea to optimists, who think that ours is a balanced existence: all that I report is that I haven't confirmed it.

But the body of a girl, and the body of a crow--

And, going over files of newspapers, I came upon this:

The body of a woman, found in the River Dee, near the town of Eccleston (London Daily Express, June 12, 1911). And nearby was found the body of another woman. One of these women was a resident of Eccleston: the other was a visitor from the Isle of Man. They had been unknown to each other. About ten o'clock, morning of June 10th, they had gone out from houses in opposite parts of the town.

New York American, Oct. 20, 1929--"Two bodies found in desert mystery." In the Coachella desert, near Indio, California, had been found two dead men, about Too yards apart. One had been a resident of Coachella, but the other was not identified. "Authorities believe there was no connection between the two deaths."

In the New York Herald, Nov. 26, 1911, there is an account of the hanging of three men, for the murder of Sir Edmund Berry Godfrey, on Greenberry Hill, London. The names of the murderers were Green, Berry, and Hill. It does seem that this was only a matter of chance. Still, it may have been no coincidence, but a savage pun mixed with murder. New York Sun, Oct. 7, 1930--arm of William Lumsden, of Roslyn, Washington, crushed under a tractor. He was the third person, in three generations, in his family, to lose a left arm. This was coincidence, or I shall have to come out, accepting that there may be "curses" on families. But, near the beginning of a book, I don't like to come out so definitely. And we're getting away from our subject, which is Bodies.

"Unexplained drownings in Douglas Harbor, Isle of Man." In the London Daily News, Aug. 19, 1910, it was said that the bodies of a young man and of a girl had been found in the harbor. They were known as a "young couple," and their drowning would be understandable in terms of a common emotion, were it not that also there was a body of a middle-aged man "not known in any way connected with them."

London Daily Chronicle, Sept. 10, 1924--"Near Saltdean, Sussex, Mr. F. Pender, with two passengers in his sidecar, collided with a post, and all were seriously injured. In a field, by the side of the road, was found the body of a Rodwell shepherd, named Funnell, who had no known relation with the accident."

An occurrence of the 14th of June, 1931, is told of, in the Home News (Bronx) of the 15th. "When Policeman Talbot, of the E. 126th St. station, went into Mt. Morris Park, at 10 A.M., yesterday, to awaken a man apparently asleep on a bench near the 124th St. gate, he found the man dead. Dr. Patterson, of Harlem Hospital, said that death had probably been caused by heart failure." New York Sun, June 15--that soon after the finding of this body on the bench, another dead man was found on a bench nearby.

I have two stories, which resemble the foregoing stories, but I should like to have them considered together.

In November, 1888 (St. Louis Globe-Democrat, Dec. 20, 1888), two residents of Birmingham, Alabama, were murdered, and their bodies were found in the woods. "Then there was such a new mystery that these murder-mysteries were being overlooked." In the woods, near Birmingham, was found a third body. But this was the body of a stranger. "The body lies unidentified at the undertaker's rooms. No one who had seen it can remember having seen the man in life, and identification seems impossible. The dead man was evidently in good circumstances, if not wealthy, and what he could have been doing at the spot where his body was found is a mystery. Several persons who have seen the body are of the opinion that the man was a foreigner. Anyway he was an entire stranger in this vicinity, and his coming must have been as mysterious as his death."

I noted these circumstances, simply as a mystery. But when a situation repeats, I notice with my livelier interest. This situation is of local murders, and the appearance of the corpse of a stranger, who had not been a tramp.

Philadelphia Public Ledger, Feb. 4, 1892--murder near Johnstown, Pa.--a man and his wife, named Kring, had been butchered, and their bodies had been burned. Then, in the woods, near Johnstown, the corpse of a stranger was found. The body was well-dressed, but could not be identified. Another body was found--"well-dressed man, who bore no means of identification."

There is a view by which it can be shown, or more or less demonstrated, that there never has been a coincidence. That is, in anything like a final sense. By a coincidence is meant a false appearance, or suggestion, of relations among circumstances. But anybody who accepts that there is an underlying oneness of all things, accepts that there are no utter absences of relations among circumstances--

Or that there are no coincidences, in the sense that there are no real discords in either colors or musical notes--

That any two colors, or sounds, can be harmonized, by intermediately relating them to other colors, or sounds.

And I'd not say that my question, as to what the disappearance of one Ambrose could have to do with the disappearance of another Ambrose, is so senseless. The idea of causing Ambrose Small to disappear may have had origin in somebody's mind, by suggestion from the disappearance of Ambrose Bierce. If in no terms of physical abduction can the disappearance of Ambrose Small be explained, I'll not say that that has any meaning, until the physicists intelligibly define what they mean by physical terms.

3

IN days of yore, when I was an especially bad young one, my punishment was having to go to the store, Saturdays, and work. I had to scrape off labels of other dealers' canned goods, and paste on my parent's label. Theoretically, I was so forced to labor to teach me the errors of deceitful ways. A good many brats are brought up, in the straight and narrow, somewhat deviously.

One time I had pyramids of canned goods, containing a variety of fruits and vegetables. But I had used all except peach labels. I pasted the peach labels on peach cans, and then came to apricots. Well, aren't apricots peaches? And there are plums that are virtually apricots. I went on, either mischievously, or scientifically, pasting the peach labels on cans of plums, cherries, string beans, and succotash. I can't quite define my motive, because to this day it has not been decided whether I am a humorist or a scientist. I think that it was mischief, but, as we go along, there will come a more respectful recognition that also it was scientific procedure.

In the town of Derby, England--see the Derby Mercury, May 15, and following issues, 1905--there were occurrences that, to the undiscerning, will seem to have nothing to do with either peaches or succotash. In a girls' school, girls screamed and dropped to the floor, unconscious. There are readers who will think over well-known ways of peaches and succotash, and won't know what I am writing about. There are others, who will see "symbolism" in it, and will send me appreciations, and I won't know what they're writing about. In five days, there were forty-five instances of girls who screamed and dropped unconscious. "The girls were exceedingly weak, and had to be carried home. One child had lost strength so that she could not even sit up." It was thought that some unknown, noxious gas, or vapor, was present: but mice were placed in the schoolrooms, and they were unaffected. Then the scientific explanation was "mass psychology." Having no more data to work on, it seems to me that this explanation is a fitting description. If a girl fainted, and, if, sympathetically, another girl fainted, it is well in accord with our impressions of human nature, which sees, eats, smells, thinks, loves, hates, talks, dresses, reads, and undergoes surgical operations, contagiously, to think of forty-three other girls losing consciousness, in involuntary imitativeness. There are mature persons who may feel superior to such hysteria, but so many of them haven't much consciousness.

In the Brooklyn Eagle, Aug. 1, 1894, there is a story of "mass psychology." In this case, too, it seems to me that the description fits--maybe. Considering the way people live, it is natural to them to die imitatively. There was, in July, 1894, a panic in a large vineyard, at Collis, near Fresno, California. Somebody in this vineyard had dropped dead of "heart failure." Somebody else dropped dead. A third victim had dropped and was dying. There wasn't a scientist, with a good and sticky explanation, on the place. It will be thought amusing: but the people in this vineyard believed that something uncanny was occurring, and they fled. "Everybody has left the place, and the authorities are preparing to begin a searching investigation." Anything more upon this subject is not findable. That is the usual experience after an announcement of a "searching investigation."

If something can't be described any other way, it's "mass psychology." In the town of Bradford, England, in a house, in Columbia Street, 1st of March, 1923, there was one of those occasions of the congratulations, hates, malices, and gaieties, and more or less venomous jealousies that combine in the state that is said to be merry, of a wedding party. The babble of this wedding party suddenly turned to delirium. There were screams, and guests dropped to the floor, unconscious. Wedding bells--the gongs of ambulances--four persons were taken to hospitals.

This occurrence was told of in the London newspapers, and, though strange, it seemed that the conventional explanation fitted it. Yorkshire Evening Argus--published in Bradford--March 3, 1923--particulars that make for restiveness against any conventional explanation--people in adjoining houses had been affected by this "mysterious malady." Several names of families, members of which had been overcome, unaccountably, were published--Downing, Blakey, Ingram.

If people, in different houses, and out of contact with one another--or not so circumstanced as to "mass" their psychologies--and all narrowly localized in one small neighborhood, were similarly affected, it seemed clear that here was a case of common exposure to something that was poisonous, or otherwise injurious. Of course an escape of gas was thought of; but there was no odor of gas. No leakage of gas was found. There was the usual searching investigation that precedes forgetfulness. It was somebody's suggestion that the "mysterious malady" had been caused by fumes from a nearby factory chimney. I think that the wedding party was the central circumstance, but I don't think of a factory chimney, which had never so expressed itself before, suddenly fuming at a wedding party. An Argus reporter wrote that the Health Officers had rejected this suggestion, and that he had investigated, and had detected no unusual odor in the neighborhood.

In this occurrence at Bradford, there was no odor of gas. I have noted a case in London, in which there was an odor of gas; nevertheless this case is no less mysterious. In the Weekly Dispatch (London), June 12, 1910, it is called "one of the most remarkable and mysterious cases of gas poisoning that have occurred in London in recent years." Early in the morning of June 10th, a woman telephoned to a police station, telling of what she thought was an escape of gas. A policeman went to the house, which was in Neale Street (Holborn). He considered the supposed leakage alarming, and rapped on doors of another floor in the house. There was no response, and he broke down a door, finding the occupants unconscious. In two neighboring houses, four unconscious persons were found. A circumstance that was considered extraordinary was that between these two houses was one in which nobody was affected, and in which there was no odor of gas. The gas company sent men, who searched for a leak, but in vain. Fumes, as if from an uncommon and easily discoverable escape of gas, had overcome occupants of three houses, but according to the local newspaper (the Holborn Guardian) the gas company, a week later, had been unable to discover its origin. In December, 1921, there was an occurrence in the village of Zetel, Germany (London Daily News, Jan. 2, 1922). This was in the streets of a town. Somebody dropped unconscious: and, whether in an epidemic of fright, accounted for in terms of "mass psychology," or not, other persons dropped unconscious. "So far no light has been thrown on the mystery." It was thought that a "current of some kind" had passed over the village. This resembles the occurrence at El Paso, Texas, June 19, 1929 (New York Sun, Dec. 6, 1930). Scores of persons, in the streets, dropped unconscious, and several of them died. Whatever appeared here was called a "deadly miasma." And the linkage goes on to the scores of deaths in a fog, in the Meuse Valley, Belgium, Dec. 5, 1930--so that one could smoothly and logically start with affairs in a girls' school, and end up with a meteorological discussion.

Lloyd's Weekly News (London) Jan. 17, 1909--story from the Caucasian city of Baku. M. Krassil-rukoff, and two companions, had gone upon a hunting trip, to Sand Island, in the Caspian Sea. Nothing had been heard from them, and there was an investigation. The searchers came upon the bodies of the three men, lying in positions that indicated that they had died without a struggle. No marks of injuries; no disarrangement of clothes. At the autopsy, no trace of poison was found. "The doctors, though they would not commit themselves to an explanation, thought the men had been stifled."

The Observer (London) Aug. 23, 1925--"A mysterious tragedy is reported from the Polish Tatra mountains, near the health-resort of Zakopane. A party, composed of Mr. Kasznica, the Judge of the Supreme Court, his wife, a twelve-year-old son, and a young student of Cracow University, started in fine weather for a short excursion in the neighboring mountains. Two days later, three of them were found dead."

Mrs. Kasznica was alive. She told that all were climbing, and were in good condition, when suffocation came upon them. "A stifling wind," she thought. One after another they had dropped unconscious. The post-mortem examinations revealed nothing that indicated deaths by suffocation, nor anything else that could be definitely settled upon. "Some newspapers suggest a crime, but so far the case remains a mystery."

There have been cases that have been called mysterious, though they seem explicable enough in known circumstances of human affairs. See a story in the New York World-Telegram, March 9, 1931--about thirty men and women at work in the Howard Clothes Company factory, Nassau Street, Brooklyn-- sudden terror and a panic of these people, to get to the street. The place was filled with a pungent, sickening odor. In the street, men and women collapsed, or reeled, and wandered away, in a semi- conscious condition. Several dozen of them were carried into stores, where they were given first-aid treatment, until ambulances arrived.

The phenomenon occurred in the second floor of the Cary Building, occupied by the clothing company. Nobody in any other part of the building was affected. All gas fixtures in the factory were intact. No gas bomb was found. Nothing was found out. But, considering many crimes of this period, the suspicion is strong that in some way, as an expression of human hatred, of origin in industrial troubles, a volume of poisonous gas had been discharged into this factory.

And it may be that, in terms of revenges, we are on the track of a general expression, even if we think of a hate that could pursue people far up on a mountainside.

In hosts of minds, today, are impressions that the word "eerie" means nothing except convenience to makers of crossword puzzles. There are gulfs of the unaccountable, but they are bridged by terminology. Four persons were taken from a wedding party to hospitals. Well, if not another case of such jocularity as mixing brick-has with confetti, it was ice cream again, and ptomaine poisoning. There is such a satisfaction in so explaining, and showing that one knows better than to sound the p in ptomaine, that probably vast holes of ignorance always will be bridged by very slender pedantries. Asphyxiation has seduced hosts of suspicions that would be resolute against such a common explanation as "gas poisoning."

New York Sun, May 22, 1928--story from the town of Newton, Mass. In this town, a physician was, by telephone, called to the home of William M. Duncan. There was nobody to meet him at the front door, but he got into the house. He called, but nobody answered. There seemed to be nobody at home, but he went through the house. He came to a room, upon the floor of which were lying four bodies. There was no odor of gas, but the doctor worked over the four, as if upon cases of asphyxiation, and they revived, and tried to explain. Duncan had gone to this room, and, upon entering it, had dropped, unconscious.

Wondering at what was delaying him, his wife had followed, and down she had fallen. One of his sons came next, and, upon entering this room, had fallen to the floor. The other son, by chance, went to this room, and felt something overcoming him. Before losing consciousness, he had staggered to the telephone.

The doctor's explanation was "mass psychology."

It is likely that readers of the Sun were puzzled, until they came to this explanation, and then--"Oh, of course! mass psychology."

There is a continuity of all things that makes classifications fictions. But all human knowledge depends upon arrangements. Then all books--scientific, theological, philosophical--are only literary. In Scotland, in the month of September, 1903, there was an occurrence that can as reasonably be considered a case of "mass psychology," as can be some of the foregoing instances: but now we are emerging into data that seem to be of physical attacks. There will be more emerging. One can't, unless one be hopelessly, if not brutally, a scientist, or a logician, tie to any classification. The story is told in the Daily Messenger (Paris) Sept. 13, 1903.

In a coal mine, near Coalbridge, Scotland, miners came upon the bodies of three men. There was no coal gas. There was no sign of violence of any kind. Two of these men were dead, but one of them revived.

He could tell, enlighteningly, no more than could any other survivor in the stories of this group. He told that his name was Robert Bell, and that, with his two cousins, he had been walking in the mine, when he felt what he described as a "shock." No disturbance had been felt by anyone else in the mine. Though other parts of this mine were lighted by electricity, there was not a wire in this part. There was, at this point, a deadly discharge of an unknown force, just when, by coincidence, three men happened to be passing, or something more purposeful is suggested.

Down in the dark of a coal mine--and there is a seeming of the congruous between mysterious attacks and surroundings. Now I have a story of a similar occurrence at a point that was one of this earth's most crowded thoroughfares. See the New York Herald, Jan. 23, 1909. John Harding, who was the head of a department in John Wanamaker's store, was crossing Fifth Avenue, at Thirty-third Street, when he felt a stinging sensation upon his chest. There was no sign of a missile of any kind. Then he saw, nearby, a man, who was rubbing his arm, looking around angrily. The other man told Harding that something unseen had struck him. If this occurrence had been late at night, and, if only two persons were crossing Fifth Avenue, at Thirty- third Street; and if a force of intensity enough to kill had struck them, the explanation, upon the finding of the bodies, would probably be that two men had, by coincidence, died in one place, of heart failure. At any rate, see back to the case of the bodies on benches of a Harlem park. No reporter of the finding of these bodies questioned the explanation that two men, sitting near each other, had died, virtually simultaneously, of heart failure, by coincidence.

We emerge from seeming attacks upon more than one person at a time, into seemingly definitely directed attacks upon single persons. New York Herald Tribune, Dec. 4, 1931--Ann Harding, film actress, accompanied by her secretary, on her way, by train, to Venice, Florida. There came an intense pain in her shoulder. Miss Harding could not continue traveling, and left the train, at Jacksonville. A physician examined her, and found that her shoulder was dislocated. The secretary was mystified, because she had seen the occurrence of nothing by which to explain, and Miss Harding could offer no explanation of her injury.

Upon Dec. 7, 1931--see the New York Times, Dec. 8, 1931--the German steamship Brechsee arrived at Horsens, Jutland. Captain Ahrenkield told of one of his sailors, who had been unaccountably wounded. The man had been injured during a storm, but he seemed to have been singled out by something other than stormy conditions. The captain had seen him, wounded by nothing that was visible, falling to the deck, unconscious. It was a serious wound, four inches long, that had appeared upon the sailor's head, and the captain had sewed it with ordinary needle and thread.

In this case, unaccountable wounds did not appear upon several other sailors. Suppose, later, I tell of instances in which a number of persons were so injured. Mass psychology?

4

NOT a bottle of catsup can fall from a tenement-house fire-escape, in Harlem, without being noted--not only by the indignant people downstairs, but--even though infinitesimally--universally—maybe

Affecting the price of pajamas, in Jersey City: the temper of somebody's mother-in-law, in Greenland; the demand, in China, for rhinoceros horns for the cure of rheumatism--maybe--

Because all things are inter-related--continuous--of an underlying oneness--

So then the underlying logic of the boy--who was guilty of much, but was at least innocent of ever having heard of a syllogism--who pasted a peach label on a can of string beans.

All things are so inter-related that, though the difference between a fruit and what is commonly called a vegetable seems obvious, there is no defining either. A tomato, for instance, represents the merging-point. Which is it--fruit or vegetable?

So then the underlying logic of the scientist--who is guilty of much, but also is very innocent--who, having started somewhere with his explanation of "mass psychology," keeps right on, sticking on that explanation. Inasmuch as there is always a view somewhere, in defense of anything conceivable, he must be at least minutely reasonable. If "mass psychology" applies definitely to one occurrence, it must, even though almost imperceptibly, apply to all occurrences. Phenomena of a man alone on a desert island can be explained in terms of "mass psychology"--inasmuch as the mind of no man is a unit, but is a community of mental states that influence one another. Inter-relations of all things-- and I can feel something like the hand of Emma Piggott reaching out to the hand, as it were, of the asphyxiated woman on the mountainside. John Doughty and bodies on benches in a Harlem park-- as oxygen has affinity for hydrogen. Rose Smith--Ambrose Small--the body of a shepherd named Funnell--

Upon the morning of April 10, 1893, after several men had been taken to a Brooklyn hospital, somebody's attention was attracted to something queer. Several accidents, in quick succession, in different parts of the city would not be considered strange, but a similarity was noted. See the Brooklyn Eagle, April 10, 1893.

Then there was a hustle of ambulances, and much ringing of gongs--

Alex. Burgman, Geo. Sychers, Lawrence Beck, George Barton, Patrick Gibbons, James Meehan, George Bedell, Michael Brown, John Trowbridge, Timothy Hennessy, Philip Oldwell, and an unknown man--

In the course of a few hours, these men were injured, in the streets of Brooklyn, almost all of them by falling from high places, or by being struck by objects that fell from high places.

Again it is one of my questions that are so foolish, and that may not be so senseless--what could the fall of a man from a roof, in one part of Brooklyn, have to do with a rap on the sconce, by a flower pot, of another man, in another part of Brooklyn?

In the town of Colchester, England--as told in Lloyd's Daily News (London) April 30, 1911--a soldier, garrisoned at Colchester, was, upon the evening of April 24th, struck senseless. He was so seriously injured that he was taken to the Garrison Hospital. Here he could give no account of what had befallen him. The next night, to this hospital, was taken another seriously injured soldier, who had been "struck senseless by an unseen assailant." Four nights later, a third soldier was taken to this hospital, suffering from the effects of a blow, about which he could tell nothing.

I have come upon a case of the "mass psychology" of lace curtains. About the last of March, 1892--see the Brooklyn Eagle, April 19, 1892--people who had been away from home, in Chicago, returned to find that during their absence there had been an orgy of curtains. Lace curtains were lying about, in lumps and distortions. It was a melancholy prostration of virtues: things so flimsy and frail, yet so upright, so long as they are supported. Bureau drawers had been ransacked for jewelry, and jewelry had been found. But nothing had been stolen. Strewn about were fragments of rings and watches that had been savagely smashed.

There are, in this account, several touches of the ghost story. There are many records of similar wanton, or furious, destructions in houses where poltergeist disturbances were occurring. Also there was mystery, because the police could not find out how this house had been entered.

Then came news of another house, which, while the dwellers were away, had been "mysteriously entered." Lace curtains, in rags, were lying about, and so were remains of dresses that had been slashed. Jewelry and other ornaments had been smashed. Nothing had been stolen.

So far as the police could learn, the occupants of these houses had no common enemy. A rage against lace curtains is hard to explain, but the hatred of somebody, whose windows were bare, against all finery and ornaments, is easily understandable. Soon after rages had swept through these two houses, other houses were entered, with no sign of how the vandal got in, and lace curtains were pulled down, and there was much destruction of finery and ornaments, and nothing was stolen.

New York Times, Jan. 26, 1873--that, in England, during the Pytchley hunt, Gen. Mayow fell dead from his saddle, and that about the same time, in Gloucestershire, the daughter of the Bishop of Gloucestershire, while hunting, was seriously injured; and that, upon the same day, in the north of England, a Miss Cavendish, while hunting, was killed. Not long afterward, a clergyman was killed, while hunting, in Lincolnshire. About the same time, two hunters, near Sanders' Gorse, were thrown, and were seriously injured.

In one of my incurable, scientific moments, I suggest that when diverse units, of, however, one character in common, are similarly affected, the incident force is related to the common character. But there is no suggestion that any visible hater of fox-hunters was traveling in England, pulling people from saddles, and tripping horses. But that there always has been intense feeling, in England, against fox-hunters is apparent to anyone who conceives of himself as a farmer--and his fences broken, and his crops trampled by an invasion of red coats--and a wild desire to make a Bunker Hill of it.

In the New York Evening World, Dec. 26, 1930, it was said that Warden Lewis E. Lawes, of Sing Sing Prison, had been ill. The Warden recovered, and, upon Christmas morning, left his room. He was told that a friend of his, Maurice Conway, who had come to visit him, had been found dead in bed. Upon Christmas Eve, Keeper John Hyland had been operated upon, "for appendicitis," and was in a serious condition in Ossining Hospital. In the same hospital was Keeper John Wescott, who also had been stricken "with appendicitis." Keeper Henry Barrett was in this hospital, waiting to be operated upon "for hernia."

Probably the most hated man in the New York State Prison Service was Asael J. Granger, Head Keeper of Clinton Prison, at Dannemora. He had effectively quelled the prison riot of July 22, 1929. Upon this Christmas Day, of 1930, in the Champlain Valley Hospital, Plattsburg, N. Y., Granger was operated upon "for appendicitis." Two days later he died. About this time, Harry M. Kaiser, the Warden of Clinton Prison, was suffering from what was said to be "high blood pressure." He died, three months later (New York Herald Tribune, March 24, 1931).

The London newspapers of March, 1926, told of fires that had

simultaneously broken out in several parts of Closes Hall, the residence of Captain B. Heaton, near Clitheroe, Lancashire. The fires were in the woodwork under the roof, and were believed to have been caused by sparks from the kitchen stove. These fires were in places that were inaccessible to any ordinary incendiary: to get to them, the firemen had to chop holes in the roof. Nothing was said of previous fires here. Maybe it is strange that sparks from a kitchen stove should simultaneously ignite remote parts of a house, distances apart.

A fire in somebody's house did not much interest me: but then I read of a succession of similars. In three months, there had been ten other mansion fires. "Scotland Yard recently made arrangements for all details of mansion fires to be sent to them, in order that the circumstances might be collated, and the probable cause of the outbreaks discovered."

April 2, 1926--Ashley Moor, a mansion near Leominster, destroyed by fire.

Somebody, or something, was burning mansions. How it was done was the mystery. There was a scare, and probably these houses were more than ordinarily guarded: but so well-protected are they, ordinarily, that some extraordinary means of entrance is suggested. In no report was it said that there was any evidence of how an incendiary got into a house. No theft was reported. For months, every now and then there was a mansion fire. Presumably the detectives of Scotland Yard were busily collating.

The London newspapers, of November 6th, told of the thirtieth mansion fire in about ten months. There were flaming mansions, and there were flaming utterances, in England.

Sometimes I am a collector of data, and only a collector, and am likely to be gross and miserly, piling up notes, pleased with merely numerically adding to my stores. Other times I have joys, when unexpectedly coming upon an outrageous story that may not be altogether a lie, or upon a macabre little thing that may make some reviewer of my more or less good works mad. But always there is present a feeling of unexplained relations of events that I note; and it is this far-away, haunting, or often taunting, awareness, or suspicion, that keeps me piling on.

Or, in a feeling of relatability of seemingly most incongruous occurrences that nevertheless may be correlated into the service of one general theme, I am like a primitive farmer, who conceives that a zebra and a cow may be hitched together to draw his plow.

But isn't there something common about zebras and cows?

An ostrich and a hyena.

Then the concept of a pageantry--the ransack of the jungles for creatures of the widest unlikeness to draw his plow--and former wild clatters of hoofs and patters of paws are the tramp of a song--here come the animals, two by two.

Or John Doughty, three abreast with the dead men of a Harlem park, pulling on my theme--followed by the forty-five schoolgirls of Derby--and the fish dealer's housemaid, with her arms full of sponges and Turkish towels--followed by burning beds, most suggestively associated with her, but in no way that any conventional thinker can explain.

Or the mansion fires in England, in the year 1926--and, in a minor hitch-up, I feel the relatability of two scenes:

In Hyde Park, London, an orator shouts: "What we want is no king and no law! How we'll get it will be, not with ballots, but with bullets!"

Far away in Gloucestershire, a house that dates back to Elizabethan times unaccountably bursts into flames.

5

    "GOOD morning!" said the dog. He disappeared in a thin, greenish vapor.

I have this record, upon newspaper authority.

It can't be said--and therefore will be said--that I have a marvelous credulity for newspaper yarns.

But I am so obviously offering everything in this book, as fiction. That is, if there is fiction. But this book is fiction in the sense that Pickwick Papers, and The Adventures of Sherlock Holmes, and Uncle Tom's Cabin; Newton's Principia, Darwin's Origin of Species, Genesis, Gulliver's Travels, and mathematical theorems, and every history of the United States, and all other histories, are fictions. A library-myth that irritates me most is the classification of books under "fiction" and "non-fiction." And yet there is something about the yarns that were told by Dickens that sets them apart, as it were, from the yarns that were told by Euclid. There is much in Dickens' grotesqueries that has the correspondence with experience that is called "truth," whereas such Euclidian characters as "mathematical points" are the vacancies that might be expected from a mind that had had scarcely any experience. That dog-story is axiomatic. It must be taken on faith. And, even though with effects that sometimes are not much admired, I ask questions.

It was told in the New York World, July 29, 1908--many petty robberies, in the neighborhood of Lincoln Avenue, Pittsburgh--detectives detailed to catch the thief. Early in the morning of July 26th, a big, black dog sauntered past them. "Good morning!" said the dog. He disappeared in a thin, greenish vapor. There will be readers who will want to know what I mean by turning down this story, while accepting so many others in this book.

It is because I never write about marvels. The wonderful, or the never-before-heard-of, I leave to whimsical, or radical, fellows. All books written by me are of quite ordinary occurrences.

If, say sometime in the year 1847, a New Orleans newspaper told of a cat, who said: "Well, is it warm enough for you?" and instantly disappeared sulphurously, as should everybody who says that; and, if I had a clipping, dated sometime in the year 1930, telling of a mouse, who squeaked: "I was along this way, and thought I'd drop in," and vanished along a trail of purple sparklets; and something similar from the St. Helena Guardian, Aug. 17, 1905; and something like that from the Madras Mail, year 1879--I'd consider the story of the polite dog no marvel, and I'd admit him to our fold.

But it is not that I take numerous repetitions, as a standard for admission--

The fellow who found the pearl in the oyster stew--the old fiddle that turned out to be a Stradivarius-- the ring that was lost in a lake, and then what was found when a fish was caught--

But these often repeated yarns are conventional yarns.

And almost all liars are conventionalists.

The one quality that the lower animals have not in common with human beings is creative imagination. Neither a man, nor a dog, nor an oyster ever has had any. Of course there is another view, by which is seen that there is in everything a touch of creativeness. I cannot say that truth is stranger than fiction, because I have never had acquaintance with either. Though I have classed myself with some noted fictionists, I have to accept that the absolute fictionist never has existed. There is a fictional coloration to everybody's account of an "actual occurrence," and there is at least the lurk somewhere of what is called the "actual" in everybody's yarn. There is the hyphenated state of truth-fiction. Out of dozens of reported pearls in stews, most likely there have been instances; most likely once upon a time an old fiddle did turn out to be a Stradivarius; and it could be that once upon a time somebody did get a ring back fishwise.

But when I come upon the unconventional repeating, in times and places far apart, I feel--even though I have no absolute standards to judge by--that I am outside the field of ordinary liars.

Even in the matter of the talking dog, I think that the writer probably had something to base upon. Perhaps he had heard of talking dogs. It is not that I think it impossible that detectives could meet a dog, who would say: "Good morning!" That's no marvel. It is "Good morning!" and disappearing in the thin, greenish vapor that I am making such a time about. In the New York Herald Tribune, Feb. 21, 1928, there was an account of a French bulldog, owned by Mrs. Mabel Robinson, of Bangor, Maine. He could distinctly say: "Hello!" Mrs. J. Stuart Tompkins, tot West 85th Street, New York, read of this animal, and called up the Herald Tribune, telling of her dog, a Great Dane, who was at least equally accomplished.

A reporter went to interview the dog, and handed him a piece of candy. "Thank you!" said the dog.

In the city of Northampton, England--see Lloyd's Weekly News (London), March 2, 1912--a detective chased a burglar, who had entered a hardware store. The burglar got away. The detective went back, and got into the store. There were objects hanging on hooks, overhead. "By coincidence," just as the detective passed under one of them, it fell. It was a scythe-blade. It cut off his ear. Now I am upon familiar ground; there are suggestions in this story that correlate with suggestions in other stories.

"A bank in Blackpool was robbed, in broad daylight, on Saturday, in mysterious circumstances"--so says the London Daily Telegraph, Aug. 7, 1926. It was one of the largest establishments in town--the Blackpool branch of the Midland Bank. At noon, Saturday, while the doors were closing, an official of the Corporation Tramways Department went into the building, with a bag, which contained L800, in Treasury notes. In the presence of about twenty-five customers, he placed the bag upon a counter. Then the doorman unlocked the front door for him to go out, and then return with another amount of money, in silver, from a motor van. The bag had vanished from the counter. It was a large, leather bag. Nobody could, without making himself conspicuous, try to conceal it. Nobody wearing a maternity cloak was reported.

In the afternoon, in a side street, near the bank, the bag was found, and was taken to a police station. But the lock on it was peculiar and complicated, and the police could not open it. An official of the Tramways Department was sent for. When the Tramways man arrived with the key, no money was found in the bag. If a bag can vanish from a bank, without passing the doorman, I record no marvel in telling of money that vanished from a bag, though maybe the bag had not been opened.

Well, then, there's nothing marvelous about it, if from a locked drawer of Mrs. Bradley's bureau, money disappeared. New York Times, Feb. 28, 1874--Mrs. Lydia Bradley, of Peoria, Ill., "mysteriously robbed." There were other occurrences; and they, too, were anything but marvelous. Pictures came down from the walls, and furniture sauntered about the place. Stoves slung their lids at people. Such doings have often been reported from houses, in the throes of poltergeist disturbances. There are many records of pictures that couldn't be kept hanging on walls. Chairs and tables have been known to form in orderly fashion, three or four abreast, and parade. In Mrs. Bradley's home, the doings were in the presence of the housemaid, Margaret Corvell. So the girl was suspected, and one time, in the midst of pranks by things that are ordinarily so staid and settled, somebody held her hands. While her hands were held, a loud crash was heard. A piano, which up to that moment had been behaving itself properly, joined in. But the girl was accused. She confessed to everything, including the stealing of the money, except whatever had occurred when her hands were held. There are dozens of poltergeist cases, in which the girl--oftenest a young housemaid--has confessed to all particulars, except things that occurred while she was held, tied, or being knocked about. Ignoring these omissions, accounts by investigators end with the satisfactory explanation that the girl had confessed.

In the Home News (Bronx, N. Y.), Sept. 25, 1927, is a story of "ghost-like depredations." In the town of Barberton, Ohio, lived an uncatchable thief. I call attention to an element often of openness, often of defiance, that will appear in many of our stories. It is as if there are criminals, and sometimes mischievous fellows, who can do unaccountable things and delight in mystifying their victims, confident that they cannot be caught. For ten years the uncatchable thief of Barberton had been operating, periodically. In some periods, as if to show off his talents, he returned to the same house half a dozen times. In January, 1925, the police of London were in the state of mind of the rest of us, when we try to solve crossword puzzles that have been filled in with alleged Scotch dialect, obsolete terms, and names of improbable South American rodents. Somebody was playing a game, unfairly making it difficult. The things that he did were what a crossword author would call "vars." He was called the "cat burglar." Since his time, many minor fellows have been so named. The newspapers stressed what they called this criminal's uncanny ability to enter houses, but I think that the stress should have been upon his knowledge of just where to go, after entering houses. Whether he had the property of invisibility or not, residents of Mayfair reported losses of money and jewelry that could not be more mystifying if an invisible being had come in through doors or windows without having to open them, and had strolled through rooms, sizing up the lay of things. He was called the "cat burglar," because there was no conventional way of accounting for his entrances, except by thinking that he had climbed up the sides of houses--always knowing just what room to climb to--climbing with a skill that no cat has ever had. Sometimes it was said that marks were seen on drain pipes and on window sills. Just so long as the police can say something, that is accepted as next best to doing something. Of course, in this respect, I'd not pick out any one profession.

The "cat burglar" piled up jewelry that would satisfy anybody's dream of expensive junk, and then he vanished, maybe not in a thin, greenish vapor, but anyway in an atmosphere of the unfair mystification of crosswords that have been made difficult with "vars" and "obs." Perhaps marks were found on drain pipes and on window sills. But only logicians think that anything has any exclusive meaning. If I had the power of invisibly entering houses, but preferred to turn off suspicions, I'd make marks on drain pipes and window sills. Everything that ever has meant anything has just as truly meant something else.

Otherwise experts, called to testify, at trials, would not be the fantastic exhibits that they so often are.

New York Evening Post, March 14, 1928--people in a block of houses, in the Third District of Vienna, terrorized. They were "haunted by a mysterious person," who entered houses, and stole small objects, never taking money, doing these things just to show what he could do. Then, from dusk to dawn, the police formed in a cordon around this block, and at approaches to it stationed police dogs. The disappearances of small objects, of little value, continued. There were stories of this "uncanny burglar or maniac" having been seen, "running lizardwise along moonlit roofs." My own notion is that nothing was seen running along roofs. There was such excitement that the "highest authorities" of Vienna University offered their mentalities for the help of the baffled policemen and their dogs. I wish I could record an intellectual contest between college professors and dogs; there might be some glee for my malices. There are probably many college professors, who at times read of strange crimes, and sympathize with civilization, because they had not taken to detective work. However, nothing more was said of the professors who offered to help the cops and the dogs. But there was a challenge here, and I am sorry to note that it was not accepted. It would have been a crowning show-off, if this perhaps occult sportsman had entered the homes of some of these "highest authorities," and had stolen from them whatever it is by which "highest authorities" maintain their authority, or had robbed them of their pants. But he did not rise to this opportunity. After we have more data, it will be my expression that probably he could not practice outside this one block of houses.

However, he got into a house in which lived a policeman, and he went to the policeman's bedroom. He touched nothing else, but stole the policeman's revolver.

Upon the afternoon of June 18, 1907, occurred one of the most sensational, insolent, contemptible, or magnificent thefts in the annals of crime, as viewed by most Englishmen; or a crime not without a little interest to Americans. On a table, on the lawn back of the grandstand, at Ascot, the Ascot Cup was upon exhibition, 13 inches high, and 6 inches in diameter; 20-carat gold; weight 68 ounces. The cup was guarded by a policeman and by a representative of the makers. The story is told, in the London Times, June 19th. Presumably all around was a crowd, kept at a distance by the policeman, though, according to the standards of the Times, in the year 1907, it was not dignified to go into details much. From what I know of the religion of the Turf, in England, I assume that there was a crowd of devotees, looking worshipfully at this ikon.

It wasn't there.

About this time, there were a place and a time and a treasure that were worthy the attention of, or that were a challenge to, any magician. The place was Dublin Castle. Outside, day and night, a policeman and a soldier were on duty. Within a distance of fifty yards were the headquarters of the Dublin metropolitan police; of the Royal Irish Constabulary; the Dublin detective force; the military garrison. It was at the time of the Irish International Exhibition, at Dublin. Upon the 10th of

July, King Edward and Queen Alexandra were to arrive to visit the Exhibition. In a safe in the strong room of the Castle had been kept the jewels that were worn by the Lord Lieutenant, upon State occasions. They were a barbaric pile of bracelets, rings, and other insignia, of a value of $250,000.

And of course. They had disappeared about the time of the disappearance of the Ascot Cup: sometime between June firth and July 6th.

All investigations came to nothing. For about twenty-four years nothing new came out. Then, according to a dispatch from London to the New York Times, Sept. 6, 1931, there was a report of attempted negotiations with the Dublin authorities, or an offer by which, "under certain conditions," the jewels would be returned. If this rumor were authentic, the remarkable part is that the various jeweled objects had not been broken up, but for twenty-four years had been kept intact. This is the look of the stunt.

But what I am worrying about is the big dog who said "Good morning!" and disappeared in a thin, greenish vapor. I am not satisfied with my explanation of why I rejected him. Considering some of my acceptances, it seems so illogical to turn down the dog who said "Good morning!"--except that only to the purist, or the scholar, can there be either the logical or the illogical. We have to get along with the logical-illogical, in our existence of the hyphen. Everything that is said to be logical is somewhere out of agreement with something, and everything that is said to be illogical is somewhere in agreement with something.

I need not worry about the big dog who said "Good morning!" If, considering some of my acceptances, I inconsistently turn him down, I am consistent with something else, and that is the need in every mind to turn down something--the need in every mind that believes, or accepts anything, to consider something else silly, preposterous, false, evil, immoral, terrible--taboo. It is not necessary that we should all agree in being revolted, shocked, or contemptuous. Some of us take Jehovah, and some of us take Allah, to despise, or to be amused with. To give it limits within which to seem to be, and to give it contrasts by which to seem to be, every mind must practice exclusions.

I draw my line at the dog who said "Good morning!" and disappeared

in a thin, greenish vapor. He is a symbol of the false and arbitrary and unreasonable and inconsistent-- though of course also the reasonable and consistent--limit, which everybody must somewhere set, in order to pretend to be.

You can't fool me with that dog-story.

6

CONSERVATISM is our opposition. But I am in considerable sympathy with conservatives. I am often lazy, myself.

It's evenings, when I'm somewhat played out, when I'm likely to be most conservative. Everything that is highest and noblest in my composition is most pronounced when I'm not good for much. I may be quite savage, mornings: but, as my energy plays out, I become nobler and nobler, and lazier, and conservativer. Most likely my last utterance will be a platitude, if I've been dying long enough. If not, I shall probably laugh.

I like to read my Evening Newspaper comfortably. And it is uncomfortably that I come upon any new idea, or suggestion of the new, in an Evening Newspaper. It's a botheration, and I don't understand it, and it will cost me some thinking--oh, well, I'll clip it out, anyway.

But where are the scissors? But they aren't. Has anybody a pin? Nobody has. There was a time when one could maneuver over to the edge of a carpet, without having to leave one's chair, and pull up a tack. But everybody has rugs, nowadays. Oh, well, let it go.

Something in a newspaper about a mysterious hair-clipper. This is a new department of data, though hair-stealing links with other mysterious thefts. Where's a pin? Oh, well, there's nothing in particular about this matter of hair-clipping. A petty thief stole hair to sell, of course. Vague suggestions hanging over from reading of various phases of "black magic"--but, if there is a market for human hair, hair- clippers are accounted for--still--

And so I could go on, every now and then, for many years, feeling a haunt of a new idea, but feeling more comfortable, if doing nothing about it. But, daytimes, I go to Libraries, and, if several times, close together, something that is new to me, in newspapers, attracts my attention, I get the power somewhere to make a note of it.

These vague, new ideas that flutter momentarily in every mind--sometimes they're as hard to catch as is the moment they flutter in. It's like trying to pin a butterfly without catching it. They're gone. They can't develop, because one doesn't, or can't, note them, and collect notes. We'd all be somewhat enlightened--if that would be any good to us--were it not for easy chairs. Where's a pin? Hereafter I'm going to have a pet porcupine around the house. One can't learn much and also be comfortable. One can't learn much and let anybody else be comfortable.

Two cases of hairdressers' windows broken, and women's switches stolen. Probably to sell to other hairdressers.

I noted this, just as an oddity: London Daily Chronicle, July 9, 1913--Paris--wealthy engineer, named Leramgourg, arrested. "At Leramgourg's residence, the police found locks of hair of 94 women."

I put this item with others upon freaks of collectors. In Oklahoma City, July, 1907, somebody collected ears. Bodies of three men--ears cut off. In April, 1913, a collector, who was known as Jack the Slipper- snatcher, operated in the subways of New York City. Girl going up the steps of a subway exit--one foot up from a step--the snatch of her slipper--

The fantastic, or the amusing--but it is as close to the appalling as is the beautiful to the hideous--
The murderer of the Conners child, in New York, in July, 1916, hacked hair from his victim.

I have only two records of male victims of hair-clippers. I conceive that once upon a time abundant whiskers were tempting. Where do manufacturers of false whiskers get their material? Both of these victims were children. There was a case of three gypsy women, who waylaid a boy, aged eight, and cut off his hair. That they were gypsies may be of occult suggestion, but this could be simply the theft of something that could be sold.

A case is told of, in the People (London), Jan. 23, 1921. The residents of Glenshamrock Farm, Anchenleck, Ayrshire, Scotland, awoke one morning to find that during the night a burglar had made off with various articles. There were screams from the bedroom of a young female member of the household. Upon awakening, she had learned that her hair had been cut off. I say that this case was told of--but a case of what? And, in the New York Sun, March 7, 1928--a case of what? An old man had entered the home of Angelo Nappi, 83 1/2 Garside Street, Newark, N. J., and had cut off the hair of his three little daughters.

Old age and youth--male and female--there is the haunt, in stories of hair-clippers, of something that is not of hair-selling. If Jack the Slipper-snatcher were in the second-hand business, he'd have maneuvered girls into having both feet in the air. I take a story from the Medium and Daybreak, Dec. 13, 1889. It was copied from the Brockville (Ontario) Daily Times, November 13th. There were doings in the home of George Dagg, a farmer, living in the Township of Clarendon, Province of Quebec, Canada. With Dagg lived his wife, two young children, and a little girl, aged 11, Dina McLean, who had been adopted from an orphan asylum. The report from which I quote was the result of investigations by Percy Woodcock. I know that that sounds fictitious, but just the same Percy Woodcock was a well-known painter. Also Mr. Woodcock was a spiritualist. It could be that he colored as much on paper as on canvas.

The first of the "uncanny" occurrences--as they are so persistently called by persons who do not realize how common they are--was upon September 15th. Windowpanes broke. There were unaccountable fires--as many as eight a day. Stones of unknown origin were thrown. A large stone struck one of the children, and "strange to say, it did not hurt her in the least"

And I give my opinion that, in comments upon my writings, my madness has been over-emphasized. Of course I couldn't pass any alienist's examination--but could any alienist? But when I come upon a detail like this of stones striking people harmlessly, in an Ontario newspaper, and have noted the same detail in a story in a Constantinople newspaper, and have come upon it in newspapers of Adelaide, South Australia, and Cornwall, England, and other places--and when I note that it is no standardized detail of ghost stories, so that probably not one of the writers had ever heard of anything of the kind before--I'd consider myself sane and reasonable in giving heed to this, if there were sanity and reasonableness.

"One afternoon, little Dina felt her hair, which hung in a long braid down her back, suddenly pulled, and, on crying out, the family found her braid almost cut off, simply hanging by a few hairs. On the same day, the little boy said that something had pulled his hair all over. Immediately it was seen by his mother that his hair, also, had been cut off, in chunks, as it were, all over his head."

Woodcock told of a voice that was heard. This is an element that does not appear in the great majority of cases of poltergeist disturbances. His story is of conversations that were carried on between him and an invisible being. There was a feud between the Daggs and neighbors named Wallace; and "the voice" accused Mrs. Wallace of having sent him, or her, or it, or whatever, to persecute the Daggs. Most of the time, the house was crowded. When this accusation was heard, a number of farmers went to the home of the Wallaces, and returned with Mrs. Wallace. The story is that "the voice" again accused Mrs. Wallace, but then made statements that were so inconsistent that it was not believed. It was an obscene voice, and Mr. Woodcock was shocked. He reasoned with it, pointing out that there were farmeresses present. And "the voice" was ashamed of itself. It repented. It sang a hymn and departed.

I take something from the Religio-Philosophical Journal, Oct. 4, 1873, and following issues, as copied from the Durand (Wisconsin) Times, and other newspapers. Home of Mr. Lynch, 14 miles from Menomonie, Wisconsin--had moved from Indiana, a few years before, and was living with a second wife and the four children of the first wife. She had died shortly before he had moved. Lynch went to town one day, and returned with a dress for his wife. Soon afterward, this dress was found in the barn, slashed to shreds. Objects all over the house vanished. Lynch bought another dress. This was found, in the barn, cut down to fit one of the children. Eggs rose from tables, teacups leaped, and a pan of soft soap wandered from room to room. One of the children, a boy, aged six, was thought to be playing tricks, because phenomena centered around him. Nobody lambasted him until he confessed, but he was tied in a chair--teacups as lively as ever.

There was the usual openness. No midnight mysteries of a haunted house. Sightseers were arriving in such numbers that there was no room in the house for them. Several hundred of them lounged outside, sitting on fences or leaning against anything that would hold them up, ready for a dash into the house, at any announcement of doings.

"One day one of the children, named Rena, was standing close to Mrs. Lynch. Her hair was sheared off, close to her scalp, and vanished."

There have been single instances, and there have been hair-clipping scares that were attributed to "mass psychology." Also I have noted cases in which girls were accused of having cut off their own hair, hoping to take up some newspaper space. My only reason for doubt is the satisfactory endings of these accounts, with statements that the girls had confessed.

There were accounts in the London newspapers, of Dec. 2 and 10, 1922, of a scare in places east and west of London. In a street, in Uxbridge, Middlesex, a woman found that her braid had been cut off. She had been aware of no such operation, but remembered that, in a crowd, her hat had been pushed over her eyes. According to the stories, women were terrorized by "a vanishing man." "Disappeared as if by magic." It is an uncatchable again, a defiant fellow, operating openly, as if confident that he could not be caught. Note that these are not ghost stories. They are stories of human beings, who seemed to have ghostly qualities, or powers. Dorris Whiting, aged 17, approaching her home, in the village of Orpington, saw a man, leaning on the gate. As she was passing him, he grabbed her, and cut off her hair. The girl screamed, and her father and brother ran to her. They searched, but the clipper was unfindable. A maid, employed by Mrs. Glanfield, of Crofton Hall, Orpington, was pounced upon by a man, who hacked off a handful of her hair. He vanished. There was excitement in Orpington, at the end of a bus route. A girl exclaimed that much of her hair had been cut off. Merely this does not seem mysterious; it seems that a deft fellow could have done this without being seen by the other passengers. But other girls were saying whatever girls say when they discover that their hair has been cut off. At Enfield, a girl named Brand, employed as a typist, at the Constitutional Club, was near the club house, one morning, about eight o'clock, when a man grabbed her and cut off her hair. "No trace of him was found, though the search was taken up a minute after the outrage."

I have noted occurrences in London, which look as if there was a desire, not generally for hair, or for anybody's hair, but for the hair, and then more hair, of one victim. See the Kensington (London) Express, Aug. 23, 5907. Twice a girl's hair had been clipped. In a London street, she felt a clip, the third time. The girl accused a man. He was arrested and was arraigned at the Mansion House. Neither the girl nor anybody else had seen him as a clipper, but he had "walked sharply away," and when accused had run. Nothing was said of either scissors or hair in any quantity found in his possession. The hair that had been cut off was not found. But "there was some hair on his jacket," and he was found guilty and was fined.

I have record of another case of "mass psychology." It is my expression that the description "mass psychology" does partly apply to it, just as would "horizontal ineptitude," or "metacarpal iridescence," or any other idea, or combination of ideas, apply, to some degree, to anything. In an existence of the hyphen, it is impossible to be altogether wrong--or right. This is why it is so hard to learn anything. It is hard to overcome that which cannot be altogether wrong with that which cannot be altogether right. I look forward to the time when I shall refuse to learn another thing, having accumulated errors enough. In the Spiritualist, July 21, 1876, was published a story of "mass excitement" in Nanking and other cities of China. Uncatchables, who could not even be seen, were cutting off the pigtails of Chinamen, and there was a panic. More of the story was told, but I preferred to take accounts from a local newspaper. I give details, as I found them, in various issues of the North China Herald, from May 20 to Sept. 16, 1876.

Panic in Nanking and other towns, and its spread to Shanghai-- people believed that invisibles were cutting off their pigtails. It was said that, regard this of the invisibles as one would, there was no doubt that a number of pigtails had been cut off, and that great alarm existed, in consequence. "Many Chinamen have lost their tails, and we can hardly admit that the imaginary spirits are real men with steel shears, for it could hardly happen that someone would not be detected, before this, in the act of cutting. The most likely explanation is that the agents, whoever they may be, operate by means of some potent acid."

Panic spreading to Hangchow--"Numerous cases are reported, but few of them are authentic." "The cases are increasing daily."

In the streets of Shanghai, men, fearing attacks behind, were holding their pigtails in front of them. Quack doctors were advertising charms. Probably the reputable physicians, devoted to their own incantations, were indignant about this. The Military Commandant stationed soldiers in various parts of the city. "Suffice it to mention that, amongst much that is untrustworthy, there seem good grounds for believing that some children have actually lost part of their tails."

Sellers of charms suspected of cutting off pigtails, to stimulate business--mischievous children suspected--missionaries accused, and anti-Christian placards appearing in public places--rumors of drops of ink thrown in people's faces, "by invisible agencies," and people so treated dying--inhabitants of Woosin and Soochow mad with terror--the lynching of suspected persons--arrests and torture. People had suspended work, and had organized into guards. At Soo-chow broke out "the crushing mania," or a belief that at night people were crushed in their beds. The beating of gongs was taken up so that the supply ran out, and anybody who wanted a gong had to wait for one to be made. The standardized way of telling of such a scare is to elaborate upon the extremities at the climax of the excitement, and to ignore, or slightingly to touch upon, the incidents that preceded. There was a panic, or a mania, in China. Perhaps there was. I have no Chinaman's account. For all I know, some Chinaman may have sent an account to his newspaper, of us, beating gongs, during the parrot-disease scare, of the year 1929, having seen a janitor knocking off dust from the cover of an ash can. There was considerable excitement that was the product of delusions: nevertheless it does seem acceptable that there were cases of mysterious hair-clipping.

7

RABID vampires--and froth on their bloody mouths. See the New York Times, Sept. 5, 1931--rabies in vampire bats, reported from the island of Trinidad. Or a jungle at night--darkness and dankness, tangle and murk--and little white streaks that are purities in the dark--pure, white froths on the bloody mouths of flying bats--or that there is nothing that is beautiful and white, aglow against tangle and dark, that is not symbolized by froth on a vampire's mouth.

I note that it is ten minutes past nine in the morning. At ten minutes past nine, tonight, if I think of this matter--and can reach a pencil, without having to get up from my chair--though sometimes I can scrawl a little with the burnt end of a match--I shall probably make a note to strike out those rabid bats, with froth on their bloody mouths. I shall be prim and austere, all played out, after my labors of the day, and with my horse powers stabled for the night. My better self is ascendant when my energy is low. The best literary standards are affronted by those sensational bats. I now have a theory that our existence, as a whole, is an organism that is very old--a globular thing within a starry shell, afloat in a super-existence in which there may be countless other organisms--and that we, as cells in its composition, partake of, and are ruled by, its permeating senility. The theologians have recognized that the ideal is the imitation of God. If we be a part of such an organic thing, this thing is God to us, as I am God to the cells that compose me. When I see myself, and cats, and dogs losing irregularities of conduct, and approaching the irreproachable, with advancing age, I see that what is ennobling us is senility. I conclude that the virtues, the austerities, the proprieties are ideal in our existence, because they are imitations of the state of a whole existence, which is very old, good, and beyond reproach. The ideal state is meekness, or humility, or the semi-invalid state of the old. Year after year I am becoming nobler and nobler. If I can live to be decrepit enough, I shall be a saint.

It may be that there are vampires other than vampire bats. I have wondered at the specialization of appetite in the traditional stories of vampires. If blood be desired, why not the blood of cattle and sheep? According to many stories there have been unexplained attacks upon human beings; also there have been countless outrages upon other animals.

Possibly the remote ancestors of human beings were apes, though no evolutionist has made clear to me reasons for doubting the equally plausible theory that apes have either ascended, or descended, from humans. Still, I think that humans may have evolved from apes, because the simians openly imitate humans, as if conscious of a higher state, whereas the humans who act like apes are likely to deny it when criticized. Slashers and rippers of cattle may be throw-backs to the ape-era. But, though it is said that, in the Kenya Colony, Africa, baboons sometimes mutilate cattle, I'd not say that the case against them has been made out. London Daily Mail, May 18, 1925--that, for some years, an alarming epidemic of sheep-slashing and cattle-ripping had been breaking out, in the month of April, on Kenya stock ranches. Natives were blamed, but then it was learned that their cattle, too, had been attacked. Then it was said to be proved that chacma baboons were the marauders. Possibly the baboons, too, were unjustly blamed. Then what? The wounds were long, deep cuts, as if vicious slashes with a knife; but it was explained that baboons kill by ripping with their thumbnails.

The most widely known case of cattle-mutilation is that in which was involved a young lawyer, George Edalji, son of a Hindu, who was a clergyman in the village of Wyrley, Staffordshire, England. The first of a series of outrages occurred upon the night of Feb. 2, 1903. A valuable horse was ripped. Then, at intervals, up to August 27th, there were mutilations of horses, cows, and sheep.

Suspicion was directed to Edalji, because of anonymous letters, accusing him.

After the mutilation of a horse, August 27th, Edalji was arrested. The police searched his house, and, according to them, found an old coat, upon which were bloodstains. In the presence of Edalji's parents and his sister, the police said that there were horse hairs upon this coat. The coat was taken to the police station, where Dr. Butler, the police surgeon, examined it, reporting that upon it he had found twenty-nine horse hairs. The police said that shoes worn by Edalji exactly fitted tracks in the field, where the horse had been mutilated. They learned that the young man had been away from home, that night, "taking a walk," as told by him. The case against Edalji convinced a jury, which found him guilty, and he was sentenced to seven years, penal servitude.

I now have a theory that our existence is a phantom--that it died, long ago, probably of old age--that the thing is a ghost. So the unreality of its composition--its phantom justice and make-believe juries and incredible judges. There seems to be a ghostly justice surviving in the old spook, having the ghost's liking for public appearances, at times. Let there be publicity enough, and Justice prevails. In a Dreyfus case, when the attention of the world is attracted, Justice, after much delay, and after a fashion, appears.

Probably in the prison with Edalji were other prisoners who had been sent there, about as he had been sent. They stayed there. But Sir Arthur Conan Doyle, with much publicity, took up Edalji's case. In his account, in Great Stories of Real Life, Doyle says that when the police inspector found the old coat, upon which, according to him, there were horse hairs, Mrs. Edalji and Miss Edalji examined it and denied that there was a horse hair upon it: that Edalji's father said: "You can take the coat. I am satisfied that there is no horse hair on it." Doyle's statements imply that somewhere near the police station was a stable. As to the statement that Edalji's shoes exactly fitted tracks in the field, where the horse was ripped, Doyle says that the outrage occurred just outside a large colliery, and that hundreds of excited miners had swarmed over the place, making it impossible to pick out any one track. Because of Doyle's disclosures-- so it is said--or because of the publicity, the Government appointed a Committee to investigate, and the report of this Committee was that Edalji had been wrongfully convicted.

Sometimes slashers of cattle have been caught, and, when called upon to explain, have said that they had obeyed an "irresistible impulse." The better-educated of these unresisting ones transform the rude word "slasher" into "vivisectionist," and, instead of sneaking into fields at night, work at regular hours, in their laboratories. There are persons who wonder at the state of mind of the people in general, back in times when the torture of humans was sanctioned. The guts of a man were dragged out for the glory of God. "Abdominal exploration" of a dog is for the glory of Science. The state of mind that was, and the state of mind that is, are about the same, and the unpleasant features of anything are glossed over, so long as mainly anything is glorious.

According to a reconsideration, by the English Government, in the Edalji case, the slasher of cattle, of Wyrley, remained uncaught. In the summer of 1907, in the same region, again there was slashing. Aug. 22, 1907--a horse mutilated, near Wyrley. It was said that blood had been found on the horns of a cow, and that the horse had been gored. Five nights later, two horses, in another field, were slashed so that they died. September 8--horse slashed, at Breenwood, Staffordshire. A young butcher, named Morgan, was accused, but he was able to show that he had been in his home, at the time. For about a month injuries to horses continued to be reported. They had been injured "by barbed wires," or "by nails projecting from fences."

8

SOME time in the year 1867, a fishing smack sailed from Boston. One of the sailors was a Portuguese, who called himself "James Brown." Two of the crew were missing, and were searched for. The captain went into the hold. He held up his lantern, and saw the body of one of these men, in the clutches of "Brown," who was sucking blood from it. Nearby was the body of the other sailor. It was bloodless. "Brown" was tried, convicted, and sentenced to be hanged, but President Johnson commuted the sentence to life imprisonment. In October, 1892, the vampire was transferred from the Ohio Penitentiary to the National Asylum, Washington, D. C., and his story was re-told in the newspapers. See the Brooklyn Eagle, Nov. 4, 1892.

Ottawa Free Press, Sept. 17, 1910--that, near the town of Galazanna, Portugal, a child had been found dead, in a field. The corpse was bloodless. The child had been seen last with a man named Salvarrey. He was arrested, and confessed that he was a vampire.

See the New York Sun, April 14, 1931, for an account of the murders of nine persons, all but one of them females, which in the year 1929 terrorized the people of Dusseldorf, Germany. The murderer, Peter Kurten, was caught. At his trial, he made no defense, and described himself as a vampire.

I have a collection of stories of children, upon whom, at night, small wounds appeared. Rather to my own wonderment, considering that I am a theorist, I have not jumped to the conclusion that these stories are data of vampires, but have thought the explanation of rat bites satisfactory enough. But, in the Yorkshire Evening Argus, March 13, 1924, I came upon a rat story that seems queer. Inquest upon the death of Martha Senior, aged 68, of New Street, Batley. "On the toes and fingers were a lot of wounds that rather suggested rat bites." It was said that these little wounds could have had nothing to do with the woman's death, which, according to the coroner, was from valvular heart disease. The only explanation acceptable to the coroner was that, before the police took charge of the body, the woman must have been dead considerable time, during which rats mutilated the corpse. But Mrs. Elizabeth Lake, a neighbor, testified that she had found Mrs. Senior lying on the floor, and that Mrs. Senior had told her that she was dying. This statement meant that the woman had been attacked by something, before dying. The coroner disposed of it by saying that the woman must have been dead considerable time, before the body was found, and that Mrs. Lake was mistaken in thinking that Mrs. Senior had spoken to her.

The fun of everything, in our existence of comedy-tragedy--and I was suspicious of the story of terrorized Chinamen, as told by English reporters, because it was a story of panic that omitted the jokes-mania without the smile. Every fiendish occurrence that gnashes its circumstances, and sinks its particulars into a victim, wags a joke. In June, 1899, there was, in many parts of the U. S. A., much amusement. Something, in New York City, Washington, and Chicago, was sending people to hospitals. I don't recommend the beating of a gong to drive away a hellish thing: but I think that that treatment is as enlightened as is giving to it a funny name. Hospitals of Ann Arbor, Mich.; Toledo, Ohio; Rochester, N. Y.; Reading, Pa.--

"The kissing bug," it was called.

The story of the origin of the "kissing bug" scare-joke in that, upon the 19th of June, 1899, a Washington newspaper man, hearing of an unusual number of persons, who, at the Emergency Hospital, had applied for treatment for "bug bites," investigated, learning of "a very noticeable number of patients," who were suffering with swellings, mostly upon their lips, "apparently the result of insect bites." According to Dr. L. O. Howard, writing in Popular Science Monthly, 56-31, there were six insects, in the United States, that could inflict dangerous bites, or punctures, but all of them were of uncommon occurrence. So Dr. Howard rejected the insect-explanation. In his opinion there had arisen a senseless scare, like those of former times, in southern Europe, when hosts of hysterical persons imagined that tarantulas had bitten them.

This is "mass psychology" again--or the Taboo-explanation. To the regret of my contrariness, it is impossible for me utterly to disagree with anybody. I think with Dr. Howard that the "kissing bug" scare was like the tarantula scares. But it could be that some of those people of southern Europe did not merely imagine that something was biting them. If somebody should like to write a book, but is like millions of persons who would like to write books, but fortunately don't know just what to write books about, I suggest a study of scares, with the idea of showing that they were not altogether hysteria and mass psychology, and that there may have been something to be scared about. New York Herald, July 9--names and addresses of persons, who upon one day (8th of July) had either scared their bodies into producing swellings, or had been bitten by something that the scientists refused to believe existed. And people who were bitten captured insects. Entomological News, September, 1899--some of these insects, which were sent to the Academy of Natural Sciences of Philadelphia, were house flies, bees, beetles, and even a butterfly. There are wings of vampires that lull with scientific articles. See Taboo, as represented by Dr. E. Murray-Aaron, writing in the Scientific American, July 22, 1899--nothing but sensation-mongering from Richmond, Va., to Augusta, Me.

There was a sensational horse, in Cincinnati. His jaw swelled. Would a child, aged. four, be too young for "mass psychology"? I suppose not. I am not denying that there was much mass psychology in this. Cedar Falls, Iowa--a four-year-old child bitten. Trenton, N. J.--Helen Lersch, two years old, bitten--died. Bay Shore, L. I.--a child, aged two, bitten.

Later, I shall give instances of sizeable wounds that have appeared upon people: but, in this chapter, I am considering tiny punctures that may not have been either rat bites or insect stings. An account, in the Chicago Tribune, July 11, 1899, is suggestive of traditional vampire stories. A woman had been bitten. "The marks of two small incisors could be seen."

I don't know whether I am of a cruel and bloodthirsty disposition, or not. Most likely I am, but not more so than any other historian. Or, conforming to the conditions of our existence, I am amiable-bloodthirsty. In my desire for vampires, which is not in the least a queer desire, inasmuch as I have a theory that there are vampires, I was not satisfied with the "kissing bug": what I wanted was an account of hospital cases, not in the summer time. The insect-explanation, even though it was not upheld by Taboo, is too much at home, in the summer time. I needed an account, not in the summer time, to fill out my collection of data. Any collector will understand how pleased I was to come upon--London Daily Mail, April 20, 1920--an account of human suffering. "A number of people in country places have been bitten by some mysterious creature with a very poisonous fang. It is rare for any sort of poisonous bite or sting to occur before summer, and as a rule the culprit is known. This spring doctors have attended case after case, where the swellings have been sudden and severe, though there is little sign of the bite, itself." I have record of several winter time cases. See La Nature, (Supplement) Jan. 16, 1897--that, while filling a stove with coal, in a house in the Rue de la Tour, Paris, a concierge had felt a stinging sensation upon his arm, which swelled. He was taken to a hospital, where he died. People in the house said that they had seen gigantic wasps entering the house by way of stovepipes.

But the most mysterious of cases of insect bites, or alleged insect bites, is that of the small wound that led to the death of Lord Carnarvon, if be accepted that his death, and the deaths of fourteen other persons, were in any especial way related to the opening, or the violation, of the tomb of Tut-Ankh- Amen. Lord Carnarvon was stung by what was supposed to be an insect. What was said to be blood poisoning set in. What was said to be septic pneumonia followed.

The stories of the "kissing bug" differ from vampire stories, in that victims were painfully wounded. But there was an occurrence in Upper Broadway, New York City, May 7, 1909, that may be more in agreement. It seems possible that a woman could, in a street crowd, viciously jab several persons with a hat pin, without being detected: but it does seem unlikely that she could enjoy such a stroll, jabbing at least five men and a woman, before being interfered with. A Broadway policeman learned that upon somebody a small wound, as if made by a hat pin, had appeared. Four other men and a woman joined the crowd and showed that they had been similarly wounded. The policeman arrested, as the cause of the excitement, a woman, who told that her name was Mary Maloney, and gave a false address.

Perhaps she had no address. She may have been guilty, but perhaps she was shabby. If somebody must be arrested, it is wise to pick out one who does not look very self-defensive. "Plead guilty and you'll get off with a light sentence." It is dangerous to be anywhere near any scene of crime, considering the way detectives pick up "suspects," even an hour or so later, obviously arguing that when somebody commits a crime, he hangs around to be suspected.

I have never been jabbed with a hat pin, but I have sat on pointed things, and my responses were so energetic that I suspect that at least six persons were not jabbed with a hat pin, before the jabber was caught. See data to come, that indicate that people may be--by some means at present not understood--wounded, and not know it until later. Also that a woman was accused makes me doubt that the marauder was caught. Women don't do such things. I have a long list of Jacks, ranging from the rippers and stranglers to the egg throwers and the ink squirters: but Mary Maloney is the only alleged Jill in my collection. Women don't do such things. They have their own deviltries.

Upon Dec. 4, 1913, Mrs. Wesley Graff, who sat in a box, in the Lyric Theatre, New York City, felt something scratching her hand. She felt a pain like the sting of a wasp, and, staggering from her chair, fainted, first accusing a young man near her. The manager of the theater held the young man, and called the police. Policemen searched, and found, on the floor, a common darning needle. It was their theory that the young man was a white slaver, who by means of a hypodermic injection, had sought to render a victim insensible, probably having waiting outside, a cab, to which he, explaining that he was her companion, would carry her. There were marks upon Mrs. Graff's arm, but it seems that they were not made by a darning needle.

With the idea that the needle might be tipped with a drug, the police sent it to a chemist. To my astonishment, I record that he reported that he had found neither drug, nor poison, on it. A strange circumstance is that, at this place, where a woman was wounded somewhat as if by a darning needle, was found this darning needle, which was suggestive of a commonplace explanation.

Then arose the story that a gang of white slavers was operating in the city. But in the newspapers were published interviews with physicians, who stated that they knew of no drug by which women could be affected so as to make them easily abductable, because the pain of an injection would give minutes of warning, before a victim could be rendered helpless. But it may be that something, or somebody, was abroad, mysteriously wounding women. In the Brooklyn Eagle, December 6, it was said that, in a period of two weeks, the Committee of Fourteen, of New York City, had heard a dozen complaints of mysterious, minor attacks upon women, and had investigated, but had been unable to learn anything definite in any case.

See back to the story of the Chicago woman, and "marks of two small incisors." Upon Mrs. Graff's arm were two little punctures. December 29--girl named Marian Brindle said that something had stung her. Upon her arm were two little punctures. It may be that, in the period of the scare in New York City, the first occurrence of which was in November, 1913, a vampire was abroad. It could be that we pick up the trail more than a year before this time. In October, 1912, Miss Jean Milne, aged 67, was living alone in her home, in West Ferry, Dundee, Scotland. London Times, Nov. 5, 1912--the finding of her body. The woman had been beaten, presumably with a poker, which was found, according to the account in the Times: but it was said that, though she had been struck on the head, her skull was not fractured: so her death was not altogether accounted for. There was more of this story, in the London Weekly Dispatch, Nov. 24, 1912. Upon this body were found perforations, as if having been made by a fork.

Late at night, Feb. 2, 1913, the body of a woman was found on the tracks of the London Underground Railway, near the Kensington High-street station. The body had been run over, and the head had been cut off. The body was identified as that of Miss Maud Frances Davies, who, alone, had been traveling around the world, and, earlier in the day, had, upon a ship train, arrived in London. She had friends and relatives in South Kensington, and presumably she was on her way to visit them. But the explanation at the inquest (London Times, Feb. 6, 1913) was that she had probably committed suicide by placing her neck upon a rail.

"Dr. Townsend said that over the heart he found a number of small, punctured wounds, over a dozen of which had penetrated the muscles; and one had entered the ventricle cavity of the heart. These punctures had been caused in life, with a sharp instrument, such as a hat pin. They were not enough to cause death, but had been made a few hours previously."

Upon December 29th, of this year, 1913, a woman, known as "Scotch Dolly," was found dead in her room, 18 Etham Street, S. E., London. A man, who had lived with her, was arrested, but was released, because he was able to show that, before the time of her death, he had left the woman. Her face was bruised, but she had seldom been sober, and the man, Williams, before leaving her, had struck her. The verdict was that she had died of heart failure, "from shock."

Upon one of this woman's legs was found a series of 38 little, double wounds. They were not explained. "The Coroner: 'Have you ever had a similar case, yourself?' Dr. Spilsbury: 'No: not exactly like this.'"

9

UPON April 16, 1922, a man was taken to Charing Cross Hospital, London, suffering from a wound in his neck. It was said that he would tell nothing about himself, except that, while walking along a turning, off Coventry Street, he had been stabbed. Hours later, another man, who had been wounded in the neck, entered the hospital. He told, with a foreign accent, that in a turning, off Coventry Street, he had been so wounded. He signed his name in the hospital register as Pilbert, but would, it was said, give no other information about the assault upon him. Late in the day, another wounded man was taken to this hospital, where, according to the records, he refused to tell anything about what had befallen him, except that he had been stabbed in the neck while walking along a turning off Coventry Street.

In the pockets of one of these men were found racing slips. The police explained that probably all of them were victims of a turf-feud.

It is, considering many other data, quite thinkable that, instead of refusing to tell how they had been wounded, these men were unable to tell, but that this inability was so mysterious that the hospital authorities recorded it as refusal. See the London Daily Express, April 17, and the People, April 23, 1922.

In a London hospital, there is small chance for an unconventional record, and probably in no London newspaper would have been published any reporter's notion of the lurk of an invisible and murderous thing, in a turning, off Coventry Street. But, in the London Daily Mail, Sept. 26, 1923, there was an account of something like this, but far away. It was a facetious account. Murderous things always have, somewhere, been regarded humorously. Or fondly. No address was published, or probably this one would have received letters from women, wanting to marry it. The story was that, in September, 1923, there was a Mumiai scare in India. Mumiais are invisibles that grab people. They have no sense of the mystic: don't dwell in enchanted woods, nor feel out for victims from old towers, or ruins; no valuation for midnight. In daylight, in the streets of cities, they grab people. Coolies, in the city of Lahore, believed that a Mumiai was abroad. There was a panic in Lahore, and it fed upon screams of rickshaw men, who thought that they were grabbed.

Probably the Daily Mail published this story, because of wavelets of gratification that arose from it, at London breakfast tables. It is usually thought that the value of coolies is only in their willingness to work for a few cents a day: but I have a notion that they have another function; or that, if it were not for coolies, and their silly superstitions that give the rest of us some sense of superiority to keep going on, millions of the rest of us would lie down and die of chagrin. Sometime I shall develop a theory of Evolution in aristocratic terms, showing that things probably made of themselves oysters and lions and hyenas, just for the thrill of gratification in being able to say that at least they weren't elephants, or worms, or human beings. I know how it is, myself, and have compensations, in thinking of silly, credulous people who believe that a dog ever said "Good morning!" and disappeared in a thin, greenish vapor.

Away back in the year 1890, the Japanese were coolies. Then they showed such talents for slaughter that now they are respected everywhere. But, in the year 1890, the Japanese were supposed to be little more than a nation of artists. A story of a panic in Japan was something to smile smugly about. I take a story from the Religio-Philosophical Journal, May 17, 1890, as copied from the newspapers. People in Japan thought that, sometimes in the streets, and sometimes in their houses, an invisible thing was attacking them. They thought that upon persons were appearing wounds, each a slash about an inch long. They thought that, at the time of an attack, little pain was felt.

Possibly a Jap, educated according to what is supposed to be an education, having his ideas as to the identity and geographical distribution of coolies, has looked over files of American newspapers, and has come upon accounts of a series of occurrences in New York City, in the winter of 1891-92, and has been amused to note the mystery that New York reporters infused into their accounts of woundings of men, in the streets of New York. The reporters told of a "vanishing man." The assassin "disappeared marvelously." As noted, in the New York Sun, Jan. 14, 1892, five men had been stabbed by an unknown assailant. There were other attacks. The police were blamed, and in the downtown precincts of the city, the most important order, each day, was to catch the stabber.

January 17--"Slasher captured." The police were out to get him, and one of them got an unterrifying-looking little fellow, named Dowd. It was said that he had been caught, stabbing a man.

In the mixture of all situations, it is impossible to be unable to pick out grounds for reasonably believing, or disbelieving, anything. Say that it is our preference to believe--or to accept--that it was not the "marvelously disappearing" slasher who was caught, but somebody else who would do just as well. Then we note that, twenty minutes earlier, another policeman had caught a man, who had, this policeman said, seized somebody, and was about to stab him. Or June, 1899--and two men were out to catch the "kissing bug"--and one of them caught a beetle, and the other nabbed a butterfly. The policeman of the first arrest was ignored: the captor of Dowd was made a roundsman.

Dowd pleaded not guilty. He said that he had had nothing to do with the other assaults, and had drawn a knife only in this one case, which had been a quarrel. His lawyer pleaded not guilty, but insane. He was found insane, and was sent to the asylum for insane criminals, at Auburn, N. Y.

The outrages in New York stopped. Brooklyn Eagle, March 12, 1892--dispatch from Vienna, Austria-- "This city continues to be shocked by mysterious murders. The latest victim is Leopold Buchinger, who was stabbed to the heart by an undetected assassin, in one of the most public places in Vienna. This makes the list of such tragedies five in number, and there is a growing feeling of terror among the public."

Say that it's an old castle, hidden away in a Balkan forest--and somebody was wounded, at night-- but, as if lulled by a vampire's wings, felt no pain. This would be only an ordinarily incredible story. In November, 1901, a woman told a policeman, of Kiel, Germany, that, while walking in a street in Kiel, she learned that she had been unaccountably wounded. She had felt no pain. She could not explain.

The police probably explained. If a doctor was consulted, he probably explained learnedly.

Another woman--about thirty women--"curious and inexplicable attacks." Then men were similarly injured. About eighty persons, openly, in the streets, were stabbed by an uncatchable--an invisible--or it may be the most fitting description to say that, upon the bodies of people of Kiel, wounds appeared. See the London Daily Mail, Dec. 7, 1901--"The extraordinary thing about the mystery is that some marvelously sharp instrument must have been used, because the victims do not seem to know that they are wounded, until several minutes after an attack."

And yet I think that something of an explanation of these Jacks is findable in every male's recollections of his own boyhood--the ringing of door bells, just to torment people--stretching a string over sidewalks, to knock off hats--other, pestiferous tricks. It is not only "just for fun"; there is an engagement of the imagination in these pranks. It will be my expression that, when the more powerful and more definite imagination of an adult human similarly engages and concentrates, phenomena that will be considered beyond belief, or acceptance, by readers who do not realize of what common occurrence they are, develop.

We have had stories of series of accidents, and perhaps my suspicion that they were not mere coincidences has been regarded at least tolerantly. I have data of three automobile accidents that occurred at times not far apart; and, as to this series, I note a seeming association with minor attacks upon other automobiles, and upon people, that suggests the doings of one criminal. If so, he will have to be called occult, whether we take readily to, or are much repelled by, that term. Upon the night of April 9, 1927, Alexander Nemko and Pearl Devon were motoring through Hyde Park, London, when their car dashed down an incline, and plunged into the Serpentine. The car sank in fifteen feet of water. Though terrified and drowning, Nemko had his wits with him, so that he opened the door of the car, and dragged his companion to the surface, and, with her, swam ashore.

There was nothing in the lay of the land by which to explain. The newspapers noted that there had never been an accident here before. "The steering gear apparently failed," was Nemko's attempt to explain. Perhaps it is queer that right at this point, so near a body of water, the steering gear failed: but, considered by itself, as mysteries usually are considered, there is little that can be said against Nemko's way of explaining.

Two nights later, a taxicab plunged into the Thames, at Walton.

The passenger swam ashore, but the driver was, it seems, drowned. His body was dredged for but was not found. The passenger, who must have been jostled past having any clear remembrance of what occurred, explained that, at the brink of the river, the rear wheels of the car had dropped into a deep rut, and that the car had jolted into the river.

Upon May 3rd--see the London Evening Standard, May 6--William Farrance and Beatrice Villes, of Linomroad, Clapham, London, were driving near Tunbridge Wells, when the car suddenly plunged toward a hedge, at the left of the road. Farrance succeeded in forcing the car to the right. Again something drove it toward the hedge. Farrance was powerless to stop it, and it broke through the hedge, overturning, killing the girl.

A schoolgirl, Beryl de Meza, was shot by somebody unknown and unseen, while playing in the street, near her home, at Hampstead, London.

At Sheffield, there was an occurrence that was atrocious, but that may not be uncanny, but that attracts my attention because of the fiendishness of something else with which it associates. At the Soho Grinding Works, it was found, morning of April 29th, that grinding wheels had been chipped, and that belting had been stripped from pulleys. Nails had been driven, points upward, in chairs upon which the grinders sat. Tools had been thrown into motors, and currents had been turned on, causing much damage. All this looks like sabotage, malicious but scarcely "fiendish": but in a building next door there had been doings that are so describable. Chickens had been tortured: combs cut off, legs broken, the head of one burned: others mutilated, and their injuries smeared with white paint.

London Evening Standard, May 5--"Mystery of four shooting affairs." A boy, playing in Mitcham

Park, London, was shot in the head, by an air gun, it was thought, though no air gun pellet was found. At Tooting Bec-common, an "air gun pellet"--though it was not said that an air gun pellet was found-- passed through the windshield of a motor car. In Stamford two men were shot by an unknown assailant. London Sunday Express, May 8--Mr. George Berlam, of Leigh-on-Sea, motoring on the road from London to Southend--he heard a report, and his windshield was splintered. In accounts of the punctured windshield, at Tooting Bec-common, the driver of the car was quoted as saying that he had heard a report, and at the same time a laugh, "though nobody was about, at the time."

Wounds have appeared upon people. Usually the explanation is that they were stabbed. Objects have been mutilated. Windowpanes and automobile windshields have been pierced, as if by bullets, but by bullets that could not be found. Such were the doings of the "phantom sniper of Camden" (N. J.). He appeared first, in November, 1927: but the first clipping that I have, relating to him, is from the New York Evening Post, Jan. 26, 1928--a store window pierced by a bullet--the eighth reported occurrence. Later, the stories were definitely of a "phantom sniper" and his "phantom bullets."

New York Herald Tribune, Feb. 9, 1928--Collingswood, N. J., February 8--"The 'phantom sniper,' if it was the work of South Jersey's mysterious marksman, scored his most sensational attack tonight when a window in the home of William T. Turnbull was shattered by what appeared to be a charge of shot.

"Police at first believed it an attempted assassination, but, as in all the other cases, no missile was found. "Turnbull, a Philadelphia stockbroker, and a former president of the Collingswood Borough Council, who was seated near the window, reading, was spattered with glass. He said that an automobile had stopped in front of the house a few minutes before. The absence of any grains of shot added to the mystery."

I have sent letters of enquiry to all persons mentioned in the various reports. I have received not one answer. It may be preferable to some readers to think that there are no such persons. Still, I note that not one of these letters was dead-lettered back to me.

The attacks continued until Feb. 28, 1928. Windowpanes and windshields of automobiles were pierced by something that made no report of a gun, and that was unfindable. Something, or somebody, who was unseen, caused excitement in half a dozen towns from Philadelphia to Newark. Even if I could persuade myself that I am over-fanciful in my own notions, the seemingly veritable stories of the existence of a missile-less gun would be interesting. Authorities in Jersey towns, noting the range of the malefactor, were especially watchful of motorists: but it is my notion that he had no need for anything on wheels in which to do his traveling. I noticed a similar range, in the doings in England, in April and May, 1927.

Snipings by the "Camden phantom" were the show-off, and nobody was injured by him: but a more harmful fellow operated in Boston, beginning about Nov. 1, 1930. I think that these sportsmen, who possibly are sentimental opponents to the shooting of game birds and deer, and practice their cruelties in ways that seem to them less condemnable, divide into the unoccult, and into more imaginative fellows who have found out how to practice occultly. In Boston, a noiseless weapon was used, but, this time, in two weeks, two men and a woman were seriously injured, and bullets of small caliber were removed from their wounds. These attacks so alarmed people that policemen, armed with riot guns, lined the roads south of Boston, with orders to catch the "silent sniper." The attacks continued until about the middle of February, 1931. Nobody was caught. In this period (Nov. 12, 1931) a dispatch to the newspapers, from Bogota, Colombia, told of a "puzzling crime wave." In the hospitals were forty-five persons, suffering with stab wounds. "The police were unable to explain what appeared to be a general attack, but they arrested more than 200 persons."

Another occurrence of "phantom bullets," in the State of New Jersey, was told of, in the New York Herald, Feb. 2, 1916. Mr. and Mrs. Charles F. Repp, of Glassboro, N. J., had been fired upon by "phantom bullets." This was a special attack upon one house. There were sounds of breaking glass, and bullet holes were found in windowpanes, but nothing beyond the windowpanes was marked. It is such a circumstance as was told of in accounts of the "Camden sniper." It is as if somebody fired, not only with a missile-less gun, or with invisible bullets, but as if with intent only to perforate windows, and with the effects controlled by, and limited by, his intent. Consequently, instead of thinking of a shooting at windowpanes, I tend simply to think that holes appeared in window glass. Nobody in the house was injured, but Mr. and Mrs. Repp were terrified and they fled. Members of the Township Committee investigated, and they reported that, though no bullets were findable, the windows "were broken much as a window usually is, when a bullet crashes through it."

That's the story. Of witnesses, I. C. Soddy and Howard R. Moore were mentioned. I sent letters of enquiry to all persons whose names were given, and received not one reply. There are several ways of explaining. One is that it is probable that persons who have experiences such as those told of in this book, receive so many "crank letters" that they answer none. Dear me--once upon a time, I enjoyed a sense of amusement and superiority toward "cranks." And now here am I, a "crank," myself. Like most writers, I have the moralist somewhere in my composition, and here I warn--take care, oh, reader, with whom you are amused, unless you enjoy laughing at yourself.

It seemed to me doubtful that a woman could go along Upper Broadway, and jab, with a hat pin, five men and a woman, before being caught. There has been a gathering of suggestions of not ordinary woundings. In Lloyd's Weekly News (London) Feb. 21, 1909, there was an account of a panic in Berlin. Many women, in the streets of the city, had been stabbed. It was said that the assailant had been seen, and he was described as "a young man, always vanishing." If he was seen, he is another of the "uncatchables." In this newspaper of February 23, it was said that 73 women had been stabbed, all except four of them not seriously. We have had data that suggest the existence of vampires, other than humans of the type of the Portuguese sailor: but the brazen and serialized--sometimes murderous, but sometimes petty--assaults upon men and women are of a different order, and seem to me to be the work of imaginative criminals, stabbing people to make mystery, and to make a stir. I feel that I can understand their motives, because once upon a time I was an imaginative criminal, myself. Once upon a time I was a boy. One time, when I was a boy, I caught a lot of flies. There was nothing of the criminal, nor of the malicious, in what I did, this time, but it seems to give me an understanding of the "phantom" stabbers and snipers. I painted the backs of the flies red, and turned them loose. There was an imaginative pleasure in thinking of flies, so bearing my mark, attracting attention, causing people to wonder, spreading far, appearing in distant places, so marked by me.

In some of our stories there is much suggestion that there was no "vanishing man"--that wounds appeared upon people, as appeared--or as it was said to have appeared--a wound on the head of a sailor.

See back to the story told by the captain of the Brechsee. Or that wounds appeared upon people, and that the victims, examined by the police, were more or less bullied into giving some kind of description of an assailant. However, some of the stories of the "vanishing man" look as if he, too, may be. There may be several ways of doing these things. Early in the year 1907, a "vanishing man" was reported from the town of Winchester, England. I take from the Weekly Dispatch (London), Feb. 10, 1907. Women of Winchester were complaining of an "uncatchable," who was committing petty assaults upon them, such as rapping their hands. "A mysterious feature of the affair is that the man disappears, as if by magic."

The "phantom stabber" of Bridgeport, Conn., appeared first Feb. 20, 1925, and the last of his attacks, of which I have record, was upon June 1, 1928. That was a long time in which to operate uncaught. In the daytime, mostly, though sometimes at night, girls were stabbed: in the streets; in such public places as a department store, and the entrance of a library. Descriptions of the assailant were indefinite. In almost all instances the wounds were not serious. One of the stories, as told in the New York Herald Tribune, Aug. 27, 1927, is typical of the circumstances of publicity, or of the confidence of an assailant that he could not be caught. If my stories will be regarded as ghost stories, a novelty about them is the eeriness of crowded thoroughfares--a lurk near Coventry Street, London, and the sneak of an invisible in Broadway, New York. I expect sometime to hear of a haunted subway, during rush hours. Edgar Allan Poe would say of me that I'm no artist, and don't know how to infuse atmosphere. One would think that I had never heard of the uncanniness of dark nights in lonely places. Some of the stories are of desperate plays for notoriety. I have a story now, not of doings in a graveyard, but in a department store. Bridgeport, Conn.--staged on a staircase, with an audience of hundreds of persons, there was a very theatrical performance. A review of this melodrama was published in the Herald Tribune--

"The stabber who has terrorized Bridgeport for the last thirty months suddenly appeared this afternoon and claimed his twenty-third victim in a crowded down-town department store. The victim was Isabelle Pelskur, fourteen, 539 Main Street, messenger girl employed in the D. M. Read store. The girl was stabbed in the store where she is employed.

"The stabbing occurred at 4:50, just two minutes before closing time of the store. Already some of the store doors had been locked, and the large crowd of shoppers were being ushered from the store. The employees were leaving their counters, and the victim had started up the stairs from the arcade side of the first floor to the women's dressing room.

"The girl had scarcely ascended more than half a dozen steps when she was attacked by the assailant who lunged his sharp blade into her side, causing a severe wound."

He got away. Nobody reported having seen him escaping. The girl could give only a "meager" description of him.

10

RELATIVELY to the principles of modern science, werewolves cannot be. But I know of no such principle that is other than tautology or approximation. It is myth-stuff. Then, if relatively to a group of phantoms, werewolves cannot be, there are at least negative grounds for thinking that they are quite likely.

Relatively to the principles, or lack of principles, of ultra-modern science, there isn't anything that can't be, even though also it is not clear how anything can be.

So my acceptance, or pseudo-conclusion, is that werewolves are quite likely-unlikely.

Once upon a time, when minds were dosed with the pill-theory of matter, werewolves were said to be physically impossible. Very little globes were said to be the ultimates of matter, and were supposed to be understandable, and people thought they knew what matter is. But the pills have rolled away. Now we are told that the ultimates are waves. It is impossible to think of a wave. One has to think of something that is waving. If anybody can think of crime, virtue, or color, independent of somebody who is criminal, virtuous, or colored, that thinker--or whatever--may say that he knows what he is talking about, in denying the existence of anything, upon physical grounds. To say that the "ultimate waves" are electrical comes no closer to saying something. If there is no definition of electricity better than that of saying that it is a mode of motion, we're not enlighteningly told that the "ultimate waves" are moving motions.

My suspicion is that we've got everything reversed; or that all things that have the sanction of scientists, or that are in agreement with their myths, are ghosts: and that things called "ghosts," are, because they are not in agreement with the spooks of science, the more nearly real things. I now suspect that the spiritualists are reversedly right--that there is a ghost-world--but that it is our existence--that when spirits die they become human beings.

I now have a theory that once upon a time, we were real and alive, but departed into this state that we call "existence"--that we have carried over with us from the real existence, from which we died, the ideas of Truth, and of axioms and principles and generalizations--ideas that really meant something when we were really alive, but that, of course, now, in our phantom-existence--which is demonstrable by any X-ray photograph of any of us--can have only phantom-meaning--so then our never-ending, but always frustrated, search for our lost reality. We come upon chimera and mystification, but persistently have beliefs, as retentions from an experience in which there were things to believe in. I'd not say that all of us are directly ghosts: most of us may be the descendants of the departed from a real existence, who, in our spook-world, pseudo-propagated.

Once upon a time--but in our own times--there were two alleged marvels that were sources of uncommon contempt, or amusement, to scientists: they were the transformation of elements into other elements, and the transformation of human animals into other animals.

The history of science is a record of the transformations of contempts and amusements.

I think that the idea of werewolves is most silly, degraded, and superstitious: therefore I incline toward it respectfully. It is so laughable that I am serious about this.

Marauding animals have often unaccountably appeared in, or near, human communities, in Europe and the United States. The explanation of an escape from a menagerie has, many times, been unsatisfactory, or has had nothing to base upon. I have collected notes upon these occurrences, as teleportations, but also there may be lycanthropy.

Nobody has ever been finally reasonable, and it is impossible for me to be absolutely unreasonable. I can tell no yarn that is wholly a yarn, if it be my whim, or inspiration, to come out for the existence of werewolves.

What is there that absolutely sets apart the story of a man who turned into an ape, or a hyena, from the story of a caterpillar that became a butterfly? Or rascals who almost starve to death, and then learn to take on the looks of philanthropists? There are shabby young doctors and clergymen, who turn so sleek, after learning the lingo of altruists, that they have the appearance of very different animals. Or the series of portraits of Napoleon Bonaparte--and so much of his mind upon classical models--and the transformation of a haggard young man into much resemblance to the Roman Emperor Augustus.

It is a matter of common belief that men have come from animals called "lower," not necessarily from apes, though the ape-theory seems to fit best, and is the most popular. Then why not that occasionally a human sloughs backward? Data of reversions, not of individuals, but of species, are common in biology.

I have come upon many allusions to the "leopard men" and the "hyena men" of African tribes, but the most definite story that I know of is an article by Richard Bagot, in the Cornhill Magazine, October, 1918, upon the alleged powers of natives of Northern Nigeria to take on the forms of lower animals. An experience attributed to Capt. Shott, D. S. O., is told of. It is said that raiding hyenas had been wounded by gun-traps, and in each case had been traced to a point where the hyena tracks had ceased, and had been succeeded by human footprints, leading to a native town. A particular of the traditional werewolf story is that when a werewolf is injured, the injury appears upon a corresponding part of the human being of its origin. Bagot told of Capt. Shott's experience, alleged experience, whatever, with "an enormous brute" that had been shot, and had made off, leaving tracks that were followed. The hunters came to a spot where they found the jaw of the animal, lying in a pool of blood. The tracks went on toward a native town. The next day a native died. His jaw had been shot away.

There have been many appearances of animals that were unexplained--anyway until I appeared upon the horizon of this field of data. It seems to me that my expressions upon Teleportations are somewhat satisfactory in most of the cases--that is, that there is a force, distributive of forms of life and other phenomena that could switch an animal, say from a jungle in Madagascar to a back yard somewhere in Nebraska. But theories of mine are not so godlike as to deny any right of being for all other theories. I'd not be dogmatic and say positively that once upon a time a lemur was magically transported from Africa to Nebraska: possibly somebody in Lincoln, Nebraska, had been transformed into a lemur, or was a werelemur.

Whatever the explanation may be, the story was told, in the New York Sun, Nov. 12, 1931. Dr. E. R. Mathers, of Lincoln, Nebraska, had seen a strange, small animal in his yard, acting queerly. The next day he found the creature dead. The body was taken to Dr. I. H. Blake, of the University of Nebraska, who identified it as that of an African lemur, of the Galaga group. A lemur is a monkey-like animal, with a long snout: size about that of a monkey.

I wrote to Dr. Mathers about this, and, considerably to my surprise, because mostly my "crank" letters are very properly ignored, received an answer, dated Nov. 21, 1931. Dr. Mathers verified the story. The lemur, stuffed and mounted, is now upon exhibition in the museum of the State University, at Lincoln. Where it had come from had not been learned. There was no story of an escape, anywhere, that could match this appearance in a back yard. Accounts had been spread-headed, with illustrations, in the Lincoln State Journal, October 23rd, and in the Sunday State Journal, October

25th: but not even in some other back yard had this animal been seen, according to absence of statements. I neglected to ask whether, at the time of the appearance of the lemur, the disappearance of any resident of Lincoln was reported.

Suppose, at a meeting of the National Academy of Sciences, 1 should read a paper upon the transformation of a man into a hyena. There would be only one way of doing that. I recommend it to unrecognized geniuses, who can't otherwise get a hearing. It would have to be a hold-up.

But, without having to pull a gun, at the meeting of the N. A. S., at New Haven, Conn., Nov. 18, 1931, Dr. Richard C. Tolman suggested that energy may be transforming into matter.

If one can't think of a man transforming into a hyena, let one try to think of the motions of a thing turning into a thing.

My expression is that, in our existence of the hyphen, or of intermediateness between so-called opposites, there is no energy, and there is no matter: but that there is matter-energy, manifesting in different degrees of emphasis one way or the other: That it is not thinkable that energy could turn into matter: but that it is thinkable that energy-matter could, by a difference of emphasis, turn into matter-energy--

Or that there is no man who is without the hyena-element in his composition, and that there is no hyena that is not at least rudimentarily human--or that at least it may be reasoned that, by no absolute transformation, but by a shift of emphasis, a man-hyena might turn into a hyena-man.

The year 1931--and there were everywhere, but most notably in the U. S. A., such shifts, or reversions, from the state that is called "civilization," that there was talk of repealing laws against carrying weapons, and of the arming of citizens to protect themselves, as if such cities as New York and Chicago were frontier towns. Out of policemen--in all except physical appearance--had come wolves that had preyed upon nocturnal women. There were chases of savages through the streets of New York City.

Jackals on juries picked up bits from kills by bigger beasts, and snarled their jackal-verdicts.

New York Times, June 30, 1931--"Police at Mineola hunt apelike animal--hairy creature, about four feet tall."

Out of judges had come swine.

County Judge W. Bernard Vause found guilty of using the mails to defraud, and sentenced to six years in Atlanta Penitentiary. Federal Judge Grover M. Moscowitz was censured by the House of Representatives. The Magistrates, who, facing charges of corruption, resigned, were Mancuso, Ewald, McQuade, Goodman, Simpson. Vitale was removed. Crater disappeared. Rosenbluth went away, for his health's sake.

And, near Mineola, Long Island, a gorilla was reported. The first excitement was at Lewis & Valentine's nursery--story told by half a dozen persons--an ape that had come out of the woods, had looked them over, and had retreated. It seems that the police heard of "mass psychology": so they had to explain less learnedly. Several days later, they were so impressed with repeating stories that a dozen members of the Nassau County Police Department were armed with shotguns, and were assigned to ape-duty.

No circus had appeared anywhere near Mineola, about this time; and from neither any Zoo, nor from anybody's smaller menagerie, had the escape of any animal been reported. Ordinarily let nothing escape, or let nothing large, wild, and hairy appear, but let it be called an ape, anyway--and, upon the rise of an ape-scare, one expects to hear of cows reported as gorillas: trees, shadows, vacancies taking on ape-forms. But--New York Herald Tribune, June 27th--Mrs. E. H. Tandy, of Star Cliff Drive, Malverne, reported something as if she had not heard of the ape-scare. She called up the. police station, saying that there was a lion in her back yard. The policeman, who incredulously received this message, waited for another policeman to return to the station, and share the joke. Both waited for the arrival of a third disbeliever. The three incredulous policemen set out, several hours after the telephone call, and by that time there wasn't anything to. disturb anybody's conventional beliefs, in Mrs. Tandy's back yard.

There was no marauding. All the stories were of a large and hairy animal that was appearing and disappearing--

And appearing and disappearing in the vast jungles not far from Mineola, Long Island, were skunks that were coming from lawyers. Some of them were caught and rendered inoffensive by disbarment. There was a capture of several dozen medical hyenas, who had been picking up livings in the trains of bootleggers. It could be that an occurrence, in New Jersey, was not at all special, but represented a general slump back toward a state of about simian development. There was an examination of applicants for positions in the schools of Irvington. In mathematics, no question beyond arithmetic was asked: in spelling, no unusual word was listed. One hundred and sixteen applicants took the examination, and all failed to pass. The average mark was 31.5. The creep of jungle-life stripped clothes from people. Nudists appeared in many places. And it was not until later in the year, that the staunchest opponent of disclosures spoke out, in the name of decency, or swaddling--or when Pope Pius XI refused to receive Mahatma Gandhi, unless he'd put on pants.

Upon the 29th of June, the ape-story was taken so seriously, at Mineola, that Police Captain Earle Comstock ordered out a dozen special motor patrols, armed with revolvers and sawed-off shotguns, with gas and ball ammunition, led by Sergeant Berkley Hyde. A posse of citizens was organized, and it was joined by twenty nurserymen, who were armed with sickles, clubs, and pitchforks. Numerous footprints were found. "The prints seemed to be solely those of the hind feet, and were about the size and shape of a man's hand, though the thumb was set farther back than would be the case with a man's hand." However, no ape was seen. As to prior observations, Policeman Fred Koehler, who had been assigned to investigate, reported statements by ten persons.

The animal disappeared about the last of June. Upon July 18th, it was reported again, and by persons who were out of communization with each other. It was near Huntington, L. I. A nurseryman, named Stockman, called up the police, saying that members of his family had seen an animal, resembling a gorilla, running through shrubbery. Then a farmer, named Bruno, three miles away, telephoned that he had seen a strange animal. Policemen went to both places, and found tracks, but lost them in the woods. The animal was not reported again.

And I suppose I shall get a letter from somebody in Long Island, asking me not to publish his name, unless I consider that positively necessary, but assuring me that, of all the theorists, who had tried to explain the Ape of Mineola, only I have insight and penetration

Or an impulse that had come upon him, in June, 1931, to climb trees, and to chatter, and to pick over the heads of his neighbors--and then blankness. He had awakened from a trance, and had found on his carpet tracks of "thumbed footprints." A peculiar, greenish mud. He had gone to Lewis and Valentine's nursery, and there he had seen a patch of this mud, which was not known to exist anywhere else.

And, if I don't take seriously this letter that I shall probably receive from somebody in Long Island, it will be because probably also I shall hear from somebody else, telling me that above all he shrinks from notoriety, but that personal considerations must be swept aside for the sake of science--that, as told in the newspapers, somebody had slung a brick, hitting the retreating ape, and that he had been unable to sit down next morning.

But the germination of a new idea, I'm feeling. I have wondered about occultly stealing a money-bag from a bank. But that is so paltry, compared with abilities, not considered occult, by which respectable operators steal banks. Or psychically dislocating somebody's shoulder, in a petty revenge--whereas, politically, and upon the noblest of idealistic principles, whole nations may be dislocated. But, when it comes to the Miracle of Mineola, I feel the stirrings of Usefulness--

Or the makings of a new religion--founded as solidly as any religion ever has been founded--

All ye who are world-weary--unsatisfied with mere nudism, which isn't reverting far enough--unsatisfied with decadence in creeds and politics of today, which conceivably might be more primitive--conceiving that, after all, the confusion in the sciences isn't blankness, and that the cave-arts are at least scrawling something--all ye who are craving a more drastic degeneration--and a possible answer to your prayer--

"Make me, oh, make me, an ape again!"

What I need, to keep me somewhat happy, and to some degree interested in my work, is opposition. If lofty and academic, so much the better: if sanctified, I'm in great luck. I suspect that it may be regrettable, but, though I am much of a builder, I can't be somewhat happy, as a writer, unless also I'm mauling something. Most likely this is the werewolf in my composition. But the science of physics, which, at one time, was thought forever to have disposed of werewolves, vampires, witches, and other pets of mine, is today such an attempted systematization of the principles of magic, that I am at a loss for eminent professors to be disagreeable to. Upon the principles of quantum mechanics, one can make reasonable almost any miracle, such as entering a closed room without penetrating a wall, or jumping from one place to another without traversing the space between. The only reason why the exponents of ultra-modern mechanics are taken more solemnly than I am is that the reader does not have to pretend that he knows what I am writing about, There are alarmed scientists, who try to confine their ideas of magic to the actions of electronic particles, or waves: but, in the Physical Review, April, 1931, were published letters from Prof. Einstein, Prof. R. C. Tolman, and Dr. Boris Podolsky that indicate that this refinement cannot be maintained. Prof. Einstein applies the Principle of Uncertainty not only to atomic affairs, but to such occurrences as the opening and shutting of a shutter on a camera.

There can be no science, or pretended science, except upon the basis of ideal certainty. Anything else is to some degree guesswork. As a guesser, I'll not admit my inferiority to any scientist, imbecile, or rabbit. The position today of what is said to be the science of physics is so desperate, and so confused, that its exponents are trying to incorporate into one system both former principles and the denial of them. Even in the anaemia and frazzle of religion, today, there is no worse state of desperation, or decomposition. The attempt to take the principle of uncertainty--or the principle of un-principledness--into science is about the same as would be an attempt by theologians to preach the word of God, and also include atheism in their doctrines.

As an Intermediatist, I find the principle of uncertainty unsatisfactorily expressed. My own expressions are upon the principled-unprincipled rule-misrule of our pseudo-existence by certainty-uncertainty--

Or, whereas it seems unquestionable that no man has ever been transformed into a hyena, we can be no more than sure-unsure about this.

About the first of January, 1849, somebody, employed in a Paris cemetery, came upon parts of a human body, strewn on the walks. Up in the leafless trees dangled parts of a body. He came to a new-made grave, from which, during the night, had been dug the corpse of a woman. This corpse had been torn to pieces, which, in a frenzy, had been scattered. For details, see Galignani's Messenger (Paris) March 10, 23, 24, 1849.

Several nights later, in another Paris cemetery, there was a similar occurrence. The cemeteries of Paris were guarded by men and dogs, but the ghoul eluded them, and dug up bodies of women. Upon the night of March 8th, guards outside the cemetery of St. Parnasse saw somebody, or something, climbing a wall of the cemetery. Face of a wolf, or a clothed hyena--they could give no description. They fired at it, but it escaped.

Near a new-made grave, at St. Parnasse, they set a spring-gun. It was loaded with nails and bits of iron, for the sake of scattering. One morning, later in March, it was found that, during the night, this gun had discharged. Part of a soldier's uniform that had been shot away was found.

A gravedigger heard of a soldier, who had been taken to a Paris hospital, where he had told that he had been shot by an unknown assailant. It was said that he had been wounded by a discharge of nails and bits of iron.

The soldier's name was Francis Bertrand. The suspicion against him was considered preposterous. He was a young man of twenty-five, who had advanced himself to the position of Sergeant-Major of Infantry. "He bore a good name, and was accounted a man of gentle disposition, and an excellent soldier."

But his uniform was examined, and the fragment of cloth that had been found in the cemetery fitted into a gap in the sleeve of it.

The crime of the ghoul was unknown, or was unrecognized, in French law. Bertrand was found guilty, and was sentenced to imprisonment for one year, the maximum penalty for the only charge that could be brought against him. Virtually he could explain nothing, except that he had surrendered to an "irresistible impulse." But there is one detail of his account of himself that I especially notice. It is that, after each desecration, there came to him another irresistible impulse." That was to make for shelter--a hut, a trench in a field, anywhere--and there lie in, a trance, then rising from the ghoul into the soldier. I have picked up another item. It is from the San Francisco Daily Evening Bulletin, June 27, 1874-- "Bertrand the Ghoul is still alive: he is cured of his hideous disease, and is cited as a model of gentleness and propriety."

11

DAMN the particle, but there is salvation for the aggregate. A gust of wind is wild and free, but there are handcuffs on the storm.

During the World War, no course of a single bullet could have been predicted absolutely, but any competent mathematician could have written the equations of the conflict as a whole.

This is the attempt by the theologians of science to admit the Uncertainty Principle, and to cancel it. Similarly reason the scientists of theology:

The single records of the Bible may not be altogether accurate, but the good, old book, as a whole, is Immortal Truth.

Says Dr. C. G. Darwin, in New Conceptions of Matter:

"We cannot say exactly what will happen to a single electron, but we can confidently estimate the probabilities. If an experiment is carried out, with a thousand electrons, what was a probability for one, becomes nearly a certainty. Physical theory confidently predicts that the millions of millions of electrons in our bodies will behave even more regularly, and that to find a case of noticeable departure from the average, we should have to wait for a time quite fantastically longer than the estimated age of the universe."

This reasoning is based upon the scientific delusion that there are final bodies, or wholes.

Arthur B. Mitchell, of 472 McAllister Avenue, Utica, N. Y., goes out for the evening. It can't be said exactly what will happen to a single cell of Mr. Mitchell's composition, but every wink of an eye, or scratch of an ear, of this body, as a whole, can be foretold.

But now we have a change of view, as to this body that had been regarded as a whole. Now Mr. Mitchell is regarded as one of many units in this community known as Utica. Now the admission is that Mr. Mitchell's conduct may be slightly irregular, but the contention is that the politics of Utica, as a whole, is never a surprise.

But surprising things, in Utica, are reported. Well, Utica is only one of the many communities that make up the State of New York. But the State of New York--

My own expression is that ours is an intermediate existence, poised, or fluctuating back and forth between two unrealizable extremes that may be called positiveness and negativeness; a hyphenated state of goodness-badness, coldness-heat, equilibrium-inequilibrium, certainty-uncertainty. I conceive of our existence as an organism in which positivizing and negativizing manifestations, or conflicts, are metabolic. Certainty, or regularity, exists to a high degree, in the movements of the planets, but not absolutely, because of small, un-formulable digressions: and negativeness exists to a high degree, in the freaks of a cyclone, though not absolutely, because a still more frenzied state of eccentricity can always be thought of.

My expression is that there are things, beings, and events that conform strikingly to regularized generalizations, but that also there are outrageous, silly, fiendish, bizarre, idiotic, monstrous things, beings, and events that illustrate just as strikingly universal imbecility, crime, or unformulability, or fantasy.

In the London newspapers, last of March, 1908, was told a story, which, when starting off, was called "what the coroner for South Northumberland described as the most case that he had ever investigated." The story was of a woman, at Whitley Bay, near Blyth, England, who, according to her statement, had found her sister, burned to death on an unscorched bed. This was the equivalence of the old stories of "spontaneous combustion of human bodies." It was said that the coroner was at first puzzled by this story; but that he learned that the woman who told it had been intoxicated, and soon compelled her to admit that she had found her sister, suffering from burns, in another part of the house, and had carried her to her bedroom.

But, in my experience with Taboo, I have so many notes upon coroners, who have seen to it that testimony was what it should be; and so many records of fires that, according to all that is supposed to be known of chemical affinity, should not have been, that, between what should and what shouldn't, I am so confused that all that I can say about a story of a woman who burned to death on an unscorched bed is that it is possible-impossible.

Looking over data, I note a case that has no bearing on the story of the burning woman on the unscorched bed, but that is a story of strange fires, or of fires that would be strange, if stories of similar fires were not so common. It is a case that interests me, because it aligns with the stories of Emma Piggott and John Doughty. There was an occurrence, and it was followed by something else that seems related: but, in terms of common knowledge, it cannot be maintained that between the first occurrence and the following occurrences there was relationship. Most of the story was told in the London Times, Aug. 21, 1856: but, whenever it is possible for me to do so, I go to local newspapers for what I call data. I take from various issues of the Bedford Times and the Bedford Mercury.

Upon the 12th of August, 1856, a resident of Bedford, named Moulton, was absent from home. He was upon a business trip to Ireland. At home were Mrs. Moulton and the housemaid, Anne Fennimore. To fumigate the house, the girl burned sulphur, in an earthenware jar, on the floor. The burning sulphur ran out on the floor, and set the house afire. This fire was put out. About an hour later, a mattress was found burning, in another room. But the fire from the sulphur had not extended beyond one room, and this mattress was in another part of the house. Smoke was seen, coming from a chest. Later, smoke was seen coming from a closet, and in it linen was found burning. Other isolated fires broke out. Moulton was sent for, and returned, upon the evening of the 16th. He took off damp clothes, and threw them on the floor. Next morning these clothes were found afire. Then came a succession of about forty fires, in curtains, in closets, and in bureau drawers. Neighbors and policemen came in, and were soon fearful for their safety. Not only objects around them flamed: so flamed their handkerchiefs.

There were so many witnesses, and so much talk in the town, that there was an investigation. Considering that nobody was harmed, it seems queer to read that the investigation was a coroner's inquest: but the coroner was the official who took up the investigation. Witnesses told of such occurrences as picking up a pillow and setting it down--pillow flaming. There was an attempt to explain, in commonplace terms: but nothing that could suggest arson was found, and Moulton had insured neither the house nor the furniture. The outstanding puzzlement was that an ordinary fire seemed to be in some way related to the fires that followed it, but in no way that could be defined. The verdict of the jury was that the fire from the burning sulphur was accidental, but that there was no evidence to show what had caused the succeeding fires.

This story attracted attention in London. After the first account, in the Times, there was considerable correspondence. At the inquest, two physicians had given their opinion that the sulphur fire must have been the cause of the other fires--or that inflammable, sulphurous fumes had probably spread throughout Moulton's house. But the jury had refused to accept this explanation, because of testimony that chairs and sofas that had been carried out into the yard, had flamed. The fires were in a period of five days, and it is probable that in that length of time any permeation by fumes would have been detected. In the discussion in the Times it was pointed out that sulphurous fumes are oxides and are not inflammable.

However, I come to another fire, and maybe I'll explain this one.

It was upon the night of Jan. 21, 1909. Upon this night, a small-town woman exasperated a New York hotel clerk. Perhaps I explain her unusual behavior by thinking that, having come from a small town, she started picturing the dangers of the big city, and let her imaginings become an obsession. The woman was Mrs. Mary Wells Jennings, of Brewster, N. Y. Place--the Greek Hotel, 30 E. 42nd Street. See the Brooklyn Eagle, Jan. 22, 1909. Mrs. Jennings asked the night clerk to change her room, saying that she feared fire. The clerk assigned her to another room. Not long afterward-- wouldn't he let her have another room? So another room. Again she annoyed the clerk. Room changed again. A few hours later, in an unoccupied room, where, during alterations, paints were stored, a fire broke out.

St. Louis Globe-Democrat, Dec. 16, 1889.--"In some mysterious way, a fire started in the mahogany desk in the center of the office of the Secretary of War, at Washington, D. C. Several official papers were destroyed, but it was said that they were of no especial value, and could be replaced. Secretary Proctor cannot understand how the fire originated, as he does not smoke, and keeps no matches about his desk."

It may be that there have been other cases, in which, "in some mysterious way" have been destroyed papers that were of no especial value, and could be replaced. Upon Sept. 16, 1920, London newspapers told of three fires that had broken out simultaneously in different departments of the Government Office, in Tothill Street, Westminster, London. It was not said that papers of no especial value had been destroyed, but it was said that these simultaneous fires had not been explained. London Sunday Express, May 2, 1920--"Upon the night of April 28, fire of mysterious origin broke out at the War Office, Constantinople, where the archives are stored. The iron doors were locked, and it was impossible to gain entrance to the building until afternoon. Many important documents were destroyed."

The body of a girl--and the body of a crow--and a newspaper correspondent's vague feeling of an unknown relationship--A woman who was away from home

Upon the night of April 6, 1919--see the Dartford (Kent) Chronicle, April 7--Mr. J. Temple Thurston was alone in his home, Hawley Manor, near Dartford. His wife was abroad. Particulars of the absence of his wife, or of anything leading to the absence of his wife, are missing. Something had broken up this home. The servants had been dismissed. Thurston was alone.

At 2:40 o'clock, morning of April 7th, the firemen were called to Hawley Manor. Outside Thurston's room, the house was blazing: but in his room there was no fire. Thurston was dead. His body was scorched: but upon his clothes there was no trace of fire.

12

FROM the story of J. Temple Thurston, I pick up that this man, with his clothes on, was so scorched as to bring on death by heart failure, by a fire that did not affect his clothes. This body was fully clothed, when found, about three o'clock in the morning. Thurston had not been sitting up, drinking. There was no suggestion that he had been reading. It was commented upon, at the inquest, as queer, that he should have been up and fully clothed about three o'clock in the morning. The verdict, at the inquest, was of death from heart failure, due to inhaling smoke. The scorches were large red patches on the thighs and lower parts of the legs. It was much as if, bound to a stake, the man had stood in a fire that had not mounted high.

In this burning house, nothing was afire in Thurston's room. Nothing was found--such as charred fragments of nightclothes--to suggest that, about three o'clock, Thurston, awakened by a fire elsewhere in the house, had gone from his room, and had been burned, and had returned to his room, where he had dressed, but had then been overcome.

It may be that he had died hours before the house was afire. It has seemed to me most fitting to regard all accounts in this book, as "stories." There has been a permeation of the fantastic, or whatever we think we mean by "untrueness." Our stories have not been realistic. And there is something about the story of J. Temple Thurston that, to me, gives it the look of a revised story. It is as if, in an imagined scene, an author had killed off a character by burning, and then, thinking it over, as some writers do, had noted inconsistencies, such as a burned body, and no mention of a fire anywhere in the house--so then, as an afterthought, the fire in the house--but, still, such an amateurish negligence in the authorship of this story, that the fire was not explained.

To the firemen, this fire in the house was as unaccountable as, to the coroner, was the burned body in the unscorched clothes. When the firemen broke into Hawley Manor, they found the fire raging outside Thurston's room. It was near no fireplace; near no electric wires that might have crossed. There was no odor of paraffin, nor was there anything else suggestive of arson, or of ordinary arson. There had been no robbery. In Thurston's pockets were money and his watch. The fire, of unknown origin, seemed directed upon Thurston's room, as if to destroy, clothes and all, this burned body in the unscorched clothes. Outside, the door of this room was blazing, when the firemen arrived.

We have had other stories of unaccountable injuries. According to them, men and women have been stabbed, but have not known until later that they were wounded. There was no evidence to indicate that Thurston knew of his scorched condition, tried to escape, or called for help.

There are stories of persons who have been found dead, with bullet wounds, under clothing that showed no sign of the passage of bullets. The police-explanation has been of persons who were killed, while undressed, and were then dressed by the murderers. New York Times, July I, 1872--mysterious murder, at Bridgeport, Conn., of Capt. Colvocoresses--shot through the heart--clothes not perforated. Brooklyn Eagle, July 8, 1891--Carl Gros found dead, near Maspeth, L. I.--no marks in the clothes to correspond with wounds in the body. Man found dead in Paris, Feb. 14, 1912--bullet wound--no sign of bullet passing through clothes.

I have come upon so many stories of showers of stones that have entered closed rooms, leaving no sign of entrance in either ceilings or walls, that I have not much sense of strangeness in the idea that bullets, or a knife, could pierce a body, under uncut clothes. There are stories of bullets that have entered closed rooms, without disturbing the materials of walls or ceilings.

Dispatch, dated March 3, 1929, to the San Francisco Chronicle--clipping sent to me by Miriam Allen de Ford, of San Francisco--"Newton, N. J.--The county prosecutor's office here is baffled by the greatest mystery in its history. For days a rain of buckshot, at intervals, has been falling in the office of the Newton garage, a small room, with one door and one window. There are no marks on the walls or ceiling, and there are no holes in the room, through which the shot could enter."

About two years later, being not very speedy in getting around to this, I wrote to the County Prosecutor, at Newton, and received a reply, signed by Mr. George R. Vaughan--"This occurrence turned out to be a hoax, perpetrated by some local jokesters."

There is a story, in the Charleston (S. C.) News and Courier, Nov. 12, 1886, not of bullets falling in a closed room, but, nevertheless, of unaccountable bullets--two men in a field, near Walterboro, Colleton Co., S. C.--small shot falling around them. They thought that it was a discharge from a sportsman's gun, but the rain of lead continued. They gathered specimens, which they took to the office of the Colleton Press.

Religio-Philosophical Journal, March 6, 1880--copying from the Cincinnati Inquirer--that, at Lebanon, Ohio, people of the town were in a state of excitement: that showers of birdshot were falling from the ceiling of John W. Lingo's hardware store. A committee had been appointed, and according to its report, the phenomenon was veritable: slow-falling volleys of shot, not of the size of any sold in the store, were appearing from no detectable point of origin. There was another circumstance, and it may have had much to do with the phenomenon: about five years before, somebody, at night, had entered this store, and had been shot by Lingo, escaping without being identified.

In the R. P. J., April 24, 1880, a correspondent, J. H. Marshall, wrote, after having read of the Lingo case, of experiences of his, in the summer of 1867. Bullets fell in every room in his house, forcefully, but not with gunshot velocity--large birdshot--broad daylight--short intervals, and then falls that lasted an hour or more. Many bullets appeared, but when Marshall undertook to gather them, he could never find more than half a dozen. About the same time raps were heard.

How bullets could enter closed rooms is no more mysterious than is the howness of Houdini's escape from prison cells, though, according to all that was supposed to be known of physical confinements, that was impossible. In Russia, Houdini made, from a prison van, an escape that involved no expert knowledge, nor dexterity, in matters of locks. He was put into this van, and the door was soldered. He appeared outside, and the police called it an unfair contest, because, so to pass through solid walls, he must have been a spirit. Anyway, this story is told by Will Goldston, President of the Magicians' Club (London).

I have a story of a horse that appeared in what would, to any ordinary horse, be a closed room. It makes one nervous, maybe. One glances around, and would at least not be incredulous, seeing almost any damned thing, sitting in a chair, staring at one. I'd like to have readers, who consider themselves superior to such notions, note whether they can resist just a glance. The story of the horse was told in the London Daily Mail, May 28, 1906. If anyone wants to argue that it is all fantasy and lies, I think, myself, that it is more comfortable so to argue. One morning, in May, 1906, at Furnace Mill, Lambhurst, Kent, England, the miller, J. C. Playfair, went to his stable, and found horses turned around in their stalls, and one of them missing. It is common for one who has lost something, to search in all reasonable places, and then, in desperation, to look into places where not at all reasonably could the missing thing be. Adjoining the stable, was a hay room: the doorway was barely wide enough for a man to enter. Mr. Playfair, unable to find a trace of his missing horse, went to the hay room doorway, probably feeling as irrational as would somebody, who had lost an elephant, peering into a kitchen closet. The horse was in the hay room. A partition had to be knocked down to get him out.

There were other occurrences that could not be. Heavy barrels of lime, with nobody perceptibly near them, were hurled down the stairs. This was in the daytime. Though occasionally I do go slinking about, at night, with our data, mostly ours are sunlight mysteries. The mill was an isolated building, and nobody--at least nobody seeable--could approach it unseen. There were two watchdogs. A large water butt, so heavy that to move it was beyond human strength, was overthrown. Locked and bolted doors opened. I mention that the miller had a young son.

About the middle of March, 1901--that a woman was stabbed to death, in a fiction--or in a scene like an imagined scene that did not belong to what we call "reality." The look of the story of Lavinia Farrar is that it, too, was "revised," and by an amateurish, or negligent, or in some unknown way hampered, "author," who, in an attempt to cover up his crime, bungled--or that this woman had been killed inexplicably, in commonplace terms, and that, later, means were taken, but awkwardly, or almost blindly, and only by way of increasing the mystery, to make the murder seem understandable in terms of common human experience.

Cambridge (England) Daily News, March 16, 1901--that Lavinia Farrar, aged 72, a blind woman, "of independent means," had been found dead on her kitchen floor, face bruised, nose broken. Near the body was a blood-stained knife, and there were drops of blood on the floor. The body was dressed, and, until the post-mortem examination, no wound to account for the death was seen. At the inquest, two doctors testified that the woman had been stabbed to the heart, but that there was no puncture in her garments of which there were four. The woman, undressed, could not have stabbed herself, and then have dressed, because death had come to her almost instantly. A knife could not have been inserted through openings in the garments, because their fastenings were along lines far apart.

A knife was on the floor, and blood was on the floor. But it seemed that this blood had not come from the woman's wound. This wound was almost bloodless. Only one of her garments, the innermost, was blood-stained, and only slightly. There had been no robbery. The jury returned an open verdict.

Upon the evening of March 9, 1929--see the New York Times, March 10, 11, 1929--Isidor Fink, of 4 East 132nd Street, New York City, was ironing something. He was the proprietor of the Fifth Avenue Laundry. A hot iron was on the gas stove. Because of the hold-ups that were of such frequent occurrence at the time, he was afraid; the windows of his room were closed, and the door was bolted. A woman, who heard screams, and sounds as if of blows, but no sound of shots, notified the police. Policeman Albert Kattenborn went to the place, but was unable to get in. He lifted a boy through the transom. The boy unbolted the door. On the floor lay Fink, two bullet wounds in his chest, and one in his left wrist, which was powder-marked. He was dead. There was money in his pockets, and the cash register had not been touched. No weapon was found. The man had died instantly, or almost instantly.

There was a theory that the murderer had crawled through the transom. A hinge on this transom was broken, but there was no statement, as to the look of this break, as indicating recency, or not. The transom was so narrow that Policeman Kattenborn had to lift a boy through it. It would have to be thought that, having sneaked noiselessly through this transom, the murderer then, with much difficulty, left the room the same way, instead of simply unbolting the door. It might be thought that the murderer had climbed up, outside, and had fired through the transom. But Fink's wrist was powder-burned, indicating that he had not been fired at from a distance. More than two years later, Police Commissioner Mulrooney, in a radio-talk, called this murder, in a closed room, an "insoluble mystery."

13

IF a man was scorched, though upon his clothes there was no sign of fire, it could be that the woman of Whitley Bay, who told of having found her sister burned to death on an unscorched bed, reported accurately. If the woman confessed that she had lied, that ends the mystery, or that stimulates interest. The statement that somebody, operated upon by the police, or by a coroner, confessed, has the meaning that has a statement that under pressure an apple produces cider. However, this analogy breaks down. I have never heard of an apple that would, if properly pressed, yield cider, if wanted; or ginger ale, if required; or home brew, all according to what was wanted.

Once upon a time, when mine was an undeveloped suspiciousness, and I'd let dogmatists pull their pedantries over my perceptions, I nevertheless collected occasional notes upon what seemed to me to be unexplained phenomena. I don't do things mildly, and at the same time much enjoy myself in various ways: I act as if trying to make allness out of something. A search for the unexplained became an obsession. I undertook the job of going through all scientific periodicals, at least by way of indexes, published in English and French, from the year 1800, available in the libraries of New York and London. As I went along, with my little suspicions in their infancies, new subjects appeared to me--something queer about some hailstorms--the odd and the unexplained in archaeological discoveries, and in Arctic explorations. By the time I got through with the "grand tour," as I called this search of all available periodicals, to distinguish it from special investigations, I was interested in so many subjects that had cropped up later, or that I had missed earlier, that I made the tour all over again--and then again had the same experience, and had to go touring again--and so on--until now it is my recognition that in every field of phenomena--and in later years I have multiplied my subjects by very much shifting to the newspapers--is somewhere the unexplained, or the irreconcilable, or the mysterious--in unformulable motions of all planets; volcanic eruptions, murders, hailstorms, protective colorations of insects, chemical reactions, disappearances of human beings, stars, comets, juries, diseases, cats, lampposts, newly married couples, cathode rays, hoaxes, impostures, wars, births, deaths.

Everywhere is the tabooed, or the disregarded. The monks of science dwell in smuggeries that are walled away from event-jungles. Or some of them do. Nowadays a good many of them are going native. There are scientific dervishes who whirl amok, brandishing startling statements; but mostly they whirl not far from their origins, and their excitements are exaggerations of old-fashioned complacencies.

Because of several cases that I have noted, the subject of Fires attracted my attention. One reads hundreds of accounts of fires, and many of them are mysterious, but one's ruling thought is that the unexplained would be renderable in terms of accidents, carelessness, or arson, if one knew all the circumstances. But keep this subject in mind, and, as in every other field of phenomena, one comes upon cases that are irreconcilables.

Glasgow News, May 20, 1878--doings in John Shattock's farmhouse, near Bridgewater. Fires had started up unaccountably. A Superintendent of Police investigated and suspected a servant girl, Ann Kidner, aged 12, because he had seen a hayrick flame, while she was passing it. Loud raps were heard. Things in the house, such as dishes and loaves of bread, moved about. The policeman ignored whatever he could not explain, and arrested the girl, accusing her of tossing lighted matches. But a magistrate freed her, saying that the evidence was insufficient.

There is a story of "devilish manifestations," in the Quebec Daily Mercury, Oct. 6, 1880. For two weeks, in the Hudson Hotel, in the town of Hudson, on the Ottawa River, furniture had been given to disorderly conduct: the beds had been especially excitable. A fire had broken out in a stall in the stable. This fire was quenched, but another fire broke out. A priest was sent for, and he sprinkled the stable with holy water. The stable burned down.

There are several recorded cases of such. fires ending with the burning of buildings; but a similarity that runs through the great majority of the stories is of fires localized in special places, and not extending.

They are oftenest in the presence of a girl, aged from 12 to 20; but seldom do they occur at night, when they would be most dangerous. It is a peculiarity. See back to the case of the fires in the house in Bedford. It seems that, if those fires had been ordinary fires, the house would have burned down. The cases are of fires, in unscorched surroundings.

New Zealand Times, Dec. 9, 1886--copying from the San Francisco Bulletin, about October 14-- that Willie Brough, 12 years old, who had caused excitement in the town of Turlock, Madison Co., Cal., by setting things afire, "by his glance," had been expelled from the Turlock school, because of his freaks. His parents had cast him off, believing him to be possessed by a devil, but a farmer had taken him in, and had sent him to school. "On the first day, there were five fires in the school: one in the center of the ceiling, one in the teacher's desk, one in her wardrobe, and two on the wall. The boy discovered all, and cried from fright. The trustees met and expelled him, that night." For another account, see the New York Herald, Oct. 16, 1886. Setting fire to teacher's desk, or to her wardrobe, is understandable, and would have been more understandable to me, when I was 12 years old; but in terms of no known powers of mischievous youngsters, can there be an explanation of setting a ceiling, or walls, afire. It seems to me that no yarn- spinner would have thought of any such particular, or would have made his story look improbable with it, if he had thought of it. I have other accounts in which similar statements occur. This particular of fires on walls is unknown in standardized yarns of uncanny doings. If writers of subsequent accounts probably had never heard of Willie Brough, it is improbable that several of them could invent, or would invent, anything so unlikely. It seems that my reasoning is that, under some circumstances, if something is highly unlikely, it is probable. John Stuart Mill missed that.

Upon the 6th of August, 1887, in a little, two-story frame house, in Victoria Street, Woodstock,

New Brunswick, occupied by Reginald C. Hoyt, his wife, four children of his own, and two nieces, fires broke out. See the New York World, Aug. 8, 1887. Within a few hours, there were about forty fires. They were fires in un-scorched surroundings. They did not extend to their surroundings, because they were immediately put out, or because some unknown condition limited them. "The fires can be traced to no human agency, and even the most skeptical are staggered. Now a curtain, high up and out of reach, would burst into flames, then a bed quilt in another room: a basket of clothes on a shed, a child's dress, hanging on a hook."

New York Herald, Jan. 6, 1895--fires in the home of Adam Colwell, 84 Guernsey Street, Greenpoint, Brooklyn--that, in 20 hours, preceding noon, January 5th, when Colwell's frame house burned down, there had been many fires. Policemen had been sent to investigate. They had seen furniture burst into flames. Policemen and firemen had reported that the fires were of unknown origin. The Fire Marshal said: "It might be thought that the child Rhoda started two of the fires, but she cannot be considered guilty of the others, as she was being questioned, when some of them began. I do not want to be quoted as a believer in the supernatural, but I have no explanation to offer, as to the cause of the fires, or of the throwing around of the furniture." Colwell's story was that, upon the afternoon of January 4th, in the presence of his wife and his step- daughter Rhoda, aged 16, a crash was heard. A large, empty, parlor stove had fallen to the floor. Four pictures, fell from walls. Colwell had been out. Upon his return, while hearing an account of what had occurred, he smelled smoke. A bed was afire. He called a policeman, Roundsman Daly, who put out the fire, and then, because of unaccountable circumstances, remained in the house. It was said that the Roundsman saw wallpaper, near the shoulder of Colwell's son Willie start to burn. Detective Sergeant Dunn arrived. There was another fire, and a heavy lamp fell from a hook. The house burned down, and the Colwells, who were in poor circumstances, lost everything but their clothes. They were taken to the police station.

Captain Rhoades, of the Greenpoint Precinct, said: "The people we arrested had nothing to do with the strange fires. The more I look into it, the deeper the mystery. So far I can attribute it to no other cause than a supernatural agency. Why, the fires broke out under the very noses of the men I sent to investigate."

Sergeant Dunn--"There were things that happened before my eyes that I did not believe were possible."

New York Herald, January 7--"Policemen and firemen artfully tricked by a pretty, young girl."

Mr. J. L. Hope, of Flushing, L. I., had called upon Captain Rhoades, telling him that Rhoda had been a housemaid in his home, where, between November 19 and December 19, four mysterious fires had occurred. "Now the Captain was sure of Rhoda's guilt, and he told her so." "She was frightened, and was advised to tell the truth."

And Rhoda told what she was "advised" to tell. She "sobbed" that she had started the fires, because she did not like the neighborhood in which she lived, and wanted to move away: that she had knocked pictures from the walls, while her mother was in another part of the house, and had dropped burning matches into beds, continuing; her trickeries after policemen, detectives, and firemen had arrived. The Colwells were poor people, and occupied only the top floor of the house that burned down. Colwell, a carpenter, had been out of work two years, and the family was living on the small wages of his son.

Insurance was not mentioned.

The police captain's conclusion was that the fires that had seemed "supernatural" to him, were nat-urally accounted for, because, if when Rhoda was in Flushing, she set things afire, fires in her own home could be so explained. Rather than to start a long investigation into the origin of the fires in Flushing, the police captain gave the girl what was considered sound and wholesome advice. And--though it seems quaint, today--the girl listened to advice. "Pretty, young girls" have tricked more than policemen and firemen. Possibly a dozen male susceptibles could have looked right at this pretty, young girl, and not have seen her strike a match, and flip it into furniture; but no flip of a match could set wallpaper afire.

The case is like the case of Emma Piggott. Only to one person's motives could fires be attributed: but by no known means could she have started some of these fires.

Said Dr. Hastings H. Hart, of the Russell Sage Foundation, as reported in the newspapers, May 10, 1931: "Morons for the most part can be the most useful citizens, and a great deal of the valuable work being done in the United States is being done by such mentally deficient persons."

Dr. Hart was given very good newspaper space for this opinion, which turned out to be popular. One can't offend anybody with any statement that is interpreted as applying to everybody else. Inas-much as my own usefulness has not been very widely recognized, I am a little flattered, myself. To deny, ridicule, or reasonably explain away occurrences that are the data of this book, is what I call useful. A general acceptance that such things are would be unsettling. I am an evil one, quite as was anybody, in the past, who collected data that were contrary to the orthodoxy of his time. Some of the most useful work is being done in the support of Taboo. The break of Taboo in any savage tribe would bring on perhaps fatal disorders. As to the taboos of savages, my impressions are that it is their taboos that are keeping them from being civilized; that, consequently, one fetish is worth a hundred missionaries.

I shall take an account of "mysterious fires" from the St. Louis Globe-Democrat, Dec. 19, 1891. I shall go on to quote from a Canadian newspaper, with the idea of supporting Dr. Hart's observa-tions. Reporters, scientists, policemen, spiritualists--all have investigated phenomena of "poltergeist girls" in ways essentially the same as the way of a Canadian newspaper man--and that has been to pick out whatever agreed with their preconceptions, or with their mental deficiencies, or their social usefulness, and to disregard everything else.

According to the story in the Globe-Democrat, there had been "extraordinary" occurrences in the home of Robert Dawson, a farmer, at Thorah, near Toronto, Canada. In his household were his wife and an adopted daughter, an English girl, Jennie Bramwell, aged 14. Adopted daughters, with housemaids, are attracting my attention, in these cases. The girl had been ill. She had gone into a trance, and had exclaimed: "Look at that!" pointing to a ceiling. The ceiling was afire. Soon the girl startled Mr. and Mrs. Dawson by pointing to another fire. Next day many fires broke out. As soon as one was extinguished, another started up. While Mrs. Dawson and the girl were sitting, facing a wall, the wallpaper blazed.

Jennie Bramwell's dress flamed, and Mrs. Dawson's hands were burned, extinguishing the fire. For a week, fires broke out. A kitten flamed. A circumstance that is unlike a particular in the Bedford case, is that furniture carried outside, and set in the yard, did not burn.

An account, in the Toronto Globe, November 9, was by a reporter, who was a person of usefulness. He told of the charred patches of wallpaper, which looked as if a lighted lamp had been held to the places.

Conditions were miserable. All furniture had been moved to the yard. The girl had been sent back to the orphan asylum, from which she had been adopted, because the fires had been attributed to her. With her departure, phenomena had stopped. The reporter described her as "a half-witted girl, who had walked about, setting things afire." He was doubtful as to what to think of the reported flaming of a kitten, and asked to see it. He wrote that it was nothing but a kitten, with a few hairs on its back slightly singed. But the chief difficulty was to explain the fire on the ceiling, and the fires on the walls. I'll not experiment, but I assume that I could flip matches all day at a wall, and not set wallpaper afire. The reporter asked Mrs. Dawson whether the girl had any knowledge of chemistry. According to him, the answer was that this little girl, aged 14, who had been brought up in an orphan asylum, was "well-versed in the rudiments of the science." Basing upon this outcome of his investigations, and forgetting that he had called the well-versed, little chemist "half- witted," or being more sophisticated than I seem to think, and seeing no inconsistency between scientific knowledge and imbecility, the useful reporter then needed only several data more to solve the mystery. He enquired in the town, and learned that the well-versed and half-witted little chemist was also "an incorrigible little thief." He went to the drug store, and learned that several times the girl had been sent there on errands. The mystery was solved: the girl had stolen "some chemical," which she had applied to various parts of Dawson's house.

Occurrences of more recent date. Story in the London Daily Mail, Dec. 13, 1921, of a boy, in Budapest, in whose presence furniture moved. The boy was about 13 years of age. Since about his 12th birthday, fires had often broken out, in his presence. Alarmed neighbors, or "superstitious" neighbors, as they were described, in the account, had driven him and his mother from their home. It was said that, when he slept, flames flickered over him, and singed his pillow.

In the New York Times, Aug. 25, 1929, was published a story of excitement upon the West Indian island of Antigua. This is a story that reverses the particulars of some of the other stories. It is an account of a girl whose clothes flamed, leaving her body unscorched. This girl, a Negress, named Lily White, living in the village of Liberta, flamed, while walking in the streets. However, at home, too, the clothes of this girl often burst into flames. She became dependent upon her neighbors for something to wear. When she was in bed, sheets burned around her, seemingly harmlessly to her, according to the story.

Early in March, 1922, an expedition, composed of newspaper reporters and photographers, headed by Dr. Walter Franklin Prince, arrived at a deserted house that was surrounded by snow banks out of which stuck the blackened backs, legs, and arms of burned furniture. The newspapers had told of doings in this house, near Antigonish, Nova Scotia, and had emphasized the circumstance that, "in the dead of winter," Alexander MacDonald and his family had been driven from their home, by "mysterious fires," unaccountable sounds, and the meanderings of crockery. The phenomena had centered around Mary Ellen, MacDonald's adopted daughter. With the idea that the house was haunted, the expedition entered, and made itself at home, everybody quick on the draw for note paper or camera. Mostly, in poltergeist cases, I see nothing to suggest that the girls--boys sometimes--are mediums, or are operated upon by spirits; the phenomena seem to be occult powers of youngsters. In MacDonald's house, the investigators came upon nothing that suggested the presence of spirits. Mary Ellen and her father, or father by adoption, were induced to return to the house, but nothing occurred. Usually, in cases of poltergeist girls, phenomena are not of long duration. Dr. Prince interviewed neighbors, and recorded their testimony that dozens of fires had broken out, in this girl's presence: but more striking than any testimony by witnesses was the sight, outside this house, of the blackened furniture, sticking out of snow banks.

New York Sun, Feb. 2, 1932--a dispatch from Bladenboro, North Carolina. "Fires, which apparently spring from nowhere, consuming the household effects of C. H. Williamson, here, have placed this community in a state of excitement, and continue to burn. Saturday a window shade and curtain burned in the Williamson home. Since then fire has burst out in five rooms. Five window shades, bed coverings, tablecloths, and other effects have suddenly burst into flames, under the noses of the watchers.

Williamson's daughter stood in the middle of the floor, with no fire near. Suddenly her dress ignited. That was too much, and household goods were removed from the house."

In the New York Sun, Dec. 1, 1882, is an account of the occult powers of A. W. Underwood, a Negro, aged 24, of Paw, Michigan. The account, copied from the Michigan Medical News, was written by Dr. L. C. Woodman, of Paw. It was Dr. Woodman's statement that he was convinced that Underwood's phenomena were genuine. "He will take anybody's handkerchief, and hold it to his mouth, rub it vigorously, while breathing on it, and immediately it bursts into flames, and burns until consumed. He will strip, and will rinse out his mouth thoroughly, and submit to the most rigorous examination to preclude the possibility of any humbug, and then by his breath, blown upon any paper, or cloth, envelop it in flames. He will, while out gunning, lie down, after collecting dry leaves, and by breathing on them start a fire."

In the New York Sun, July 9, 1927, is an account of a visit by Vice-President Dawes, to Memphis, Tennessee. In this city lived a car-repairer, who was also a magician. "He took General Dawes' handkerchief, and breathed upon it, and it caught fire."

Out of the case of the Negro who breathed dry leaves afire, I conceive of the rudiments of a general expression, which I expect to develop later. The phenomena look to me like a survival of a power that may have been common in the times of primitive men. Breathing dry leaves afire would, once upon a time, be a miracle of the highest value. I speculate how that could have come about. Most likely there never has been human intelligence keen enough to conceive of the uses of fire, in times when uses of fire were not of conventional knowledge. But, if we can think of our existence as a whole--perhaps only one of countless existences in the cosmos--as a developing organism, we can think of a fire-inducing power appearing automatically in some human beings, at a time of its need in the development of human phenomena. So fire-geniuses appeared. By a genius I mean one who can't avoid knowledge of fire, because he can't help setting things afire.

I think of these fire-agents as the most valuable members of a savage community, in primitive times: most likely beginning humbly, regarded as freaks; most likely persecuted at first, but becoming established, and then so overcharging for their services that it was learned how, by rubbing sticks, to do without them--so then their fall from importance, and the dwindling of them into their present, rare occurrence--but the preservation of them, as occasionals, by Nature, as an insurance, because there's no knowing when we'll all go back to savagery again, degrading down to an ignorance of even how to start fires--so then a revival of the fire- agents, and civilization starting up again--only again to be overthrown by wars and grafts, doctors, lawyers, and other racketeers; corrupt judges and cowardly juries--starting down again, perhaps this time not stopping short of worms. Occasionally I contribute to the not very progressive science of biology, and, as I explain atavistic persons in societies, I now make suggestions as to vestigial organs and structures in human bodies--that the vestigial may not be merely a relic, but may be insurance--that the vestigial tail of a human being is no mere functionless retention, but is a provision against times when back to the furry state we may go, and need means for wagging our emotions. Conceive of a powerful backward slide, and one conceives of the appearance, by only an accentuation of the existing, of hosts of werewolves and wereskunks and werehyenas in the streets of New York City.

Mostly our data indicate that occasional human beings have the fire-inducing power. But it looks as if it were not merely that, in the presence of the Negress, Lily White, fires started: it looks as if these fires were attacks upon her. Men and women have been found, burned to death, and explanations at inquests have not been satisfactory. There are records of open, and savage, seizures, by flames, of people.

Annual Register, 1820-13--that Elizabeth Barnes, a girl aged 10, had been taken to court, accused by John Wright, a linen draper, of Foley-place, Mary-le-bon, London, of having repeatedly, and "by some extraordinary means," set fire to the clothing of Wright's mother, by which she had been burned so severely that she was not expected to live. The girl had been a servant in Wright's household. Upon January 5th an unexplained fire had broken out. Upon the 7th, Mrs. Wright and the girl were sitting by the hearth, in the kitchen. Nothing is said, in the account, of relations between these two. Mrs. Wright got up from her chair, and was walking away, when she saw that her clothes were afire. Again, upon January 12th, she was, with the girl, in the kitchen, about eight feet from the hearth, where "a very small fire" was burning. Suddenly her clothes flamed. The next day, Wright heard screams from the kitchen, where his mother was, and where the girl had been. He ran into the room, and found his mother in flames. Only a moment before had the girl left the kitchen, and this time Wright accused her. But it was Mrs. Wright's belief that the girl had nothing to do with her misfortunes, and that "something supernatural" was assailing her. She sent for her daughter, who arrived, to guard her. She continued to believe that the girl could have had nothing to do with the fires, and went to the kitchen, where the girl was, and again "by some unknown means, she caught fire." "She was so dreadfully burned that she was put to bed." When she had gone to sleep, her son and daughter left the room--and were immediately brought back by her screams, finding her surrounded by flames. Then the girl was told to leave the house. She left, and there were no more fires. This seemed conclusive, and the Wrights caused her arrest. At the hearing, the magistrate said that he had no doubt that the girl was guilty, but that he could not pronounce sentence, until Mrs. Wright should so recover as to testify.

In Cosmos, 3-6-242, is a physician's report upon a case. It is a communication by Dr. Bertholle to the Societe Medico-Chirurgicale:

That, upon the 1st of August, 1869, the police of Paris had sent for Dr. Bertholle, in the matter of a woman, who had been found, burned to death. Under the burned body, the floor was burned, but there was nothing to indicate the origin of the fire. Bedclothes, mattresses, curtains, all other things in the room, showed not a trace of fire. But this body was burned, as if it had been the midst of flames of the intensity of a furnace. Dr. Bertholle's report was technical and detailed: left arm totally consumed; right hand burned to cinders; no trace left of internal organs in the thorax, and organs in the abdomen unrecognizable. The woman had made no outcry, and no other sound had been heard by other dwellers in the house. It is localization, or specialization, again--a burned body in an almost unscorched room.

Upon the night of Dec. 23, 1916--see the New York Herald, Dec. 27, 28, 1916--Thomas W. Morphey, proprietor of the Lake Denmark Hotel, seven miles from Dover, N. J., was awakened by moaning sounds. He went down the stairs, and found his housekeeper, Lillian Green, burned and dying. On the floor under her was a small, charred place, but nothing else, except her clothes, showed any trace of fire. At a hospital, the woman was able to speak, but it seems that she could not explain. She died without explaining.

One of my methods, when searching for what I call data, is to note, in headlines, or in catalogues, or indexes, such clue-words, or clue-phrases, as I call them, as "mystery solved," or an assurance that something has been explained. When I read that common sense has triumphed, and that another superstition has been laid low, that is a stimulus to me to be busy--

Or that story of the drunken woman, of Whitley Bay, near Blyth, who had told of finding her sister burned to death on an unscorched bed, and had recanted. Having read that this mystery had been satisfactorily explained, I got a volume of the Blyth News.

The story in the local newspaper is largely in agreement with the story in the London newspapers: nevertheless there are grounds for doubts that make me think it worthwhile to re-tell the story.

The account is of two retired schoolteachers, Margaret and Wilhelmina Dewar, who lived in the town of Whitley Bay, near Blyth. In the evening of March 22, 1908, Margaret Dewar ran into a neighbor's house, telling that she had found her sister, burned to death. Neighbors went to the house with her. On a bed, which showed no trace of fire, lay the charred body of Wilhelmina Dewar. It was Margaret's statement that so she had found the body, and so she testified, at the inquest. And there was no sign of fire in any other part of the house.

So this woman testified. The coroner said that he did not believe her. He called a policeman, who said that, at the time of the finding of the body, the woman was so drunk that she could not have known what she was saying. The policeman was not called upon to state how he distinguished between signs of excitement and terror, and intoxication. But there was no accusation that, while upon the witness stand, this woman was intoxicated, and here she told the same story. The coroner urged her to recant. She said that she could not change her story.

So preposterous a story as that of a woman who had burned to death on an unscorched bed, if heeded, or if permitted to be told, would be letting "black magic," or witchcraft, into English legal proceedings. The coroner tried persistently to make the woman change it. She persisted in refusing. The coroner abruptly adjourned the inquest until April 1st.

Upon April 1st, Margaret Dewar confessed. Any reason for her telling of a lie, in the first place, is not discoverable. But there were strong reasons for her telling what she was wanted to tell. The local newspaper was against her. Probably the coroner terrified her. Most likely all her neighbors were against her, and hers were the fears of anybody, in a small town, surrounded by hostile neighbors. When the inquest was resumed, Margaret Dewar confessed that she had been inaccurate, and that she had found her sister burned, but alive, in a lower part of the house, and had helped her up to her room, where she had died. In this new story, there was no attempt to account for the fire; but the coroner was satisfied. There was not a sign of fire anywhere in the lower part of this house. But the proper testimony had been recorded. Why Margaret Dewar should have told the story that was called a lie was not inquired into. There are thousands of inquests at which testimonies are proper stories.

Madras Mail, May 13, 1907--a woman in the village of Manner, near Dinapore--flames that had consumed her body, but not her clothes--that two constables had found the corpse in a room, in which nothing else showed signs of fire, and had carried the smoldering body, in the unscorched clothes, to the District Magistrate. Toronto Globe, Jan. 28, 1907--dispatch from Pittsburgh, Pa.-- that Albert Houck had found the body of his wife, "burned to a crisp," lying upon a table--no sign of fire upon the table, nor anywhere else in the house. New York Sun, Jan. 24, 1930--coroner's inquiry, at Kingston, N. Y., into the death of Mrs. Stanley Lake. "Although her body was severely burned, her clothing was not even scorched."[

14

THE story of the "mad bats of Trinidad" is that the discoverer of them had solved a mystery of many deaths of human beings and cattle. "Dr. Pawan, a Trinidad scientist, had discovered that the infection had been caused by mad vampire bats, affected by rabies, which they transmitted in a new form of insidious hydrophobia."

But the existence of hydrophobia is so questionable, or of such rare occurrence, even in dogs, that the story of the "mad bats of Trinidad" looks like some more of the sensationalism in science that is so obtrusive today, and compared with which I am, myself, only a little wild now and then. It is probable that the deaths of human beings and cattle, in Trinidad, have not been accounted for. Once upon a time the explanation would have been "witchcraft." Now it's "rabid vampires." The old hag on her broomstick is of inferior theatrical interest, compared with the insane blood-sucker.

The germ-theory of diseases is probably like all other theories, ranging from those of Moses and Newton and Einstein and Brother Voliva down, or maybe up, or perhaps crosswise, to mine, or anybody else's.

Many cases may be correlated under one explanation, but there must be exceptions. No pure, or homogeneous, case of any kind is findable: so every case is variously classifiable. There have been many cases of ailments and deaths of human beings that have not been satisfactorily explained in the medical terms that are just now fashionable, but that will probably be out of style, after a while. Nowadays one is smug with what one takes for progress, thinking of old-time physicians prescribing dried toads for ailments. Here's something for the enjoyment of future smugness. Newspapers of Jan. 14, 1932-- important medical discovery--dried pigs' stomachs, as a cure for anaemia. I now have a theory of what is called evolution, in terms of fashions--that somewhere, perhaps on high, there is a Paris--where, once upon a time, were dictated the modes in bugs and worms, and then the costumes of birds and mammals; grotesquely stretching the necks of giraffes, and then quite as unreasonably reacting with a repentance of hippopotami; passing on to a mental field of alternating extravagances and puritanisms, sometimes neat and tasteful, but often elaborate and rococo, with religions, philosophies, and sciences, imposing upon the fashion-slaves of this earth the latest thing in theories.

In the New York Sun, Jan. 17, 1930, Dr. E. S. Godfrey, of the New York State Department of Health, told, in an interview, of mysterious deaths on a vessel. In a period of four years, twenty-seven officers and men had been stricken by what was called "typhoid fever." Taking his science from the Sunday Newspapers, which had full-paged the story of "Typhoid Mary," a scientific detective, with his microscope, boarded this vessel, and of course soon announced that he had "tracked down" one of the sailors, as a "typhoid carrier." Such sleuthing has become a modernized witch-finding. There are, in New York State, today, persecutions that are in some cases as deadly as the witchcraft-persecutions of the past. "There are 188 women and 90 men recorded as typhoid-carriers, in New York State." Why there should be twice as many women as men is plain enough: the carrier-finders, with "Typhoid Mary" in mind, probably went looking for women. It may be a matter of difficulty, or it may be impossible, in times of general unemployment, for somebody in the grocery or dairy business, to change into some other occupation: but these 278 "typhoid-carriers," tracked down by medical Sherlock Holmeses, who had read of "typhoid-carriers," are prohibited from working in food-trades, and have to report to district health officers once every three months. But this is for the protection of the rest of us. But that is what the witch-finders used to say. Chivalry can't die, so long as there is tyranny: every tyrant has been much given to protecting somebody or something. It is one of the blessings of our era that we are tormented by so many abominations, enormities, and pestiferous, smaller botherations that we can't concentrate upon the germ-scares that the medical "finders" would spread, if it were not for so much competition. They did spread, with some success, with their parrot-scare, in the year 1929. Abandoned parrots, in their cages, were found, frozen to death, in parks and doorways. Probably the psittacosis scare, of 1929, did not become the hysteria of former scares, because lay-alarmists were checked by their inability to pronounce the name of it.

There must be something the matter with the germ-theory of diseases, or the nursing and medical professions would not be so overcrowded. There must be something the matter with the germ-theory of diseases, if there is something the matter with every theory. I looked up the case of "Typhoid Mary." With the preconceptions of everybody who looks up cases, I went looking for something to pick on. It was impossible for me to fail to find what I wanted to consider a case of injustice, if ours is an existence of justice-injustice. I of course found that the case of "Typhoid Mary" as a germ-carrier was not made out so clearly as the "finders" of today suppose.

In the year 1906, it was noted that in several homes, in New York City, where Mary had been employed as a cook, there had been illnesses that were said to be cases of typhoid fever. The matter was investigated, according to what was supposed to be scientific knowledge, in the year 1906. The germ- theory of diseases was the dominant idea. Not a thought was given to relations between this woman and her victims. Had there been quarrels, before illnesses of persons, living in the same house with her, occurred? What was the disposition of the woman? There are millions of men and women, with long hours and little pay, who may, in their states of mind, be more dangerous than germs. There are cooks with grievances, as well as cooks with germs. But Mary's malices were not examined. It was "found" that, though immune herself, she was a distributor of typhoid bacilli. For three years she was "detained" in a hospital, by the public health officials of New York City.

And then what became of Mary's germs? According to one examination, she had them. According to another examination, she hadn't them. At the end of three years, Mary was examined again, and, according to all tests, she hadn't them. She was released, upon promising to report periodically to the Board of Health.

Probably because of lively impressions of "detention," Mary did not keep her promise. Under various aliases, she obtained work as a cook.

About five years later, twenty-five persons, in the Sloane Maternity Hospital, New York City, were stricken with what was said to be typhoid fever. Two of them died. See the Outlook, 109-803. And Mary was doing the cooking at the hospital. The Public Health officials "detained" her again, following their conclusion that they said was obvious. I know of hosts of cases that are obvious one way, and just as apparent some other way; conclusive, according to one theorist, and positively established, according to opposing theorists.

She had them, when, to support a theory, she should have them. She hadn't them, when her own support, as "detained," was becoming expensive. She had--she hadn't--But it does seem that in some way this woman was related to the occurrence of illnesses, sometimes fatal.

Of all germ-distributors, the most notorious was Dr. Arthur W. Waite, who, in the year 1916, was an embarrassment to medical science. In his bacteriological laboratory, he had billions of germs. Waite planned to kill his father-in-law, John E. Peck, 435 Riverside Drive, New York City. He fed the old man germs of diphtheria, but got no results. He induced Peck to use a nasal spray, in which he had planted colonies of the germs of tuberculosis. Not a cough. He fed the old man calomel, to weaken his resistance. He turned loose hordes of germs of typhoid, and then tried influenza. In desperation, he lost all standing in the annals of distinctive crimes, and went common, or used arsenic. The old-fashioned method was a success. One's impression is that, if anything, diets and inhalations of germs may be healthful.

It is not that I am attacking the germ-theory of diseases as absolute nonsense. I do not attack this theory as absolute nonsense, because I conceive of no theory that is more than partly nonsensical. I have some latitude. Let the conventionalists have their theory that germs cause diseases, and let their opponents have their theory that diseases cause germs, or that diseased conditions attract germs. Also there is room for dozens of other theories. Under the heading "Invalidism," I have noted 43 cases of human beings who were ill, sometimes temporarily, and sometimes dying, at the time of uncanny--though rather common--occurrences in their homes. No conventional theory fits these cases. But the stories, as collected by me, are only fragments.

One day, in July, 1890, in the home of Mr. Piddock, in Haferroad, Clapham, London--see the London Echo, July 16, 1890--the daughter of this household was dying. Volleys of stones, of origin that could not be found out, were breaking through the glass of the conservatory. It is probable that not a doctor, in London, in the year 1890--nor in the year 1930--if what is known as a reputable physician--would admit any possibility of relationship between a dying girl and stones that were breaking windows.

But why should any doctor, in London, in the year 1890, or any other year, accept the existence of any relation between a bombardment of a house and a girl's dying condition? He would be as well-justified in explaining that there was only coincidence, as were early paleontologists in so explaining, when they came upon bones of a huge body, and, some distance away, a relatively small skull--explaining that the skull only happened to be near the other bones. They had never heard of dinosaurs. If many times they came upon similar skulls associating with similar other bones, some of them would at least refuse any longer to believe in mere coincidence; but the more academic ones, affronted by a new thought, would continue in their thought-ruts, decrying all reported instances as yarns, fakery, imposture, nonsense.

The dying girl--showers of stones--

New York Sun, Dec. 22, 30, 1883--that, in a closed room in a house in Jordan, N. Y., in which a man was dying, stones were falling.

In the home of Alexander Urquhart, Aberdeen, Scotland, there was an invalid boy. Stories of doings in this house were told in London newspapers, early in January, 1920. The boy was simply set down as "an invalid boy," and presumably doctors were not mystified by his ailment. Nobody was recorded as suspecting anything but coincidence between whatever may have been the matter with him, and phenomena that centered around him, as he lay in his bed. It was as if he were bombarded by unseen bombs. Explosive sounds that shook the house occurred over his bed, and, according to reports by policemen, the bed was violently shaken. Policemen reported that objects, in the boy's room, moved--

London Daily News, Jan. 10, 1920--"Aberdeen ghost laid low--prosaic explanation for strange sounds-- nothing but a piece of wood that the wind had been knocking against a side of the house." That probably convinced the London readers who preferred something like the "mice-behind-the-baseboards" conclusion to such stories. But the Glasgow Herald, of the 13th, continued to tell of "thumping sounds that shook the house and rattled the dishes."

The data are protrusions from burials. The body of a girl--the body of a crow. Somebody. dying-- and hostile demonstrations that cannot conventionally be explained. But if there were connecting circumstances, they are now undiscoverable. It is said that there is a science of comparative anatomy, by which, given any bone of an animal, the whole skeleton can be reconstructed. So stated, this is one of the tall stories of science. The "father" of the science of comparative anatomy never reconstructed anything except conventionally. The paleontologists have reconstructed crowds of skeletons that are exhibited as evidences of evolution: but Cuvier not only never reconstructed anything new, but is now notorious as a savage persecutor of evolutionists. There cannot be reconstruction, unless there be a model. We may have a comparative anatomy of our fragmentary circumstances, if we can fit the pieces to a situation-model. And it may be that we are slowly building that. Of course anything of the nature of old-fashioned, absolute science is no dream of mine.

From the Port of Spain (Trinidad) Mirror, and the Port of Spain Gazette, I take a story of phenomena that began Nov. 12, 1905, in Mrs. Lorelhei's boarding house, in Queen Street, Port of Spain. The house was pelted with stones. A malicious neighbor was suspected, but then, inside the house, there were occurrences that, at least physically, could be attributed to nobody. Objects were thrown about. Chairs fell over, got up, and whirled. Out of a basket of potatoes, flew the potatoes. Stones fell from unseen points of origin, in rooms. A doctor was quoted as saying that he had seen some of these doings. He had been visiting a girl, who, in this house, was ill.

In the Religio-Philosophical Journal, July 15, 1882, as copied from the New York Sun, there is a boardinghouse story. Mrs. William Swift's boardinghouse, 52 Willoughby Street, Brooklyn--the occupant of the back parlor was ill. Raps were heard. Several times appeared a floating, vaporous body, shaped like a football. Upon the ailing boarder, the effect of this object was like an electric shock.

In the Religio-Philosophical Journal, March 31, 1883, and the New York Times, March 12, 1883, there are accounts of the bewitchment of the house, 33 Church Street, New Haven, Conn. Tramping sounds-- objects flying about. A woman in this house was ill. While she was preparing medicine in a cup, the spoon flew away. Sounds like Hey, diddle, diddle! Then it was as if an occult enemy took a shot at her. An unfindable bullet made a hole in a glass.

In the Bristol (England) Mercury, Oct. 12, 1889, and in the Northern Daily Telegraph, Oct. 8, 1889, are accounts of loud sounds of unknown origin in a house in the village of Hornington, near Salisbury. Here a child, Lydia Hewlett, aged nine, "was stricken with a mysterious illness, lying in bed, never speaking, never moving, apparently at death's door." It was said that this child had incurred the enmity of a gypsy, whom she had caught stealing vegetables in a neighbor's garden.

One of the cases of "mysterious family maladies," accompanied by poltergeist disturbances, was reported by the Guernsey Star, March 5, 1903. In the home of a resident of the island of Guernsey, Mr. B. Collinette, several members of the family were taken ill. Things were flying about.

Early in the year 1893--as told in the New York World, Feb. 17, 19, 1896--an elderly man, named Mack, appeared, with his invalid wife, and his daughter Mary, in the town of Bellport, Long Island, N. Y., and made of the ground floor of their house a little candy store. The account in the World is of a starting up of persecutions of this family that were attributed to hostility of other storekeepers, and to dislike "probably because of their thrift." Stones were thrown at the house "by street gamins." Several boys were arrested, but there was no evidence against them. At the time of one of the bombardments, Mary was on the porch of the house. A big dog appeared. He ran against her, knocking her down, injuring her spine so that she was a cripple the rest of her life. All details of this story are in terms of persecutions by neighbors: in the terms of the telling, there is no suggestion of anything occult. Unidentified persons were throwing stones.

The terrified girl took to her bed. Stones thumped on the roof above her, throwing her into spasms of fright. In one of these convulsions, she died. Missing in this story is anything relating to Mack's experiences before arriving in Bellport. His daughter was crippled, and died of fright. He arrived with an invalid wife. In his biography of the Bishop of Zanzibar (Frank, Bishop of Zanzibar)--I take from a review in the London Daily Express, Oct. 27, 1926--Dr. H. Maynard Smith, Canon of Gloucester, tells of poltergeist persecutions, near the mission station, at Weti. Clods of earth, of undetectable origin, were bombarding a house in which lived a man and his wife. Clods fell inside the house. The bishop investigated, and he was struck by a clod. Inside the house, he saw a mass of mud appear on a ceiling. The door was open, but this point on the ceiling was in a position that could not be hit by anyone throwing anything from outside. There was no open window.

The bishop came ceremoniously the next morning, and solemnly exorcised the supposed spirit.

That these stories indicate the existence of spirits is what I do not think. But it seems that the bishop made an impression. The mud-slinging stopped. But then illness came upon the woman of this house.

Upon the night of Aug. 9, 1920, as told in the London Daily Mail, Aug. 19, 1920, a shower of small stones broke the windows in the top floor of Wellington Villa, Grove-road, South Woodford, London, occupied by Mr. H. T. Gaskin, an American, the inventor of the Gaskin Life Boat. There were many showers of stones of undetectable origin. Upon the night of the 13th, policemen took positions in the house, in the street, on roofs, and in trees. The upper floor of the house was bombarded with stones, but where they came from could not be found out. Night of the 14th--a procession. Forty policemen, some of them local, and some of them from Scotland Yard, marched down Grove-road, and went up on roofs, or climbed into trees. Volleys of stones arrived, but the forty policemen learned no more than had the smaller numbers of the preceding investigations. Nevertheless it seems that they made an impression. Phenomena stopped.

The patter of stones--and policemen on roofs, and policemen in trees, and the street packed with sightseers--and this is a spot of excitement--but it has no environment. I can pick up no trace of relations between anybody in this house and anybody outside.

In one of the rooms lay an invalid. Mr. Gaskin was suffering from what was said to be sciatica. In an interview he said that he could not account for the attack upon him, or upon the house: that, so far as he knew, he had no enemy.

In some of these cases, I have tried to dig into blankness. I have shoveled vacancy. I have written to Mr. and Mrs. Gaskin, but have received no answer. I have looked over the index of the London Times, before and after August, 1920, with the idea of coming upon something, such as a record of a law case, or some other breeder of enmity, in which Mr. Gaskin might have been involved, but have come upon nothing.

15

NOW I have a theory that our existence is a hermaphrodite--

Or the unproductivity of it, in the sense that the beings, and seas, and houses, and trees, and the fruits of trees, its "immortal truths," and "rocks of ages" that it seems to produce are only flutters that seem to be real productions to us, because we see them very slow-motioned.

My interpretation of theology is that, though mythologically much confused, it is an awareness of the wholeness of one existence--perhaps one of countless existences in the cosmos--and that its distortions are founded upon intuitive knowledge of the unproductive state of this one existence, as a whole--and so its visions of a divine sterility, which are illustrated with figures of blonde hermaphrodites. Of course there are stray legends of male angels, but such stories are symbols of the inconsistency that co-exists with the consistency of all things phenomenal--

I'd be queried, if I should say, of the consummation of any human romance, that it is parthenogenetic: but humanity, regarded as a whole, is sustained by self-fertilization. Except for occasional, vague stories of external enrichments, there are no records of invigorations imparted to the human kind from gorillas, hyenas, or swine. Elephants fertilize elephants. I conceive of no bizarre, little love story, with a fruitful outcome, of the attractions of a rhinoceros to a humming bird. Though I have a venerable, little story-- account sent to me by Mr. Ernest Doerfler, Bronx, N. Y.--of an eighteenth-century scientist, whose theory it was that human females can be pollinated, and who experimented, by exposing a buxom female to the incidence of the east wind, and of course was successful in establishing his theory, I have no other datum of human and vegetable unions: so this reported occurrence must be considered one of the marvels from which this book of not uncommon events holds aloof.

The parthenogenetic triumphs of the human intellect are circular stupidities. The mathematicians, in their intuitions of the state of a whole, have represented what to the devout is divinity, with the circle, which, to them the "perfect figure," symbolizes getting nowhere.

Much of the argument in this book will depend upon our acceptance that nothing in our existence is real. The Whole may be Realness. Out of its phenomena, it may be non-phenomenally producing offspring-realnesses. That is not our present subject. But up comes the question: If nothing phenomenal is real, is everything phenomenal really unreal? But, if I accept that nothing is real, in phenomenal existence, I cannot accept that anything, therein, is really unreal. So my acceptance, in accordance with our general philosophy of the hyphen, is that all things perceptible to us are real-unreal, varying from the direction of one extreme to the other, according to whatever may be the degree of their appearance of individuality. If anybody has the notion that he is a real being--and by realness I mean individuality, or call it entity, or unrelatedness--let him try to tell why he thinks he exists, in a real sense. Recall the most celebrated of the parthenogenetic attempts to make this demonstration:

I think: therefore I am.

Why do I think?

Because I am.

Why am I?

Because I think.

The noblest triumphs of the human intellect are about as sublime as would be the description of a house in terms of its roof, whereas the description would be equally sublime, if in terms of the cellar, or the bathroom. That is Newtonism--or a description of things in terms of one of their aspects, or gravitation. It is Darwinism--a description of all life in terms of selection, one of its aspects. Gravitation is only another name for attraction. Sir Isaac Newton's contribution to the glories of human knowledge is that an apple falls because it drops. All living things are selected by environment, said Darwin. Then, according to him, when he shifted aspects, all things constituting living environment are selected.

Darwinism--that selection selects.

The materialists explain all things, except what they deny, or disregard, in terms of the material. The immaterialists, such as the absolute and the subjective idealists, explain all things in terms of the immaterial. My expression is in terms of the continuity of the material and the immaterial--or that one of these extremes is only an accentuation on one side, and the other only an accentuation on the other side, of the hyphenated state of the material-immaterial.

I am a being who thinks: therefore I am a being who thinks. In this circular stupidity there is a simple unity that commends it to conventional lovers of the good, the true, and the beautiful.

I do not think. I have never had a thought. Therefore something or another. I do not think, but thoughts occur in what is said to be "my" mind--though, instead of being "in" it, they are it--just as inhabitants do not occur in a city, but are the city. There is a governing tendency among these thoughts, just as there is among people in any community, or as there is in the movements of the planets, or in the arrangements of cells constituting a plant, or an animal. So far as goes any aware-ness of "mine," "I" have no soul, no self, no entity, though at times of something like a harmoniza-tion of "my" elements, "I" approximate to a state of unified being.

When I see--as for convenience "I" shall say, even though there is no I that is other than a very im-perfectly co-ordinated aggregation of experience-states, sometimes ferociously antagonizing one an-other, but mostly maintaining a kind of civilization--but when I see that my thoughts are ruled by tendencies, such as to harmonize, organize, or co-ordinate: that they tend to integrate, segregate, nu-cleate, equilibrate--I am conscious of mere mechanical processes that mean no more in the arrange-ments of my ideas than they mean in the arrangements of my bones. I'd no more think of offering my ideas as immortal truth than I'd think of publishing X-ray photographs of my bones, as eternal. But the organizing tendency implicit in all things--along with the disorganizing tendency implicit in all things--has admirably expressed itself in the design that is my skeleton. I think so. I have no rea-son to think that my skeleton is in any way inferior to anybody else's skeleton. I feel that if I could arrange my ideas with the art that has arranged my bones, I'd have, for writing a book, the justifica-tion that all writers feel the need of, trying to excuse themselves for writing books.

But I do not think that mechanism is all that there is in our existence. Only the old-fashioned abso-lutist conceives, or says he conceives, of our existence as absolutely mechanical. There is an individ-uality in things that is not of mechanical relations, because individuality is unrelatedness. I conceive of our existence as positive-negative, or as mechanical-immechanical.

But my methods are the largely mechanical methods of everybody, and of everything, that harmo-nizes, or organizes. One of these methods is classification. I am impelled to arrange my materials un-der headings--quite as a wind arranges fallen leaves, of various sizes, into groups--as a magnet makes selections from a pile of various things. So, again, when I see that my thoughts are coerced by con-ventional processes, I can think of my thoughts as nothing but the products of coercions. I'd not do these slaves the honor of believing them. They impose upon me only to the degree of temporary acceptance of some of them.

Merely thoughtfully, or only intellectually, I have made a collection of notes, under the classification of "Explosions." Some of the occurrences look as if explosive attacks, of an occult order, have been made upon human beings; or as if psychic bombs have been thrown invisibly at people, or at their property.

In the New York Tribune, Jan. 7, 1900, there is an account of poltergeist disturbances in a house, in Hyde Park, Chicago. According to the now well-known ways of chairs and tables, at times, these things hopped about, or moved with more dignity. It was as if into this house stole an invisible but futile assassin. See back to accounts of visible but futile bullets. Time after time there was a sound like the discharge of a revolver. It was noted that this firing always occurred "at about the height of a man's shoulder." In a booklet, A Disturbed House and its Relief, Ada M. Sharpe tells of a seeming psychic bombardment of her home in Tackley, Oxen, England. Beginning upon April 24, 1905, and continuing three years, at times, detonations, as if of exploding bombs, were heard in this house. Upon the 1st of May, 1911 (Lloyd's Weekly News, July 30; Wandsworth Borough News, July 21) unaccountable fires broke out in the house of Mr. J. A. Harvey, 356 York-road, Wandsworth, London. Preceding one of these fires, there were three explosions of unknown origin. In January, 1892 (Peterborough Advertiser, Jan. 10, 1892), a house in Peterborough, England, occupied by a family named Rimes, was repeatedly shaken, as if bombed, and as if bombed futilely. Nobody was injured, and there was no damage.

In the Religio-Philosophical Journal, Dec. 25, 1880--copied from the Owatonna (Minn.) Review-- there is a story maybe of a psychic bomb that was tossed through the wall of a house, in Owatonna, penetrating the wall, without leaving a sign of its passage through the material. It was in a house occupied by a family named Dimant. There had been petty persecutions by an uncatchable: such as persistent ringing of the doorbell. One evening, members of this family were in one of the rooms, when something exploded. Mrs. Dimant was knocked insensible. Fragments of a cylindrical glass object were found. But no window had been open, and there had been no other way by which, by known means, this object could have entered this house.

I note something of agreement between notions that are now developing--notions that will be called various names, one of which is not "practical"--and experiments by inventors that are attempts to be very practical. It is said that by means of "rays" inventors have been able to set off distant explosives. If by other means, or by subtler "rays," explosions at a distance can be made to occur, whatever the practical ones are trying to do may be far more effectively accomplished--if the data of this chapter do mean that there have 'peen explosions that were the products of means, or powers, that are at present mysterious.

There are stories of brilliantly luminous things that are called "globe lightning" that have appeared in houses, and have moved about, before exploding, as if guided by intelligence of their own, or as if directed by a distant control. These stories are easily findable in books treating of lightning and the freaks of lightning. I pick out an account from a periodical. There seems to be no relation with lightning. In the English Mechanic, 90-140, Col. G. T. Plunket tells of an experience, in July, 1909, in his home, in Wimbledon, London. He and his wife were sitting in one of their rooms, when his wife saw a luminous thing moving toward them. It went to a chair, upon the back of which it seemed to rest, for a moment. It exploded. Col. Plunket did not see this thing, but he heard the explosion. As to the lightning-explanation, he writes that it was a fine evening.

London Daily Mail, July 23, 1925--"Explosion riddle--mystery of a boy's wounds." "Injured by a mysterious explosion, which occurred in his mother's house, at Riverhall-street, South Lambeth, S.W., yesterday morning, Charley Orchard, 5, was conveyed to hospital in a serious condition. He was hurt on the face and chest, and some of his fingers were blown away.

"His mother had just called him to breakfast when the explosion occurred.

"Neighbors who heard the report of the explosion thought there was an outbreak of fire and summoned the fire brigade.

"An all-day search failed to discover the cause of the explosion."

The London newspapers, Sept. 26, 1910, told of a tremendous, unexplained explosion in a house in Willesden, London. I take from the local newspaper, the Willesden Chronicle, September 30--"a fire of a most mysterious character ... absolutely no cause can be assigned for the outbreak, which was followed by a terrific explosion, completely wrecking the premises." But in no account is it made clear that first there was a fire, and that the explosion followed. A policeman, standing on a nearby corner, saw this house, 71 Walm-lane, Willesden, flame and burst apart. "Windows and doors in the back of the house were blown a distance of 60 feet." "On examination of the premises, it was found that the two gas meters under the stairs had been shut off: so it was evident that the explosion was not caused by gas. Representatives of the Salvage Corps and of the Home Office investigated, but could conclude nothing except that chemicals, or petrol, might have exploded."

The occupants of this house, named Reece, were out of town, week-ending. Mr. Reece was communicated with, and it was his statement that there had been nothing in the house that could have exploded.

Willesden Chronicle, October 7--"Mystery cleared up. A charred sofa in the drawing room and other evidence reveal the cause of the outbreak." Before leaving the house, Saturday morning (September 24th), Mr. Reece, while smoking a pipe, had leaned over this sofa, and sparks from his pipe had fallen upon it. For 36 hours a fire, so caused, had smoldered, before bursting into flames. There were two standard spirit lamps in the room. In the fire, they must have exploded simultaneously.

The writer of this explanation picked the remains of a sofa out of a wreck of charred furniture. He leaned Reece over the sofa, because that would make his explanation work out as it should work out. Reece made no such statement, and he was not quoted. The explosion of two spirit lamps could do much damage, but this explosion was tremendous. The house was wrecked. The walls that remained standing were in such a toppling condition that the ruins were roped off.

The jagged walls of this wrecked house are more of our protrusions from vacancy. We visualize them in an environment of blankness. Somewhere there may have been a witch or a wizard.

Upon June 13, 1885, a resident of Pondicherry, Madras, India, was sitting in a closed room, when a mist appeared near him. At the same time there was a violent explosion. This man, M. Andre, sent an account to the French Academy. I take from a report, in L'Astronomie, 1886-310. M. Andre tried to explain in conventional terms, mentioning that at the time the weather was semi-stormy, and that an hour later rain fell heavily.

In times still farther back, the mist would have been told of, as the partly materialized form of an enemy, who had expressed his malices explosively. In times, still somewhere in the future, this may seem the most likely explanation.

Or the mist was something like the partly visible smoking fuse of an invisible bomb that had been discharged by a distant witch or wizard. And that does not seem to me to be much more of a marvel than would be somebody's ability to blow up a quantity of dynamite, though at a distance, and with no connecting wires.

In the New York Herald Tribune, Nov. 29, 1931, there is an account of the doings of Kurt Schimkus, of Berlin, who had arrived, in Chicago, to demonstrate his ability to discharge, from a distance, explosives, by means of what he called his "anti-war rays." According to reports from Germany, Schimkus had so exploded submarine mines and stores of buried cartridges. Herr Schimkus will have success and renown, I think: he knows that nothing great and noble and of benefit to mankind has ever been accomplished without much lubrication. He announced that slaughter was far-removed from his visions: that he was an agency for peace on earth and good will to man, because by exploding an enemy's munitions, with his "anti-war rays," he would make war impossible. Innocently, myself, I speculate upon the possible use of "psychic bombs," in blowing up tree stumps, in the cause of new pastures.

In the New York Herald Tribune, March 25, 1931, there is a story of an explosion that may have been set off by "rays" that at present are not understood. It is the story of the explosion that wrecked the sealing ship, Viking, off Horse Island, north of New Brunswick. It reminds me of the woman, who, in the New York hotel, feared fire. This ship was upon a moving picture expedition. Varrick Frissell, film producer, aboard this vessel, started to think of the kegs of powder aboard, and he became apprehensive. He started to make a warning sign to hang on the door of the powder room. Just then the ship blew up.

New York Herald Tribune, Dec. 13, 1931--an account of disasters to two wives of a man--not a datum of his relations, or former relations, with anybody else. In the year 1924, illness was upon the wife of W. A. Baker, an oil man, who lived in Pasadena, California. It was said that her affliction was cancer. She was found, hanged, in her home. It was said that despondency had driven her to suicide. In the year 1926, Baker married again. Upon the night of Dec. 12, 1931, there was an explosion, somewhere under the bed of the second Mrs. Baker, or in the room underneath. The bed was hurled to the ceiling, and Mrs. Baker was killed. It was a tremendous explosion, but nobody else in the house was harmed.

Bomb experts investigated. They concluded that no known explosive had been used. They said that there had been no escape of gas. "The full force of the explosion seemed concentrated almost beneath Mrs. Baker's room."

In the years 1921-22, and early in the year 1923, there were, in England and other countries, explosions of coal such as had never occurred before. There was a violent explosion in a grate in a house in Guildford, near London, which killed a woman, and knocked down walls of the house (London Daily News, Sept. 16, 1921). There were other explosions of coal, during this year, but in 1922 attention was attracted by many instances.

In this period there was much disaffection among British coal miners. There was a suspicion that miners were mixing dynamite into coal. But, whether we think that the miners had anything to do with these explosions, or not, suspicions against them, in England, were checked by the circumstances that no case of the finding of dynamite in coal was reported, and that there were no explosions of coal in the rough processes of shipments.

There came reports from France. Then stoves, in which was burned British coal, were blowing up in France, Belgium, and Switzerland. The climax came about the first of January, 1923, when in one day there were several of these explosions in Paris, and explosions in three towns in England.

About the first of January, 1921, Mr. T. S. Frost, of 8 Ferristone road, Hornsey, London, bought a load of coal. In his home were three children, Gordon, Bertie, and Muriel. I take data from the London newspapers, but especially from the local newspapers, the Hornsey Journal, and the North Middlesex Chronicle. In the grates of this house, coal exploded. Also, coal in buckets exploded. A policeman was called in. He made his report upon coal that not only exploded, but hopped out of grates, and sauntered along floors, so remarkable that an Inspector of Police investigated. According to a newspaper, it was this Inspector's statement that he had picked up a piece of coal, which had broken into three parts, and had then vanished from his hands. It was said that burning coals leaped from grates, and fell in showers in other rooms, having passed through walls, without leaving signs of this passage. Flatirons, coal buckets, other objects "danced." Ornaments were dislodged, but fell to the floor without breaking. A pot on a tripod swung, though nobody was near it. The phenomena occurred in the presence of one of the boys, especially, and sometimes in the presence of the other boy.

There has been no poltergeist case better investigated. I know of no denial of the phenomena by any investigator. One of the witnesses was the Rev. A. L. Gardiner, vicar of St. Gabriel's, Wood Green, London. "There can be no doubt of the phenomena. I have seen them, myself." Another witness was Dr. Herbert Lemerle, of Hornsey. Dr. Lemerle told of a clock that mysteriously vanished. Upon the 8th of May, a public meeting was held in Hornsey, to discuss the phenomena.

In the newspapers there was a tendency to explain it all as mischief by the children of this household. The child, Muriel, terrified by the doings, died upon April 1st. The boy, Gordon, frightened into a nervous breakdown, was taken to the Lewisham Hospital.

The coal in all these cases was coal from British coal mines. The newspapers that told of these explosions told of the bitterness and vengefulness of British coal miners, enraged by hardships and reduced wages, uncommon in even their harsh experiences--

Or see back--

There's a shout of vengefulness, in Hyde Park, London--far away, in Gloucestershire, an ancient mansion bursts into flames.

16

BUT why this everlasting attempt to solve something?--whereas it is our acceptance that, in a final sense, there is, in phenomenal affairs, nothing--or that there is only the state of something-nothing--so that all problems are only soluble-insoluble--or that most of the social problems we have, today, were at one time conceived of as solutions of preceding problems--or that every Moses leads his people out of Egypt into perhaps a damn sight worse--Promised Lands of watered milk and much-adulterated honey-- so why these everlasting attempts to solve something?

But to take surgical operations upon warders of Sing Prison, and the loss of rectitude by lace curtains, and the vanishing man of Berlin; "Typhoid Mary," and a Chinese hair-clipper, and explosions of coal, and bodies on benches in a Harlem Park--

Robert Browning's conception was to take three sounds, and make, not a fourth, but a star.

Out of seven colors, not to lay on daubs, but to paint a picture. Out of seven million Americans, Russians, Germans, Irishmen, Italians, and on, so long as geography holds out, not to pile a population, but to organize--more or less--into New York City.

Sulphur and lava in a barren plain, and a salty block of stone, shaped roughly like a woman--signs of erosion on rocks far above water-level--a meteor that had set a bush afire--the differences of languages of peoples--and all the other elements that organized into Genesis.

Data of variations and heredity and adaptations; of multiplications and of checks and of the doctrine of Malthus; of acquired characters and of transmissions--and they organized into The Origin of Species--

Just as, once upon a time, minerals that had affinity for one another came together and took on geometrical appearances. But a crystal is not supposed to be either a prohibition or an anti-prohibition argument. I know of a crystal of quartz that weighs several hundred pounds. But it has not been mistaken for propaganda--

Or all theories--theological, scientific, philosophical--and that they represent the same organizing process--but that self-conscious theorists, instead of recognizing that thought-forms were appearing in their minds, as in wider existence have appeared crystalline constructions, have believed that it was immortal Truth that they were conceiving.

Oxygen and sulphur and carbon--

Or Emma Piggott and Ambrose Small and Rose Smith--

Or let's have just a little, minor expression, or organization, a small composition, arranging the data of poltergeist girls. The elements of this synthesis are moving objects, fires, girls in strange surroundings, youth and the atavism of youth.

Case of Jennie Bramwell--she was an adopted daughter. The Antigonish girl was an adopted daughter. See the Dagg case--adopted daughter. "Adoption" is a good deal of a disguise for getting little girls to work for not much more than nothing. It is not so much that so many poltergeist girls have been housemaids and "adopted daughters," as that so many of them have been not in their own homes; lost and helpless youngsters; under hard taskmasters, in strange surroundings--

Or the first uncertain and precarious appearances of human beings upon this earth--and a need for them, and a fostering, a nurturing, a protection, far different from conditions in these swarming times, when the need is for eliminations--

A lost child in primordial woods--and the value of her, which no genius, king, or leveler of kings, has today--

That objects moved in her presence--fruits of trees that came down from the trees and set themselves beside her--the shaking of bushes that cast, to her, berries--then night and coldness--faggots, joining twigs, and dancing around her--heaping--the crackling of flames to warm her--

Or that, to this day, grotesque capers of chairs, the antics of sofas, and the seeming wantonness of flames are survivals of co-operations that once upon a time moved even the trees, when a child was lost in a forest.

The old mathematicians had this aesthetic appraisal of their thoughts: they wrought theorems and calculi "for elegance," and were scornful of uses. But virtually everything that they produced "for elegance" was put to work by astronomers, navigators, surveyors. I assemble, compositionally, what I call data: but I am much depressed, perhaps, fearing that they have meaning outside themselves, or may be useful.

There is, upon this earth, today, at least one artist. Prof. Albert Einstein put together, into what he called one organic whole, such a diversity of elements as electro-magnetic waves and irregularities in the motions of the planet Mercury; the fall of a stone from a train to an embankment, the geometry of hyper-space, and accelerated co-ordinate systems, and Lorentz transformations, and the displacements of stars during eclipses--

And the exploitation of everything by something, or, more or less remotely, by everything else--the need of astronomers for Einsteinism, because it was so encouragingly unintelligible, whereas school-boys were beginning to pick Newtonism to pieces--and in the year 1918 it was announced that the useful Einstein had predicted displacements of stars, according to his theory, and that his predictions had been confirmed.

For purposes of renewed confirmation--or maybe in innocence of trying to confirm anything, or at least not consciously intending to observe whatever was wanted--an expedition was sent by Lick Observatory to report upon the displacement of stars during the solar eclipse of October, 1922. The astronomers of this expedition agreed that the displacements of stars confirmed Einstein, the Prophet. Einstein was said to be useful, and, in California, school children, dressed in white, sang unto him kindred unintelligibilities. In New York, mounted policemen roughly held back crowds from him, just as he, to make his system of thoughts, had clubbed many astronomical data into insensibility. He had taken into his system of thoughts irregularities of the planet Mercury, but had left out irregularities of the planet Venus. Crowds took him into their holiday-making, but omitted asking what it was all about.

Upon June 12, 1931, Prof. Erwin Freundlicher reported to the Physics Association of Berlin that, according to his observations, during the eclipse of May 9, 1929, stars were not displaced, as, according to Einstein, they should be--or that, outside itself, Einsteinism is meaningless.

There was no excitement over this tragedy, or comedy, because this earth's intellectuals, mostly, take notice only when they're told to take notice; and to orthodoxy it seemed wisest that this earth's thinkers should not think about this. Prof. Freundlicher's explained the astronomers of the Lick expedition, quite as I explain all astronomers. He gave his opinion that they had confirmed Einstein because "they had left out of consideration observations that did not fit in with the results that they wanted to obtain." If there be much more of such agreements with me, I shall have to hunt me some new heresies. For an account of Prof. Freundlicher's report, see the New York Herald Tribune, June 14, 1931.

Outside itself Einsteinism has no meaning.

As a worthless thing--As an unrelated thing its state is that of which artists have dreamed, in their quest for absoluteness--the dream of "art for art's sake."

Up to Dec. 6, 1931, I thought of Prof. Einstein's theories as almost alone, or as representing almost sublime worthlessness. But New York Times, Dec. 6, 1931--scientists of the University of California,

experimenting upon an admixture of phosphorus in the food of swine, were developing luminous pigs. "Just what they will be good for has not yet been announced."

Mine is a dream of being not worth a displaced star to anybody. I protest that with the elements of this book my only motive is compositional--but comes the suspicion that I protest too much.

There has been a gathering of suggestions--that there are subtler "rays" than anything that is known in radioactivity, and that they may be developed into usefulness. The Ascot Cup and the Dublin jewels-- and, if they were switched away by a means of transportation now not commonly known, a common knowledge may be developed to enormous advantage in commercial and recreational and explorative transportations.

In the period of my writing of this book, Californian scientists were trying to make pigs shine at night. Another scientist, who could not yet announce much usefulness, was feeding skimmed milk to huckleberries. For all I know one of us may revolutionize something or another.

17

LONDON Daily Chronicle, March 30, 1922--"It is incredible, but nothing has been heard of Holding."

For three weeks a search had been going on--cyclists, police, farmers, people from villages.

At half past ten o'clock, morning of the 7th of March, 1922, Flying Officer B. Holding had set out from an aerodrome, near Chester, England, upon what was intended by him to be a short flight in Wales. About eleven o'clock, he was seen, near Llangollen, Wales, turning back, heading back toward Chester

Holding disappeared far from the sea, and he disappeared over a densely populated land. One of my jobs was that of looking over six London newspapers for the years 1919-1926, and it is improbable that anything was learned of what became of Holding, later, without my knowing of it. I haven't a datum upon which to speculate, in the Holding mystery: but now I have a story of two men, whose track on land stopped as abruptly as stopped Holding's track in the sky: and this time I note an additional circumstance. The story of these men is laid in a surrounding of hates of the intensity of oriental fanaticism.

Upon July 24, 1924, at a time of Arab hostility, Flight-Lieutenant W. T. Day and Pilot Officer D. R. Stewart were sent from British headquarters, upon an ordinary reconnaissance over a desert in Mesopotamia.

According to schedule, they would not be absent more than several hours. I take this account from the London Sunday Express, Sept. 21, 28, 1924.

The men did not return, and they were searched for. The plane was soon found, in the desert. Why it should have landed was a problem. "There was some petrol left in the tank. There was nothing wrong with the craft. It was, in fact, flown back to the aerodrome." But the men were missing. "So far as can be ascertained, they encountered no meteorological conditions that might have forced them to land." There were no marks to indicate that the plane had been shot at. There may be some way, at present very exclusively known, of picking an aeroplane out of the sky. According to the rest of this story, there may be some such way of picking men out of a desert.

In the sand, around the plane, were seen the footprints of Day and Stewart. "They were traced, side by side, for some forty yards from the machine. Then, as suddenly as if they had come to the brink of a cliff, the marks ended."

The landing of the plane was unaccountable. But, accepting that as a minor mystery, the suggested explanation of the abrupt ending of the footprints was that Day and Stewart had been captured by hostile Bedouins, who had brushed away all trails in the sand, starting at the point forty yards from the plane. But hostile Bedouins could not be thought of as keeping on brushing indefinitely, and a search was made for a renewal of traces.

Aeroplanes, armored cars, and mounted police searched. Rewards were offered. Tribal patrols searched unceasingly for four days. Nowhere beyond the point where the tracks in the sand ended abruptly, were other tracks found. The latest account of which I have record is from the London Sunday News, March 15, 1925--mystery of the missing British airmen still unsolved.

London Evening News, Sept. 28, 1923--"Second-Lieut. Morand, while at shooting practice, at Gadaux, France--himself firing at a target on the ground, while a sergeant piloted the machine--suddenly fell back, calling to the pilot to land, as he had been wounded. It was found that he had a serious wound in his shoulder, and he was taken to Bordeaux, by the hospital aeroplane." It was said that he had been shot. "But no clue has been found, as to the origin of the shot."

I especially notice this case, because it was at a time of other "accidents" to French fliers. The other "accidents" were different, in that they did not occur in France, and in that they were not shootings. I know of no case that in all particulars I can match with the disappearance of Day and Stewart: but there are records of airmen who, flying over a land where the sight of them directed hate upon them, were unaccountably picked out of the sky.

In this summer of 1923, French aviators told of inexplicable mishaps and forced landings, while flying over German territory. The instances were so frequent that there arose a belief that, with "secret rays," the Germans were practicing upon French aeroplanes. From a general impression of an existence of rationality-irrationality, we can conceive that the Germans were practicing upon French aeroplanes something that they were most particularly endeavoring to keep secret from France--if they had any such powers. But I think that they had not--or that officially they had not. There may have been a hidden experimenter, unknown to the German authorities.

An article upon this subject was published in the London Daily Mail, Sept. 1, 1923. "Two theories have been put forward. One is that by a concentration of wireless rays the magneto of the aeroplane may be affected; and another is that a new ray, which will melt certain metals, has been discovered. In this connection it is notable that most of the forced landings of the French aeroplanes, when flying from Strasbourg to Prague, have taken place in the vicinity of a German aerodrome, near Furth." It was said that for some time, at the German wireless station at Nauen, there had been experiments upon directional wireless, with the object of sending out rays, concentrated along a certain path, as the beams of a searchlight are directed. The authorities at Nauen denied that they had knowledge of anything that could have affected the French aeroplanes, in ways reported, or supposed. Automobiles can be stopped, by wireless control, if they be provided with special magnetos: otherwise not. Sir Oliver Lodge was quoted, by the Daily Mail, as saying that he knew of no rays that could stop a motor, unless specially equipped. Professor A. M. Low's opinion was that someday distant motors may be stopped--"I feel confident that, in 50 or 60 years' time, such a thing will be possible." Prof. Low said that he knew of laboratory experiments in which, over a distance of two feet, rays of sufficient power to melt a small coil of wire had been transmitted. But, as to the reported "accidents" in Germany, Prof. Low said: "There is a wide difference between transmitting such a power over a distance of a foot or two, and a distance of one or two thousand yards."

In the Daily Mail, April 5, 1924, was an account of invisible rays, which had been discovered by Mr. H. Grindell-Mathews, powerful enough, under laboratory conditions, to stop the engine of a motorcycle at a distance of fifty feet.

Of course high among virtues are the honorable lies of Governments. Whether virtuously said, or accurately reported, I don't know: but it is said, or reported, that, in the year 1929, the British

Government spent $500,000 investigating alleged long-distance "death-rays," and developed nothing that was effective. It is said, or reported, that the Italian navy gave opportunity to an inventor to demonstrate what he could do with "death-rays," but that his demonstrations came to nothing. We have no data for thinking that, in the year 1929, any Government was in possession of a secret of long- distance "death-rays." The forced landings of French aeroplanes, in the summer of 1923, remain unexplained.

There may be powerful rays that are not electromagnetic. French aviators may have been brought to earth by no power that is called "physical"--though I know of no real demarcation between what is called physical and what is called mental. See back to the series of "mysterious attacks," in England, in April and May, 1927. Three times, as if acted upon by an unknown influence, automobiles behaved unaccountably.

Our data are upon "accidents" that have not been satisfactorily explained. There have been occurrences that were similar to effects that inventors are, by mechanical means, striving for, in the cause of military efficiencies. And these experimenters are practical persons. It may be that we are on the track of a subtler slaughter. It looks as if a lonely possessor of a secret, such as is called "occult," operated wantonly, or in the malicious exercise of a power, upon automobiles, in England, in the months of April and May, 1927. He was a criminal. But I am a practical thinker, and a useful citizen, on the track of much efficiency, which will be at the disposal of God's second choice of people--which I think we must be, judging by the afflictions that are upon us, at this time of writing--a power that would, by this great nation, be used only righteously, if anybody could ever distinguish between righteousness and exploitation and tyranny. One of the engaging paradoxes of our existence--which strip mathematics of meaning--is that a million times a crime is patriotism. I am unable to conceive that a power to pick planes out of the sky would be so terrible as to stop war, because up comes the notion that counter-operations would pick the pickers. If we could have new abominations, so unmistakably abominable as to hush the lubricators, who plan murder to stop slaughter--but that is only dreamery, here in our existence of the hyphen, which is the symbol of hypocrisy.

New York Times, Oct. 25, 1930--that about forty automobiles had been stalled, for an hour, on the road, in Saxony, between Risa and Wurzen.

About forty chauffeurs were probably not voiceless, in this matter; and, if the German Government were experimenting with "secret rays," that was some more of its public secrecy. In the Times, October 27, was quoted the mathematician and former Premier of France, Paul Painleve--"No experiment thus far conducted would permit us to credit such a report, nor give any prospect of seeing it accomplished in the near future."

Upon May 26, 1925--see the London Daily Mail, May 28, 1925--at Andover, Hampshire, England, a corporal of the R.A.F., making a parachute practice jump, was killed by a fall of 1,900 feet from an aeroplane. There is not a datum for thinking that there was anything to this occurrence that aligns it with other occurrences told of in this chapter. But there is an association. About the time of the accident, or whatever it was that befell this man, and at the same place, Flight Sergeant Frank Lowry, and Flying Officer John Kenneth Smith, pilot, were in an aeroplane, making wireless tests. They had been in the air about fifteen minutes, when Smith, having called to his companion, without hearing from him, looked around, and saw smoke coming from the back cockpit, and saw Lowry in a state of collapse.

Lowry was dead. "Flight-Lieut. Cyril Norman Ellen said that there was nothing in the machine likely to kill a man, and that Lowry must have come in contact with an electric current in the air. No similar case has been reported."

In the Daily Mail, Oct. 14, 1921, a writer (T. Gifford) tells of a scene of "accidents," at a point on a road in Dartmoor. This story is like the account of the series of "accidents" to automobiles, in England, in April and May, 1927, except that the "accidents" were strictly localized.

The story told by Gifford is that one day in June, 1921, a doctor, riding on his motor-cycle, with his two children in a side-car, suddenly, at this point, on the Dartmoor road, called to the children to jump. The machine swerved, and the doctor was killed. Several weeks later, at this place, a motor coach suddenly swerved, and several passengers were thrown out. Upon Aug. 26, 1924 a Captain M.--for whom I apologize--it is not often that a Mr. X. or a Captain M. appears in these records-- was, at this point on the road, thrown from his motor-cycle. Interviewed by Gifford, he told, after evasions, that something described by him as "invisible hands" had seized upon his hands, forcing the machine into the turf.

More details were published in the Daily Mail, October 17, of this year. The scene of the "accidents" was on the road, near the Dartmoor village of Post Bridge. In the first instance, the victim was Dr. E. H. Helby, Medical Officer of Princetown Prison.

In Light, Aug. 26, 1922, a correspondent noted another "accident" at this point. Details of the fourth "accident" were told, in the London Sunday Express, Sept. 12, 1926. The victim was traveling on his motor-cycle. "He was suddenly and violently unseated from his mount, and knew no more until he regained consciousness in a cottage, to which he had been carried, after a collapse." The injured man could not explain.

18

I RECORD that, once upon a time, down from the sky came a shower of virgins.

Of course they weren't really virgins. I can't accept the reality of anything, in such an indeterminate existence as ours.

See the English Mechanic, 87-436--a shower of large hailstones, at Remiremont, France, May 26, 1907. Definitely upon some of these objects were printed representations of the Virgin of the Hermits.

It used to be the fashion, simply and brusquely to deny such a story, and call it a device of priestcraft: but the tendency of disbelievers, today, is not to be so free and monotonous with accusations, and to think that very likely unusual hailstones did fall, at Remiremont, and that out of irregularities or discolorations upon them, pious inhabitants imagined pictorial representations. I think, myself, that the imprints upon these hailstones were of imaginative origin, but in the sense that illustrations in a book are; and were not simply imagined by the inhabitants of Remiremont, any more than are some of the illustrations of some books only smudges that are so imaginatively interpreted by readers that they are taken as pictures.

The story of the hailstones of Remiremont is unique in my records. And a statement of mine has been that our data are of the not extremely uncommon. But, early in this book, I pointed out that any two discordant colors may be harmonized by means of other colors; and there are no data, thinkable by me, that cannot be more or less suavely co-ordinated, if smoothly doctored; or that cannot be aligned with the ordinary, if that be desirable.

I am a Jesuit. I shift aspects from hailstones with pictures on them, to pictures on hailstones--and go on with stories of pictures on other unlikely materials.

According to accounts--copied from newspapers--in the Spiritual Magazine, n.s., 7-360, and in the Religio-Philosophical Journal, March 29, 1873, there was more excitement in Baden-Baden, upon March 12, 1872, than at Remiremont. Upon the morning of this day, people saw pictures that in some unaccountable way had been printed upon windowpanes of houses, with no knowledge by occupants as to how they got there. At first the representations were crosses, but then other figures appeared. The authorities of Baden ordered the windows to be washed, but the pictures were indelible. Acids were used, without effect. Two days later, crosses and death's heads appeared upon window glass, at Rastadt.

The epidemic broke out at Boulley, five leagues from Metz. Here, because of feeling, still intense from the Franco-Prussian War, the authorities were alarmed. Crosses and other religious emblems appeared upon windowpanes--pictures of many kinds--death's heads, eagles, rainbows. A detail of Prussian soldiers was sent to one house to smash a window, upon which was pictured a band of French Zouaves and their flags. It was said that at night the pictures were invisible. But the soldiers did not miss their chance: they smashed a lot of windows, anyway. Next morning it looked as if there had been a battle. In the midst of havoc, the Zouaves were still flying their colors.

This story, I should say, then became a standardized newspaper yarn. I have a collection of stories of pictures appearing upon window glass that were almost busily told in American newspapers, after March, 1872, not petering out until about the year 1890.

But it cannot be said that all stories told in the United States, of this phenomenon, or alleged phenomenon, were echoes of the reported European occurrences, because stories, though in no such profusion as subsequently, had been told in the United States before March, 1872. New York Herald, Aug. 20, 1870--a representation of a woman's face, appearing upon window glass, in a house in Lawrence, Mass. The occupant of the house was so pestered by crowds of sightseers that, not succeeding in washing off the picture, he removed the window sash. Human Nature, June, 1871--copied from the Chicago Times--house in Milan, Ohio, occupied by two tenants, named Horner and Ashley. On windowpanes appeared blotches, as if of water mixed with tar, or crude oil--likenesses of human faces taking form in these places. New York Times, Jan. 18, 1871--that, in Sandusky and Cincinnati, Ohio, pictures of women had appeared upon windowpanes.

Still, it might be thought that there was one origin for all the stories, and that that was the spirit-photograph controversy, which, in the early eighteen-seventies, was a subject of intense beliefs and disbeliefs, in both Europe and America. A point that has not been taken up, in this controversy, which continues to this day, even after the fateful spread of knowledge of double exposure, is whether the human imagination can affect a photographic plate. I incline to the idea that almost all spirit- photographs have been frauds, but that a few may not have been--that no spirits were present, but that, occasionally, or very rarely, a quite spookless medium has, in a profound belief in spirits, engendered, out of visualizations, something wraith-like that has been recorded by a camera. Against the explanation that stories of pictures on windowpanes probably had origin in the spirit-photograph craze, I mention that similar stories were told centuries before photography was invented. For an account of representations of crosses that appeared, not upon window glass, but upon people's clothes, as told by Joseph Grunpech, in his book, Speculum Naturalis Coelestis, published in the year 1508, see Notes and Queries, April 2, 1892.

"After the death of Dean Vaughan, of Llandaff, there suddenly appeared on a wall of the Llandaff Cathedral, a large blotch of dampness, or minute fungi, formed into a life-like outline of the dean's face" (Notes and Queries, Feb. 8, 1902).

Throughout this book, my views, or preconceptions, or bigotries, are against spiritual interpretations, or assertions of the existence of spirits, as independent very long from human bodies. However, I think of the temporary detachability of mentalities from bodies, and that is much like an acceptance of the existence of spirits. My notion is that Dean Vaughan departed, going where any iceberg goes when it melts, or where any flame goes when it is extinguished: that intense visualizations of him, by a member of his congregation, may have pictorially marked the wall of the church.

According to reports, in the London Daily Express, July 17 and 30, and in the Sunday Express, Aug. 12, 1923, it may be thought, by anybody so inclined to think, that, in England, in the summer of 1923, an artistic magician was traveling, and exercising his talents. Somebody, or something, was perhaps impressing pictures upon walls and pillars of churches. The first report was that, on the wall of Christ Church, Oxford, had appeared a portrait of the famous Oxford cleric, Dean Liddell, long dead.

Other reports came from Bath, Bristol, and Uphill, Somerset. At Bath--in the old abbey of Bath--the picture was of a soldier, carrying a pack. The Abbey authorities scraped off this picture, but the portrait, at Oxford, was not touched.

There is a description, in T. P.'s and Cassell's Weekly (London), Sept. 11, 1926, of the portrait on the wall of Christ Church, Oxford, as seen three years later. It is described as "a faithful and unmistakable likeness of the late Dean Liddell, who died in the year 1898." "One does not need to call in play any imaginative faculty to reconstruct the head. It is set perfectly straight upon the wall, as it might have been drawn by the hand of a master artist. Yet it is not etched; neither is it sketched, not sculptured, but it is there plain for all eyes to see."

And it is beginning to look as if, having started somewhat eccentrically with a story of virgins, we are making our way out of the marvelous. Now accept that there is a very ordinary witchcraft, by which, under the name of telepathy, pictures can be transferred from one mind to another, and there is reduction of the preposterousness of stories of representations on hailstones, window glass, and Other materials. We are conceiving that human beings may have learned an extension of the telepathic process, so as to transfer pictures to various materials. So far as go my own experiences, I do not know that telepathy exists. I think so, according to many notes that I have taken upon vagrant impressions that come and go, when my mind is upon something else. I have often experimented. When I incline to think that there is telepathy, the experiments are convincing that there is. When I think over the same experiments, and incline against they, they indicate that there isn't.

New York Sun, Jan. 16, 1929--hundreds of persons standing, or kneeling, at night, before the door of St. Ann's Roman Catholic Church, in Keansburg, N. J. They saw, or thought they saw, on the dark, oak door, the figure of a woman, in trailing, white robes, emitting a glow. The pastor of the church, the Rev. Thomas A. Kearney, was interviewed. "I don't believe that it is a miracle, or that it has anything to do with the supernatural. As I see it, it is unquestionably in the outline of a human figure, white-robed, and emitting light. It is rather like a very thin motion picture negative that was under-exposed, and in which human outlines and detail are extremely thin. Yet it seems to be there." Or pictures on hailstones--and wounds that appeared on the bodies of people. In the name of the everlasting If, which mocks the severity of every theorem in every textbook, and is not so very remote from every datum of mine, we can think that, by imaginative means, at present not understood, wounds appeared upon people in Japan, and Germany, and in a turning, off Coventry Street, London, if we can accept that in some such way, pictures ever have appeared upon hailstones, windowpanes, and other places. And we can think that pictures have appeared upon hailstones, windowpanes, and other places, if we can think that wounds have appeared upon people in Japan and other places. Ave the earthworm!

It is my method not to try to solve problems--so far as the solubility-insolubility of problems permits--in whatever narrow specializations of thought I find them stated: but, if, for instance, I come upon a mystery that the spiritualists have taken over, to have an eye for data that may have bearing, from chemical, zoological, meteorological, sociological, or entomological sources--being unable to fail, of course, because the analogue of anything electrical, or planetary, is findable in biological, ethical, or political phenomena. We shall travel far, even to unborn infants, to make hailstones reasonable.

I have so many heresies--along with my almost incredible credulities--or pseudo-credulities, seeing that I have freed my mind of beliefs--that, mostly, I cannot trace my infidelities, or enlightenments, back to their sources. But I do remember when first I doubted the denial by conventional science of the existence of prenatal markings. I read Dr. Weismann's book upon this subject, and his arguments against the possibility of pre-natal markings convinced me that they are quite possible. And this conversion cost me something. Before reading Dr. Weismann, I had felt superior to peasants, or the "man in the street," as philosophers call him, whose belief is that pregnant women, if frightened, mark their offsprings with representations of rats, spiders, or whatever; or, if having a longing for strawberries, fruitfully illustrate their progeny, and were at one time of much service to melodrama. I don't know about the rats and the strawberries, but Dr. Weismann told of such cases as that of a woman with a remarkable and distinctive disfigurement of an ear, and of her similarly marked offspring. His argument was that thousands of women are disfigured in various ways, and that thousands of offsprings are disfigured, and that it is not strange that in one case the disfigurement of an offspring should correspond to the disfigurement of a parent. But so he argued about other remarkable cases, and left me in a state of mind that has often repeated: and that is with the idea that much mental development is in rising down to the peasants again.

If there can be pre-natal markings of bodies, and, as I interpret Dr. Weismann's denials, there can be, and, if they be of mental origin, my mind is open to the idea that other--and still more profoundly damned stories of strange markings--may be similarly explained. If a conventional physician is scornful, hearing of a human infant, pre-natally marked, I'd like to hear his opinion of a story that I take from the London Daily Express, May 14, 1921. Kitten, born at Nice, France--white belly distinctly marked with the gray figures, 1921--the mother cat had probably been looking intently at something, such as a , so dated. "Or reading a newspaper?" said scornful doctor would ask, pointing out that, if I think there are talking dogs, it is only a small "extension," as I'd call it, to think of educated cats keeping themselves informed upon current events.

London Sunday News, Aug. 3, 1926--"Dorothy Parrot, 4-year-old child of R. S. Parrot, of Winget Mill, Georgia, was marked by a red spot on her body. Out of this spot formed three letters, R. I. C. Doctors cannot explain."

London Daily Express, Nov. 17, 1913--phenomena of a girl, aged 12, of the village of Bussus-Bus-Suel, near Abbeville, France. If asked questions, answers appeared in letterings on her arms, legs, and shoulders. Also upon her body appeared pictures, such as of a ladder, a dog, a horse.

In September, 1926, a Rumanian girl, Eleonore Zegun, was taken to London, for observation by the National Laboratory for Psychical Research. Countess Wassilko-Serecki, who had taken the girl to London, said, in an interview (London Evening Standard, Oct. 1, 1926), that she had seen the word Dracu form upon the girl's arm. This word is the Rumanian word for the Devil.

Or the Handwriting on the Wall--and why don't I come out frankly in favor of all, or anyway a goodly number of, the yarns, or the data, of the Bible? The Defender of Some of the Faith is clearly becoming my title.

In recent years I have noted much that has impressed upon my mind the thought that religionists have taken over many phenomena, as exclusively their own--have colored and discredited with their emotional explanations--but that someday some of these occurrences will be rescued from theological interpretations and exploitations, and will be the subject-matter of--

New enlightenments and new dogmas, new progresses, delusions, freedoms, and tyrannies.

I incline to the acceptance of many stories of miracles, but think that these miracles would have occurred, if this earth had been inhabited by atheists. To me, the Bible is folklore, and therefore is not pure fantasy, but comprises much that will be rehabilitated. But also to me the Bible is non-existent. This is in the sense that, except in my earlier writings, I have drawn a dead-line, for data, at the year 1800. I may, upon rare occasions, dip farther back, but my notes start in the year 1800. I shall probably raise this limit to 1850, or maybe 1900. I take for a principle that our concern is not in marvels. It is in repetitions, or sometimes in almost the commonplace. There is no desirability in going back to antiquity for data, because, unless phenomena be appearing now, they are of only historical interest. At present, there is too much history.

Handwritings on walls--I have several accounts: but, if anybody should be interested enough to look up this phenomenon for himself, he will find the most nearly acceptable record in the case of Esther Cox, of Amherst, Nova Scotia. This case was of wide notoriety, and, of it, it could be said that it was well-investigated, if it can be supposed that there ever has been a case of anything that has been more than glanced at, or more than painstakingly and profoundly studied, simply to confirm somebody's theory.

If I should tell of a woman, who, by mental picturings, not only marked the body of her unborn infant, but transformed herself into the appearance of a tiger, or a lamppost, or became a weretiger, or a werelamppost--or of a magician, who, beginning with depicting forest scenes on window glass, had learned to transform himself into a weredeer, or a weretree--I'd tell of a kind of sorcery that used to be of somewhat common occurrence.

I have a specimen. It is a Ceylon leaf insect. It is a wereleaf. The leaf insect's likeness to a leaf is too strikingly detailed to permit any explanation of accidental resemblance.

There are butterflies, which, with wings closed, look so much like dried leaves that at a distance of a few feet they are indistinguishable from dried leaves. There are tree hoppers with the appearance of thorns; stick insects, cinder beetles, spiders that look like buds of flowers. In all instances these are highly realistic portraitures, such as the writer, who described the portrait of Dean Liddell, on the church wall, would call the handiwork of a master artist.

There have been so many instances of this miracle that I now have a theory that, of themselves, men never did evolve from lower animals: but that, in early and plastic times, a human being from somewhere else appeared upon this earth, and that many kinds of animals took him for a model, and rudely and grotesquely imitated his appearance, so that, today, though the gorillas of the Congo, and of Chicago, are only caricatures, some of the rest of us are somewhat passable imitations of human beings.

The conventional explanation of the leaf insect, for instance, is that once upon a time a species of insects somewhat resembled leaves of trees, and that the individuals that most closely approximated to this appearance had the best chance to survive, and that in succeeding generations, still higher approximations were still better protected from their deceived enemies.

An intelligence from somewhere else, not well-acquainted with human beings--or whatever we are--but knowing of the picture galleries of this earth, might, in Darwinian terms, just as logically explain the origin of those pictures--that canvases that were daubed on, without purpose, appeared; and that the daubs that more clearly represented something recognizable were protected, and that still higher approximations had a still better chance, and that so appeared, finally, highly realistic pictures, though the painters had been purposeless, and with no consciousness of what they were doing Which contrasts with anybody's experience with painters, who are not only conscious of what they're doing, but are likely to make everybody else conscious of what they're so conscious of.

It is not merely that hands of artists have painted pictures upon canvas: it is that, upon canvas, artists have realized their imaginings. But, without hands of artists, strikingly realistic pictures and exquisite modelings have appeared. It may be that for crosses on windowpanes, emblems on hailstones, faces on church walls, pre-natal markings, the stigmata, telepathic transferences of pictures, and leaf insects we shall conceive of one expression.

To the clergyman who told the story of the hailstones of Remiremont, the most important circumstance was that, a few days before the occurrence, the Town Council had forbidden a religious procession, and that, at the time of the fall of the hailstones, there was much religious excitement in Remiremont.

English Mechanic, 87-436--story told by Abbe Gueniot, of Remiremont:

That, upon the afternoon of the 26th of May, 1907, the Abbe was in his library, aware of a hailstorm, but paying no attention to it, when a woman of his household called to him to see the extraordinary hailstones that were falling. She told him that images of "Our Lady of the Treasures" were printed on them.

"In order to satisfy her, I glanced carelessly at the hailstones, which she held in her hand. But, since I did not want to see anything, and moreover could not do so, without my spectacles, I turned to go back to my book. She urged: 'I beg of you to put on your glasses.' I did so, and saw very distinctly on the front of the hailstones, which were slightly convex in the center, although the edges were somewhat worn, the bust of a woman, with a robe that was turned up at the bottom, like a priest's cope. I should, perhaps, describe it more exactly by saying that it was like the Virgin of the Hermits. The outline of the images was slightly hollow, as if they had been formed with a punch, but were very boldly drawn. Mlle. Andre asked me to notice certain details of the costume, but I refused to look at it any longer. I was ashamed of my credulity, feeling sure that the Blessed Virgin would hardly concern herself with instantaneous photographs on hailstones. I said: 'But do you not see that these hailstones have fallen on vegetables, and received these impressions? Take them away: they are no good to me.' I returned to my book, without giving further thought to what had happened. But my mind was disturbed by the singular formation of these hailstones. I picked up three in order to weigh them, without looking closely. They weighed between six and seven ounces. One of them was perfectly round, like balls with which children play, and had a seam all around it, as though it had been cast in a mold."

Then the Abbe's conclusions:

"Savants, though you may try your hardest to explain these facts by natural causes, you will not succeed." He thinks that the artillery of heaven had been directed the impious Town Council. However people with cabbages suffered more than people with impieties.

"What appeared most worthy of notice was that the hailstones, which should have been precipitated to the ground, in accordance with the laws of acceleration of falling bodies, appeared to have fallen from a height of but a few yards." But other, or unmarked, hailstones, in this storm, did considerable damage. The Abbe says that many persons had seen the images. He collected the signatures of fifty persons who asserted that they had been witnesses.

I notice several details. One is the matter of a hailstone with a seam around it, as if it had been cast in a mold. This looks as if some hoaxer, or pietist--who was all prepared, having prophetic knowledge that an extraordinary shower of big hailstones was coming--had cast printed lumps of ice in a mold. But accounts of big hailstones, ridged or seamed, are common. Another detail is something that I should say the Abbe Gueniot had never before heard of. The detail of slow-falling objects is common in stories of occult occurrences, but, though for more than ten years I have had an eye for such reports, in reading of hundreds, or thousands, of hailstorms, I know of only half a dozen records of slow- falling hailstones.

In the English Mechanic, 87-507, there is more upon this subject. It is said that, according to the newspapers of Remiremont, these "prints" were inside the hailstones, and were found on surfaces of hailstones that had been split: that 107 persons had given testimony to the Bishop of Sainte-Die; and that several scientists, one of whom was M. de Lapparent, the Secretary of the French Academy, had been consulted. The opinion of M. de Lapparent was that lightning might have struck a medal of the Virgin, and might have reproduced the image upon the hailstones.

I have never come upon any other supposition that there can be manifold reproductions of images, or prints, by lightning. The stories of lightning-pictures are mostly unsatisfactory, because most of them are of alleged pictures of leaves of trees, and, when investigated, turn out to be simply forked veinings, not very leaf-like. There is no other record, findable by me, of hailstones said to be pictorially marked by lightning, or by anything else. It would be much of coincidence, if, at a time of religious excitement in Remiremont, lightning should make its only known, or reported, pictures on hailstones, and make those pictures religious emblems. But that the religious excitement did have much to do with the religious pictures on hailstones, is thinkable by me.

19

THE astronomers are issuing pronouncements upon what can't be seen with telescopes. The physicists are announcing discoveries that can't be seen with microscopes. I wonder whether anybody can see any meaning in an accusation that my stories are about invisibles.
I am a sensationalist.
And it is supposed that modern science, which is supposed to be my chief opposition, is remote from me and my methods.
In December, 1931, Dr. Humason, of Mount Wilson Observatory, announced his discovery of two nebulae that are speeding away from this earth, at a rate of 15,000 miles a second. There was a race. Prof. Hubble started it in the year 1930, with announced discoveries of nebulae rushing away at--oh, a mere two or three thousand miles a second. In March, 1931, somebody held the record with an 8,000- mile nebula. At this time of writing, Dr. Humason is ahead.
When a tabloid newspaper reporter announces speedy doings by more or less nebulous citizens, as "ascertained" by him, by methods that did not necessarily indicate anything of the kind, his performance is called sensationalism.
It is my statement that Dr. Hubble and Dr. Humason are making their announcements, as inferences from a method that does not necessarily indicate anything of the kind.
In the New York Herald Tribune, Jan. 6, 1932, Dr. Charles B. Davenport, of the department of genetics, in Carnegie Institution, received only four inches of space for one of those scares that used to be spread- headed--unknown disease that may wipe out all humanity. "Sometime in the future our boasted skyscrapers may become inhabited by bats, and the safe deposit vaults of our cities become the caves of wild animals." The unknown disease is antiquated sensationalism. I look back at my own notion of the appearance of werethings in the streets of New York--
I now have a little story that pleases me, not so much because I think that I at least hold my own with my professorial rivals, but because, with it, I exercise some of those detective abilities that all of us, even professional detectives, possibly, are so sure we have. I reconstruct, according to my abilities, an incident that occurred somewhere near the city of Wolverhampton, England, about the first of December, 1890. The part of the story of which I have no record--that is the hypothetical part--is that, at this time, somewhere near Wolverhampton, lived a tormented young man. He was a good young man. Not really, of course, if nothing's real. But he approximated. Though for months he had not gone traveling, he was obsessed with a vividly detailed scene of himself, behaving in an unseemly manner to a female, in a railway compartment. There was another mystery. Somebody had asked him to account for his absence, somewhere, about the first of December, whereas he was convinced that he had not been absent--and yet--but he could make nothing of these two mysteries.

Upon the Thursday before the 6th of December, 1890--see the Birmingham Daily Post, December 6--a woman was traveling alone, in a compartment of a train from Wolverhampton to Snow Hill. According to my reconstruction, she began to think of stories of reprehensible conduct by predatory males to females traveling alone in railway compartments.

The part of the story that I take from the Birmingham Post is that when a train went past Soho Station, a woman fell from it. She gave her name as Matilda Crawford, and said that a young man had insulted her.

An odd detail is that it was not her statement that she had leaped from the train, but that the insulting young man had pushed her through a window.

In the next compartment had sat a detective. At an inquiry, he testified that--at least so far as went his observations upon visible entrances and exits--there had been nobody but this woman in this compartment.

In the New York Herald Tribune, Jan. 23, 1932, was published an explanation, by Dr. Frederick B. Robinson, president of New York City College, of some of us sensationalists:

"'Professors have not scored so well in making good appearances from the publicity standpoint,' Dr. Robinson said. 'Living sheltered lives,' he added, 'they yearn for public notice and sometimes get it at the expense of their college. Surely a great New England institution was not elevated in public esteem when one of its professors of English engaged in a series of publicity-stunts, the first of which was to give solemn advice to young men to be snobs.'"

At a meeting of the American Chemical Society, at Buffalo, N. Y., Sept. 3, 1931, Dr. William Engleback told of cases in which, by the use of glandular extracts, the height of dwarfed children had been increased an inch or two. For the announcement of this mild little miracle, he received several inches of newspaper space. New York Times, Dec. 16, 1931--meeting of the Institute of Advanced Education, at the Roerich Museum, New York--something more like a miracle. I measured. Dr. Louis Berman got eleven inches of newspaper space. Dr. Berman's announcement was that the sorcerers of his cult--the endocrinologists--would breed human beings sixteen feet high.

Meeting of the American Association for the Advancement of Science, in New Orleans, December, 1931-

-report upon the work of Dr. Richard P. Strong, of the Harvard Medical School, in the matter of the filaria worms that infest human bodies--and an attempt to make it more interesting. That an ancient mystery had been solved--Biblical story of the fiery serpents at last explained. There's no more resemblance between these tiny worms and the big fiery things that--we are told--grabbed people, than between any caterpillar and a red-hot elephant. But that the filaria worms had been "identified" as the fiery monsters of antiquity was considered a good story, and was given much space in the newspapers. However, see an editorial, not altogether admiring, in the New York Herald Tribune, Jan. 5, 1932.

Still, I do, after a fashion, hold my own. New York Sun, Oct. 9, 1931--that, shortly after the Civil War,

Captain Neil Curry sailed from Liverpool to San Francisco. The vessel caught fire, about 1,500 miles off the west coast of Mexico. The Captain, his wife, and two children, and thirty-two members of the crew took to three small boats, and headed for the mainland. Then details of suffering for water.

"Talk of miracles!" In the midst of the ocean, they found themselves in a volume of fresh water.

I note the statement that Capt. Curry discovered fresh water around the boats, not by a disturbance of any kind, but because of the green color of it, contrasting with the blue of the salt water.

I wrote to Capt. Curry, who at the time of my writing was living in Emporia, Kansas, and received an answer from him, dated Oct. 21, 1931, saying that the story in the Sun was accurate except as to the time; that the occurrence had been in the year 1881.

Here is something, both very different and strikingly similar, which I take from Dr. Richardson's Journal, as quoted by Sir John Franklin, in his Narrative of a Journey to the Polar Sea, p. 157--a story of a young Chipewyan Indian. His wife had died, and he was trying to save his new-born child. "To still its cries, he applied it to his breast, praying earnestly to the great Master of Life, to assist him. The force of the powerful passion by which he was actuated produced the same effect in his case, as it has done in some others, which are recorded: a flow of milk actually took place from his breast."

Intensest of need for water--and it may be that, to persons so suffering, water has been responsively transported. But there have been cases of extremest need for water to die by. One can think of situations in which more frenziedly have there been prayers for water, for death, than ever for water to live by.

New York Sun, Feb. 4, 1892--that, after the burial of Frances Burke, of Dunkirk, N. Y., her relatives, suspecting that she had been in a trance, had her body exhumed. The girl was found dead in a coffin that was full of water. It was the coroner's opinion that she had been buried alive, and had been drowned in her coffin. No opinion as to the origin of the water was published.

20

THE importance of the invisible--

That I'd starve to death, in the midst of eatables, were it not for the invisible means of locomotion by which I go and get them, and the untouchable and unseeable processes by which I digest them--

That every stout and determined materialist, arguing his rejection of the unseeable and the untouchable, lives in a phantom existence, from which he would fade away were it not for his support by invisibles--

The heat of his body--and heat has never been seen.

His own unseeable thoughts, by which he argues against the existence of the invisible.

Nobody has ever seen steam. Electricity is invisible. The science of physics is occultism. Experts in the uses of steam and electricity are sorcerers. Mostly we do not think of their practices as witchcraft, but we have an opinion upon what would have been thought of them, in earlier stages of the Dark Age we're living in.

Or by the "occult," or by what is called the "supernatural," I mean something like an experience that I once saw occur to some acquaintances of mine.

A neighbor had pigeons, and the pigeons loafed on my window sill. They were tempted to come in, but for weeks, stretched necks, fearing to enter. I wished they would come in. I went four blocks to get them sunflower seeds. Though I will go thousands of miles for data, it is most unusual for me to go four blocks--it's eight blocks, counting both ways--for anybody. One time I found three of them, who had flown through an open window, and were upon the frame of a closed window. I went to them slowly, so as not to alarm them. It seems that I am of a romantic disposition, and, if I take a liking to anybody, who seems female, like almost all birds, I want her to perch on my finger. So I put out a finger. But all three birds tried to fly through the glass. They could not learn, by rebuffs, but kept on trying to escape through the glass. If, back in the coop, these pigeons could have told their story, it would have been that they were perched somewhere, when suddenly the air hardened. Everything in front was as clearly visible as before, but the air had suddenly turned impenetrable. Most likely the other pigeons would have said: "Oh, go tell that to the sparrows!"

There is a moral in this, and it applies to a great deal in this book, which is upon the realization of wishes. I had wished for pigeons. I got them. After the investigation by the three pioneers all of them came in. There were nine of them. It was the unusually warm summer of 1931, and the windows had to be kept open. Pigeons on the backs of chairs. They came up on the table, and inspected what I had for dinner. Other times they spent on the rug, in stately groups and processions, except every now and then,, when they were not so dignified. I could not shoo them out, because I had invited them. Finally, I did get screens: but it takes. weeks to be so intelligent. So the moral is in the observation that, if you wish for something, you had better look out, because you may be so unfortunate as to get it. It is better to be humble and contented with almost nothing, because there's no knowing what something may do to you. Much is said of the "cruelty of Nature": but, when a man is denied his "heart's desire," that is mercy.

But I am suspicious of all this wisdom, because it makes for humility and contentment. These thoughts are community-thoughts, and tend to suppress the individual. They are corollaries of mechanistic philosophy, and I represent revolt against mechanistic philosophy, not as applying to a great deal, but as absolute.

Nevertheless, by the "occult," or the "supernatural," I do not mean that I think that it is altogether exemplified by the experience of the pigeons. In our existence of law-lawlessness, I conceive of two magics: one as representing unknown law, and the other as expressing lawlessness--or that a man may fall from a roof, and alight unharmed, because of anti-gravitational law; and that another man may fall from a roof, and alight unharmed, as an expression of the exceptional, of the defiance of gravitation, of universal inconsistency, of defiance of everything.

London Times, October--

Oh, well, just as an exception of our own--never mind the data, this time--take my word for it that I could cite instances of remarkable falls, if I wanted to.

It looks to me as if, for instance, some fishes climb trees, as an expression of lawlessness, by which there is somewhere an exception to the generalization that fishes must be aquatic. I think that Thou Shalt Not was written on high, addressed to fishes. Whereupon a fish climbed a tree. Or that it is law that hybrids shall be sterile--and that, not two, but three, animals went into a conspiracy, out of which came the okapi. There is a "law" of specialization. Evolutionists make much of it. Stores specialize, so that dealers in pants do not sell prunes. But then appear drugstores, which sell drugs, books, soups, and mouse traps.

I have had what I think is about the average experience with magic. But, except in several periods, I have taken notes upon my experiences: and most persons do not do this, and forget. We forget so easily that I have looked over notes, and have come upon details of which I had no remembrance.

From records of my own experiences, I take an account of a series of small occurrences, several particulars of which are of importance to our general argument.

I was living in London--39 Marchmont Street, W. C. 1. I was gathering data, in the British Museum Library. In my searches, I had noted instances of pictures falling from walls, at times of poltergeist disturbances: but I note here that my data upon physical subjects, such as earthquakes and auroral beams and lights on dark parts of the moon were about five to one, as compared with numbers of data upon matters of psychic research. Later, the preponderance shifted the other way. The subject of pictures falling from walls was in my mind, but it was much submerged by other subjects and aspects of subjects. It was so inactive in my mind that, when I was told of several pictures that had fallen from walls in our house, I put that down to household insecurities, and paid no more attention.

The abbreviations in the notes are A, for my wife; Mrs. M., for the landlady; E, the landlady's daughter; the C's the tenants upstairs.

According to me, this is not the unsatisfactoriness of so many stories about a Mr. X, or a Mrs. Y., because, according to me, only two of us, whom I identify, were more than minor figures: also we may suspect that, of these two, one was rather more central than the other--according to me.

However, also, I suspect that, if E should tell this story, I'd be put down, much minored, as Mr. F. A and I occupied the middle floor, which was of two rooms, one of them used by us as a kitchen, though it was furnished to rent as a furnished room.

March 1, 1924--see Charles Fort's Notes, Letter E, Box 27--"I was reading last night, in the kitchen, when I heard a thump. Sometimes I am not easily startled, and I looked around in a leisurely manner, seeing that a picture had fallen, glass not breaking, having fallen upon a pile of magazines in a corner. Two lace curtains at sides of window. Picture fell at foot of left curtain. Now, according to my impression, the bottom of the right-hand curtain was vigorously shaken, for several seconds, an appreciable length of time after the fall of the picture.

"Morning of the 12th--find that one of the brass rings, on the back of the picture frame, to which the cord was attached, had been broken in two places--metal bright at the fractures.

"A reminded me that, in the C's room, two pictures had fallen recently."

I have kept this little brass ring, broken through in one place, and the segment between the breaks, hanging by a metal shred at the point of the other break. The picture was not heavy. The look is that there had been a sharp, strong pull on the picture cord, so doubly to break this ring.

"March 18, 1924--about 5 P.M., I was sitting in the corner, where the picture fell. There was a startling, crackling sound, as if of window glass breaking. It was so sharp and loud that for hours afterward I had a sense of alertness to dodge missiles. It was so loud that Mrs. C., upstairs, heard it."

But nothing had broken a windowpane. I found one small crack in a corner, but the edges were grimy, indicating that it had been made long before.

"March 28, 1924--This morning, I found a second picture--or the fourth, including the falls in the rooms upstairs--on the floor, in the same corner. It had fallen from a place about three feet above a bureau, upon which are piled my boxes of notes. It seems clear that the picture did not ordinarily fall, or it would have hit the notes, and there would have been a heartbreaking mess of notes all over the floor."

Oh, very. Sometimes I knock over a box of notes, and it's a job of hours to get them back in their places. I don't know whether it has any meaning, but I think about this: the accounts of pictures falling from walls, which were among these notes.

"The glass in the picture was not broken. This time, the cord, and not a ring, was broken. I quickly tied the broken cord, and put the picture back. I suppose I should have had A for a witness. Partly I did not want to alarm her, and partly I did not want her to tell, and start a ghost-scare centering around me." I would have it that, in some unknown way, I was the one who was doing this. I'd like to meet Mrs. C., sometime, and perhaps listen to her hint that she has psychic powers, and hint that she was the one who went around psychically, knocking down pictures in our house.

The cord of this second, or fourth, picture was heavy and strong. It was beyond my strength to break a length of it. But something had broken this strong cord. I looked at the small nail in the wall. It showed no sign of strain.

Of course I was reasoning about all this. Said I: "If, when this house was furnished, all the pictures were put up about the same time, their cords may all weaken about the same time." But a ring broke, one of the times. Upstairs, one of the pictures had fallen in a kitchen, and the other in a living room, where conditions were different. Smoke in a kitchen has chemical effects upon picture cords.

"April 18, 1924--A took a picture down from the kitchen wall, to wash the glass--London smoke. The picture seemed to fall from the wall into her hands. A said: 'Another picture cord rotten.' Then: 'No: the nail came out.' But the cord had not broken, and the nail was in the wall. Later, that day, A said: 'I don't understand how that picture came down.'"

There was nothing resembling a "scare" in the house. There were no discussions. I think that there was an occasional laughing suggestion--"Must be spooks around." I had three or four reasons for saying nothing about the matter to anybody.

"July 26, 1924--Heard a sound downstairs. Then Fannie called up: 'Mrs. Fort, did you hear that? A picture fell right off the wall.'"

I go on with my account, or with the mistake that I am making. Just so long as I gave the New York Something or Another, or the Tasmanian Whatever, for reference, that was all very well. But now I tell a story of my own, and everybody who hasn't had pictures drop from walls, in his presence, will resent pictures falling from walls, because of my occult powers.

There are several notes that may indicate a relation between my thoughts upon falling pictures, and then, later, a falling picture.

"Oct. 22, 1924--Yesterday, I was in the front room, thinking casually of the pictures that fell from the walls. This evening, my eyes bad. Unable to read. Was sitting, staring at the kitchen wall, fiddling with a piece of string. Anything to pass away time. I was staring right at a picture above corner of bureau, where the notes are, but having no consciousness of the picture. It fell. It hit boxes of notes, dropped to floor, frame at a corner broken, glass broken."

There was another circumstance. I remember nothing about it. The notes upon it are as brief as if I had not been especially impressed by something that I now think was one of the strangest particulars--that is, if by indicating that I had searched for something, I meant that I had searched thoroughly.

"The cord was broken several inches from one of the fastenings on back of picture. But there should have been this fastening, a dangling piece of cord, several inches long. This missing. I can't find it."

"Night of Sept. 28-29, 1925--a picture fell in Mrs. M's room." Note the lapse of time.

I am sorry to record that a note, dated Nov. 3, 1926, is missing. As I remember it, and according to allusions, in notes of November 4th, it was only a remark of mine that for more than a year no picture had fallen.

"Nov. 4, 1926--This is worth noting. Last night, I noted about the pictures, because earlier in the evening, talking over psychic experiences with France and others, I had mentioned falling pictures in our house.

Tonight, when I came home, A told me of a loud sound that had been heard, and how welcome it was to her, because it had interrupted E, in a long, tiresome account of the plot of a moving picture. Later, A exclaimed: 'Here's what made the noise!' She had turned on the light, in the front room, and on the floor was a large picture. I had not mentioned to A that yesterday my mind was upon falling pictures. I took that note after she had gone to bed. I looked at the picture--cord broken, with frayed ends. I have kept a loop of this cord. The break is under a knot in it. Nov. 5--I have not strongly enough emphasized A's state of mind, at the time of the fall of the picture. E's long account of a movie had annoyed her almost beyond endurance, and probably her hope for an interruption was keen." Here is an admission that I did not think, or suspect, that it was I, who was the magician, this time.

In October, 1929, we were living in New York, or, anyway, in the Bronx. I do not have pictures on walls, in places of my own. I can't get the pictures I'd like to have: so I don't have any. I haven't been able to get around to painting my own pictures, but, if I ever do, maybe I'll have the right kind to put up.

"October 15, 1929--I was looking over these notes, and I called A from the kitchen to discuss them. I note that A had been doing nothing in the kitchen. She had just come in: had gone to the kitchen to see what the birds were doing. While discussing those falling pictures, we heard a loud sound. Ran back, and found on the kitchen floor a pan that had fallen from a pile of utensils in a closet."

"Oct. 18, 1930--I made an experiment. I read these notes aloud to A, to see whether there would be a repetition of the experience of Oct. 15, 1929. Nothing fell."

"Nov. 19, 1931--tried that again. Nothing moved. Well, then, if I'm not a wizard, I'm not going to let anybody else tell me that he's a wizard."

21

I LOOKED at a picture, and it fell from a wall.

The diabolical thought of Usefulness rises in my mind. If ever I can make up my mind to declare myself the enemy of all mankind, then shall I turn altruist, and devote my life to being of use and of benefit to my fellow-beings.

Everything that is of slavery, ancient and modern, is a phenomenon of usefulness. The prisons are filled with unconventional interpreters of uses. If it were not for uses, we'd be free of lawyers. Give up the idea of improvements, and that is an escape from politicians.

Do unto others as you would that others should do unto you, and you may make the litter of their circumstances that you have made of your own. The good Samaritan binds up wounds with poison ivy. If I give anybody a coin, I hand him good and evil, just as truly as I hand him head and tail. Whoever discovered the uses of coal was a benefactor of all mankind, and most damnably something else. Automobiles, and their seemingly indispensable services--but automobiles and crime and a million exasperations. There are persons who think they see clear advantages in the use of a telephone--then the telephone rings.

If, by looking at it, a picture can be taken down from a wall, why could not a house be pulled down, by still more intently staring at it?

If, occultly, mentally, physically, however, a house could be pulled down, why could not a house be put up, by concentrating upon its materials?

Now visions of the Era of Witchcraft--miracles of invisible bricklaying, and marvels of masonry without masons--subtle uses and advantages that will merge both A. D. and B. C. into one period of barbarism, known as B. W.--

But the factories and labors and laborers--everything else that is now employed in our primitive ways of buildings houses. Unemployment and starvation and charity-- political disturbances--the outcry against putting the machines out of work. There is no understanding any messiah, inventor, discoverer, or anybody else who is working for betterment, except by recognizing him as partly a fiend.

And yet, in one respect, I am suspicious of all this wisdom. The only reason that it is not conventional mechanistic philosophy is that the conventionalist is more subdued. But, if to every action there is a reaction that is equal and opposite, there is to every advantage, or betterment, an equal disadvantage, or worsement. This view--except as quantitatively expressed--seems to me to be in full agreement with my experiences with advantages and uses and betterments: but, as quantitatively expressed, it is without authority to me, because I cannot accept that ever has any action-reaction been cut in two, its parts separated, and isolated, so that it could be determined what either part was equal to.

I looked at a picture, and it fell from a wall.

Once upon a time, Dr. Gilbert waved a wand that he had rubbed with the skin of a cat, and bits of paper rose from a table. This was in the year 1, of Our Lord, Electricity, who was born as a parlor-stunt.

And yet there are many persons, who have read widely, who think that witchcraft, or the idea of witchcraft, has passed away.

They have not read widely enough. They have not thought widely enough. What idea has ever passed away? Witchcraft, instead of being a "superstition of the past," is of common report. I look over my data for the year 1924, for instance, and note the number of cases, most of them called "poltergeist disturbances," that were reported in England. Probably in the United States more numerously were cases reported, but, because of library facilities, I have especially noted phenomena in England. Cases of witchcraft and other uncanny occurrences, in England, in the year 1924, were reported from East Barnet, Monkton, Lymm, Bradford, Chiswick, Mount-sorrel, Dudley, Hayes, Maidstone, Minster Thanet, Epping, Grimsby, Keighley, and Clyst St. Lawrence.

In the year 1927. New York Herald Tribune, Aug. 12, 1927--Fred Koett and his wife compelled to move from their home, near Ellenwood, Kansas. For months this house had been bewitched--pictures turned to the wall--other objects moving about--their pet dog stabbed with a pitch fork, by an invisible. New York Herald Tribune, Sept. 12, 1927--Frank Decker's barn, near Fredon, N. J., destroyed by fire. For five years there had been unaccountable noises, opening and shutting doors, and pictures on walls swinging back and forth. Home News (Bronx), Nov. 27, 1927--belief of William Blair, County Tyrone, Ireland, that his cattle were bewitched. He accused a neighbor, Isabella Hazelton, of being a witch--"witch" sued him for slander--L5 and costs.

My general expression is against the existence of poltergeists as spirits--but that the doings are the phenomena of undeveloped magicians, mostly youngsters, who have no awareness of their powers as their own--or, in the cases of mischievous, or malicious, persecutions, are more or less consciously directed influences by enemies--or that, in this aspect, "poltergeist disturbances" are witchcraft under a new name. The change of name came about probably for two reasons: such a reaction against the atrocities of witchcraft-trials that the existence of witches was sweepingly denied, so that continuing phenomena had to be called something else; and the endeavor by the spiritualists to take over witchcraft, as evidence of the existence of "spirits of the departed."

If witches there be, there must of course be some humorous witches. The trail of the joke crosses our accounts of the most deadly occurrences. In many accounts of poltergeist disturbances, the look is more of mischief than of hate for victims. The London Daily Mail, May 1, 1907, is responsible for what is coming now:

An elderly woman, Mme. Blerotti, had called upon the Magistrate of the Ste. Marguerite district of Paris, and had told him that, at the risk of being thought a madwoman, she had a complaint to make against somebody unknown. She lived in a flat, in the Rue Montreuil, with her son and her brother. Every time she entered the flat, she was compelled by some unseen force to walk on her hands, with her legs in the air. The woman was detained by the magistrate, who sent a policeman to the address given. The policeman returned with Mme. Blerotti's son, a clerk, aged 27. "What my mother has told you, is true," he said. "I do not pretend to explain it. I only know that when my mother, my uncle, and myself enter the flat, we are immediately impelled to walk on our hands." M. Paul Reiss, aged fifty, the third occupant of the flat, was sent for. "It is perfectly true," he said. "Everytime I go in, I am irresistibly impelled to walk around on my hands." The concierge of the house was brought to the magistrate. "To tell the truth," he said, "I thought that my tenants had gone mad, but as soon as I entered the rooms occupied by them, I found myself on all fours, endeavoring to throw my feet in the air."

The magistrate concluded that here was an unknown malady. He ordered that the apartments should be disinfected.

There used to be a newspaper story of the "traveling needle." People perhaps sat on needles, though they thought it more dignified to report that needles had entered their bodies by way of their elbows. Then, five, ten, twenty, years later, the needles came out by way of distant parts. We seldom hear of the "traveling needle," nowadays: so I think that most--not all--of these old stories were newspaper yarns. I was interested in these stories, as told back in the eighteen-eighties and nineties, but never came upon one that seemed to me to be authentic, or to offer material much to speculate upon. I took suggestion from the method of "black magic," of piercing, with a needle, the heart, or some other part, of an image of a proposed victim, and, according to beliefs, succeeded in affecting a corresponding part of a human being

An inquest, in the Shoreditch (London) Coroner's court, Nov. 14, 1919--a child, Rosina Newton, aged thirteen months, had died. A needle was found in her heart. "There was no skin-wound to show where it had entered the body." It was the short life of this child that attracted my attention. The parents had no remembrance of any injury to her, such as that of a needle entering her body.

It seems unlikely that anybody so intensely hated this infant as to concentrate upon a desire for her death: but I have stories that may indicate the doing of harm to children as vengeance upon parents. And in the annals of "black magic" often appears the sorcerer, who obtains something of the belongings, or of the body, of a victim, to secure a contact, or a sense of contact. Parings of fingernails are recommended, but the procuring of a lock of the victim's hair is supposed to be most effective. There may be psychic hounds, who, from a belonging, pick up a scent, and then maintain, and operate along, a path, or a current, between themselves and their victims. In such terms, of harm, or of possession, may be understandable the hair-clippers of our records.

There is a strange story, in the Times of India (Bombay), Aug. 30, 1928. A part of this story that does not seem so very strange to me is that three times a new-born infant of a Muslim woman, of Bhonghir, had been "mysteriously and supernaturally" snatched away from her. The strange part is that the police, though they had explained that these disappearances were only ordinary, or "natural," kidnappings, had gone to the trouble of taking this woman, who for the fourth time was in a state of expectation, to the Victoria Zenana Hospital, at Secunderabad; and that the hospital authorities had gone to the trouble and expense of assigning her to a special ward, where special nurses watched her, night and day. The fourth infant arrived, and this one, so surrounded by test-conditions, did not mysteriously vanish: so it was supposed to be demonstrated that the three disappearances were ordinary kidnappings. The explanation that occurs to one is that, though it was not mentioned in the Times of India, there was probably a scare, at Bhonghir, and that this demonstration was made to allay it.

Just how, by ordinary, or "natural," means, anybody could, time after time, without being seen, snatch a new-born infant from a woman, was not inquired into. All such "demonstrations" start with the implied assumption that there is not witchcraft, and then show that there is not witchcraft. That is, there is no consideration for the thought that a witch might exist and might fear to practice so publicly as in a hospital ward. The "demonstration" was that there was not witchcraft in a hospital ward, and that therefore there is not witchcraft. Many of our data are of most public, or daring, or defiant occurrences: but it is notable that they stop--mostly, though not invariably--when public attention is aroused.

Sometimes they stop, and then renew periodically.

About the first of May, 1922, Pauline Picard, a Breton child, aged 12, disappeared from her home on a farm, near Brest, France. I take this account from various issues of the Journal des Debats (Paris), May and June, 1922. Upon May 26th, a cyclist, passing Picard's farm, saw something in a field, not far from the road. He investigated. He came upon Pauline's naked and headless body. At the roadside were found her clothes. It was noted that they were "neatly folded."

The body was decomposed. Hands and feet, as well as head, were missing. This body, visible from the road, was found at a point half a mile from the Picard farmhouse.

It seems most likely that, if it was seen by a passing cyclist, it could not long have been lying so conspicuous, but unseen, by members of the Picard family. Nevertheless, that it had so lain was the opinion that was accepted at the inquest. It was said that the child must have wandered from home, and, returning, must have died of exhaustion; and that the body had been defaced by rats and foxes. This story of the wandering child, dying of exhaustion, half a mile from her home, was given plausibility by the circumstances that once before Pauline had wandered far, and that she had been affected mentally. At least, she had disappeared, and had been found far away.

Upon April 6th, of this year 1922, Pauline disappeared. Several days later, a child was found wandering in the streets of Cherbourg. The Picards were notified, and, going to Cherbourg, identified this child as Pauline, who, however, did not recognize them, being in a state of lapsed consciousness, or amnesia. If Pauline Picard, aged 12, had made this journey afoot, or by means that are called "natural," between a farm near Brest and Cherbourg, in a state of amnesia, which it seems would somewhere be noted, but had not been reported, she had gone, unreported, a distance by land of about 230 miles.

Twice Pauline Picard disappeared. The first disappearance was not an ordinary runaway, or was not an ordinary kidnaping, because something had profoundly affected this child mentally. I have notes upon more than a few cases of persons who have appeared, as if they had been occultly transported, or at any rate have appeared in places so far from their homes that they were untraceable, and were amnesiatics. An expression for which I should like to find material is that, three times, in distant parts of India, "wolf children" were reported, after the times of disappearance of the infants of Bhonghir. The official explanation of the second disappearance and the death of Pauline Picard bears the marks of dictation by Taboo. If the body of this child had been also otherwise mutilated, the explanation of defacement by rats and foxes would be more nearly convincing: but something, or somebody, had, as if to prevent identification, removed, without other mutilations, hands and feet and head--and also, contradictorily, had placed the body in a conspicuous position, as if planning to have it found. The verdict at the inquest required belief that this decomposed body had lain, conspicuous, but unseen, for several weeks, in this field. There is a small particular that adds to the improbability. It seems that the clothes—also conspicuous by the roadside--had not been lying there, for several weeks, subject to the disturbing effects of rains and wind. They were "neatly folded."

It is as if somebody had removed head, hands, and feet from this body, and had stripped the clothes from it, so that it could not be identified; and had placed the clothes nearby, so that it could be identified.

A field--the dismembered body of a child--a farmhouse nearby. But I can pick up no knowledge of relations with environment. Friendly neighbors--or a neighbor with a grudge--all around is vacancy. A case that was called "unparalleled" was told of, in the New York newspapers, April 30, 1931. Here, too, the surroundings are blankness: in the usual way the story was told, as an unrelated thing. Perhaps, somewhere nearby, brooding over a crystal globe, or some other concentration-device, was the origin of a series of misfortunes.

Early in April, 1931, Valentine Minder, of Happauge, Long Island, N. Y., was suffering with what was said to be mastoiditis. His eight children were stricken with what was said to be measles, and then, one after another, in a period of eight days, the eight children were taken ill with mastoiditis, and were removed to a hospital. The circumstance, because of which these cases were called "unparalleled," is that mastoiditis was supposed to be not contagious.

These cases, which, if "unparalleled," were mysterious, were a culmination of a series of misfortunes. About two years before, Minder's home had burned down. Then came his illness, a loss of vitality, the loss of his job, and a state of destitution. Toward the end of 1930, Mrs. Minder was stricken with an indefinable illness, and became an invalid.

So far as was known, mastoiditis is not contagious. Out of many cases of family maladies, misfortunes, and fatalities, I pick one in which it seems that even more decidedly there is no place for the idea of contagion. Of course there is a place for the idea of coincidence. That is one square peg that fits into round holes and octagonal holes; dodecagonal holes, cracks, slits, gaps--or seems to, so long as whether it does or doesn't is not enquired into. London Daily Chronicle, Nov. 3, 1926--that Mr. A. C. Peckover, the well-known violinist, one of the examiners to the Royal College of Music, had at the home of his sister, in Skipton, awakened one morning, to find himself blind. He was taken to the Bradford Eye and Ear Hospital. Here was his father, who, almost simultaneously, had been stricken with blindness.

In the matter of the deaths that followed the opening of Tut-Ankh-Amen's tomb, it is my notion that, if "curses" there be, they lose their vitality, anyway after several thousand years--

Or that a tomb was violated, and that funerals followed--by the deadly magic of no mummy, but of a living Egyptian--that, somewhere in Egypt, a sense of desecration became an obsession, from which came "rays," or a more personal and searching vengeance.

I wonder why the "wealthy farmer" appears in so many records of more or less uncanny doings. Perhaps any farmer who becomes wealthy, so becomes by sharp practices, and has enemies, whose malices against him demonstrate. In November, 1890, the household of Stephen Haven, a wealthy farmer, living near Fowlerville, Michigan, was startled by cries, one night. Haven was found at the bottom of a deep well. He had walked in his sleep. Two months later, he was again missing from his bedroom, was searched for, and was found, standing, with the water up to his neck, in Silver Lake. Other members of the family were alarmed and alert. They heard slight sounds, one night--Haven was found, fast asleep, trying to set the house afire. Another time--and a thud was heard. The man, asleep, had tried to hang himself.

According to the story, as told in the Brooklyn Eagle, Nov. 18, 1892, Haven had finally been found dead at night. He had fallen from the upper-story doorway of his barn.

See back to occurrences in Sing Prison, in December, 1930. New York Herald Tribune, Jan. 18, 1932--"Warden Lewis E. Lawes fell this evening on the sleet-covered steps of his home, at the prison, and his right arm was broken in three places."

In matters of witchcraft, my general expression--as I say, to signify that neither as to anything in this book, nor anywhere else, have I beliefs--my general expression upon poltergeist girls is not that they are mediums, controlled by spirits, but that effects in their presence are phenomena of their own powers, or talents, or whatever: but that there are cases in which it seems to me that youngsters were mediums, or factors, not to spirits, but to living human beings, who had become witches, or wizards, by their hates--or that, in some cases, sorcery, unless so involuntarily accomplished, cannot operate. See back to the Dagg case--here there seemed to be a girl's own phenomena, and also the presence of another being, who was invisible. The story was probably largely a distortion. The story was that there was a feud--that a "voice" accused a neighbor, Mrs. Wallace, of having sent it into the Dagg home. If this woman could invisibly transport herself into somebody else's home, for purposes of malice and persecution, we'd not expect her to accuse herself--but there is such an element in a hate, as a sense of dissatisfaction with injuring an enemy, unless the victim knows who's doing it. Also the accusation was soon confused into an acquittal.

I have noted a case of occurrences in a shop, in London, which I tell of, mostly because it has highly the look of authenticity. Not a girl but a boy was present. I'd think that the doings were his own phenomena, were it not for the circumstance of "timing." By "timing," in this case, I mean the occurrence of phenomena upon the same days of weeks. The phenomenon of "timing," or the occurrences of doings, about the same time each day, appears in many accounts of persecutions by invisibles, for which I have found no room, in this book.

London Weekly Dispatch, Aug. 18, 1907--disturbances in the stationery shop of Arthur Herbert George, 20 Butte Street, South Kensington, London, according to Mr. George's sworn statement, before the Commissioner for Oathes, at 85 Gloucester-road, South Kensington. George and his assistant, a boy, or a young man, aged 17, saw books and piles of stationery slide unaccountably from shelves. Everything that they replaced fell again, so that they could make no progress, trying to restore order. No vibration, no force of any kind, was felt. Two electric lamps in the window toppled over. Then there was livelier action: packages of note paper flew around, striking George and his assistant several times. George shut the door, so that customers should not come in and be injured. The next day boxes of stationery and bottles of ink were flying around, and four persons were struck. To this statement was appended an affidavit by an antique dealer, Sidney Guy Adams, 23 Butte Street, testifying that he had seen heavy packages of note paper flying around, and that he had been struck by one of them. In the Weekly Dispatch, September 1, it was said that there had been a repetition of the disturbances, upon the same days of the week (Wednesday, Thursday, and Friday) as the days of former phenomena. The damage to goods amounted to about L10.

Upon May 31, 1905, Englishmen--in a land where reported witchcraft is of common occurrence-- were startled. This tabooed subject had been brought up in Parliament. A member of the House of Commons had told of a case of witchcraft, and had asked for an investigation.

See back to "mysterious thefts." Accept data and implications of almost any of the succeeding groups of stories, and "cat burglars," and other larcenous practitioners, become thinkable as adepts in skills that are not describable as "physical."

Dean Forest Mercury, May 26, 1905--that L50 had been stolen from a drawer in the home of John Markey near Blakeney (Dean Forest). The disappearance of this money was considered unaccountable.

Just why, I could not find out, because the influence of Taboo smothered much, in this case. The members of this household could not explain how this money could have vanished, and brooding over the mystery made them "superstitious." They asked a woman, who, according to her reputation, had much knowledge of witchcraft, to investigate. Then came occurrences that made them extremely, hysterically, insanely "superstitious." It was as if an invisible resented the interference. Soon after the arrival of this woman-- Ellen Haywood--something went through this house, smashing windows, crockery, and other breakables.

That is about all that I can pick up from the local newspaper, and from other newspapers published in the neighborhood.

Markey's daughter broke down, with terror. There is only this record: no particulars of her experiences. Without detail, or comment, it is told that Markey's granddaughter became insane. Both women were removed, one to a hospital, and the other to an asylum. Markey's wife ran screaming from the house, and hid in the forest. A Police Inspector came from Gloucester, and organized a search for her; but she was not found. For three days, without food or shelter, she hid. Then she returned, telling that she had seen the searchers, but had been in such a state of terror--by whatever was censored out of the records-

-that she had been afraid to come out of hiding. Markey's son became violently insane, smashing furniture, and seriously injuring himself, crying out that the whole family was bewitched. He, too, was taken to an asylum.

There was a demand for an inquiry into this case, and it was voiced in the House of Commons. It was voiced against Taboo. There is no more to tell.

I have notes upon another case that looks like resentment against an intrusion--if a woman died, but not in an epileptic fit, as alleged. There were accounts in the London newspapers, but I take from a local newspaper, the Wisbech Advertiser, Feb. 27, 1923, home of Mr. Scrimshaw, at Gorefield, near Wisbech. Other members of Scrimshaw's household were his mother, aged 82, and his daughter, Olive, aged 16. The phenomena were in the presence of this girl. First, Mrs. Scrimshaw's lace cap rose from her head.

Then a washstand crashed to the floor. Objects, such as books, dishes, a water filter, fell to the floor. There was much smashing of furniture and crockery. Names of neighbors, who witnessed these unconventionalities, are John Fennelow, T. Marrick, W. Maxey, and G. T. Ward. A piano that weighed 400 pounds moved from place to place. Police-constable Hudson was a witness of some of the phenomena. As to a suggestion that, for any reason of notoriety, or hoaxing, Scrimshaw could be implicated, it was noted that the damage to furniture amounted to about L140.

A woman--Mrs. J. T. Holmes--who, sometime before, had been accused of witchcraft, went to this house, and practiced various incantations to exorcise the witch, or the evil spirit, or whatever. She died suddenly. It was said that she was subject to fits, and had died in one of her convulsions. Whether his decision related to Taboo, or not, the coroner decided not to hold an inquest.

Upon Dec. 12, 1930--see the Home News (Bronx), Dec. 22, 1930--a resident of the Bronx, Elisha Shamray--who had changed his name from Rayevsky--opened a pharmaceutical laboratory, in Jackson Street, lower East Side, New York. During the night he died. His brother, Dr. Charles Rayevsky, came from Liberty, N. Y., to arrange for the funeral. He died a week later. The next night, the third of these brothers, Michael Shamray, Tremont Ave., Bronx, was on his way to arrange for the second funeral. He was struck by an automobile, and was killed.

In August, 1927, Wayne B. Wheeler was the general counsel of the Anti-saloon League of America. Upon August 13th, an oil stove exploded, in his home, and his wife was killed. Later, his father dropped dead. Upon the 5th of September, Wheeler died.

New York Sun, Feb. 3, 1932--Mount Vernon, Ohio, February 3--"Fear that the mysterious illness which has killed three young brothers may strike again in the same family gripped surviving members of the household, today."

Upon the 24th of January, Stanley Paazig, aged 9, died in the home of his parents, on a farm, near Mount Vernon. Upon the 31st, Raymond, aged 8, died. Marion, aged 6, died, February 2nd.

The State Health Department had been unable to identify the malady. "Chemists spent twenty-four hours making tests of the youngest victim's blood, without finding a trace of poison."

22

BELIEF in God--in Nothing--in Einstein--a matter of fashion--

Or that college professors are mannequins, who doll up in the latest proper things to believe, and guide their young customers modishly.

Fashions often revert, but to be popular they modify. It could be that a re-dressed doctrine of witch-craft will be the proper acceptance. Come unto me, and maybe I'll make you stylish. It is quite possible to touch up beliefs that are now considered dowdy, and restore them to fashionableness. I conceive of nothing, in religion, science, or philosophy, that is more than the proper thing to wear, for a while.

"Typhoid Mary"--I doubt her germs--or I suspect that she was more malicious than germy. But nobody else--at least so far as go the published accounts--which could not be expected to go very far back in the years 1906-14--thought of ignoring her germs, and of bottling her "rays." For my own suspicion that this was a case of witchcraft, I shall, for a while, probably be persecuted, by an amused tolerance, but, if back in the year 1906, anybody had given his opinion that "Typhoid Mary" was a witch, he'd have been laughed at outright.

Nobody accused "Typhoid Mary," except properly. According to the demonology of her era, she was distributing billions of little devils. Her case is framed with the unrecorded. As to her relations with her victims, I have nothing upon which to speculate.

The homes of dying men and women have been bombarded with stones of undetectable origin. Nobody was accused. We have had data of unexplained explosions, and data of seeming effects of "rays," not physical, upon motors. To me it is thinkable that a distant enemy could, invisibly, make an oil stove explode, and kill a woman, and then--if by means other than any known radioactivity, aero-planes ever have been picked from the sky--pick from existence other members of her family. The explosion of the oil stove is simply a bang, such as cartoonists sometimes draw, with a margin of vacancy.

But there have been cases of persons who were accused of witchcraft.

This statement--like every other statement, issuing from the Supreme Court of the United States of America, from a nursery, from a meeting of the Amer. Assoc. Ad. Sci., or from the gossip of imbeciles-- means whatever anybody wants it to mean. One interpretation is that superstitious people have attributed various misfortunes, which were probably due to their own ignorance and incompetence, to the malice of neighbors. At any rate, these cases are sketches of relations with environment, and so far we have been in a garden of evil, in which blossomed deaths and destructions, without visible stems, and without signs of the existence of roots.

New York Evening World, Sept. 14, 1928--Michael Drouse, a farmer, living near Bruce, Wis., who shot and fatally wounded John Wierzba, forty-four, told Sheriff Dobson that he did it because Wierzba had bewitched his cows. New York Times, Sept. 8, 1929--action by the Rye (N. Y.) National Bank against Leland Waterbury, of Poundridge, for recovery of properties, which the bank alleged had been taken from its client, Howard I. Saires, by "evil eye" methods. "The case has come to be known as the 'Westchester witchcraft case'." New York Times, Oct. 9, 1930--charges of sorcery brought against Henry Dorn, of Janesville, Wis. "After a member of the State Board of Medical Examiners listened to the charges of sorcery, he said that he was convinced that they were unfounded." Dorn's sister had accused him of "casting spells of sickness" upon members of her household.

So that case was disposed of.

I am not given to fortune-telling. I dislike the idea of fortune-telling, so called, or termed more pretentiously. But I do think that anybody could tell the fortune of any member of any State Board of Medical Examiners, who would say, of any charge of sorcery, that he was convinced that it was well-founded.

There were other charges against Dorn. They remind one of accusations in old-time witchcraft trials That Dorn had caused apples to rot on trees, cows to go dry, and hens to cease laying.

Opponents to the idea of witchcraft are much influenced by their inability to conceive how anybody could make apples rot; inability to visualize the process of drying a cow, or entering into the organism of a hen, and stopping her productions. And science does not tell them how this could be done. So.

Also they cannot conceive how something makes apples grow, or why they don't rot on trees; how the milk of a cow is secreted, or why she shouldn't be dry; how the egg of a hen develops. And science does not tell them.

It's every man for himself, and save who can--and damnation is in accepting any messiah's offers of salvation. We're told too much, and we're told too little. We rely. And for two pins--having had experiences by which I am pretty well assured that nobody ever has two pins, when they're called for--I'd finish this book, as a personal philosophy, or for myself, alone, and then burn it. It's everybody for himself, or he isn't anybody.

It's every thinker for himself. He can be told of nothing but surfaces. Theological fundamentalists say, rootily, they think, that all things have makers--that God made all things. Then what made God? even little boys ask. Space is curved, and behind space, or space-time, there is nothing, says Prof. Einstein. Also may he be construed as saying that it is only relatively to something else that anything can be curved.

Throughout this book there is a permeation that may be interpreted as helplessness and hopelessness-- absence of anything in science more than approximately to rely on--solaces and reassurances of religion, but any other religion would do as well--all progresses returning to their points of origin--philosophies only intellectual dress-making--

But, if it's every man for himself, it is my expression that out of his illusion that he has a self, he may develop one.

In records of witchcraft trials, often appears the statement that the accused person was seen, at the time of doings, in a partly visible, or semi-substantial, state. In June, 1880, at High Easter, Essex, England (London Times, June 24, 1880), there were poltergeist disturbances in the home of a family named Brewster. Furniture wandered. A bed rocked. Brewster saw, or thought he saw, a shadowy shape, which he recognized as that of his neighbor, Susan Sharpe. He and his son went to the home of the woman, and dragged her to a pond. They threw her into the pond, to see whether she would sink or float. But, though once upon a time, this was the scientific thing to do, fashions in science had changed. Brewster and his son were arrested, and were bound over to keep the peace--just as should be any woman, who, during rush hours in the subway, should appear in a hoop skirt.

A case that was a blend of ancient accusations and modern explanations was reported in the London Evening News, July 14, 1921--that is, "mysterious illnesses" attributed to the doings of an enemy, but an attempt to explain materialistically. Residents of a house in Putney had, in the London South Western Court, accused their neighbor, Frank Gordon Hatton, of "administering poisonous fumes down their chimney." Saying that the complainants had failed to prove their case, the magistrate dismissed the charge.

If anybody could have a sane idea as to what he means by insanity, he might know what he is thinking about, by bringing in this convenient way of explaining unconventional human conduct. Whatever insanity is supposed to be, it cannot so satisfactorily be applied as the explanation of two persons' beliefs relatively to one set of circumstances. According to newspaper accounts of a murder, in July, 1929, Eugene Burgess, and his wife, Pearl, went insane together, upon the same subject. It was their belief that, when Burgess's mother died, in the year 1927, she had been "willed to death" by a neighbor, Mrs. Etta Fairchild. It was their belief that this woman had cast illness upon their daughter. They killed Mrs. Fairchild. In an account, in the New York Sun, Oct. 16, 1929, Mrs. Burgess is described: "Belying the comparison to the ignorant peasant women, who have stood for trial for similar crimes, for hundreds of years, Mrs. Burgess looks like a prosperous clubwoman."

These are accounts of accusations of witchcraft, by persons, against other persons, according to their superstitions, or perceptions. Now there will be accounts of cases in which there are suggestions of witchcraft to me, according to my ignorance, or enlightenment.

Chicago Tribune, Oct. 14, 1892--marvelous--though not at all extraordinary--doings in the home of Jerry Meyers, a farmer, living near Hazelwood, Ohio. Meyers had been absent from his home, driving his wife to the railroad station. When he returned, he heard a hysterical story from his niece, Ann Avery, of Middletown, Ohio, who was visiting him. Soon after he and Mrs. Meyers had left the house stones were thrown at her, or fell around her. Objects in the house moved toward her. Mr. Meyers was probably astonished to hear this, but what he wanted was his dinner. The girl went to the barn to gather eggs. On her way back, stones fell around her. Whether Meyers got his dinner, or not, he got a gun. Neighbors had heard of the doings. Stationed around the house were men with shotguns: but stones of unknown origin continued to bombard the house. Ann Avery fled back to her home in Middletown. Phenomena stopped.

In this case of the girl who was driven from her uncle's home, the circumstance that I pick out as significant is that assailments by stones began soon after Mrs. Meyers left the house. It was said that she had gone to visit friends, in the village of Lockland. Of course hospitalities often are queer, but there is a good deal of queerness in the hospitality of somebody who would go visiting somewhere else, while her husband's niece was visiting in her home.

About the last of November, 1892, in the town of Hamilton, Ontario, a man was on his way to a railroad station. In a cell, in a prison, in Fall River, Massachusetts, sat a woman.

Henry G. Trickey was, in Hamilton, on his way to a railroad station. In the Fall River jail was Lizzie Borden, who was accused of having murdered her parents.

In August, 1892, Trickey, a reporter of the Boston Globe, had written what was described as a "scandalous article" about Lizzie Borden. The Globe learned that the story was false, and apologized. Trickey was indicted.

He went to Canada. This looks as if he had fled from prosecution.

Lizzie Borden sat in her cell. There may have been something more deadly than an indictment, from which there was no escape for Trickey. While boarding a train, at Hamilton, he fell, and was killed.

In the town of Eastbourne, Sussex, England, in April, 1922, John Blackman, a well-known labor leader, was committed to prison, under a maintenance order, for arrears due to his wife. The judge who committed him died suddenly. When Blackman was released, he still refused to pay so back he went to prison. The judge who sent him back "died suddenly." He continued to refuse to pay, and twice again was re-committed to prison, and each time the judge in his case "died suddenly." See Lloyd's Sunday News (London), Oct. 14, 1923.

Upon Nov. 29, 1931, there was an amateur theatrical performance in the home of Miss Phoebe Bradshaw, 106 Bedford Street, New York City. Villain--Clarence Hitchcock, 23 Grove Street, New York. Wronged husband--John L. Tilker, 1976 Belmont Avenue, Bronx. Tilker was given a cap pistol. Also he carried a loaded revolver of his own, for which he had a permit. When the time came, Tilker, with his own revolver, fired at Hitchcock, shooting him in the neck. "He was apparently new at play-acting, and in his excitement fired his own revolver, instead of the dummy."

Hitchcock lay dying in St. Vincent's Hospital. Soon something occurred to Tilker. He was taken to the Willard Parker Hospital, suffering from what was said to be scarlet fever. Hitchcock died, Jan. 17, 1932. See the New York Herald Tribune, Jan. 18, 1932.

New York Evening Journal, Feb. 6, 1930--"Two bitter women enemies are teetering on the verge of death, today, one of them 'doing satisfactorily,' while the other is weaker, and in a highly critical condition. Both are sufferers from cancer. They are Mrs. Frances Stevens Hall and her most hated opponent, in the famed Hall-Mills trial, Jane Gibson, whose testimony was used in an effort to send Mrs. Hall to the electric chair."

Upon the 8th of February, Jane Gibson died.

In the Fall of 1922, Mrs. Jane Gibson was a sturdy woman-farmer. It was her accusation that, upon the night of the murder of Dr. Edward Hall and Elinor Mills, Sept. 14, 1922, she had seen Mrs. Hall bending over the bodies. So she testified. She returned to her home, and soon afterward was stricken. At the re- trial, in November, 1926, she repeated the accusation, though she had to be carried on a cot into the court room. "Most of her days since that time were spent in the hospital."

23

DEAD men in a Harlem park--and houses are torn by explosions, of unknown origin--the sneak of an invisible clipper of hair--vampires and murder--theatrically a girl is stabbed, on a staircase, in the presence of a large audience--the internal organs of a woman are burned into unrecognizability--And the stoutest opponents of witchcraft, one with persecutions, and the other with denials, have been religion and science--

And more power to them, for it--

Except that witchcraft is appalling.

In our existence of the hyphen, the appalling can be only one view of a state that combines the direst and the most desirable. Religion is belief in a supreme being. Science is belief in a supreme generalization. Essentially they are the same. Both are the suppressors of witchcraft, and I shall take up these oppositions together. But, in a state of realness-unrealness, there cannot be real opposition. In our existence of the hyphen, what is called opposition is only one view of the state of opposition-stimulation.

There is no way of judging anything, except by its manifestations. Just as much as it has been light, religion has been darkness. Today it is twilight. In the past it was mercy and charity and persecution and bloody, maniacal, sadistic hatred--hymns from chapels and screams from holy slaughterhouses--aspirations going up from this earth, with smoke from burning bodies. I can say that from religion we have never had opposition, because there never has been religion--that is that religion never has existed, as apart from all other virtues and vices and blessings and scourges--that, like all other alleged things, beings, or institutions, religion never has, in a final sense, had identity. An atheist, of zeal, may be thought of as religious. Or I can take the unmonistic view, and accept that there is, or used to be, religion, just as, practically, I ignore that all things and beings of my daily experiences are so bound up with one another that they have not identities, and go about my daily affairs as if things and beings really were entities.

New York Sun, March 26, 1910--eruption of Mt. Etna--people of Borelli praying--the oncoming lava. The molten flood moved onward toward a shrine. Here the praying ones concentrated. The lava reached the shrine, and suddenly changed its course.

New York Times, July 27, 1931--"A revival of the ancient rain dance of Northern Saskatchewan Indians, despite the ban by the government agents, is reported to have occurred recently. Fields were parched and cattle were suffering when Chief Buffalo Bow, head of the File Hills Reserve, decided to invoke the Great Spirit. The forty-eight-hour dance, led by six singers in relays, centered about a great tree, on the bark of which a petition for aid had been carved. The Great Spirit seemed to answer, for soon after the mystic rites had been performed, the rain began and continued for two days, July 14 and 15, bringing relief all over Saskatchewan."

If, according to the views of the majority of the inhabitants of this earth, both Jehovah and the Great Spirit are myths, lava, if it would not have changed its course anyway, and rain, if it would not have fallen anyway, were influenced by witchcraft, if there be witchcraft. My general situation is that of any mathematician. Consider any of his theorems. The parallelogram of forces. In the textbooks, this demonstration works out--if the incident forces be without irregularities--if resistances be unchanging--if the body acted upon be changeless--if the student has no awareness of the changes and the irregularities that are everywhere.

In the London Daily Chronicle, July 7, 1924, was reported a case of an English girl who had come back from Lourdes, cured, she thought. It is not often that the doctors will have anything to do with one of these cases; but it was arranged to investigate this case. At the Hospital of St. John and St. Elizabeth, St. John's Wood, London, the girl was examined by 50 doctors. She had gone, with a nurse, to Lourdes. The nurse was questioned, and testified that the girl's hand had been covered with sores, from blood poisoning, and that she had been cured, at Lourdes. The diseased condition of the girl, when she arrived in Lourdes, was certified by three doctors, of Lourdes. The sores had disappeared, but some contraction of the hand remained. The official decision of the 50 doctors, who were not of Lourdes, was: "On the evidence submitted, the cure is not proven."

I should like to come upon a record of the opinions of 50 drivers of hansom cabs, as to automobiles, when automobiles were new and uncertain, but were of some slight menace to the incomes of hansom cabbies.

In the New York World-Telegram, July 24, 1931, there is a story of a boy, who, at the Medical Center Hospital, New York, was cured of paralysis by the touch of a bit of bone of St. Anne, taken to the hospital from the Church of St. Anne, 110 East 12th Street, New York City. The boy was the son of Hugh F. Gaffney, 348 East 18th Street, New York City.

If, according to the views of the majority of the inhabitants of this earth, there is no more divinity at Lourdes, or at 110 East 12th Street, than anywhere else, there are reasons for thinking that it is witchcraft that is practiced at these places.

The function of God is the focus. An intense mental state is impossible, unless there be something, or the illusion of something, to center upon. Given any other equally serviceable concentration-device, prayers are unnecessary. I conceive of the magic of prayers. I conceive of the magic of blasphemies. There is witchcraft in religion: there may be witchcraft in atheism.

In the New York Evening World, Sept. 19, 1930, is an account of joy in Naples: the shouts of crowds, and the ringing of church bells. In the Cappella del Tesora Cathedral had been displayed the phial containing the "blood of St. Januarius." It had boiled.

It is my notion that, if intenser than the faith in Naples, had been a desire for a frustration of this miracle, the "blood of St. Januarius" might have frozen.

Upon the 5th of March, 1931--see the New York Herald Tribune, March 6th-15,000 worshipers were kneeling, at a pontifical high mass, in the Municipal Plaza, at San Antonio, Texas. Considering the intense antagonism to Catholicism in Mexico, at this time, one thinks of the presence of some of this feeling in San Antonio. From a palm tree, the topmost tuft fell into the kneeling congregation. Six persons were taken to the hospital.

My general expression is that some of the reported phenomena that are called "miracles" probably have occurred, but have been arbitrarily taken over by the religionists, though they are the exclusive properties of priests no more than of traveling salesmen--that scientists have been repelled by the reported phenomena, because of a fear of contamination from priestcraft--but that any scientist who preaches the "ideals of science," and also lets fear of contaminations influence him is as false to his preachments as ever any priest has been.

See the New York Herald Tribune, Dec. 6, 1931--an account of the opening, in Goa, Portuguese India, of the coffin of Saint Francis Xavier.

"A special emissary, sent by Pope Pius XI, led the ceremonial procession, in which marched three archbishops, fifteen bishops, and hundreds of other members of the clergy. A throng of ten thousand persons heard the papal mass and benediction, in the Church of Don Jesus.

"The congregation passed before the coffin, and kissed the dead saint's feet."

But there have been scientists, especially medical scientists, who, in spite of contaminations, have not been held back from investigations.

In January, 1932, the New York newspapers told that many miracles had been reported in Goa.

There is no opposition, as sheer, to witchcraft, by religionists. It is competition.

24

OUR only important opposition is, not science, but a belief that we are in conflict with science. This is an old-fashioned belief.

There is nothing told of in this book that is more of an affront to old-time dogmas than is the theory of the Nobel Prize-winner, Dr. Bohr, that the sun is "deriving" its energy from nowhere.

The quantum theory is a doctrine of magic. The idea of playing leapfrog, without having to leap over the other frog, is simply another representation of the idea of entering a closed room without passing through the walls. But there is a big difference between "authoritative pronouncements" and my expressions. It is the difference between sub-atomic events and occurrences in boarding houses. The difference is in many minds--unlike my mind, to which all things are phenomena, and to which all records are, or may be, data--in which electrons and protons are dignified little things, whereas boarders and tramps on park benches can't be taken solemnly. Charles Darwin was similarly received when, in the place of academic speculations upon evolution, he treated of bugs and bones and insides of animals.

Not, of course, that I mean anything by anything.

Quantum-magic is a doctrine of discontinuity. So it seems to be opposed to my expressions upon hyphenation, which seem to be altogether a philosophy of continuity. But I have indicated that also I hyphenate in another "dimension." I conceive of all phenomena as representing continuity in one "dimension," and as representing discontinuity in another "dimension"--that is, all phenomena as inter- dependent and bound up with one another, or continuous, and at the same time so individualized that nothing is exactly like anything else, or that everything is alone, or discontinuous. I conceive of our existence as one organic state, or being, that is an individual, or that is unrelated to anything else, such as other existences, in the cosmos, its state of oneness expressing in the continuity of its internal phenomena, and its state of individuality, or apartness from everything else in the cosmos, expressing in a permeation of that individuality, or discontinuity, throughout its phenomena. Of course, if the word cosmos means organized universality, I misuse the word here. For various reasons I let it stand.

There are hosts of persons, who consider themselves up-to-date, or ahead of that; who bandy arguments in the latest, scientific lingo, and believe anything that they're told to believe of electrons, but would be incapable of extending an idea from electrons to boarders--even though they argue that every boarder is only a composition of electrons--and go right on thinking of affairs, in general, in old- fashioned, materialistic terms.

Well; then, in old-fashioned terms, what had I this morning for breakfast?

I think: therefore I had breakfast.

If no line of demarcation can be drawn between one's breakfasts and one's thoughts, or between a cereal and a cerebration, this is the continuity of the material and the immaterial. If there is no material, as absolutely differentiated from the immaterial, what becomes of any opposition from what may still survive of what is called materialistic science?

"Science is systematized and formulated knowledge."

Then anybody who has systematized and formulated knowledge enough to appear, on time, at the breakfast table, is, to that degree, a scientist. There are scientific dogs. Most of them have a great deal of systematized and formulated knowledge. Cats and rabbits and all those irritating South American rodents that were discovered by cross-word puzzle-makers are scientists. A magnet scientifically picks out and classifies iron filings from a mass of various materials. Science does not exist, as a distinguishable entity.

Our data have been upon witchcraft in love affairs; in small-town malices, and occasional murders of no importance. According to the phantom, materialistic science, there is no witchcraft. In the monistic sense, I agree. Witchcraft is so bound up with other "natural forces," that it cannot be picked out, as having independent existence. But, in terms of common illusions, I accept that there is witchcraft; and, just for the sake of seeming to have opposition, which makes for more interest, I pretend that there is science.

Stars and planets and ultra-violet radiations from the sun--paleolithic and neolithic inter-relationships, and zymotic multiplications, and tetrahedronic equilaterality--

And the little Colwell girl, who kept the firemen busy--and a kid named "Rena" got a haircut--there was a house in which a pan of soft soap wandered from room to room--a woman alone in a compartment of a railway train, and then maybe she wasn't alone

The disdain of any academic scientist--if among the sensationalists of today, there survive an academic scientist--for what I call the data of witchcraft--

And now my subject is witchcraft in science.

In the year 1913, the German scientist, Emil Abderhalden, announced his discovery of the synthesis of inorganic materials into edible substances. It was said that to avoid all uncertainties--this back in those supreme old days when all scientists were certain--this announcement had been long-delayed. But experiments had been successes. Dogs fed upon synthetic foods had gained weight astonishingly, as compared with dogs that had been fed ordinary meals. Reports were much tabulated. Statistics--very statistical. Then came the War. If Dr. Abderhalden, or anybody else in Germany, could out of muds of various kinds have produced those alleged meals, perhaps we'd all be fighting to this day. As it is, we have had a rest, and can do the necessary breeding, before again starting up atrocities. So, at least for the sake of vigorous new abominations, it seems to be just as well that some of the widely advertised scientific successes aren't so successful.

But the dogs got fat.

There is scarcely an annual meeting of any prominent scientific association at which are not made, by eminent doctors and professors, announcements of great discoveries that, by long and careful experimentation, constructive and eliminative tests, and guards against all possible sources of error, have been established. A year or so later, these boons to suffering humanity are forgotten.

Almost always these announcements are not especially questioned, and bring no confusion upon their sponsors. There is much "scientific caution." A scientist doesn't know but that he may make an announcement, himself, someday. But about the middle of July, 1931, Professor Wilhelm Gluud, of the Westphalian University of Munster, was not received with the usual "caution." Prof. Gluud announced--these Professors never merely say anything--that synthetic albumen could be produced from coal. This dreamery was attacked, and later, in July, Prof. Gluud admitted that he had been "premature" in his announcement.

But something had convinced a scientist, of international reputation, so that he had risked that reputation by making his announcement.

So one inclines to think.

If he had made no experiments, and had simply and irresponsibly squawked into publicity, we have some more monism, and can draw no line between a Westphalian Professor and any Coney Island "barker." But, if he did make experiments, and, if, in spite of later developments, which showed that, according to chemical principles, success was impossible, he nevertheless had reasons to believe that some of his experiments were successes, these successes that agreed with his theory were realizations of his imaginings.

About the same time (July, 1931) another scientist was embarrassed. The Russian physiologist, Pavlov, had announced that he had taught white mice to respond to a bell, at meal time--

But now see here!

Just how disdainful should persons who put in their time ringing dinner bells for mice be of others who collect accounts of meandering pans of soft soap?

It was Pavlov's statement, or "announcement," that he had taught white mice to respond to a bell, at meal time, and that a second generation of white mice had been keener in so responding. This improvement was supposed to represent cumulative hereditary influences.

But Sir Arthur Thompson, of Aberdeen, Scotland, made an announcement.

And now see here, again! I should like to hear Sir Arthur's opinion upon the dignity of such subjects as "the vanishing man," and stones that were pegged at a farmer's niece. He, too, had been ringing dinner bells for animals.

Thompson's announcement was that he had noted no improved teachableness in a second generation of white mice. Whereupon Pavlov withdrew his announcement, saying that he must have been deceived by his assistant.

This is becoming a stock-retreat. Before he shot himself, in August, 1925, Prof. Kammerer, accused of having faked, with India ink, what he called acquired characters on the feet of toads, explained that he had been betrayed by an assistant.

I conceive that, though Pavlov retreated before a "higher authority," his white mice may have been keener in a second generation, though nobody else's white mice would have been of any improved discernment in a fifteenth generation--and that, though biologically, nuptial pads could not possibly appear upon the feet of Prof. Kammerer's toads--

Pictures on hailstones--a face on a cathedral wall--and an insect takes on the appearance of a leaf--

That it may be that a man did not altogether deceive himself and others, but that faint markings did appear upon the feet of toads, as responses to his theory--but in all the uncertainty and the evanescence of the incipient--that, convinced that he was right, Prof. Kammerer may have supplemented faint markings with India ink, just to tide over, at a time of enquiry--then exposure-- suicide.

The story of cancer-cure announcements is a record of abounding successes in the treatment of cancerous dogs, cats, chickens, rats, mice, and guinea pigs--followed by appeals to the public for funds for the study of the unknown causes, and the still undiscovered cure for cancer. Look over the records of cancerous growths that, according to triumphant announcements have been absorbed, or stopped, in mice and guinea pigs, and try to think that all were only deliberate deceptions. My good-bad opinion of human nature won't stand it. But, if some of these experiments were the successes they were said to be, and if the treatments are now repudiated or forgotten, these successes were re-alized imaginings. I know of nothing in science that has the look of better establishment than that there have been some cures of cancer, under radium-treatment. But, in the year 1930, the British Radium Commission issued a warning that the use of radium had not been established as a cancer-cure. The look to me is that, in all the earnestness and charlatanry; devotion to ideals, and fakery, and insincerity; exploitations and duperies of this cult, some cures, as if by the use of radium, have occurred; but that applications of soft soap, if subject to an equal intensity of thought, would have done just as well--

Which brings us to the appalling unnecessity of vivisection, if experiments upon the animals of a toy Noah's Ark, to cure them of their splinters, would be just as enlightening, if anything can be con-strued into meaning anything that anybody wants it to mean--in. an existence in which there is not meaning, but meaning-meaninglessness.

And--not wanting to write three or four hundred pages upon this subject--I shall not go much into records of professorial rascals, or faithful and devoted scientists, who have exploited, or have tried to minister unto, the desire of old codgers to caper. I take from the New York Evening Post, April 12, 1928, an account of "discoveries of major importance to the science of rejuvenation," as announced, in Berlin, by Professor Steinach, to the annual Congress of German Surgeons. Professor Steinach's announcement was that he had discovered the secret of rejuvenation in uses of the pituitary gland. If any reader isn't quite sure where the pituitary is, I remind him that it is connected with the fundibu-lum. It is in a part of the body that is most profoundly engaged in sex-relations. It is in the brain.

Dr. Steinach announced that, with twelve injections of pituitary serum, in senile rats, he had "re-stored their failing appetites, induced a new growth of hair, rejuvenated all bodily functions, and had generally transformed ailing, or half-dead, creatures into youthful animals."

There is witchcraft in science--

If bald old rats have turned young and hairy--if dogs, fed on coal-products, have astonishingly fat-tened-- if tens of thousands of mice and guinea pigs have magically gone fat, or gone thin, in the presence of experimenters--

If, in not all these cases has the treacherous, or perhaps kindhearted, assistant slipped, say, a brisk and hairy young rat into the place of a decrepit old codger; or has not, in secret rascality, or benevolence, meatily supplemented the fare of dogs supposed to be thriving upon coal-products--

If not in all these cases have eminent trappers lied snares for dollars.

My pseudo-conclusion, or acceptance--which is as far as I can go, in the fiction that we're living--is that some of these announcements have been pretty nearly faithful reports of occurrences; and that, by witchcraft, or in response to intense desires of experimenters, senile rats have lost the compensa-tions of old age, and have suffered again the tormenting restlessness of youth--all this by witchcraft, and not by injections that in themselves could have no more of a rejuvenating effect upon either rats or humans than upon mummies.

But, if Prof. Steinach, by witchcraft, or by the effects of belief, did grow hair upon the bald skin of a rat-- to say nothing of the more frolicsome effects of his practices--how comes it that he was not equally successful with the human subjects of his sorcery? Today the Steinach treatment stands discredited.

Especially destructive have been Dr. Alexis Carrel's attacks upon it. It may be that the Professor's own greed defeated him. It may be that he failed because he dissipated his sorcery among many customers.

25

IF I can bridge a gap--

Then that, in a moment of religious excitation, an inhabitant of Remiremont, focusing upon a point in the sky, transferred a pictorial representation from his mind to hailstones--

The turning, off Coventry Street--streets in Japan, Kiel, Berlin, New York City--other places--and that wounds, as imagined by haters of people, have appeared upon the bodies of people--

Or the story of the sailor aboard the steamship Breeshe, in December, 1931--and that it was during a storm--and that in the mind of somebody else aboard this vessel a hate pictured this man, as struck by lightning, and that upon his head appeared a wound, as pictured.

The gap, or the supposed gap, is the difference, or the supposed, absolute difference, between the imagined and the physical.

Or, for instance, the disappearance of Ambrose Small, of Toronto--and it was just about what his secretary, who had embezzled from him, probably wished for, probably unaware that an inventory would betray him. A picturization, in the secretary's mind, of his employer, shooting away to Patagonia, to Franz Josef Land, or to the moon--so far away that he could never get back--but could the imagined realize? Or why didn't I keep track, in the newspapers of December, 1919, for mention of the body of a man, washed up on a beach of Java, scarcely decipherable papers in the pockets indicating that the man was a Canadian? Are the so-called asteroids bodies of people who have been witched away into outer space?

Rose Smith--that when she was released from prison, her visualizations crept up behind her former employer, and killed him? According to some viewpoints, I might as well try to think of a villain, in a moving picture, suddenly jumping from the screen, and attacking people in the audience. I haven't tried that, yet.

Case of Emma Piggott--and the fires in the home of her employers were just about what the girl, alarmed by the greediness of her thefts, may have wished for. Also there are data that may mean that, because of experiences unknown to anybody else, this girl knew that, from a distance, she could start fires.

There is an appearance of affinity between the Piggott case and the fires in the house in Bedford. There was a sulphur fire that was ordinary. It was followed by a series of fires that were, at least according to impressions in Bedford, extraordinary. In no terms of physics, nor of chemistry, was an explanation possible; yet investigators felt that a relationship of some kind did exist. The relationship may have existed in the mind of Anne Fennimore. After the sulphur fire had been put out, she may have started fearing fires, especially in the absence of the only male member of the household. Her fear may have realized.

Story of the Colwell girl--here, too, fires in a house seem to have related to a girl's mental state--or that the fires were related to her desire to move to another house. Having the not uncommon experience of learning how persuasive are police captains, she "listened to advice," and confessed to effects, in terms of ordinary incendiarism, though, according to reports by firemen and policemen, some of the fires could not have been produced by flipping lighted matches.

In the case of Jennie Bramwell, there is no knowing what were the feelings of this girl, who had been "adopted," probably to do hard farm-work. If she, too, had nascently the fire-inducing power, which manifested under the influence of desire, or emotion, I think of her, in the midst of drudgery, wishing destruction upon the property of her exploiters, and fires following. At any rate, the story of the little Barnes girl, which quite equals anything from the annals of demonology, is very suggestive--or the smolder of hate, in the mind of a child, for an exploiter--and flames leaped upon a woman.

There is a particular in the case of Emma Piggott that makes it different from the other cases. In the other cases, fires broke out in the presence of girls. But, according to evidence, Emma Piggott was not in the house wherein started the fires for which she was accused. Then this seems to be a case of distance- ignition, or of distance-witchcraft. I'd not say that invisibly starting a fire, at a distance, by means of mental rays, is any more mysterious than is the shooting-off of distant explosives by means of rays called physical, which nobody understands.

I am bringing out:

That, as a "natural force," there is a fire-inducing power;

That, mostly, it appears, independently of wishes, or of the knowledge, of the subjects, but that sometimes, conformably to wishes, it is used--

That everything that I call witchcraft is only some special manifestation of transformations, or transportations, that, in various manifestations, are general throughout "Nature."

The "accidents" on the Dartmoor road--or that somewhere near this road lived a cripple. That his mind had shaped to his body--or that somewhere near this road lived somebody who had been injured by a motor car, and lay on his bed, or sat in his invalid's chair, and radiated against the nearby road a hate for all motorists, sometimes with a ferocity, or with a directness, that knocked cars to destruction.

Or Brooklyn, April to, 1893--see back to the supposed series of coincidences--man after man injured by falling from a high place, or being struck by a falling object--or that somewhere in Brooklyn was somebody who had been crippled by a fall, and, brooding over what he considered a monstrous injustice that had so singled him out, radiated influences that similarly injured others.

See back to the account of what occurred to French aeroplanes, flying over German territory. Tracks in the sands of a desert. Occurrences, about Christmas Day, 1930, in Sing and Dannemora Prisons--or a prisoner in a punishment-cell--and nothing to do in the dark, except to concentrate upon vengefulness. I think that sometimes, coming from dungeons, there are stinks of hates that can be smelled. It was a time that for almost everybody else was a holiday.

Tracks that stopped, in a desert--or the tracks of a child that stopped, on a farm, in Brittany--the story of Pauline Picard:

Or the hate of a neighbor for the Picards, and vengeance by teleporting their offspring--the finding of Pauline in Cherbourg--again her disappearance--

That this time the body of the child was mutilated and stripped, so that it could not be identified, and was transported to some lonely place, where it decomposed--

But a change of purpose, or a vengefulness that required that the parents should know--transportation to the field, of this body, which probably could not be identified--transportation of the "neatly folded" clothes, so that it could be identified.

In the matter of the two bodies on benches in a Harlem park, I have another datum. I think I have. The dates of June 14 and June 16 are close together, and Mt. Morris Park and Morningside Park are not far apart--

Or a man who lived in Harlem, in June, 1931--and that he was a park bencher--about whom I can say nothing except that his trousers were blue, and that his hat was gray. Something may have sapped him, pursued him, driven him into vagrancy--

But that he probably had the sense of localization, as to benches, that everybody has in so many ways, such as going to the same seat, or as near as possible to the same seat, upon every visit to a moving picture theater--that every morning he had sat on a particular bench, in Mt. Morris Park--

But that, upon the morning of June 14th, because of a whim, suspicion, or intuitive fear, he went to Morningside Park instead--

That somebody else sat on his particular bench--that there something that was an intensification of the experiences of John Harding and another man, when crossing Fifth Avenue, at Thirty-third Street--to the man who was sitting on this particular bench, and to another man upon a nearby bench--

But that, two days later, the trail of the intended victim was picked up--

Home News (Bronx) June 17, 1931--that, in Morningside Park, morning of the 16th, a policeman noticed a man--blue trousers and gray hat--seemingly asleep on a bench. The man was dead. "Heart failure."

At a time of intensely bitter revolts by coal miners against their hardships, there were many coal explosions, but in grates and stoves, and not in shipments. No finding of dynamite in coal was reported. If in coal there is storage of radiations from the sun, coal may be absorbent to other kinds of radiations-- or a savagely vengeful miner's hope for future harm in every lump he handled. If, in the house in Hornsey, there were not only coal-explosions, but also poltergeist doings, we note that these phenomena occurred only in the presence of the two boys of the household--or especially one of these boys. Between the occultism of adolescence and the occultism of lumps of coal, surcharged with hatreds, there may have been rapport.

That, somewhere near the town of Saltdean, Sussex, September, 1924, somebody hated a shepherd, and stopped the life of him, as have been stopped the motions of motors--and that the place remained surcharged with malign vibrations that affected somebody else, who came along, in a sidecar. The wedding party at Bradford--and the gaiety of weddings is sometimes the bubbling of vitriol--or that, from a witch, or a wizard, so made by jealousy, mental fumes played upon this house, and spread to other houses. At the same time, there are data that make me think that volumes of deadly gases may be occultly transported. And a young couple, walking along a shore of the Isle of

Man--that, from a state of jealousy, witchery flung them into the harbor, and that somebody who stepped into the area of this influence was knocked after them. See back to the story of a room in a house in Newton, Massachusetts. See other cases of "mass psychology." See a general clearing up--

If I can bridge the gap between the subjective and the objective, between what is called the real and what is called the unreal, or between the imaginary and the physical.

When, in our philosophy of the hyphen, we think of neither the material nor the immaterial, but of the material-immaterial, accentuated one way or the other in all phenomena; when we think of the imaginary, as deriving from material sustenance, or, instead of transforming absolutely, only shifting accentuation, we accept that there is continuity between what is called the real and what is called the unreal, so that a passage from one state to the other is across no real gap, or is no absolute jump. If there is no realness that can be finally set apart from unrealness--in phenomenal being--my term of the "realization of the imaginary," though a convenience is a misnomer. Maybe the word trans-mediumization, meaning the passage of phenomena from one medium of existence to another, is not altogether too awkward, and is long and important-enough-looking to give me the appearance of really saying something. I mean the imposition of the imaginary upon the physical. I mean, not the action of mind upon matter, but the action of mind-matter upon matter-mind.

Theoretically there is no gap. But very much mine are inductive methods. We shall have data. Not that I can more than really-unreally mean anything by that. The interpretations will be mine, but the data will be for anybody to form his own opinions upon.

Granting that the gap has not been disposed of, inductively, I reduce it to two questions:

Can one's mind, as I shall call it, affect one's own body, as I shall call it?

If so, that is personal witchcraft, or internal witchcraft.

Can one's mind affect the bodies of other persons and other things outside?

If so, that is what I shall call external witchcraft.

26

HATES and malices--murderous radiations from human minds--

Or the flashes and roars of a thunderstorm--

And there has been the equivalence of picking strokes of lightning out of the sky, and harnessing them to a job.

A house afire--or somebody boils an egg.

Devastation or convenience--

Or what of it, if I bridge a gap?

I take it that the story of Marjory Quirk is only an extreme instance of cases of internal, or personal, witchcraft that, today, are commonly accepted. London Daily Express, Oct. 3, 1911--inquest upon the body of Marjory Quirk, daughter of the Bishop of Sheffield. The girl had been ill of melancholia. In a suicidal impulse she drank, from a cup, what she believed to be paraffin. She was violently sick. She died. "There had been no paraffin in the cup. There was no trace of it in her mouth or throat."

New York Herald Tribune, Jan. 30, 1932--Boston, Jan. 29--"Nearly half a hundred students and physicians living in Vanderbilt Hall of the Harvard Medical School have experienced mild cases of what apparently was paratyphoid, it was learned today. The first thirty of the group fell ill two weeks ago, following a fraternity dinner, at which Dr. George H. Bigelow, state health commissioner, discussed 'food poisoning.' A few days later twenty more men reported themselves ill. The food was prepared at the hall.

"Today state health officers started an examination of kitchen help in the belief that one of the employees may be a typhoid carrier. College authorities said they did not believe the food itself was at fault, but were inclined to think the subject of Dr. Bigelow's address may have influenced some of the diners to diagnose mere gastronomic disturbances more seriously. All of the students have recovered."

To say that fifty young men had gastronomic disturbances is to say much against conditions of health in the Harvard Medical School. To say that the subject of illness may have induced illness is to say that there was personal, or internal, witchcraft, usually called auto-suggestion. See back to "Typhoid Mary" and other probable victims of carrier-finders. To say that there may have been a carrier among the kitchen help is to attribute to him, and is to say that it was only by coincidence that illnesses occurred after a talk upon illnesses. It's a hell of a way, anyway, to have dinner with a lot of young men, and talk to them about food-poisoning. Hereafter Dr. Bigelow may have to buy his own dinners. If he tells shark- stories, while bathing, he'll do lonesome swimming.

Physiologists deny that fright can turn one's hair white. They argue that they cannot conceive how a fright could withdraw the pigmentation from hairs: so they conclude that all alleged records of this phenomenon are yarns. Say it's a black-haired person. The physiologists, except very sketchily, cannot tell us how that hair became black, in the first place. Somewhere, all the opposition to the data of this book is because the data are not in agreement with something that is not known.

There have been many alleged instances. See the indexes of Notes and Queries, series 6, 7, 10. I used to argue that Queen Marie Antoinette's deprivation of cosmetics, in prison, probably accounted for her case. Now that my notions have shifted, that cynicism has lost its force to me. Mostly the instances of hair turning white, because of fright, are antiques, and can't be investigated now. But see the New York Times, Feb. 8, 1932:

Story of the sinking of a fishing schooner, by the Belgian steamship, Jean Jadot--twenty-one members of the crew drowned--six of them saved, among them Arthur Burke, aged 52.

"Arthur Burke's hair was streaked with gray before the collision, but was quite gray when Burke landed yesterday, at Pier 2, Erie Basin."

It may be that there have been thousands, or hundreds of thousands, of cases in which human beings have died in violent convulsions that were the products of beliefs--and that, also, merciful, but expensive, science has saved a multitude of lives, with a serum that has induced contrary beliefs--just as that serum, if injected into the veins of somebody, suffering under the oppressive pronouncement that twice two are four, could be his salvation by inducing a belief that twice two are purple, if he should want so to be affected--

Or what has become of hydrophobia?

In the New York Telegram, Nov. 26, 1929, was published a letter from Gustave Stryker, quoting Dr. Mathew Woods, of Philadelphia, a member of the Philadelphia County Medical Society. Dr. Woods had better look out, unless he's aiming at cutting down expenses, such as dues to societies. Said Dr. Woods:

"We have observed with regret numerous sensational stories, concerning alleged mad dogs and the terrible results to human beings bitten by them, which are published from time to time in the newspapers.

"Such accounts frighten people into various disorders, and cause brutal treatment of animals suspected of madness, and yet there is on record a great mass of testimony from physicians asserting the great rarity of hydrophobia even in the dog, while many medical men of wide experience are of the opinion that if it develops in human beings at all it is only upon extremely rare occasions, and that the condition of hysterical excitement in man, described by newspapers as 'hydrophobia,' is merely a series of symptoms due usually to the dread of the disease, such a dread being caused by realistic newspapers and other reports, acting upon the imaginations of persons scratched or bitten by animals suspected of rabies.

"At the Philadelphia dog pound where on an average more than 6,000 vagrant dogs are taken annually, and where the catchers and keepers are frequently bitten while handling them, not one case of hydrophobia has occurred during its entire history of twenty-five years, in which time, about 150,000 have been handled."

My own attention was first attracted, long ago, when I noticed, going over files of newspapers, the frequency of reported cases of hydrophobia, a generation or so ago, and the fewness of such reports in the newspapers of later times. Dogs are muzzled, now--in streets, in houses they're not. Vaccines, or powdered toads, caught at midnight, in graveyards, would probably cure many cases, but would not reduce the number of cases in dogs, if there ever have been cases of hydrophobia in dogs.

In the New York Times, July 4,:931, was published a report by M. Roeland, of the Municipal Council of Paris:

"It will be noticed that rabies has almost entirely disappeared, although the number of dogs has increased. From 166,917 dogs in Paris, in 1924, the number had risen, in 1929, to 230,674. In spite of this marked increase, only ten cases of rabies in animals were observed. There were no cases of rabies in man."

Sometimes it is my notion that there never has been a case of hydrophobia, as anything but an instance of personal witchcraft: but there are so many data for thinking that a disease in general is very 'much like an individual case of the disease, in that it runs its course and then disappears--quite independently of treatment, whether by the poisoned teat of a cow, or the dried sore of a mummy--that I suspect that once upon a time there was, to some degree, hydrophobia. When I was a boy, pitted faces were common. What has become of smallpox? Where are yellow fever and cholera? I'm not supposed to answer my own questions, am I? But serums, say the doctors. But there are enormous areas in the Americas and Europe, where vaccines have never penetrated. But they did it, say the doctors.

Eclipses occur, and savages are frightened. The medicine men wave wands--the sun is cured--they did it.

The story of diseases reads like human history--the rise and fall of Black Death--and the appearance and rule of Smallpox--the Tubercular Empire--and the United Afflictions of Yellow Fever and Cholera. Some of them passed away before serums were thought of, and in times when sanitation was unpopular.

Several hundred years ago there was a lepers' house in every good-sized city in England. A hundred years ago there had not been much of what is called improvement in medicine and sanitation, but leprosy had virtually disappeared, in England. Possibly the origin of leprosy in England was in personal witchcraft--or that if the Bible had never devastated England, nobody there would have had the idea of leprosy--that when wicked doubts arose, the nasty suspicions of people made them clean. So it may be that once upon a time there was hydrophobia: but the indications are that most of the cases that are reported in these times are sorceries wrought by the minds of victims upon their bodies.

A case, the details of which suggest that occasionally a dog may be rabid, but that his bites are dangerous only to a most imaginatively excited victim, is told of, in the New York Herald Tribune, Nov. 16, 1931. Ten men were bitten by a dog. "The dog was killed, and was found to have the rabies." The men were sailors aboard the United States destroyer, J. D. Edwards, at Cheefoo, China. One of these sailors died of hydrophobia. The nine others showed no sign of the disease.

In such a matter as a fright turning hair gray, it is probable that conventional scientists mechanically, unintelligently, or with little consciousness of the whyness of their opposition, deny the occurrences, as unquestioning obediences to Taboo. My own concatenation of thoughts is--that, if one's mental state can affect the color of one's hair, a mental state may in other ways affect one's body--and then that one's mental state may affect the bodies of others--and this is the path to witchcraft. It is not so much that conventional scientists disregard, or deny, what they cannot explain--if, in anything like a final sense, nothing ever has been, or can be, explained. It is that they disregard or deny, to clip concatenations that would lead them from concealed ignorance into obvious bewilderment. Every science is a mutilated octopus. If its tentacles were not clipped to stumps, it would feel its way into disturbing contacts. To a believer, the effect of the contemplation of a science is of being in the presence of the good, the true, and the beautiful. But what he is awed by is Mutilation. To our crippled intellects, only the maimed is what we call understandable, because the unclipped ramifies away into all other things. According to my aesthetics, what is meant by the beautiful is symmetrical deformation. By Justice--in phenomenal being--I mean the appearance of balance, by which a reaction is made to look equal and opposite to an action--so arbitrarily wrought by the clip and disregard of all ramifications of the action--expressing in the supposed condign punishment of a man, regardless of effects upon other persons. This is the arbitrary basis of the mechanical theory of existence--the idea that an action can be picked out of a maze of interrelationships, as if it were a thing in itself. Some wisdom of mine is that if a man is dying of starvation he cannot commit a crime. He is good. The god of all idealists is Malnutrition. If all crimes are expressions of energy, it is unjust to pick on men for their crimes. A higher jurisprudence would indict their breakfasts. A good cook is responsible for more evil than ever the Demon Rum has been: and, if we'd all sit down and starve to death, at last would be realized Utopia.

My expression is that, if illnesses, physical contortions, and deaths can be imposed by the imaginations of persons upon their own bodies, we may develop the subject-matter of a preceding chapter, with more striking data--

Or the phenomenon of the stigmata--

Which, considered sacred by pietists, is aligned by me with hydrophobia.

This phenomenon is as profoundly damned, in the views of all properly trained thinkers, as are crucifixes, sacraments, and priestly vestments. As to its occurrence, I can quote dozens of churchmen, of the "highest authority," but not one scientist, except a few Catholic scientists.

Over and over and over--science and its system--and theology and its system--and the fights between interpretations by both--and my thought that the freeing of data from the coercions of both, may, or may not, be of value. Once upon a time the religionists denied, or disregarded much that the scientists announced. They have given in so disastrously, or have been licked so to a frazzle, that, in my general impression of controversies that end up in compromises, this is defeat too nearly complete to be lasting. I conceive of a return-movement--open to freethinkers and atheists--in which many of the data of religionists--scrubbed clean of holiness--will be accepted.

As to the records of stigmatics, I omit the best-known, and most convincingly reported, of all the cases, the case of the French girl, Louise Lateau, because much has been published upon her phenomena, and because accounts are easily available.

In the newspapers of July, 1922--I take from the London Daily Express, July 10th--was reported the case of Mary Reilly, aged 20, in the Home of the Sisters of the Good Shepherd, Peekskill, N. Y. It was said that intermittently, upon her side, appeared a manifestation in the form of a cross of blood. Mostly the appearances are of the "five wounds of Christ," or six, including marks on the forehead. For an account of the case of Rose Ferron, see the New York Herald Tribune, March 25, 1928. According to this story, Rose Ferron, aged 25, of 86 Asylum Street, Woonsocket, R. I., had, since March 17, 1916, been a stigmatic, wounds appearing upon her hands, feet, and forehead. The hysterical condition of this girl--in both the common and the medical meaning of the term--is indicated by the circumstance that for three years she had been strapped to her bed, with only her right arm free.

At this time of writing, I have, for four years, been keeping track of the case of Theresa Neumann, the stigmatic girl of Konnersreuth, Germany: and, up to this time, there has been no exposure of imposture. See the New York Times, April 8, 1928--roads leading to her home jammed with automobiles, carriages, motorcycles, vans, and pilgrims on foot. Considering the facilities--or the facilities, if nothing goes wrong--of modern travel, it is probable that no other miracle has been so multitudinously witnessed. A girl in bed--and all day long, the tramp of thousands past her. Whether admission was charged, I do not know. The story of this girl agrees with the stories of other stigmatics: flows of blood, from quick-healing wounds, and phenomena on Fridays. It was said that medical men had become interested, and had "demanded" Theresa's removal to a clinic, where she could be subjected to a prolonged examination, but that the Church authorities had objected. This is about what would be expected of Church authorities: and that the medical men, unable to have their own way, then disregarded the case is something else that is about what would be expected.

My expression is that, upon stigmatic girls have appeared wounds, similar to the alleged wounds of a historical, and therefore doubtful, character, because this melodrama is most strikingly stimulative to the imagination--but that an atheistic girl--if there could be anything for an atheistic girl to be equally imaginatively hysterical about-- might reproduce other representations upon her body. In the Month, 134-249, is an account of Marie- Julie Jahenny, of the village of La Fraudais (Loire-Inferieure), France, who, upon March 21, 1873, became a stigmatic. Upon her body appeared the "five wounds." Then, upon her breast appeared the picture of a flower. It is said that for twenty years this picture of a flower remained visible. According to the story, it was in the mind of the girl before it appeared upon her body, because she predicted that it would appear. One has notions of the possible use of indelible ink, or of tattooing. That is very good. One should have notions.

If a girl drinks a liquid that would harm nobody else, and dies, can a man inflict upon himself injuries that would kill anybody else, and be unharmed?

There is a kind of stigmatism that differs from the foregoing cases, in that weapons are used to bring on effects: but the wounds are similar to the wounds of stigmatic girls, or simply are not wounds, in an ordinary, physical sense. There is an account, in the Sphinx, March, 1893, of a fakir, Soliman Ben Aissa, who was exhibiting in Germany; who stabbed daggers into his cheeks and tongue, and into his abdomen, harmlessly, and with quick-healing wounds.

Such magicians are of rare occurrence, anyway in the United States and Europe: but the minor ones who eat glass and swallow nails are not uncommon.

But, if in Germany, or anywhere else, in countries that are said to be Christian, any man ever did savagely stab himself in the abdomen, and be unhurt, and repeat his performances, how is it that the phenomenon is not well-known and generally accepted?

The question is like another:

If, in the Theological Era, a man went around blaspheming, during thunderstorms, and was unhurt, though churches were struck by lightning, how long would he remain well-known?

In March, 1920, a band of Arab dervishes exhibited in the London' music halls. In the London Daily News, March 12, 1920, are reproduced photographs of these magicians, showing them with skewers that they had thrust through their flesh, painlessly and bloodlessly.

Taboo. The censor stopped the show.

For an account of phenomena, or alleged phenomena, of the Silesian cobbler, Paul Diebel, who exhibited in Berlin, in December, 1927, see the New York Times, Dec. 18, 1927. "Blood flows from his eyes, and open wounds appear on his chest, after he has concentrated mentally for six minutes, it is declared. He drives daggers through his arms and legs, and even permits himself to be nailed to a cross, without any suffering, it is said. His manager asserts that he can remain thus for ten hours. His self- inflicted wounds, it is declared, bleed or not, as he wishes, and a few minutes after the knife or nails are withdrawn all evidence of incisions vanishes."

The only thing that can be said against this story is that it is unbelievable.

New York Herald Tribune, Feb. 6, 1928--that, in Vienna, the police had interfered with Diebel, and had forbidden him to perform. It was explained that this was because he would not give them a free exhibition, to prove the genuineness of his exhibitions. "In Munich, recently, he remained nailed to a cross several hours, smoking cigarettes and joking with his audience."

After April 8, 1928--see the New York Times of this date--I lose track of Paul Diebel. The story ends with an explanation. Nothing is said of the alleged crucifixions. The explanation is a retreat to statements that are supposed to be understandable in commonplace terms. I do not think that they are so understandable. "Diebel has disclosed his secret to the public, saying that shortly before his appearance, he scratched his flesh with his fingernails, or a sharp instrument, being careful not to cut it. On the stage by contracting his muscles, those formerly invisible lines assumed blood-red hue and often bled."

I have heard of other persons, who have "disclosed" trade secrets.

Upon March 2, 1931, a man lay, most publicly, upon a bed of nails. See the New York Herald Tribune, March 3, 1931. In Union Square, New York City, an unoriental magician, named Brawman, from the unmystical region of Pelham Bay, in the Bronx, gave an exhibition that was staged by the magazine, Science and Invention. This fakir from the Bronx lay upon a bed of 1,200 nails. In response to his invitation, ten men walked on his body, pressing the points of the nails into his back. He stood up, showing deep, red marks made by the nails. These marks soon faded away.

I have thought of leaf insects as pictorial representations wrought in the bodies of insects, by their imaginations, or by the imaginative qualities of the substances of their bodies--back in plastic times, when insects were probably not so set in their ways as they now are. The conventional explanation of protective colorations and formations has, as to some of these insects, considerable reasonableness. But there is one of these creatures--the Tasmanian leaf insect--that represents an artistry that so transcends utility that I considered the specimen that I saw, in the American Museum of Natural History, misplaced: it should have been in the Metropolitan Museum of Art. This leaf insect has reproduced the appearance of a leaf down to such tiny details as serrated edges. The deception of enemies, or survival-value, has had nothing thinkable to do with some of the making of this remarkable likeness, because such minute particulars as serrations would be invisible to any bird, unless so close that the undisguisable insect- characteristics would be apparent.

I now have the case of what I consider a stigmatic bird. It is most unprotectively marked. Upon its breast it bears betrayal--or it is so conspicuously marked that one doubts that there is much for the theory of protective coloration to base upon, if conspicuously marked forms of life survive everywhere, and if many of them cannot be explained away, as Darwin explained away some of them, in terms of warnings.

It is a story of the sensitiveness of pigeons. I have told of the pigeons with whom I was acquainted. One day a boy shot one, and the body lay where the others saw it. They were so nervous that they flew, hearing trifling sounds that, before, they'd not have noticed. They were so suspicious that they kept away from the window sills. For a month they remembered.

The bleeding-heart pigeon of the Philippines--the spot of red on its breast--or that its breast remembered--

Or once upon a time--back in plastic times when the forms and plumages of birds were not so fixed, or established, as they now are--an ancestral pigeon and her mate. The swoop of a hawk--a wound on his breast--and that sentiment in her plumage was so sympathetically moved that it stigmatized her, or reproduced on offsprings, and is to this day the recorded impression of an ancient little tragedy.

A simple red spot on the breast of a bird would not be conceived of, by me, as having any such significance. It is not a simple, red spot, only vaguely suggestive of a wound, on the breast of the bleeding-heart pigeon of the Philippines. The bordering red feathers are stiff, as if clotted. They have the appearance of coagulation.

Conceiving of the transmission of a pictorial representation, by heredity, is conceiving of external stigmatism, but of internal origin. If I could think that a human being's intense mental state, at the sight of a wound, had marked a pigeon, that would be more of a span over our gap. But I have noted an observation for thinking that the sight of a dead and mutilated pigeon may intensely affect the imaginations of other pigeons. If anybody thinks that birds have not imagination, let him tell me with what a parrot of mine foresees what I am going to do to him, when I catch him up to some of his mischief, such as gouging furniture. The body of a dead and mutilated companion prints on the minds of other pigeons: but I have not a datum for thinking that the skeleton, or any part of the skeleton, of a pigeon, would be of any meaning to other pigeons. I have never heard of anything that indicates that in the mind of any other living thing is the mystic awe that human beings, or most human beings, have for bones--

Or a moth sat on a skull--

And that so it rested, with no more concern than it would feel upon a stone. That a human being came suddenly upon the skull, and that, from him, a gush of mystic fright marked the moth--The Death's Head Moth.

On the back of the thorax of this insect is a representation of a human skull that is as faithful a likeness as ever any pirate drew. In Borneo and many other places, there is not much abhorrence for a human skull: but the death's head moth is a native of England.

Or the death's heads that appeared upon windowpanes, at Boulley--except that perhaps there were no such occurrences at Boulley. Suppose most of what I call data may be yarns. But the numbers of them-- except, what does that mean? Oh, nothing, except that some of our opponents, if out in a storm long enough, might have it dawn on them that it was raining.

If I could say of any pictorial representation that has appeared on the wall of a church that it was probably not an interpretation of chance arrangements of lights and shades, but was a transference from somebody's mind, then from a case like this, of the pretty, the artistic, or of what would be thought of by some persons as the spiritual, and a subject to be treated reverently, would flow into probability a flood of everything that is bizarre, malicious, depraved, and terrifying in witchcraft-- and of course jostles of suggestions of uses.

In this subject I have had much experience. Long ago, I experimented. I covered sheets of paper with scrawls, to see what I could visualize out of them; tacked a sheet of wrapping paper to a ceiling, and smudged it with a candle flame; made what I called a "visualizing curtain," which was a white window shade, covered with scrawls and smudges; went on into three dimensions, with boards veneered with clay. It was long ago--about 1907. I visualized much, but the thought never occurred to me that I marked anything. It was my theory that, with a visualizing device, I could make imaginary characters perform for me more vividly than in my mind, and that I could write a novel about their doings. Out of this idea I developed nothing, anyway at the time. I have had much experience with visualizations that were, according to my beliefs, at the time, only my own imaginings, and I have had not one experience--so recognized by me--of ever having imaginatively marked anything. Not that I mean anything by anything.

There is one of these appearances that many readers of this book may investigate. Upon Feb. 23, 1932, New York newspapers reported a clearly discernible figure of Christ in the variegations of the sepia-toned marble of the sanctuary wall of St. Bartholomew's Church, Park Avenue and Fiftieth Street, New York City.

In the New York Times, Feb. 24, 1932, the rector of the church, the Rev. Dr. Robert Norwood, is quoted:

"One day, at the conclusion of my talk, I happened to glance at the sanctuary wall and was amazed to see this lovely figure of Christ in the marble. I had never noticed it before. As it seemed to me to be an actual expression on the face of the marble of what I was preaching, 'His Glorious Body,' I consider it a curious and beautiful happening. I have a weird theory that the force of thought, a dominant thought, may be strong enough to be somehow transferred to stone in its receptive state."

In 1920, a censor stopped a show: but, in 1930, the Ladies' Home Journal published William Seabrook's story--clipping sent to me by Mr. Charles McDaniel, East Liberty P.O., Pittsburgh, Pa. There was a performance in the village of Doa, in the Ben-Hounien territory of the French West African Colony.

It is a story of sorceries practiced by magicians, not upon their own bodies, but upon the bodies of others.

"There were the two living children close to me. I touched them with my hands. And there equally close were the two men with their swords. The swords were iron, three-dimensional, metal, cold and hard.

And this is what I now saw with my eyes, but you will understand why I am reluctant to tell of it, and that I do not know what seeing means:

"Each man, holding his sword stiffly upward with his left hand, tossed a child high in the air with his right, then caught it full upon the point, impaling it like a butterfly on a pin. No blood flowed, but the two children were there, held aloft, pierced through and through, impaled upon the swords.

"The crowd screamed now, falling to its knees. Many veiled their eyes with their hands, and others fell prostrate. Through the crowd the jugglers marched, each bearing a child aloft, impaled upon his sword, and disappeared into the witch doctor's inclosure."

Later Seabrook saw the children, and touched them, and had the impression that he would have, looking at a dynamo, or at a storm at sea, at something falling from a table, or at a baby crawling-- that he was in the presence of the unknown.

27

THE twitch of the legs of a frog--and Emma Piggott swiped a powder puff.

The mysterious twitchings of electrified legs--and unutterable flutterings in the mind of Galvani. His travail of mental miscarriages--or ideas that could not be born properly. The twitch of trivialities that were faint and fantastic germinations in the mind of Galvani--the uninterpretable meanings of far- distant hums of motors--these pre-natal stirrings of aeroplanes and transportation systems and the lighting operations of cities--

Twitch of the legs of a frog--

A woman, from Brewster, N. Y., annoyed a hotel clerk.

My general expression is that all human beings who can do anything, and dogs that track unseen quarry, and homing pigeons, and bird-charming snakes, and caterpillars who transform into butter-flies, are magicians. In the lower--or quite as truly the higher, considering them the more aristocratic and established--forms of being, the miracles are standardized and limited: but human affairs are still developing, and "sports," as the biologists call them, are of far more frequent occurrence among humans. But their development depends very much upon a sense of sureness of reward for the pains, travail, and discouragements of the long, little-paid period of apprenticeship, which makes questionable whether it is ever worthwhile to learn anything. Reward depends upon harmonization with the dominant spirit of an era.

Considering modern data, it is likely that many of the fakirs of the past, who are now known as saints, did, or to some degree did, perform the miracles that have been attributed to them. Miracles, or stunts, that were in accord with the dominant power of the period were fostered, and miracles that conflicted with, or that did not contribute to, the glory of the Church, were discouraged, or were savagely suppressed. There could be no development of mechanical, chemical, or electric miracles--

And that, in the succeeding age of Materialism--or call it the Industrial Era--there is the same state of subservience to a dominant, so that young men are trained to the glory of the job, and dream and invent in fields that are likely to interest stockholders, and are schooled into thinking that all magics, except their own industrial magics, are fakes, superstitions, or newspaper yarns.

I am of the Industrial Era, myself; and, even though I can see only advantages-disadvantages in all uses, I am very largely only a practical thinker--

Or the trail of a working witchcraft--and we're on the scent of utilities--

Or that, if a girl, in the town of Derby, set a house afire, by a process that is now somewhat understandable, a fireman could, if he had a still better understanding, have put out that fire without moving from his office. If the mechanism of a motor can invisibly be stopped, all the motors of the world may, without the dirt, crime, misery, and exploitation of coal-mining, be started and operated. If Ambrose Small was wished so far away that he never got back--though that there is magic in a mere wish, or in a mere hope, or hate, I do not think--the present snails of the wheels and planes may be replaced by instantaneous teleportations. If we can think that quacks and cranks and scientists of highest repute, who have announced successes, which were in opposition to supposed medical, physical, chemical, or biological principles--which are now considered impostures, or errors, or "premature announcements"--may not in all cases have altogether deceived themselves, or tried to deceive others, we--or maybe only I--extend this suspicion into mechanical fields.

Now it is my expression that all perpetual motion cranks may not have been dupes, or rascals--that they may have been right, occasionally--that their wheels may sometimes have turned, their marbles rolled, their various gimcracks twirled, in an excess of reaction over action, either because sometimes will occur exceptions to any such supposed law as "the conservation of energy," or because motivating "rays" emanated from the inventors--

That sometimes engines have run, fueled with zeals--but have, by such incipient, or undeveloped, witchcraft, operated only transiently, or only momentarily--but that they may be forerunners to such a revolution of the affairs of this earth, as once upon a time were flutters of the little lids of teakettles--

A new era of new happiness and new hells to pay; ambitions somewhat realized, and hopes dashed to nothing; new crimes, pastimes, products, employments, unemployments; labor troubles, or strikes that would be world-wide; new delights, new diseases, disasters such as had never before been heard of--

In this existence of the desirable-undesirable.

Wild carrots in a field--and to me came a dissatisfaction with ham and cabbage. That was too bad: there isn't much that is better. My notion was that probably all around were roots and shoots and foliages that might be, but that never had been, developed eatably--but that most unlikely would be the cultivation of something new to go with ham, in the place of cabbage, because of the conventionalized requirements of markets. But once upon a time there were wild cabbages and wild beets and wild onions, and they were poor, little incipiencies until they were called for by markets. I think so. I don't know. At any rate, this applies to wild fruits.

There are sword swallowers and fire eaters, fire breathers, fire walkers; basket tricksters, table tilters, handcuff escapers. There is no knowing what development could do with these wild talents: but Help Wanted if for—Reasonable and confidential accntnts; comptometer oprs., fire re-ins., exp., Christian; sec'ys, credit exp.,advance, Chris.; P & S expr.; fast sandwich men; reception men, 35-45, good educ., ap. tall, Chris.--

But I do think that one hundred years ago an advertisement for a fast sandwich man would have looked as strange as today would look an advertisement for "poltergeist."

Against all the opposition in the world, I make this statement--that once I knew a magician. I was a witness of a performance that may someday be considered understandable, but that, in these primitive times, so transcends what is said to be the known that it is what I mean by magic.

When the magician and I were first acquainted, he gave no sign of occult abilities. He was one of the friendliest of fellows, but that was not likely to endear him to anybody, because he was about equally effusive to everybody. He had frenzies. Once he tore down the landlord's curtains. He bit holes into a book of mine, and chewed the landlord's slippers.

The landlord got rid of him. This was in London. The landlord took him about ten miles away, and left him, probably leaping upon somebody, writhing joy for anybody who would notice him. He was young.

It was about two weeks later. Looking out a front window, I saw the magician coming along, on the other side of the street. He was sniffing his way along, but went right on past our house, without recognizing it. He came to a point where he stopped and smelled. He smelled and he smelled. He crossed the street, and came back, and lay down in front of the house. The landlord took him in, and gave him a bone.

But I cannot accept that the magician smelled his way home, or picked up a trail, taking about two weeks on his way. The smelling played a part, and was useful in a final recognition: but smelling indiscriminately, he could have nosed his way, for years, through the streets of London, before coming to the right scent.

New York Sun, April 24, 1931--an account, by Adolph Pizaldt, of Allentown, Pa., of a large, mongrel magician, who had been taken in a baggage car, a distance of 340 miles, and had found his way back home, in a week or so. New York Herald Tribune, July 4, 1931--a curly magician, who, in Canada, had found his way back home, over a distance of 400 miles.

New York Herald Tribune, Aug. 13, 1931--The Man They Could Not Drown--

"Hartford, Conn., Aug. 12--Angelo Faticoni, known as 'The Human Cork,' because he could stay afloat in water for fifteen hours with twenty pounds of lead tied to his ankles, died on August 2 in

Jacksonville, Fla., it became known here today. He was seventy-two years old.

"Faticoni could sleep in water, roll up into a ball, lie on his side, or assume any position asked of him. Once he was sewn into a bag and then thrown headforemost into the water, with a twenty-pound cannonball lashed to his legs. His head reappeared on the surface soon afterward, and he remained motionless in that position for eight hours. Another time he swam across the Hudson tied to a chair weighted with lead.

"Some years ago he went to Harvard to perform for the students and faculty. He had been examined by medical authorities who failed to find support for their theory that he was able to float at such great lengths by the nature of his internal organs, which they believed were different from those of most men.

"Faticoni had often promised to reveal the secret of how he became 'The Human Cork,' but he never did."

There are many accounts of poltergeist-phenomena that are so obscured by the preconceptions of witnesses that one can't tell whether they are stories of girls who had occult powers, or of invisible beings, who, in the presence of girl-mediums, manifested. But the story of Angelique Cottin is an account of a girl, who, by an unknown influence of her own, acted upon objects in ways like those that have been attributed to spirits. The phenomena of Angelique Cottin, of the town of La Perriere, France, began upon Jan. 15, 1846, and lasted ten weeks. Anybody who would like to read an account of this wild, or undeveloped, talent, that is free from interpretations by spiritualists and anti-spiritualists, should go to the contemporaneous story, published in the Journal des Debats (Paris) February, 1846.

Here are accounts by M. Arago and other scientists. When Angelique Cottin went near objects, they bounded away. She could have made a perpetual motion machine whiz. She was known as the Electric Girl, so called, because nobody knew what to call her. When she tried to sit in a chair, there was low comedy. The chair was pulled away, or, rather, was invisibly pushed away. There was such force here that a strong man could not hold the chair. A table, weighing 60 pounds, rose from the floor, when she touched it. When she went to bed, the bed rocked.

And I suppose that, in early times of magnetic investigations, people who heard of objects that moved in the presence of a magnet, said--"But what of it?"

Faraday showed them.

A table, weighing 60 pounds, rises a few feet from the floor--well, then, it's some time, far ahead, in the Witchcraft Era--and a multi-cellular formation of poltergeist-girls is assembled in the presence of building materials. Stone blocks and steel girders rise a mile or so into their assigned positions in the latest sky-prodder. Maybe. Tall buildings will have their day, but first there will have to be a show-off of what could be done.

I now have a theory that the Pyramids were built by poltergeist-girls. The Chinese Wall is no longer mysterious. Every now and then I reconstruct a science. I may take up neo-archaeology sometime. Old archaeology, with its fakes and guesses, and conflicting pedantries, holds out an invitation for a ferocious and joyous holiday.

Human hopes, wishes, ambitions, prayers and hates--and the futility of them--the waste of millions of trickles of vibrations, today--unorganized forces that are doing nothing. But put them to work together, or concentrate mental ripples into torrents, and gather these torrents into Niagara Falls of emotions-- and, if there isn't any happiness, except in being of use, I am conceiving of cataractuous happiness--

Or sometime in the Witchcraft Era--and every morning, promptly at nine o'clock, crowds of human wishers, dignified under the name of transmediumizers, arrive at their wishing stations, or mental power-houses, and in an organization of what are now only scattered and wasted hopes and hates concentrate upon the running of all motors of all cities. Just as they're all nicely organized and pretty nearly satisfied, it will be learned that motors aren't necessary.

In one way, witchcraft has been put to work: that is that wild talents have been exhibited, and so have been sources of incomes. But here is only the incipiency of the stunt. In August, 1883, in the home of Lulu Hurst, aged 15, at Cedarville, Georgia, there were poltergeist disturbances. Pebbles moved in the presence of the girl, things vanished, crockery was smashed, and, if the girl thought of a tune, it would be heard, rapping at the head of her bed. In February, 1884, Lulu was giving public performances. In New York City, she appeared in Wallack's Theatre. It could be that a girl, aged 15, if competently managed, was able to deceive everybody who went up on the stage. She at least made all witnesses think that, when a man, weighing 200 pounds, sat in a chair, she, by touching the chair, made it rise and throw him to the floor--

And I am very much like an Indian, of long ago; an Indian, thinking of the force of a waterfall; unable to conceive of a waterwheel; simply thinking of all this force that was making only a little spectacle-Or the state of melancholy into which I am perhaps cast, thinking that a little poltergeist girl, if properly trained, could make all witnesses believe that she raised building materials forty or eighty stories, by simply touching them--thinking that nobody is doing anything about this--

Except that I am not clear that anything would be gained by it--or by anything else.

Lulu Hurst either had powers that far transcended muscular powers, or she had talents of deception far superior to the abilities of ordinary deceivers. Sometimes she tossed about zoo-pound men, or made it look as if she did; and sometimes she placed her hands on a chair, and five men either could not move that chair, or were good actors, and earned whatever the confederates of stage magicians were paid, at that time.

In November, 1891, Mrs. Annie Abbott, called the Little Georgia Magnet, put on a show, in the Alhambra Music Hall, London. She weighed about 98 pounds, and, if she so willed it, a man could easily lift her.

The next moment, six men, three on each side of her, grasping her by her elbows, could not lift her. When she stood on a chair, the six men could not, when the chair was removed, prevent her from descending to the floor. If anybody suggests that, when volunteers were called for from the audience, it was the same six who responded, at every performance, I think that that is a pretty good suggestion. Because of many other data, it hasn't much force with me; but, in these early times of us primitives, almost any suggestion has value. I take these accounts from Holms' Facts of Psychic Science. I have theme from other sources, also.

In September, 1921, Mary Richardson gave performances, at the Olympic Music Hall, Liverpool. Easily lifted one moment--the next moment, six men--same six, maybe--could not move her. By touching a man, she knocked him flat. It is either that she traveled with a staff of thirteen comedians, whose stunt it was to form in a line, pretending their utmost to push her, but seeming to fail comically, considering the size of her, or that she was a magician.

It is impossible to get anywhere by reasoning. This is because--as can be shown, monistically--there isn't anywhere. Or it is impossible to get anywhere, because one can get everywhere. I can find equally good reasons for laughing, or for being serious, about all this. Holms tells that he was one of those in the audience, who, though not taking part, went up on the stage; and that he put his hand between Mrs. Richardson and the leader of the string of thirteen men, who were almost dislocating one another's shoulder blades, pushing their hardest against her, and that he felt no pressure. So he was convinced not that she resisted pressure, but that pressure could not touch her.

Suppose it was that pressure could not touch her. Could blows harm her? Could bullets touch her? Did Robert Houdin have this power, when he faced an Arab firing squad, and is the story of the substituted blanks for bullets only just some more of what Taboo is telling everywhere? One untouchable man could own the world--except that he'd have a weakness somewhere, or, in general, could be no more than the untouchable-touchable. But he could add to our bewilderments by making much history before being touched. Well, then, if there are magicians, why haven't magicians seized upon political powers? I don't know that they haven't.

It may be the secret of fire-walking--or that wizards walk over red-hot stones, unharmed, because they do not touch the stones. However, for some readers, it is more comfortable to disbelieve that anybody ever has been a fire-walker. For an uncomfortable moment, read an account, in Current Literature, 32- 98--exhibition by a Hawaiian fire-walker, at Honolulu, Jan. 19, 1901. The story is that this wizard walked on stones of "a fierce, red glow," with flames spouting from burning wood, underneath; walking back and forth four times.

There is a muscular strength of men, and it may be that sometimes appears a strength to which would apply the description "occult," or "psychic." In the New York Herald Tribune, Jan. 24, 1932, was reported the death of Mrs. Betsy Anna Talks, of 149 Fourteenth Road, Whitestone, Queens, N. Y.--who had often performed such feats as carrying a barrel of sugar, weighing 400 pounds--had carried, under each arm, a sack of potatoes, whereas, in fields, usually two men lug one sack--had impatiently watched two men, clumsily moving a 550-pound barrel of salt, in a cart, and had taken it down for them.

There are "gospel truths," and "irrefutable principles," and "whatever goes up must come down," and "men are strong and women are weak"--but somewhere there's a woman who takes a barrel of salt away from two men. But we think in generalizations, and enact laws in generalizations, and "women are weak," and, if I should look it up, I'd be not at all surprised to learn that Mrs. Talks was receiving alimony.

I now recall another series of my own experiences with what may be my own very wild talents. I took no notes upon the occurrences, because I had decided that note-taking would make me self-conscious. I do not now take this view. I was walking along West Forty-second Street, N. Y. C., when the notion came to me that I could "see" what was in a show window, which, some distance ahead, was invisible to me. I said to myself: "Turkey tracks in red snow." It should be noted that "red snow" was one of the phenomena of my interests, at this time. I came to the window, and saw track-like lines of black fountain pens, grouped in fours, one behind, and the three others trifurcating from it, on a background of pink cardboard.

At last I was a wizard!

Another time, picking out a distant window, invisible to me--or ocularly invisible to me--I said "Ripple marks on a sandy beach." It was a show window. Several men were removing exhibits from it, and there was virtually nothing left except a yellow-plush floor covering. Decoratively, this covering had been ruffled, or given a wavy appearance.

Another time--"Robinson Crusoe and Friday's footprints." When I came to the place, I saw that it was a cobbler's shop, and that, hanging in the window, was a string of shoe soles.

I'm sorry.

I should like to hear of somebody, who would manfully declare himself a wizard, and say--"Take it or leave it!" I can't do this, because I too well remember other circumstances. Maybe it's my timidity, but I now save myself from the resentment, or the mean envy, of readers, who say, of a distant store window, "popular novels," and it's pumpkins. My experiments kept up about a month. Say that I experimented about a thousand times. Out of a thousand attempts, I can record only three seemingly striking successes, though I recall some minor ones. Throughout this book, I have taken the stand that nobody can be always wrong, but it does seem to me that I approximated so highly that I am nothing short of a negative genius. Nevertheless, the first of these experiences impresses me. It came to me when, so far as I know, I was not thinking of anything of the kind, though sub-consciously I was carrying much lore upon various psychic subjects.

These things may be done, but everybody who is interested has noticed the triviality and the casualness of them. They--such as telepathic experiences--come and go, and then when one tries to develop an ability, the successes aren't enough to encourage anybody, except somebody who is determined to be encouraged.

Well, then, if wild talents come and go, and can't be developed, or can't be depended upon, even people who are disposed to accept that they exist, can't see the good of them.

But accept that there are adepts: probably they had to go through long periods of apprenticeship, in which, though they deceived themselves by hugely over-emphasizing successes, and forgetting failures, they could not impress any parlor, or speakeasy, audience. I have told of my experiments of about a month. It takes five years to learn the rudiments of writing a book, selling gents' hosiery, or panhandling.

Everybody who can do anything got from the gods, or whatever, nothing but a wild thing. Read a book, or look at a picture. The composer has taken a wild talent that nobody else in the world believed in; a thing that came and went and flouted and deceived him; maybe starved him; almost ruined him--and has put that damn thing to work.

Upon Nov. 29, 1931, died a wild talent. It was wild of origin, but was of considerable development. See the New York Herald Tribune, Nov. 30, 1931. John D. Reese had died in his home, in Youngstown, Ohio. Mr. Reese was a "healer." He was not a "divine healer." He means much to my expression that the religionists have been permitted to take unto themselves much that is not theirs exclusively. Once we heard only of "divine healers." Now there are "healers." It is something of a start of a divorcement that may develop enormously. Sometime I am going to loot the records of saints, for suggestions that may be of value to bright atheists, willing to study and experiment. "Reese had never studied medicine. The only instruction he had ever received was from an aged healer, in the mountains of Wales, when he was a boy. Physicians could not explain his art, and, after satisfying themselves that he was not a charlatan, would shrug, and say simply that he had 'divine power'." But Reese never described himself as a "divine healer," and, though by methods no less divine than those of the Salvation Army and other religious organizations, he made a fortune out of his practices, he was associated with no church. He was about thirty years old when he became aware of his talent. One day, in the year 1887, a man in a rolling mill fell from a ladder, and was injured. It was "a severe spinal sprain," according to a physician. "Mr. Reese stooped and ran his fingers up and down the man's back. The man smiled, and while the physician and the mill hands gaped in wonder, he rose to his feet, and announced that he felt strong again, with not a trace of pain. He went back to work, and Mr. Reese's reputation as a healer was spread abroad."

Then there were thousands of cases of successful treatments. Hans Wagner, shortstop of the Pittsburgh Pirates, was carried from the baseball field, one day: something in his back had snapped, and it seemed that his career had ended. He was treated by Reese, and within a few days was back short-stopping.

When Lloyd George visited the United States, after the War, he shook hands so many times that his hand was twisted out of shape. Winston Churchill, in a later visit, had what was said to be an automobile accident, and said that he was compelled to hold his arm in a sling. But Lloyd George was so cordially greeted that he was maimed. "Doctors said that only months of rest and massage could restore the cramped muscles." "Reese shook hands with the statesman, pressed gently, and then harder, disengaged their hands with a wrench, and Lloyd George's hand was strong again."

One of the most important particulars in this story of a talent, or of witchcraft, that was put to work, is that probably it was a case of a magician who was taught. Reese, when a boy, received instructions in therapeutic magic, and then, in the stresses of making a living, forgot, so far as went the knowledge of his active consciousness. But it seems that sub-consciously a development was going on, and suddenly, when the man was thirty-two years of age, manifested.

My notion is that wild talents exist in the profusion of the weeds of the fields. Also my notion is that, were it not for the conventions of markets, many weeds could be developed into valuable, edible vegetables. The one great ambition of my life, for which I would abandon my typewriter at any time-- well, not if I were joyously setting down some particularly nasty little swipe at priests or scientists--is to say to chairs and tables, "Fall in! forward! march!" and have them obey me. I have tried this, as I don't mind recording, because one can't be of an enquiring and experimental nature, and also be very sensible. But a more unmilitary lot of furniture than mine, nobody has. Most likely, for these attempts, I'll be hounded by pacifists. I should very much like to be a wizard, and be of great negative benefit to my fellow beings, by doing nothing for anybody. And I have had many experiences that lead me to think that almost everybody else not only would like to be a wizard, but at times thinks he is one. I think that he is right. It is monism that if anybody's a wizard, everybody is, to some degree, a wizard.

One time--spring of 1931--my landlord received some chicks from the country, and put them in an enclosure at the end of the yard. They grew, and later I thought it interesting, listening to the first, uncertain attempts of two of them to crow. It was as interesting as is watching young, human males trying to take on grown-up ways. But then I thought of what was ahead, at four o'clock, or there-abouts, mornings. I'm a crank about sleeping, because at times I have put in much disagreeable time with insomnia. I worried about this, and I spoke about it.

There was not another sound from the two, little roosters. At last!

Months went by. Confirmation. I was a wizard.

One day in October, the landlord's son-in-law said to me: "There hasn't been a sound from them since."

I tried not to look self-conscious.

Said he: "Last May, one day, I was looking at them, and I said, in my own mind: 'If we lose tenants on account of you, I'll wring your necks.' They never crowed again."

Again it's the Principle of Uncertainty, by which the path of a particle cannot be foretold, and by which there's no knowing who stopped the roosters. Well, we're both--or one of us is--very inferior in matters of magic, according to a story that is told of Madame Blavatsky. The little bird of a cuckoo clock annoyed her. Said she: "Damn that bird! shut up!" The cuckoo never spoke again.

By the cultivation of wild talents, I do not mean only the learning of the secret of the man they could not drown, and having the advantage of that ability, at times of shipwreck--of the man they could not confine, so that enormous would be the relief from the messiahs of the legislatures, if nobody could be locked up for failure to keep track of all their laws--of the woman they could not touch, so that there could be no more automobile accidents--of myself and the roosters--though just here my landlord's son- in-law will read scornfully--so that all radios can be stopped immediately after break-fast, and all tenors and sopranos forever.

Only the secret of burning mansions in England; appearances of wounds on bodies, or of pictures on hailstones; bodies on benches of a Harlem park; strange explosions, and forced landings of aero-planes, and the case of Lizzie Borden

Those are only specializations. If all are only different manifestations of one force, or radio-activity, transmediumization, or whatever, that is the subject for research and experiment that may develop--New triumphs and new disasters; happiness and miseries--a new era, in which people will think back, with contempt, or with horror, at our times, unless they start to think a little more keenly of their own affairs.

In the presence of a poltergeist girl, who, so far as is now knowable, exerts no force, objects move.

But this is a book of no marvels.

In the presence of certain substances, which so far as is now knowable, exert no force, other sub-stances move, or transmute into very different substances.

This is a common phenomenon, to which the chemists have given the name catalysis.

All around are wild talents, and it occurs to nobody to try to cultivate them, except as expressions of personal feelings, or as freaks for which to charge admission. I conceive of powers and the uses of human powers that will someday transcend the stunts of music halls and seances and sideshows, as public utilities have passed beyond the toy-stages of their origins. Sometimes I tend to thinking con-structively--or batteries of witches teleported to Nicaragua, where speedily they cut a canal by dis-solving trees and rocks--the tumults of floods, and then magic by which they cannot touch houses--cyclones that smash villages, and then cannot push feathers. But also I think that there is nothing in this subject that is more reasonable than is the Taboo that is preventing, or delaying, development. I mean that semi-enlightenment that so earnestly, and with such keen, one-sided foresight fought to suppress gunpowder and the printing press and the discovery of America. With the advantages of practical witchcraft would come criminal enormities. Of course they would be somewhat adapted to. But I'd not like to have it thought that I am only an altruist, or of the humble mental development of a Utopian, who advocates something, as a blessing, without awareness of it as also a curse. Every folly, futility, and source of corruption of today, if a change from affairs primordial, was at one time preached as cure and salvation by some messiah or another. One reason why I never pray for anything is that I'm afraid I might get it.

Or the uses of witchcraft in warfare--

But that, without the sanction of hypocrisy, superintendence by hypocrisy, the blessing by hypocrisy, nothing ever does come about--

Or military demonstrations of the overwhelming effects of trained hates--scientific uses of destructive bolts of a million hate-power--the blasting of enemies by disciplined ferocities--

And the reduction of cannons to the importance of fire crackers--a battleship at sea, or a toy boat in a bathtub--

The palpitations of hypocrisy--the brass bands of hypocrisy--the peace on earth and good will to man of hypocrisy--or much celebration, because of the solemn agreements of nations to scrap their battleships and armed aeroplanes--outlawry of poison gases, and the melting of cannon--once it is recognized that these things aren't worth a damn in the Era of Witchcraft--

But of course not that witchcraft would be practiced in warfare. Oh, no: witchcraft would make war too terrible. Really, the Christian thing to do would be to develop the uses of the new magic, so that in the future a war could not even be contemplated.

Later: A squad of poltergeist girls--and they pick a fleet out of the sea, or out of the sky--if, as far back as the year 1923, something picked French aeroplanes out of the sky--arguing that some nations that renounced fleets as obsolete would go on building them just the same.

Girls at the front--and they are discussing their usual not very profound subjects. The alarm--the enemy is advancing. Command to the poltergeist girls to concentrate--and under their chairs they stick their wads of chewing gum.

A regiment bursts into flames, and the soldiers are torches. Horses snort smoke from the combustion of their entrails. Reinforcements are smashed under cliffs that are teleported from the Rocky Mountains.

The snatch of Niagara Falls--it pours upon the battlefield. The little poltergeist girls reach for their wads of chewing gum.

28

THAT everything that is desirable is not worth having--that happiness and unhappiness are emotional rhythms that are so nearly independent of one's circumstances that good news or bad news only stimulate the amplitude of these waves, without affecting the ratio of ups to downs--or that one might as well try to make, in a pond, waves that are altitudes only as to try to be happy without suffering equal and corresponding unhappiness.

But, so severely stated, this is mechanistic philosophy.

And I am a mechanist-immechanist.

Sometimes something that is desirable is not only not worth having, but is a damn sight worse than that.

Is life worth living? Like everybody else, I have many times asked that question, usually deciding negatively, because I am most likely to ask myself whether life is worth living at times when I am convinced it isn't. One day, in one of my frequent, and probably incurable, scientific moments, it occurred to me to find out. For a month, at the end of each day, I set down a plus sign, or a minus sign, indicating that, in my opinion, life had, or had not, been worth living, that day. At the end of the month, I totaled up, and I can't say that I was altogether pleased to learn that the pluses had won the game. It is not dignified to be optimistic.

I had no units by which to make my alleged determinations. Some of the plus days may have been only faintly positive, and, here and there, one of the minus days may have been so ferociously negative as to balance a dozen faintly positive days. Of course I did attempt gradations of notation, but they were only cutting pseudo-units into smaller pseudo-units. Also, out of a highly negative, or very distressing, experience, one may learn something that will mean a row of pluses in the future. Also, some pluses simply mean that one has misinterpreted events of a day, and is in for much minus--

Or that nothing--a joy or a sorrow, the planet Jupiter, or an electron-- can be picked out of its environment, so as finally to be labeled either plus or minus, because as a finally identifiable thing it does not exist--or that such attempted isolations and determinations are only scientific.

I have picked out witchcraft, as if there were witchcraft, as an identifiable thing, state, or activity. But, if by witchcraft, I mean phenomena as diverse as the mimicry of a leaf by a leaf-insect, and illnesses in a house where "Typhoid Mary" was cooking, and the harmless impalement, on spears, of children, I mean, by witchcraft in general, nothing that can be picked out of one commonality of phenomena. All phenomena are rhythmic, somewhere between the metrical and the frenzied, with final extremes unreachable in an existence of the metrical-unmetrical. The mechanical theory of existence is as narrowly lopsided as would be a theory that all things are good, large, or hot. It is Puritanism. It is the text-book science that tells of the clock-work revolutions of the planet Jupiter, and omits mention of Jupiter's little, vagabond moons, which would be fired from any job, in human affairs, because of their unpunctualities--and omits mention that there's a good deal the matter with the clock-work of most clocks. Mechanistic philosophy is a dream of a finality of exact responses to stimuli, and of absolute equivalences. Inasmuch as the advantages and disadvantages of anything can no more be picked out, isolated, identified, and quantitatively determined, than can the rise of a wave be clipped from its fall, it is only scientific dreamery to say what anything is equal and opposite to.

And, at the same time, in the midst of a submergence in commonality, there is a permeation of all phenomena by an individuality that is so marked that, just as truly as all things merge indistinguishably into all other things, all things represent the unmergeable. So then there is something pervasive of every action and every advantage that makes it alone, incommensurable, and incomparable with a reaction, or a disadvantage.

Our state of the hyphen is the state of the gamble. Go to no den of a mathematician for enlightenment. Try Monte Carlo. Out of science is fading certainty as fast as ever it departed from theology. In its place we have adventure. Accepting that there is witchcraft, in the sense in which we accept that there is electricity, magnetism, or life, the acceptance is that there is no absolute poise between advantages and disadvantages.

Or that practical witchcraft, or the development of wild talents, might be of such benefits as to draw in future records of human affairs the new dividing line of A. W. and B. W.--or might be a catastrophe that would drive all human life back into Indians, or Zulus, or things furrier--

If by any chance the evils of witchcraft could compare with, or beat to an issue, the demoralizations of law, justice, business, sex, literature, education, pacifism, militarism, idealism, materialism, which at present, are incomprehensibly not yet equal and opposite to stabilizations that are saving us from, or are denying us, the jungles--

Or let all persons of foresight, if of sedentary habits, shift positions occasionally, so as not to suppress too much their vertebral stubs that their descendants may need as the bases of more graceful appendages.

But my own expression is that any state of being that can so survive its altruists and its egotists, its benefactors and its exploiters, its artists, gunmen, bankers, lawyers, and doctors would be almost immune to the eviler magics of witchcraft, because it is itself a miracle.

29

STUNTS of sideshows, and the miracles of pietists, and the phenomena of spiritualistic medium--

Or that the knack that tips a table may tilt an epoch.

Or much of the "parlor magic" of times gone by, and now it is industrial chemistry. And Taboo, by which earlier experimenters in the trained forces of today were under suspicion as traffickers with demons.

I take for a pseudo-principle, by which I mean a standard of judgment that sometimes works out, and sometimes doesn't work out--which is as near to wisdom as I can arrive, in an existence of truth-nonsense--that, someday to be considered right, is first to be unholy. It is out of blasphemy that new religions arise. It is by thinking things that schoolboys know better than to think that discoveries are made. It is because our visions are not delirious enough, or degraded, or nonsensical enough, that all of us are not prophets. Let any thoughtful, properly trained man, who has had all the benefits of an academic education, predict--at least, then, we know what won't be. We have, then, at our command, a kind of negative clairvoyance--if we know just where to go for an insight into what won't be.

The trail of a working witchcraft--but, if we are traffickers with demons, the traffic isn't much congested, at present. Someday almost every particular in this book may look quaint, but it may be that the principle of putting the witches to work will seem as sound as now seems the employment of steam and electric demons. Our instances of practical witchcraft have been practical enough, so long as they were paying attractions at exhibitions, but the exhibition implies the marvel, or what people regard as the marvel, and the spirit of this book is of commonplaceness, or of coming commonplaceness--or that there isn't anything in it, except of course its vagaries of theories and minor interpretations, that won't someday be considered as unsensational as the subject-matters of textbooks upon chemistry and mechanics. My interest is in magic, as the daily grind--the miracle as a job--sorceries as public utilities.

There is one manifestation of witchcraft that has been put to work. It is a miracle with a job. Dowsing.

It is commonly known as water-divining. It is witchcraft. One cannot say that, because of some unknown chemical, or bio-chemical, affinity, a wand bends in a hand, in the presence of underground water. The wand bends only in the hand of a magician.

It is witchcraft. So, though there are scientists who are giving in to its existence, there are others, or hosts of others, who never will give in. Something about both kinds of scientists was published in Time, Feb. 9, 1931. It was said that Oscar E. Meinzer, of the U. S. Geological Survey, having investigated dowsers, had published his findings which were that "further tests ... of so-called 'witching' for water, oil, or other minerals, would be a misuse of public funds." Also it was shown that conclusions by Dr. Charles Albert Browne, of the U. S. Department of Agriculture, disagreed with Mr. Meinzer's findings. "On a large sugar-beet estate, near Magdeburg, Dr. Browne saw one of Germany's most famed dowsers at work. Covering his chest with a padded leather jacket, the dowser took in his hands a looped steel divining rod, and began to pace the ground. Suddenly the loop shot upward, and hit him a sharp blow on the chest. Continuing, he charted the outlines of an underground stream. Then, using an aluminum rod, which he said was much more sensitive, he estimated the depth of the stream. A rod of still another metal indicated that the water was good to drink. When Dr. Browne tried to use the rod, himself, he could get no chest blows, unless the dowser was holding one end. Dr. Browne then questioned German scientists. The majority answered that, with all humbuggery discounted, a large number of successes remained, which could not be accounted for by luck or chance." For queer places--or for places in which scientists of not so far back would have predicted that such yokelry as dowsing would never be admitted--see Science, Jan. 23, 1931, or the Annual Report of the Smithsonian Institution, 1928, p. 325. Here full particulars of Dr. Browne's investigation are published.

The Department of Public Works, of Brisbane, Queensland, Australia, has employed a dowser, since the year 1916 (Notes and Queries, 150-235). New York Times, July 26, 1931--two Australian states were employing dowsers.

I don't know that I mean much by that. The freaks and faddists who get themselves employed by governments make me think that I am not very convincing here. But I have no record of a dowser with a political job before the year 1916: and, wherever I got all this respectfulness of mine for the job, it is the entrance of magic into the job that I am bent upon showing.

In the London Observer, May 2, 1926, it is said that the Government of Bombay was employing an official water diviner, who, in one district of scarcity of water, had indicated about fifty sources of supply, at forty-seven of which water had been found. The writer of this account says that members of one of the biggest firms of well-boring engineers had informed him that they had successfully employed dowsers in Wales, Oxfordshire, and Surrey.

In Nature, Sept. 8, 1928, there is an account, by Dr. A. E. M. Geddes, of experiments with dowsers.

Dr. Geddes' conclusion is that the faculty of water-divining is possessed by some persons, who respond to at present unknown, external stimuli.

It is not that I am maintaining that out of the mouths of babes, and from the vaporings of yokels, we shall receive wisdom--but that sometimes we may. Peasants have believed in dowsing, and scientists used to believe that dowsing was only a belief of peasants. Now there are so many scientists who believe in dowsing that the suspicion comes to me that it may be only a myth, after all.

In the matter of dowsing, the opposition that Mr. Meinzer represents is as understandable as is the opposition that once was waged by priestcraft against the system that he now represents. Let in, against the former dominant, data of raised beaches, or of deposits of fossils, and each intruder would make a way for other iniquities. Now, relatively to the Taboo of today, let in any of the occurrences told of in this book, and by its suggestions and affiliations, or linkages, it would make an opening for an irruption.

Very largely, dowsing, or witchcraft put to work, has been let in.

30

IT has been my expression that, for instance, African fakirs achieved the harmless impalement of children by a process that would ordinarily be called imposing the imaginary upon the physical, but that is called by me imposing the imaginary-physical upon the physical-imaginary. I think that this is the conscious power and method of adepts: but I think that in the great majority of our stories, effects have been wrought unconsciously, so far as went active awareness, by witches and wizards. I am impressed more with an experience of my own than with any record of other doings. I looked, or stared, at a picture on a wall. Somewhere in my mind were many impressions of falling pictures. But I was not actively thinking of falling pictures. The picture fell from the wall.

See back to the Blackman case--the four judges, who "died suddenly." It was Blackman, who called attention to these deaths. Why? Vanity of the magician? I think that more likely these victims were removed by a wizardry of Blackman's of which he was unconscious. I think that if a man so earnestly objected to paying alimony that, instead, he went to jail four times, he'd overlook his judges and take a shorter cut, on behalf of his income, if he consciously reasoned about it.

It would seem that visualizations have had nothing to do with many occurrences told of in this book. Still, by a wild talent I mean something that comes and goes, and is under no control, but that may be caught and trained. Also there are cases that look very much like controlled uses of visualizations upon physical affairs. In this view, I have noted an aspect of doings that is a support for our expression upon transmediumization.

The real, as it is called, or the objective, the external, the material, cannot be absolutely set apart from the subjective, or the imaginary: but there are quasi-attributes of the imaginary. There have been occurrences that I think were transmediumizations, because I think that they were marked by indications of having carried over, from an imaginative origin, into physical being, or into what is called "real life," the quasi-attributes of their origin.

A peculiarity of fires that are called--or that used to be called--"spontaneous combustions of human bodies," is that the fires do not communicate to surrounding objects and fabrics, or that they extend only to a small degree around. There are stories of other such fires, which cannot be "real fires," as compared with fires called "real." In the St. Louis Globe-Democrat, October 2, or about Oct. 2, 1889, there is a story of restricted fires, said to have occurred in the home of Samuel Miller, upon a farm, six miles west of Findlay, Ohio. A bed had burst into flames, burning down to a heap of ashes, but setting nothing else afire, not even scorching the floor underneath. The next day, "about the same time in the afternoon," a chest of clothes flamed, and was consumed, without setting anything else afire. The third day, at the same time, another bed, and nothing but the bed burned. See back to the fires in the house in Bladenboro, N. C., February, 1932. A long account of these fires, from a San Diego (Cal.) newspaper, was sent to me by Margaret M. Page, of San Diego. In it one of the phenomena considered most remarkable was that fires broke out close to inflammable materials that were unaffected by the flames. Names of several witnesses--Mayor J. A. Bridger, of Bladenboro, J. B. Edwards, a Wilmington health officer, and Dr. S. S. Hutchinson, of Bladenboro.

It is as if somebody had vengefully imagined fires, and in special places had localized fires, according to his visualizations. Such localizing, or focusing, omitting surroundings, is a quasi-attribute of all visualizations. One vividly visualizes a face, and a body is ignored by the imagination. Let somebody visualize a bed afire, and exhaust his imaginative powers in this specialization: I conceive of the bed burning, as imagined, and nothing else burning, because nothing else was included in the mental picture that transmediumized, it having been taken for granted, by the visualizer, that, like a fire of physical origin, this fire would extend. It seems to me to be only ordinarily impossible to understand the burning of a woman on an unscorched bed as the "realization" of an imagined scene in which the burning body was pictured, with neglect of anything else consuming.

See back to unsatisfactory attempts to attribute punctures of windowpanes and automobile shields, to a missile-less weapon. The invisible bullets stopped short, after penetrating glass. If we can think of an intent, more mischievous than malicious, that was only upon shooting through glass, and that gave no consideration to subsequent courses of bullets, we can think of occurrences that took place, as visualized, and as restricted by visualizations.

Doings in closed rooms--but my monism, by which I accept that all psychical magic links some-where with more or less commonplace physical magic.

New York Times, June 18, 1880--Rochester, N. Y.--a woman dead in her bed, and the bed post hacked as if with a hatchet. It was known that nobody had entered this room. But something had killed this woman, leaving no sign of either entrance or exit.

It was during a thunderstorm, and the woman had been killed by lightning.

The man of one of our stories--J. Temple Thurston--alone in his room--and that a pictorial representation of his death by fire was enacting in a distant mind--and that into the phase of existence that is called "real" stole the imaginary--scorching his body, but not his clothes, because so was pictured the burning of him--and that, hours later, there came into the mind of the sorcerer a fear that this imposition of what is called the imaginary upon what is called the physical bore quasi-attributes of its origin, or was not realistic, or would be, in physical terms, unaccountable, and would attract attention-- and that the fire in the house was visualized, and was "realized," but by a visualization that in turn left some particulars unaccounted for.

Lavinia Farrar was a woman of "independent means." Hosts of men and women have been shot, or stabbed, or poisoned, because of their "independent means." But that Mrs. Farrar was thought to death--or that upon her, too, out of the imaginary world in somebody's mind, stole a story--that it made of her, too, so fictional a being that of her death there is no explanation in ordinary, realistic terms.

That here, too, there was an after-thought, or an after-picturation, which, by way of attempted explanation, "realized" a knife and blood on the floor, but overlooked other details that made this occurrence inexplicable in terms of ordinary murders--or that this woman had been stabbed in the heart, through unpunctured clothes, because it was, with the neglect of everything else, the wound in the heart that had been visualized.

The germ of this expression is in anybody's acceptance that a stigmatic girl can transfer a wound, as pictured in her mind, into appearance upon her body. The expression requires that there may be external, as well as personal, stigmatism.

It seems to me to be as nearly unquestionable as anything in human affairs goes, that there have been stigmatic girls. There may have been many cases of different kinds of personal stigmatism.

There are emotions that are as intense as religious excitation. One of them is terror.

The story of Isidor Fink is a story of a fear that preceded a murder. It could be that Fink's was a specific fear, of somebody whom he had harmed, and not a general fear of the hold-ups that, at the time, were so prevalent in New York City. According to Police Commissioner Mulrooney, it was impossible, in terms of ordinary human experience, to explain this closed-room murder.

Or Isidor Fink, at work in his laundry--and his mind upon somebody whom he had injured--and that his fears of revenge were picturing an assassination of which he was the victim--that his physical body was seized upon by his own picturization of himself, as shot by an enemy.

31

IN February, 1885, in an English prison, there was one of the dream-like occurrences that the materialists think are real. But every character concerned in it was fading away, so that now there is probably no survivor. From time to time repairs had to be made, because the walls of the prison were dissolving. By way of rusts, the iron bars were disappearing.

Upon February 23, 1885--as we say, in terms of our fanciful demarcations--just as if a 23rd of February, which is only relative to rhythms of sunshine, could be a real day--just as if one could say really where a January stops and a February begins--just as if one could really pick a period out of time, and say that there ever was really a year 1885--

Early in what is called a morning of what is so arbitrarily and fancifully called the 23rd of February, 1885, John Lee, in his cell, in the penitentiary, at Exeter, England, was waiting to be hanged.

In the yard of a prison of stone, with bars of iron, John Lee was led past a group of hard and motionless witnesses, to the scaffold. There were newspaper men present. Though they probably considered it professional to look as expressionless as stones, or bars of iron, there was nothing in Lee's case to be sentimental about. His crime had been commonplace and sordid. He was a laborer, who had lived with an old woman, who had a little property, and, hoping to get that, he had killed her. John Lee was led past a group, almost of minerals. It was a scene of the mechanism and solidity of legal procedure, as nearly real as mechanism and solidity can be.

Noose on his neck, and up on the scaffold they stood him on a trap door. The door was held in position by a bolt. When this bolt was drawn, the door fell--

John Lee, who hadn't a friend, and hadn't a dollar--

The Sheriff of Exeter, behind whom was Great Britain.

The Sheriff waved his hand. It represented Justice and Great Britain.

The bolt was drawn, but the trap door did not fall. John Lee stood with the noose around his neck. It was embarrassing. He should be strangling. There is something of an etiquette in all things, and this was indecorum. They tinkered with the bolt. There was no difficulty, whatsoever, with the bolt: but when it was drawn, with John Lee standing on the trap door, the door would not fall.

Something unreasonable was happening. Just what is the procedure, in the case of somebody, who is standing erect, when he should be dangling? The Sheriff ordered John Lee back to his cell.

The people in this prison yard were not so stolid. They fluttered, and groups of them were talking it over. But there was no talk that could do John Lee any good. This was what is called stern reality. The Sheriff did not flutter. I have a note upon him, twenty years later: he was in trouble with a religious sect of which he was a member, because he ordered his beer by the barrel. He was as solid as beer and beef and the British Government.

The warders looked into the matter thoroughly--except that there wasn't anything to look into. Every time they drew back the bolt, with John Lee out of the way, the door fell, as it should fall. One of the warders stood in Lee's place, where, instead of placing the noose around his neck, he clung to the rope. The bolt was drawn, the door fell, as it should fall, and down dropped the warder, as he should drop.

There was a woman they could not push. A man they could not crucify. The man they could not drown. There was the man they could not imprison. The dog they could not lose.

John Lee was led back to the scaffold. The witnesses did not know whether to be awed or not. But, after all, it was just one of those things that nobody could explain, but that could not happen again--

Or that to a college professor it could not--to anybody educated in the first principles of mechanics and physics it could not--that, to anybody, not an untutored laboring man, but committed to unquestioning belief in everything that a professor of physics would say in maintaining that the trap door would have to fall--

The bolt was drawn.

The trap door would not fall.

John Lee stood unchangeable.

That when, the first time, John Lee was led past these newspaper men, and town officials, and others who had been invited to the ceremony, any one of them could have overstepped any line that all were told to toe would have been little short of inconceivable. But a doctor, whose professional appearance was much faded, interceded. Others were shaky. The Sheriff said that John Lee had been sentenced to be hanged, and that John Lee would be hanged.

They had done everything thinkable. Any suggestions? Somebody suggested that rains might have swollen the wooden door, causing friction. There had been, in all tests, no friction: but, by way of taking every possible precaution, a warder planed the edges of the door. They experimented, and, every time, the door fell, as it should fall.

They stood him on the scaffold again.

The door would not fall.

This scene of an attempted execution dissolved, like a dream-picture. The newspaper men faded away, or burst away. The newspaper men ran out into the streets of Exeter. In the streets, they ran, shouting the news of the man who could not be hanged. The Sheriff, who had tried hard to be a real Sheriff, went to pieces. He'd do this about it, and then he'd do that about it, and then "Take him away!" He communicated with the Home Secretary. There was something about all this that so shook the Home Secretary that he authorized a delay.

The matter was debated in the House of Commons, where some of the members denounced a proposed defeat of justice by superstition. Nevertheless the execution was not attempted again. Lee's sentence was commuted to life-imprisonment, but he was released in December, 1907. His story was re-told in the newspapers of that time. I take from Lloyd's Weekly News (London) Jan. 5, 1908. I have tried to think of a conventional explanation, in the case of John Lee. All attempts fail. He hadn't a dollar.

There may be some commonplace explanation that I have not thought of: but my notion is that the explanation that I have thought of will someday be considered as commonplace as are now regarded the impenetrable mysteries of electricity and radio-activity.

32

   IT'S the old controversy--the action of mind upon matter. But, in the philosophy of the hyphen, an uncrossable gap is disposed of, and the problem is rendered into thinkable terms, by asking whether mind-matter can act upon matter-mind.

I am beginning to see whence all my specialization, not much short of hypnotization, upon magic, as the job. Just why am I so bent upon cooping people into multicellular formations, and setting batteries of disciplined sorcerers at work, bewitching into useful revolutions all the motors of the world?

As to the job, and anything that is supposed to be not a job, there is only the state of job-recreation, or recreation-job. I have cut out of my own affairs very much of so-called recreation, simply because I feel that I cannot give to so-called enjoyments the labors that they exact. I'd often like to be happy, but I don't want to go through the equivalence of digging a ditch, or of breaking stones, to enjoy myself. I have seen, by other persons, very labored and painful efforts to be happy. So then I am so much concerned with the job, because, though it hyphenates, there isn't anything else.

Probably it will be some time before any college professor, of whatever we think we mean by importance, will admit that, by witchcraft, or by the development of what are now only wild talents, all the motors of this earth may be set going and kept at work. But "highest authority" no longer unitedly opposes the more or less remote possibility of such operations. See an interview, with Dr. Arthur H. Compton, Professor of Physics, at the University of Chicago, published in the New York Times, Jan. 3, 1932. Said Dr. Compton: "The new physics does not suggest a solution of the old question of how mind acts on matter. It does definitely, however, admit the possibility of such action, and suggests where the action may take effect."

I don't know that I am much more of a heretic, myself. In my stories, I have admitted possibilities, and I have made suggestions.

But the difference is that the professors will not be concrete, and I give instances. Dr. Compton's views are ripe with the interpretation that transportation systems, and the lighting of cities, and the operation of factories may someday be the outcome of what he calls the "action of mind on matter," or what I'd call mechanical witchcraft. But toyers with abstractions falter, the moment one says-- "For instance?"

The fuel-less motor, which is by most persons considered a dream, or a swindle, associates most with the name of John Worrell Keely, though there have been other experimenters, or impostors, or magicians. The earliest fuel-less motor "crank" of whom I have record is John Murray Spear, back in the period of 1855, though of course various "cranks" of all ages can be linked with this swindle, dream, or most practical project. The latest, at this writing, is a young man, Lester J. Hendershot, of Pittsburgh, Pa. I take data from the New York Herald Tribune, Feb. 27-March 10, 1928. It was Henderson's statement that he had invented a motor that operated by deriving force from "this earth's magnetic field." Nobody knows what that means. But Hendershot was backed by Major Thomas Lanphier, U. S. Army, commandant of Selfridge Field, Detroit. It was said that at tests of Selfridge Field, a model of the "miracle motor" had invisibly generated power enough to light two 110-volt lamps, and that another had run a small sewing machine. Major Lanphier stated that he had helped to make one of these models, which were of simple construction, and that he was sure that there was nothing fraudulent about it.

This espousal by Major Lanphier array, considering that to orthodox scientists it was the equivalence of belief in miracles, seem extraordinary: but it seems to me that the attacks that were made upon Hendershot were more extraordinary--or significant. It would seem that, if a simple, little contrivance, weighing less than ten pounds, were a fraud, the mechanics of Selfridge Field, or anywhere else, could determine that in about a minute, especially if they had themselves made it, under directions. If the thing were a fraud, it would seem that it would have to be obviously a fraud. Who'd bother? But Dr. Frederick Hochstetter, head of the Hochstetter Research Laboratory, of Pittsburgh, went to New York about it. He hired a lecture room, or a "salon," of a New York hotel, telling reporters that he had come to expose a fraud, which would be capable of destroying faith in science for 1,000 years. If so, even to me this would not be desirable. I should like to see faith in science destroyed for 20 years, and then be restored for a while, and then be knocked flat again, and then revive--and so on, in a healthy alternation. Dr. Hochstetter exhibited models of the motor. They couldn't generate the light of a 1-volt firefly. They couldn't stitch a fairy's breeches. Dr. Hochstetter lectured upon what he called a fraud. But the motive for all this? Dr. Hochstetter explained that his only motive was that "pure science might shine forth untarnished."

It was traveling far, and going to trouble and expense to maintain the shine of a purity, the polish of which was threatened by no more than a youngster, of whom most of the world had never heard before. What I pick up is that there must have been an alarm that was no ordinary alarm, somewhere. I pick up that at tests, in Detroit, in Hendershot's presence, his motors worked; that, in New . York, not in his presence, his motors did not work.

Then came the denouement, by which most stories of exposed impostors end up, or are said to end up.

Said Dr. Hochstetter--dramatically, I suppose, inasmuch as he was much worked up over all this--he had discovered that concealed in one of the motors was a carbon pencil battery.

Just about so, in the literature of Taboo, end almost all stories of doings that are "alarming." There is no chance for a come-back from the "exposed impostor." He is shown as sneaking off-stage, in confusion and defeat. But some readers are having a glimmer of what I mean by taking so much material from the newspapers. They get statements from "exposed impostors." They ridicule and belittle, and publish much that is one-sided, but they do give the chance for the come-back.

Came back Hendershot:

That Dr. Hochstetter was quite right in his accusation, but only insofar as it applied to an incident of several years before. In his early experiments Hendershot, having no assurance of the good faith of visitors, had stuck into his motor various devices "to lead them away from the real idea I was working on." But, in the tests at Selfridge Field there had been no such "leads," and there had been no means of concealments in motors that mechanics employed by Major Lanphier had made.

Two weeks later, Hendershot dropped out of the newspapers. Perhaps a manufacturer of ordinary motors bought him off. But he dropped out by way of a strange story. It is strange to me, because I recall the small claims that were made for the motor--alleged power not sufficient to harm anybody--only enough to run a sewing machine, or to light lamps with 220 volts. New York Herald Tribune, March 10, 1928--that Lester J. Hendershot, the Pittsburgh inventor of the "miracle motor," was a patient in the Emergency Hospital, Washington, D. C. It is said that, in the office of a patent attorney, he was demonstrating his "fuel-less motor," when a bolt estimated at 2,000 volts shot from it, and temporarily paralyzed him.

It was Hendershot's statement that his motor derived force from "this earth's magnetic field." It is probable that, if the motor was driven by his own magic, he would, even if he knew this, attribute it to something else. It is likely that spiritualistic mediums--or a few of them--have occult powers of their own: but they attribute them to spirits. Probably some stage-magicians have occult powers: but, in a traditional fear of persecutions of witchcraft, they feel that it is safer to say that the hand is quicker than the eye. "Divine healers" and founders of religions have been careful to explain that their talents were not their own.

In November, 1874, John Worrell Keely exhibited, to a dozen well-known Philadelphians, his motor. They were hard-headed business men--as far as hard heads go--which isn't very far--but they were not dupes and gulls of the most plastic degree. They saw, or thought they saw, this motor operate, though connected in no way with any conventionally recognized source of power. Some of these witnesses considered the motor worth backing. Keely, too, explained that something outside himself was the moving force, but nobody has ever been able to explain his explanations. Unlike Hendershot's simple contrivance, Keely's motor was a large and complicated structure. The name of it was formidable. When spoken of familiarly, it was a vibratory generator, but the full name of the monster was the Hydro- pneumatic-pulsating-vacue-engine. A company was organized, and, after that, everything was very unsatisfactory, except to Keely. There was something human about this engine-- just as any monist, of course thinks there is to everything--such as rats and trees and people. It was like so many promising young men, who arrive at middle age, still promising, and go to their graves, having, just before dying, promised something or another. It can't be said that the engine worked. The human-like thing had talents, and was capable of sensational stunts, but it couldn't earn a dollar. That is, at an honest day's toil, it could not, but with its promises it brought tens of thousands of dollars to Keely. It is said that, though he lived well, he spent much of this money in experiments. Here, too, just what I suspect--though don't have it that I think I'm the only one who has had this idea-- was just what was not asserted. That his motor moved responsively to a wizardry of his own, was just what Keely never said. It could be that it was a motivation of his own, but that he did not know it.

Mesmer, in his earlier phases, believed that he wrought cures with magnets, and he elaborated very terminological theories, in terms of magnets, until he either conceived, or admitted, that his effects were wrought by his own magic.

I should like to have an opinion upon fuel-less engines, from an official of General Motors, to compare with what the doctors of Vienna and Paris thought of Mesmer.

For eight years there was faith: but then (December, 1882), there was a meeting of disappointed stockholders of the Keely Motor Co. In the midst of protests and accusations, Keely announced that, though he would not publicly divulge the secret of his motor, he would tell everything to any representative of the dissatisfied ones. A stockholder named Boekel was agreed upon. Boekel's report was that it would be improper to describe the principle of the mechanism, but that "Mr. Keely had discovered all that he had claimed." There is no way of inquiring into how Mr. Boekel was convinced. Considering the billions of human beings who have been "convinced" by bombardments of words and phrases beyond their comprehension, I think that Mr. Boekel was reduced to a state of mental helplessness by flows of a hydro-pneumatic-pulsating-vacue terminology; and that faithfully he kept his promise not to explain, because he had not more than the slightest comprehension of what it was that had convinced him.

But I do not think that any character of Mr. Keely's general abilities has ever practiced successfully without the aid of religion. Be good for a little while, and you shall have everlasting reward. Keely was religious in preaching his doctrine of goodness: benefits to mankind, releases from enslavement, spare time for the cultivation of the best that is in everybody, promised by his motor--and in six months the stock will be quoted at several times its present value. I haven't a notion that John Worrell Keely, with a need for business, and a throb for suffering humanity, was any less sincere than was General Booth, for instance.

In November, 1898, Keely died. Clarence B. Moore, son of his patron, Mrs. Bloomfield Moore-- short tens of thousands of dollars in his inheritance, because of Keely and his promises--rented Keely's house, and investigated. According to his findings, Keely was "an unadulterated rascal."

This is too definite to suit my notions of us phenomena. The unadulterated, whether of food we eat, or the air we breathe, or of idealism, or of villainy, is unfindable. Even adultery is adulterated. There are qualms and other mixtures.

Moore said that he had found the evidences of rascality. The motor was not the isolated mechanism that, according to him, the stockholders of the Keely Motor Co. had been deceived into thinking it was: he had found an iron pipe and other tubes, and wires that led from the motor to the cellar. Here was a large, spherical, metallic object. There were ashes.

Imposture exposed--the motor had been run by a compressed air engine, in the cellar.

Anybody who has ever tried to keep a secret twenty-four hours, will marvel at this story of an impostor who, against all the forces of revelation, such as gas men, and coal men, and other persons who get into cellars--against inquisitive neighbors, and, if possible, even more inquisitive newspaper men--against disappointed stockholders and outraged conventionalists--kept secret, for twenty-four years, his engine in the cellar.

It made no difference what else came out. Taboo had, or pretended it had, something to base on. Almost all people of all eras are hypnotics. Their beliefs are induced beliefs. The proper authorities saw to it that the proper belief should be induced, and people believed properly.

Stockholders said that they knew of the spherical object, or the alleged compressed air engine in the cellar, because Keely had made no secret of it. Nobody demonstrated that by means of this object, the motor could be run. But beliefs can be run. So meaningless, in any sense of organization, were the wires and tubes, that I think of Hendershot's statement that he had complicated his motor with "leads," as he called them.

Stones that have fallen in houses where people were dying--the rambles of a pan of soft soap--chairs that have moved about in the presence of poltergeist girls--

But, in the presence of John Worrell Keely, there were disciplined motions of a motor. For twenty-four years there were demonstrations, and though there was much of a stir-up of accusations, never was Keely caught helping out a little. There was no red light, nor semi-darkness. The motor stood in no cabinet. Keely's stockholders were of a superior intelligence, as stockholders go, inasmuch as many of them investigated, somewhat, before speculating. They saw this solemn, big contrivance go around and around. Sometimes they saw sensational stunts. The thing tore thick ropes apart, broke iron bars, and shot bullets through a twelve- inch plank. I conceive that the motivation of this thing was a wild talent--an uncultivated, rude, and unreliable power, such as is all genius in its infancy--

That Keely operated his motor by a development of mere "willing," or visualizing, whether consciously; or not knowing how he got his effects--succeeding spasmodically sometimes, failing often, according to the experience of all pioneers--impostor and messiah.

Justifying himself, in the midst of promises that came to nothing, because he could say to